Nanostructured Carbon Materials for Catalysis

RSC Catalysis Series

Titles in the Series:

1: Carbons and Carbon Supported Catalysts in Hydroprocessing
2: Chiral Sulfur Ligands: Asymmetric Catalysis
3: Recent Developments in Asymmetric Organocatalysis
4: Catalysis in the Refining of Fischer–Tropsch Syncrude
5: Organocatalytic Enantioselective Conjugate Addition Reactions: A Powerful Tool for the Stereocontrolled Synthesis of Complex Molecules
6: *N*-Heterocyclic Carbenes: From Laboratory Curiosities to Efficient Synthetic Tools
7: *P*-Stereogenic Ligands in Enantioselective Catalysis
8: Chemistry of the Morita–Baylis–Hillman Reaction
9: Proton-Coupled Electron Transfer: A Carrefour of Chemical Reactivity Traditions
10: Asymmetric Domino Reactions
11: C-H and C-X Bond Functionalization: Transition Metal Mediation
12: Metal Organic Frameworks as Heterogeneous Catalysts
13: Environmental Catalysis Over Gold-Based Materials
14: Computational Catalysis
15: Catalysis in Ionic Liquids: From Catalyst Synthesis to Application
16: Economic Synthesis of Heterocycles: Zinc, Iron, Copper, Cobalt, Manganese and Nickel Catalysts
17: Metal Nanoparticles for Catalysis: Advances and Applications
18: Heterogeneous Gold Catalysts and Catalysis
19: Conjugated Linoleic Acids and Conjugated Vegetable Oils
20: Enantioselective Multicatalysed Tandem Reactions
21: New Trends in Cross-Coupling: Theory and Applications
22: Atomically-Precise Methods for Synthesis of Solid Catalysts
23: Nanostructured Carbon Materials for Catalysis

How to obtain future titles on publication:
A standing order plan is available for this series. A standing order will bring delivery of each new volume immediately on publication.

For further information please contact:
Book Sales Department, Royal Society of Chemistry, Thomas Graham House, Science Park, Milton Road, Cambridge, CB4 0WF, UK
Telephone: +44 (0)1223 420066, Fax: +44 (0)1223 420247
Email: booksales@rsc.org
Visit our website at www.rsc.org/books

Nanostructured Carbon Materials for Catalysis

Philippe Serp
Laboratoire de Chimie de Coordination UPR CNRS 8241, Composante ENSIACET, Université de Toulouse, Toulouse, France
Email: philippe.serp@ensiacet.fr

Bruno Machado
Laboratoire de Chimie de Coordination UPR CNRS 8241, Composante ENSIACET, Université de Toulouse, Toulouse, France
Email: bruno.machado@ensiacet.fr

RSC Catalysis Series No. 23

Print ISBN: 978-1-84973-909-2
PDF eISBN: 978-1-78262-256-7
ISSN: 1757-6725

A catalogue record for this book is available from the British Library

Published by The Royal Society of Chemistry,
Thomas Graham House, Science Park, Milton Road,
Cambridge CB4 0WF, UK

Registered Charity Number 207890

For further information see our web site at www.rsc.org

Printed in the United Kingdom by CPI Group (UK) Ltd, Croydon, CR0 4YY, UK

Foreword

Carbon plays a well-established and important role in catalysis for a wide range of applications, both as a support material and as catalyst in its own right. This role is growing for two independent but equally important reasons. Firstly, a number of emerging liquid-phase processes under demanding conditions and more specifically aqueous phase biomass conversion, call for a chemical stability of supports that surpasses that of metal oxide materials. Secondly, the almost explosive development of new carbon nanostructures over the last decades enables control of carbon materials at multiple length scales in very new ways. For a long period of time activated carbons, and carbon black to a lesser extent, have dominated catalysis applications. With the advanced control of nanostructured carbons in materials, *inter alia* carbon nanofibers, carbon nanotubes, carbon onions and most recently graphene, an unsurpassed control over carbon as catalyst or as support comes within reach. It seems, therefore, very timely that the current book authored by Bruno Machado and Philippe Serp appears. Both authors have contributed over the years in a very significant way to the utilization and scientific understanding of carbon nanomaterials in catalysis and can be considered as world-leading experts. What also makes this book special to me is its comprehensive nature. It starts from a description of the allotropes of carbon and ends with the engineering considerations of carbon utilization such as shaping, safety aspects and life cycle analysis. In between it discusses adsorption, surface chemistry, carbon as catalyst, emplacement of metals and grafting onto carbon supports, catalysis with supported metals for hydrogenation, and extending into oxidation catalysis, polymerization and photo-catalysis. Energy conversion and storage including carbon in electrodes and batteries is described as well. I consider

RSC Catalysis Series No. 23
Nanostructured Carbon Materials for Catalysis
By Philippe Serp and Bruno Machado

Published by the Royal Society of Chemistry, www.rsc.org

this book a 'must have' and even a 'must read' for many that work in the area of (carbon in) catalysis.

Krijn de Jong
Professor of Inorganic Chemistry and Catalysis
Utrecht University

Preface

Different types of carbon materials, such as activated carbon, graphite, and carbon black, have been successfully used over the last century as catalysts or catalyst supports in the chemical industry or for energy or environmental applications. Even though significant research and development has been made during this period concerning the knowledge of carbon surface chemistry, carbon materials were often considered as "complex and poorly understood" by the catalysis community. Low mechanical and thermal stability, together with narrow microporosity and inconsistent quality of activated carbons are usually pointed out as main disadvantages, limiting their applications. Pure graphite and carbon black often suffer from irregular pore structure and insufficient anchoring sites for the active phase. Highly competitive carbon materials for catalysis should therefore combine a controlled porosity and surface chemistry, and good thermal and mechanical properties with an acceptable stability and price. If such a material did not previously exist, the arrival of nanotechnologies in the 1980s, with the discovery of fullerenes in 1985, and the identification of carbon nanotubes in 1991 has permitted a huge boost of carbon chemistry and physics. Last but not least, the production, isolation, identification and characterization of graphene in 2010 have significantly contributed to the fact that nowadays carbon materials have achieved an important place in the scientific community. Yearly, a huge number of scientific publications are published on nanostructured carbon materials, and in the last thirty years a significant part of them is devoted to their use in catalysis.

Though the use of carbon materials for catalysis is nowadays a well-recognized field of research (with dedicated congresses and books) the field of nanostructured carbon materials for catalysis has undergone explosive growth in recent years. We have witnessed the emergence of many novel approaches to synthesis and synthetic design, control of size, morphology,

RSC Catalysis Series No. 23
Nanostructured Carbon Materials for Catalysis
By Philippe Serp and Bruno Machado

Published by the Royal Society of Chemistry, www.rsc.org

as well as novel applications of nanostructured catalytic carbon materials. Hence, we feel that there is a need to combine in a single platform the various properties of nanostructured carbon materials, particularly those related to catalysis such as surface chemistry and adsorption, the details of catalyst preparation, and how and where these novel materials are being used in catalysis. We believe that a fundamental understanding of the role of carbon surface properties should lead to a systematic procedure for the design of catalytic materials with improved performances. This is why we have introduced this book.

In the present book, we want to give the reader a comprehensive overview of what a nanostructured carbon is, and to rationalize the advantages of these catalytic materials regarding their activity, selectivity and stability. Thus, each chapter will provide a critical overview of a specific domain through relevant examples of the literature. In this sense, this book is the first to introduce the concepts and main achievements, while covering the main aspects of nanostructured carbon material for catalysis, both as a support and a metal-free catalyst. It is our hope that this book will not only prove suitable for self-study and teaching purposes, but will also inspire further research and discovery, thus setting the standard in the field of nanostructured carbon materials for catalysis for the years to come.

Philippe Serp
Bruno Machado
Toulouse

Contents

RSC Catalysis Series No. 23
Nanostructured Carbon Materials for Catalysis
By Philippe Serp and Bruno Machado

Published by the Royal Society of Chemistry, www.rsc.org

CHAPTER 1

Carbon (Nano)materials for Catalysis

1.1 Introduction

Both organic and inorganic carbons play a key role in catalysis. Organic molecules form the huge and very complex discipline of organic chemistry, and they are, in most catalytic applications, the substrates and the products of the process under consideration. In homogeneous catalysis, carbon is often the main constituent of the organic ligands surrounding the metallic center. In enzymatic catalysis it constitutes the backbone of the active species. In heterogeneous catalysis, carbon materials are unique catalyst supports, allowing the anchoring of the active phase, and can also act as catalysts or catalyst poisons (carbon deposits) by themselves.

The physical and chemical properties of carbon materials, such as their tunable porosity and surface chemistry, make them suitable for application in many catalytic processes. Traditionally, carbon materials have been used as supports for catalysts in heterogeneous catalytic processes, although their use as catalysts on their own is becoming more and more common.[1] Although several kinds of carbon materials have been studied, activated carbon (AC) and carbon black (CB) are the most commonly used carbon supports. The typically large surface area and high porosity of activated carbon catalysts favor the dispersion of the active phase over the support and increase its resistance to sintering at high metal loadings. The pore size distribution can be adjusted to suit the requirements of several reactions. The surface chemistry of carbon catalysts influences their performance as catalysts and catalyst supports. Carbon materials are normally hydrophobic and they usually show a low affinity towards

RSC Catalysis Series No. 23
Nanostructured Carbon Materials for Catalysis
By Philippe Serp and Bruno Machado

Published by the Royal Society of Chemistry, www.rsc.org

polar solvents, such as water, and a high affinity towards solvents such as acetone. Although their hydrophobic nature may affect the dispersion of the active phase over the carbon support, the surface chemistry of carbon materials can easily be modified, for example by oxidation, to increase their hydrophilicity and favor ionic exchange. Apart from an easily tailorable porous structure and surface chemistry, carbon materials present other advantages: (i) metals on the support can be easily reduced; (ii) the carbon structure is resistant to acids and bases; (iii) the structure is stable at high temperatures (even above 1023 K under inert atmosphere); (iv) porous carbon catalysts can be prepared in different physical forms, such as granules, cloth, fibers, pellets, *etc.*; (v) the active phase can be easily recovered; and (vi) the cost of conventional carbon supports is usually lower than that of other conventional supports, such as alumina and silica. Nevertheless, carbon supports also present some disadvantages: they can be easily gasified, which makes them difficult to use in high temperature hydrogenation and oxidation reactions, and their reproducibility can be poor, especially activated carbon-based catalysts, since different batches of the same material can contain varying ash amounts. In this introductory chapter, we will: (i) briefly introduce the main carbon and graphite (nano)materials relevant to catalysis, (ii) present the main application of carbon and graphite materials in catalysis, and (iii) highlight the possible perspectives of using nanocarbons in catalysis.

1.2 Carbon (Nano)materials

The capability of a chemical element to combine its atoms to form such polymorphs is not unique to carbon. Other elements in the fourth column of the periodic table (silicon, germanium, and tin) also have this characteristic. However carbon is unique in the number and the variety of its polymorphs. These allotropes are composed entirely of carbon but have different physical structures and, exclusively for carbon, have different names: graphite, diamond, lonsdaleite, and fullerene, among others. Additionally, carbon as a solid denotes all natural and synthetic substances consisting mainly of atoms of the element carbon, such as single crystals of diamond and graphite, as well as the full variety of carbon and graphite materials. A result of this diversity is that the carbon terminology can be confusing for the non-specialist. The terminology used so far is mainly based on technological tradition and on the standardized characterization methods derived from decades of industrial experience. As a consequence, for many years, carbon science was a very specialized field, considered by many to be too complicated. More recently, carbon science has gained high visibility with the discovery of fullerenes in 1985 and the first HR-TEM observations of carbon nanotubes (CNTs) in 1991. This visibility has been further heightened by the 1996 Nobel Prize in Chemistry awarded to R. F. Curl, H. Kroto and R. E. Smalley for their discovery of fullerenes and the 2010 Nobel Prize in Physics awarded to A. Geim and

K. Novoselov for ground-breaking experiments regarding the two-dimensional material graphene (denoted SG for single layer graphene). Because of the increasing interdisciplinary importance of this group of materials in science and technology, it is obvious that clear definitions of the corresponding terms are required. In order to clarify the terminology, we will refer to the recommended terminology for the description of carbon as a solid (IUPAC Recommendations 1995). In the following part of this chapter we will briefly review the main carbon and graphite (nano)materials relevant to catalysis and their production processes.

1.2.1 Activated Carbons

The term activated carbon (also known as activated charcoal) defines a group of materials with highly developed internal surface area and porosity, and hence a large capacity for adsorbing chemicals from gases and liquids.[2] The adsorption on the surface is essentially due to Van der Waals or London dispersion forces. This force is strong over short distances, equal between all carbon atoms and not dependent on external parameters such as pressure or temperature. Thus, adsorbed molecules will be held most strongly where they are surrounded by the most carbon atoms. The area presenting a high density of graphitic basal structural units will favor a high adsorption. High temperature treatment (>1500 K) of AC can favor the adsorption sites by increasing the density of "π-sites" present on partly graphitized structure.[3,4] Almost all precursors containing a high fixed carbon content can potentially be activated. The most commonly used raw materials are coal, coconut shells, wood (both soft and hard),[5] peat and petroleum based residues[6,7] or agricultural residues.[8,9] Most carbonaceous materials do have a certain degree of porosity and an internal surface area in the range of 10–15 $m^2\ g^{-1}$. There are many activation methods to produce an AC that can basically be categorized into two: physical and chemical activation. In the same way, the manufacture of AC involves two main stages, the carbonization of the starting material and the activation of the resulting char. The choice of the activation method is dependent upon the starting material and whether a low or high density, powdered or granular carbon is desired.

Carbonization. The first step is a thermal treatment of the raw material that implies dehydration (eqn (1.1)) and where most of the non-carbon elements, such as dust and volatile substances, are eliminated by heating the source under anaerobic conditions. The aim of the carbonization stage is to conserve the carbonaceous structure of the material, which is achieved by burning off the material at a range of temperatures from 673 to 1123 K. The char is constituted when carbon atoms regroup themselves into sheets forming rigid and dense clusters of microcrystals, each one consisting of several layers of graphitic planes. Each atom inside one stack is bonded to four adjacent carbon atoms. Thus, the carbon atoms on the edges of the planes have a high adsorption potential available. The internal structure is neither homogeneous nor regular; leaving free interstices that constitute the porosity of the char. These interstices may be filled or blocked by disorganized carbon resulting from deposition and decomposition of tars making them not always accessible and reducing the porosity.

$$C_x(H_2O)_y \rightarrow C(s) + yH_2O \textit{ Carbonization} \tag{1.1}$$

There are three stages in the carbonization process. First of all, is the loss of water in the 373–473 K range. The second stage is the primary pyrolysis, which takes place in the 473–773 K temperature range and is characterized by a large generation of gases and tars due to the elimination of volatile matter and tars, causing a remarkable reduction of weight. This also forms the basic structure of the char. The third step is the consolidation of char structure in the range of temperatures from 773 to 1123 K, with a very small weight loss. The small reduction in bulk density with increasing carbonization temperature, coupled with the large reduction in weight (especially in the second stage) implies a contraction of the precursor material; such contraction continues above 773 K, since the density slightly increases in the third stage temperature range.

Physical activation. The objective of the activation process is to enhance the pore structure. The physical activation is the partial gasification of the char with steam, carbon dioxide and air, or a mixture of these, at temperatures around 1073–1473 K (lower for air). Once the non-desirable materials have been removed, the char is accessible. The main change is an increase in pore volume exposing the crystallites to the action of the activating agent (oxidizing gases) for further development of porosity with increasing burn-off. The oxidizing gases employed, steam or CO_2, are reactive agents that react with the raw material and also remove volatile material from the solid (eqn (1.2) and (1.3). For a given temperature, the reactivity with steam is larger than that with carbon dioxide. Many authors have shown that the most important variables in the gasification process from the point of view of porosity development are:

- the choice of activating agent
- the final burn-off temperature reached
- the presence of inorganic impurities that catalyze or inhibit the gasification reaction.

$$C(s) + 2H_2O \rightarrow CO_2 + 2H_2 \textit{ Steam activation} \left(\Delta H = +75 \text{ kJ mol}^{-1}\right) \tag{1.2}$$

$$C(s) + CO_2 \rightarrow 2CO \textit{ Activation by } CO_2 \left(\Delta H = +159 \text{ kJ mol}^{-1}\right) \tag{1.3}$$

Chemical activation. During chemical activation, carbonization and activation are accomplished in a single step by carrying out thermal decomposition of the raw material impregnated with certain chemical agents such as H_3PO_4, H_2SO_4, HNO_3, NaOH, KOH or $ZnCl_2$. In this way, a carbonized product with a well-developed porosity may be obtained in a single operation. The activating agents employed function as dehydrating agents that influence pyrolytic decomposition, inhibiting the formation of tar and thereby enhancing the yield of carbon. The yield and properties of AC depend on the impregnation conditions, such as impregnation ratio (weight of activating reagent/ weight of carbon precursor), time of pre-drying of impregnated materials, as well as pyrolysis conditions, such as temperature, soaking time (period of time that the sample and chemical are in contact) and atmosphere. All these process variables vary with the type of carbon precursor and the activating agent. The temperatures used in chemical activation are lower than that used in the physical activation process. As a result, the development of a porous structure is better controlled in the case of chemical activation. In spite of the abovementioned advantage, the chemical method has its own inherent drawbacks, such as the need for washing of the product to remove the residual inorganic material.

After activation, the carbon will have acquired an internal surface area between 700 and 1200 $m^2\ g^{-1}$, depending on the processing conditions. The internal surface area must be accessible to the passage of reactants for adsorption. Thus, it is necessary that an AC has not only a highly developed internal surface but accessibility to that surface *via* a network of pores of different diameters. All ACs contain micropores, mesopores, and macropores within their structures but the relative proportions vary considerably according to the raw material. In general, it can be said that macropores are of little value in their surface area, except for the adsorption of unusually large molecules and are, therefore, usually considered as an access point to micropores. Mesopores do not generally play a large role in adsorption, except in particular carbons where the surface area attributable to such pores is appreciable (usually 400 $m^2\ g^{-1}$ or more). Thus, it is the micropore structure of an activated carbon that plays an effective role in adsorption. It is, therefore, important that activated carbon is not classified as a single product but rather a range of products suitable for a variety of specific applications. The ash content is an important physical characteristic of AC. This is the inorganic, commonly inert, amorphous and unusable part present in the activated carbon. This ash comes initially from the raw material. The lower the ash content, the better the activated carbon. The practical limit for the level of ash content allowed in the activated carbon varies from 2 to 5 wt%. Since these impurities on the AC surface often result in undesirable side reactions, AC will be preferably derived from raw materials with a high degree of purity. In those cases where additional purity is required, AC can be acid washed before use. Other important physical characteristics of AC with respect to catalysis are: the hardness, the apparent density, the moisture, the pH_{PZC}† value and the particle size distribution. Powdered activated carbon is made up of crushed or ground carbon particles, 95–100% of which will pass through a designated mesh sieve or sieves. Granular activated carbon can be either in the granular form or extruded. The use of AC as a catalyst by itself dates from more than one century ago, when G. Lemoine first reported that carbon is an effective catalyst for the decomposition of hydrogen peroxide solutions, and for the decomposition of ethanol in acetaldehyde and dihydrogen.[10] This subject has been recently reviewed by Figueiredo *et al.*[11] and it is briefly discussed in Chapters 5 and 6. The use of AC as a support for catalysis has been the subject of several books and review articles[12–15], and it is also discussed in Chapters 7, 8 and 9.

1.2.2 Graphite and Graphitized Materials

Hardly any of the vast amounts of natural or manufactured graphite is currently used in catalyst supports. This is mainly due to the low surface area of graphite of only 10–50 $m^2 g^{-1}$. Additionally, the relatively low reaction ability of natural graphite towards the activation agents (steam, oxygen, carbon dioxide) can create problems in the preparation of porous carbon supports. But graphite materials are able to form graphite intercalated compounds, which can sufficiently (by hundreds of times) increase their volume after a high temperature treatment, allowing a significant expansion of the material along the crystallographic *c*-axis to occur.[16] This unique property of intercalated graphite was used for manufacturing thermally expanded graphite by chemical and electrochemical methods with sulfuric and nitric acids.[17] Such a material, developing a specific surface area of 100–300 $m^2 g^{-1}$ has been used as a catalyst support.[18–21] High surface area graphite can also be obtained by high-energy ball-milling with an increase of surface area from 10 to 600 $m^2 g^{-1}$.[22] High surface area graphite (HSAG) consists of crystalline graphite structures of small particle sizes, with medium surface areas and whose mesopores are the principal contribution to their porosity. These materials also exhibit high reactivity, specifically because of the unsaturated valences at the edges of the graphitic layers. They have been mainly used as supports for catalysts for various reactions, including the oxygen reduction reaction in fuel cells,[23] NO reduction,[24] wet air oxidation,[25] hydrogenations,[26] hydrodechlorination[27] and decomposition of NH_3.[28]

†The existence of surface functional groups on carbon materials cause both negatively and positively charged surface sites to exist in aqueous solution, depending on the pH. At some pH, the carbon surface charge will be zero; called the point of zero charge (pH_{PZC}). At $pH_{PZC} < pH$, the carbon surface becomes negatively charge, favorable for the adsorption of cationic substances. Adsorption of anionic substances will be favored at $pH_{PZC} > pH$.

1.2.3 Glass-like Carbon

Glass-like (also called glassy or vitreous) carbon derives its name from exhibiting fracture behavior similar to glass, from having a disordered structure over large dimensions (although it contains a graphitic microcrystalline structure), and because it is a hard shiny material, capable of being highly polished. Glass-like carbon is an example of a non-graphitizing carbon, which is a carbon that cannot be transformed into crystalline graphite even at temperatures of 3273 K and above. Glass-like carbon is a macroisotropic microcrystalline solid, with a density approximately 60% of that for monocrystalline graphite and thus suggests high porosity. However, the practically complete absence of opened pores causes its chemical inertness even to active oxidizers. The structure of glass-like carbon has long been a subject of debate.[29,30] Early structural models assumed that both sp^2- and sp^3-bonded atoms were present, but it is now known that glass-like carbon is 100% sp^2 and contains a high proportion of fullerene related structures.[31] While certain ambiguities still exist about the precise molecular structure of glass-like carbon, this model does take into account the relatively low density of glass-like carbon, *ca.* 1.5 g cm^{-3} compared with 2.3 g cm^{-3} for graphite (which suggests the existence of voids), the impermeability of glass-like carbon to gases (which suggest that these voids are not connected), and the isotropic conductivity observed for this material. Glass-like carbon is the most important carbon material for use as an electrode.[32] It is particularly useful in electrochemical applications because of its low electrical resistivity, impermeability to gases, high thermal stability, and because it has the widest potential range observed for carbon electrodes. In addition to very high thermal stability, its distinguishing properties include its extreme resistance to chemical attack: it has been demonstrated that the rates of oxidation of glassy carbon in oxygen, carbon dioxide or water vapor are lower than those of any other carbon. It is also highly resistant to attack by acids. While normal graphite is reduced to a powder by a mixture of concentrated sulfuric and nitric acids at room temperature, glassy carbon is unaffected by such treatment.

The current approach used for the conventional preparation of glassy carbon solids involves the careful pyrolysis of one of a variety of polymeric materials including poly(vinyl chloride), poly(vinylidene chloride), cellulose, phenolic resin, poly(furfuryl alcohol), and poly(acrylonitrile). The processing requirements for the preparation of glassy carbon are quite demanding.[33] The chemical transformations that occur on the formation of glassy carbon are, of course, dependent on the starting polymer. In general, thermal treatment to *ca.* 773 K results in the loss of small molecules, such as carbon monoxide, carbon dioxide, hydrogen chloride, hydrogen cyanide, water, and small hydrocarbons. Above 973 K, the loss of hydrogen results in the formation of carbon–carbon bonds and a highly cross-linked structure. Heat treatments up to 2273–3273 K are usually performed to produce the final material. The extreme thermal requirements for these syntheses yield glassy carbon with

<0.5% of any element other than carbon, independent of the precursor material. Reticulated vitreous carbon is an open-pore foam material of honeycomb structure composed solely of vitreous carbon. The structure of reticulated vitreous carbon is achieved by polymerization of a resin combined with foaming agents, followed by carbonization. The result is a low volume disordered glassy porous carbon with some crystallographic order, low electrical resistance and a continuous skeletal structure.[34] The solid-phase transformation of metal-containing glass-like carbon nanoparticles into nanotubes by heating at temperatures of 1073–1273 K has been reported.[35]

Glass-like carbons have been used as electrocatalyst by themselves,[36] and as support for both noble and non-noble metal electrocatalysts.[37–40]

1.2.4 Carbon Nanostructures

1.2.4.1 Carbon Blacks

Carbon blacks, a very pure form of soot, constitute a group of materials that are characterized by having near spherical carbon particles of colloidal size, which are produced by four different processes:

- Furnace black process, in which aromatic oils (based on crude oil) are burned in a reactor, producing CB and tail-gas. After cooling, the carbon black is separated from the tail-gas, densified, and processed into pellets.
- Thermal black process, in which natural gas is burned in a reactor to produce CB and tail-gas, and is then processed in a fashion similar to the furnace black process.
- Acetylene black process, which is similar to the thermal black process, except that acetylene is the raw material used.
- Lampblack process, in which an aromatic oil (based on coal tar) is heated in a flat cast-iron pan to produce CB.

Over 90–95% of carbon black produced is of the furnace type and thermal black follows at a distant second place. Commercial carbon black materials generally contain more than 97% elemental carbon with variable amounts of oxygen, hydrogen and sulfur. The different carbon black materials are characterized by their primary particles size, their aggregation and agglomeration status as well as their impurity profile.[41]

The products made by each process have unique characteristics. During production, the colloidal carbon particles coalesce into chemically fused aggregates and agglomerates (groups of aggregates) with varying morphologies. Their fundamental properties vary with feedstock and manufacturing conditions, and they are usually classified according to their method of preparation or intended application. The key properties for CB are considered to be fineness (primary particle size), structure (aggregate size/shape), porosity, and surface chemistry. Typically, the average primary particle diameter of commercial CB materials ranges from 10 to 100 nm, while the average aggregate/agglomerate size is in the range 100–800 nm or above. Carbon black is initially formed as roughly spherical primary particles, which, in most cases, rapidly form aggregates. An aggregate is a chain of primary carbon particles that are permanently fused together in a random branching structure. The aggregate may consist of between a few to hundreds of spherical particles (or, as in the case of thermal black, primarily single spheres rather than chains). Accordingly, on the basis of their primary particle size, all CB materials are considered as nano-structured materials. It is believed that carbon black particles form in three distinct phases. The first stage, known as particle inception, involves homogeneous reactions between hydrocarbon species that combine into larger aromatic layers and eventually condense out of the vapor phase to form nuclei. In the second stage, called growth, two processes occur: nuclei coalescence and surface deposition, the latter process being responsible for most of the mass increase of the primary particles. Finally, in the chain formation stage, relatively large spheroidal particles become joined together, without coalescing, to form long chains. Many aspects of this mechanism are poorly understood, in particular the nature of the initial nucleus.

The longstanding problem of the detailed structure of CB particles remains unresolved.[42] A number of points can be made with reasonable confidence, however. Although it is true that the basic structural units that make up the particles are arranged concentrically around the center, HRTEM images suggest that the units are more extensive, and more defective, than the rather perfect fragments proposed in some models. Detailed TEM studies show that fullerene-like structures are probably present, but are not responsible for the sphericity of the particles.

The BET surface-area of CB covers a wide range *i.e.*, from few tens for acetylene or thermal blacks to greater than 1000 $m^2\ g^{-1}$. The porosity in CB also varies from mild surface pitting to the actual hollowing out of particles. Additional porosity is also created by the intra- and inter-aggregate voids that are formed between the small, fused, primary carbon particles (see Figure 2.3 in Chapter 2) . The surface area of CB is generally considered as being more accessible than other forms of high surface-area carbon.

Graphitized carbon black is another support material of interest to catalyst manufactures. This high surface material (Figure 1.1) is obtained by recrystallization of the spherical CB particles at 2773–3273 K.

The partially crystallized material possesses well-ordered domains. The degree of graphitization is determined by process temperature. Highly conductive CBs are characterized by a high structure (*i.e.*, aggregates with a highly branched, open structure), high porosity, small particle size, and a chemically clean (oxygen free) surface. The conductivity of carbon blacks is typically in the range 10^{-1} to 10^2 $(\Omega\ cm)^{-1}$ and is influenced by the relative ability of electrons to jump the gap between closely-spaced aggregates (electron tunneling) and by graphitic conduction *via* touching aggregates. CBs are conventional support for fuel cells electrocatalysis.[43,44]

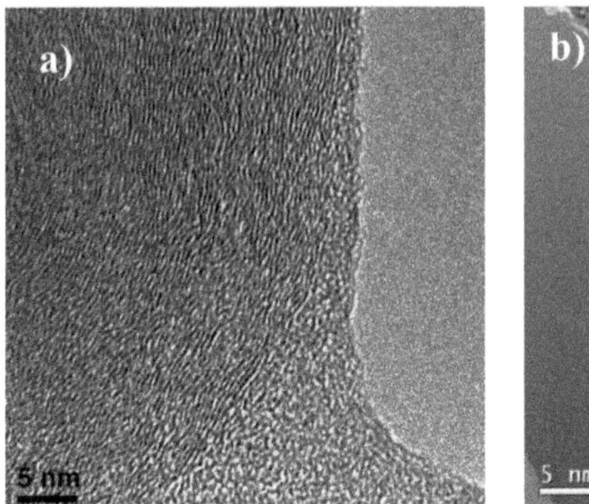

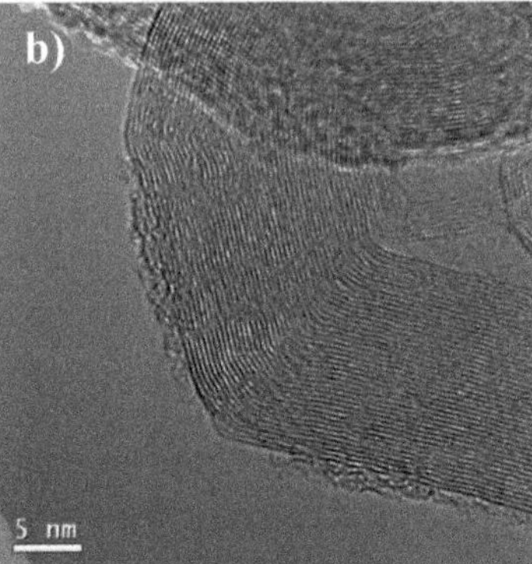

Figure 1.1 Conventional HRTEM micrographs of (a) untreated CB particles; and (b) CB particles heat treated at 2973 K.

1.2.4.2 Carbon Nanotubes and Nanofibers

Carbon nanotubes (CNTs) and nanofibers (CNFs) are classically produced by the catalytic decomposition of certain hydrocarbons.[45] By careful manipulation of various parameters, it is possible to generate nanostructures in assorted conformations and also to control their crystalline order (Figure 1.2).

CNTs can be either single-walled (SWCNTs) or multi-walled (MWCNTs). Most SWCNTs have a diameter of close to 1 nanometre, with a tube length that can be many millions of times longer. The structure of a SWCNT can be conceptualized by wrapping a one-atom-thick layer of graphite called graphene into a seamless cylinder (Figure 1.3). The way the graphene sheet is wrapped is represented by a pair of indices (n,m). The integers n and m denote the number of unit vectors ($\mathbf{a}_1$ and $\mathbf{a}_2$) along two directions in the honeycomb crystal lattice of graphene. Thus, the crystalline structure of SWCNTs can be characterized by the (chiral) roll-up vector **R** within a corresponding graphite sheet plane, where $\mathbf{R} = m\mathbf{a}_1 + n\mathbf{a}_2$, with vectors $\mathbf{a}_1$ and $\mathbf{a}_2$ describing the base unit of the graphite sheet plane. The roll-up vector results from the difference between the two grid points of the graphite lattice that coincide in a single grid point of the nanotube on the imaginary roll-up. If $m = 0$, the nanotubes are called zigzag nanotubes, and if $n = m$, the nanotubes are called armchair nanotubes. Otherwise, they are (incorrectly) called chiral. For a given (n,m) nanotube, if $n = m$, the nanotube is metallic; if $n - m$ is a multiple of 3, then the nanotube is semiconducting with a very small band gap, otherwise the nanotube is a moderate semiconductor.

MWCNTs consist of multiple rolled layers (concentric tubes) of graphene. The interlayer distance in MWCNTs is close to the distance between graphene layers in graphite, approximately 3.4 Å. Each individual shell can be described as a SWCNT, which can be metallic or semiconducting. Because of statistical probability and restrictions on the relative diameters of the individual tubes, one of the shells, and thus the whole MWCNT, is usually a zero-gap metal.

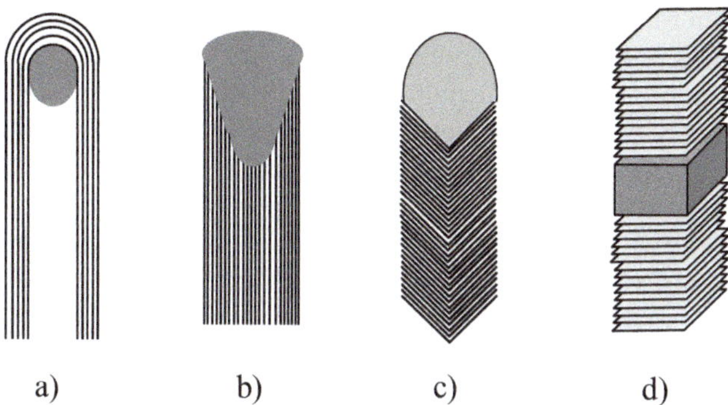

Figure 1.2 Different carbon nanostructures produced by catalytic chemical vapor deposition: (a) multi-walled carbon nanotubes; (b) ribbon-carbon nanofibers (r-CNFs); (c) fishbone-carbon nanofibers (h-CNFs); and (d) platelet-carbon nanofibers (p-CNFs).

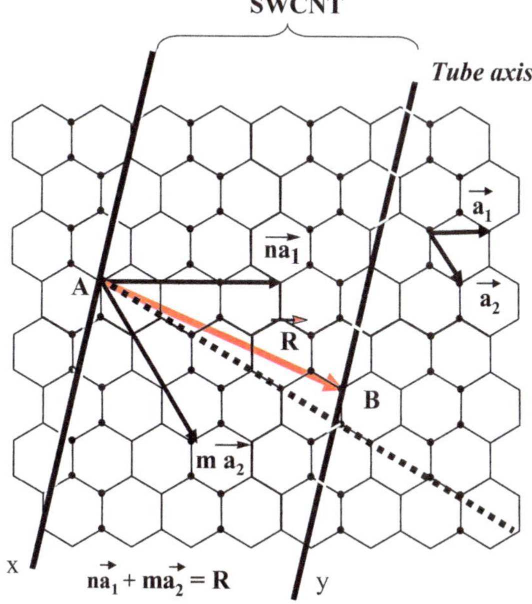

Figure 1.3 Structure of a SWCNT from its chiral vector.

There are generally three types of CNFs (Figure 1.2): the herringbone (h-CNFs) or fishbone, in which the graphene layers are stacked obliquely with respect to the fiber axis; the platelet (p-CNFs), in which the graphene layers are perpendicular to the fiber axis; and the ribbon (r-CNFs), in which the graphene layers are parallel to the growth axis.[46] The specific surface area and porosity of CNFs can be significantly modified by an activation process that removes the most reactive carbon atoms from the structure, increasing the surface area and porosity.

Recently, there is an increasing interest in the application of CNTs or carbon nanofibers (CNFs) as supports for catalysis since the nanoscale tubular morphology of these materials can offer a unique combination of low electrical resistivity and high porosity in a readily accessible structure.[47,48] Thus, CNTs represent an interesting alternative to conventional supports for a number of reasons,[49–52] including: (i) their high purity that eliminates self-poisoning; (ii) their impressive mechanical properties, high electrical conductivity and thermal stability; (iii) the high accessibility of the active phase and the absence of any microporosity (for MWCNT and CNFs), thus eliminating diffusion and intra-particle mass transfer in the reaction media; (iv) the possibility for macroscopic shaping of the support; (v) the possibility of tuning the specific metal–support interactions, which can directly affect the catalytic activity and selectivity; and (vi) the possibility of confinement effects in their inner cavity. Additionally, compared to conventional supports, CNTs have a high flexibility for the dispersion of the active phase since it is possible to: (i) modulate their specific surface area (10–250 $m^2\ g^{-1}$ for CNFs, 50–500 $m^2\ g^{-1}$ for MWCNTs and 400–900 $m^2\ g^{-1}$ for SWCNTs) or their internal diameter (5–100 nm for MWCNTs); (ii) easily chemically functionalize their surfaces; (iii) change their chemical composition (*e.g.* nitrogen- or boron-doped CNTs); and (iv) deposit the catalytic phase either on their external surface or in their inner cavity.

1.2.4.3 Graphene and Few-layer Graphene

Graphene is a material consisting of a 2-D layer of sp^2 hybridized carbon atoms, the parent material of CNTs. It can be produced by micromechanical cleavage of graphite, liquid phase exfoliation resulting in dispersed graphene flakes, or chemical vapor deposition. From a theoretical point of view, it provides the ultimate two-dimensional model of a catalytic support. Its unique physical, chemical and mechanical properties are outstanding, and could allow the preparation of composite-materials with unprecedented characteristics. Even though the use of a single graphene (SG) sheet as a catalytic support has not yet been reported, some promising results have already been obtained with few-layer graphene (FLG).[53] Chemical oxidation of graphite results in exfoliated sheets, called graphene oxide (GO). The utilization of graphene-based materials as a two-dimensional catalyst support with the possibility to harness its redox properties opens up new opportunities for designing next-generation catalysts.[54–56] Incorporation of catalyst particles onto SG or FLG with a good distribution can provide great versatility in carrying out catalytic processes, and is a new member of the carbon-supported catalysts that currently attract special research efforts.

Nanodiamonds. The nanodiamonds also referred to as ultra-disperse diamonds are particles in the 2–5 nm size range. Nanoscale diamond particles were first produced by detonation in the USSR in the 1960s, but they remained essentially unknown to the rest of the world until the end of the

1980s. Typically, to synthesize nanodiamonds, explosives with a negative oxygen balance are detonated in a closed metallic chamber in an atmosphere of N_2, CO_2 and liquid or solid H_2O. After detonation, diamond-containing soot is collected from the bottom and the walls of the chamber. During detonation, the pressure and temperature rise instantaneously, reaching a point that falls within the region of liquid carbon clusters of 1–2 nm in size. As the temperature and pressure decrease along the isentrope, carbon atoms condense into nanoclusters, which further coalesce into larger liquid droplets and crystallize. When the pressure drops below the diamond–graphite equilibrium line, the growth of diamond is replaced by the formation of graphite. Today, nanodiamonds can be produced by a variety of techniques including the detonation technique, laser ablation, high-energy ball milling of high-pressure high-temperature (HPHT) diamond microcrystals, plasma-assisted chemical vapor deposition (CVD), autoclave synthesis from supercritical fluids, chlorination of carbides, ion irradiation of graphite, electron irradiation of carbon 'onions' and ultrasound cavitation, with the first three of these methods being used commercially.[57] While diamond is metastable relative to graphite under atmospheric pressure, at the nanoscale, diamond particles less than 5 nm are more stable than graphite. For example, Barnard compared the energies of graphite, diamond, and fullerene on the basis of density functional theory to conclude that diamond could be the stable phase of carbon clusters in the size range of 1.9–5.2 nm.[58] Nanodiamond is often described as a crystalline diamond core with a perfect diamond lattice surrounded by an amorphous shell with a combination of sp^2/sp^3 bonds or onion-like graphite shell. The surface of sp^3 clusters must be either stabilized through termination with functional groups or reconstructed into sp^2 carbon. Therefore, in addition to size and shape, the stability of carbon nanoparticles also depends on their surface terminations. Even though theoretical studies to date have considered both spherical and polyhedral nanodiamond particles, high resolution transmission electron microscope (HRTEM) images of nanodiamond have confirmed the latter. Barnard *et al.* using first principles computer simulations, have shown how the polyhedral shape affects the stability of nanodiamond particles.[58] In particular, they considered nanodiamond particles of octahedral, cuboctahedral, and cubic morphologies up to 2 nm in diameter and discussed the shape dependent stability of nanodiamond particles. Today, although boron-doped diamond has been used by many researchers as a conductive support in electrocatalysis, especially in environmental applications,[59] few studies are devoted to the use of this material in catalysis.[60–65]

1.2.4.4 Carbon Onions

Carbon onions are multilayered spherical nanostructures made up of nested fullerene cages.[66] They were discovered in 1992 by Ugarte,[67] and until quite recently they were seldom studied, mainly because of their unavailability and lack of convenient methods for their preparation and separation. By focusing an electron beam on a sample of amorphous carbon, Ugarte was able to observe the formation of carbon onion *in situ*. Under the electron beam, the amorphous carbon graphitizes and begins to curl, and after sufficient time, the graphitic carbon closes on itself, forming an onion. The curving and closure occurs in order to minimize the surface energy of the newly formed edge planes of graphite, which is about 30 × that of the basal plane. A variety of methods have been used to make carbon onions, including vacuum annealing of nanodiamonds, graphite arc discharge underwater, flash pyrolysis, ion implantation into copper and silver, vapor deposition, stainless steel autoclave reactions, and laser irradiation. Most of these methods lead to low yields and/or many by-products.[68] Given the easier functionalization and higher thermal stability of the carbon onions prepared from nanodiamonds, they are the most obvious choice for studying the potential applications of these multi-shelled fullerene structures. This class of nanocarbons has been referred to by multiple names including carbon onions, carbon nano-onions, multilayer fullerenes, onion-like fullerenes and onion-like carbon (OLC). Those names cover all kinds of concentric shells, from nested fullerenes to small (<100 nm) polyhedral nanostructures. Recently, the transformation of detonation nanodiamond particles into nested fullerene-like carbon structures at temperatures above 800 K has been explored as a way to make the nanographitic structures. In the literature, these multilayered cages obtained by annealing of detonation nanodiamond have been referred to as OLC, mainly to distinguish them from the ideal carbon onions consisting of layers of enclosed fullerene molecules of different sizes.[69] Carbon onions represent one of the least studied carbon nanomaterials. However, because of their unique 0-D structure, small (<10 nm) diameter, high electrical conductivity, and relatively easy dispersion compared to 1-D nanotubes and 2-D graphene, OLCs are seeing a large increase in attention for energy storage applications.[68] As far as the fullerene family is concerned, some attempts to use them as catalyst support or metal free catalyst have been reported.[70–74]

1.3 Carbon Materials as Catalysts

1.3.1 Reactions

The catalytic application of carbon materials can be traced back to the use of AC in the treatment of waste water and gas. Today, most of the reactions that are catalyzed by carbon catalysts can be classified into one of the following groups: (i) oxidation-reduction; (ii) hydrogenation-dehydrogenation; (iii) combination with halogens; and (iv) decomposition.[11] There are also examples of the catalysis of dehydration, isomerisation and polymerisation reactions. Some of the most relevant reactions catalyzed by carbon catalysts are summarized in Table 1.1.

It has been proved industrially that AC displays good catalytic performance in the dechlorination and desulfation of the waste gases of industrial processes. AC catalysts have been used for a long time in the production

Table 1.1 Reactions catalyzed by carbon catalysts.

General classification	Example
Oxidation–reduction	$SO_2 + ½O_2 \rightarrow SO_3$
	$NO + ½O_2 \rightarrow NO_2$
	deNOx
	$2H_2S + O_2 \rightarrow S_2 + 2H_2O$
	$C_6H_5C_2H_5 + ½O_2 \rightarrow C_6H_5C_2H_3 + H_2O$
	Toxin oxidation (creatinine)
	Oxidation of industrial effluents (oxalic acid)
Hydrogenation–dehydrogenation	$RX + H_2 \rightarrow RH + HX$ (X = Cl, Br)
	$HCOOH \rightarrow CO_2 + H_2$
	$CH_3CHOHCH_3 \rightarrow CH_3COCH_3 + H_2$
Combination with halogens	$H_2 + Br_2 \rightarrow 2HBr$
	$CO + Cl_2 \rightarrow COCl_2$ (phosgene)
	$C_2H_4 + 5Cl_2 \rightarrow C_2Cl_6 + 4HCl$
	$SO_2 + Cl_2 \rightarrow SO_2Cl_2$
	$C_6H_5CH_3 + Cl_2 \rightarrow C_6H_5CH_2Cl + HCl$
Decomposition	$2H_2O_2 \rightarrow 2H_2O + O_2$
	$CH_4 \rightarrow C + 2H_2$
Dehydration, isomerization, and polymerisation	$HCOOH \rightarrow H_2O + CO$
	$3C_2H_2 \rightarrow C_6H_6$
	α-Olefins → poly(α-olefins)
	α-Oxime → β-oxime
Emerging applications	Oxygen reduction reaction in fuel cells[77–80] or for water oxidation[81]
	Carbon Molecular sieves for shape selectivity reactions[82]
	Catalytic wet air oxidation of organic pollutants[83]
	Gasification of organics and biomass[84]
	CO_2 reforming of methane[85]
	Methylamines synthesis[86]

of phosgene[75] and sulfur halides.[76] Anhydrous chlorine gas is reacted with high-purity CO in the presence of an AC catalyst producing phosgene, some unwanted by-products, and considerable heat. The production process is continuous with the raw materials carefully metered and excess heat removed.

The production of SO_2Cl_2 by the reaction of chlorine with SO_2 in the presence of a carbon catalyst is a well-known process. Typically, chlorine and sulfur dioxide are dissolved in a solvent as sulfuryl chloride before contact with the catalyst. Thionyl chloride is typically made by treating SO_2 with chlorine and sulfur dichloride in the presence of a carbon catalyst. Carbon catalysts are known to degrade during such processes. A variety of AC has also been found to catalyze a highly selective reaction between phosgene and formaldehyde to produce dichloromethane and CO_2.[87] Another important industrial application of carbon catalysts is in fuel gas cleaning. The dry desulfurization, denitrification and air toxic removal processes using activated carbon were originally researched and developed during the 1960s by Bergbau Forschung in Germany. In these processes, the active carbon acts simultaneously as an adsorbent and as a catalyst in the temperature range 383–443 K, *i.e.*, under conditions where the material is stable in the presence of oxygen.[88] From 1992, Mitsui Mining developed a technology to produce the activated carbon used in the dry $DeSO_x$/$DeNO_x$/Air Toxic removal process based on their own metallurgical coke manufacturing technology. These tests have proven the system's capability by removing over 99% of the SO_x. Another important use of carbon catalysts for environmental cleaning is the removal of halogen from halogen-containing compounds. Two basic approaches can be used: (i) gas phase catalyzed oxidation of the halogen-containing compounds to carbon dioxide and the corresponding halogen-containing acid[89] and (ii) catalytic dehalogenation.[90] Activated carbon is also used as an oxidation catalyst during the production of glyphosate, an easy degradable herbicide that is produced through oxidative decarboxylation of *n*-phosphonomethyliminodiacetic acid. This reaction, taking place in liquid phase at a temperature higher than 353 K, involves air as the oxidant and a specific AC as catalyst, giving high conversion yields.

The oxidative dehydrogenation of ethylbenzene (ODE), an exothermic reaction, is an alternative route to the production of styrene that has the main advantage the fact that conversions are not limited by equilibrium and can be carried out at lower temperatures (623–723 K). During the last three decades, a group of catalysts was reported, such as alumina and various metal oxides and phosphates, that showed an activity and selectivity in the ODE to styrene comparable to the iron catalyst. Evidence was gradually accumulated that the active sites were not located on the initial catalyst surface, but on a carbonaceous over-layer. This carbonaceous over-layer is initially deposited on the surface, and it is one reason why the catalysts exhibited induction periods in their activities. Thus, a large variety of carbon or graphite materials,[1,91] including nitrogen-doped ones[92] have been studied in the literature for ODE. It is noteworthy that most of the effort into oxidative dehydrogenation reactions was focussed on the conversion of ethylbenzene, and only few works have looked at the conversion of alkanes.[93] One possible reason is that the intermediate products obtained in the ODH of ethylbenzene is much more stable than those in the ODH of alkanes, as the radical is stabilized by the delocalized π bonds. The catalytic performance of various carbon catalysts are shown in Table 1.2. These and other catalytic applications are reviewed in more detail in Chapters 5 and 6.

1.3.2 Influence of Carbon Properties on Catalysis

In the late 1960s, R. W. Coughlin suggested in a landmark paper that the catalytic activity and selectivity of carbon catalysts was related to their surface chemistry and electronic properties.[103] Indeed, carbon materials can exhibit the properties of a conductor, semi-conductor or insulator, depending on the methods of pre-treatment and preparation. By applying the right treatment, the catalytic properties of carbon can be adjusted to suit a specific application. Thermal and graphitization treatments would favor metallic behavior;

Table 1.2 Catalytic performances of carbon catalysts in the oxidative dehydrogenation of light alkanes.

Catalysts	Reactants	Products	T/K	Conv. (%)	Selec. (%)	References
Coal	Butane	Butene, butadiene	973	~40	~17	94
CNTs	Propane	Propene	773	42	40	95
CNTs	Propane	Propene	673	6	25	96
P-doped CNTs	Butane	Butene	723	20	60	97,98
N-doped CNTs	Propane	Propene	673	10	20–60	92
AC	*iso*-Butane	*iso*-Butene	648	25	60	99
P-doped carbon	*iso*-Butane	*iso*-Butene	673	2.1	90	100
Fullerene-like carbon	*iso*-Butane	*iso*-Butene	673	0.3–10	50–80	101
FLG	*iso*-Butane	*iso*-Butene	673	1–4	40–60	102

oxidation would tend to localize π-electrons and lead to semi conductivity; and other treatments can result in highly disordered carbon with insulator-type behavior. Moreover, in view of the strong electronic anisotropy of graphitic carbon, a broad spectrum of crystalline properties can be possible within a given carbon catalyst, which may explain the poor selectivity that is sometimes observed with carbon catalysts. More recently, the influence of heteroatoms (N, B, P) on the catalytic activity of carbons was interpreted in terms of semiconductor properties.[104] The insertion of nitrogen atoms into the graphite lattice lowers the band gap, leading to higher electron mobility and lowering the electron work function at the carbon/fluid interface. Thus, N-doped carbons exhibit enhanced catalytic activity in electron transfer reactions, in the same way as semiconducting metal oxide catalysts. In many cases, the effect of nitrogen-doping of carbons on their catalytic properties might be due in fact to two overlapping effects, catalysis by basic surface sites and by electron donation. More recently, the heteroatom-doping of carbon nanomaterials (*e.g.*, CNTs, FLG, mesoporous carbon) to induce intramolecular charge transfer has been shown to be a promising approach to the development of metal-free, carbon-based electrocatalysts.[77–80,105–108] Nitrogen-doped carbon nanomaterials show an even higher electrocatalytic activity and better long-term operation stability than that of commercially available Pt-based electrodes for oxygen reduction in fuel cells.[108] DFT calculations have shown that the active catalytic sites on single nitrogen doped graphene have either high positive spin density or high positive atomic charge density. The nitrogen doping introduces asymmetry spin density and atomic charge density, making it possible for N-graphene to show high electrocatalytic activities for the oxygen reduction reaction.[109] Based on first principles method, the catalytic property of nitrogen-doped graphene has been investigated for the oxygen reduction reaction.[110] It is revealed that nitrogen clusters, other than the isolated one, are the most efficient catalytic sites for oxygen reduction. Clusters with three or four nitrogen atoms are found to be optimal. A first-principles study was also conducted to investigate the effect of N-doping in CNTs on oxygen reduction reaction.[111] The results show that the catalytic activity of Pt-doped CNTs is higher than that of N-doped CNTs, as revealed from the cleaved bond length of O–O. However, N-doped CNTs outperform both pristine and Pt-doped CNTs on the specific thermal capacity, which can yield a higher structural stability in the ORR operation. These results show that, theoretically, catalytic properties similar or even superior to platinum can be obtained.

Later on, the catalytic activity of carbon materials came to be associated with their surface area. However, in several cases, the correlation was not found when using activated carbons with different surface areas and pore size distribution for a given catalytic reaction.[13] This means that the catalytic performances of carbon materials should also be related to the chemical nature of the surface of carbon catalysts, as suggested by Coughlin in 1969.[103] The investigation on the relationship between the chemical properties of carbon materials and the catalytic activity has been studied for several decades. Generally, two approaches have been widely followed in the surface chemistry of carbon: one is a "solid state chemistry" approach and the other is an "organic surface groups" approach. The former focuses on the crystalline microstructure of carbon materials, whereas the latter focuses on the organic character of the surface groups. In the "solid state chemistry" approach, the defects on the surface of carbon materials are considered as active sites since the edge-side carbon atoms are more reactive. The "organic surface groups" approach deals with the nature and the functionality of surface complexes of oxygen and other heteroatoms chemisorbed at the surface defects. Obviously, the combination of both approaches could lead to a deeper insight of the real reaction process taking place on the surface of carbon materials. For instance, the dependence of the chemical nature of AC on the raw material and preparation history was always observed, suggesting that the microstructure should be a key factor for the reaction activity. On the other hand, AC annealed in H_2 exhibited no activity for the dehydration and dehydrogenation of alcohols, while the oxidation treatment by nitric acid considerably increased the activity of the same carbon by two orders of magnitude, suggesting that the catalytic activity should be attributed to the surface functionalities.[112] The influence of heat-treatment on surface functionality stability and thus concentration on the catalytic activity should also be carefully considered.[113] Table 1.3 provides an overview of the various reactions that can be catalyzed with carbon, together with the type of surface chemistry required or the nature of the active sites, where they have been identified.

The redox couple quinone–hydroquinone was found to be involved in the oxidative dehydrogenation of hydrocarbons,[93] while carboxylic acid groups are the active sites for the dehydration of alcohols.[112] Carbon materials have also been used in environmental catalysis for the removal of SO_x, NO_x and H_2S from gaseous streams.[131] Basic carbons were the most active in these processes,

Table 1.3 Various reactions catalyzed by carbon materials, and the type of surface chemistry required or the nature of the active sites.

Reactions	Surface chemistry/Active sites	References
Oxidative dehydrogenation	Quinones	114
Dehydration of alcohols	Carboxylic acids	115
	$C-O-PO_3$ and $C-PO_3$	116
Dehydrogenation of alcohols	Lewis acids and basic sites	117
NO_x reduction (SCR with NH_3)	Acidic oxides + basic sites	118,119
NO oxidation	Basic sites, vacancies	120
SO_2 oxidation	Basic sites, pyridinic-N6	121
H_2S oxidation	Basic sites	122
Dehydrohalogenation	Pyridinic nitrogen sites	123
Liquid phase		
Hydrogen peroxide reactions	Basic sites	124,125
Catalytic ozonation	Basic sites	126,127
Catalytic wet air oxidation	Basic sites	128–130

particularly N-doped carbons.[132–134] Linear correlations between the catalytic activity and the concentration of pyridinic groups have been reported in the oxidation of SO_2[121] and for the reduction of NO.[135,136] Thus, if we consider nitrogen-doped carbon materials, in many cases the effect of nitrogen-doping of carbons on their catalytic properties might be due to two overlapping effects: electron donation[104] and by catalysis by basic surface sites.[137] Activated carbon has also a high catalytic activity for the oxidative desulfurization of diesel fuels with hydrogen peroxide.[138,139] The oxidation of organic compounds in liquid effluents is another environmental application of carbon catalysts, using air, oxygen, ozone or hydrogen peroxide as oxidants.[126,139–146] The reaction mechanisms in the liquid phase are more complex; nevertheless, the following general conclusions can be drawn: (i) the reaction mechanisms involve free radical species; (ii) basic carbons are the best catalysts; and (iii) oxidation of the organic compounds may occur both in the liquid phase (homogeneous reaction) and on the catalyst surface (heterogeneous reaction). For methane decomposition or reforming, the activity seems to be related to the BET area and oxygen surface groups, specifically those desorbed as CO in TPD experiments.[147] The steam reforming reactions of biomass materials in supercritical water on AC carbon catalyst has also been reported.[148]

Of course, to be complete, a correlation between carbon materials properties and their performances as catalyst should take into account the nature, the concentration and the accessibility of the active sites. Thus, such studies should use a series of carbon materials of the same nature and origin, which are produced with very similar textural properties and different amounts of surface functional groups. Concerning the nature of the active sites, the basal planes are not very reactive; therefore, we may expect to find active sites essentially at the edges of the graphene layers, where the unsaturated carbon atoms may chemisorb oxygen, nitrogen or sulfur, originating surface groups such as those discussed in Chapter 4. These groups can act as active sites in various acid–base or redox reactions. The concentration of these active sites depends to a large extent on the microcrystalline structure of the carbon material. Small crystallites can expose more edges; therefore more surface groups can be formed, while the role of the basal planes and the π electrons will become more important with large crystallites. Thus, the orientation of the graphene layers and the ratio between the prismatic and basal plane areas may affect the reactivity.[149–151] As far as the accessibility of the active sites is concerned, it will be determined by the pore sizes, especially in the case of microporous materials, such as AC, for which pore diffusion limitations become important in liquid,[152] supercritical[153] and gas[154] phase reactions. Notably, intra-particular diffusion limitations in the carbon can also operate.[155] Therefore, deactivation and diffusion phenomena will in general affect more strongly the performance of microporous carbons. As a result, there has been a need to develop mesoporous carbon catalysts such as CNTs, CNFs, aerogels, xerogels and templated carbons. Finally, as technical carbon materials may comprise a significant amount of impurities, such as sulfur or various oxides, modification of catalytic activity is sometime observed.

From this brief overview on the use of carbon materials as catalyst, it appears that the use of conventional carbons such as AC covers a wide range of reactions. The main properties that control carbon material catalytic activity are related to: (i) their electronic properties, (ii) their surface chemistry, (iii) their porosity, and (iv) their thermal stability. While AC, CB or glass-like carbon have long been used as catalysts for certain chemical and electrochemical processes, the recent availability, environmental acceptability, corrosion resistance, and unique surface properties of carbon nanomaterials have made them ideal candidates for metal-free catalysis. Thus, CNTs, nanodiamonds, and FLG offer new opportunities for the development of advanced carbon-based catalysts with much improved catalytic performance. Additionally, the introduction of surface heteroatoms, such as nitrogen[156] and/or sulfur,[157–162] into these carbon nanomaterials could further cause electron modulation to provide desirable electronic structures for many catalytic processes of practical significance. Consequently, considerable effort has recently been directed toward the development of metal-free carbon nanomaterials for various catalytic processes (see Chapters 5 and 6), involving either oxidation or reduction reactions.[163] We can particularly emphasise the use of CNTs for oxidative dehydrogenation of aromatic hydrocarbons and alkanes as well as ORRs, particularly in alkaline medium.

1.4 Carbon as a Catalyst Support

Among their many interesting applications, carbon and graphite materials have been considered over the last decades for their utilization in several processes involving heterogeneous catalytic reactions. Most of these catalysts consist of metals or metallic compounds supported on several materials, the role of which is not only to maintain the catalytic phase in a well dispersed state but also affect the catalytic activity, by means of direct participation in any of the steps of the reaction mechanism, or by favoring the interactions between active phase and support. This participation and their interaction with the active phase make catalyst supports more than just simple active phase carriers. Some of the properties desirable in a support are its inertness towards unwanted reactions, stability under regeneration and reaction conditions, adequate mechanical properties, tunable surface area, porosity, and physical form, *i.e.* the possibility of being manufactured in granulates or conformates of different size and shape to suit different chemical reactor configurations. Only a few out of a wide range of potential materials totally fulfil this set of desirable properties and combine them in an optimal way. To date the most important carbon supports from an industrial point of view are AC and CB. The main reason for the success of those materials is their commercial availability and variety of different grades, so that the final catalyst can be tailored to the end user's requirements. Other carbon supports like carbon aerogels and CNTs are the focus of modern catalytic research, but so far have not been used in commercial processes.

Carbon supported metal catalysts are employed in a number of applications including hydrodesulfurization of petroleum, hydrodenitrogenation, dehydrohalogenation, hydrogenation of CO, hydrogenation of halogenated nitroaromatics compounds and nitro compounds, hydrogenation of unsaturated fatty acids, hydrogenation of alkenes and alkynes, oxidation of organic compounds and organic pollutants, and for fuel cells.[12–15,45,47,48,146,164–170] However, the number of industrial processes using carbon or graphite materials as catalyst or catalyst support is still relatively limited. The large scale synthesis of vinyl acetate and vinyl chloride, and the desulfurization of natural gas on AC impregnated with ZnO, CuO, or Fe_2O_3 are important technical applications.[12] Additionally, in the *Catalytic Reaction Guide* published by Johnson Matthey, of 69 reactions of industrial interest catalyzed by noble metals, 50 used carbon materials as catalyst support (*e.g.* fatty acid hydrogenation, selective nitrobenzene hydrogenations, reductive alkylation, hydrogenation of dinitrotoluene, butanediol synthesis, purified terephthalic acid, *etc.*).[165]

In the past, the lack of fundamental understanding of many aspects of the use of carbon in catalysis and the irreproducibility of the results obtained when employing AC supported catalysts caused a limited application of carbon as catalyst and more so as catalyst support. But the continuous studies to better understand all aspects of the physical and chemical characteristics of carbon material, especially for AC (surface area and porosity) and, even more so, the possibility of controlling the surface chemistry of such materials is the origin of much important research carried out in industrial chemistry over the last decade.

1.4.1 Role of Carbon Properties in Performances of the Support

In heterogeneous catalysis, carbon materials have been used for a long time because they can be used directly as catalysts, and moreover, they can satisfy most of the properties desired for a suitable support. It has been shown that although the surface area and the porosity may be very important in the preparation and properties of the corresponding catalysts, the role of carbon surface chemistry is also extremely important. The number and strength of the surface groups influence the apparent acidity or basicity of the carbon surface.[171] Moreover, as already stated, a few percent of inorganic matter from the organic precursor (*e.g.* coal or wood) could be present on carbons and this fact is considered important for their performances, especially when they are used as catalyst supports. However, although the catalytic effect is mainly a result of the chemical properties of the active phase, the dispersion and the local distribution of the active phase across the carbon support, as well as the interaction between active phase and support, are significantly important.[150] These are just the aspects of carbon supports that make them so attractive for heterogeneous catalysis. These parameters can be modified to satisfy any specific requirement during catalyst preparation making the surface not only physically but also chemically accessible to the precursor and diminishing the deactivation by sintering. It seems obvious that the future is promising once it is understood that not only the surface area and porosity of carbon materials are the key for their uses, and that both physical and chemical surface properties of carbon support have to be taken into account when designing a solid catalyst.[172]

1.4.1.1 Surface Area and Porosity

Both high surface area and a well-developed porosity are very important for achieving a high dispersion (fraction of metal atoms that are on the surface of the support in relation to the total metal loading) of the active phase in the catalyst. Carbon materials, especially AC, exhibit surface areas significantly higher than other conventional oxide catalyst supports (Table 1.4). However, a great part of this surface area may be due to microporosity and, in this case, it may not be available to precursors or reactants.

Many studies report the effect of porosity and surface area on metal dispersion and catalytic activity. As far as metal dispersion is concerned, it is difficult to separate the influence of surface area from that of surface chemistry.[173] Indeed, irrespective of the model applied to describe the adsorption properties of metal-precursors during ion adsorption and impregnation, all authors agree with the fact that the surface composition of carbon plays a crucial role in adsorption of metal ions. Often, oxygen groups are introduced *via* oxidation of the surface of the carbon (see Chapter 4) to enhance the adsorption of the metal in order to obtain a high dispersion and a high metal-loading. But the role of basic sites, being either oxygen-containing groups such as chromene- and pyrone-like structures or π-sites, is also claimed to be of importance.[172]

An interesting study shows the separate influence of porosity, surface chemistry and ash contents on preparation of MoS_2/AC catalysts.[174] The effect of these properties on the preparation of the catalysts is as follows: (i) increasing surface area and porosity of the carbon facilitates the loading

Table 1.4 Main features of conventional catalyst supports.

Carrier	Features
Activated carbon	Surface area: 800–1500 m^2 g^{-1} Heat stability
Alumina	Surface area: 100–300 m^2 g^{-1} Type α, γ, η are often used as supports Reasonable price Heat resistance Alkali resistance
Silica	Surface area: 200–600 m^2 g^{-1}
Zeolite	Surface area: 350–900 m^2 g^{-1} Type A, X, Y Mordenite, Erionite, ZSM-5 are often used as supports. High controllability of pore size.
Titania	Surface area: 40–100 m^2 g^{-1}
Magnesia	Surface area: 50–200 m^2 g^{-1} Basicity Strong adsorption of carbon dioxide and water in air.

of Mo; (ii) ash content has a relatively small (but significant) negative effect in the adsorption of the metal precursor; and (iii) oxygen surface groups of the carbon support increase the adsorption of the metal precursor. For the preparation of Pd/AC catalysts by deposition/precipitation method on AC of different textural properties but similar surface chemistry, it was shown that the prime parameter that determines the Pd particle size is the extent of AC surface in the meso- and macropores.[175] Mesopore volume and mean mesopore size have also been identified as being important parameters to control Pt[176–178] and Ru[179] particle size and dispersion on carbon materials. It was found that the presence of ultra-small mesopores (diameters < 3 nm) in the support results in the formation of ultra-fine ruthenium particles. Ordered uniform porous carbon frameworks with pore sizes in the range of 10 to 1000 nm were used to prepare bimetallic PtRu electrocatalysts.[180] Such supports resulted in much improved catalytic activity for methanol oxidation due to their high surface areas, large pore volumes, and three-dimensionally interconnected uniform pore structures, which allow a higher degree of dispersion of the catalysts and efficient diffusion of reagents.

Studies on confinement effects in carbon-based catalysts have recently appeared.[49,181,182] The confinement effects that influence chemical reactions can be classified into three groups: (i) shape-catalytic effects, *i.e.*, the effect of the shape of the confining material and/or the reduced dimensionality of the porous space, (ii) physical (or "soft") effects including the influence of dispersion and electrostatic interactions with the confining material, and (iii) chemical (or "hard") effects, *i.e.*, interactions that involve significant electron rearrangement, including the formation and breaking of chemical bonds with the confining material.[183] The latter is usually considered to be the actual catalytic effect, and it is the one that has the most obvious influence on the reaction rates, as it alters the reaction mechanism. However, the first and second types of effects can also have a strong influence on both the rates and equilibrium yields, as has been shown in several recent theoretical calculations and experimental studies.[181] Multi-walled CNTs, which present a high mesoporosity, appear to be ideal candidate as support to investigate confinement effects in catalysis.[150]

However, in some cases, the high surface area of the carbon support may be detrimental if the active phase is confined in narrow micropores, which are not accessible to the reactant molecules. This is especially important in processes where large molecules are involved, as in the treatment of petroleum feedstocks, and in liquid phase reactions in which diffusion of reactants and/or products may be hindered by the narrow porosity.

The shape of the pores of the carbon support can also play an important role in the catalytic process. Laine *et al.* have used AC as support for NiMo catalysts for HDS.[184,185] On the basis of their results, these authors suggested that the narrow slit-shaped pores in AC are able to lower the vapor pressure of sulfur to such an extent that they create a driving force for sulfur transfer from the active compound to the micropores, this process forming active vacancies in the metal sulfide. This so called "sink effect" was not observed in microporous silica, whose pores are not slit-shaped. Shape selectivity has also been reported for a variety of microporous carbon, such as for hydrogenation,[186–190] methanol decomposition,[191] or the Fischer–Tropsch reaction.[192]

It appears that a large surface area formed by accessible pores is important to obtain highly dispersed and active catalysts. The possibility to confine the active phase in well-defined mesopores such as the ones present in CNTs opens perspectives for unexpected reactivity due to confinement effects. However, there are some other carbon characteristics that have to be taken into account in order to explain the catalytic behavior of carbon supported catalysts. One of the most important is the surface chemistry.

1.4.1.2 Surface Chemical Properties

Apart from the effects of wetting, surface area and pore size distribution, carbon surface functionalities govern the extent of adsorption of the catalyst precursor and the extent of its reduction or conversion to active state. If most of the studies have focused on the effect of carbon surface chemistry on catalyst performances, the chemical compatibility between the surface and the precursor should also been taken into account, as stated in a landmark paper by F. J. Derbyshire *et al.* "The affinity between a particular carbon surface and the (selected catalyst) precursor will depend upon the compatibility of the two chemical structures".[193] Thus, an important factor influencing the dispersion of the active phase is the nature of the precursor. Rodríguez-Reinoso *et al.* used two different iron precursors (iron nitrate in aqueous solution and iron pentacarbonyl in organic solution) to prepare Fe/AC.[194] They obtained an increase in iron dispersion with the support surface area for the nitrate series, but a high and unaffected dispersion was found for the $[Fe(CO)_5]$ series. Similarly, for iron supported on carbon nanosphere, iron acetate yielded higher dispersion than iron nitrate, allowing better performances for Fischer–Tropsch synthesis.[195] Large differences were also reported on the dispersion and location of iron nanoparticles on AC prepared from iron acetate, iron sulfate and iron nitrate: (i) preparation from iron sulfate resulted in more external Fe particles and higher leaching; (ii) preparation from iron acetate resulted in internal Fe particles and low leaching; and preparation from iron nitrate resulted in good Fe dispersion, but high leaching.[196] In a comprehensive study of Pd/AC catalysts, V. A. Likholobov *et al.* stated: "The influence of the state of the carbon surface, *e.g.* its coverage with functional groups, on the dispersion of the active phase proved comparable, however, to the effect of the nature of the metal precursor (in their case H_2PdCl_4, $(Pd(OAc)_2$ or $([Pd(NH_3)_4](NO_3)_2)$ and other conditions used during catalyst preparation".[197] In a study on the influence of various parameters on the confinement yield of metallic nanoparticles in CNTs, Nguyen *et al.* show that the nature of the metal precursor (nitrate, chloride or organometallic) is a critical issue; whatever the metal (Co, Pd or Ru) the best yield was obtained with the nitrate precursors.[198] The compatibility between the internal surface of the CNT and the metal precursor was at the origin of this result. The nuclearity of the metal precursor may also affect the final particle size. For a given support, the higher the precursor nuclearity, the higher the final metal particle size.[198] In the case of bimetallic catalysts, an influence of the nature of the precursor has also been noticed. Devillers *et al.* studied the influence of metallic precursors on the properties of carbon-supported bismuth-promoted palladium catalysts for the selective oxidation of glucose to gluconic acid.[199] They used different precursors for Pd and Bi deposition and find that the best catalytic activities were obtained when using acetate precursors instead of classical salts, such as chlorides or nitrates. However, for Ni–W/AC catalysts used in phenol hydrodeoxygenation, it was shown that the effect of the tungsten precursor (silicotungstic, phosphotungstic, and tungstic acids) on catalytic performances was negligible.[200]

The presence of surface groups that contains heteroatoms (O, N, H) can affect the preparation of carbon-supported catalysts, as they confer to the carbon surface an acid–base and hydrophilic character. The presence of these groups can also affect the adsorption/desorption phenomena and, thus, impact the catalytic activity.

The carbon surface functional groups (SFGs) strongly impact catalyst performances. First, it provides the anchoring sites for the metal precursor. For example, –COOH groups are the sites where ion exchange (cationic precursors) or oxidative addition takes place.[150] Second, the optimum surface chemistry allows favorable electrostatic (coulombic) interaction between the support and the metal precursor.[201–205] For example, adsorption of anions occurs only when the pH of the solution containing the catalyst precursor is lower than the PZC of the carbon; and opposite is true for cationic precursors. J. R. Regalbuto *et al.* clearly show that, contrarily to oxide supports, alteration of the PZC of carbon surfaces (by oxidation) influences the adsorption of Pt complexes in a way that it is systematic and controllable.[202] Thus, while cationic metal precursors are usually used on silica, and anionic ones on alumina, both types of precursors can be used on carbon.[206] It is also worth noting that although the presence of oxygen SFGs has often been reported to have a positive impact on metal dispersion, some studies have shown the opposite.[207] Possible explanations are that: (i) on microporous supports, a fast nucleation could hinder the precursor from penetrating the porous network of the supports; or (ii) the anchoring centers decompose during catalyst activation. Indeed, the heat treatment in H_2 to reduce the catalysts may have a decisive influence in the resulting platinum dispersion, because this treatment leads to the decomposition and some redistribution of surface oxygen complexes, producing the mobility and, consequently, the agglomeration of platinum particles. Third, the optimum surface chemistry prevents excessive catalyst mobility on the support surface. Both dispersion and resistance to sintering of Pt/CB catalysts were found to be a function of the number of oxygen surface groups of the support.[208,209] For example, the relatively stable carbonyl groups appear to be quite effective for this purpose.[210] Thus, although it has been experimentally shown that the more acidic groups, such as the carboxylic ones, decrease the hydrophobic character of the carbon surface and positively impact the metal dispersion, providing anchoring sites, on the other hand, the less acidic and more thermally stable surface groups, such as the carbonyls, favor the interaction between the metal precursor or the metal particle and the carbon surface, thus minimizing sintering. However, it is often forgotten that the thermal stability of the carboxylic groups is limited to temperatures close to 673 K. Nevertheless, such temperatures are often applied for the decomposition and/or reduction of the metallic precursor. Thus, the question concerning whether these oxygen SFGs affect the final metallic dispersion, either in a positive way due to their exchange properties or in a negative way (sintering of the metal species and a loss of dispersion) as a consequence of their decomposition during catalyst pretreatments, remains open to discussion. In a study on Ru/CNT catalysts, B. Machado *et al.* have shown that carboxylic acid, carboxylic anhydride, and lactone groups act as anchoring centers for the Ru precursor, presumably as surface acetato ligands. After a high temperature treatment performed in order to remove the SFG, the Ru/CNT material can react with oxygen from air, *via* a surface reconstruction reaction, which reforms a stable surface Ru–acetato interface.[150] An optimal surface chemistry also facilitates catalyst conversion into its catalytically active state. This was nicely illustrated in the work of Gallezot *et al.*[211] and Prado-Burguete *et al.*[208,212] who reported studies on the role of surface oxygen groups on the dispersion and resistance to sintering of Pt/C catalysts. They prepared a number of CBs with similar porosities but different amounts of oxygen SFGs, which were impregnated with H_2PtCl_6 solutions. It was observed that the more acidic the groups were, the less hydrophobic was the CB surface, making the surface more accessible to the aqueous solution of the metal precursor upon impregnation. Thus, the Pt dispersion increased with an increase in the amount of oxygen SFGs. On the other hand, the less acidic and more thermally stable surface groups favored the interaction between the metal precursor or the metal particle with the carbon surface, this minimizing the Pt sintering. Similar results have been obtained in the preparation of K promoted Ru/C catalysts for ammonia synthesis, although in this case the AC was oxidized with HNO_3.[213] The authors concluded that the presence of oxygen surface groups improved the hydrophilic character of the carbon surface, thus enhancing the dispersion of K and Ru.

The presence or absence of surface functionalities can also directly affect the catalytic behavior of the active phase. The effect of acidic treatments on N_2O reduction over Ni catalysts supported on AC was systematically studied by Lu *et al.*[214] It was found that surface chemistry plays an important role in the N_2O–carbon reaction catalyzed by the Ni catalyst. HNO_3 treatment produces a significant amount of active acidic surface groups, such as carboxyl and lactone, resulting in uniform catalyst dispersion and high catalytic activity. HCl treatment of the pristine AC decreases the number of active acidic groups and increases the number of inactive groups, playing the opposite role in the catalyst dispersion and catalytic activity. A strong effect of the AC surface chemistry on the activity of Au/AC catalysts was also reported for glycerol oxidation.[215] Gold particles with similar average sizes resulted in different performances depending on the amount of oxygenated groups on the surface of the support used. Basic oxygen-free supports, characterized by a high density of free π-electrons, lead to an enhancement of the gold catalyst activity. These characteristics can easily be achieved by thermal treatments at high temperatures, which remove the oxygen-containing surface groups. The role of the AC surface chemistry was explained by considering the capability of oxygen-free supports to promote electron mobility. The mobility of the electrons from and to the gold surface promotes both adsorption and regeneration of hydroxide ions, which are necessary for the oxidation reaction. A clear effect of the oxygen SFGs created by HNO_3 oxidation was also reported for Pt/CNT catalysts involved in different probe reactions, including both steam and liquid phase reforming of hydrocarbon oxygenates and dehydrogenation of alkanes in the liquid and gas phases.[216] Compared to the pristine CNT supports, Pt dispersion is improved on HNO_3 treated supports, but the turnover frequency of aqueous phase reforming decreases by half. Higher turn-over frequencies can be obtained by removing the oxygen surface groups *via* high temperature annealing. A comparison of the results obtained in the different reactions suggests that the oxygen surface groups are only detrimental to reactions in a binary mixture with two components of different hydrophilicity due to their competitive adsorption on the catalyst supports.

Various carbon and graphite materials have been used for selective hydrogenation of α, β-unsaturated aldehydes, such as CNFs, CNTs, graphite, AC, CBs, and fullerenes. The presence or not of surface oxygenated groups also directly affect the catalytic behavior of the active phase. In a study on vapor-phase crotonaldehyde hydrogenation on Pt/AC catalysts, Coloma *et al.* used a demineralized AC oxidized with H_2O_2 to introduce oxygen surface functionalities (AC_{O_x}) and then heat treated under He at 773 K to remove the less stable surface groups ($AC_{O_x}T$).[217] The specific activity of 1% Pt/AC catalysts was high, and followed the order $Pt/AC < Pt/AC_{O_x}T < Pt/AC_{O_x}$. The selectivity to the unsaturated alcohol was low for catalyst Pt/AC and much larger for the others, especially for catalyst Pt/AC_{O_x}. Toebes *et al.* used Ru/CNF catalysts to study the influence of oxygen SFGs on the catalytic performance for the hydrogenation of cinnamaldehyde.[218] After reduction, the catalysts were heat-treated in nitrogen at different temperatures to control the number of oxygen SFGs. The overall specific activity increased by a factor of 22 after the heat treatment, which was related to the decreased number of oxygen surface groups. Cinnamyl alcohol selectivity decreases from 48 to 8% due to the enhanced rate of hydrocinnamaldehyde production with increasing heat treatment. In a similar study carried out with Pt/CNF catalysts,[219,220] the authors also found an increase in catalytic activity with increasing thermal treatment temperature, and a linear decrease in the hydrogenation activity with an increase of the number of acidic groups on the carbon nanofiber surface. They suggested that the hydrogenation process was favored by the adsorption of cinnamaldehyde on the carbon support after removal of the oxygen-containing surface complexes. Similar results have been obtained for Pt/CNT.[216,221] For bimetallic PtRu/CNT systems, the removal of oxygen SFGs by heat-treatment under inert atmosphere (up to 1273 K) induces an increase of both activity and selectivity towards cinnamyl alcohol.[222,223] The activity increase was attributed: (i) to a better adsorption of cinnamaldehyde, the CNT surface thus acting as reservoir of substrate where a high local concentration of cinnamaldehyde is readily accessible for the catalyst; and (ii) to a better diffusion of cinnamaldehyde and cinnamyl alcohol on the support. The selectivity increase was attributed to an enrichment of the surface of the particles in ruthenium after the heat-treatment. The possible favored adsorption of cinnamyl alcohol on the heat-treated surface of the catalyst during the reaction, which creates a steric effect that inhibits the hydrogenation of the C=C bond of cinnamaldehyde has also been proposed for Ir/Graphite catalysts.[224] The positive or negative impact on catalytic performances of surface oxygen groups removal after supported catalyst preparation has also been investigated for other reactions, such as enantioselective hydrogenation of methyl pyruvate over Pt/C catalysts,[225] oxidation of benzyl alcohol over Ru/CNFs,[226] hydrodechlorination of chlorobenzenes over Pd/AC, Pd/graphite and Pd/CNFs,[227] and sorbitol hydrogenolysis to glycols on Ru/CNF.[228] It is also worth mentioning that the effect of the gradual thermal decomposition of surface oxygen groups on the chemical and catalytic properties of oxidized AC has also been reported.[113,229,230]

Although less studied than oxygen, nitrogen surface groups can also influence metal dispersion and catalytic activity (see Chapters 4 and 6), as recently reviewed for CNTs.[169] The effect of nitrogen functionalities depends on the system studied. In this way, Derbyshire *et al.* obtained more active Mo catalysts for HDS by pre-nitriding the carbon support,[193] and Guerrero-Ruiz *et al.* found the same effect for Fe/C and Ru/C catalysts.[231] Pd nanoparticles have been deposited on N-doped nanocarbons and compared with undoped catalysts prepared on CB for the direct synthesis of H_2O_2.[232] The Pd on N-doped CNT gives high productivities to H_2O_2. The introduction of nitrogen into the CNT structure favors not only the dispersion of Pd but also the specific turnover. Stone-fruit AC and modified supports containing acidic oxygen (HNO_3 oxidation) and basic nitrogen groups (treatment with nitrogen-containing compounds) have been used to prepare palladium catalysts by wet impregnation.[233] The influence of the nature of the functional groups on the dispersion and oxidation state of Pd and its activity in hydrogen oxidation have been investigated. Palladium dispersion was found to increase with the basic strength of functional groups on the support. XPS studies have shown that introduction of amine surface groups results in an increased proportion of Pd^0, which is resistant to re-oxidation. Palladium catalysts supported on AC modified by diethylamine groups are found to exhibit the highest metal dispersion and greatest activity in hydrogen oxidation. In another study, ruthenium catalysts were supported on two different carbon materials, CNT and bamboo-like CNT doped with nitrogen.[234] The Ru catalysts were tested in the catalytic ammonia decomposition reaction. The catalytic activity of Ru particles was significantly improved when supported on CNTs doped with nitrogen. Better dispersions and activity have also been reported for metals such as Pt,[235–240] PtRu,[241] PtFe and PtCo,[242] PtSn,[243] Co,[244] or Fe[245] deposited on nitrogen-doped carbon for fuel cell electrocatalysis. Calculations performed on Pt/N-doped-HOPG (HOPG: highly oriented pyrolytic graphite) show that nitrogen doping at high doses likely causes agglomerated nitrogenous defect clusters.[246] For Pt clusters supported on HOPG with nitrogen defects, calculations show a greater driving force for nucleation and greater particle tethering, particularly on pyridine and pyrrole groups. The intrinsic enhancement of activity by chemical doping of the catalyst support is attributed to a down-shift in the Pt *d*-band because of charge donation from the dopant to platinum.

Improved catalyst-support interactions correlated to high substrate nitrogen content in immediate proximity to stabilized metal nanoparticles have been demonstrated by R. O'Haire *et al.* using model substrates through principal component analysis of electron energy loss spectral imaging datasets acquired on an aberration-corrected scanning transmission electron microscope.[247] This improved stability has been confirmed in other studies dealing with fuel cell electrocatalysis,[248] or Fischer–Tropsch synthesis.[249]

On the other hand, Wachowski *et al.* studied the polymerization of styrene with a $[CpTiCl_2(OC_6H_4Cl\text{-}p)]$ catalyst supported on carbon materials with different degrees of coalification, and analyzed the effect of the modification of the support by nitrogen on the efficiency of the catalytic system in polymerization of styrene.[250] It was found that the introduction of nitrogen functionalities on the carbon surface lowered the catalytic activity of these systems.

To conclude, it appears that it is more than the quantity of oxygen or nitrogen introduced in the carbon or graphite (nano)material. It is the quality (type) of these groups that controls the extent of metal precursor sorption: its distribution throughout the support, the resistance to active phase agglomeration upon drying/reduction, and ultimately the performance of carbon-supported catalysts.

1.4.1.3 Inertness

Compared to conventional oxide supports such as silica, alumina, titania, or ceria, it is clear that the carbon surface (in spite of the presence of surface hetero-atoms) is less reactive. Thus, Milone *et al.*,[251] investigated the influence of the support (SiO_2 or AC) on the hydrogenation of citronellal on Ru supported catalysts, and showed that the main products obtained on the Ru/SiO_2 catalyst were the unsaturated cyclic alcohols, produced by citronellal isomerization on the SiO_2 surface. However, the main reaction products on Ru/AC were the open chain hydrogenated products, and this was attributed to the low activity of the carbon surface towards the isomerisation reaction.

The carbon surface inertness may also significantly impact catalyst preparation. An analysis of the literature suggests that there are at least four different characteristics of carbon that can be utilized to generate metal surfaces not found on refractory oxide supports.[252]

First, on graphitic carbon many metals interact very weakly, allowing bimetallic particles to form structures identical to those anticipated for bulk materials. Of particular significance is the formation of true alloys, both in the bulk and on the (catalytic) surface of the bimetallic particles. In contrast, on conventional refractory oxide supports these same structures will not form for certain base-metal/noble-metal pairs. Instead, a preferential and strong interaction between the more "base" metal and the support generally leads to preferential segregation of that metal to the refractory oxide interface and, concomitantly, dominance of the catalytic interface by the "more noble" metal. As a result of these structural differences, the catalytic chemistry, both activity and selectivity, of some bimetallic particles supported on refractory oxides and graphitic carbons are dramatically different. One representative example is the preparation of bimetallic PtSn[253,254] or PtRu[222] catalysts for selective hydrogenation of α,β-unsaturated aldehydes. The catalytic behavior of this system is determined by at least three aspects that determine the catalytic activity and the selectivity towards the desired product: (i) the oxidation state of the promoter (Ru or Sn) in the catalyst; (ii) the possibility of formation of alloyed phases; and (iii) the extent of metal–support interactions. It is particularly interesting to obtain the alloyed phase, and this is determined by the ease of reduction of the tin species which, in turn, depends on the interaction between the tin precursor and the support. For PtSn/TiO_2 catalysts it was shown by XPS that even after reduction under hydrogen at 773 K, the majority of the tin (about 78%) was in an oxidized state, this limiting the possibility of formation of alloy phases. However, a nearly complete reduction of tin to the metallic state can be achieved by using carbon materials as supports, depending on the preparation method and on the relative amount of tin.[253,254]

Second, it is clear that it is possible to directly bond metals to unsaturated active sites on high surface-area CBs, AC, *etc*. This has been demonstrated to yield thermally stable particles of a unique structure.[255] Thus, on CB it was demonstrated that iron particles are more resistant to sintering if deposited on an oxygen-free support than on a functionalized support. However, for Pd nanoparticles deposited on single-walled CNTs, the dispersion of Pd nanoclusters was enhanced by creating defects on the nanotube walls, which lead to a stronger metal–support interaction.[256] DFT calculations have shown that the binding energy of Pd is significantly enhanced when the CNT surface is oxygen-functionalized, compared to the case of the pristine SWCNT surface. The electronic interaction of Pd atoms with oxygen at the defect sites results in a stronger bonding. These calculations were consistent with experimental measurements, since microscopy images clearly show that the functionalized SWCNT surface is much more effective than the pristine surface in anchoring Pd nanoclusters. However, strong interaction between Pd clusters and the carbon surface can be effective without functionalization, and a molecular orbital study has shown that supported Pd^0 atoms and clusters are strongly bound to unsaturated and defect surface sites of AC surface.[257] In such positions, interaction of Pd atoms with the support is much stronger than that with each other. That provides the driving force for atomic dispersity of Pd/AC catalysts. Similar conclusions were drawn from DFT calculation and TEM observations in the case of Au clusters deposited on CNTs.[258] Indeed, both experimental and theoretical studies show that surface defects are the anchoring sites of Au nanoparticles. Very weak binding is identified between Au clusters and pristine CNTs, or defective CNTs when Au clusters are located far from the vacancy. It is only when Au clusters are directly located on the defect site, that very strong interactions between them are found. The strong interaction is caused by the charge transfer from Au to the defect. This study suggests that defect of CNTs can be used for the grafting of Au nanoparticles without requiring surface oxidation. Potentially, the dispersion of Au clusters can be controlled or regulated by controlling the density of defects. Another study on Au/CNF catalysts showed that the shapes of the Au nanoparticles depended on the different interactions with the support.[259] Thus, gold nanoparticles were supported on two types of CNFs with different degrees of graphitization. TEM investigation of these materials showed that the degree of the surface graphitization strongly influences the structures of the supported gold particles. The more ordered graphitic layers of the CNF surface led to Au particles preferring to immobilize on their (111) plane, exhibiting more facet area. In contrast, disordered CNF surfaces led to a random orientation of supported particles. For Co/CNF catalysts, the interaction of precipitated cobalt oxides with the CNF support was shown to be influenced to a large extent by heat treatment in an inert atmosphere.[260] Heat treatment of the CNF at 573 K resulted in an increase in the interaction between the Co particles and the support. Under these conditions, a small amount of Co carbide and Co metal was detected by the XRD and XPS analyses. Heat treatment at 873 K resulted in a further increase in the interaction between the metal and the support leading to increasing amounts of Co carbide and Co metal.

On refractory oxides, strong interaction generally leads to the creation of complex, ionic-bonded "interface" phases.

Third, the carbon structure can be manipulated to generate shape-selective supports. This can also be done with refractory oxides, but only carbon surfaces are neutral. Thus, only on carbon will reduced metal readily form. The easier metal reduction on carbon support compared to conventional oxide supports has been evidenced in the case of Fe catalysts.[261–263] The inertness of the carbon surface facilitates the presence of the zero-valence iron in the catalyst, which is more difficult in the case of other supports, such as alumina, on which the reduction of the oxidized iron species is hindered. Vannice *et al.* carried out an extensive study of carbon-supported iron catalysts using different carbons and preparation methods, and concluded that highly dispersed Fe/C catalysts could be prepared on high surface area carbons due to the weak chemical interactions between oxidized iron precursors and the carbon surface.[261–263] There is surprisingly little research into these phenomena, suggesting there are many opportunities to create unique metal surfaces using carbon as a support.

Fourth, a characteristic of carbon is the possibility with some materials, such as CNFs, to tune the basal/edge site ratio. The interactions of metal nanoparticles with CNF surfaces can be described in terms of the binding energy of metal nanoparticles or clusters with the CNF surfaces.[151] Thus, it has been shown that unsaturated surface carbon atoms formed much stronger bonds with Pd atoms than the basal graphite plane, owing to the formation of several Pd–C bonds and to the stabilization of the π-electronic system of several nearest-neighboring aromatic rings.[257] The order of the interaction of metals with CNFs and CNTs, p-CNFs > h-CNFs > r-CNFs = CNTs, is opposite to the order of the ratios of edge sites to the basal plane. The possible reactivity of carbon towards hydrogen gas (catalyzed methanation) should also be considered. If the catalyst is submitted to high temperature treatment under H_2, stability issues should be taken into account. In that case, Machado *et al.* show that it is the basal/edge site ratio of the carbon materials that should be considered.[150] In a study on Ru/C catalyst (C = CNT, CNF, FLG), they show that the more stable catalyst (Ar/H_2–973 K) is Ru/CNT. The limited hydrogasification and CH_4 formation should result from the orientation of the graphene layers in CNTs that differ from that observed on conventional carbon supports such as AC, CB, HOPG, FLG or diamond crystallite, for which the concentration of edges is much higher.

Finally, it is worth noting that an "inert" surface means that non-functionalized carbon materials can present electron conductivity provided the size of the graphite-like crystallites is large enough. An increase in the concentration of oxygen-containing surface functionalities will result in an increase of the work function and concomitant increase in the resistance of compacted carbon powders.[264,265]

In conclusion, many parameters have to be taken into account during the rational design of a carbon supported catalyst (Scheme 1.1). The natures of the metallic precursor and of the carbon/graphitic material have to be

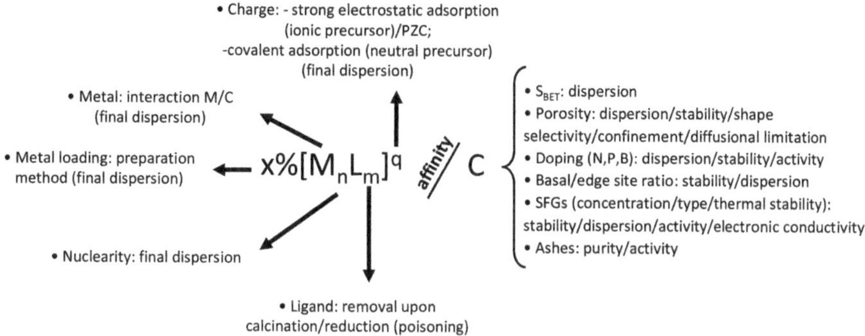

Scheme 1.1 Main parameters to evaluate the rational design of a carbon supported catalyst.

carefully considered, and their potential affinity evaluated. Although relatively good knowledge of conventional oxide support (silica, alumina, zeolites) surface chemistry has already allowed the design, at the molecular level, of supported nanoparticles and single site catalysts,[266,267] such examples are still rare for carbon materials.[268]

1.5 Concluding Remarks and Future Perspectives

Carbon is an extraordinary element that exists in a large variety of forms. Graphite, coal, soot, and diamonds are all nearly pure, naturally occurring, forms of carbon. Carbon and graphite materials can also be produced in various forms such as AC, CB, CNTs, fullerenes or graphene. Today, a large variety of carbon materials are industrially produced. The versatility of these materials and the richness and the complexity of carbon physical chemistry make the design of structurally controlled catalysts a challenging task. The description of carbon surfaces (chemistry and pore description) is a key point towards a better control of the catalytic performances. For most catalytic applications, the surface chemistry of the carbons has a major influence on the performance of the material *e.g.* dispersions in solvents, dispersion of metallic species on carbon surfaces, adsorption of reactants/products on carbons *etc.* The catalytic reactions are often carried out in meso- (CB) or microporous (AC) carbon or graphite materials, which can enhance or reduce reaction yields through a host of different effects, including an increase in the surface area per unit volume, the selective adsorption of reactants and/or products, confinement and shape-catalytic effects, among others. A fundamental understanding of the role of each one of these different effects should lead to a systematic procedure for the design of improved catalytic materials that takes advantage of all of them simultaneously. In that context, the use of carbon nanomaterials offers several advantages with respect to conventional supports such as AC, glass-like carbon or CB. Indeed, the graphene sheets can be used as basic building blocks to tune the structure of carbon nanomaterials and, thus, the properties of metal particles and their catalytic performance. Carbon nanomaterials, with their unique features and specific characteristics (see Chapter 2), can be used to manipulate the catalytic properties of nanocatalysts by tuning the properties of the support, including the orientation of the graphene sheets, as well as their diameter, crystallinity, conductivity, *etc.* Additionally, carbon nanomaterials often present high purity (avoiding self-poisoning), high mesoporosity (improving mass transfer), and high mechanical and thermal stability. Several excellent reviews cover these subjects.[45,47,48,50,53,56,72,151,269–275] Currently, carbon nanomaterials are mainly used as model catalyst supports or catalysts. Some issues should be addressed before an industrial use of these materials as catalysts or catalyst supports. First, the availability and price of these nanomaterials still remain issues for their use in catalysis. Indeed, even if CNTs are today produced on the industrial scale, the price (~50 € kg^{-1}) is still high compared to high quality CB or AC (~10 € kg^{-1}). Only a very significant improvement of the catalytic performances will justify the choice of such an advanced material. Second, micrometric agglomerates of these materials have often a low bulk density. The typical value of bulk density for industrial CNT agglomerates ranges today between 60 and 270 kg m^{-3}, compared to 20–550 kg m^{-3} for CB and 300–580 kg m^{-3} for AC. This would handicap the economic use in a reactor since the mass of catalyst per reactor volume would be low. Therefore, synthesis routes to carbon nanomaterial bodies with high bulk densities and high strength while maintaining the porosity are much desired. Some encouraging results have already been obtained in this direction.[276]

References

1. P. Serp, in *Comprehensive Inorganic Chemistry II*, ed. Editors-in-Chief R. Jan and P. Kenneth, Elsevier, Amsterdam, 2nd edn, 2013, pp. 323–369.
2. H. Marsh, F. Rodríguez-Reinoso, *Activated Carbons*, Elsevier, Oxford, 2006.
3. J. F. H. Peter, L. Zheng and S. Kazu, *J. Phys.: Condens. Matter*, 2008, **20**, 362201.
4. M. Kang, Y.-S. Bae and C.-H. Lee, *Carbon*, 2005, **43**, 1512–1516.
5. Suhas, P. J. M. Carrott and M. M. L. Ribeiro Carrott, *Bioresour. Technol.*, 2007, **98**, 2301–2312.
6. E. L. K. Mui, D. C. K. Ko and G. McKay, *Carbon*, 2004, **42**, 2789–2805.
7. J. M. Dias, M. C. M. Alvim-Ferraz, M. F. Almeida, J. Rivera-Utrilla and M. Sánchez-Polo, *J. Environ. Manage.*, 2007, **85**, 833–846.
8. O. Ioannidou and A. Zabaniotou, *Renewable Sustainable Energy Rev.*, 2007, **11**, 1966–2005.
9. Y. Chen, Y. Zhu, Z. Wang, Y. Li, L. Wang, L. Ding, X. Gao, Y. Ma and Y. Guo, *Adv. Colloid Interface Sci.*, 2011, **163**, 39–52.
10. G. Lemoine, *C. R. Acad. Sci.*, 1907, **144**, 357–358.

11. J. L. Figueiredo and M. F. R. Pereira, *Carbon Materials for Catalysis*, John Wiley & Sons, Inc., 2008, pp. 177–217.
12. H. Jüntgen, *Fuel.*, 1986, **65**, 1436–1446.
13. F. Rodríguez-Reinoso, *Carbon*, 1998, **36**, 159–175.
14. E. Furimsky, *Carbons and Carbon Supported Catalysts in Hydroprocessing*, 2008.
15. Y. Yang, K. Chiang and N. Burke, *Catal. Today*, 2011, **178**, 197–205.
16. D. D. L. Chung, *J. Mater. Sci.*, 1987, **22**, 4190–4198.
17. A. V. Yakovlev, A. I. Finaenov, S. L. Zabud'kov and E. V. Yakovleva, *Russ. J. Appl. Chem.*, 2006, **79**, 1741–1751.
18. W. Li, C. Han, W. Liu, M. Zhang and K. Tao, *Catal. Today*, 2007, **125**, 278–281.
19. P. Shi, S. Zhu, H. Zheng, D. Li and S. Xu, *Desalin. Water Treat.*, 2013, 1–8.
20. J. Bian, M. Xiao, S. J. Wang, Y. X. Lu and Y. Z. Meng, *J. Colloid Interface Sci.*, 2009, **334**, 50–57.
21. I. M. Afanasov, O. N. Shornikova, V. V. Avdeev, O. I. Lebedev, G. V. Tendeloo and A. T. Matveev, *Carbon*, 2009, **47**, 513–518.
22. H.-Q. Li, Y.-G. Wang, C.-X. Wang and Y.-Y. Xia, *J. Power Sources*, 2008, **185**, 1557–1562.
23. P. Ferreira-Aparicio, M. A. Folgado and L. Daza, *J. Power Sources*, 2009, **192**, 57–62.
24. C. Marquez-Alvarez, I. Rodriguez-Ramos and A. Guerrero-Ruiz, *Carbon*, 1996, **34**, 339–346.
25. P. Gallezot, S. Chaumet, A. Perrard and P. Isnard, *J. Catal.*, 1997, **168**, 104–109.
26. B. Bachiller-Baeza, A. Guerrero-Ruiz, P. Wang and I. Rodríguez-Ramos, *J. Catal.*, 2001, **204**, 450–459.
27. E. Díaz, S. Ordóñez, R. F. Bueres, E. Asedegbega-Nieto and H. Sastre, *Appl. Catal., B*, 2010, **99**, 181–190.
28. R. Sørensen, A. Klerke, U. Quaade, S. Jensen, O. Hansen and C. Christensen, *Catal. Lett.*, 2006, **112**, 77–81.
29. S. Otani and A. Oya, *Materials Science and Technology*, Wiley-VCH Verlag GmbH & Co. KGaA, 2006.
30. L. A. Pesin, *J. Mater. Sci.*, 2002, **37**, 1–28.
31. P. J. F. Harris, *Philos. Mag.*, 2004, **84**, 3159–3167.
32. R. L. McCreery, *Chem. Rev.*, 2008, **108**, 2646–2687.
33. V. D. Chekanova and A. S. Fialkov, *Russ. Chem. Rev.*, 1971, **40**, 413.
34. J. M. Friedrich, C. Ponce-de-León, G. W. Reade and F. C. Walsh, *J. Electroanal. Chem.*, 2004, **561**, 203–217.
35. G. Du, C. Song, J. Zhao, S. Feng and Z. Zhu, *Carbon*, 2008, **46**, 92–98.
36. K. Tammeveski, K. Kontturi, R. J. Nichols, R. J. Potter and D. J. Schiffrin, *J. Electroanal. Chem.*, 2001, **515**, 101–112.
37. Y. Gorlin, C.-J. Chung, D. Nordlund, B. M. Clemens and T. F. Jaramillo, *ACS Catal.*, 2012, **2**, 2687–2694.
38. Y. Gu, J. St-Pierre and H. J. Ploehn, *Langmuir*, 2008, **24**, 12680–12689.

39. N. L. Pocard, D. C. Alsmeyer, R. L. McCreery, T. X. Neenan and M. R. Callstrom, *J. Am. Chem. Soc.*, 1992, **114**, 769–771.
40. W. M. Costa, W. S. Cardoso, E. P. Marques, C. W. B. Bezerra, A. A. P. Ferreira, C. Song, J. Zhang and A. L. B. Marques, *J. Braz. Chem. Soc.*, 2013, **24**, 651–656.
41. M.-J. Wang, C. A. Gray, S. R. Reznek, K. Mahmud and Y. Kutsovsky, *Encyclopedia of Polymer Science and Technology*, John Wiley & Sons, Inc., 2002.
42. P. J. F. Harris, *Crit. Rev. Solid State Mater. Sci.*, 2005, **30**, 235–253.
43. M. Carmo, A. R. dos Santos, J. G. R. Poco and M. Linardi, *J. Power Sources*, 2007, **173**, 860–866.
44. E. Antolini, *Appl. Catal., B*, 2009, **88**, 1–24.
45. K. P. De Jong and J. W. Geus, *Catal. Rev.*, 2000, **42**, 481–510.
46. Y. Kim, T. Hayashi, M. Endo and M. Dresselhaus, in *Springer Handbook of Nanomaterials*, ed. R. Vajtai, Springer Berlin Heidelberg, 2013, pp. 233–262.
47. P. Serp, M. Corrias and P. Kalck, *Appl. Catal., A*, 2003, **253**, 337–358.
48. J. H. Bitter, *J. Mater. Chem.*, 2010, **20**, 7312–7321.
49. P. Serp and E. Castillejos, *ChemCatChem.*, 2010, **2**, 41–47.
50. P. Serp, *Carbon Materials for Catalysis*, John Wiley & Sons, Inc., 2008, pp. 309–372.
51. E. Castillejos and P. Serp, *Carbon Nanotubes and Related Structures*, Wiley-VCH Verlag GmbH & Co. KGaA, 2010, pp. 321–347.
52. A. K. Mark, *Dekker Encyclopedia of Nanoscience and Nanotechnology*, Taylor & Francis, 2nd edn, 2009, pp. 2723–2736.
53. B. F. Machado and P. Serp, *Catal. Sci. Technol.*, 2012, **2**, 54–75.
54. C. Huang, C. Li and G. Shi, *Energy Environ. Sci.*, 2012, **5**, 8848–8868.
55. B. Garg and Y.-C. Ling, *Green Mater.*, 2013, **1**, 47–61.
56. J. Pyun, *Angew. Chem., Int. Ed.*, 2011, **50**, 46–48.
57. V. N. Mochalin, O. Shenderova, D. Ho and Y. Gogotsi, *Nat. Nano.*, 2012, **7**, 11–23.
58. A. Barnard, in *Synthesis., Properties and Applications of Ultrananocrystalline Diamond*, ed. D. Gruen, O. Shenderova and A. Vul', Springer Netherlands, 2005, vol. 192, pp. 25–38.
59. J. M. Peralta-Hernández, K. I. B. Eguiluz, A. Hernández-Ramírez, J. L. Guzmán-Mar, L. Hinojosa-Reyes, C. A. Martínez-Huitle and G. R. Salazar-Banda, *Int. J. Electrochem.*, 2012, 154316.
60. O. V. Turova, E. V. Starodubtseva, M. G. Vinogradov, V. I. Sokolov, N. V. Abramova, A. Y. Vul and A. E. Alexenskiy, *Catal. Commun.*, 2011, **12**, 577–579.
61. J. Zhang, D. S. Su, R. Blume, R. Schlögl, R. Wang, X. Yang and A. Gajović, *Angew. Chem., Int. Ed.*, 2010, **49**, 8640–8644.
62. G. P. Bogatyreva, M. A. Marinich, E. V. Ishchenko, V. L. Gvyazdovskaya, G. A. Bazalii and N. A. Oleinik, *Phys. Solid State*, 2004, **46**, 738–741.
63. Y. Cao, X. Luo, H. Yu, F. Peng, H. Wang and G. Ning, *Catal. Sci. Technol.*, 2013, **3**, 2654–2660.

64. E. A. Tveritinova, I. I. Kulakova, Y. N. Zhitnev, A. V. Fionov, A. Lund, W. Chen, I. Buyanova and V. V. Lunin, *Russ. J. Phys. Chem. A*, 2012, **86**, 26–31.
65. L. La-Torre-Riveros, R. Guzman-Blas, A. E. Méndez-Torres, M. Prelas, D. A. Tryk and C. R. Cabrera, *ACS Appl. Mater. Interfaces*, 2012, **4**, 1134–1147.
66. J. K. M and Y. Gogotsi, *Interface*, 2013, **22**, 61–66.
67. D. Ugarte, *Nature*, 1992, **359**, 707–709.
68. M. E. Plonska-Brzezinska and L. Echegoyen, *J. Mater. Chem. A*, 2013, **1**, 13703–13714.
69. V. L. Kuznetsov, A. L. Chuvilin, Y. V. Butenko, I. Y. Mal'kov and V. M. Titov, *Chem. Phys. Lett.*, 1994, **222**, 343–348.
70. T. Braun, M. Wohlers, T. Belz, G. Nowitzke, G. Wortmann, Y. Uchida, N. Pfänder and R. Schlögl, *Catal. Lett.*, 1997, **43**, 167–173.
71. N. F. Goldshleger, *Fullerene Sci. Technol.*, 2001, **9**, 255–280.
72. B. Coq, J. Marc Planeix and V. Brotons, *Appl. Catal., A*, 1998, **173**, 175–183.
73. L. Pacosová, C. Kartusch, P. Kukula and J. A. van Bokhoven, *ChemCatChem.*, 2011, **3**, 154–156.
74. L. L. Lazarus and R. L. Brutchey, *Dalton Trans.*, 2010, **39**, 7888–7890.
75. W. Schneider and W. Diller, *Ullmann's Encyclopedia of Industrial Chemistry*, Wiley-VCH Verlag GmbH & Co. KGaA, 2000.
76. H.-D. Lauss and W. Steffens, *Ullmann's Encyclopedia of Industrial Chemistry*, Wiley-VCH Verlag GmbH & Co. KGaA, 2000.
77. Y. Shao, J. Sui, G. Yin and Y. Gao, *Appl. Catal., B*, 2008, **79**, 89–99.
78. L. Feng, Y. Yan, Y. Chen and L. Wang, *Energy Environ. Sci.*, 2011, **4**, 1892–1899.
79. M. Zhang and L. Dai, *Nano Energy*, 2012, **1**, 514–517.
80. V. J. Watson, C. Nieto Delgado and B. E. Logan, *Environ. Sci. Technol.*, 2013, **47**, 6704–6710.
81. Y. Zhao, R. Nakamura, K. Kamiya, S. Nakanishi and K. Hashimoto, *Nat. Commun.*, 2013, **4**, 2390.
82. D. Lafyatis, R. Mariwala, E. Lowenthal and H. Foley, *Expanded Clays and Other Microporous Solids*, ed. M. Occelli and H. Robson, Springer US, 1992, pp. 318–332.
83. F. Stüber, J. Font, A. Fortuny, C. Bengoa, A. Eftaxias and A. Fabregat, *Top. Catal.*, 2005, **33**, 3–50.
84. D. C. Elliott, *Biofuels., Bioprod. Biorefin.*, 2008, **2**, 254–265.
85. Q. Song, R. Xiao, Y. Li and L. Shen, *Ind. Eng. Chem. Res.*, 2008, **47**, 4349–4357.
86. M. Pérez-Mendoza, M. Domingo-García and F. J. López-Garzón, *Appl. Catal., A*, 2002, **224**, 239–253.
87. T. Anthony Ryan and M. H. Stacey, *Fuel.*, 1984, **63**, 1101–1106.
88. H. Kühl. H. Jüntgen, in *Chemistry & Physics of Carbon*, ed. T. P. A. Kruck, Marcel Dekker, New York, 1989, vol. 22, pp. 145–195.
89. S. C. Petrosius and R. S. Drago, *J. Chem. Soc., Chem. Commun.*, 1992, 344–345.

90. M. Farcasiu, S. C. Petrosius and E. P. Ladner, *J. Catal.*, 1994, **146**, 313–316.
91. N. V. Qui, P. Scholz, T. F. Keller, K. Pollok and B. Ondruschka, *Chem. Eng. Technol.*, 2013, **36**, 300–306.
92. C. Chen, J. Zhang, B. Zhang, C. Yu, F. Peng and D. Su, *Chem. Commun.*, 2013, **49**, 8151–8153.
93. S. Viviane, H. O. Steven and L. Chengdu, *Novel Materials for Catalysis and Fuels Processing*, American Chemical Society, 2013, vol. 1132, pp. 247–258.
94. F. J. Maldonado-Hódar, L. M. Madeira and M. F. Portela, *Appl. Catal., A*, 1999, **178**, 49–60.
95. Z.-J. Sui, J.-H. Zhou, Y.-C. Dai and W.-K. Yuan, *Catal. Today*, 2005, **106**, 90–94.
96. A. Rinaldi, J. Zhang, B. Frank, D. S. Su, S. B. Abd Hamid and R. Schlögl, *ChemSusChem.*, 2010, **3**, 254–260.
97. J. Zhang, X. Liu, R. Blume, A. Zhang, R. Schlögl and D. S. Su, *Science*, 2008, **322**, 73–77.
98. X. Liu, B. Frank, W. Zhang, T. P. Cotter, R. Schlögl and D. S. Su, *Angew. Chem., Int. Ed.*, 2011, **50**, 3318–3322.
99. J. de Jesús Díaz Velásquez, L. M. C. Suárez and J. L. Figueiredo, *Appl. Catal., A*, 2006, **311**, 51–57.
100. V. Schwartz, H. Xie, H. M. Meyer Iii, S. H. Overbury and C. Liang, *Carbon*, 2011, **49**, 659–668.
101. C. Liang, H. Xie, V. Schwartz, J. Howe, S. Dai and S. H. Overbury, *J. Am. Chem. Soc.*, 2009, **131**, 7735–7741.
102. V. Schwartz, W. Fu, Y.-T. Tsai, H. M. Meyer, A. J. Rondinone, J. Chen, Z. Wu, S. H. Overbury and C. Liang, *ChemSusChem.*, 2013, **6**, 840–846.
103. R. W. Coughlin, *Prod. Res. Dev.*, 1969, **8**, 12–23.
104. V. V. Strelko, V. S. Kuts and P. A. Thrower, *Carbon*, 2000, **38**, 1499–1503.
105. S. Chen, J. Bi, Y. Zhao, L. Yang, C. Zhang, Y. Ma, Q. Wu, X. Wang and Z. Hu, *Adv. Mater.*, 2012, **24**, 5593–5597.
106. K. Ai, Y. Liu, C. Ruan, L. Lu and G. Lu, *Adv. Mater.*, 2013, **25**, 998–1003.
107. K. Gong, F. Du, Z. Xia, M. Durstock and L. Dai, *Science*, 2009, **323**, 760–764.
108. L. Qu, Y. Liu, J.-B. Baek and L. Dai, *ACS Nano.*, 2010, **4**, 1321–1326.
109. L. Zhang and Z. Xia, *J. Phys. Chem. C*, 2011, **115**, 11170–11176.
110. Y. Feng, F. Li, Z. Hu, X. Luo, L. Zhang, X.-F. Zhou, H.-T. Wang, J.-J. Xu and E. G. Wang, *Phys. Rev. B*, 2012, **85**, 155454.
111. C. H. San and C. W. Hong, *J. Electrochem. Soc.*, 2012, **159**, K116–K121.
112. F. Carrasco-Marín, A. Mueden and C. Moreno-Castilla, *J. Phys. Chem. B*, 1998, **102**, 9239–9244.
113. G. S. Szymański, Z. Karpiński, S. Biniak and A. Świątkowski, *Carbon*, 2002, **40**, 2627–2639.
114. J. L. Figueiredo and M. F. R. Pereira, *Catal. Today*, 2010, **150**, 2–7.
115. C. Moreno-Castilla, F. Carrasco-Marín, C. Parejo-Pérez and M. V. López Ramón, *Carbon*, 2001, **39**, 869–875.

116. J. Bedia, R. Barrionuevo, J. Rodríguez-Mirasol and T. Cordero, *Appl. Catal., B*, 2011, **103**, 302–310.
117. G. S. Szymański and G. Rychlicki, *Carbon*, 1993, **31**, 247–257.
118. H. Teng, Y.-T. Tu, Y.-C. Lai and C.-C. Lin, *Carbon*, 2001, **39**, 575–582.
119. G. S. Szymański, T. Grzybek and H. Papp, *Catal. Today*, 2004, **90**, 51–59.
120. I. Mochida, Y. Kawabuchi, S. Kawano, Y. Matsumura and M. Yoshikawa, *Fuel*, 1997, **76**, 543–548.
121. E. Raymundo-Piñero, D. Cazorla-Amorós and A. Linares-Solano, *Carbon*, 2003, **41**, 1925–1932.
122. F. Adib, A. Bagreev and T. J. Bandosz, *Langmuir*, 1999, **16**, 1980–1986.
123. C. Sotowa, Y. Watanabe, S. Yatsunami, Y. Korai and I. Mochida, *Appl. Catal., A*, 1999, **180**, 317–323.
124. L. B. Khalil, B. S. Girgis and T. A. M. Tawfik, *J. Chem. Technol. Biotechnol.*, 2001, **76**, 1132–1140.
125. R. S. Ribeiro, A. M. T. Silva, J. L. Figueiredo, J. L. Faria and H. T. Gomes, *Carbon*, 2013, **62**, 97–108.
126. M. Sánchez-Polo, U. von Gunten and J. Rivera-Utrilla, *Water Res.*, 2005, **39**, 3189–3198.
127. J. Rivera-Utrilla and M. Sánchez-Polo, *Langmuir*, 2004, **20**, 9217–9222.
128. C. Aguilar, R. García, G. Soto-Garrido and R. Arriagada, *Appl. Catal., B*, 2003, **46**, 229–237.
129. R. P. Rocha, J. P. S. Sousa, A. M. T. Silva, M. F. R. Pereira and J. L. Figueiredo, *Appl. Catal., B*, 2011, **104**, 330–336.
130. M. Santiago, F. Stüber, A. Fortuny, A. Fabregat and J. Font, *Carbon*, 2005, **43**, 2134–2145.
131. J. N. Armor, *Appl. Catal., B*, 1992, **1**, 221–256.
132. B. Stöhr, H. P. Boehm and R. Schlögl, *Carbon*, 1991, **29**, 707–720.
133. J. P. Boudou, *Carbon*, 2003, **41**, 1955–1963.
134. F. Sun, J. Liu, H. Chen, Z. Zhang, W. Qiao, D. Long and L. Ling, *ACS Catal.*, 2013, **3**, 862–870.
135. T. Suzuki, T. Kyotani and A. Tomita, *Ind. Eng. Chem. Res.*, 1994, **33**, 2840–2845.
136. M.-C. Huang and H. Teng, *Carbon*, 2003, **41**, 951–957.
137. H.-P. Boehm, *Carbon Materials for Catalysis*, John Wiley & Sons, Inc., 2008, pp. 219–265.
138. G. Yu, S. Lu, H. Chen and Z. Zhu, *Energy Fuels.*, 2004, **19**, 447–452.
139. L. C. A. Oliveira, C. N. Silva, M. I. Yoshida and R. M. Lago, *Carbon*, 2004, **42**, 2279–2284.
140. M. E. Suarez-Ojeda, F. Stüber, A. Fortuny, A. Fabregat, J. Carrera and J. Font, *Appl. Catal., B*, 2005, **58**, 105–114.
141. A. Fortuny, J. Font and A. Fabregat, *Appl. Catal., B*, 1998, **19**, 165–173.
142. J. Ma, M.-H. Sui, Z.-L. Chen and L.-N. Wang, *Ozone: Sci. Eng.*, 2004, **26**, 3–10.
143. U. Jans and J. Hoigné, *Ozone: Sci. Eng.*, 1998, **20**, 67–90.

144. P. C. C. Faria, J. J. M. Órfão and M. F. R. Pereira, *Appl. Catal., B*, 2008, **79**, 237–243.
145. A. Georgi and F.-D. Kopinke, *Appl. Catal., B*, 2005, **58**, 9–18.
146. S. Navalon, A. Dhakshinamoorthy, M. Alvaro and H. Garcia, *ChemSusChem.*, 2011, **4**, 1712–1730.
147. B. Fidalgo and J. Á. MenÉNdez, *Chin. J. Catal.*, 2011, **32**, 207–216.
148. X. Xu, Y. Matsumura, J. Stenberg and M. J. Antal, *Ind. Eng. Chem. Res.*, 1996, **35**, 2522–2530.
149. T.-J. Zhao, W.-Z. Sun, X.-Y. Gu, M. Rønning, D. Chen, Y.-C. Dai, W.-K. Yuan and A. Holmen, *Appl. Catal., A*, 2007, **323**, 135–146.
150. B. F. Machado, M. Oubenali, M. Rosa Axet, T. Trang Nguyen, M. Tunckol, M. Girleanu, O. Ersen, I. C. Gerber and P. Serp, *J. Catal.*, 2014, **309**, 185–198.
151. J. Zhu, A. Holmen and D. Chen, *ChemCatChem.*, 2013, **5**, 378–401.
152. A. Villa, N. Janjic, P. Spontoni, D. Wang, D. S. Su and L. Prati, *Appl. Catal., A*, 2009, **364**, 221–228.
153. T. Nunoura, G. Lee, Y. Matsumura and K. Yamamoto, *Ind. Eng. Chem. Res.*, 2003, **42**, 3522–3531.
154. S. Yao, C. Yang, Y. Tan and Y. Han, *Catal. Commun.*, 2008, **9**, 2107–2111.
155. M. Sánchez-Polo, R. Leyva-Ramos and J. Rivera-Utrilla, *Carbon*, 2005, **43**, 962–969.
156. H. Wang, T. Maiyalagan and X. Wang, *ACS Catal.*, 2012, **2**, 781–794.
157. Z. Liu, H. Nie, Z. Yang, J. Zhang, Z. Jin, Y. Lu, Z. Xiao and S. Huang, *Nanoscale*, 2013, **5**, 3283–3288.
158. C. H. Choi, M. W. Chung, Y. J. Jun and S. I. Woo, *RSC Adv.*, 2013, **3**, 12417–12422.
159. F. Sun, J. Wang, H. Chen, W. Qiao, L. Ling and D. Long, *Sci. Rep.*, 2013, **3**, 2823.
160. Z. Yang, Z. Yao, G. Li, G. Fang, H. Nie, Z. Liu, X. Zhou, X. a. Chen and S. Huang, *ACS Nano.*, 2011, **6**, 205–211.
161. R. Wang, D. C. Higgins, M. A. Hoque, D. Lee, F. Hassan and Z. Chen, *Sci. Rep.*, 2013, 3.
162. J. Y. Shimin Kang and J. Chang, *Int. Rev. Chem. Eng.*, 2012, **5**, 133–144.
163. D. Yu, E. Nagelli, F. Du and L. Dai, *J. Phys. Chem. Lett.*, 2010, **1**, 2165–2173.
164. E. Auer, A. Freund, J. Pietsch and T. Tacke, *Appl. Catal., A*, 1998, **173**, 259–271.
165. V. Arunajatesan, B. Chen, K. Möbus, D. J. Ostgard, T. Tacke and D. Wolf, *Carbon Materials for Catalysis*, John Wiley & Sons, Inc., 2008, pp. 535–572.
166. V. Calvino-Casilda, A. J. López-Peinado, C. J. Durán-Valle and R. M. Martín-Aranda, *Catal. Rev.*, 2010, **52**, 325–380.
167. W. Zhang, P. Sherrell, A. I. Minett, J. M. Razal and J. Chen, *Energy Environ. Sci.*, 2010, **3**, 1286–1293.
168. S. Tang, G. Sun, J. Qi, S. Sun, J. Guo, Q. Xin and G. M. Haarberg, *Chin. J. Catal.*, 2010, **31**, 12–17.

169. L. Mabena, S. Sinha Ray, S. Mhlanga and N. Coville, *Appl. Nanosci.*, 2011, **1**, 67–77.
170. E. Bailón-García, F. Maldonado-Hódar, A. Pérez-Cadenas and F. Carrasco-Marín, *Catalysts*, 2013, **3**, 853–877.
171. J. L. Figueiredo, *J. Mater. Chem. A*, 2013, **1**, 9351–9364.
172. J. H. Bitter and K. P. de Jong, *Carbon Materials for Catalysis*, John Wiley & Sons, Inc., 2008, pp. 157–176.
173. M. L. Toebes, J. A. van Dillen and K. P. de Jong, *J. Mol. Catal. A: Chem.*, 2001, **173**, 75–98.
174. A. Martí'n-Gullón, C. Prado-Burguete and F. Rodríguez-Reinoso, *Carbon*, 1993, **31**, 1099–1105.
175. A. Cabiac, T. Cacciaguerra, P. Trens, R. Durand, G. Delahay, A. Medevielle, D. Plée and B. Coq, *Appl. Catal., A*, 2008, **340**, 229–235.
176. M. B. Dawidziuk, F. Carrasco-Marín and C. Moreno-Castilla, *Carbon*, 2009, **47**, 2679–2687.
177. Y. Liu, C. Ji, W. Gu, J. Jorne and H. A. Gasteiger, *J. Electrochem. Soc.*, 2011, **158**, B614–B621.
178. M. E. Gálvez, A. C. Alegre, D. Sebastián, R. Moliner and M. J. Lázaro, *Int. J. Electrochem.*, 2012, 267893.
179. Z. Kowalczyk, S. Jodzis, W. Raróg, J. Zieliński, J. Pielaszek and A. Presz, *Appl. Catal., A*, 1999, **184**, 95–102.
180. G. S. Chai, S. B. Yoon, J.-S. Yu, J.-H. Choi and Y.-E. Sung, *J. Phys. Chem. B*, 2004, **108**, 7074–7079.
181. X. Pan and X. Bao, *Nanomaterials in Catalysis*, Wiley-VCH Verlag GmbH & Co. KGaA, 2013, pp. 415–441.
182. X. Pan and X. Bao, *Acc. Chem. Res.*, 2011, **44**, 553–562.
183. F. Goettmann and C. Sanchez, *J. Mater. Chem.*, 2007, **17**, 24–30.
184. J. Laine, F. Severino and M. Labady, *J. Catal.*, 1994, **147**, 355–357.
185. J. Laine, M. Labady, F. Severino and S. Yunes, *J. Catal.*, 1997, **166**, 384–387.
186. D. L. Trimm and B. J. Cooper, *J. Chem. Soc. D.*, 1970, 477–478.
187. S. Kogan, M. V. Landau, M. Herskowitz and J. E. Koresh, in *Studies in Surface Science and Catalysis*, ed. J. Barbier, J. Barrault, C. Bouchoule, D. Duprez, G. Pérot, M. Guisnet and C. Montassier, Elsevier, 1993, vol. 78, pp. 353–359.
188. P. C. L'Argentière, M. E. Quiroga, D. A. Liprandi, E. A. Cagnola, M. C. Román-Martínez, J. A. Díaz-Auñón and C. Salinas-Martínez de Lecea, *Catal. Lett.*, 2003, **87**, 97–101.
189. M. Peer, A. Qajar, B.-P. M. Holbrook, R. Rajagopalan and H. C. Foley, *Carbon*, 2013, **57**, 485–497.
190. F. Glenk, T. Knorr, M. Schirmer, S. Gütlein and B. J. M. Etzold, *Chem. Eng. Technol.*, 2010, **33**, 698–703.
191. K. Miura, J. i. Hayashi, T. Kawaguchi and K. Hashimoto, *Carbon*, 1993, **31**, 667–674.
192. D. S. Lafyatis and H. C. Foley, *Chem. Eng. Sci.*, 1990, **45**, 2567–2574.

193. F. J. Derbyshire, V. H. J. de Beer, G. M. K. Abotsi, A. W. Scaroni, J. M. Solar and D. J. Skrovanek, *Appl. Catal.*, 1986, **27**, 117–131.
194. F. Rodriguez-Reinoso, C. Salinas-Martinez de Lecea, A. Sepulveda-Escribano and J. D. Lopez-Gonzalez, *Catal. Today*, 1990, **7**, 287–298.
195. H. Xiong, M. Moyo, M. A. M. Motchelaho, L. L. Jewell and N. J. Coville, *Appl. Catal., A*, 2010, **388**, 168–178.
196. F. M. Duarte, F. J. Maldonado-Hódar and L. M. Madeira, *Appl. Catal., A*, 2013, **458**, 39–47.
197. M. Gurrath, T. Kuretzky, H. P. Boehm, L. B. Okhlopkova, A. S. Lisitsyn and V. A. Likholobov, *Carbon*, 2000, **38**, 1241–1255.
198. T. Trang Nguyen and P. Serp, *ChemCatChem.*, 2013, **5**, 3595–3603.
199. M. Wenkin, R. Touillaux, P. Ruiz, B. Delmon and M. Devillers, *Appl. Catal., A*, 1996, **148**, 181–199.
200. S. Echeandia, P. L. Arias, V. L. Barrio, B. Pawelec and J. L. G. Fierro, *Appl. Catal., B*, 2010, **101**, 1–12.
201. L. D'Souza, J. R. Regalbuto and J. T. Miller, *J. Catal.*, 2008, **254**, 157–169.
202. X. Hao, L. Quach, J. Korah, W. A. Spieker and J. R. Regalbuto, *J. Mol. Catal. A: Chem.*, 2004, **219**, 97–107.
203. S. Lambert, N. Job, L. D'Souza, M. F. R. Pereira, R. Pirard, B. Heinrichs, J. L. Figueiredo, J.-P. Pirard and J. R. Regalbuto, *J. Catal.*, 2009, **261**, 23–33.
204. Y. Verde, G. Alonso, V. Ramos, H. Zhang, A. J. Jacobson and A. Keer, *Appl. Catal., A*, 2004, **277**, 201–207.
205. X. Yu and S. Ye, *J. Power Sources*, 2007, **172**, 133–144.
206. L. Jiao, Y. Zha, X. Hao and J. R. Regalbuto, in *Studies in Surface Science and Catalysis*, ed. M. Devillers, D. E. De Vos, S. Hermans, P. A. Jacobs, J. A. Martens, E.M. Gaigneaux and P. Ruiz, Elsevier, 2006, vol. 162, pp. 211–218.
207. M. C. Román-Martínez, D. Cazorla-Amorós, A. Linares-Solano, C. S. -M. n. De Lecea, H. Yamashita and M. Anpo, *Carbon*, 1995, **33**, 3–13.
208. C. Prado-Burguete, A. Linares-Solano, F. Rodríguez-Reinoso and C. S.-M. de Lecea, *J. Catal.*, 1989, **115**, 98–106.
209. E. Antolini, *J. Mater. Sci.*, 2003, **38**, 2995–3005.
210. F. Coloma, A. Sepulveda-Escribano, J. L. G. Fierro and F. Rodriguez-Reinoso, *Langmuir*, 1994, **10**, 750–755.
211. D. Richard and P. Gallezot, in *Studies in Surface Science and Catalysis*, ed. P. Grange, P. A. Jacobs, B.Delmon and G. Poncelet, Elsevier, 1987, vol. 31, pp. 71–81.
212. C. Prado-Burguete, A. Linares-Solano, F. Rodriguez-Reinoso and C. S.-M. De Lecea, *J. Catal.*, 1991, **128**, 397–404.
213. W. Han, H. Liu and H. Zhu, *Catal. Commun.*, 2007, **8**, 351–354.
214. Z. H. Zhu, S. Wang, G. Q. Lu and D. K. Zhang, *Catal. Today*, 1999, **53**, 669–681.
215. E. G. Rodrigues, M. F. R. Pereira, X. Chen, J. J. Delgado and J. J. M. Órfão, *J. Catal.*, 2011, **281**, 119–127.
216. X. Wang, N. Li, J. A. Webb, L. D. Pfefferle and G. L. Haller, *Appl. Catal., B*, 2010, **101**, 21–30.

217. F. Coloma, A. Sepúlveda-Escribano, J. L. G. Fierro and F. Rodríguez-Reinoso, *Appl. Catal., A*, 1997, **150**, 165–183.
218. M. L. Toebes, F. F. Prinsloo, J. H. Bitter, A. J. van Dillen and K. P. de Jong, *J. Catal.*, 2003, **214**, 78–87.
219. M. L. Toebes, Y. Zhang, J. Hájek, T. Alexander Nijhuis, J. H. Bitter, A. Jos van Dillen, D. Y. Murzin, D. C. Koningsberger and K. P. de Jong, *J. Catal.*, 2004, **226**, 215–225.
220. M. L. Toebes, T. Alexander Nijhuis, J. Hájek, J. H. Bitter, A. Jos van Dillen, D. Y. Murzin and K. P. de Jong, *Chem. Eng. Sci.*, 2005, **60**, 5682–5695.
221. A. Solhy, B. F. Machado, J. Beausoleil, Y. Kihn, F. Gonçalves, M. F. R. Pereira, J. J. M. Órfão, J. L. Figueiredo, J. L. Faria and P. Serp, *Carbon*, 2008, **46**, 1194–1207.
222. J. Teddy, A. Falqui, A. Corrias, D. Carta, P. Lecante, I. Gerber and P. Serp, *J. Catal.*, 2011, **278**, 59–70.
223. H. Vu, F. Gonçalves, R. Philippe, E. Lamouroux, M. Corrias, Y. Kihn, D. Plee, P. Kalck and P. Serp, *J. Catal.*, 2006, **240**, 18–22.
224. J. P. Breen, R. Burch, J. Gomez-Lopez, K. Griffin and M. Hayes, *Appl. Catal., A*, 2004, **268**, 267–274.
225. M. A. Fraga, M. J. Mendes and E. Jordão, *J. Mol. Catal. A: Chem.*, 2002, **179**, 243–251.
226. T. Tang, C. Yin, N. Xiao, M. Guo and F.-S. Xiao, *Catal. Lett.*, 2009, **127**, 400–405.
227. C. Amorim and M. A. Keane, *J. Chem. Technol. Biotechnol.*, 2008, **83**, 662–672.
228. L. Zhao, J. Zhou, H. Chen, M. Zhang, Z. Sui and X. Zhou, *Korean J. Chem. Eng.*, 2010, **27**, 1412–1418.
229. A. Pigamo, M. Besson, B. Blanc, P. Gallezot, A. Blackburn, O. Kozynchenko, S. Tennison, E. Crezee and F. Kapteijn, *Carbon*, 2002, **40**, 1267–1278.
230. M. F. R. Pereira, J. J. M. Órfão and J. L. Figueiredo, *Appl. Catal., A*, 1999, **184**, 153–160.
231. A. Guerrero-Ruiz, I. Rodriguez-Ramos, F. Rodriguez-Reinoso, C. Moreno-Castilla and J. D. López-González, *Carbon*, 1988, **26**, 417–423.
232. S. Abate, R. Arrigo, M. E. Schuster, S. Perathoner, G. Centi, A. Villa, D. Su and R. Schlögl, *Catal. Today*, 2010, **157**, 280–285.
233. V. Z. Radkevich, T. L. Senko, K. Wilson, L. M. Grishenko, A. N. Zaderko and V. Y. Diyuk, *Appl. Catal., A*, 2008, **335**, 241–251.
234. F. R. García-García, J. Álvarez-Rodríguez, I. Rodríguez-Ramos and A. Guerrero-Ruiz, *Carbon*, 2010, **48**, 267–276.
235. B. Choi, H. Yoon, I.-S. Park, J. Jang and Y.-E. Sung, *Carbon*, 2007, **45**, 2496–2501.
236. T. Maiyalagan, B. Viswanathan and U. V. Varadaraju, *Electrochem. Commun.*, 2005, **7**, 905–912.
237. C.-H. Hsu, H.-M. Wu and P.-L. Kuo, *Chem. Commun.*, 2010, **46**, 7628–7630.
238. F. Hasché, T.-P. Fellinger, M. Oezaslan, J. P. Paraknowitsch, M. Antonietti and P. Strasser, *ChemCatChem.*, 2012, **4**, 479–483.

239. T. Zhou, H. Wang, J. Key, S. Ji, V. Linkov and R. Wang, *RSC Adv.*, 2013, **3**, 16949–16953.
240. X. Tuaev, J. P. Paraknowitsch, R. Illgen, A. Thomas and P. Strasser, *Phys. Chem. Chem. Phys.*, 2012, **14**, 6444–6447.
241. Z. Liu, F. Su, X. Zhang and S. W. Tay, *ACS Appl. Mater. Interfaces*, 2011, **3**, 3824–3830.
242. B. P. Vinayan and S. Ramaprabhu, *Nanoscale*, 2013, **5**, 5109–5118.
243. W. Xizhang, X. Hua, Y. Lijun, W. Huakai, Z. Pengyuan, Q. Xintai, W. Yangnian, M. Yanwen, W. Qiang and H. Zheng, *Nanotechnology*, 2011, **22**, 395401.
244. H. T. Chung, J. H. Won and P. Zelenay, *Nat. Commun.*, 2013, **4**, 1922.
245. K. Parvez, S. Yang, Y. Hernandez, A. Winter, A. Turchanin, X. Feng and K. Müllen, *ACS Nano.*, 2012, **6**, 9541–9550.
246. T. Holme, Y. Zhou, R. Pasquarelli and R. O'Hayre, *Phys. Chem. Chem. Phys.*, 2010, **12**, 9461–9468.
247. S. Pylypenko, A. Borisevich, K. L. More, A. R. Corpuz, T. Holme, A. A. Dameron, T. S. Olson, H. N. Dinh, T. Gennett and R. O'Hayre, *Energy Environ. Sci.*, 2013, **6**, 2957–2964.
248. Y. Zhang, J. Ge, L. Wang, D. Wang, F. Ding, X. Tao and W. Chen, *Sci. Rep.*, 2013, **3**, 2771.
249. H. J. Schulte, B. Graf, W. Xia and M. Muhler, *ChemCatChem.*, 2012, **4**, 350–355.
250. L. Wachowski, W. Skupiński and M. Hofman, *Appl. Catal., A*, 2006, **303**, 230–233.
251. C. Milone, C. Gangemi, R. Ingoglia, G. Neri and S. Galvagno, *Appl. Catal., A*, 1999, **184**, 89–94.
252. J. Phillips, J. Weigle, M. Herskowitz and S. Kogan, *Appl. Catal., A*, 1998, **173**, 273–287.
253. F. Coloma, A. Sepúlveda-Escribano, J. L. G. Fierro and F. Rodríguez-Reinoso, *Appl. Catal., A*, 1996, **136**, 231–248.
254. F. Coloma, A. Sepúlveda-Escribano, J. L. G. Fierro and F. Rodríguez-Reinoso, *Appl. Catal., A*, 1996, **148**, 63–80.
255. A. A. Chen, M. A. Vannice and J. Phillips, *J. Phys. Chem.*, 1987, **91**, 6257–6269.
256. T. Prasomsri, D. Shi and D. E. Resasco, *Chem. Phys. Lett.*, 2010, **497**, 103–107.
257. I. Efremenko and M. Sheintuch, *J. Catal.*, 2003, **214**, 53–67.
258. Y.-A. Lv, Y.-H. Cui, X.-N. Li, X.-Z. Song, J.-G. Wang and M. Dong, *Phys. E.*, 2010, **42**, 1746–1750.
259. D. Wang, A. Villa, D. Su, L. Prati and R. Schlögl, *ChemCatChem.*, 2013, **5**, 2717–2723.
260. M. M. Keyser and F. F. Prinsloo, in *Studies in Surface Science and Catalysis*, ed. B. H. Davis and M. L. Occelli, Elsevier, 2007, vol. 163, pp. 45–73.
261. H. J. Jung, P. L. Walker Jr. and A. Vannice, *J. Catal.*, 1982, **75**, 416–422.

262. J. W. Niemantsverdriet, A. M. Van der Kraan, W. N. Delgass and M. A. Vannice, *J. Phys. Chem.*, 1985, **89**, 67–72.
263. A. Chen, M. Kaminsky, G. L. Geoffroy and M. A. Vannice, *J. Phys. Chem.*, 1986, **90**, 4810–4819.
264. D. Pantea, H. Darmstadt, S. Kaliaguine and C. Roy, *Appl. Surf. Sci.*, 2003, **217**, 181–193.
265. F. Maillard, P. A. Simonov and E. R. Savinova, *Carbon Materials for Catalysis*, John Wiley & Sons, Inc., 2008, pp. 429–480.
266. J. M. Thomas, R. Raja and D. W. Lewis, *Angew. Chem., Int. Ed.*, 2005, **44**, 6456–6482.
267. V. Dal Santo, F. Liguori, C. Pirovano and M. Guidotti, *Molecules*, 2010, **15**, 3829–3856.
268. C. Avelino, C. Patricia, B. Mercedes, J. S. Maria, N. Javier, Y. Miguel José, L. Eduardo, P. Alvaro, M. A. López-Quintela, B. David, M. Ernest, G. Gemma and M. Alvaro, *Nat. Chem.*, 2013, **5**, 775–781.
269. D. S. Su, S. Perathoner and G. Centi, *Chem. Rev.*, 2013, **113**, 5782–5816.
270. A. Schaetz, M. Zeltner and W. J. Stark, *ACS Catal.*, 2012, **2**, 1267–1284.
271. D. Haag and H. Kung, *Top. Catal.*, 2014, **57**, 762–773.
272. B. Garg and Y.-C. Ling, *Green Mater.*, 2013, **1**, 47–61.
273. X. Zhou, J. Qiao, L. Yang and J. Zhang, *Adv. Energy Mater.*, 2014, **4**, 1301523.
274. Q. Li, R. Cao, J. Cho and G. Wu, *Adv. Energy Mater.*, 2014, **4**, 1301415.
275. G. Blanita and M. D. Lazar, *Micro Nanosys*, 2013, **5**, 138–146.
276. M. K. van der Lee, A. J. van Dillen, J. W. Geus, K. P. de Jong and J. H. Bitter, *Carbon*, 2006, **44**, 629–637.

CHAPTER 2

Classification, Structure and Bulk Properties of Nanostructured Carbon Materials

2.1 Classification

The ground state electronic configuration of the carbon atom is ($1s^2$) ($2s^2 2p_x 2p_y$). The electrons in the innermost shell of a carbon atom constitute the electronic core, which is sufficiently compact to allow the outer valence electrons to organize themselves to hybridize, so as to form linear (sp), planar (sp^2), or tetrahedral bonds (sp^3) with the electrons of neighbouring atoms. With these three hybrid orbitals, sp^3, sp^2, and sp, carbon atoms show a variety of combination of chemical bonds. For kinetic and mainly thermodynamic issues, the sp^2 (or $sp^{2+\delta}$) and sp^3 carbon forms are the predominant ones for inorganic carbon materials; the sp hybridization being limited to scarce surface carbon species.

The inorganic carbon family has been classified by Inagaki and Feyu into four basic forms: diamond, graphite, fullerene and carbyne.[1] The nanocarbon family outlined in Figure 2.1 is based on two major characteristics: the type of carbon atom hybridization and the characteristic size. Starting with a description of the bonding nature, this scheme analyzes how different classes of carbon networks are formed with increasing characteristic size of the carbon structure. The two first circles of the figure consist of organic (macro)molecules, and the hierarchy of carbon materials can be described as an extension of organic molecular species to inorganic all-carbon materials

RSC Catalysis Series No. 23
Nanostructured Carbon Materials for Catalysis
By Philippe Serp and Bruno Machado

Published by the Royal Society of Chemistry, www.rsc.org

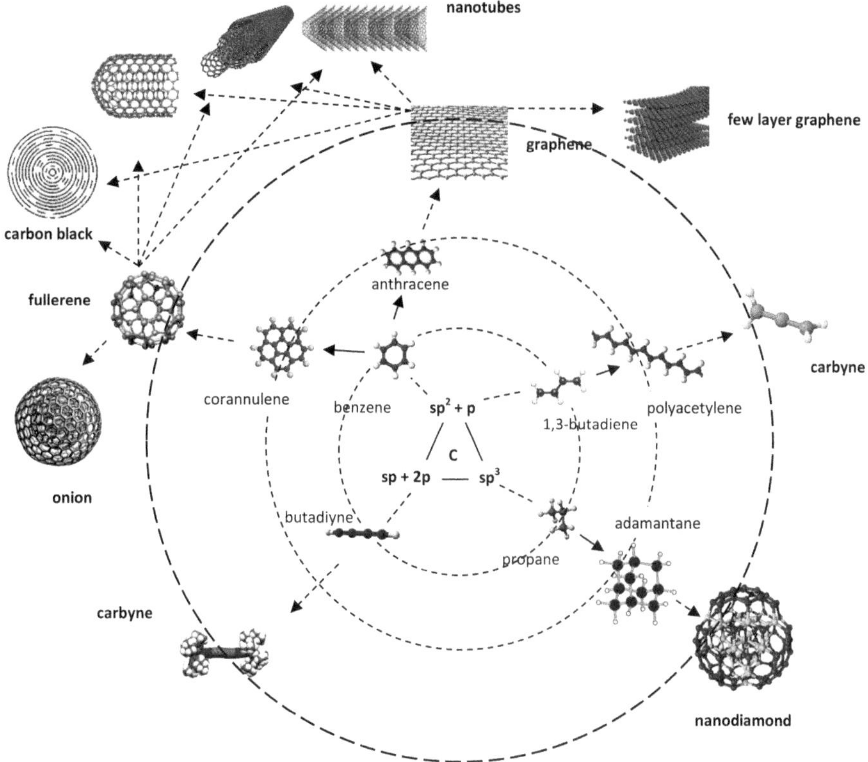

Figure 2.1 Classification of carbon nanomaterials according to their hybridization.

through a wide variety of carbon entities in the nanoscopic size range. The third circle is the basic structural units in the carbon nanoworld: fullerenes, graphene, carbyne and nanodiamond clusters. The existence of carbyne or linear acetylenic carbon of chemical structure $-(C{\equiv}C)_n-$ with sp orbital hybridization as a carbon allotrope is still controversial.[2] It should be noted that topological similarities between organic species and inorganic materials are emphasized here, not the synthesis of the units.

The next structural level, with a corresponding increase of the characteristic sizes, can be considered as consisting of assemblies of the structural units, ranging from simple forms, such as multi-walled CNT (MWCNT) or few layer graphene (FLG), to more complicated carbon architectures, such as carbon black. A recent study suggests that such a complex chemistry can occur in a simple candle flame, where four known carbon forms (diamond, graphitic, fullerenic and amorphous particles) have been identified.[3] From this scheme, it is obvious that one of the approaches to classify carbon nanostructures is based on combination of the type of hybridization of carbon bonds within the structure and characteristic size of the structure.

Graphene and FLG consist in pure sp^2 nanocarbon. It is worth noting that for those nanomaterials (as well as for carbon nanofibers), edge effects will be increasingly important in catalysis as the size of the graphene flakes decreases. The edge morphology tremendously affects not only the chemical reactivity but also the edge stabilities, which determine the band gap properties, magnetic and optical behavior of graphene. The structure of graphene edges has been intensively studied,[4–8] and will be discussed in Section 2.2. Nanodiamonds can consist of pure sp^3 nanocarbon. Whereas the bare (non-functionalized) surfaces of cubic crystals exhibit structures similar to bulk diamond, the surfaces of octahedral, cuboctahedral and spherical clusters show a transition from sp^3 carbon to sp^2 carbon.[9] The surface of sp^3 clusters must be either stabilized through termination with functional groups or reconstructed into sp^2 carbon.

Aromatic bonding of carbon results in the formation of sheets, which may vary in extent and even possess significant curvature or even closure, as in the case of nanotubes and fullerenes. Because of the curvature of the surface, fullerene and also small diameter CNT hybridization falls between graphite (sp^2) and diamond (sp^3) and these new carbon allotropes are therefore of intermediate, and perhaps variable hybridization. Computer calculations and experimental evidences indicate that curvature, introduced through different kinds of 'defects', produces stable atomic arrangements. These 'defects', which can be pentagons, heptagons, rings with more atoms, or more complex architectures, change the electronic, magnetic, and mechanical properties of the structure.

A classification of carbon nanomaterials based upon the scheme and degree of preferred orientation of anisotropic layers has also been proposed, as illustrated in Figure 2.2.[1] Since these textures are constructed by fundamental nano-sized structural units, they are called nano-textures. First, random and oriented nano-textures are differentiated, and then the latter classified by the orientation scheme: parallel to the reference plane (planar orientation) or around the reference point (point orientation). The extreme case of planar orientation, *i.e.* perfect orientation with large-sized planes, is few layer graphene. Axial orientation of layers is found in some fibrous carbon materials as CNTs, some nanofilaments and vapor-grown carbon fibers; the fibrous macro-structure morphology is possible because of this axial orientation scheme. Note that this is not the case for some carbon nanofibers, which present a fibrous texture and graphene layers presenting an angle, or even being perpendicular, to the reference axis (see Section 1.2.4). In point orientation, concentric and radial alignments also have to be differentiated.

The extremes of concentric point orientation are the family of fullerenes. They can also be found in carbon onions and in CB particles formed by minute hexagonal carbon layers. Radial alignment of layers to form a spheroidal macrostructure is found in carbon spherules, which are formed from a mixture of polyethylene and polyvinylchloride by pressure carbonization.[10]

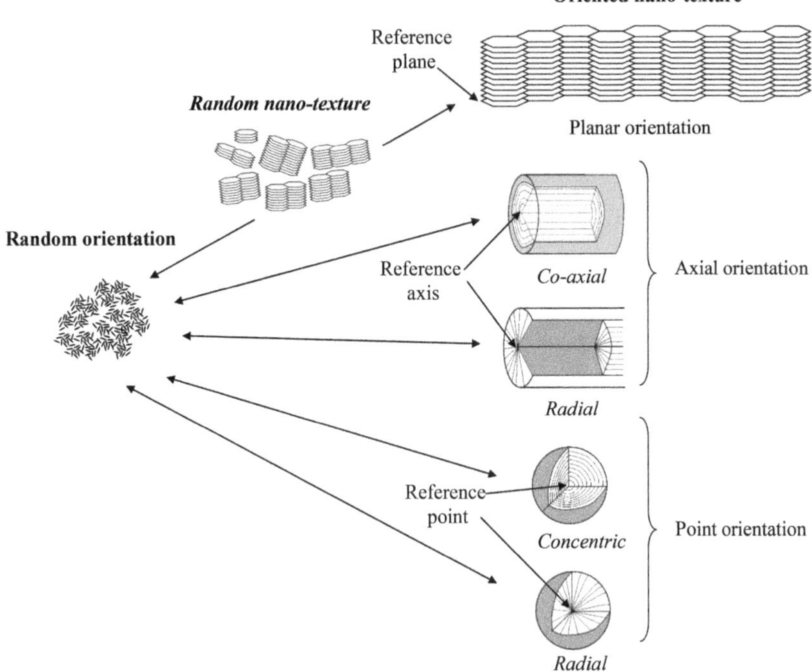

Figure 2.2 Classification of nano-textures in carbon nanomaterials. Adapted from ref. 1.

The structure of mesophase spheres is close to the radial point orientation scheme, but in their centres the orientation of layers is not radial. The texture with random orientation occurs in soot and some carbon nanodots derived from soot oxidation.[11,12]

The arrangement of carbon nanomaterials in 2-D or 3-D-macrostructures (Figure 2.3) is also important for catalytic applications, since this arrangement will dictate material porosity and impact the adsorption phenomena (see Chapter 3).

Although the smallest indivisible unit of carbon black is the aggregate, in the TEM images aggregates appear to be formed by spherical particles, which are fused together. Aggregates connect through Van der Waals forces into networks called agglomerates. The internal structure of aggregates is not well understood.[13]

The ability to assemble carbon nanomaterials into macrostructures that exhibit the properties displayed by the carbon nanomaterials on the nanoscale is the key to utilizing them in practical applications, such as chemical sensors, or catalyst supports. This issue will be discussed in Chapter 10.

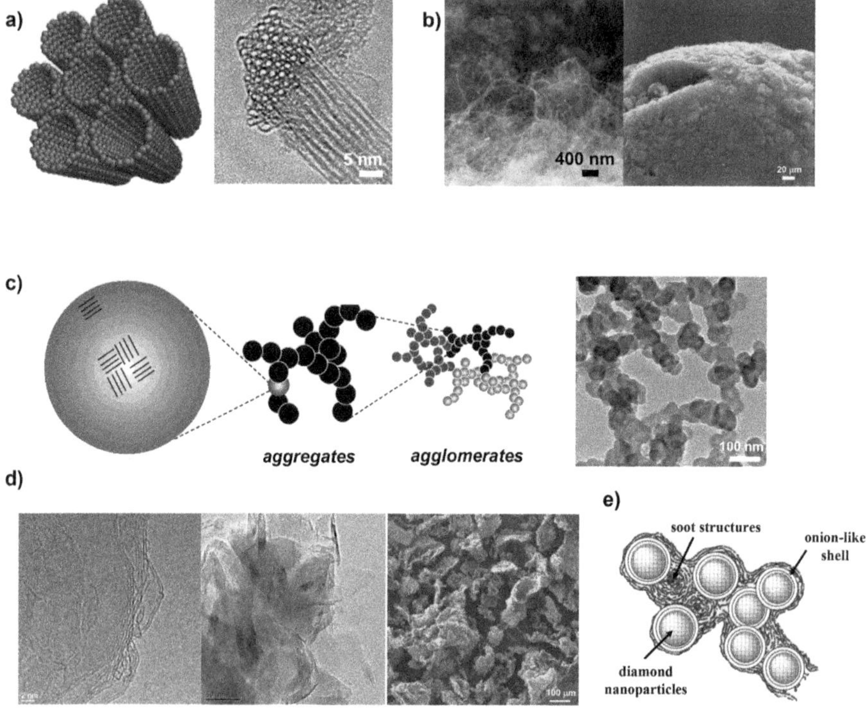

Figure 2.3 Macrostructure of (a) SWCNTs; (b) MWCNTs; (c) CB; (d) FLG; and (e) nanodiamonds.

2.2 Structure of Graphene and FLG: sp^2 Carbon Nanomaterials

The single-layer graphene (SG) is a one-atom thick sheet of hexagonally arranged sp^2-bonded carbon atoms. With the impressive progress in research on graphene mono- and bilayers, recent attention has also turned to graphene few layer counterparts. In FLG, the crystallographic stacking of the individual graphene sheets provides an additional degree of freedom. The bond is covalent (σ bond) and has a short length (0.141 nm) and high strength (524 kJ mol^{-1}). The hybridized fourth valence electron is paired with another delocalized electron of the adjacent plane by a much weaker *van der Waals bond* (a secondary bond arising from structural polarization) of only 7 kJ mol^{-1} (π bond). The spacing between the layer planes is relatively large (0.335 nm) or more than twice the spacing between atoms within the basal plane and approximately twice the van der Waals radius of carbon. The distinct lattice symmetries associated with different stacking orders of FLG have been predicted to strongly influence the electronic properties of FLG, including the band structure, interlayer screening, magnetic state, and spin–orbit coupling.[14] For graphene trilayers, two stable crystallographic configurations are predicted: ABA and ABC stacking order (Figure 2.4). In the absence

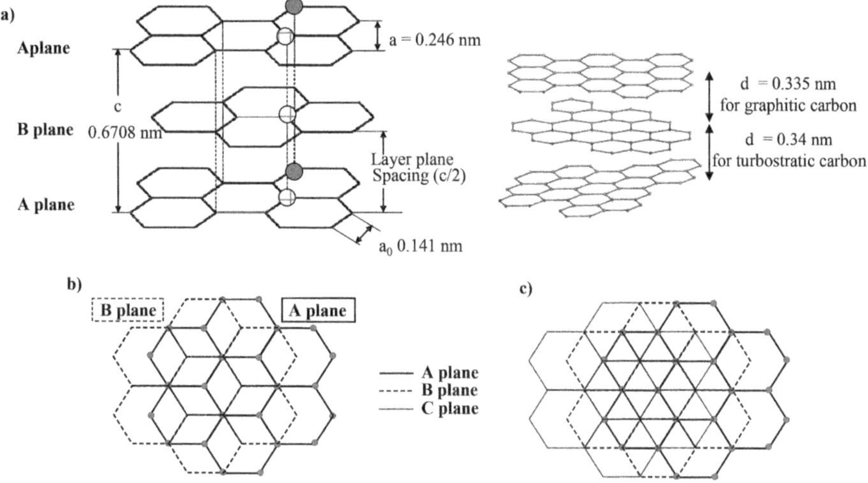

Figure 2.4 Lattice structure of trilayer graphene with (a) ABA (turbostratic structure is shown for comparison on the right side) and (b, c) ABC stacking sequence. The grey and black dots represent carbon atoms in the A and B sub-lattices of the graphene honeycomb structure.

of direct evidence of ABC stacking order in trilayers, ABA stacking order has generally been presumed in most studies of exfoliated materials, as this structure is believed to be slightly more thermodynamically stable than the other. The graphene planes in FLG are often arranged in the so-called ABAB Bernal stacking sequence. Recent studies, however, indicate distinct properties for these two types of graphene trilayers. ABA stacked trilayers are semimetals with an electrically tunable band overlap, while ABC-stacked trilayers are predicted to be semiconductors with an electrically tunable band gap.[14] The crystal lattice parameters, *i.e.*, the relative position of its carbon atoms (along the orthohexagonal axes) are: $a = 0.246$ nm and $c = 0.6708$ nm.

In turbostratic FLG there is no stacking order between adjacent layers; due to rotational stacking faults between the different graphene planes. This results in an increase in the interlayer spacing, which can increase from 0.3354 nm to more than 0.345 nm. If the value exceeds this, the structure exfoliates. These multilayered structures are of interest since introduction of rotational stacking faults into AB-Bernal stacked graphene causes decoupling of adjacent graphene layers. As a result the dispersion relation close to the K-point is altered from a parabolic (AB) to linear band behavior (rotational stacking fault). Thus, electronic properties typical of isolated SG can be observed in bilayer and multilayer graphene structures.[15] Graphene sheets with a relative rotation between them give rise to Moiré patterns in scanning tunneling microscopy (STM) or transmission electron microscopy (TEM) images. Raman spectroscopy played also an important role in the structural characterization of graphitic materials, in particular providing valuable information about stacking of the graphene layers.[14,16]

The quasi-crystalline domains in carbon nanomaterials, usually referred to as graphitic crystallites, nanocrystallites, or Basic Structural Units (BSU) are characterized by the distance between graphene planes (d_{002}) and the average sizes of the crystallites, L_a and L_c, where L_a and L_c are the crystallite sizes in the layer plane and in its normal direction, respectively (Figure 2.5). Quantitative characterization of the microstructure can be performed by measuring BSU parameters, including the average distance of the interlayer spacing (d_{002}), the stacking layer length (L_a) and the layer thickness (L_c) from the 002 lattice fringe images. These important parameters can be calculated from XRD patterns.

The d_{002} values are determined from the (002) diffraction peak positions. The interlayer spacing for graphitic stacking is known to be 0.3354 nm and that for turbostratic stacking is reported to be about 0.344 nm. Therefore, the observed interlayer spacing d_{002} is an average value depending on the relative ratio of graphitic to turbostratic stacking, and so decreases gradually to the value of 0.3354 nm with structural improvement by heat treatment.

From XRD data of d_{002}, a graphitization index (g_P) can be derived by applying the following eqn (2.1):

$$g_p = \frac{0.3440 - \mathrm{d}_{002}}{0.3440 - 0.3354} \tag{2.1}$$

It is used to characterize quantitatively the degree of similarity between carbon material and a perfect single crystal of graphite.

The dimensions of carbon crystallites are determined from the analysis of X-ray diffraction line broadening. According to the Scherrer formula, L_c is equal to:

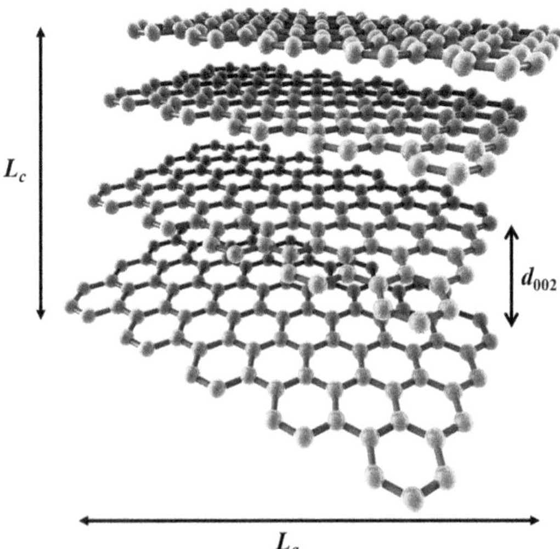

Figure 2.5 Crystallographic parameters d_{002}, L_a and L_c of a graphite crystallite.

$$L_c = 0.89\lambda / (B\cos\theta_c) \quad (2.2)$$

where λ is the X-ray wavelength, B is the angular width (radians) of the (002) diffraction peak at half-maximum intensity (corrected for instrumental broadening) and θ_c is the Bragg angle for reflection (002). The layer dimension L_a is calculated by use of the eqn (2.3):

$$L_a = 1.84\lambda / (B\cos\theta_a) \quad (2.3)$$

where B and θ_a correspond to reflection (100) (often labelled (10) for two dimensional structures).

The edge shape of graphene is also of great interest for adsorption, since the chirality and morphology of the edges are known to determine the electronic properties.[17] The edge morphology significantly affects the chemical reactivity and also the edge stability, which determine the band gap properties, and magnetic and optical behavior of graphene. Graphene edges can be configured as either zigzag or armchair based on morphology, open or closed depending on whether they are decorated with fully coordinated bonds or not, or folded/unfolded upon edge terminations (Figure 2.6). Graphene, with its sp^2-hybridized carbon atoms packed in a hexagonal lattice, has edges that consist of two different configurations referred to as "armchair" and "zigzag". An edge may also be comprised of a combination of both configurations. Carbon atoms of the zigzag edge have an unpaired electron, which is active to combine with other reactants (carbene like chemistry). The carbon atoms of the armchair edge side are less reactive because of a triple covalent bond between the two open edge

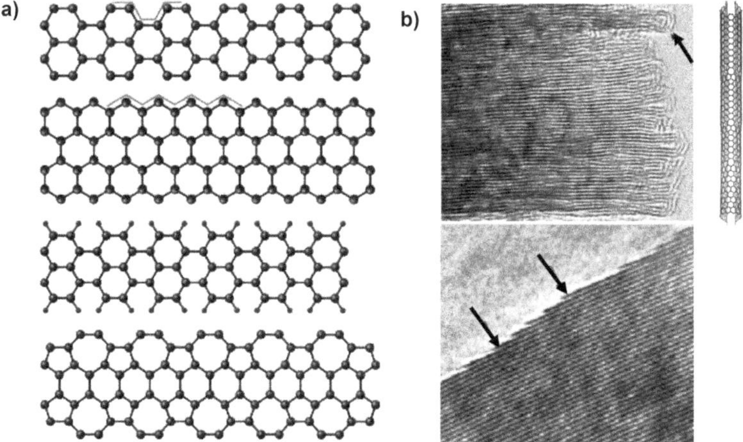

Figure 2.6 (a) Schematic diagram showing graphene ribbons with different edge structures. From top: 'armchair' edges and 'zigzag' edges (outlined in grey), armchair edges terminated with hydrogen atoms, and zigzag edges reconstructed with pentagons and heptagons. (b) From top: closed (folding) edges, and open edges.

carbon atoms of each edge hexagonal ring. Therefore, the chemically reactive sites in carbon materials, rather than being (as is too often assumed) either the very unstable unadulterated radicals formed by H removal from graphene edges or the very stable hydrogen-saturated graphene edges, are proposed to be an isolated carbene-type zigzag carbon atom and a carbyne-type armchair pair of carbon atoms.[18] Energetically, the zigzag edges are more stable than armchair edges. Therefore, the interconversion of zigzag edges to the armchair configuration is difficult due to their greater stability, which requires migration of the edge atoms to be transformed. Therefore, the reactivity sequence of the edge carbon atoms should be: carbene zigzag carbon > armchair carbon > zigzag carbon. The concentration of the edge sites at the periphery of graphene nanodomains is significant compared to the basal plane sites. Even though many studies claim that the zigzag edges are the most stable configuration, there are some studies contradicting this fact. Indeed, the edge configuration can strongly be altered by the chemical environment, which varies the edge energies and the edge forces. Edge-to-edge forces therefore become a dominant effect once the edges are terminated with functional groups involving oxygen, hydrogen, nitrogen, and boron using oxygen gas plasma, hydrogenation, and doping with either nitrogen or boron. In principle, the impact of the chemical functionalization determines the edge states, which can be easily tailored through either covalent or non-covalent bonding.

Consequently, structural edge modification or reconstruction is likely to occur upon either edge passivation or termination. More interestingly, etched defective sites as well as etched holes behave as hosts for molecules to adsorb, especially for gas species, which alter the electronic and electrical properties of graphene. Although much progress has been made in understanding and distinguishing the edge configuration and types of edge defects and their locations, there is significantly more to be explored given the highly facile properties of the material. This area still remains challenging due to the limitations of characterization techniques for such atomic scale understanding, and thus requires more involvement of researchers for further investigations.

2.3 Structure of Diamond Nanoparticles: sp^3 Carbon Nanomaterials

Diamond is a relatively simple substance in the sense that its structure and properties are essentially isotropic, in contrast to the pronounced anisotropy of graphite. However, unlike graphite, it has several crystalline forms and polytypes. Each diamond tetrahedron combines with four other tetrahedrals to form strongly-bonded, three-dimensional and entirely covalent crystalline structures. Diamond has two such structures, one with a cubic symmetry (the more common and stable) and one with a hexagonal symmetry found in nature as the mineral lonsdaleite (Figure 2.7). Cubic diamond is by far the more common structure. The covalent link between the carbon atoms of

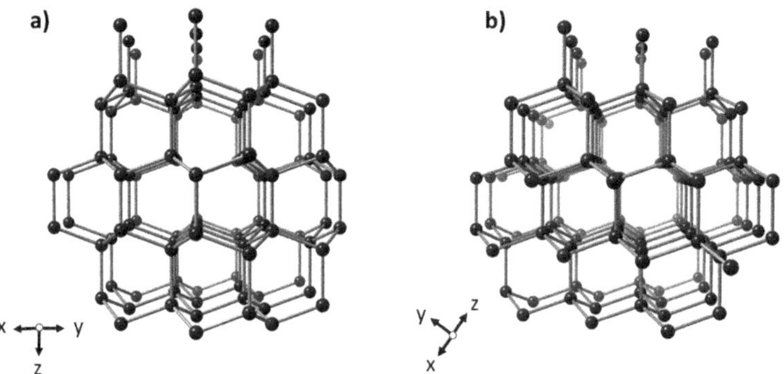

Figure 2.7 (a) Hexagonal diamond structure, the diagram highlights the ABAB... sequence of the hexagonal diamond structure; and (b) cubic diamond structure, the diagram highlights the ABCABC... sequence of the cubic diamond structure.

diamond is characterized by a small bond length (0.154 nm) and high bond energy of 7.37 eV. Each diamond unit cell has eight atoms.

In diamond all four outer electrons of each carbon atom are 'localized' between the atoms in covalent bonding. The movement of electrons is restricted and diamond does not conduct electric current.

Diamond nanoparticles are not small-sized diamond crystals. Thus, the structure of nanodiamonds has been explained within the model of icosahedral nanoparticles with a shell structure.[19] The nearest environment of all the atoms involved in the core of the particle consists of slightly distorted tetrahedra. As the particle size increases, the structure of the outer shells becomes more and more different from the diamond structure and transforms gradually into a graphene structure. Fragments of diamond and graphite are coherently joined together into a single shell structure. While the core of diamond nanoparticles is well characterized and there are little doubts about the presence of a diamond cage, the structure of nanodiamond surfaces is more controversial, as it is much more difficult to characterize them experimentally. Do diamond nanoparticles have a fullerenic, onion-shell, or disordered graphitic surface as has been widely concluded or assumed? Is there an amorphous graphitic component, or are they fully sp^3-hybridized with a hydrogenated surface? Are there dangling bonds at the surface? Are there many OH groups at the diamond surface, or is the O–H vibration in Raman spectroscopy just from adsorbed water? Is the nitrogen at the surface or in the core?[20] The presence of hydrogen atoms at the surface appears to be dependent on preparation conditions. If hydrogen is not present at the surface, both computer simulations and experiments suggest that graphitic like reconstructions and onion-like structures appear at the surface of nanodiamonds (see also Chapter 3, Section 3.2.2).[9] In particular, first principles calculations have found fullerene-like reconstructions in the 2–4 nm size range, leading to nanoparticles that have a hybrid structure these nanoparticles have been called bucky-diamonds.[9] The sp^2 content is massively dependent on the production and purification of a particular diamond sample. In the case of detonation diamond, oxygen containing groups are usually present on the particle surface.[21]

2.4 Structure of Fullerene and Related Materials: $sp^{2+\delta}$ Hybridization

2.4.1 Fullerenes, Onions and Carbon Black

Unlike graphite or diamond, fullerenes are not a single material, but a family of molecular geodesic structures in the form of cage-like spheroids, consisting of a network of five-membered rings (pentagons) and six-membered rings (hexagons). The bucky-ball molecule follows Euler's Theorem, which specifies that any convex closed-caged structure can be made up of any number of hexagons but must include exactly 12 pentagons in order to provide the appropriate curvature necessary to close the cage. Fullerenes can have a variable number of hexagons (m), with the general composition: C_{20+2m}. Although C_{20} is theoretically possible, it is a highly unlikely structure due to the fact that two pentagons do not go together well structurally. This is due to added strain on the geometry. Fullerenes with less than 60 carbon atoms are rare. The next smallest fullerene is C_{70}, a rugby-ball shaped molecule that is commonly found in soot. Other fullerenes were discovered with more and fewer carbon atoms then in the C_{60}, ranging from 28 up into the hundreds, though C_{60} remains the easiest to produce.

In order to account for the bonding of the carbon atoms of a fullerene molecule, the hybridization must be a modification of the sp^3 hybridization of diamond and sp^2 hybridization of graphite. In this case the σ orbital no longer contains all of the s-orbital character and the π orbital is no longer of purely p-orbital character, as they are in graphite. Unlike the sp^3 or sp^2 hybridizations, the hybridization in fullerene is not fixed but has variable characteristics depending on the number of carbon atoms in the molecule.[22] This number varies from twenty for the smallest geometrically (but not thermodynamically) feasible fullerene C_{20}, to infinity for graphite (which could be considered as the extreme case of all the possible fullerene structures). It determines the size of the molecule as well as the angle θ of the basic pyramid of the structure (the common angle to the three σ-bonds). The number of carbon atoms, the pyramidization angle (θ – 90°), and the nature of the hybridization are related and this relationship (in this case the s character in the π-orbital) is given in Figure 2.8.

The bond lengths of the fullerenes are reported as 0.145 ± 0.0015 nm for the bonds between five- and six-membered rings, and 0.140 ± 0.0015 nm for the bond between the six-membered rings. The C_{60} has a calculated diameter of 0.710 ± 0.007 nm. The rehybridization plays an important role in determining the electronic structure of the fullerene's family and it is the combination

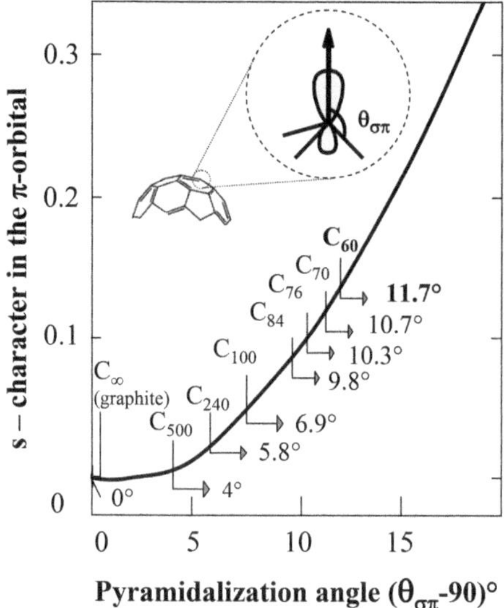

Figure 2.8 Hybridization of fullerene molecules as a function of pyramidization angle ($\theta_{\sigma\pi}$– 90°). $\theta_{\sigma\pi}$ is the common angle of the three σ bonds. Adapted from ref. 22.

of topology and rehybridization that together account for the possible specific reactivity of all the curved sp^2 nanostructures.

The atomic structure of carbon onions can be view as multilayer carbon nanoparticles consisting of closed curved graphite surfaces inserted into each other. These particles are distinguished by a variety of shapes and are referred to as Russian matreshkas, onion-like fullerenes, multilayer fullerenes, *etc.* The spherical particles known to date, which have been synthesized by various methods, can be conventionally classified into three groups according to their shell shape: spherical, spheroidal, and icosahedral.[23] The evolution of the carbon onion structure from spherical to polyhedral is correlated with changes in the sp^3/sp^2 ratio. Icosahedral fullerenes, whose topology is controlled only by pentagons and hexagons with more than two hundred atoms, appear in the form of spherical and polyhedral fullerenes along the twofold and fivefold axes, respectively. Spherical shells are isomers of the icosahedral fullerene, and their topology is enriched by seven- and eight-membered cycles, which makes their surface very close to spherical. Spherical and icosahedral nanoparticles have been studied both theoretically using various models and experimentally, and Figure 2.9 shows the results obtained on a tetrahedral carbon onion.[24]

Concerning carbon black particles, it appears that the longstanding problem of their detailed structure remains unresolved.[25] A number of points can be made with reasonable confidence, however. The first models put forward by Franklin in 1951 based on X-ray diffraction analysis (Figure 2.10a),[26] and Heidenreich *et al.*[27] in 1968 based on a detailed investigation of high resolution phase contrast images (Figure 2.10b) seems to be overly simplistic.

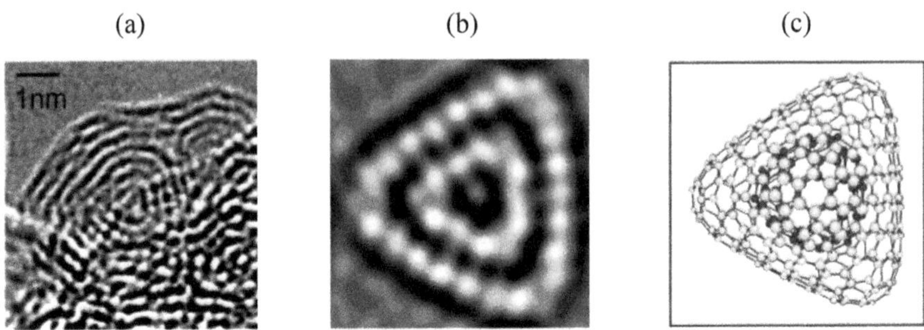

Figure 2.9 (a) HREM image of a tetrahedral carbon onion; (b) calculated HREM images of C84@C276; and (c) proposed structure models of C84@C276.

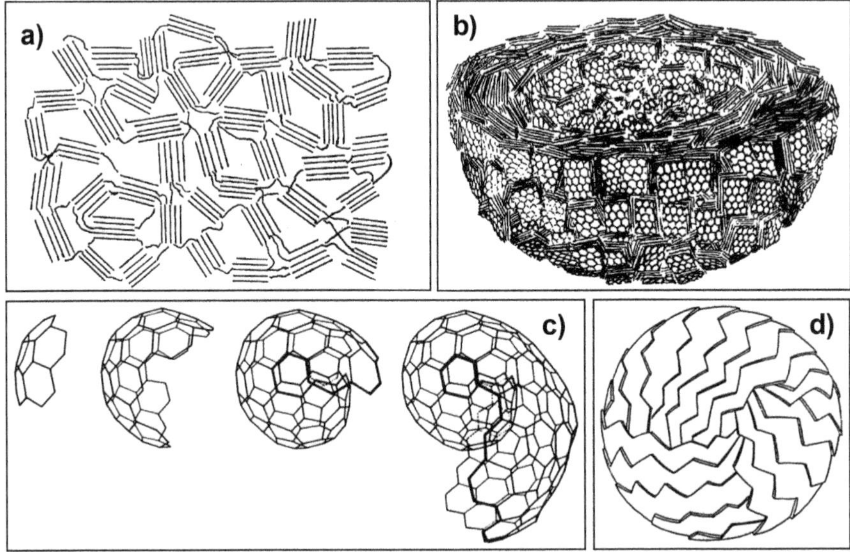

Figure 2.10 (a) R.E. Franklin model of carbon black structure; (b) Hess-Ban-Heidenreich model of carbon black structure; (c) illustration of Kroto–McKay icospiral growth model; and (d) Donnet–Custodéro model of carbon black structure.

Although it is true that the basic structural units that make up the particles are arranged concentrically around the centre, recent HRTEM observations suggest that the units are more extensive, and more defective, than the rather perfect fragments in the Heidenreich model.

In 1986, Kroto, Smalley, and their co-workers put forward a new mechanism for soot formation, the icospiral growth mechanism,[28,29] which was later refined by Kroto and McKay.[30] The mechanism was based on the "pentagon-road" model of fullerene formation (Figure 2.10c). The essential element of the pentagon-road model is the incorporation of pentagonal rings into a growing carbon network, driven by the need to eliminate dangling bonds. If the pentagons occur in the "correct" positions then C_{60} and other fullerenes will form, but in general closed structures will not be obtained. Kroto and Smalley suggested that if the growing shell fails to close, it would tend to curl around itself like a nautilus shell, as illustrated in Figure 2.10c. In the refined model, Kroto and McKay suggested that the pentagons in a new layer would tend to form directly above those in the previous layer in an epitaxial manner, resulting in 12 columns of pentagonal rings. They argued that the spiralling structure would become increasingly faceted as it grew larger, in the same way that giant fullerenes are expected to be much more faceted than small ones. There also seems to be little evidence to support the icospiral growth model of Kroto, Smalley *et al.* This model predicts that CB particles should have a relatively ordered structure, rather than the disordered structure that is observed experimentally. Detailed TEM studies show that fullerene-like structures are probably present (presence of non-hexagonal rings),[31] but are not responsible for the sphericity of the particles. Instead, the roundness of the particles is believed to be a consequence of energetics, coupled with surface mass growth *via* a radical addition mechanism. This is an area where ultra-high resolution TEM would be of great value.

Donnet and Custodéro have presented STM measurements apparently indicating that carbon black particles are made up of overlapping scales, as illustrated in Figure 2.10d.[32] This model is reminiscent of the Kroto–McKay icospiral growth model, and Donnet and Custodéro believe that their observations provide support for the view that the 'seed' for soot particle growth may be a fullerene-like icospiral.

Finally, Homan[33] and later on Osawa[34] proposed that both soot particles and fullerenes have common precursors: very reactive particles called aromers for *aro*matic oligo*mers* (Figure 2.11).

These aromers are obtained by oligomerization of polycyclic aromatic hydrocarbons (PAHs), and contrarily to PAHs, are not planar.[34] These observations have lead Osawa *et al.* to propose a unified mechanism of soot and fullerene formation (Figure 2.11).[34] Once aromers are formed, the products undergo reversible spiroid–onion transformation involving in-plane rearrangements. Continuous rearrangement process leads to the thermodynamically most stable $60n^2$ ($n = 1, 2, 3, \ldots$) configuration of the infinite carbon nano-onion series with C_{60} ($n = 1$) in the center. Thus, concomitant formation of soot particles and fullerenes can be understood within the framework of a single collective mechanism. Single-layer fullerenes are formed only when oligomerization of PAH happened to lead to closure of a spherical polyhedral solid during the first spiral turn. Even in such a case, the total number of carbon atoms will be adjusted by the one-layer version of onion–spiroid transformation. Otherwise, continuous addition of small carbon radicals, such as C_2, PAH, and aromer radicals onto the radical terminal at the outermost layer of spiroids lead to primary particles of soot.

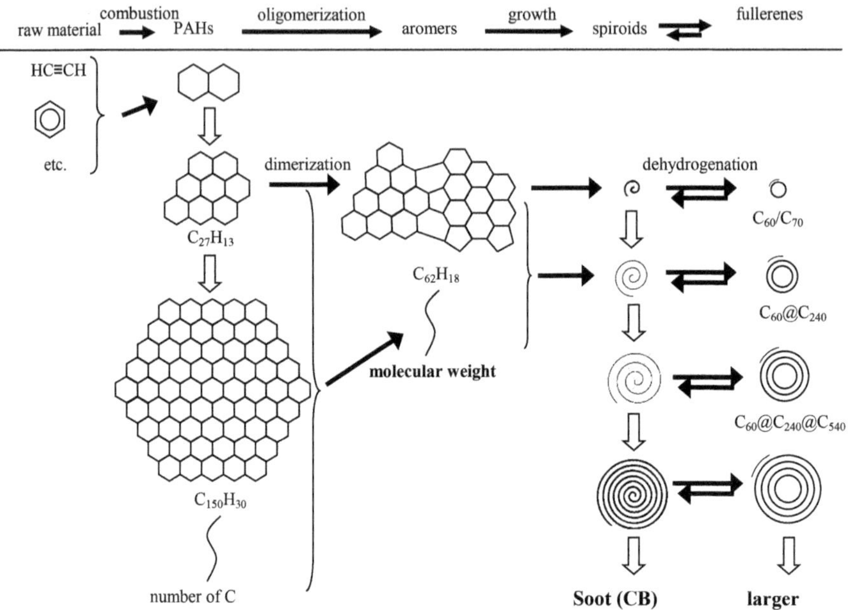

Figure 2.11 Unified mechanism of formation/growth of soot, C_{60} and nano-onions proposed by Osawa *et al.* (some PAHs and only one aromer out of numerous structures are depicted as examples). Adapted from ref. 34.

2.4.2 Carbon Nanotubes

In graphite the four outer electrons of carbon form three sp^2 hybridized σ bonds and one π orbital, which gives the conduction band six Fermi points and a linear dispersion around each of them. If a graphene sheet is rolled up into a structure like a CNT the orbital structure of carbon is altered, because the bond length between carbon atoms decreases and the bond angle changes. The σ and π orbitals are no longer perpendicular to each other. Overlap of the π orbitals is introduced. As a consequence the parts of the π orbitals inside and outside of a nanotube rearrange, in a way that the outer contribution is larger than the inner one (see Figure 2.12a). The curvature induces a mixed state of σ and π orbitals, called rehybridization. Consequently, the π-orbital is always inclined by the hybridization angle $\delta = a/(2\sqrt{3}d)$ (a the length of the lattice unit vector and d the diameter) relative to the normal direction to the tube's surface.[35] This tilting angle is strongly dependent on CNT diameter and helicity (chirality).[36] Curvature in the nanotube introduces misalignment of π-orbitals within the graphene sheet (Figure. 2.12b). The π-orbitals of a nanotube are not pointed directly towards the central axis of the nanotube, and some adjacent carbon π-orbitals have a misalignment angle, φ, between them. The π-orbitals of adjacent carbon atoms in a (10,10) and (5,5) nanotube have a π-orbital misalignments for the

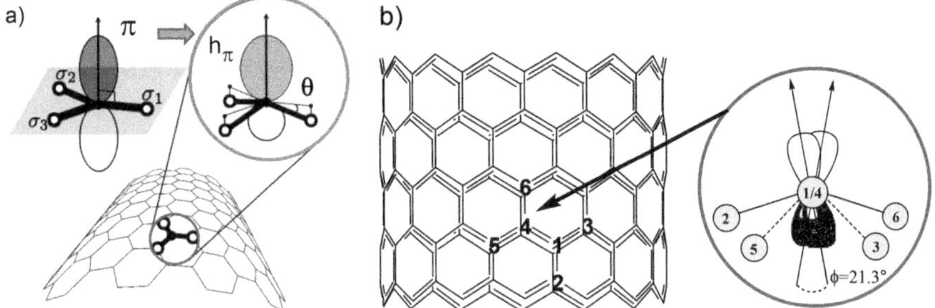

Figure 2.12 (a) Diagram showing the π orbital in planar graphene and its change into h_π under bending together with the pyramidalization angle θ. (b) The π-orbital misalignment angles ϕ along the C_1–C_4.

two different C–C-bonds of $\varphi = 0°$ and 10.4° for (10,10) and 0° and 21.3° for (5,5), respectively.

The rehybridization can also be explained by a mixture of sp^2 and sp^3 orbitals. While single sp^2 and sp^3 orbitals are saturated, the mixed state contains unsaturated orbitals. The mixing of a tetravalent sp^3 orbital to a trivalent sp^2 orbital leaves one hybrid orbital free for binding. The higher the ratio of the sp^3 contribution the more free bonds exist and the higher the reactivity. As an index for local reactivity the pyramidalization angle θ is used (see Figure 2.12a), which is sufficient for describing the curvature-induced shift in sp^2 hybridization. θ is the angle between the σ and π orbitals minus 90° ($\theta = 0°$ for the sp^2 hybridization). The tetrahedral sp^3 orbital has a pyramidalization angle of 19.5°. This angle changes depending on the local mixture of sp^2 and sp^3 orbitals. The degree of hybridization $sp^{2+\eta}$ could be obtained, η being a number between 0 and 1. The comparison between different pyramidalization angles allows the comparison of reactivity, where a higher angle results in a larger reactivity. Thus, the degree of hybridization is highly curvature dependent, so that it is even possible to increase the reactivity solely by bending a CNT.

To summarize, strain in non-defective nanotube sidewalls is manifested in pyramidalization and π-orbital misalignment, which can be relieved by the addition of an atom or functional group to the CNT external surface. As the nanotube diameter increases, both the pyramidalization angle and π-orbital misalignment angle decrease, lowering the chemical reactivity of the C–C-bonds, eventually approaching planar graphite for very large CNT diameters.

Finally, the CNT convex (outer) surface is chemically reactive because the convex arrangement of pyramidalized sp^2 carbon atoms is correctly disposed for the formation of chemical bonds with reagent species. For the same reason, the concave (inner) surface should be more inert and can withstand the presence of highly reactive species encapsulated within the nanotubes.

2.5 Bulk Properties of Carbon Nanomaterials

This section will be devoted to the description of the main bulk properties of carbon nanomaterials relevant for catalysis. The physical properties of carbon vary widely with the allotropic form. The physical and chemical surface properties, which are related to adsorption, an important elemental step for both catalyst preparation and catalytic reactions, will be developed in Chapter 3.

The bulk properties of carbon and graphite materials reflect their structure, and particularly their degree of anisotropy between the *a* and *b* axes (very strong aromatic C–C bond), and the *c* direction, perpendicular to the basal plane (very weak 'van der Waals' or π–π interactions). For diamond, an isotropic material, the sp^3 tetrahedral arrangement of the C atoms is responsible of its specific properties: super hard, electrical insulator with extremely high thermal conductivity. X-ray diffraction and Raman spectroscopy[37] are the most powerful single tools for characterizing the structural properties of carbon and graphite materials.[38] From its earliest applications to carbon materials, Raman spectrometry has usually been considered as an alternative to X-ray diffraction (XRD), with respect to which it offers as main advantages a higher surface selectivity (with a sampling depth estimated to be about 100 nm) and the possibility of sampling reduced areas of the surface. Both advantages have found applications in the study of carbons.[38]

Typical bulk property values of nanostructured carbon and graphite materials of interest for catalysis are given in Table 2.1. The properties of activated carbon are also given for comparison purposes. The wide range of values reported for a given property of these nanomaterials arise either from a difference in measurement methods or from the origin of the material (presence of defects, functional groups, number of layers, *etc.*).

The electronic or electrical properties of carbons and graphite materials are of practical interest in electrocatalysis. The anisotropy and its degree of replication are responsible for the behavior, from good conductors (FLG, some CNTs) to semi-conductors (some SWCNTs) and to insulators (nanodiamonds). For single-walled CNTs, theoretical calculations predicted that they could exhibit either metallic or semiconducting behavior depending only on diameter and helicity (see Section 1.2.4.2). This ability to display fundamentally distinct electronic properties without changing the local bonding was experimentally confirmed by STM, which is able to resolve simultaneously the atomic lattice and the electronic density of states of a material. The electronic structure properties of bulk carbon materials are influenced by the so-called crystallographic defects. There are three major groups of crystallographic defects: point defects, planar defects and bulk defects (see Section 3.2.4). Understanding the influence of crystallographic defects on the electronic structure properties of carbon-based materials is a necessary requirement if these materials are to be used in electrocatalysis.

The thermal properties often follow the electronic properties quite closely and are of importance for thermal management of a given catalytic reaction, both on the macro- and the micro-scale. Carbon allotropes and their derivatives occupy a unique place in terms of their ability to conduct heat.[39] The

Table 2.1 Physical bulk properties of various carbon and graphite nanomaterials.

Property	Graphene/FLG	Nanodiamond	AC	CB	CNF	MWCNT	SWCNT
d_{002}/nm	0.339–0.34[a]	—	>0.344	0.35–0.37[b]	0.34–0.36	0.339–0.348	—
L_c/nm	<3[a]	—	<5	1–4	>5	—	—
L_a/nm	>100[a]	—	<5	1–5	5–50	—	—
Electrical resistivity (μΩm)	0.1–0.4[c]	10^{17}–10^{24}	10^3–10^6	10^2	0.5–20	0.6–20	0.1–35
ab-direction	10–>40[c]						
c-direction							
Thermal conductivity/W m^{-1} K^{-1}		700–3300	0.15–0.5	1–150			
ab-direction	3000–5000[c]				10–1100	50–200[d]	1000–6600
c-direction	6–80[c]						
Young modulus/TPa	1	0.4–1.2	—	—	0.1–0.6	0.1–0.95	0.1–1.5
Tensile strength/GPa	5–130[c]	0.5–1.4	—	—	15–30	2–60	4–50
Packing (bulk) density/g cm^3	0.03–2[e]	0.09–3[f]	0.2–1	0.1–0.55	1.4–2.2	0.1–0.3	0.1–0.2

[a]For FLG.
[b]Values ranging between 0.345 and 0.339 nm can be obtained upon graphitization between 1773 and 3773 K, respectively.
[c]The lower values are for FLG.
[d]The lower value for a powder of defective MWCNTs.
[e]0.03 to 1 for FLG powder and >1 for films.
[f]0.09–0.7 for nanodiamond powders.

room-temperature thermal conductivity of carbon materials span an extraordinary large range – of over five orders of magnitude – from the lowest in amorphous carbons to the highest in graphene and carbon nanotubes. For SG and FLG the thermal conductivity decreases as the number of atomic layers increases. However, it still remains very high at 1300 W m^{-1} K^{-1} for graphene containing four atomic layers. The thermal conductivity of graphene-based materials can be tuned over a wide range by the introduction of disorder or edge roughness. Contrary to the common assumption of the dominant effect of the contact conductance, the contribution of the finite thermal conductivity of individual nanotubes is found to control the value of the conductivity at material densities and CNT lengths typical for real materials.[40]

Carbon and graphite materials such as graphite and glassy carbon absorb light over a wide energy range, at least from deep UV to radio frequencies. Crystalline diamond is optically transparent from its band gap at *ca.* 220 nm into the infrared, but impurities and defects can result in weak absorption throughout the visible region. Graphene is in many ways the ultimate thin film, only one atomic layer thick, and has photonic properties of high interest. Noteworthy is that for pristine, unbiased graphene an impressive 2.3% of incident visible light is absorbed. Optical properties of interest for catalytic applications are particularly those that reveal contributions of the inter-band transitions of π-electrons, because these can, in principle, shed light on the electron-donating or electron-accepting properties of carbons.

Resistance to attrition and crushing are also important parameters in catalysis since it will strongly impact the filtration behavior of supports as powdered or granular carbon-based materials. Thus, if carbon's surface chemistry is duly utilized it is not necessary to use as support a carbon adsorbent with very high surface area and consequent poor mechanical resistance. In that respect, fibrous carbons offer advantages not only in terms of mechanical properties but also better control of transport and other characteristics that can ensure maximum accessibility of catalytically active sites, and not only their maximum concentration.

Another additional issue is of special relevance here, because of the impact of bulk properties on surface properties: the effect of heteroatom incorporation into the carbon structure (doped-carbon) or between the graphene layers (intercalated carbons). Heteroatom doping (*e.g.*, boron, sulfur, phosphorous, and nitrogen) of graphitic carbon lattices, either during the production process or by a post-treatment, affects various physicochemical properties of sp^2 carbon materials.[41,42] These modifications strongly depend on the type of dopants, concentrations and their location within the graphene systems. In particular, doping with boron or nitrogen has received growing attention because significant changes in hardness, electrical conductivity (additional electronic states around the Fermi level), and chemical reactivity have been theoretically predicted and experimentally observed.[43,44] Thus, the use of these materials as catalyst support or directly as catalyst is of great interest (see Chapter 6).[45,46] The possibility to dope sp^3 carbon materials such as diamond with heteroatoms has also been reported.[47,48] Because of its large band-gap of more than 5 eV undoped diamond is electrically insulating. But as with other large band-gap materials it can be made conducting by doping it with specific elements. Currently, in most cases boron is used as doping agent, and a p-semiconductor results. If phosphorus or nitrogen are used a n-semiconductor is produced.

Through physical, chemical or electrochemical methods, ions or compounds can be intercalated between the graphene layers. Intercalation provides to the host material a means for controlled variation of many physical properties over wide ranges.[49] Because the free carrier concentration of graphite host is very low ($\sim 10^{-4}$ free carriers per atom at room temperature), intercalation with different chemical species and concentrations enables a wide variation of the free carrier concentration and thus of the electrical, thermal and magnetic properties of the host material. By separating the graphite layers through intercalation with alkali metals followed by exfoliation with aqueous solvents, thin graphite nanoplatelets can be formed with potentially high surface area.[50]

References

1. M. Inagaki, F. Kang, *Carbon Materials Science and Engineering : from Fundamental to Applications*, Tsinghua University Press, 2006.
2. H. W. Kroto, *Chem. World*, 2010, **7**, 37.
3. Z. Su, W. Zhou and Y. Zhang, *Chem. Commun.*, 2011, **47**, 4700–4702.
4. M. C. Acik and J. Yves, *Jpn. J. Appl. Phys.*, 2011, **50**, 070101.
5. X. Zhang, J. Xin and F. Ding, *Nanoscale*, 2013, **5**, 2556–2569.
6. L. R. Radovic and B. Bockrath, *J. Am. Chem. Soc.*, 2005, **127**, 5917–5927.
7. X. Jia, J. Campos-Delgado, M. Terrones, V. Meunier and M. S. Dresselhaus, *Nanoscale*, 2011, **3**, 86–95.
8. S. Rotkin and Y. Gogotsi, *Mater. Res. Innovations*, 2002, **5**, 191–200.
9. V. N. Mochalin, O. Shenderova, D. Ho and Y. Gogotsi, *Nat Nanotechnol.*, 2012, **7**, 11–23.
10. V. G. Pol, M. Motiei, A. Gedanken, J. Calderon-Moreno and M. Yoshimura, *Carbon*, 2004, **42**, 111–116.
11. V. Fernandez-Alos, J. K. Watson, R. v. Wal and J. P. Mathews, *Combust. Flame*, 2011, **158**, 1807–1813.
12. H. Li, Z. Kang, Y. Liu and S.-T. Lee, *J. Mater. Chem.*, 2012, **22**, 24230–24253.
13. T. Ungár, J. Gubicza, G. Ribárik, C. Pantea and T. W. Zerda, *Carbon*, 2002, **40**, 929–937.
14. C. H. Lui, Z. Li, Z. Chen, P. V. Klimov, L. E. Brus and T. F. Heinz, *Nano Lett.*, 2010, **11**, 164–169.
15. S. Shallcross, S. Sharma, E. Kandelaki and O. A. Pankratov, *Phys. Rev. B.*, 2010, **81**, 165105.
16. M. A. Pimenta, G. Dresselhaus, M. S. Dresselhaus, L. G. Cancado, A. Jorio and R. Saito, *Phys. Chem. Chem. Phys.*, 2007, **9**, 1276–1290.
17. M. Acik and Y. J. Chabal, *Jpn. J. Appl. Phys.*, 2011, **50**, 070101.

18. L. R. Radovic, *J. Am. Chem. Soc.*, 2009, **131**, 17166–17175.
19. V. Y. Shevchenko, A. E. Madison and G. S. Yur'ev, *Glass Phys. Chem.*, 2006, **32**, 261–266.
20. X. Fang, J. Mao, E. M. Levin and K. Schmidt-Rohr, *J. Am. Chem. Soc.*, 2009, **131**, 1426–1435.
21. A. Krueger, *J. Mater. Chem.*, 2008, **18**, 1485–1492.
22. R. C. Haddon, *Acc. Chem. Res.*, 1992, **25**, 127–133.
23. O. E. Glukhova, A. I. Zhbanov and A. G. Rezkov, *Phys. Solid State*, 2005, **47**, 390–396.
24. I. Narita, T. Oku, K. Suganuma, K. Hiraga and E. Aoyagi, *J. Mater. Chem.*, 2001, **11**, 1761–1762.
25. P. J. F. Harris, *Crit. Rev. Solid State Mater. Sci.*, 2005, **30**, 235–253.
26. R. E. Franklin, *Proc. R. Soc. Lond., A.*, 1951, **209**, 196–218.
27. W. M. Hess, L. L Ban and R. D. Heidenreich, *J. Appl. Cryst.*, 1968, **1**, 1–19.
28. Q. L. Zhang, S. C. O'Brien, J. R. Heath, Y. Liu, R. F. Curl, H. W. Kroto and R. E. Smalley, *J. Phys. Chem.*, 1986, **90**, 525–528.
29. H. W. Kroto, *Comp. Math. Appl.*, 1989, **17**, 417–423.
30. H. W. Kroto and K. McKay, *Nature*, 1988, **331**, 328–331.
31. F. Cataldo, *Carbon*, 2002, **40**, 157–162.
32. J. B. Donnet and E. Custodero, *Carbon*, 1992, **30**, 813–815.
33. K.-H. Homann, *Angew. Chem., Int. Ed.*, 1998, **37**, 2434–2451.
34. M. O. a. E. Ōsawa, in *Carbon Nanotechnology*, ed. L. Dai, Elsevier, Dordrecht, 2006, pp. 127–151.
35. A. Kleiner and S. Eggert, *Phys. Rev. B.*, 2001, **64**, 113402.
36. Y. Ouyang, J.-C. Peng, H. Wang and Z.-H. Peng, *Chin. Phys. B.*, 2008, **17**, 3123–3127.
37. R. Saito, M. Hofmann, G. Dresselhaus, A. Jorio and M. S. Dresselhaus, *Adv. Phys.*, 2011, **60**, 413–550.
38. A. Cuesta, P. Dhamelincourt, J. Laureyns, A. Martinez-Alonso and J. M. D. Tascon, *J. Mater. Chem.*, 1998, **8**, 2875–2879.
39. A. A. Balandin, *Nat. Mater.*, 2011, **10**, 569–581.
40. A. N. Volkov and L. V. Zhigilei, *Appl. Phys. Lett.*, 2012, **101**, 043113–043115.
41. J. P. Paraknowitsch and A. Thomas, *Energy Environ. Sci.*, 2013, **6**, 2839–2855.
42. D. Jana, C.-L. Sun, L.-C. Chen and K.-H. Chen, *Prog. Mater. Sci.*, 2013, **58**, 565–635.
43. R. Lv and M. Terrones, *Mater. Lett.*, 2012, **78**, 209–218.
44. S.-S. Yu and W.-T. Zheng, *Nanoscale*, 2010, **2**, 1069–1082.
45. Y. Zhao, R. Nakamura, K. Kamiya, S. Nakanishi and K. Hashimoto, *Nat Commun.*, 2013, **4**, 2390.
46. L. Mabena, S. Sinha Ray, S. Mhlanga and N. Coville, *Appl. Nanosci.*, 2011, **1**, 67–77.
47. A. Kraft, *Int. J. Electrochem. Sci.*, 2007, **2**, 355–385.
48. V. Pichot, O. Stephan, M. Comet, E. Fousson, J. Mory, K. March and D. Spitzer, *J. Phys. Chem. C.*, 2010, **114**, 10082–10087.
49. M. Inagaki, *Solid State Ionics*, 1996, **86–88**, Part 2, 833–839.
50. L. M. Viculis, J. J. Mack, O. M. Mayer, H. T. Hahn and R. B. Kaner, *J. Mater. Chem.*, 2005, **15**, 974–978.

CHAPTER 3

A Molecular View of Adsorption on Nanostructured Carbon Materials

3.1 Introduction

Nanostructured carbon materials are among the most prominent nanoscale materials studied today. The confinement to one or two dimensions and the high symmetry of these materials lead to new physical properties and many potential applications, including adsorption, a fundamental step of catalysis.[1,2] Fullerenes, CNTs and SG represent a class of carbon nanomaterials with 0-D, 1-D and 2-D structures and no bulk counterpart. Their unique structure gives rise to spectacular properties. The hollow 1-D structure of CNTs and the ultimate 2-D structure of graphene are especially attractive for adsorption-related applications. As for any material, adsorption on nanostructured carbon materials is related to their surface physical and chemical properties. Surface area and porosity are important physical properties that influence the quality, utility, and handling of nanostructured carbon materials, which often must be carefully engineered to perform specific functions. Differences in the surface area and porosity of nanostructured carbon materials, which otherwise may have the same physical dimensions, can greatly influence their adsorption performances and therefore have a pronounced impact on their applications. Therefore, it is critically important that these characteristics be accurately determined and controlled.[3] The surface chemistry of nanostructured carbon materials is governed by basal and edge carbon atoms,[4,5] as well as by the presence of defects (*i.e.*, structural carbon vacancies, non-aromatic rings).[6] These imperfections and defects along the edges of graphene layers are the most

RSC Catalysis Series No. 23
Nanostructured Carbon Materials for Catalysis
By Philippe Serp and Bruno Machado

Published by the Royal Society of Chemistry, www.rsc.org

active sites owing to high densities of unpaired electrons. Indeed, the zigzag sites are carbene-like, and the armchair sites are carbyne-like.[6] Heteroatoms, such as oxygen, hydrogen, nitrogen, and sulfur can be chemisorbed, leading to stable surface compounds, and resulting in a complex surface chemistry (see Chapter 4), if compared to classical oxides (Al_2O_3, SiO_2, TiO_2). Among the heteroatoms that can be introduced, nitrogen has been particularly studied since it provides basicity to the surface (see Chapter 6).[7,8] If most of the adsorption sites for chemisorption on sp^2 and $sp^{2+\delta}$ carbon nanomaterials are on the basal planes (inner or outer surface in the case of CNTs), the much higher activity of the heterogeneous groups can result in significant effects on the overall adsorption capacity. The exact mechanism by which the heterogeneous surface affects the adsorption is of great current interest. Computational approaches at the molecular scale can provide microscopic insight into adsorption behavior from the bottom-up, complement and secure correct interpretation of experimental results, and are imperative to new material design and advanced technological innovation.

3.2 Adsorption Sites

The interaction of nanostructured carbon materials with their environment, and in particular with gases or liquids adsorbed either on their internal (CNTs) or external surfaces, is attracting increasing attention. Many of the intriguing properties of SG are a consequence of its 2-D electronic band structure. Electrons in SG exhibit a characteristic linear dispersion relation between energy and momentum near the K point of the Brillouin zone. Interlayer coupling in FLG leads to a dramatic change of this electronic structure, with the emergence of hyperbolically dispersing bands.[9] Furthermore, the electronic properties of FLG are predicted to be highly sensitive to crystallographic stacking sequence. Thus, even though several physical properties are shared between graphene and its few-layers, recent theoretical and experimental advances demonstrate that each specific thickness of FLG is a material with unique physical properties.[10] To date, the role of stacking has, however, been overlooked in experimental studies dealing with adsorption on FLG with crystalline order. CNTs are usually associated with other nanotubes in bundles, fibers, films, papers, and so on, rather than as a single entity. Each of these associations has a specific range of porosities that determines its adsorption properties.

3.2.1 Adsorption Sites on Graphene and FLG: sp^2 Carbon Nanomaterials

The structure of SG and FLG has been presented in Chapter 2 (Section 2.2). On considering adsorption on SG and FLG, the phenomena may basically occur on two kinds of surfaces: prismatic (edge) surfaces and basal plane surfaces (Figure 3.1a). Ideal (= defect-free and contaminant-free) basal plane surfaces

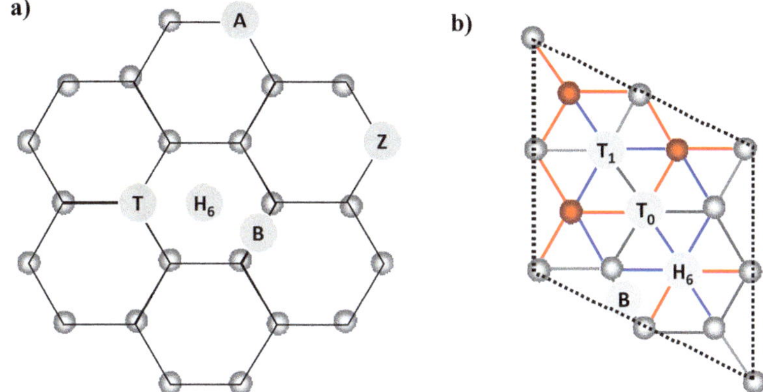

Figure 3.1 Adsorption sites on SG and FLG: (a) hollow H_6, bridge B, top T, A armchair, and Z zigzag sites; and (b) for FLG, lattice structure of ABC-stacked graphene trilayer; grey/red/blue indicate links on the top/middle/bottom layers, and $H_6/B/T_0/T_1$ distinguish the four adsorption sites for adsorption on graphene: hollow center (H_6), bridge center (B), and two top sites (T_0 and T_1) in a 2 × 2 graphene supercell.

are homogeneous and "smooth" and consist only of carbon atoms. In contrast, the prismatic surfaces are heterogeneous and "rough" and apart from carbon contain various surface groups (mostly oxygen-containing). The ratio of the basal to the prismatic surfaces is an important factor in determining the adsorption performances of a (nano)carbon material. Gas adsorption data, such as those, which are used for determination of the Brunett–Emmet–Teller (BET) surface area of solids, can be used to estimate the absolute and relative extents of "basal plane surface" area and "non-basal plane surface" area of graphitic materials.[11,12] For defect-free SG or FLG, the different possible adsorption sites above graphene are top, T (top of a carbon atom); bridge, B (above a C–C bond); hollow, H_6 (above the center of hexagons) sites. Interestingly, there are actually two types of T site (Figure 3.1b): one (marked as T_0) with the adsorbed species above not only a carbon atom of the uppermost graphene but also a carbon atom of the second graphene layer; the other (marked as T_1) with the adsorbed species located above a carbon atom in the uppermost graphene layer and a hexagon center in the second graphene layer.

3.2.2 Adsorption Sites on Diamond Nanoparticles: sp^3 Carbon Nanomaterials

When considering adsorption on diamond surfaces, three crystallographic faces should be considered, namely the (111), (110), and (100) (Figure 3.2).[13] Normally, a diamond crystal is mainly bordered by stable (100) and (111) surfaces. The (100) surface reconstructs into an energetically favorable (2 × 1, dimer-based reconstructions) structure, while the (111) substrate exhibits the Pandey chain configuration.[14–16] The (110) face, which is found

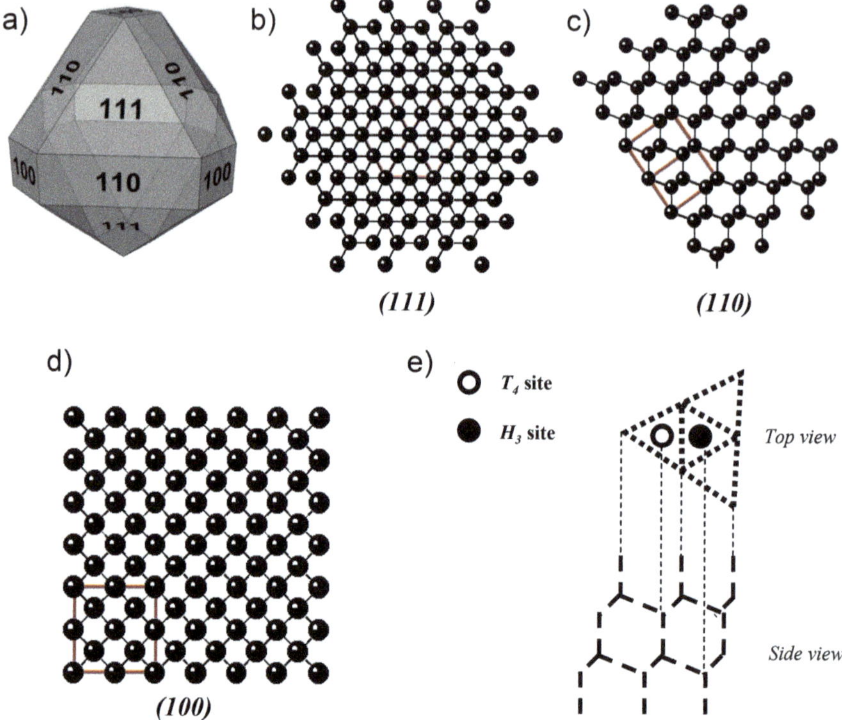

Figure 3.2 (a) Three dimensional cubic diamond crystal, with (b) (111), (c) (110) and (d) (100) faces; and (e) top view and side view of the (111) surface of diamond structure.

to be the most rapidly growing phase during chemical vapor deposition (CVD) growth assumes a stable (1 × 1) configuration.[14] There are two kinds of adsorption sites, called T_4 and H_3 on the (111) surface of the diamond crystal structure (Figure 3.2e).

It has been reported that physico-chemical properties of diamond surfaces are closely related to the chemisorbed species on the surface.[17] Hydrogen chemisorption on a chemical vapor deposition-grown diamond surface is well-known to be important for stabilizing diamond surface structures with sp^3 hybridization. It is generally accepted that at complete or near complete hydrogen coverage the diamond (111) surface has a bulk-terminated structure, with the dangling bonds or radicals terminated by hydrogen atoms (Figure 3.3). It has been shown that hydrogenation of (100) diamond surfaces on nanocrystalline diamonds serves to produce a more bulk-diamond-like surface structure, and reduces localized structural differences related to nanocrystal corners and edges.[18] Importantly, hydrogen-terminated diamond exhibits a high surface conductivity.[19] Many reports have suggested that a H-chemisorbed structure is necessary to provide a negative electron affinity condition on the diamond surfaces. The monohydride terminated (100)–(2 × 1) surface is found to have

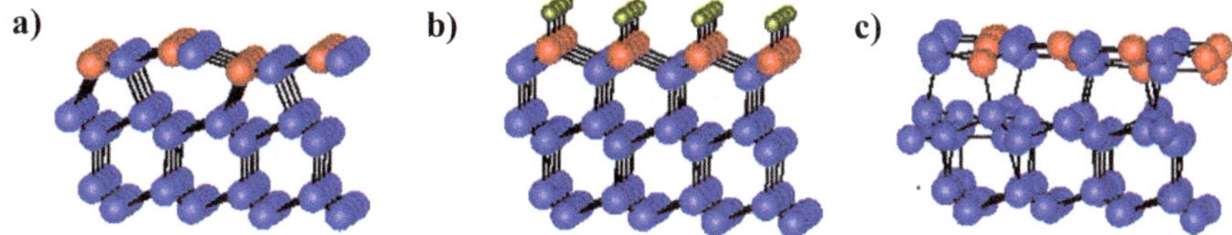

Figure 3.3 Side views of various C(111) surfaces (the larger blue and red spheres represent C atoms, and the smaller spheres represent H atoms. The red C atoms are radical sites in the bulk-terminated (1 × 1) structure): (a) bulk terminated structure, (b) structures from all *p*-bonded chain reconstructions, and (c) H-terminated structures.

negative electron affinity, while the corresponding bare surface exhibits positive electron affinity.[20] It was also reported that the negative electron affinity condition could change to a positive electron affinity by oxidation of the H-chemisorbed diamond surfaces. Oxidized diamond surfaces usually show characteristics completely different from those of the H-chemisorbed diamond surfaces. The unique electron affinity condition, or the surface potential, is strongly related to the chemisorbed species on diamond surfaces. The differences between O-terminated and H-terminated diamond surfaces include electrical conductivity, hydrophobicity and hydrophilicity, negative and positive electron affinities and fluorescence.

On the other hand, the clean surfaces exhibit gap levels in their electronic density of states due to dangling bonds, which disappear upon termination of the substrates with hydrogen atoms. The clean (111) diamond surface undergoes reconstruction with a π-bonded chain structure (Figure 3.3b). In this case the radicals that would be found on the clean bulk terminated surface are eliminated in favor of chains of sp^2-hybridized carbon atoms. DFT results indicate that dangling bonds on the nanodiamond surfaces play an important role in the graphitization process (bucky-diamond formation), and the orientation of the dangling bonds on different nanodiamond surfaces determines whether there will be a graphitization process or not.[21] Thermal treatment of the hydrogen-free (111) diamond surfaces can lead to graphitization of a surface region. It is known that hydrogenation of the nanodiamond surfaces prevents such graphitization by the elimination of the dangling bonds thus stabilizing the crystal structure. The structure of the C(111) surface at intermediate H coverage, however, is largely unknown. Similar consideration can be made for the two other faces (110), and (100).

3.2.3 Adsorption Sites on Fullerene and Related Materials: $sp^{2+\delta}$ Hybridization

As the adsorption on individual fullerenes or small aggregates of fullerenes has remained largely unexplored because of technical challenges,[22,23] last but not least, this section will focus on adsorption sites on CNTs. Because the electronic structure of CNTs is highly dependent on the nanotube characteristics, such as tube diameter and helicity, it is expected that they will also affect the adsorption.[24–26] Although CNT closed or opened tips do not constitute the main part of the surface area, a different adsorption is expected with respect to the tube walls, since pentagons or edges are present, respectively.[27,28] As the length of the CNT increases, the charges of the middle and top (or bottom) carbon layers decrease. The larger the length and section of a SWCNT, the stronger the adsorption between SWCNT and the gas.[29] Additionally, the possibility to adsorb species on the convex (exohedral adsorption) or concave (endohedral adsorption) surface should be taken into account.

The nature and strength of the adsorption is directly correlated to carbon hybridization. Strain in non-defective nanotube sidewalls is manifested in pyramidalization and π-orbital misalignment (Chapter 2, Section 2.4.2), which can be relieved by the addition of an atom or functional group to the CNT external surface. Even when they have the same diameter, SWCNT has a much smaller pyramidalization angle than fullerenes because of the cyclic bending (instead of spherical as in fullerene). As the nanotube diameter increases, both the pyramidalization angle and π-orbital misalignment angle decrease, lowering the chemical reactivity of the C–C bonds, eventually approaching planar graphite for very large CNT diameters. Finally, the CNT convex (outer) surface is chemically reactive because the convex arrangement of pyramidalized sp^2 carbon atoms is correctly disposed for the formation of chemical bonds with reagent species. For the same reason, the concave (inner) surface should be more inert and can withstand the presence of highly reactive species encapsulated within the nanotubes.[30,31] The same tendency is true and even more pronounced for fullerenes,[32] for which different adsorption sites have been identified (Figure 3.4). Of course, distortions of CNTs will accentuate this phenomenon, suggesting that local reactivity can be significantly enhanced.[33]

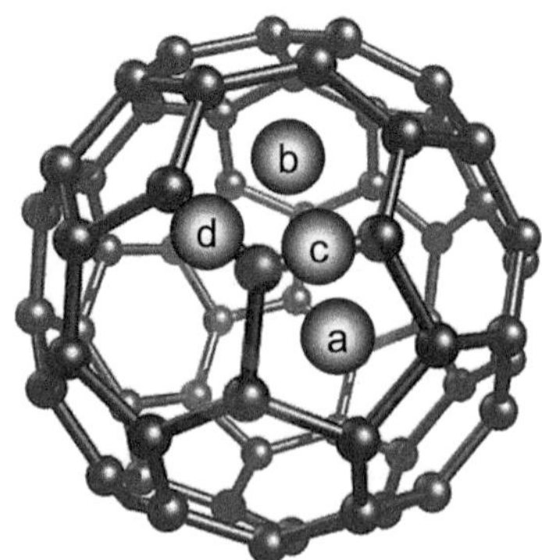

Figure 3.4 Identified adsorption sites on fullerenes.

3.2.4 Presence of Defects

Defects in three-dimensional crystals are referred to as *intrinsic* when the crystalline order is perturbed without the presence of foreign atoms. The latter are denoted as impurities and constitute *extrinsic* defects. In macroscopic crystalline materials, intrinsic defects have different dimensionalities. Point defects, typically vacancies or interstitial atoms, are zero-dimensional, whereas the concept of dislocations is based on one dimensional line of defects. Grain boundaries or stacking faults extend in two dimensions, while inclusions and voids have a finite size in all three dimensions. Foreign atoms may exist as zero-dimensional defects when they substitute individual atoms of the crystal or are located on interstitial sites. Agglomerations of foreign atoms can extend to more dimensions. The reduced dimensionality of nanostructured carbon materials decreases the number of possible defect types. Therefore, in graphene-based materials, the concept of zero-dimensional point defects is quite similar to bulk crystals, but line defects play a different role. Truly three dimensional defects do not exist in graphene. It is well-known that defects are not always stationary and that their migration can have an important influence on the properties of a defective crystal. In graphene-based materials, each defect has certain mobility parallel to the graphene plane. Several studies have indicated that defective sites are of great importance to the adsorption process.[34–39] The presence of these defects in the structure creates highly reactive sites for the adsorption of various species. This means that the overall chemical reactivity of carbon materials depends strongly on the number of surface defects present in the sample.

3.2.4.1 Point Defects

Single and Multiple Vacancies. The simplest defect in any graphene-based material is the missing lattice atom (Figure 3.5a). The calculated migration barrier for a single vacancy (SV) in graphene is about 1.3 eV. This already allows a measurable migration slightly above room temperature (373–473 K). Double vacancies (DV) can be created either by the coalescence of two SVs or by removing two neighboring carbon atoms. On such defects the initial three-fold symmetry is broken and a vacancy is formed.

Removing a carbon atom leaves a vacant space with three dangling carbon bonds in the network.[40] The triplet state is slightly more stable than the singlet state.[41] These three dangling bonds are unstable and undergo recombination to make a chemical bond between two of them forming a pentagon ring and one remaining dangling bond in a nonagon ring.[42] The related single occupied molecular orbital (MO), given in Figure 3.6, clearly indicates the propensity of one carbon atom of the defect to be very reactive. Worthy of note is that such MOs and the singlet-triplet quasi-degeneracy point to the chemistry of carbenes.

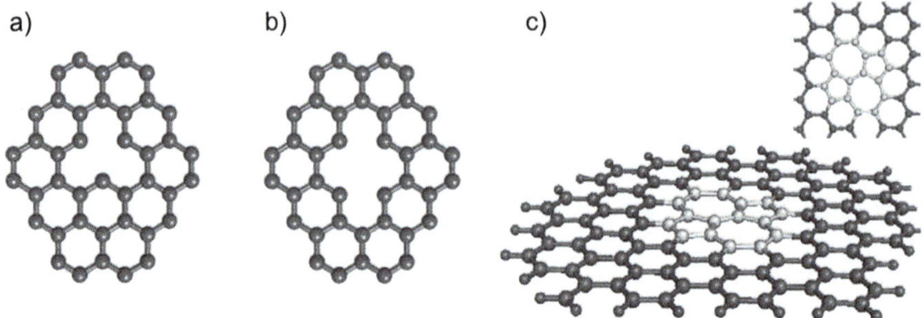

Figure 3.5 Point defects in the graphene system. Mono-vacancy (a); di-vacancy (b) and (c) Stone–Wales defect.

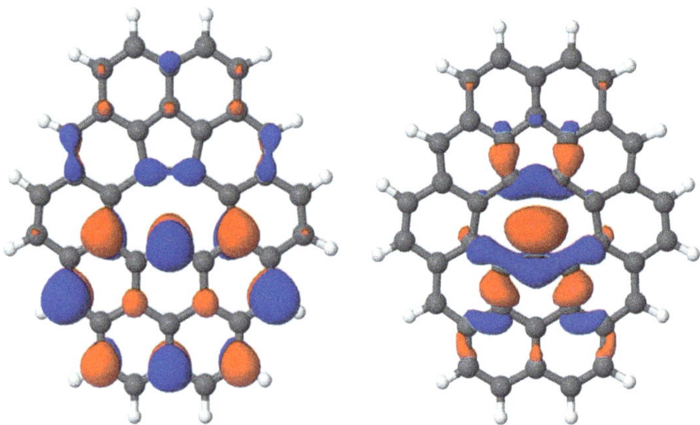

Figure 3.6 Single occupied MOs in the lowest triplet state for a vacancy on a graphene sheet. Adapted from ref. 41.

For CNTs, the formation of a vacancy and its orientation depends on the radius and chirality of nanotubes. Such a disturbance causes significant changes in their transport, magnetic, mechanical, and optical properties.[43,44] On the other hand, the dangling bond can also participate in the interaction with other molecules as well as it can be the center of CNT functionalization.

Isolated vacancies are the most studied defect in diamond, both experimentally and theoretically.[45] Its most important practical property is optical absorption, like in the color centers, which gives diamond a green or sometimes even green-blue color (in pure diamond).[46] In nanodiamonds, vacancies have been much less studied than in graphene-based materials,[47] except for nitrogen vacancy defects.[48]

Stone–Wales Defects. As mentioned above, one of the unique properties of the graphene lattice is its ability to reconstruct by forming non hexagonal rings. The simplest example is the Stone–Wales (SW) defect, which does not involve any removed or added atoms. Four hexagons are transformed into two pentagons and two heptagons [SW(5,5-7,7) defect] by rotating one of the C–C bonds by 90°, as shown in Figure 3.5c. Such rotation creates two pairs of pentagons and heptagons opposite to each other around the rotated C–C bond. For this reason, Stone–Wales defects are also known as a 5-7-7-5 defects.

Other rings such as squares, octagons, nonagons, and decagons can also be observed at certain stages of the surface reconstruction of CNTs, but they are found to rearrange by the Stone–Wales mechanism, thus leading the structure to be mainly constituted of five-, six-, and seven-membered rings.[49] The pairs of pentagons and heptagons create a local disturbance (electrons are more concentrated around the double bonds involved in the defect instead of participating in the delocalized electron cloud above the graphene as usual), which serves as a reactive center for adsorption of various atoms, nanoparticles (NPs), or molecules.[50–52] Carbon atoms belonging to the tips are more reactive that those belonging to the sidewall of the nanotube, even in the presence of a Stone–Wales defect.[53] Among the carbons making up the tips, the activity of each pyramidal structure is strongly correlated to its local curvature angle.

Carbon Adatoms. Interstitial atoms, as they appear in three-dimensional crystals, do not exist in graphene. This is because placing an atom to any in-plane position, for example, in the center of a hexagon, would require a prohibitively high energy. Rather than straining the local structure in two dimensions, additional atoms use the third dimension. The energetically favored position is the bridge configuration (B-site, on top of a carbon–carbon bond, as shown in Figure 3.1.[54] When a carbon atom interacts with a perfect graphene layer, it induces a change in the hybridization of the carbon atoms in the layer. Some degree of sp^3-hybridization can appear locally so that two new covalent bonds can be formed between the adatom and the underlying atoms in the graphene plane. The small energy difference of about 0.3 eV between the local and global minima indicates that adatoms migrate easily over the graphene surface.

On diamond(001) surfaces the C adatom migrates preferably along the dimer rows, and an anisotropic diffusion of a carbon adatom on diamond(001) surfaces was suggested.[55] Adatoms or vacancies play a less pronounced role in stabilizing the diamond(111) surface.[56] This is probably due to the large energy loss upon bond stretching for diamond and may explain why no rough surface reconstructions with adatoms and holes or walls have been found for diamond.

Foreign Adatoms. The effect of a heteroatom on the properties of graphene depends on the bonding between the atom and graphene. If the bond is weak, only physisorption due to van der Waals interaction occurs. If the interaction is stronger, covalent bonding between the foreign atom and the nearest carbon atoms leads to chemisorption. Various bonding configurations, normally corresponding to high symmetry positions such as on top of a carbon atom (*T*-site), on top of the center of a hexagon (H_6-site), or the bridge position (*B*-site) are possible. Foreign atoms can also be incorporated into graphene or diamond as *substitutional* impurities. In this case, the impurity atom replaces one or two carbon atoms. Boron[57] or nitrogen[58] serve as the natural dopants in carbon structures since they have one electron less or more, respectively, and roughly the same atomic radius. Sulfur and phosphorus have just recently entered the world of carbon synthesis.[59] For these two latter elements, the caused effects are attributed more to the formation of structural defect sites or favorable spin densities throughout the carbons, as here the electronegativities of carbon and the dopants are closer to each other.

Boron is by far the most widely used doping agent to produce conducting diamond electrodes.[57,60] Boron doping leads to a p-type semiconductor. At low doping levels, the diamond acts as an extrinsic semiconductor, and at high doping levels the material acts as a semi-metal. But other dopants are also possible that provide n-type conductivity: nitrogen,[61] phosphorus[62] and sulfur.[63] By examining the potential energy surface for nitrogen and boron substitution in diamond nanocrystals along a variety of crystallographically non-equivalent substitution paths and extremum sites, Barnard *et al.* have predicted that neither nitrogen nor boron are likely to be stable as 'dopants' within nanodiamond, unless the particles are sufficiently large (>2 nm) so as to be effectively 'bulk-like'.[64]

Carbon nanotube doping has drawn substantial interest due the increased reactivity of doped nanotubes with other molecules.[65] The most important and, at the same time, the most abundant heteroatom that affects the use of CNTs is oxygen. It is usually either chemisorbed on the surface or arranged in the form of functional groups analogous to those existing in organic chemistry. Since those groups increase the reactivity of the otherwise hydrophobic carbon surface, oxidation often opens the door to further modifications *via* interactions of specific chemicals with the groups and their reactions on the surface. The boron and nitrogen serve as acceptors and donors of electrons in the CNTs. Since their atomic radii are similar to the carbon atom, they create a small perturbation in the nanotube structure in comparison to the perfect one. The nitrogen in CNTs can be seen as regular defects, which change the chemical behavior of the tubes. The reactivity of N-doped CNTs can be estimated to be higher than un-doped ones of similar diameters. Theoretical calculations predict a localization of the unpaired electrons around the nitrogen-defect in the semiconducting heteronanotubes.[65] Other heteroatoms such as hydrogen and sulfur can also be chemisorbed over a large number of imperfections and defects leading to stable surface compounds.[66]

Similarly, the N-doped graphene shows different properties compared with the pristine graphene. For instance, the spin density and charge distribution of carbon atoms will be influenced by the neighbor nitrogen dopants, which induce the "activation region" on the graphene surface.[8] This kind of activated region can participate in catalytic reactions or anchor the metal NPs used in the catalytic reaction. Moreover, after nitrogen doping in the monolayer graphene, the Fermi level shifts above the Dirac point, and the density of state near the Fermi level is suppressed; thus, the band gap between the conduction band and the valence band will be opened. Much larger atoms such as transition metal impurities have also received particular attention due to their ability to inject charge into the electron system of graphene.

In C_{60}, the dopant atoms may be located: (i) outside the cage, producing fulleride salts; (ii) inside the cage, producing a sort of super-atom; or (iii) as part of the cage itself, replacing one or more of the carbon atoms in the cage network.[67] Azafullerenes, in which one or more cage carbons are substituted by N atoms show tunable chemical and physical properties.[68]

3.2.4.2 One-dimensional Defects

Dislocation-like Defects. One dimensional defects have been observed in several experimental studies of graphene. Generally, these line defects are tilted boundaries separating two domains of different lattice orientations with the tilt axis normal to the plane. Graphene domain boundaries and the interconnection between FLG sheets are important considerations when working to optimize adsorption properties or reactivity. There are two primary ways one can envisage two separate graphene domains being connected: either by direct atomic bonding at the interface of the two graphene sheets to create a discrete atomic domain boundary, or by one sheet overlapping another to form a bilayer boundary region that is not atomically discrete and relies on interlayer van der Waals forces to hold the two domains together.[69] These shall be referred to as the atomic interface and the overlap interface, respectively. Recent investigations of monolayer graphene domain boundaries reported the existence of atomically bonded discrete interfaces. These domain boundaries were found to consist of pentagonal, hexagonal, and heptagonal carbon rings connected in a chain, allowing for two graphene domains of different lattice directions to bond together. These studies and observations illustrate the complexity of the FLG system, with both atomic domain boundaries and small regions of turbostratic graphene, originating from overlap at domain intersections, in an otherwise well-ordered AB Bernal stacked graphitic system.

Dislocations are very common defects in both natural as well as chemical vapor deposition-grown diamond. The two major types of dislocations are the *glide* set, in which bonds break between layers of atoms with different indices (those not lying directly above each other) and the *shuffle* set, in which the breaks occur between atoms of the same index.[70,71] The dislocations produce dangling bonds, which introduce energy levels into the band gap, enabling the absorption of light. Few studies are related to dislocations in nanodiamonds.[72]

Defects at the Edges of a Graphene Layer. Each graphene layer is terminated by edges with the edge atom being either free or passivated with hydrogen or oxygen atoms. The simplest edge structures are the armchair and the zigzag orientation. They can reconstruct and any other direction in between these two can be imagined. However, the zigzag and armchair orientations seem to be preferred, possibly because they minimize the number of dangling bonds at the edge. Defective edges can appear because of local changes in the reconstruction type, or because of sustained removal of carbon atoms from the edges. Under these conditions, armchair edges can be transformed to zigzag edges. An intermediate structure can be considered as a defective edge. A simple example of an edge defect is the removal of one carbon atom from a zigzag edge. This leads to one pentagon in the middle of a row of hexagons at the edge. Other edge reconstructions result in different combinations of pentagons and heptagons at the edge.

3.3 Physisorption on Nanostructured Carbon Materials

Depending on the intensity of the interaction between one or more adsorbates and a given substrate, a qualitative distinction can be usually made between *physisorption* and *chemisorption*. Physisorption results from the presence of van der Waals attractive forces due to fluctuating dipole moments between the adsorbate and the substrate (no charge transfer or electrons shared between atoms). Typically, this process involves energies in the order of 50–500 meV per atom. In the case of chemisorption the adsorption energy (E_{Ads}) of the adatom can rise to a few eV per atom, and a strong "chemical bond" is formed between the adsorbate and the substrate.

The adsorption of various species on nanostructured carbon surfaces is usually studied by first-principles calculations using density functional theory. Due to their well-defined surface, graphene and CNTs have been by far the most studied materials. For fullerenes, most of the works deal with their adsorption on surfaces. For carbon blacks, onion and nanodiamonds, their poorly understood surface explain the relatively low amount of studies. Accordingly, we will concentrate our analysis to adsorption on CNTs and graphene, and will broaden the discussion to other nanostructured carbons when possible.

The binding energy, tube/graphene–adsorbate distance, and charge transfers are generally investigated.[73,74] Atomistic modelling of gas adsorption should be treated differently depending on the specific phenomena involved, either physical adsorption or chemical bonding. Sometimes, however, the classification of the studied phenomena in terms of physical or chemical adsorption is quite difficult due to the occurrence of strong polar interactions or weak charge transfer that make the classification of the case under study uncertain. In these cases, the calculation of energetic quantities, such as the activation energy or the adsorption enthalpy, may help to get a clearer scenario, because it is expected that physical adsorption exhibits significantly lower adsorption enthalpy values than the ones involved in chemical bonds.[75] The physical adsorption is believed to act as a step in the transition towards a chemisorption state. This transition may result in the splitting of the molecule and adsorption of individual atoms, commonly identified as dissociative chemisorption.

3.3.1 Physisorption of Gases on Nanostructured Carbon Materials

Considering gas adsorption over the nanostructured carbon materials, there have been several attempts to explain the adsorption mechanism of different gases. In general terms, physisorption is believed to take place over a perfect surface. The presence of imperfections, such as defects (see Section 3.2.4) often produces stronger interactions between the gas adsorbate and the surface, thus yielding a chemisorptive mechanism. It has also been demonstrated that doping the graphene structure with metallic atoms could be an efficient way to improve the sensitivity of nanostructured carbon materials to some gas molecules (see Section 3.4.2).[73]

3.3.1.1 Physisorption on Individual Nanostructured Carbon Materials

Although most of the studies deal with a perfect surface of individual CNTs, it is worth noting that some works have been devoted to the comparison of the reactivity of the tip of CNTs with the rest of its surface. Although the contribution of the tips to adsorption in terms of specific surface area is negligible, it can be relevant at very low coverages.[76] DFT calculations have shown that O_2 and H_2O behave differently when they are adsorbed on a SWCNT tip.[77] The tip-molecule distance is 2.722 Å for O_2, and 2.488 Å for H_2O, and the binding energies are −0.51 eV and −0.12 eV, for O_2 and H_2O respectively, indicating that O_2 and H_2O are weakly adsorbed on the tip. H_2O molecules can be adsorbed on both closed and open-ended SWCNTs, and O_2 molecules can be adsorbed on a closed SWCNT. Quite differently, O_2 molecules are dissociatively chemisorbed on an open-ended zigzag SWCNT.

The adsorption of hydrogen is perhaps one of the most debated in terms of mechanism. Based on first-principles theoretical calculations[78] and DFT calculations,[79] it has been reported that on pristine SWCNTs, the dissociative adsorption of H_2 molecules is improbable due to high energy barriers of about 2–3 eV. The dissociative chemisorption weakens carbon–carbon bonds, and the concerted effect of many incoming molecules with sufficient kinetic energies can lead to the scission of the nanotube. A similar situation occurs on graphene, where chemisorption of molecular hydrogen presents rather high barriers estimated around 1.5 eV,[80] because it requires the dissociation of H_2 (dissociative adsorption). Thus, molecular physisorption is predicted to be the most stable adsorption state. The physisorption energies outside the CNTs are approximately 0.07 eV, and larger inside, reaching a value of 0.17 eV inside a

(5,5) SWCNT. The H_2 binding energy in graphene was theoretically evaluated in the range of 0.01–0.06 eV.[81,82] On the other side, H atoms have low-energy barriers, less than 0.3 eV. A DFT study of the interaction between atomic hydrogen and (5,5) and (10,0) SWCNT has shown a weak chemisorption on the outer wall.[83] A geometry relaxation produces a binding energy of about −1.5 eV. When SWCNTs are completely covered by hydrogen, the binding energy is enlarged by about 0.3 eV. From this study, a high sensitivity of the electronic structure to the presence of an H atom is reported. On graphene, commonly accepted values for H binding energy and chemisorption barriers are ~0.7 and ~0.3 eV, respectively. Theoretical studies have shown, in particular, that adsorption of the first H atom locally modifies the graphene structure favoring further H binding, with a collective stabilization effect.[84,85] The interaction of atomic hydrogen and low-energy hydrogen ions with sp^2-bonded carbon was investigated on the surfaces of C_{60} multilayer films, SWCNTs and graphite (0001).[86] These three materials have been chosen to represent sp^2-bonded carbon networks with different local curvatures and closed surfaces (*i.e.* no dangling bonds). It was found that the energy barrier for hydrogen adsorption decreases with increasing local curvature of the carbon surface. Whereas in the case of C_{60} and SWCNTs, hydrogen adsorption can be achieved by exposure to atomic hydrogen, the hydrogen adsorption on graphite (0001) requires H^+ ions of low kinetic energy. The binding strength of molecular hydrogen on either positively or negatively charged fullerenes can be dramatically enhanced to 0.18–0.32 eV.[87] The enhanced binding is delocalized in nature, surrounding the whole surface of a charged fullerene, and is attributed to the polarization of the hydrogen molecules by the high electric field generated near the surface of the charged fullerene.

Storage of fuels other than H_2 has also been studied. Ganji *et al.* studied the role of defects on the interaction of methane with SWCNTs by using DFT based calculations.[88] The energy values and H–C binding distances obtained from the *ab initio* calculations were typical of a physisorption process. The obtained results revealed a considerable increase in the adsorption binding energy of the order of 156% due to the presence of structural defects in CNTs, which affects the methane storage capacity in CNTs. Daykova *et al.* used first-principles simulations to study the adsorption of CH_4 on a perfect graphene sheet.[89] They found that CH_4 occupies the center of the hexagon and interaction between the methane molecule and graphene induces a dipole moment in CH_4. The dipole moment is *ca.* 1.5 Debye if only one bond of CH_4 faces towards the graphene surface, and it is much smaller, *i.e.*, *ca.* 0.5 Debye, if three bonds point at the surface. Calculated adsorption energies of 118–281 meV have been reported for CH_4 adsorption on fullerenes.[90] An endohedral methane complex of a fullerene derivative was synthesized by insertion of a methane molecule through the opening of an open-cage C_{60} derivative.[91] The trapped methane was confirmed by NMR spectroscopy and mass spectrometry. Both methane carbon and protons show remarkable up-field shifts in NMR, characteristic of a chemical species in a fullerene cage. CH_4 protons appear as one equivalent signal in the ^1H NMR spectrum, suggesting that even methane can rotate in a C_{60} cage.

Physisorption of noble gases is widely employed to study the structural and theoretical aspects of surface adsorption because the inertness of the noble gases typically excludes the possibility of chemical and polar interactions with the surface. For helium, the difference of energy between the most favorable adsorption site (H_6-site) and the less favorable one (top site T) characterizes surface roughness; its value, of the order of 3.45×10^{-3} eV, is comparable to that on planar graphite.[92] The energy barriers for displacements of He atoms along the nanotube walls are of the order of 1.21×10^{-3} eV and the minimum energy of He–He interaction is -0.86×10^{-3} eV. Adsorption of noble gases (Ar, Kr, Xe) on metallic and semi-conducting CNTs was investigated using the van der Waals density functional.[93] No difference was found in the adsorption between the metallic and semi-conducting nanotubes, indicating that the adsorption energies for rare gases on CNTs are not strongly influenced by differences in the electronic structure of the nanotubes. Molecular dynamics simulation has been used to study the adsorption isotherms of noble gases on open ended SWCNTs.[94] It was shown that adsorption occurs both inside and outside an open ended SWCNT in the pressure range (0.1–2.5 MPa). The interior coverage of 0.06 Xe–C and 0.08 Kr–C were reported at saturation conditions. The monolayer coverage on the exterior surface is 0.23 Xe–C and 0.25 Kr–C. Grand canonical Monte Carlo calculations performed on Xe adsorption on SWCNT, at low pressure shows a similar tendency.[95] However, in that case and at very low pressure, simulations indicate that adsorption takes place primarily on the inside of the nanotubes. Interstitial and external adsorptions were found to be negligible in comparison with adsorption inside the nanotubes. The results also indicate that the curvature of the nanotube does not substantially perturb the adsorption potential from that of a graphene sheet. The adsorption of Xe on graphene was systematically investigated by *ab initio* MP2 calculations using Dunning's correlation-consistent basis sets.[96] Polycyclic aromatic hydrocarbon (*i.e.*, coronene) was used to model the graphene surface. The equilibrium distance of Xe at the hollow site was calculated to be 3.56 Å, which was in excellent agreement with the available experimental value of 3.59 ± 0.05 Å. The corresponding binding energy at the hollow site was calculated as −142.9 meV, whereas the binding energies at the bridge and on-top sites were −130.8 and −127.4 meV, respectively.

Zhao *et al.* studied the adsorption of various gas molecules (NO_2, O_2, NH_3, CH_4, CO_2, H_2 and N_2) on SWCNTs using a first principles method.[97] The equilibrium tube–molecule distance, adsorption energy, and charge transfer for various molecules on (10,0), (17,0) and (5,5) SWCNTs were calculated, as shown in Table 3.1. The results show that most of the studied molecules (except for NO_2 and O_2) are charge donors with small charge transfer (0.01–0.035e) and weak binding (≤0.2 eV). For these molecules, the adsorption can be identified as physisorption. For O_2 and NO_2, it shows that they both are charge acceptors with large charge transfer and adsorption energies. It also demonstrated that there is no clear dependence of adsorption on the tube size and chirality.

Table 3.1 Equilibrium tube-molecule distance (d), adsorption energy (E_a), and charge transfer (Q) of various molecules on (10,0), (17,0), and (5,5) individual SWCNTs. Adapted from ref. 97.

	NO_2	O_2	NH_3	CH_4	CO_2	H_2	N_2
(10,0) SWCNT							
d Å	1.93	2.32	2.99	3.17	3.20	2.81	3.32
E_a/meV	797	509	149	190	97	113	164
Q/eV	−0.061	−0.128	0.031	0.027	0.016	0.014	0.008
Site	*T*	*A*	*T*	*H*	*H*	*H*	*H*
(5,5) SWCNT							
d Å	2.16	2.46	2.99	3.33	3.54	3.19	3.23
E_a/meV	427	306	162	122	109	84	123
Q/eV	−0.071	−0.142	0.033	0.022	0.014	0.016	0.011
Site	*T*	*A*	*T*	*H*	*H*	*H*	*H*
(17,0) SWCNT							
d Å	2.07	2.50	3.00	3.19	3.23	2.55	3.13
E_a/meV	687	487	127	72	89	49	157
Q/eV	−0.089	−0.096	0.027	0.025	0.015	0.012	0.006
Site	*T*	*A*	*T*	*H*	*H*	*H*	*H*

Leenaerts *et al.* investigated the adsorption of H_2O, NH_3, CO, NO_2 and NO on a graphene substrate using first-principles calculations.[98] They found that the adsorbates were physically adsorbed on the pristine graphene and that the charge transfer between the adsorbates and graphene was almost independent on the adsorption site. However, charge transfer did strongly depend on the orientation of the adsorbate with respect to the graphene surface. In addition, charge transfer was interpreted in light of two different mechanisms: for NO_2 on graphene, it is mainly due to the position of the lowest unoccupied molecular orbital (LUMO) below the Dirac point, whereas for all the other adsorbates it was caused by the mixing of the highest occupied molecular orbital (HOMO) or LUMO orbitals with the orbitals of graphene; the strength of the hybridization being deduced from the geometrical orientation of the HOMO and LUMO orbitals with respect to the graphene surface. The NO molecule physisorbs on a pure C_{60} with a binding energy of about −6 meV. The presence of a Sc atom on the C_{60} surface considerably enhances the capacity of C_{60} to accept electrons.[99] The population analysis reveals that a charge transfer from the Sc atom to C_{60} has occurred. Also, it was found that the dissociative chemisorption of the NO molecule on Sc/C_{60} is accompanied with a charge transfer from Sc to the NO molecule.

Infrared spectroscopy has proved to be a very sensitive technique to study different adsorption sites on SWCNTs.[100] For gas molecules adsorbed onto SWCNT surfaces, there is a dispersive interaction that causes a softening of the bonds in the molecule, seen as a downshift in IR modes (Table 3.2).

Table 3.2 Effects of interactions with nanotubes on IR spectra of small molecules.

Molecules	Non interacting I/cm^{-1}	Endohedral I/cm^{-1}	Exohedral I/cm^{-1}	Endohedral shift I/cm^{-1}
CO	$\nu = 2143$ (gas)	$\nu = 2135$	$\nu = 2140$	$\Delta = -8$
NO_2	$\nu_1 = 1868$ (gas)	$\nu_1 = 1853$	—	$\Delta = -15$
NO_2	$\nu_{SS} = 1318$ (gas)	$\nu_{SS} = 1026$	$\nu_{SS} = 1026$	$\Delta = -292$
CO_2	$\nu = 2349$ (gas)	$\nu = 2329$	$\nu = 2342$	$\Delta = -20$
SF_6	$\nu_3 = 947$ (gas)	$\nu_3 = 927$	—	$\Delta = -20$
CF_4	$\nu_3 = 1282$ (gas)	$\nu_3 = 1247$	$\nu_3 = 1267$	$\Delta = -35$
NH_3	$\nu_1 = 3336$ (gas)	$\nu_1 = 3156$	$\nu_1 = 3205$	$\Delta = -180$

This effect has been well-documented and has been used to distinguish between exohedral and endohedral adsorption, by the magnitude of the shift. This difference was shown with clarity and supported by calculations for CO adsorbed on SWCNTs. The ν(C–O) mode of endohedral CO showed a downshift Δ of 8 cm^{-1} and the exohedral CO showed a downshift of 3 cm^{-1}, compared to the gas phase. For physisorbed molecules, the higher coordination of endohedral sites causes a greater downshift for endohedrally adsorbed gas molecules, compared to the surface of bundle (Table 3.2).

The adsorption of CO_2 by zigzag and armchair SWCNTs of different diameters (0.47–1.085 nm) has been studied using DFT.[101] Different binding sites have been considered, namely, in the interior (side-on and end-on binding modes) and on the surface (parallel or perpendicular) of the nanotube. Calculations predict larger interaction energies for interior than exterior adsorption. Calculations predict Gibbs energies of binding between SWCNT and CO_2 of up to 395 meV, with the strongest binding observed for a zigzag (10,0), compared to armchair (6,6) (356 meV) and chiral (8,4) (300 meV) CNT.[102] Furthermore, the diffusion of CO_2 from the outside to the interior of a (5,5) SWCNT is an energetically barrierless and favourable process.[101] The interplay between CO_2/SWCNT and CO_2/CO_2 interactions when more than one CO_2 molecule is inside the tube was also investigated, showing interesting cooperativity effects for SWCNTs with large diameters.[101] Doping of the (10,0) tube with nitrogen increases the Gibbs energies of binding of CO_2 by *ca.* 130 meV, but slightly reduced binding is found when (6,6) and (8,4) SWCNTs are doped in similar fashion.[102] Exohedral doping of an Fe atom onto the SWCNT surface affects the adsorption energy of the quadrupolar CO_2 molecule inside the CNT of 20–30%.[103] DFT calculations give a range of shifts for the asymmetric stretching mode from about −6 to −20 cm^{-1} for internally bound CO_2, and a range from −4 to −16 cm^{-1} for externally bound CO_2 at low coverage.[104] Infrared measurements and DFT calculations indicate that two families of CO_2 adsorption sites are present. One family, exhibiting a shift of about −20 cm^{-1} is assigned to internally bound CO_2 molecules in a parallel configuration. The second family exhibits a shift of about −7 cm^{-1} and the site location and configuration for these species is still ambiguous.

3.3.1.2 Physisorption on CNT Bundles

During the growth process, CNTs usually aggregate into bundles because of Van der Waals interactions. Thus, it is most appropriate to discuss adsorption on CNT samples not in terms of adsorption on individual nanotubes but in terms of adsorption on the exterior or interior surfaces of such agglomerates. Most of the work on gas adsorption has been analyzed in terms of the homogeneous model of the bundle. This model considers that bundles are constituted by infinitely long CNTs of the same diameter, packed into perfect arrays. An alternative model, the heterogeneous bundle, has tubes of different diameters constituting the bundles. Diameter mismatch leads to the appearance of packing defects in the bundles, which give rise to interstitial channels with diameters larger than those found in the homogeneous bundles. Theoretical calculations have predicted that the molecule adsorption on the surface or inside of the SWCNT bundle is stronger than that on an individual tube.[97] A similar situation exists for MWCNTs, where adsorption could occur either on or inside the tube or between aggregated MWCNTs.[105]

It has also been shown that the curvature of the graphene sheets constituting the CNT walls can result in a lower heat of adsorption with respect to that on a planar graphitic surface due to rehybridization.[106] Additionally, DFT calculations have shown that nitrogen adsorption energies are stronger on armchair than on zigzag SWCNTs.[107] These considerations raise a number of questions with respect to the accessible sample surface area and pore size distribution (PSD), the preferred binding sites, the uptake capacity as well as the strength (chemi- or physisorption) of the corresponding adsorbate–substrate interaction.

The structure of the pores and PSD of CNTs are largely dictated by the type of CNT. The pores in CNTs show a wide range of sizes and shapes. The pores are classified by their sizes usually into three groups: (i) macropores having an average diameter of more than 50 nm, (ii) mesopores with a diameter of 2–50 nm, and (iii) micropores having an average diameter of less than 2 nm. The latter are further divided into super-micropores (0.7–2.0 nm) and ultra-micropores of diameter less than 0.7 nm. For CNTs, which can be either microporous (SWCNTs) or mesoporous (MWCNTs), one should consider two types of pores: the well-defined inner cavity, cylindrical endo-cavities or α-pores,[108] and the inter-tube voids, interstitial pores or β-pores,[109] that will both contribute to the PSD. In addition to the well-known α- and β-pores, two types of newly defined pores with diameters of 2–10 and 8–100 nm have been evidenced for SWCNT arrays, inter-bundle (packing of bundles) pores and inter-array (bundle arrays) pores.[110] For MWCNTs, it was shown than the aggregated (inter-tube) pores are much more important for adsorption than the endo-cavities.[105] It is worth noting that, since it is difficult to describe the complex secondary structures of such materials by the methods of Euclidean geometry, the method of fractal geometry has been applied.[111]

An important problem to solve when undertaking adsorption studies on nanotubes is the identification of the adsorption sites. For SWCNT bundles, there are four distinct sites for gas molecules to be adsorbed onto (shown in Figure 3.7): (i) the external surface of the bundle; (ii) the groove formed at the contact between adjacent tubes on the outside of the bundle; (iii) the interior pore of individual tubes; (iv) the interstitial channel formed between three adjacent tubes within the bundle. The gas adsorption on these sites is decided by the binding energy of the gas molecule as well as the site availability. Some of these sites may not be available for certain gases because of the gas molecule dimension and the site diameter. The interior pore is only accessible when the SWCNT is uncapped or has defects on the tube walls. For MWCNTs, adsorption can occur in the aggregated pores, inside the tube or on the external walls, as already discussed; in this latter case the presence of defects, as incomplete graphene layers, has to be taken into consideration. Although adsorption between the walls has been proposed in the case of hydrogen adsorption in herringbone-type graphite nanofibers, it is unlikely to occur in the case of MWCNTs due, for many molecules, to steric effects and should not prevail for small molecules due to the long diffusion path.

Considering closed-end SWCNTs, simple molecules can be adsorbed onto the walls of the outer nanotubes of the bundle and preferably on the external grooves. In the first stages of adsorption (corresponding to the most reactive sites for adsorption), adsorption or condensation in the interstitial channels of the SWCNT bundles depends on the size of the molecule

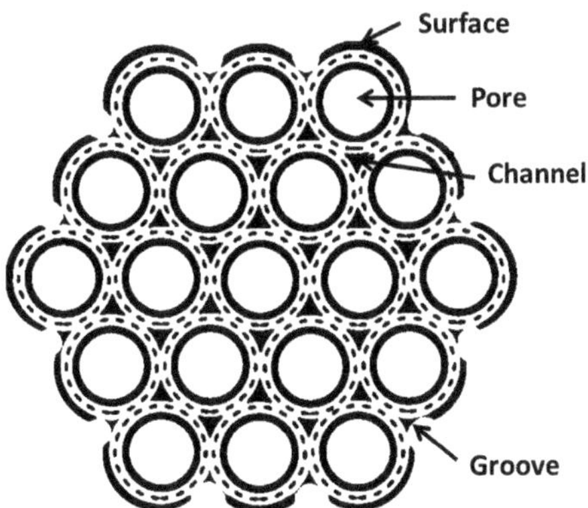

Surface: E_B = 0.062 eV; σ = 783 m^2/g
Pore: E_B = 0.049 eV; σ = 483 m^2/g
Channel: E_B = 0.119 eV; σ = 45 m^2/g
Groove: E_B = 0.089 eV; σ = 22 m^2/g

Figure 3.7 Schematic structure of a SWCNT bundle showing the available sites for gas adsorption. Dashed line indicates the nuclear skeleton of the nanotubes. Binding energies (EB) and specific surface area contributions (σ) for H_2 adsorption on these sites are indicated. Adapted from ref. 73.

(and/or on the SWCNT diameters) and on their interaction energies.[112–115] Opening of the tubes favors gas adsorption of small molecules (including O_2, N_2 and H_2) within the inner walls.[116–118] It was found that the adsorption of nitrogen on open-ended SWCNT bundles is three times larger than that on closed-ended SWCNT bundles.[119] For hydrogen and other molecules like CO, CH_4 or CF_4, computational methods have shown that, for open SWCNTs, groove sites are energetically more favourable than surface sites.[97,115,120–122]

Neutron diffraction experiments performed by Bienfait *et al.* showed the adsorption of CO_2, and other gas molecules, on the groove sites of SWCNTs bundles.[123] Yim *et al.* combined infrared spectra with local density approximation DFT to investigate CO_2 adsorption at different graphitic surfaces.[124] They found that CO_2 adsorption was preferred on the grooves and interstitial sites of bundles of SWCNTs.

Zhao *et al.* studied the adsorption of various gas molecules (Ar, N_2, H_2, O_2, H_2O, CO_2, CH_4, NO_2 and NH_3) in the SWCNT bundles and on the bundle surface using first principles method.[97] They found that the adsorption energy and charge transfer of the gases in the interstitial and groove sites of the tube bundle were considerably larger than those on the surface sites. The pore site was also energetically more favorable than the surface site. The enhancement of molecule adsorption on the groove and interstitial sites can be understood by the increased number of carbon nanotubes interacting with the molecule. Experimental results on the adsorption of NO_2 and NH_3 on SWCNTs were reported by Ellison *et al.* TPD data confirmed that NH_3 and NO_2 adsorbed molecularly and that NO_2 was slightly more strongly bound than NH_3.[121] In addition, the vibrational data were consistent with molecules that are adsorbed in interstitial channels in nanotube bundles. Rawat *et al.* measured the adsorption of xenon on purified SWCNTs for coverage in the first layer,[125] and found two rounded substeps adsorption data. This indicated the existence of at least two distinct groups of binding sites for adsorption in the monolayer: grooves and the external surface. The binding energy on the strongest binding sites for Xe on the purified nanotubes was found to be 1.6 times greater than the corresponding value on planar graphite.

The trends observed in the binding energies of gases with different van der Waals radii suggest that the groove sites of SWCNTs are the preferred low coverage adsorption sites, owing to their high binding energies. Furthermore, several studies have shown that at low coverage, the binding energy of the adsorbate on SWCNTs is between 25% and 75% higher than the monolayer binding energy on graphite. The changes observed in binding energy can be related to an increase in the coordination possibilities in binding sites, such as the groove sites in SWCNT bundles.[126,127]

There are not many studies addressing adsorption sites in MWCNT aggregates. Hilding *et al.* studied the adsorption of butane on MWCNTs at room temperature.[128] They found that butane adsorbs on MWCNTs by two mechanisms: pore condensation and surface condensation. MWCNTs with smaller outside diameters adsorbed more butane, consistent with the fact that the strain in curved graphitic surfaces affects adsorption and reaction, but not in agreement with results reported for H_2 adsorption on MWCNTs of different diameters.[129] Most of the butane adsorbs on the external surface of the MWCNTs while only a small fraction of the gas condenses in the pores.[128] The role played by isolated CNTs in MWCNT aggregates has been evidenced by P.A Gauden *et al.* in a study on nitrogen adsorption.[130]

3.3.2 Physisorption of Liquids on Nanostructured Carbon Materials

The liquid–solid surface interaction can be broadly divided into two topics, namely: (i) a study of the energetics of immersion of a solid in a liquid or wetting of a liquid on a solid surface, and (ii) adsorption on a solid surface by species present in solution. To understand these adsorption phenomena, it is necessary to not only characterize the surface structure of the solid but also the nature of the interaction between the liquid and the solid. Techniques such as immersion microcalorimetry have been used for the calculation of the heat of immersion of powders in several types of liquids. In parallel, microscopic methods have been used for determining the degree of wetting of a liquid on a solid surface. The principles of these experimental methods and the thermodynamic quantity determination there from have been detailed in several monographs.[131–133] The interactive forces between the solid surface and the liquid in which it is immersed play a key role in the phenomenon of "wetting" for the liquid and "wettability" for the solid phase. When a solid is immersed in a liquid, wetting can arise due to adhesional forces between a liquid film on the solid, spreading of a liquid drop on the solid or condensation of a liquid from the vapor phase on the solid. The determination of the wettability Δ of a surface is often carried out microscopically by measuring the contact angle θ between the two surfaces that is related to the surface tension γ of the liquid by the simple relation: $\Delta = \gamma\cos\theta$. For contact angles less than 90°, the liquid is said to wet the solid surface and *vice versa* (Figure 3.8).

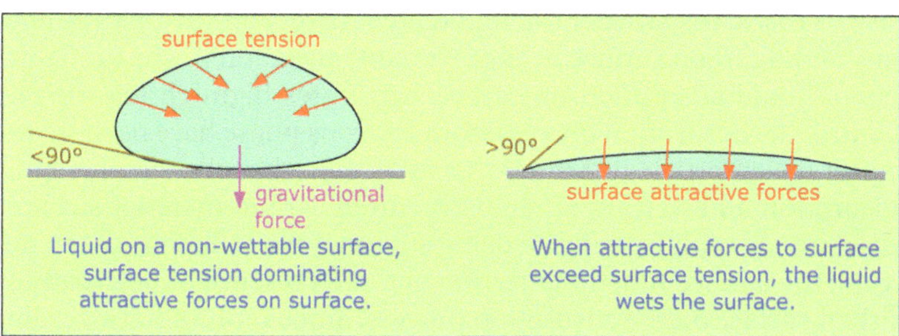

Figure 3.8 Illustration of surface wetting phenomenon.

The process of wetting being an exothermic one, wettabilities of powders can also be determined by immersion calorimetry using the relation between wettability and the heat of immersion. The advantage of this technique is that in addition to the wettability, an estimation of various other parameters such as surface polarity, porosity of the surface and the surface area of the solid can also be obtained.[131] The determination of the wettability of carbonaceous substrates such as graphite or CB is routinely carried out in view of their application in industry for oil recovery and lubrication.

The second type of adsorption, namely, adsorption on carbon substrates from species present in solution is also of great importance and a wide spectrum of adsorbates, including organic molecules used in chemical and pharmaceutical industry, biomolecules, natural organic matter (NOM) and polymers, have been investigated.[134] These processes are completely different from those involving the adsorption of gases principally due to the presence of two adsorbing species present, namely, the solute and the solvent. The overall adsorption isotherm cannot be derived from the individual adsorption isotherms of the two components. The thermodynamics and the theoretical aspects of such composite isotherms have been treated in detail.[132,133] For all practical applications, only dilute solutions have been considered where one of the components, the solvent, is in excess. In such cases, a number of assumptions about the thickness (monolayer), composition and structure of the adsorbed layer (no interaction between the adsorbed species on the substrate) can be made as a result of which an adsorption isotherm obtained by plotting the amount absorbed as a function of the concentration of the solute describes the phenomenon adequately.[131] Experimentally, this involves either the measurement of the enthalpy of adsorption of the solution by microcalorimetry or the concentration of solution after adsorption followed by sedimentation and filtration. Factors such as the pH of the solution or the presence of surface charges on the adsorbate influence the shape of the isotherm and hence give valuable information about the adsorption process. For polymers and biomolecules, steric effects may also play a role in determining the concentration of the adsorbent on the surface.[135] Even so, the mechanistic aspects of such studies is still unclear mainly due to the innate complexity of liquid adsorption, namely, non-ideal behavior due to strong intermolecular interaction when compared to gases.[136] Unlike the case of gases, the sizes of the liquid adsorbates can vary greatly and the porosity of the carbon surface needs to be suitably tailored. The fundamental aspects of liquid adsorption on carbon substrates highlighting the factors that control the adsorption process from aqueous phase have been treated in a compact review article by Moreno-Castilla.[136] In the following paragraphs, the adsorption and wetting of nanostructured carbon material surfaces by liquid adsorbates will be described, highlighting the influence of the microstructure, the surface functionalization and the presence of heteroatoms on the carbon surface on the adsorption process. Some representative examples including water, ionic liquids (ILs) and organic molecules that are liquids at 298 K are outlined; following which adsorption of species present in dilute aqueous or organic solvents on the surface of nanostructured carbon materials is highlighted using selected examples.

3.3.2.1 Physisorption of Water on Nanostructured Carbon Materials

One of the most well studied liquid adsorbates on nanostructured carbon materials is water. The existence of hydrogen bonds in water makes the fluid–fluid interaction among water molecules much stronger than that between water molecules and carbon atoms. However, at the molecular level, understanding is far from complete, with the most fundamental matter of how strong the bond is between water molecules and any carbon surface yet to be established accurately. According to the model used for the graphene surface, adsorption energy between −70 meV and −200 meV have been reported.[137]

A reasonable concern is that since CNTs are made of graphene sheets, which are hydrophobic, it may be difficult for water to be adsorbed inside nanotube. Gravimetric measurements of adsorption isotherms and kinetics of water vapor in SWCNTs were reported.[138] Adsorption was facilitated in an open configuration. Adsorption capacities of SWCNTs were found to be approximately one-half of those of AC and activated carbon fibers (ACFs). The water adsorption isotherms of SWCNTs followed type V characteristics, which are typical for a surface chemistry mediated adsorption of water. Sequential adsorption of water and organic vapor mixtures onto SWCNT bundles was studied experimentally and by grand canonical Monte Carlo simulation to elucidate the distinct interactions between select adsorbates and the nanoporous structure of SWCNTs.[139] Experimental adsorption isotherms on SWCNT bundles for hexane, methyl ethyl ketone, cyclohexane, and toluene individually mixed in carrier gases that were nearly saturated with water vapor are compared with the simulated isotherms for hexane, as a representative organic, on the external surface of the heterogeneous SWCNT bundles. From the nearly perfect overlap between the experimental and simulated isotherms, it is concluded that until near saturation only the internal pore volume of pristine SWCNT bundles fills with water. The adsorption of water vapor on the peripheral surface of the bundles remains insignificant, if not negligible, in comparison to the adsorption of water in the internal volume of the bundles. This is in contrast with the adsorption of pure hexane, which exhibits appreciable adsorption both inside the bundles and on their external surface. It is also suggested that during competitive adsorption, water molecules take precedence over small nonpolar and polar organic molecules for adsorption inside SWCNTs and leave unoccupied the hydrophobic external surface of the bundles for other more compatible adsorbates.

Interestingly, the cylindrical structure of CNTs could have a significant effect on the potential of nanotube to adsorb water. Analysis of FT-IR data for water adsorption on SWCNTs demonstrates that a small number of water molecules react with the nanotubes, forming C–O bonds, whereas a majority of the water molecules adsorb intact.[140] Water confined inside SWCNTs is of particular interest for understanding the wetting properties of the nanotubes surface. In addition, water confined in CNT cavities gives interesting information on the structure, reactivity and physical properties of confined fluids. Experimental vibrational spectroscopy studies have provided direct evidence of a water phase inside SWCNTs that exhibits an unusual form of hydrogen-bonding due to confinement.[141] Water adopts a stacked-ring structure inside CNTs, forming intra- and inter-ring hydrogen bonds. The intra-ring hydrogen bonds are bulk-like while the inter-ring hydrogen bonds are relatively weak, having a distorted geometry that gives rise to a distinct OH stretching mode. Striolo *et al.* performed grand canonical Monte Carlo simulations on the water uptake by SWCNTs and found that the calculated adsorption isotherms depended on the diameter and chirality of the nanotubes.[142] Adsorption isotherms simulated for (10,10), (12,12), (20,20) SWCNTs were found to be of type V in the IUPAC classification characterized by negligible water uptake at low pressures, sudden and complete pore filling once a threshold pressure is reached, and wide adsorption-desorption hysteresis loops. On the other hand no hysteresis was observed for water adsorption in smaller tubes (6,6) and (8,8), and the amount of water adsorbed in SWCNTs increased with increasing pressure. The authors also studied the structure of the confined "water nanotubes" and reported that water confined in (6,6) SWCNT at 298 K forms one-dimensional hydrogen-bonded chains, in which each water molecule receives one hydrogen from the preceding molecule and donates one hydrogen to the following molecule along the chain and for larger tubes, three dimensional structures such as cubic or octagonal water were predicted. Marti *et al.*[143] studied the dynamics of liquid water and its isotopes when adsorbed inside CNTs of different radii by means of molecular dynamics simulations and reported the presence of new bands in the vibrational spectra. The presence of liquid-like water at temperatures much lower than the freezing point of water was confirmed by Maniwa *et al.* using XRD where liquid-like water transformed into ice at 235 K.[144] It has been suggested that this anomalous fluid-like behavior of "water nanotubes" may explain water transport mechanisms in nominally hydrophobic regions of trans-membrane proteins. The static and dynamic behavior of liquids inside CNTs, a broad subject, which includes: the investigation of liquid entering inside the tubes, and their subsequent filling, the overall flow through tubes as well as the wetting of the nanotube walls, has been recently reviewed.[145] The continuum approximation seems to break down below 10 nm in the case of water in CNTs. A smooth liquid–gas interface (meniscus) disappears in tubes with diameter less than 8–10 nm and an anomalous behavior of water is observed in 1–7 nm CNTs.

The adsorption of water on the surface of graphene has been studied by several workers following the observation of Novoselov *et al.*,[146] who reported that traces of adsorbed water dramatically alter the mobility of FLG. Both experimental and theoretical works on graphene surfaces have been reported. The interaction energy for the water–graphite system was calculated to be of −95 meV, with the assumed geometrical structure with one hydrogen atom pointed down toward the ring system.[147] The structure with two hydrogens pointed down is predicted to be more stable, with net interaction energy of −117 meV. The structure, adsorption, electronic states, and charge transfer of small water aggregates ($5H_2O$) on the surface of a graphene layer was investigated by DFT.[148] The results show that the adsorption energy of one water molecule is primarily determined by its orientation. Although water physisorption was expected to occur in the regime of droplets, no induced impurity states close to the Fermi level of graphene interacting with small water clusters were found. Concerning the donor/acceptor tendency, the charge transfer mechanism should preferentially occur from water to graphene only when the oxygen atom is pointing toward the surface. Otherwise, and in the case of larger adsorbed clusters, charge transfers systematically occur from graphene to water. Perfect suspended graphene is rather insensitive to H_2O adsorbates, as doping requires highly oriented H_2O clusters.[149]

Gordillo *et al.* have reported interesting results about the behavior of water confined inside or close to different graphene surfaces by means of molecular dynamics simulations.[150] This allowed the calculation of the structural properties, such as atomic density profiles, molecular orientation and hydrogen bond distributions, showing a significant difference from bulk water. Important changes were also predicted for the dynamic properties such as IR spectra, translational and rotational diffusion coefficients of adsorbed water. At the carbon water interface, dielectric properties such as permittivity were found to increase up to five times that of the bulk. These changes are similar to those for confined water in CNTs.[151] Rafiee *et al.* studied the wetting of graphene sheets by water and reported that graphene sonicated in water, acetone or a combination of water and acetone showed different contact angles with water droplet ranging from super-hydrophobic to super-hydrophilic behavior.[152] The molecular arrangement of water on graphene surface has also been studied by Kimmel *et al.* who deposited thin films of H_2O or D_2O on graphene substrate at temperatures ranging from 20 to 152 K. It was found that ice layers formed on the graphene surface, showing significant wetting at 135 K.[153,154] At higher temperatures, these ice layers de-wet the surface forming ice crystals. The presence of a vacancy changed the adsorption energy of water for graphene.

C_{60} is extremely hydrophobic and its solubility in water is very low ($<10^{-12}$ g L^{-1}). However, upon extended contact with water, C_{60} is known to form negatively charged water stable aggregates (5 to 500 nm), commonly called nC_{60} or nano C_{60}. The nC_{60} has much higher solubility (*ca.* 150 mg L^{-1}) in water compared to molecular C_{60} and is known to be the most relevant form of C_{60} in the environment. The adsorption of water on CB has been calculated on several model graphitized carbon blacks using Monte Carlo simulation.[155] The results for the various surface configurations were compared with the few experimental results available in the literature. Traditionally used potential parameters provide isotherms with no similarity to those seen in experiment. Increased interaction well depths with the surface lead to more realistic isotherms but still require strongly attractive functional groups, such as phenol,[156] for qualitative agreement.

It has been shown that nanodiamond adsorbs atmospheric water quickly.[157] The water adsorbed on the surface of diamond consists of three forms: (i) the monolayer water molecules adsorbed on the sites of Lewis acid of diamond surface, such as carbon-containing species carrying positive charge; (ii) the monolayer water molecules adsorbed on the sites of Lewis base of diamond surface, such as carbon-containing species carrying negative charge, oxygen and nitrogen-containing species of diamond surface; and (iii) the adjacent adsorbed water molecules forming intermolecular hydrogen bonding. On such surfaces, the concentration of the primary adsorption centers (defects, oxygenated functional groups) determined the amount of water adsorbed.[158]

3.3.2.2 Physisorption of Ionic Liquids (ILs) on Nanostructured Carbon Materials

The synthesis, characterization and applications of CNT-IL and graphene-IL composites have been reviewed recently.[159] IL-based surfactants adsorb on the CNT surface *via* their hydrophobic long alkyl chain orienting their hydrophilic cationic imidazolium groups toward the aqueous phase. Hence, these have also been adsorbed on CNTs, fullerenes or graphene to achieve a homogeneous dispersion of the latter.[160] Several types of long- and short-chain ILs have been investigated to disperse CNTs in aqueous solvents. Surfactant-like long-chain ILs such as 1-alkyl-3-methylimidazolium bromides, butyl-α, β-bis(dodecylimidazolium bromide),[161] carbazole tailed ILs (1-*n*-(*N*-carbazole)alkyl]-3-methylimidazolium bromide)[162] and imidazolium ion-based ILs with hexadecyl alkyl chain have been found to be quite effective in dispersing CNTs in water. While CNT-IL dispersions formed by non-covalent wrapping are relatively unstable, physisorbed polymerized ionic liquids (PILs) form stable dispersions with CNTs.[163] Despite rapid advancement in applications, only a few reports exist on the fundamental adsorption characteristics of these systems. Frolov *et al.* have studied the basic mechanisms of CNT interactions with several room temperature ILs by means of fully atomistic molecular simulations, and related these interactions to the length of the alkyl cation chain on the IL, the presence of solvents and the charge on the CNT.[164] Microscopic adsorption structures of imidazolium-based ionic liquids, [mmim]PF_6, [bmim]PF_6, [mmim]Cl, and [bmim]Cl, on the surfaces of graphene plates, coronene and circumcoronene, were studied in detail by quantum chemical computation.[165] It was found that the adsorption process modifies the charges of carbon atoms of the graphite model involved in the interaction appreciably but does not modify those of the adsorbed ILs. The type of imidazolium-based ionic liquid determines the extent of charge modification of graphite model atoms.

For graphene, the physisorption of IL can be used not only to disperse but also to exfoliate graphite to form FLG.[166] It has been found that ILs are ideal for exfoliating graphite since their surface tension values match the surface energy of graphite and their ionicity helps to stabilize the exfoliated graphene sheets. ILs can adsorb on the graphene surface through non-covalent interactions of anion and/or cation with graphene. The preparation of FLG sheets, mostly with less than five layers, by direct exfoliation of pristine graphite in [bmim]NTf_2 and [bmpy]NTF_2 by sonication for one hour has been reported.[167] In addition to dispersion in solvents, physisorption of ILs can also be achieved by direct grinding of graphene or CNT with ILs giving rise to what are known as bucky gels. These have gained a lot of importance due to the ease of handling and the unique combination of the mechanical properties of CNT or graphene and the ionic conductivity of the ILs. It appears that the molecular ordering of the non-volatile ILs and the networking of CNTs by weak physical bonds give rise to the formation of the gel. Using polymerized ILs these gels can be used to prepare conducting polymer materials with enhanced mechanical properties.[168] Analogous to the other adsorbates, the adsorption of IL is also influenced by the presence of functional groups on the CNT or graphene. For graphene oxide (GO), the addition of a hydrophilic polymerized IL followed by reduction, leads to the formation of highly dispersed graphene in the aqueous phase that can be transferred into an organic phase by anion exchange.[169] Furthermore, the reducing capability of imidazolium-based polymerized IL leads to a more effective reduction of graphene oxide as confirmed by XPS.[170]

3.3.2.3 Physisorption of Ions and Molecules from Solution

The adsorption from solution of organic molecules and inorganic ions is of importance for catalysis applications. The mechanisms of metal ion adsorption on CNTs are complex and appear attributable to physical adsorption, electrostatic attraction, precipitation and chemical interactions with the surface functional groups of CNTs. The most important factors influencing these adsorptions are the pH of the solution and the nature of the functional groups on the CNT surface.[171] Among heavy metal ions, Pb^{2+}, Ni^{2+} and Cd^{2+} have been the most studied. Interestingly, the spontaneous reduction of metal ions (Au^{3+}, Pt^{2+}) on the sidewalls of carbon nanotubes has been reported.[172]

Physisorption of Aromatics. It is found that for CNTs, the adsorption of organic molecules such as benzene or toluene is enhanced in the presence of surfactants.[173,174] In general, the adsorption of organic moieties on CNTs takes place through π–π-interactions, hydrogen bonding or hydrophobic effects.[175]

The adsorption of a benzene molecule on CNTs with various diameters and chiral angles was investigated within the *ab initio* framework.[176] The physisorption of such an organic molecule is an example of non-covalent functionalization involving π-stacking interactions and corresponding to a weak binding energy. The calculations show that for small diameter tubes, the most favorable adsorption site is one type of C=C bond. The disparities between the non-equivalent bonds of a CNT are rationalized in terms of the π-orbital axis vector misalignment (see Section 2.4.2). Moreover, the curvature and the chirality effect on benzene adsorption were analyzed, showing that large diameter nanotubes are the most reactive ones.

A highly heterogeneous adsorption profile has been observed for CNTs, the adsorption involving mainly defective sites, functional groups and interstitial spaces between the grooves in bundles.[177] Various models have been proposed for the adsorption of organic molecules on CNTs.[178] For many organic molecules, multilayer adsorption followed by condensation including condensation inside CNT cavities has been observed. Molecular dynamics studies of benzene in CNTs (diameter 1.08 nm) showed that the plane of benzene is almost perpendicular to the tube axis when the molecule is near the center of the channel and parallel when near the wall of the channel.[179] The same behavior was suggested for functionalized benzene molecules inside carbon nanopores.[180] Concentration dependencies are generally not observed in aqueous solutions in contrast to that observed for organic vapors. Hysteresis is observed for several organic adsorbates that has been attributed to strong π–π coupling and capillary condensation.[181] In some cases, the adsorption of the organic molecule changes the secondary structure (for example, debundling of SWCNTs) resulting in different mechanisms for adsorption and desorption, thus giving rise to hysteresis.[182]

The structural and electronic properties of graphene upon the adsorption of benzene and naphthalene molecules have been computed.[183] The total-energy calculations suggest that for the adsorption of benzene and naphthalene on graphene, the stack (hollow) configuration is more stable than the hollow structure by an energy gain of 50 meV and 80 meV, respectively. The adsorption energy of the stack configuration for benzene/graphene and naphthalene/graphene are found to be −0.30 eV, and −0.47 eV, respectively. Both molecules adopt a planar geometry with a vertical distance of 3.52–3.54 Å above the graphene. The π-stacking interaction between various planar organic molecules, including benzene and CNTs, was investigated within the framework of *ab initio* calculations.[184] The adsorption of these molecules on the sidewall of the cylindrical carbon structure induces a small binding energy compared to conventional covalent functionalization. Such a weak interaction is found to be only physisorption and leads to minor and predictable modifications of the electronic structure. The magnitude and nature of interactions between small aromatic systems (benzene and naphthalene) and various SWCNTs were examined by MP2 theory. The π–π stacking configurations are more strongly bound than CH---π analogues.[185] There is a small preference for placement of the aromatic directly above a C=C bond center in the nanotube. All of these complexes are dominated by dispersion forces. Large nearly neutral aromatic molecules, such as naphthalene and anthracene, and small charge-transfer aromatic molecules interact more strongly with metallic SWCNTs *versus* their semiconducting counterparts.[186]

The adsorption of organic molecules on CNTs usually shows a strong dependence on the CNT morphology and is not directly related to the surface area. Liquid-phase adsorption of tetracene and phenanthrene on a SWCNT was examined.[25] Tetracene adsorption was more than six times greater than that of phenanthrene. This remarkable difference was caused by the nanoscale curvature effect of the tube surface, resulting in a difference in the amount of contact between the molecule and the tube surface.

The adsorption of neutral (poly-)aromatic, anti-aromatic, and more generally π–conjugated systems on graphene was studied as a prototypical case of π–π stacking.[187] The comparison of the dispersive *versus* electrostatic contributions to the total binding energies in the aromatic and anti-aromatic systems suggests that π–π interactions can be regarded as being prevalently dispersive in nature at large separations, whereas close to the equilibrium bonding distance, it is a complex interplay between dispersive and electrostatic Coulombic interactions. Moreover the results surprisingly indicate that the magnitude of π–π interactions normalized both per number of total atoms and carbon atoms increases significantly with the relative number of hydrogen atoms in the studied systems. The interaction of polycyclic aromatic hydrocarbon molecules with hydrogen-terminated graphene was studied using DFT.[188] The effective potential energy surfaces for the interaction of benzene, naphthalene, coronene, and ovalene with hydrogen-terminated graphene were calculated as functions of the molecular displacement along the substrate. It is shown that inclusion of the dispersive interaction, which is the most important contribution to the binding of these weakly bound systems, does not change the shape of the interaction energy surfaces or the value of the barriers to the motion of polycyclic aromatic hydrocarbon molecules on graphene.

The adsorption of simple benzene derivatives composed of a benzene ring with $-NO_2$, $-CH_3$, or $-NH_2$ functional groups on a semiconducting SWCNT was studied using DFT.[189] It was found that all of these benzene derivatives are physisorbed mainly through the interaction of the π-orbitals of the benzene ring and those of the carbon nanotube. These aromatics do not change significantly the CNT's electronic structure, and therefore only small changes in the CNT's properties are expected. The adsorption of various substituted derivatives of benzene on a graphene sheet has been computed.[190] The presence of functional groups can significantly alter the overall magnitude of π–π interactions between the adsorbed molecules and graphene by giving rise to strong medium-range interactions involving π-orbitals of the substituents. When the substituents can simultaneously permit the formation of hydrogen bonds between adsorbed molecules, it is possible to evaluate the relative contributions of hydrogen bonding and π-based interactions to the overall adsorption.

Adsorption of three aromatic organic compounds (AOCs: phenanthrene, biphenyl and 2-phenylphenol) by four types of carbonaceous adsorbents (AC, ACF, SWCNTs and MWCNTs) with different structural characteristics but similar surface polarities was examined in aqueous solutions.[191] Isotherm results demonstrated the importance of molecular sieving and micropore effects in the adsorption of AOCs by carbonaceous porous adsorbents. In the absence of the molecular sieving effect, a linear relationship was found between the adsorption capacities of AOCs and the surface areas of the adsorbents, independently of the type of adsorbent. On the other hand, the pore volume occupancies of the adsorbents followed the order of ACF > AC > SWCNT > MWCNT, indicating that the availability of adsorption sites was related to the pore size distributions of the adsorbents. ACF and AC with higher microporous volumes exhibited higher adsorption affinities to low molecular weight AOCs than SWCNT and MWCNT with higher mesopore and macropore volumes. Due to their larger pore sizes, SWCNTs and MWCNTs are expected to be more efficient in the adsorption of large size molecules. Removal of surface oxygen-containing functional groups from the SWCNT enhanced adsorption of AOCs.

The adsorption of tetraaza[14]annulene ligands (simple analogues of porphyrins and phthalocyanines) on SWCNTs is insignificantly influenced by the presence of aliphatic (methyl) or/and aromatic (benzyl) substituents on the macrocyclic molecules.[192] On the contrary, the adsorption dramatically increases in the case of metal (cobalt) complexes, resulting in substantial changes in their geometry and electronic structure. An increase in negative electrostatic potential at the exposed macrocycle side was found for Co(II) complexes. It might give rise to an enhanced reactivity toward electrophilic agents, which, along with the strong adsorption predicted, is an encouraging prerequisite for the preparation of new hybrid carbon nanotube-based catalysts.

The interaction of phenylalanine (Phe), histidine (His), tyrosine (Tyr), and tryptophan (Tryp) molecules with graphene and SWCNTs was investigated with an aim to understand the effect of curvature on the non-covalent interaction.[193] The equilibrium configurations of these complexes were found to be very similar, *i.e.*, the aromatic rings of the amino acids prefer to orient in parallel with respect to the plane of the substrates, which bears the signature of weak π–π interactions. The binding strength follows the trend: His < Phe < Tyr < Tryp. Although the qualitative trend in binding energy is almost similar between the planar graphene and CNTs structure, they differ in terms of the absolute magnitude. For the nanotube, the binding strength of these molecules is found to be weaker than the graphene sheet. Excellent correlation between the polarizability and the strength of the interaction was found: the higher the polarizability, the greater is the binding strength.

Adsorption of single-ringed N- and S-heterocyclic aromatics on SWCNTs was examined.[194] Adsorbates included pyrimidine, 2-aminopyrimidine, 4,6-diaminopyrimidine, thiophene, benzene and aniline. Adsorbents included pristine SWCNTs, oxidized SWCNTs, and nonporous graphite. Adsorption of N- and S-heterocyclic aromatics was significantly enhanced by non-hydrophobic interactions. Particularly, the $-NH_2$-substituted compounds exhibited much stronger (up to 2 orders of magnitude) adsorption affinities to oxidized SWCNTs than benzene, even though they are much less hydrophobic. The π–π coupling or electron donor–acceptor interactions are likely adsorption-enhancement mechanisms for all six compounds. The lone-pair electrons of the N-heteroatoms and the $-NH_2$ group can enable n–π electron donor–acceptor interactions with SWCNT surfaces. Lewis acid-base interactions are another significant adsorption-enhancement mechanism for the $-NH_2$-substituted compounds (and possibly for pyrimidine) on SWCNTs. For the N-heterocyclic aromatics, adsorption affinity is highly dependent on the O-functionality of the SWCNT surface and on solution pH, due to the speciation reactions of both adsorbates and SWCNT surface O-functional groups, indicating that selective adsorption of N-heterocyclic aromatics is possible by combining the surface functionality of CNTs and solution chemistry. DFT calculations suggested that aminotriazines are strongly adsorbed on graphene, in part through specific interactions of NR_2 groups with the underlying surface.[195] These interactions require distortions of NH_2 groups out of the triazine plane, which may partially compromise their ability to participate in normal intermolecular hydrogen bonds.

The adsorption of thiophene inside and outside SWCNTs and onto graphene was also computed.[196] The results indicate that thiophene adopts a nearly parallel configuration with respect to the graphene plane. The sulfur atom is 3.7 Å above the sheet, whereas the two hydrogen atoms located in carbon atoms not bonded to sulfur are 3.45 Å above it. The adsorption energy for this configuration is 385 meV. For the T-shaped configurations the potential energy surface is very flat showing different orientations with similar interaction energies. When two hydrogen atoms are positioned over a C=C bond the binding energy is 225 meV. However, when the sulfur atom is over a hexagon, the interaction energy reaches its minimum value. The vibrational frequencies of thiophene are red-shifted when it is adsorbed on graphene, with the intensity of the most prominent peak in the IR spectra increased by 34% and red-shifted by 14 cm^{-1}. For the adsorption on CNTs, the internal adsorption energies are larger than the external ones, although the former decreases rapidly as the tube radius is increased. The orientation of the thiophene molecule inside a SWCNT strongly depends on the diameter of the tube. The charge transfer between thiophene and the carbon nanostructures is minimal, thus the electronic properties are not affected by the adsorption.

The adsorption of pyridine and its derivatives on the graphene surface has been studied using DFT.[197] It has been demonstrated that the flat orientation of the adsorbate is more favorable than the normal one. Hollow adsorption site is preferred by pyridine and the most stable structure is characterized by adsorption energy of −412 meV. Substitution of hydrogen in the 4-position of pyridine yields increasing interaction between graphene and the adsorbate by up to 134 meV. Due to relatively weak interactions between studied pyridine derivatives and the substrate, caused mainly by van der Waals forces, the conical band structure of graphene stays unchanged upon adsorption.

As far as C_{60} are concerned, they behave as an electron-deficient dienophile in the Diels–Alder reactions with numerous electron-rich dienes including anthracene, pentacene, cyclopentadiene, and *o*-quinodimethane.[198] The high reactivity of C_{60} as a dienophile with non-activated dienes, such as 2,3-dimethylbutadiene (DMB) has been reported experimentally and theoretically.

Physisorption of Other Molecules. Mezgebe *et al.* compared the adsorption of hexane, iodomethane, water and formamide on SWCNTs and MWCNTs using capillary rise methods and found that the former showed a higher adsorption of polar liquids, whereas the latter showed a preference for non-polar liquids.[199] Diiodomethane showed the maximum adsorption, whereas hexane was the least adsorbed. It is also interesting to note that water adsorption curves for SWCNT showed two stages, while for MWCNT only one stage is visible, indicating that wettabilities of the two surfaces are different. The pore size compatibility with the adsorbate was found to be important for adsorption on CNT walls, especially for natural organic matter (NOM). The amount of adsorbed NOM onto CNTs increased with a rise in initial NOM concentration and with the ionic strength of the solution, but decreased with a rise in solution pH. A comparative analysis on the NOM adsorption between CNTs and granular AC has been reported. Under the same conditions, the acid-treated CNTs have the best NOM adsorption performance, followed by raw CNTs and then the AC, suggesting that the CNTs are efficient NOM adsorbents.[200,201] On the other hand, even large molecules such as enzymes were able to enter the inner pores of CNTs.[202] In some cases larger molecules can show increased adsorption on CNTs due to curvature effects, and it has been proposed that large molecules can assume twisted geometries to match the CNT surface.[25,203,204] Li *et al.* studied SWCNTs and MWCNTs as potential effective adsorbents for removal of tetracycline from aqueous solution. After normalization for adsorbent surface area, the adsorption affinity of tetracycline decreased in the order graphite/SWCNT > MWCNT > AC. The remarkably strong adsorption of tetracycline to the CNTs and to graphite can be attributed to the strong adsorptive interactions (van der Waals forces, π–π electron-donor-acceptor interactions, cation–π bonding) with the graphene surface.[205]

Using *ab initio* electronic structure calculations, Junkermeier *et al.* estimated the energy of adsorption of amine groups on graphene sheets.[206] They found that amines are found to form a semi-ionic bond of 0.778 eV on pristine graphene. Its binding is found to be modified near other defects and substituent atoms. The adsorption mechanisms of formaldehyde (H_2CO) on modified graphene, including aluminium doping, Stone–Wales defects, and a combination of these two, were investigated *via* DFT. It was found that the graphene with SW-defect is more sensitive than that of perfect graphene for detecting H_2CO molecules.[207] For graphene, π–π interactions were found to be the most important criterion for increased physisorption efficiency. The adsorption characteristics of chemically reduced graphene toward acrylonitrile, *p*-toluenesulfonic acid, 1-naphthalenesulfonic acid and methylene blue were evaluated. The results showed that organic chemicals with larger molecule size and more benzene rings possessed a higher adsorption speed and a higher maximum adsorption capacity on graphene, fluorescence spectra indicated that the adsorption of methylene blue on graphene was a π–π stacking adsorption process.[208] Similarly, graphene was found to be a good adsorbent for bisphenol A (Figure 3.9) with a maximum adsorption capacity

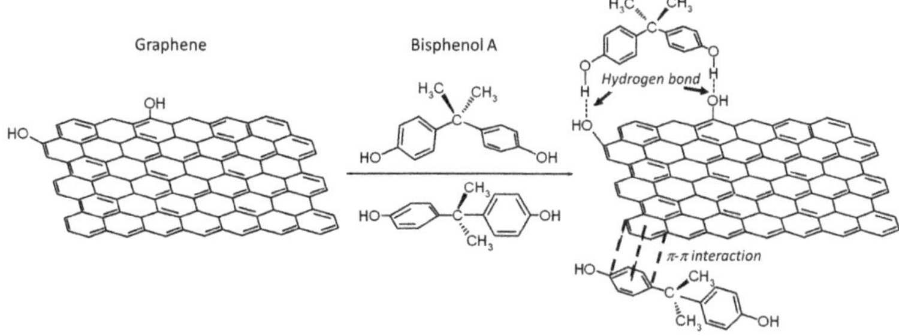

Figure 3.9 Schematic of π–π interactions and hydrogen bonding between bisphenol A and graphene. Adapted from ref. 209.

for graphene, as obtained from Langmuir adsorption isotherm.[209] Results on the adsorption of thiophene and nitrate ions show that graphene, despite its lower SSA when compared to CNTs and AC, shows a high adsorption capacity.[210,211]

The adsorption of cytosine on a graphene surface was studied using DFT with local density approximation.[212] The cytosine is physisorbed onto graphene through π–π interaction, with a binding energy around −0.39 eV. Due to the weak interaction, the electronic properties of graphene show little change upon adsorption. The cytosine/graphene interaction can be strongly enhanced by introducing metal atoms. The binding energies increase to −0.60 and −2.31 eV in the presence of Li and Co atoms, respectively.

The presence of oxygenated functional groups drastically changes the adsorption properties of CNTs and graphene. In some cases, the functional groups may increase the wettability for polar compounds, while in other cases they may block potential adsorption sites leading to reduced adsorption.[213] Hsu *et al.* have reported the adsorption and desorption of 2-propanol on oxidized CNTs. The adsorption mechanism appears to be mainly attributable to physical forces from 278 to 298 K, but seems attributable to chemical forces from 298 to 308 K.[214] Liao *et al.* discussed the sorption capacity of pristine and acid-treated MWCNTs and found that acid-treated MWCNTs become more hydrophilic due to the presence of carboxylic groups that give the surface a negative charge.[215] This negative charge reduced the adsorption of certain molecules such as resorcinol. A second effect was the surface deformation caused by functionalization that led to a decrease in pore volume and a consequent decrease in adsorption capacity. Many adsorption experiments have been carried out on graphene oxide dispersions and it was found that for methylene blue adsorption, GO was more efficient than AC or CNTs, despite the lower specific surface area. The increased activity was attributed to a combination of increased electrostatic π–π interactions between the graphene sheets and the adsorbate, as well as the hydrogen bonding that anchors the organic adsorbate on to the graphene surface.[216]

3.3.2.4 Physisorption of Solids on Nanostructured Carbon Materials

In this section, adsorption of metals, semiconductors and other inorganic and organic materials that are solids at 298 K are considered. While adsorption of metals involving chemical bonding is discussed in Section 3.4.3, in this section a few examples of physisorption of solids on nanostructured carbon material will be given.

Four limiting cases have been distinguished for the interaction of metal atoms with graphene surfaces:[217]

(a) Weak physisorption of metal atoms generally occurs when the metal atom has filled d-orbitals (in the case of transition metals such as gold) or possesses an *s,p*-like metallic structure with free-electron-like parabolic band structure (such as Pb), together with a high work function.
(b) Ionic chemisorption is characteristic of the interaction of metals of low ionization energy such as alkali metals (Li, Na, K) and alkaline earth metals (Ca, Sr, Ba). Metals with low work function lead to the injection of electrons into the conduction band of graphitic materials (n-type doping). Such charge transfer interaction with the graphitic structure largely preserves the conjugation and band structure.
(c) Covalent chemisorption of metals to graphitic systems leads to strong (destructive) rehybridization of the graphitic band structure. One such example is the formation of metal carbides by the strong interaction between graphitic surface and metals leading to metal–carbon bond formation (Ti).
(d) Covalent chemisorption of metals to graphitic systems, which is accompanied by the formation of an organometallic hexahapto(η^6)-metal bond and preserves the graphitic band structure (constructive rehybridization).

Due to the high surface area and pore volume of CNTs, colloidal particles can be anchored to the surface through electrostatic interaction between the NP surface and the chemically functionalized sites present on the support, which in the case of a nanotube are the walls, the tips and the CNT cavity. For graphene, mainly defective sites, doping sites and edges function as active sites for adsorption. The coating of solids on CNTs by physisorption and the application of coated nanotubes have been treated in many reviews[218,219] and only a few results will be highlighted, mainly concerning the deposition of metal and semiconductor NPs using self-assembly approach. The fundamentals of metal adsorption on CNTs and graphene have been treated in detail in a recent review by Machado *et al.*[220] It was found that the most stable adsorption configuration and binding energy of the adatoms on CNTs are sensitive to the type of metal element and CNT curvature, while the electronic properties depend on the valence electronic configuration of the metal atoms. The metal–CNT interaction results in considerable charge transfer accompanied by a substantial modification of the electronic states of CNTs around the Fermi energy. Among metals, noble metals used in catalysis, such as Au, Ag, Pt, Ru, Rh, and Pd, and metals used as electrodes for electronic applications, such as Ti and In, have been the most investigated. Nakada *et al.* calculated the adsorption and migration energy systematically for each adatom adsorbed using band calculation with the projector augmented wave method at the three adsorption sites on 3×3 graphene and have made general predictions on the preferential site for adsorption for the large number of elements including transition metals, alkaline earth metals and halogens.[221] Following this study, it was reported that several transition metals, such as Ti and Zr, showed a high binding energy for adsorption on graphene and hence have the potential to form viable electrodes for graphene-based devices.

Metal particles and semiconductor quantum dots have been physisorbed on the nanotube surface using suitable capping agents to provide electrostatic bonding to specified sites on the nanotube surface. The nanoscale nature of the tube provides an excellent template for a bottom-up organization of a nanocomposite. Gold nanowires were formed by the templated self-assembly of gold NPs surface coated with tetraoctylammonium bromide.[222] On heating the composite, the gold NPs could be sintered to produce wires. The strong nanotube-gold interaction is attributed to electron transfer between the gold nanoparticles to the continuum of π^* states in the nanotubes. Surface modification by AEPA (2-aminoethylphosphonic acid) and APTES (3-aminopropyltriethoxysilane) was used for the coating of CNTs with TiO_2 and SiO_2 particles.[223]

In view of its application in electronics and energy storage, metal deposition on graphene has been studied using both theoretical and experimental approaches, the former adapting the results already obtained for CNT-metal systems.[224] While clusters of metals such as Au, Pt and Pd have been adsorbed on graphene sheets and the energies of adsorption have been estimated,[225] experimental observation of such clusters have only been recently reported. For example, gold clusters on graphene have been imaged using aberration corrected HRTEM in the HAADF mode. It was observed that the gold clusters assumed a square planar arrangement and were mobile on the graphene surface.[226] Several works have studied the deposition of Pt NPs on graphene for application in oxidation–reduction reactions in electrochemical cells.[227,228] In all cases, it was observed that for efficient coating of Pt, the surface of graphene was modified to introduce oxygen containing functional groups. Similar to the case of CNTs, the adsorption of oxides and sulfide colloids on graphene surface using π–π interactions has also been reported (*e.g.*, with TiO_2, Al_2O_3, SnO_2, and clay).[229–231] For example, TiO_2 particles and graphene oxide colloids have been mixed ultrasonically followed by the reduction of GO to yield TiO_2 graphene nanocomposites. Preformed SnO_2 nanoparticles have been chemically mixed with graphene to form composite materials. Figure 3.10 shows uniform FePt alloy nanoparticles self-assembled on graphene[232] that were found to be active catalysts for the oxygen reduction reaction in 0.1M $HClO_4$.

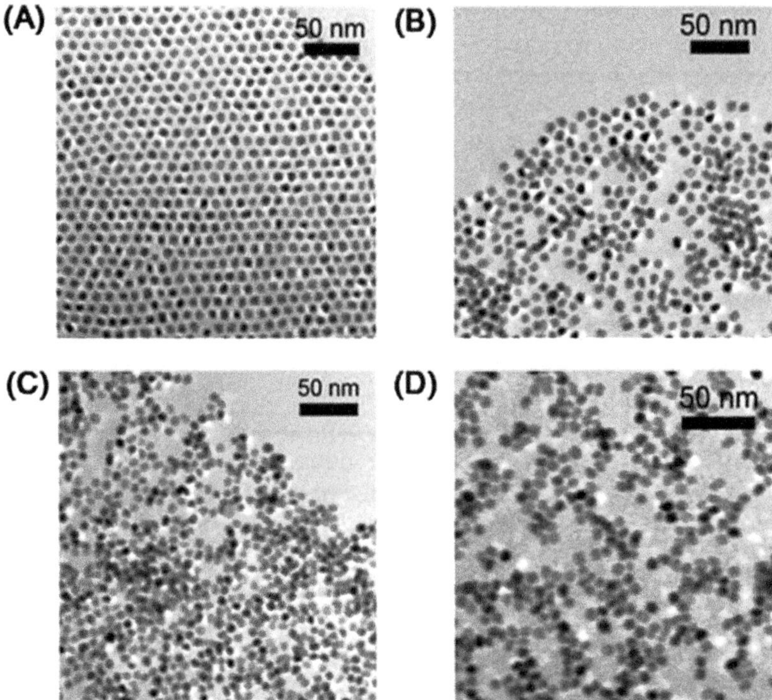

Figure 3.10 TEM images of (a) the 7 nm $Fe_{58}Pt_{42}$ NPs assembled on the amorphous carbon surface, (b) the $Fe_{58}Pt_{42}$ NPs assembled on G surface, giving G/$Fe_{58}Pt_{42}$ NPs, (c) the G/$Fe_{58}Pt_{42}$ NPs after acetic acid wash, resulting in G/$Fe_{22}Pt_{78}$ NPs, and (d) the G/$Fe_{22}Pt_{78}$ NPs annealed under Ar + 5% H_2 at 373 K for 1 h (G = graphene). Reproduced with permission from ref. 232. Copyright 2012, American Chemical Society.

3.3.2.5 Physisorption of Metals and Semiconductors on Doped Nanostructured Carbon Materials

The presence of dopants on the CNT or graphene surface has a strong effect on the dispersion and density of the deposited metal particles. The sidewalls of acid treated N-doped MWCNTs, coated with a cationic polyelectrolyte such as PDADMAC [poly(diallyldimethylammonium chloride)], were used as substrate to anchor 10 nm gold particles by dipping in a negatively charged gold colloid.[233] Well-dispersed gold particles decorated the walls of the nanotubes quite uniformly. For Pt/N-graphene used for an oxidation reaction, it has been reported that Pt nanoparticles could disperse better on the surface of N-graphene when compared to undoped graphene.[234] Groves *et al.* have studied the interaction between Pt atom and N-graphene using DFT to calculate the binding energy of Pt atom to N-graphene[235] and have reported that the presence of nitrogen doping prevents the clustering of the Pt particles by aggregation.

3.4 Chemisorption on Nanostructured Carbon Materials

3.4.1 Chemisorption of Gases on Nanostructured Carbon Materials

As mentioned earlier, defects can be seen as sites with elevated reactivity that can affect many of the material's properties. From an adsorption point of view, this can also mean that a physisorptive processes over a defect-free graphitic surface can become chemisorptive, whenever an imperfection is found. In the following sections, we will mainly discuss the interaction of various gas molecules with a modified carbon surface.

3.4.1.1 Hydrogen

Some of the most important research in this area is related to hydrogen. It has been shown that H chemisorption occurs solely on top of a C atom, which protrudes outside of the flat graphene plane changing its hybridization into a mixed sp^2–sp^3 state. An *ortho* hydrogen pair is formed for two and more H atoms.[236] Physisorbed and chemisorbed states are separated by an energy barrier, which explains why H chemisorption does not spontaneously occur at low temperatures. Binding energies lie between 0.47 eV and 1.9 eV, with a majority of data between 0.6 eV and 0.85 eV.[237,238] Hydrogenation offers the interesting possibility to manipulate both the electronic and chemical properties of graphene.[239] Both the hydrogenation and dehydrogenation process of the graphene layers are controlled by the corresponding energy barriers, which show significant dependence on the number of layers. However, contrasting results have been reported on the influence of the number of layers on graphene hydrogenation.[240,241] Transition metal doped nanocarbons can generate chemically very active sites to hydrogen, where reversible hydrogen storage is possible. Alkali metals can attract a significant amount of hydrogen molecules when they are coated onto nanostructured carbon materials, while their binding strength is not high enough to ensure reversible hydrogen storage at near-ambient conditions.[242] In spite of the numerous publications on CNTs as hydrogen storage materials there are contradictory reports on the quantity of hydrogen these carbon nanomaterials can uptake. This is mostly due to the fact that currently the exact nature of the sites on which hydrogen activation takes place before they are adsorbed on the carbon atoms of CNTs is still unkown.[243] Lopez-Corral *et al.*[244] observed that H_2 adsorption occurred preferentially on C-defective sites. During hydrogen interaction with the C-defective nanotube, the H–H bond was broken after adsorption on the parallel vacancy (adsorption energy value of 2.52 eV), representing a chemisorptive phenomenon. Gayathri *et al.* also observed a considerable increase in the adsorption binding energy of H_2 (in the order of 50%) due to the presence of structural defects in SWCNTs.[245] When alkali-CNT systems are used to adsorb some gases such as hydrogen, they form hydrides with a high degree of ionic bonding character. As a consequence, gas recovery is severely hindered, requiring a higher external energy supply to release the gas.[246,247] Among other reactions, fullerenes have demonstrated their ability to react with gaseous hydrogen *via* hydrogenation of C=C double bonds.[248] Theory predicts that a maximum of 60 hydrogen atoms can be attached both to the inside (endohedrally) and the outside (exohedrally) of the fullerene spherical surface and that a stable $C_{60}H_{60}$ isomer can be formed. The process of C_{60} hydrogenation involves formation of C–H bonds as a result of breakage of C=C double bonds and H–H bonds of molecular hydrogen to form hydrogen atoms. Although the hydrogenation reaction is exothermic, additional energy is required to break these bonds. Considering only 1,2-addition, $C_{60}H_{60}$ has the smallest reorganization energy (71 meV). Further hydrogenation does not reduce the reorganization energy. Highly hydrogenated fullerenes, especially $C_{60}H_{58}$, have very large reorganization energies because the HOMOs are localized around non hydrogenated carbon atoms. However, $C_{60}H_{60}$ has a smaller reorganization energy than other highly hydrogenated fullerenes and C_{60} because the HOMO of $C_{60}H_{60}$ is distributed over the whole molecule. Activation energies between 1 eV per H_2 and 1.7 eV per H_2 (mainly associated with atomization of molecular hydrogen) have been reported for hydrogenation,[249] and 1.6 eV/H_2 for the dehydrogenation process that re-establishes the C=C and H–H bonds.

3.4.1.2 Oxygen and Ozone

The adsorption process of other gas molecules, such as oxygen and ozone, has also been studied in detail. The O_2 molecules dissociate exothermally and chemisorb at the dangling bonds of a vacancy. When an O_2 molecule approaches dangling bonds at a top site and/or bridge site of a vacancy, charges are transferred from carbon atoms of CNTs to the oxygen molecule due to the larger electronegativity of oxygen atoms.[250] The resulting intramolecular Coulomb repulsion breaks the O–O bond exothermally without an activation barrier. This interaction keeps the oxygen atoms apart, placing them at the opposite sides of the plane. The adsorption geometry of O_2 is found to depend sensitively on the curvature of the nanotube.[251] Calculations also show that the transport properties of nanotubes are significantly affected by the presence of oxygen.[252] Grujicic *et al.* observed that in both zigzag and armchair SWCNTs, molecular oxygen is physisorbed to defect-free portions of the nanotube walls, but chemisorbed to the topological defects such as Stone–Wales.[253] The binding of molecular oxygen to a graphene sheet and to a (8,0) SWCNT was investigated by means of spin-unrestricted density-functional calculations.[254] It was found that triplet oxygen retains its spin-polarized state when interacting with graphene or the nanotube. This leads to the formation of a weak bond with essentially no charge transfer between the molecule and the sheet or tube, as one would expect for a physisorptive bond. Theoretical analysis demonstrated strengthened oxygen adsorption on defective graphene as compared to pristine graphene, resulting in trapping of the oxygen onto defects.[255] This was accompanied by significant charge transfer of up to 3e, unlike for pristine graphene. At the same time, atomic oxygen diffuses at different rates dependent on the local environment, however with relatively low barriers (mostly <1 eV), lower than for pristine graphene, thus revealing an interplay between diffusion and adsorption in this case. Interestingly, following incorporation of oxygen within defects, the morphology has shown deformation from planarity of the nanostructure, particularly with higher coverage. Oxygen atoms prefer to adsorb on both sides of the graphene sheet, while the one-side adsorption structure becomes metastable. It has been found that the structural bi-stability appears with regard to the oxygen adsorption. This bi-stability corresponds to the formation of an epoxy group or ether group, where the ether group phase is more stable than the epoxy group one.[256] On diamond surfaces, oxygen bonds in ether (C–O–C), peroxide (C–O–O–C), and carbonyl (C=O) bonding.[257] Adsorption of an O_2 molecule on a graphene sheet with and without a Stone–Wales defect was examined by the DFT.[258] The energetics have clarified that O_2 physisorption is possible for graphene with and without defects. The energy diagram of three representative O_2 adsorptions on defect-free graphene and graphene with a Stone–Wales defect is depicted in Figure 3.11. DFT calculations have also revealed significant changes in the electronic structure of graphite ribbons caused by partial oxidation of graphitic zigzag edges, transforming antiferromagnetic interactions to ferromagnetic interactions in two- and three-dimensional graphites.[259]

Picozzi *et al.* investigated ozone adsorption on CNTs by DFT.[51] They observed that the ozone adsorption process on ideal nanotubes is most likely due to physisorption, whereas when ozone adsorbs on Stone–Wales defects, a strong chemisorption occurs, leading to relevant structural relaxations and to

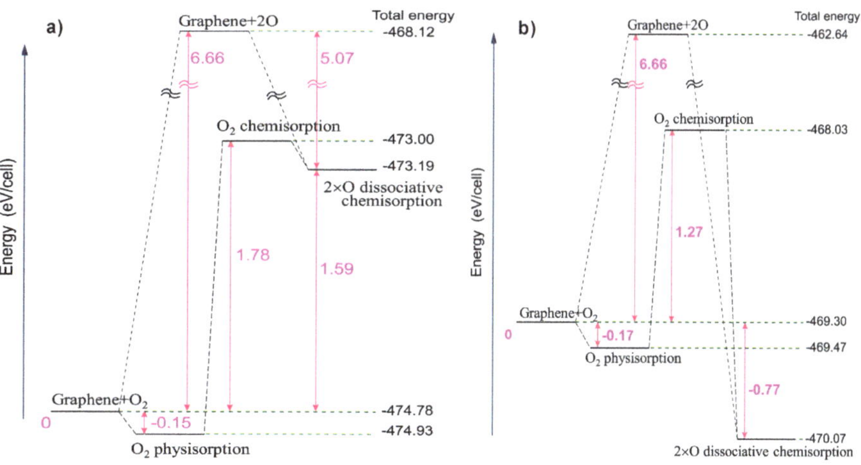

Figure 3.11 Energy diagrams of three representative O_2 adsorptions on (a) defect-free graphene, and (b) graphene with a Stone–Wales defect; physisorption, chemisorption, and dissociative chemisorption. Adapted from ref. 258.

the formation of a C=O covalent bond. A similar observation was also made by Akdim *et al.*, where some orientations of the Stone–Wales defects for both SWCNT chiralities resulted in lower formation energies, and stronger ozone adsorption.[260] Theoretical studies have shown that the binding energy of ozone on CNT is ~0.2–0.3 eV, depending on the diameter and chirality. However, contrasting results were reported by Lu *et al.*, who reported that the 1,3-cycloadditions of O_3 to the perfect site and to the Stone–Wales defect site both form primary ozonide species.[261] The exothermicity of the defect-free process is 0.68 eV greater than that involving the defect site. The activation energies for the defect-free and defect-involving processes are 0.12 and 0.77 eV, respectively. Thus far they have shown that the central 7-7 ring fusion at the Stone–Wales defect in a (5,5) SWCNT sidewall is less reactive than the defect-free sites. As the local curvature of the SW defect is far less severe than that of a perfect site, the much lower chemical reactivity at the defect site can be ascribed to the constraints of its planar local structure. Since the sidewall reactivity of perfect armchair SWCNTs decreases monotonically with an increase in tube diameter, they infer that the reactivity difference between perfect and Stone–Wales defect sites in the sidewalls of larger armchair SWCNTs should also decrease. Ozone molecules adsorb on the graphene basal plane with a binding energy of 0.25 eV, and the physisorbed molecule can chemically react with graphene to form an epoxide group and an oxygen molecule.[262] The activation energy barrier from physisorption to chemisorption is 0.72 eV, and the chemisorbed state has a binding energy of 0.33 eV. These binding energies and energy barrier indicate that the ozone adsorption on graphene is gentle and reversible. The exposure of powdered crystals of C_{60} and C_{70} fullerenes to ozone causes their immediate oxidation.[263] For CB, the reaction between ozone essentially involves two stages: the gasification of the surface to CO_2 and its functionalization with oxygenated chemical groups, mainly as –COOH but also other oxygenated chemical moieties.[264]

3.4.1.3 *NH_3 and NO_x*

More complex molecules have also been targeted in several studies, given their importance in sensor applications. Among the most investigated are NH_3 and NO_2. Chang *et al.*[265] studied the adsorption of NH_3 and NO_2 molecules on SWCNTs using DFT, and observed a small binding energy of NO_2 and NH_3 molecules to the SWCNT wall, indicating a physical and not a chemical adsorption. In sharp contrast with the case of a perfect nanotube, the adsorption of NH_3 and NO_2 at the defect site of a SWCNT is generally chemical. DFT calculations were used by Tang and Cao to investigate the effect of a Stone–Wales defect and a single vacancy on the interaction of NO_x (x = 1, 2, 3) with a semi-conducting SWCNT.[266] The adsorption of NO_x on the SWCNT with the mono vacancy defect was more favorable energetically in comparison with the NO_x adsorption on the perfect CNT as well as at the Stone–Wales defect. Such stronger interactions of NO_x with the vacancy defect were ascribed to the presence of a dangling bond (carbene-like reactivity). The adsorption of

NO_2 with the N end at the SW-defect site, which leads to the formation of C–N, is energetically more favorable than the O-down adsorption to form the C–O bond. However, the O-down adsorption of NO_2 at the single vacancy is more favorable thermodynamically, compared to the adsorption of NO_2 with the N end. The adsorption affinity order of these nitrogen oxides over the SWCNT was found to be $NO_3 > NO_2 > NO$. As for NH_3 adsorption, Turabekova *et al.* observed that the preferred carbon atom site for chemisorption of NH_3 was located in the Stone–Wales defect region.[267]

Zhang *et al.* studied the effects of Stone–Wales defects on the interactions between NH_3, NO_2 and graphene, using DFT calculations.[268] They observed that both gases interacted rather weakly with pristine graphene. Introducing Stone–Wales defects into the graphene structure had little effect on the NH_3 adsorption, but dramatically enhanced the adsorption of NO_2, causing significant deformation of the graphene sheet around the defective site. The distance between the N atom of NO_2 and the Stone–Wales defect graphene sheet was 1.56 Å, whereas the sheet–molecule distance in pristine graphene was more than 2.80 Å, indicating the chemisorption of NO_2 on the defective graphene ($E_{Ads} = -1.10$ eV), but physisorption on pristine graphene ($E_{ads} = -0.40$ eV).

In order to understand how GO sheets interact with nitrogen oxides, Tang and Cao studied the adsorption of NO_x ($x = 1, 2, 3$) and N_2O_4 on both graphene and GO (containing hydroxyl, epoxy and carbonyl functional groups) using DFT calculations.[269] They found that the adsorption of NO_x on GO was generally stronger than that on graphene. Furthermore, the GO provided various active defective sites, such as hydroxyl and carbonyl functional groups in addition to the carbon atoms near these groups, which enhance the binding energies and charge transfers for the adsorption of nitrogen oxides on these materials, inducing the chemisorption of gas molecules. The adsorption interaction of NO_3 on both graphene and GO was found to be relatively stronger than other nitrogen oxides, followed by that of NO_2. In the case of GO with hydroxyl and epoxy groups, the interaction of nitrogen oxides with GO resulted in the formation of hydrogen bonds OH•••O(N) between –OH and nitrogen oxides, as well as weak covalent bonds of C•••N, depending on the adsorption configuration of NO_2 on GO. The formation of multiple hydrogen bonds OH•••O between one NO_2 or NO_3 and GO was also assumed to be possible. For GO containing a vacancy defect decorated by the hydroxyl and carbonyl functional groups, NO_2 and NO_3 were found to be chemisorbed. The interactions of NO_2 and NO_3 with –OH were energetically more favourable than that at the carbon atom attached to –OH and with C=O. In the case of the adsorption on –OH, NO_2 and NO_3 abstracted the H atom to form nitrous acid- and nitric acid-like moieties, respectively. The adsorption of NO_2 and NO_3 at the carbon atom led to the newly formed weak covalent bonds C•••N and C•••O, respectively, as well as the hydrogen bond OH•••O.

Using different DFT methods, covalent functionalization of C_{20} fullerene with the NO_2 molecule was investigated in terms of energy, geometry, and electronic properties.[270] It was found that the molecule has strongly

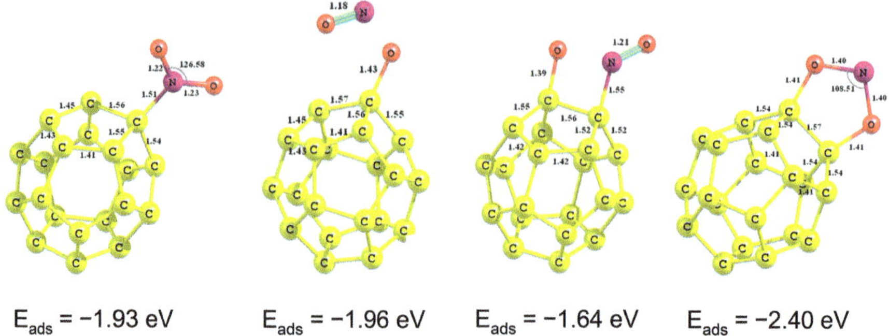

Figure 3.12 Optimized structure of NO_2 adsorption on C_{20} fullerene and their adsorption energy (B3LYP level). Reproduced with permission from ref. 270. Copyright 2013 Wiley Periodicals, Inc.

chemisorbed on the fullerene surface, and the NO_2/C_{20} configurations are stable (Figure 3.12).

The possibility of the formation of endohedral complexes between NH_3 molecules and C_{60} was investigated by using *ab initio* DFT calculations. The obtained results indicate that only one NH_3 molecule being incorporated inside the C_{60} cage can form a stable complex.[271]

Interestingly, the adsorption and reaction of NO_2 (forming NO) with either carbon black or soot held at near ambient temperatures (293–348 K) was observed.[272] It was found that initially NO_2 adsorbed very rapidly; approximately half the NO_2 adsorbed was converted to NO, leaving behind O-containing functionalities on the surface of the carbon. These adsorbed species can be used to produce CO_2 and CO by heating the carbon to a temperature above 373 K. At 293 K, the number of O-containing species, C(O), formed on the carbons was found, surprisingly, to be equal to the amount of NO_2 remaining stably adsorbed.

A theoretical study of the adsorption of nitric oxide on the (111) and (001) idealized unreconstructed surfaces of diamond has been conducted.[273] Nitric oxide is predicted to bind preferentially at the on-top position of the (111) surface. For the carbon (001) substrate, the results show that the molecular chemisorption occurs at the on-top site, which is slightly preferred over the bridge one. Dissociative chemisorption has been investigated in the case of (001) surfaces and represents the best process from a thermodynamic point of view.

3.4.1.4 Other Gases

Sanyal *et al.* studied the adsorption of various gases (O_2, CO, N_2, B_2, and H_2O) on graphene with divacancy defects with the help of *ab initio* DFT calculations.[274] They found that all molecules interacted strongly with the divacancy. In addition, a metallic behavior of the graphene layer in the presence of CO and N_2 molecules was found, with a large density of states in the vicinity of the Fermi level, suggesting an increase in the conductivity. Furthermore, they showed that N_2 atoms could dissociate in the vicinity of the defects and occupy the place where the missing C atoms of the divacancy used to be. In that way, the defective graphene structure was healed geometrically, and at the same time doped with electron states.

Hydrogen sulfide (H_2S) is one of the major environmental pollutants having its sources in natural processes. Zhang *et al.* studied the interaction between H_2S and graphene by DFT calculations and non-equilibrium Green's function formalism.[275] Based on the relatively small adsorption energy (−0.17 eV) and large graphene–molecule distance (2.58 Å), H_2S molecules were found to undergo weak physisorption on the pristine graphene. For the graphene with vacancy defects, the distance between H_2S and the defective graphene was found to decrease to 1.77 Å, and the adsorption energy was more than five times higher than that of the pristine graphene (−0.91 eV). The strong interaction was attributed to the presence of dangling bonds in the defective graphene caused by the missing carbon atom. Borisova *et al.* used DFT to study adsorption of H_2S molecules onto defective graphene as a function of vacancy concentration.[276] The calculations showed that a H_2S molecule interacts more strongly with the carbon atoms surrounding the vacancy than with the carbon atoms in a perfect arrangement. The most favorable energetically configuration was the one with the sulfur atom at the center of the vacancy. Such a configuration facilitated the covalent binding of the sulfur atom with three carbon atoms with unsaturated bonds. The chemisorption was followed by a release of the hydrogen atoms, which formed a H_2 molecule.

Analogously to the tips of nanotubes, carbon atoms present at the graphene edges also exhibit higher reactivities compared to those in the basal plane. Since the edges of graphene are very reactive, they may undergo oxidation and be terminated by oxygen atoms. When one molecule is attached to the oxygen atom of a C–O bond at the edge, the binding energies increase, indicating that graphene edges with terminating oxygen atoms play an important role in the adsorption of molecules.[277,278] Wood *et al.* used *ab initio* DFT calculations to explore the effect of chemical functionalization (presence of COOH, OH, NH_2, H_2PO_3, NO_2 and CH_3) on CO_2 and CH_4 adsorption within graphitic carbon edges.[278] They found that adsorption was a function not only of the chemical group but also of the gas geometry; polar groups (including COOH, NH_2, NO_2, and H_2PO_3) were found to be promising candidates for enhancing CO_2 and CH_4 capacity by strengthening adsorption and activating exposed edges and terraces to introduce additional binding sites.

3.4.2 Effect of Dopants on Gas Chemisorption

3.4.2.1 Adsorption of H_2, HCN, H_2S and HF

Cho and Park investigated hydrogen storage on Li-doped SWCNTs using first principle calculation based on DFT.[279] It was found that, through Li-doping, two new adsorption sites for hydrogen molecules are created in addition to the inherent three adsorptive sites which are exterior, interior and interstitial regions of pristine SWCNTs. The first site ("region 1") consisted of the nanotube sidewall whose electronic distribution status was influenced by the doped Li atoms. The second site ("region 2") existed on the positively charged Li atoms, which result from the transfer of electrons from the Li atoms to the SWCNTs. The adsorption coverage in region 1 at moderate temperature and pressures was similar to that of pristine SWCNTs, although its adsorption energy was slightly higher. As for the adsorption energy in region 2, it was approximately two times greater than that of pristine SWCNTs. A systematic increase of binding energy of hydrogen can also be achieved by moderate substitutional doping of CNTs with Mg atoms.[280] A mixed quantum mechanics/molecular mechanics model was used for investigating the nature of molecular hydrogen adsorption in pure and alkali-metal-doped SWCNTs.[281] The results demonstrate that the charge transfer from the alkali metal to the tube polarizes the H_2 molecule and this charge-induced dipole interaction is responsible for the higher hydrogen uptake of the doped tubes. It was also demonstrated that the doping of alkali metal atoms on C_{60} remarkably enhances the molecular hydrogen adsorption capacity of fullerenes, which is higher than that of other conventionally known fullerene complexes.[282]

The doping effects of B and N on atomic and molecular adsorption of H_2 in SWCNTs were investigated through DFT calculations.[283] The B-doping increases the hydrogen atomic adsorption energies both in zigzag and armchair CNTs. The B-doping forms an electron-deficient six-membered ring structure, and when hydrogen is adsorbed on top of a B atom, a coordination-like B–H bond will form. The N-doping forms an electron-rich six-membered ring structure, and decreases the hydrogen atomic adsorption energies in the N-doped SWCNT. In case of H_2 molecular adsorption, both B- and N-doping decrease the adsorption energies in SWCNTs. On N-doped CNTs the favorable hydrogen adsorption site is not on top of the nitrogen atom but on top of carbon atoms next to the nitrogen atom.[284] The adsorption of several molecular and dissociative hydrogen systems on a Pd-decorated graphene monolayer was studied using DFT.[285] The calculations show that the most favorable graphene-supported coordination structure is similar to the PdH_2 complex in vacuum, where the H–H bond is relaxed but not dissociated. Zhang *et al.* examined the interaction between H_2 and Eu-doped SWCNT using first-principles DFT calculations.[286] The results indicated that each Eu atom could adsorb five H_2 molecules. The adsorption behaviors of H_2 on pristine and Al-doped graphene were studied using DFT.[287] It was found that the physisorption of H_2 is greatly enhanced by doping Al into graphene. The doped Al varies the electronic structures of both C and H_2, causing the bands of H_2 overlapping with those of Al and C simultaneously. This induces an intensive interaction between H_2 and the Al-doped graphene. Strain-engineered adsorption of Ti on pyridinic nitrogen-doped graphene (PNG) and the hydrogen storage characteristics of Ti-decorated PNG were examined by using a first-principles approach using DFT calculations.[288] The binding energy of Ti on PNG was higher than cohesive energy of the Ti bulk, suggesting that Ti atoms prefer atomic dispersion in PNG to clustering. For this Ti-PNG system, the binding energy variation of the second and third adsorbed H_2 molecule had large values of 0.217 and 0.254 eV, respectively. Adsorption of Be atoms on B-doped graphene layer and hydrogen adsorption on the Be-decorated B-doped graphene layer were studied using DFT.[289] Be atoms are strongly adsorbed on B-doped graphene. The boron stabilizes the Be adatoms, and the average hydrogen adsorption energy is −0.289 and −0.298 eV per H_2 for single- and double-sided adsorptions, respectively. The adsorption of an H atom on pristine and B-doped graphenes has been investigated and discussed based on the DFT calculations.[290] For the adsorption of H on a top site, B-doping forms an electron-deficient structure and decreases the adsorption energy of equilibrium position about 1 eV. For the adsorption of hydrogen on top of other sites, similar results have also been found. The chemisorption processes of hydrogen adatoms on boron doped graphene sheets have also been examined by *ab initio* total energy calculations.[291] For the pristine graphene system, it was found that the binding energy of hydrogen adatoms forming H dimer-like configurations is higher by 1 eV per H atom compared with the binding energy of two isolated H monomers (E = 0.98 eV per H atom). The presence of a single substitutional boron atom increases the hydrogen binding energy by 0.9 eV per H atom compared with the one on the clean graphene sheet.

Boron-doped (8,0) SWCNTs were investigated using DFT calculations as sensor models to detect the presence of gaseous hydrogen cyanide (HCN), cyanogen chloride (CNCl) and formaldehyde (HCOH).[292,293] The replacement of a carbon by a boron atom made the system highly sensitive, by enhancing the interaction between all the molecules and B-doped SWCNT (chemisorption), in contrast with the physisorption on the pristine SWCNTs.

Zhang *et al.* studied the interaction between H_2S and doped graphene by DFT calculations and non-equilibrium Green's function formalism.[275] When one carbon atom is substituted by B or N atom in the super cell (assigned as B-graphene or N-graphene, respectively), it was found that the geometric structures of the doped graphene nanosheets had a planar configuration before and after adsorption of H_2S molecule, resulting in physisorption. In contrast, the geometry of graphene doped with Si (Si-graphene) deformed dramatically before and after adsorbing a H_2S molecule. The deviation of C–Si–C bond angles from the standard angle of sp^2 hybridization of 120° clearly indicates that the Si atom near the adsorption site has changed to a more sp^3-like hybridization. Compared to the distance of around 3.00 Å between the pristine graphene sheet and the S atom of physisorbed H_2S, the sheet–molecule distance between the Si-graphene and H_2S molecule is shortened to 2.53 Å. In addition, the E_{ads} of H_2S on Si-graphene is found to be −0.94 eV, which confirms that the Si-graphene is more favorable for H_2S adsorption than the pristine graphene. These results indicate that the H_2S molecule is chemisorbed onto the Si–graphene. Further investigation of H_2S adsorption on the surface of graphene doped with Ca, Co and Fe, showed that the H_2S molecule can bind to the surface with adsorption energies of

−0.66, −1.80 and −1.92 eV, respectively. Because absorption energy of H_2S molecule is very high, it is worth noting that desorption of the H_2S could be quite difficult.

The effect of Al doping on graphene was also addressed by Sun *et al.*,[294] by studying the adsorption of HF by first-principles calculations. They found that Al-doped graphene had a higher adsorption energy (−7.164 eV) and shorter connecting distance to the HF molecule (2.235 Å) than pristine graphene (−0.071 eV and 3.661 Å). A charge accumulation between the HF molecule and the Al-doped graphene occurred, and the Al-doped graphene gained electrons from the H atom of the HF molecule. This charge transfer would form an ionic bond between the HF molecule and the Al-doped graphene, and the HF molecule acted as an electron donor of the system. Thus, the adsorption of HF molecules on Al-doped graphene can be considered chemisorption, while physisorption takes place on pristine graphene. A principle of the enhancement of CO adsorption was developed theoretically by using DFT through doping Al into graphene.[295] The results show that the Al doped graphene has strong chemisorption of the CO molecule by forming an Al–CO bond, whereas CO adsorption onto intrinsic graphene remains weak physisorption.

3.4.2.2 Adsorption of CO, CO_2, NO_2, NH_3, H_2O and O_2

Using first-principle calculations, Peng and Cho showed that boron and nitrogen-doped SWCNTs could detect CO and H_2O.[296] The binding energies of CO and H_2O molecules with the B-doped CNTs indicated that these molecules underwent chemical adsorption, while physical adsorption occurred when CO or H_2O molecules adsorbed on the N-doped CNTs. Bai *et al.* studied the adsorption of NH_3 and NO_2 in B- or N-doped (10,0) SWCNTs using DFT.[297] NH_3 and NO_2 were only physisorbed on the (10,0) pristine SWCNTs with weak binding and little charge transfer. The N-doping did not change NH_3 adsorption in SWCNTs, but the B-doping made NH_3 chemisorption feasible, with an adsorption energy of −0.70 eV and a charge transfer of 0.31e. NO_2 was found to chemisorb on the B- or N-doped SWCNTs (adsorption energies of −1.05 eV and −0.59 eV, respectively). However, the very strong binding of NO_2 in B-doped SWCNTs makes desorption difficult.

Wang *et al.* investigated the reactivity of the intrinsic and Al-doped (8,0) SWCNT with CO using DFT calculations,[298] and observed a dramatic change on the electronic properties and geometric structures of the Al-doped SWCNTs after CO adsorption. Dai *et al.*[299] studied the effect of doped graphene (boron, nitrogen, aluminium and sulfur) on the adsorption of several other common gas molecules (NO, NO_2, NH_3, CO, CO_2, H_2O, SO_2, O_2, H_2 and N_2). First-principle calculations showed small adsorption energies and large molecule–graphene distances, pointing to no binding other than physisorption. For B-graphene, however, NO and NO_2 bind with a significant adsorption energy (E_a ~ 0.3 eV). For N-graphene, the largest binding energies (E_a ~ 0.2 eV) are found with NO_2, SO_2 and O_2, but the corresponding dopant atom–molecule distances are rather large (d > 3 Å), suggesting that a true chemical bond was not formed.

S-graphene could only bind the NO_2 molecule, with a rather large adsorption energy $E_a = -0.83$ eV, in a configuration that is similar to that of NO_2 on B-graphene. Finally, Al-graphene was the most reactive and found to bind all molecules except H_2, *via* the formation of strong Al–X (X = O, N, C) bonds of length d_{Al-X} ~2 Å or less.

Mowbray *et al.* performed a computational screening of transition metal (TM)-doped (6,6) SWCNTs on the adsorption of various gases.[300] They demonstrated, using *ab initio* calculations, that a combined Cu- and Ni-doped metallic SWCNT device may work effectively as a multifunctional sensor for both CO and NH_3. The adsorption energies shown in Table 3.3 show that the earlier TMs tend to bind the adsorbates stronger than the late TMs. DFT studies also show that Sc, Ti and V doped SWCNTs adsorb CO molecules effectively compared to pure SWCNTs.[301] The binding energy of Sc, Ti and V doped SWCNTs is 1.31, 1.81 and 0.95 eV, respectively, which is much higher than pure SWCNTs.

Tabtimsai *et al.* studied the adsorption of NO_2, NH_3, H_2O, CO_2 and H_2 gases on undoped and Zn-, Pd- and Os-doped armchair (5,5) SWCNTs using DFT calculations.[302] The bond lengths between doped metal and neighbor carbon atoms, M–C were in the order: Zn–C (2.087 Å) > Pd–C (1.956 Å) > Os–C (1.911 Å). The adsorption abilities of the gases on the Zn-, Pd- and Os-doped SWCNTs were remarkably increased as compared with the undoped SWCNT, and their adsorption abilities were: $NO_2 > NH_3 > H_2O > CO_2 > H_2$. Using the same (5,5) armchair SWCNT, Tabtimsai *et al.* investigated the Co-, Rh- and Ir-doping effect on CO_2 and NO_2 adsorption using DFT calculations.[303] Co, Rh and Ir atoms formed a pyramidal geometry with the side wall of the SWCNT (Figure 3.13), with the doping atoms occupying the top of the pyramid. The adsorption abilities on the pristine and metal-doped SWCNTs were found to be in the order: Co-doped > Ir-doped > Rh-doped >> pristine SWCNTs for CO_2 and Ir-doped > Co-doped > Rh-doped >> pristine SWCNTs for NO_2.

Gas sensing (NO_2 and NH_3) properties of platinum derivatives on SWCNTs were studied by Pannopard *et al.*[304] DFT calculations showed that Pt (both deposited and doped) could enhance adsorption and charge transfer

Table 3.3 Adsorption energies (E_{ads} in eV) for N_2, O_2, H_2O, CO, NH_3 and H_2S on a 3d transition metal-doped (6,6) SWCNTs.

	N_2	O_2	H_2O	CO	NH_3	H_2S
Ti	−0.61	−3.16	−1.09	−0.89	−1.39	−0.78
V	−0.76	−3.39	−1.08	−1.21	−1.46	−0.88
Cr	−0.57	−2.61	−0.97	−1.06	−1.35	−0.77
Mn	−0.70	−2.57	−0.89	−1.38	−1.32	−0.78
Fe	−0.77	−2.17	−0.84	−1.54	−1.31	−0.88
Co	−0.60	−1.88	−0.64	−1.23	−1.03	−0.58
Ni	−0.49	−2.00	−0.54	−1.16	−0.94	−0.69
Cu	−0.34	−1.02	−0.52	−0.97	−0.89	−0.63
Zn	−0.51	−0.90	−0.62	−1.07	−1.12	−0.67

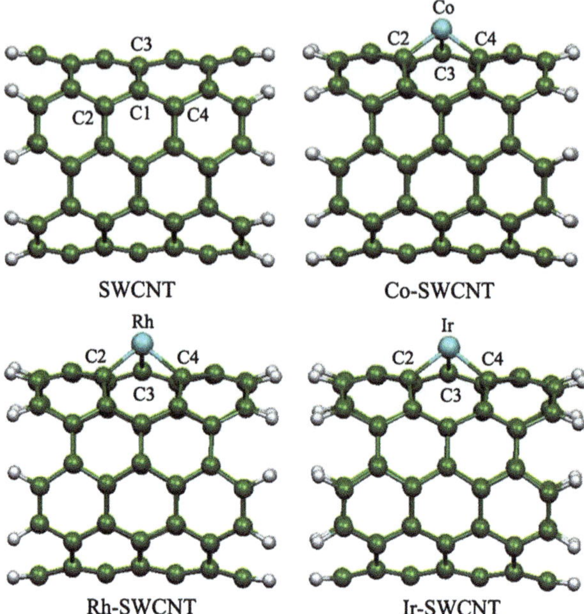

Figure 3.13 Optimized structures of pristine, Co-doped, Rh-doped and Ir-doped armchair (5,5) SWCNTs. Reprinted with permission from ref. 303. Copyright 2013 Elsevier B.V.

processes to a very large degree. For NO_2 adsorption, the Pt-doped SWCNTs showed slightly smaller binding energies and gap changes than both the pristine SWCNTs and Pt-deposited SWCNTs, whereas for NH_3, Pt-doped SWCNTs showed the largest electron transfers to the SWCNTs. A higher sensitivity of Pt-doped SWCNTs for the detection of both NO_2 and NH_3 was attributed to larger energy gap changes and a larger charge transfer. Similar results were reported for other gases such as NO, C_2H_4 and C_2H_2.[305] Zhou *et al.* investigated the adsorption sensitivity of Pd-doped (5,5) SWCNTs to small gas molecules (SO_2, CH_3OH vapour and CH_4) using DFT calculations.[306] The adsorption of SO_2 and CH_3OH onto the Pd-SWCNTs was attributed to a chemical adsorption and physical adsorption, respectively, while the interaction between CH_4 and Pd-SWCNT was an electrostatic one, and the formation of (Pd-SWCNT)-$(CH_4)^+$ complex indicated a charge transfer sensing mechanism.

Li *et al.* investigated the catalytic activity of Fe-doped graphene by the first principles method using the CO oxidation as a probe reaction.[307] The catalytic activity of the Fe-graphene system was interpreted in light of an Eley–Rideal mechanism: due to the strong hybridization between 3d states of Fe and the 2p state of O_2, the latter could be efficiently activated by Fe; then CO could be inserted into the elongated O–O bond to form a carbonate-like intermediate state CO_3 with a moderate energy barrier (0.58 eV) *via* the Eley–Rideal mechanism. Finally, CO_3 reacted with another CO to produce two CO_2 molecules also with a rather modest energy barrier (0.57 eV).

The same reaction was also used by Lu *et al.* and Song *et al.* to assess the theoretical behavior of an Au-graphene[308] and Cu-graphene[309] catalyst, respectively. The high activity of Au- and Cu-doped graphene was attributed to the electronic resonance among electronic states of CO, O_2 and metal atoms, particularly among the d states of the Au and Cu atoms and the anti-bonding $2\pi^*$ states of CO and O_2. The partially occupied d orbital of Au is localized in the vicinity of the Fermi level due to the interactions between the Au atom and graphene.

Zhou *et al.* studied the adsorption of gas molecules (O_2, CO, NO_2 and NH_3) on transition metal embedded graphene using a first principles approach based on DFT.[310] Their results showed that Ti and Au were the best in improving the chemical reactivity of graphene. Furthermore, when adsorbed on Ti or Au embedded graphene, the desorption barriers of these molecules were all quite high: close to 1 eV for NH_3 and well above 1.5 eV for other molecules, suggesting the high stability of these molecules on TM-graphene.

The reactivity of Si-doped SWCNTs towards O_2, CO_2, SO_2 and NO_2 using DFT was reported.[311] The weak binding of these molecules on CNT is due to formation of charge–dipole interactions. In case of Si-CNT, all molecules are chemisorbed to the Si–C bonds with appreciable adsorption energy and significant charge transfer. The charge density analysis reveals the formation of σ-bonds between Si and C atoms. Furthermore, the band structure and density of states clearly illustrate the creation of an extra state near the Fermi level and reduction in the band gap, which acts as a reactive center for adsorption of the molecules on Si-CNT.[312] Ao *et al.* investigated the enhancement of CO detection in Al-doped graphene using DFT calculations.[295] The results showed that CO molecules are only weakly adsorbed onto the intrinsic graphene with a small binding energy value and a large distance (d = 3.767 Å) between the CO molecule and graphene, whereas CO molecules strongly interact with the Al-doped graphene, forming an Al–CO bond (d = 1.964 Å) that introduces a large number of shallow acceptor states into the system. Physisorption of CO on pristine graphene did not alter the electron distribution of either the CO molecule or graphene, implying weak bonding characteristics. On the other hand, the chemisorption of CO on Al-doped graphene led to significant electron transfer from the graphene to the CO molecule. The electrons not only accumulated on the O atom but also on the C atom of the molecule bond with the doped Al atom. The final position of Al atom in the chemisorbed CO–Al-graphene complex was thus a direct consequence of the maximized degree of sp^3 orbital hybridization with neighboring C atoms from both the graphene layer and CO molecule.

A first principles approach was used to establish that substitutional phosphorus atoms within CNTs strongly modify the chemical properties of the surface, thus creating highly localized sites with specific affinity towards acceptor molecules.[313] Phosphorus-nitrogen co-dopants within the tubes have a similar effect for acceptor molecules, but the P–N bond can also accept charge, resulting in affinity towards donor molecules. This molecular selectivity was illustrated in CO and NH_3 adsorbed on PN-doped CNTs, O_2 on P-doped CNTs, and NO_2 and SO_2 on both P- and PN-doped CNTs. The adsorption of different chemical species onto the doped nanotubes modifies the dopant-induced localized states, which subsequently alter the electronic conductance.

The adsorption of gas molecules on P-doped graphene was theoretically studied using DFT in order to find the possibility of modulating electronic and magnetic ordering of graphene.[314] H_2, H_2O, CO_2, CO, N_2 and NH_3 molecules are physisorbed, while NO, NO_2, SO_2 and O_2 molecules are strongly chemisorbed on P-doped graphene through the formation of P–X (X = O, N, S) bonds. Dai and Yuan studied the adsorption of multiple gas molecules (N_2, O_2, H_2, H_2O, CO, CO_2, NO, NO_2, NH_3 and SO_2) on P-doped graphene using DFT.[314] When the P atom replaced a C atom in the graphene plane, the structure of the graphene underwent a significant rearrangement (due to the much larger size of the P atom with respect to C) and the P atom projects out of the graphene plane at a distance of 1.19 Å. Based on the calculated large distances from the graphene plane, H_2, H_2O, CO, CO_2, N_2 and NH_3 are considered to be physisorbed on P-doped graphene surfaces. On the other hand, O_2, NO, NO_2 and SO_2 molecules were found to strongly chemisorb onto the P-doped graphene, forming P–X (X = O, N, S) bonds with large adsorption energies (−1.0895, −0.5093, −1.8875 and −0.3216 eV, respectively).

Zhang *et al.* reported first principles simulation of the interactions between several small molecules (CO, NO, NO_2 and NH_3) and graphene sheets containing defects (single vacancies) and dopant species (B or N).[315] The calculation results suggest that the pristine graphene has weak interactions with all four gas molecules. The introduction of dopants and defects into the graphene significantly increases the molecule–graphene interaction. The minimum atom-to-atom distance between the CO and graphene is obtained when a vacancy was present (1.33 Å). This distance was close to the bond length of a C=C bond and was much shorter than that with any other type of graphene, which were 3.02 Å (pristine graphene), 2.97 Å (B-doped graphene) and 3.15 Å (N-doped graphene). The adsorption energy of CO over the defect reached −2.33 eV, which was more than one magnitude higher than that of the pristine and doped graphene. Similarly to CO, the presence of a defect on the graphene sheet also yielded the highest affinity to both NO (−3.04 eV adsorption energy and a NO–graphene distance of only 1.34 Å) and NO_2 (−3.04 eV adsorption energy a NO_2–graphene distance of only 1.42 Å), revealing the occurrence of a strong chemisorptive behavior. As for the adsorption of NH_3, this molecule showed a much stronger interaction with the B-doped graphene (−0.50 eV), which was attributed to the strong interaction between the electron-deficient B atom and the electron-donating N atom of NH_3. It was also found that the B-graphene underwent an obvious distortion upon NH_3 adsorption, indicating that the B-site is transformed from sp^2 to sp^3 hybridization. The B–N distance (1.66 Å) was very close to the B–N bond length in BH_3NH_3 (1.6576 Å), confirming the formation of a covalent bond between the

NH_3 and the B-doped graphene. DFT calculations have also shown that NH_3 is absorbed on the hollow site through the physisorption mechanism. However, the adsorption energy of NH_3 onto Al- and B-doped graphene increases, resulting in chemisorption.[316]

The interaction of CO_2 to the interior and exterior walls of pristine and nitrogen-doped SWCNTs has been studied using DFT.[102] The calculations predict Gibbs energies of binding between SWCNTs and CO_2 of up to 0.39 eV. Doping of the (10,0) tube with nitrogen increases the Gibbs energies of binding of CO_2 by *ca.* 0.13 eV. The Gibbs energy of binding of CO_2 to the exterior of the tubes is quite small compared to the binding that occurs inside the tubes. Exohedral doping of an Fe atom onto the CNT surface also affects the adsorption energy of the quadrupolar CO_2 molecule inside the CNT (20–30%).[103] Cabrera-Sanfelix studied the adsorption and reactivity of CO_2 on a vacancy defect of a graphene sheet, using DFT calculations.[317] A mechanism (Figure 3.14) to fix and remove gaseous CO_2 molecules was proposed to consist of: (i) physisorption of CO_2 on the top of the vacancy defect, with the inner axis of the molecule parallel to the surface (strong physisorption with a binding energy of −0.136 eV); (ii) formation of a lactone group with the C atoms surrounding the vacancy through a chemisorptive state (after overcoming an energy barrier of ~1.2 eV, and an energy release of about 1.4 eV); (iii) the lactone chemisorbed state shows dissociation of the CO_2 through the formation of epoxy groups (which is more stable than the preceding lactone group by ~0.614 eV); (iv) oxygen recombination from the epoxy groups and the desorption of O_2 as the global minimum (energy release of ~3.4 eV with respect to the gas phase and with an energy barrier of ~1.08 eV relative to the previous epoxy-complexes configuration), yielding a defect-free graphene sheet surface.

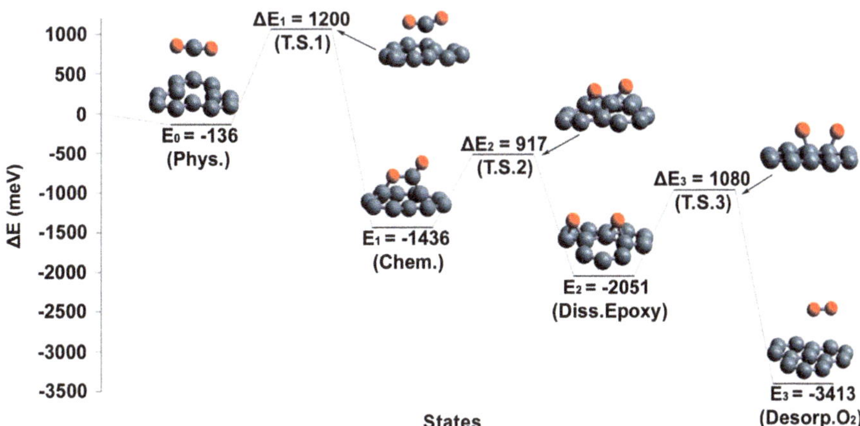

Figure 3.14 Structural and energetic scheme of the proposed global reaction path for the interaction of CO_2 with the defective graphene layer. Reprinted with permission from ref. 317. Copyright 2009, American Chemical Society.

On a graphene-C_{40} model, the adsorption energies were calculated to be 0.087 eV for the perpendicular physisorption configuration, and 0.13–0.14 eV for the parallel physisorption configuration.[318] The formation of lactone groups is the most exothermic process in CO_2 adsorption on graphene edges. The CO_2 chemisorption energies on graphene-C_{40} assuming high pressure are predicted to be 3.08–3.12 eV for the lactone systems depending on various C–O orientations. This result is in reasonably good agreement with the experimentally observed 3.25 eV. Theoretical study of sorption of CO_2 on unmodified or N-, O-, and OH-substituted graphene structures possessing one completely unsaturated edge zigzag site was reported using DFT.[319] None of the substitutions change or decrease the enthalpy in comparison with the unmodified graphene sheet. The CO_2 molecules adsorb on the edge plane surface of N-, O-, OH-containing carbon surfaces similarly or much less favorably in comparison with the unmodified surface. DFT calculations also indicate that the presence of $-NH_2$, $-H_2PO_3$, $-NO_2$, and –COOH functional groups (edge-functionalization) on graphene nanoribbons can significantly enhance CO_2 binding with respect to a hydrogen-passivated edge.[320]

The adsorption and activation of triplet O_2 on the surface of N-doped CNTs with different diameter and length were investigated.[321] The obtained results indicated that nitrogen-doping facilitated the adsorption of O_2 on CNTs. It was found that, rather than the unfavorable adsorption on pristine CNTs, the adsorption of O_2 on N-doped CNTs was obviously exothermic and the electron transition of O_2 happened in the adsorption process. The oxygen adsorbed on N-doped CNTs showed an interesting electron configuration which was similar to the active oxygen anion. DFT results show that nitrogen prefers to stay at the open-edge of the short CNTs.[322] Oxygen can adsorb and partially reduce on the carbon–nitrogen complex site and on the open carbon–carbon long bridge sites at the open-edge of the CNT. DFT calculations were used to explore reaction paths for facile oxygen dissociation on modified SWCNTs, including nitrogen doping, Stone–Wales defects, and a combination of these two.[323] It was found that oxygen dissociation is facilitated on carbon atoms neighboring a nitrogen dopant, with the dissociation barrier reduced from 2 eV to 0.68 eV. The activation barrier can be further reduced to 0.03 eV in the vicinity of a N-doped Stone–Wales defect. It has been shown that N-doping decreased the energy barrier for O_2 dissociative chemisorption on graphene from 2.39 eV (undoped case) to 1.20 eV (Figure 3.15).[324] The O_2 molecule prefers to adsorb on the graphene surface with the two O atoms close to two C atoms along the diagonal of the C_6 ring. It is found that charge transfer occurs between the O_2 molecule and the graphene sheet, which makes the adsorption energetically stable. Nitrogen doping enhances the adsorption since the N atom has one more electron than C atom. For single O atom adsorption on the graphene surface, the bridge sites are energetically most stable. However, N-doping makes the top site of the nearest neighboring C atom to the N atom the most stable adsorption site.

Dai and Yuan[325] used DFT to study the adsorption of molecular oxygen on B-, N-, Al-, Si-, P-, Cr- and Mn-doped graphene. The molecular oxygen was found to chemisorbed on Cr-, Mn-, Al-, Si- and P-doped graphene with large adsorption energies (E_{ads} in Table 3.4), and short bonds between the dopant atom and oxygen ($d_{X\text{-}O}$ in Table 3.4), indicating very stable adsorption of O_2. On the other hand, O_2 physisorbed over B- and N-doped graphene, with small adsorption energies and long distances to the graphene plane. B- and N- doped graphene retained a planar form, while other atoms project out of the graphene layer and induce a local curvature in graphene (Figure 3.16 and *h* in Table 3.4).

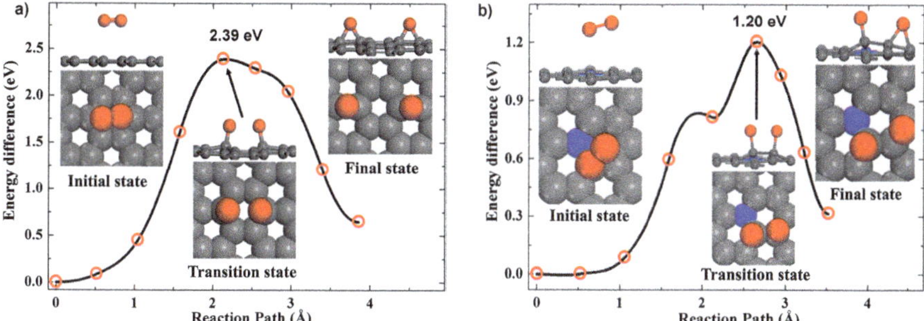

Figure 3.15 The optimized O_2 molecule dissociation reaction path and energy profile along the optimized dissociation reaction path at (a) graphene, and (b) N-doped graphene surface. Reprinted with permission from ref. 324. Copyright 2012 American Physical Society.

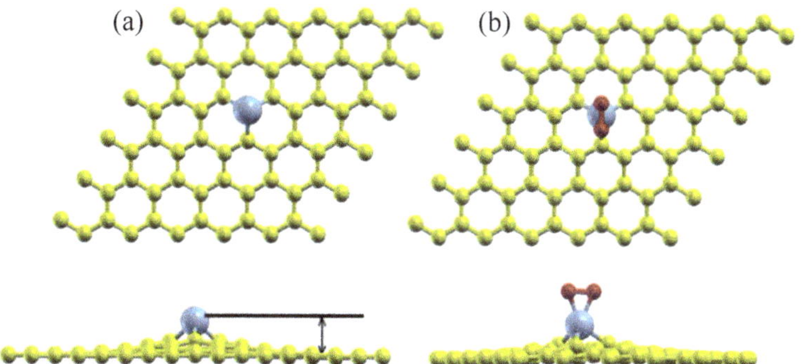

Figure 3.16 Most stable configuration of (a) Mn-doped graphene with an elevation h of 1.555 Å, and (b) O_2 on Mn-doped graphene. Adapted from ref. 325.

Table 3.4 Summary of atomic structures of the compounds, including the elevation of the dopant atoms above the graphene plane (*h*), the bond length between dopant (X) and O ($d_{X\text{-}O}$) and adsorption energy (E_{ads}).

Dopant	h Å^{-1}	$d_{X\text{-}O}$/Å	E_{ads}/eV
N	0.0000	3.3196	−0.1228
B	0.0000	3.5099	−0.0232
P	1.4591	1.6275	−1.0359
Si	1.4579	1.7109	−1.3132
Al	1.7584	1.8770	−1.5589
Mn	1.5555	1.8641	−2.0918
Cr	1.6487	1.7935	−2.6098

3.4.3 Chemisorption of Liquids and Organic Molecules on Nanostructured Carbon Materials

3.4.3.1 Adsorption of Water

The interaction of water vapor with CNTs at room temperature has been investigated using FT-IR spectroscopy and DFT calculations.[140] FT-IR data indicate that water molecules adsorb on SWCNTs at room temperature. Comparison to previous studies suggests that the water forms hydrogen-bonded structures inside the nanotubes. Analysis of the FT-IR data demonstrates that a small number of water molecules react with the nanotubes, forming C–O bonds, whereas a majority of the water molecules adsorb intact. The DFT calculations show that cleavage of an O–H bond upon adsorption to form adsorbed –H and –OH groups is energetically favorable at defect sites on nanotubes. The chemisorption of water on a (8,0) CNT is found to be very difficult, and the smallest energy barrier of the chemisorption is 2.89 eV with the H and OH dissociated from water bonding at the opposite sites of carbon hexagons.[53]

Water monomer adsorption on graphene was examined with state-of-the-art electronic structure approaches.[137] A value of −70 to −98 meV was obtained for each of the two water configurations tested (Figure 3.17a–b) with a slight preference for a structure with two hydrogens oriented towards the surface. These values are below the range obtained from calculations on cluster models. A value of -130 ± 6 meV has also been reported with two hydrogens positioned above the central ring.[326] From another theoretical study, it has been shown that the circumflex-like orientation of water is more favorable than the caron-like one (Figure 3.17c).[327] A top adsorption site is preferred by the water molecule and the most stable structure is characterized by an adsorption energy of 135 meV. So, it appears that on the perfect surface, the water physisorbs. At a vacancy site, the interaction is much more significant, with a computed binding energy of 210 meV in a weak chemisorption/strong physisorption state.[328] The H_2O sits with one H pointing down to a carbon atom, which is pulled out of the plane by 0.55 Å. From this physisorption state, dissociative chemisorption will occur after overcoming a barrier of 0.8–0.9 eV. The lowest dissociation barrier obtained is 0.47 eV along a path largely avoiding the physisorption well. The dissociation paths have an intermediate step, in which the molecule partially dissociates to H and OH. Subsequently, the chemisorbed OH stretches, breaking into O and H atoms chemisorbed on separate C atoms on the vacancy with a total exothermicity of 3.21 eV. It has also been shown that the dissociative adsorption of one water molecule at a vacancy site may lead to the formation of a "ketone-like" structure, which can then act as a nucleation center for additional water molecules.[329]

3.4.3.2 Adsorption of Alcohols and Acetone

The adsorption of methanol and ethanol on SWCNTs has been investigated using FT-IR spectroscopy and two-level ONIOM calculations.[330] Whereas methanol does not adsorb onto SWCNTs at room temperature, ethanol

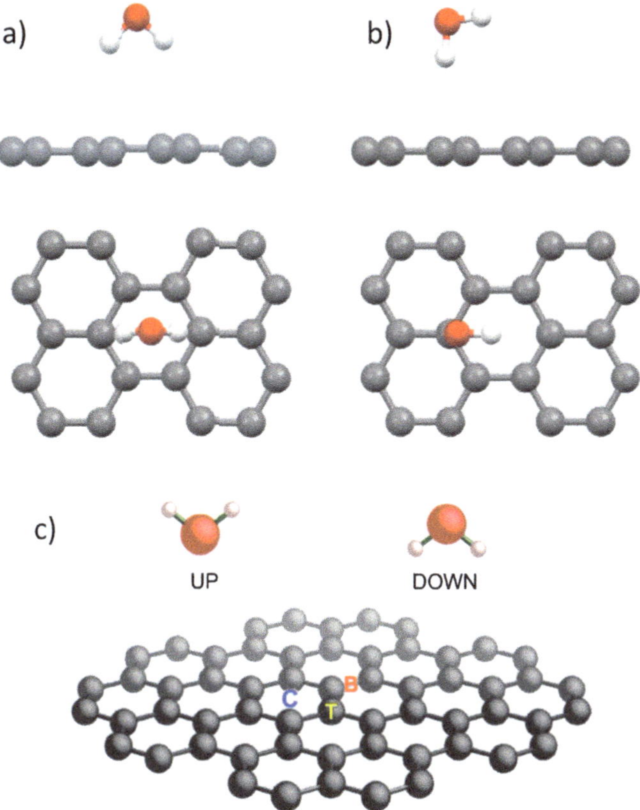

Figure 3.17 Water adsorption structures considered: (a) the two-leg structure shown from the side (top) and from above (bottom), (b) the one-leg structure shown from the side (top) and from above (bottom) from ref. 137; and (c) the H_2O-graphene arrangements are defined by the position of oxygen atom above the six-membered carbon ring (C = H6/T/B) and the water orientation (UP/DOWN). Reprinted with permission from ref. 327. Copyright 2011, Royal Society of Chemistry.

does adsorb molecularly under these conditions. The IR data show that the adsorbed ethanol is in an environment quite similar to that of liquid ethanol. Comparison to studies of clusters of ethanol indicates that adsorbed ethanol molecules likely form cyclic clusters of four or more molecules, and ONIOM calculations suggest that these clusters can form inside the largest-diameter nanotubes in the sample. Changes in the NMR chemical shifts have been reported for methanol molecules located on or in SWCNTs.[331] The results reveal that the changes in chemical shift mainly originate from the magnetic shielding induced by the delocalized π electrons. As for SWCNTs, the adsorption energies of methanol on graphene are low, typically between 200 and 350 meV.[332] In contrast, it is quite interesting to find that methanol is not only strongly chemisorbed at the zigzag edge site of defective SWCNT, but also the O–H bond of methanol is completely dissociated.[333] This suggests that the zigzag edge of SWCNT can be the active site for adsorption and activation of methanol. However, the adsorption of methanol at the armchair edge of SWCNT is rather weak, hence suggesting the crucial effect of local edge carbon atoms arrangement on the adsorption behavior of methanol on CNTs.

A combined experimental and theoretical quantification of the adsorption enthalpies of seven organic molecules (acetone, acetonitrile, dichloromethane, ethanol, ethyl acetate, hexane, and toluene) on graphene has been performed.[334] Adsorption enthalpies were measured by inverse gas chromatography and ranged from −256 meV for dichloromethane to −585 meV for toluene. The strength of interaction between graphene and the organic molecules was estimated by density functional theory, wave function theory, and empirical calculations. The calculations indicated that the interactions were governed by London dispersive forces (amounting to 60% of attractive interactions), even for the polar molecules. The results also showed that the adsorption enthalpies were largely controlled by the interaction energy. Adsorption enthalpies obtained from *ab initio* molecular dynamics were in excellent agreement with the experimental data, and follow the order: toluene > hexane > ethyl acetate > acetone > acetonitrile > ethanol > dichloromethane.

The adsorption of acetone on SWCNT at different temperatures was followed using temperature programmed desorption studies (TPD).[35] It was observed that the activation energy for desorption of acetone was much larger than the latent heat of vaporization of acetone from the liquid state (0.32 eV) and the activation energy of desorption of acetone multilayer from graphite. Following this observation, the adsorption of acetone on SWCNT was modelled using hybrid quantum mechanical and semi-empirical calculations, which showed that acetone binds preferentially to defects present in the nanotubes and was dependent on the diameter of the nanotubes The calculated binding energies showed that acetone is strongly chemisorbed on CNTs and remained bound up to high temperatures unlike its behavior on graphite. Theoretical *ab initio* calculations performed on model Ti_xO_y/CNT compounds reveal that poorly oxygenated surface Ti sites represent well-defined trapping sites for acetone chemisorption, while atomic oxygen impurities as well as oxygen-containing functional groups and more oxygenated Ti species lead to physisorbed or weakly chemisorbed configurations.[335]

3.4.4 Chemisorption of Solids on Nanostructured Carbon Materials

3.4.4.1 Alkali Metals

Rytkönen *et al.* computed the adsorption of alkali-metal – Li, Na, K, Rb, Cs – atoms, dimers, and (2 × 2) monolayers on graphene.[336] All alkaline atoms prefer to bind on the H_6-site – Li lying closest to 1.84 Å and Cs farthest 3.75 Å – from the surface. The adsorption energies range between 0.55 and 1.21 eV. The ordering of binding energies is Li > Cs ≥ Rb ≥ K > Na. Lithium has a moderate diffusion barrier of 0.21 eV, whereas the larger alkali-metals are relatively mobile with diffusion barriers of 0.02–0.06 eV. Taking into account these calculations, lithium is the only alkali-metal able to be chemisorbed on the graphene surface. The other alkali-atoms are mobile on the surface with weak binding energies. A similar observation was also made by Nakada and Ishii.[337] These authors studied the adsorption and migration energies of several adatoms, from hydrogen to bismuth, on graphene. They found that lithium (binding energy 1.36 eV with a distance to the graphene layer of 1.62 Å and migration energy of 0.3 eV) is likely to be chemisorbed on the graphene surface. The calculations of Liu *et al.* also point in this direction.[338] The interaction of lithium with graphene and the effect of the Li–Li interaction on the binding of Li to graphene has been calculated using the local density approximation to the density functional theory.[339] Lithium binds to a H_6-site at 1.64 Å from the graphene with a binding energy of 1.598 eV. The introduction of more Li atoms on the graphene sheet decreases the binding energy to 0.934 eV without changing the location and the distance between Li and graphene. There is substantial charge transfer from both lithium and carbon atoms (including those that do not surround the lithium) to a region between the Li and graphene at the minimum energy configuration. The calculations show that the Li–Li interaction is repulsive due to the charge transfer that takes place between Li and graphene, leaving a positively charged cation at the original location of the Li atom. The charge transfer remains significant even at long distances. Li adsorption on graphene for Li concentrations ranging from about 1–50% has been studied by means of DFT calculations.[340] At low adsorbant densities, a strong ionic interaction characterized by a substantial charge transfer from the adatoms to the substrate was observed. In this low concentration regime, the electronic density around the Li adatoms is well localized and does not contribute to the electronic behavior in the vicinity of the Fermi level. For larger concentrations, the formation of a chemically bound Li layer was observed, characterized by a stronger binding energy as well as a significant density of states above the Fermi level coming from both graphene and the two-dimensional Li sheet. A study of the characteristics of the bonding between lithium and carbon atoms and whether they interact *via* an sp^2 or an sp^3 hybridization was performed.[341] It was found that the carbon–lithium bond is not purely covalent but instead presents a significant ionic character. The local geometry is governed by the π-acceptor character of lithium atoms which occupy reverse positions relative to the carbon atoms as compared to the positions of hydrogen in graphene.

Concerning the adsorption of Li on CNT, it has been also observed that the increase of Li/C ratio decreases the binding energy between the lithium atom and the nanotube and increases the Li–CNT distance.[342] This can be explained by the decrease of the ionicity of the Li–CNT bond; that is the effective charges of lithium ions decrease in the order: LiC_{64} (0.57*e*), LiC_8 (0.39*e*) and LiC_4 (0.27*e*). Senami *et al.* investigated the adsorption of lithium atoms on the surface of a (12,0) CNT by using *ab initio* quantum chemical calculations to clarify the interaction between lithium atoms and between a lithium atom and the CNT.[343] The authors show that after the lithium attachment, charge is transferred from the Li atom to the CNT and the bond between the Li atom

Table 3.5 Adsorption energies per alkali atom for the Li, Na, and K adsorption on C_{60}.

System	Adsorption energy/eV
LiC_{60}	1.30
Li_2C_{60}	1.41
Li_6C_{60}	1.55
$Li_{12}C_{60}$	1.43
NaC_{60}	1.09
Na_2C_{60}	1.13
Na_6C_{60}	0.95
$Na_{12}C_{60}$	0.85
KC_{60}	1.27
K_6C_{60}	0.87
$K_{12}C_{60}$	0.81

and the CNT has ionic properties. The amount of charge transfer depends on the curvature of the CNT, being larger for the endohedral than the exohedral grafting. This explains that the *exo*-adsorption is weaker (1.70 eV) than the *endo* one (1.87 eV). The authors also observed that the repulsive force between lithium atoms strongly destabilize the system.

DFT calculations to study the behavior of potassium atoms over the surface of a C_{60} were performed.[344] The results show that the structures result from a balance between the electrostatic repulsion between the alkali atoms, due to the electronic charge transfer from the metal atoms to the C_{60} cage, *versus* the metallic bonding. The coverage by K atoms is found to be similar to that of Na atoms, with the formation of small island on the surface for $n = 12$, but contrast with that of Li atoms, which prefer to homogeneously coat the fullerene *via* pentagonal sites.[345] The adsorption energies (per alkali atom) are reported in Table 3.5.

A carbon hybrid material system consisting of SWCNTs and C_{60} has been investigated using the first-principles methods.[346] Through combining metallic SWCNTs with C_{60} of high electron affinity, the lithium adsorption energy on this hybrid system (−2.110 eV) is found to be larger than that of the pure SWCNTs (−1.720 eV). By characterizing the electronic properties of the CNT-C_{60} system, such as band structure, density of states and charge distribution, as a function of the Li adsorption in comparison with SWCNT or C_{60}, it is also found that the Li adsorption takes place preferentially on the C_{60} side due to the large adsorption energy, which imparts metallic character to the C_{60} in the CNT-C_{60} hybrid system. Investigating various adsorption sites on the CNT-C_{60} system in order to understand the adsorption mechanism of Li, it is found that Li atoms are preferentially adsorbed at every other hexagonal or pentagonal site (next nearest neighboring sites) rather than every site (nearest neighboring sites) on the hybrid system. The possibility of Li cluster formation in this CNT-C_{60} system does not seem to be high since the Li–Li binding is less favorable than the Li adsorption on the CNT-C_{60} system. *Ab initio* calculations were performed to investigate the adsorption of Li onto a clean and oxygenated diamond C(100) surface.[347] It was found that Li adopts structures on the clean C(100) surface similar to those reported for Na, K, and

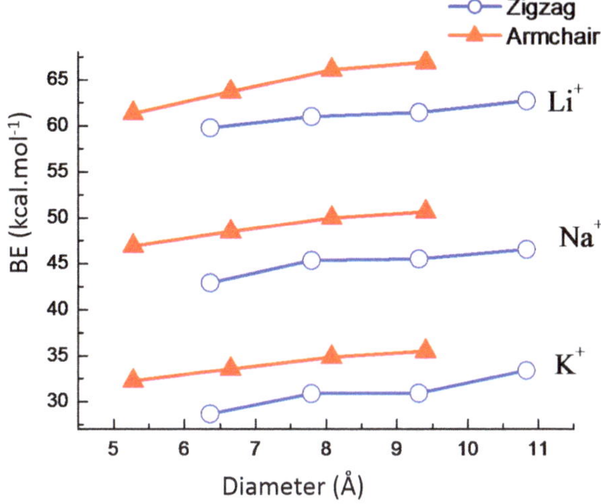

Figure 3.18 BSSE corrected B3LYP-D/6-31G*//ONIOM (M06-2X/6-31G*:PM3) binding energy (kcal mol^{-1}) of metal ions with armchair and zigzag CNTs. Reprinted with permission from ref. 349. Copyright 2012 Elsevier B.V.

Rb on diamond, though Li exhibits significantly higher binding energies in the range 2.7–3.1 eV per Li adsorbate. For the oxygenated surface, the lowest energy involving a full Li monolayer structure shows an exceptionally large work-function shift of −4.52 eV relative to the clean surface, an effect similar to that seen for Cs–O on diamond, but with a higher binding energy of 4.7 eV per Li atom.

The adsorption energy of K on graphene was estimated by DFT to be 1.3–1.6 eV.[348] It was also found that the adsorption energy of K on SWCNTs was not strongly affected by the length of the nanotube fragment, but by the nanotube surface curvature and chirality. This suggested that the adsorption of K on SWCNTs could not be described in terms of a simple classical model and that the energy of K–SWCNT interaction depended on factors related to the structure and electronic properties of SWCNTs. The binding of a series of alkali metal ions (Li^+, Na^+ and K^+) to graphene and CNTs has also been studied by first principle calculations.[349] The alkali metal ions are at distances around 1.9, 2.3 and 2.7 Å, respectively, from the surface of the carbon nanostructures and bind on the H_6-site. Li^+ shows the stronger binding energy (2.9 to 3.3 eV), followed by Na^+ with 3.0 to 2.2 eV and K^+ ion with a binding energy range of 1.3 to 2.2 eV. These values indicate that alkali metal ions interact with the carbon nanostructures by chemisorption. Regarding the effect of the curvature, all the metal ions considered exhibit stronger binding to graphene compared to CNTs. Further, the binding energy increases as the diameter increases (curvature decreases) in the case of armchair CNTs, while such clear trends did not emerge in the case of zigzag CNTs. Concerning the effect of the chirality, Figure 3.18 shows that the monocationic alkali metal ions are more stable for the armchair than for the zigzag CNTs.

3.4.4.2 Alkaline-Earth Metals

Alkaline-earth metals have attracted the attention as elements capable to form stable and uniform atomic coatings on carbon nanostructures, unlike transition metals, which tend to aggregate and form clusters.[350–355]

Liu *et al.* studied the adsorption behavior of alkaline-earth metals (Mg, Ca and Sr) on capped (5,5) and (9,0) SWCNTs by DFT to shed light on (i) the difference between adsorption on the pentagon and hexagon of capped CNTs,[356] and (ii) the effect of vacancy defects.[357] Ca and Sr are chemisorbed on the capped SWCNTs, Mg with low adsorption energies is considered as physisorbed. The authors show that there are significantly different adsorption behaviors between the different sites, being the H_6 the preferred one. The authors also show that vacancy defects have a great impact on adsorption. Charge transfers are enhanced significantly in defective CNTs, especially for Mg. The binding energies of Mg, Ca and Sr increase from 12 meV, 0.96 eV and 0.97 eV for (5,5) CNTs to 2.37 eV, 3.99 eV, 3.81 eV for defective CNTs, respectively. The presence of a defect site enhanced the binding of all alkaline-earth metals; the average increase of 2 eV on the binding energies suggest that Mg is chemisorbed when defects are introduced on the CNTs. The problem of the interaction between Be, Mg and Ca with graphene has been investigated.[358] A non-ionic adsorption of the alkaline earth elements onto graphene has been considered. It is worth mentioning that, in some cases, the ionic adsorption is more stable. In effect, in a recent work by Wang *et al.*,[359] it was shown that the interaction energy between graphene and calcium is 611 meV. The optimized structures have the alkaline earth atom over a hexagon. The binding energy between graphene and Be, Mg and Ca were estimated to be 91, 117 and 126 meV, respectively. These binding energy are much smaller than the BE between lithium and graphene. The main reason is that there exists a charge transfer between lithium and graphene, but herein we have only considered complexes in which there is no charge transfer between the alkaline earth elements and graphene. The stability of calcium adsorbates on CNTs and defective graphene was investigated using first principles calculations.[352] For ultra-narrow CNTs (3.2–5.6 Å), the effect of chirality is more important for the adsorption as compared to the diameter. The stability of Ca adsorbates on the tubes with the same chirality is enhanced when the diameter of the tubes decreases. The binding energy for a calcium atom absorbed on the (5,0) tube is about 1.4 eV higher than that on the (3,3) tube. The calcium atoms on the octagon defect of graphene are also stable. The binding energies of calcium adsorbates on narrow tubes and defective graphene are high, which indicates that the calcium monolayer should be stable without clustering.

The interactions of magnesium species (Mg, Mg^+, and Mg^{2+}) with a graphene surface have been investigated by DFT.[360] The distances of Mg atoms from the graphene surface were calculated to be 1.80 Å (Mg^{2+}), 2.16 Å (Mg^+), and 4.17 Å (Mg). The binding nature of Mg ions (Mg^{2+} and Mg^+) is caused by the charge transfer interaction, and the Mg atom interacts with the surface *via* van der Waals forces. The Mg ions can diffuse *via* the C–C bond center between hexagonal sites. The barrier heights for the diffusion of Mg^{2+} and Mg^{+} on the graphene surface were calculated to be 633 and 121 meV, respectively. The diffusion of Mg atoms proceeds without an activation barrier (only 0.3 meV). Umadevi *et al.* studied the adsorption of alkaline earth metal ions (Be^{2+}, Mg^{2+}, Ca^{2+}) on CNTs and graphene by first principles calculations.[349] The dicationic species bind to H_6-sites at distances around 1.3 Å, 1.9 Å and 2.2 Å, respectively. Be^{2+} shows the highest binding energy (14–17 eV), followed by Mg^{2+} (10–13 eV) and finally Ca^{2+} (6–8 eV). Regarding the effect of the chirality, the dicationic alkaline earth metal ions bind more strongly to the zigzag than to the armchair CNTs, except for Ca^{2+}, which shows similar binding energies on both CNTs. Thus, the alkaline-earth metal ions show an opposite behavior when compared to alkaline metal ions. In the case of dicationic metal ion complexes, the charge transfer is higher compared to that of the monocationic metal ion complexes explaining why the dicationic alkaline-earth metals show higher binding energies that the mono-cationic alkali metal ions. The binding energy is also affected by the CNTs curvature, for larger diameters the dicationic species are bound more strongly. The energy binding is similar for CNTs (14, 0) and graphene.

Both Ca and Sr can bind strongly to the C_{60} surface, and highly prefer monolayer coating.[351] The strong binding is attributed to an intriguing charge transfer mechanism involving the empty *d* levels of the metal elements. The charge redistribution, in turn, gives rise to electric fields surrounding the coated fullerenes.

3.4.4.3 d-*Metals*

The adsorption of transition metal on nanostructured carbon materials is one of the most studied due to the variety of possible applications of such systems, including catalysis.[218]

Single Metal Atoms. In general, transition metal adatoms are mainly chemisorbed on the graphene surfaces, showing a high preference to adsorb with a η^6 hollow geometry. *B* and *T*-sites are most stable for transition metals with a filled or near-filled d-shell, such as Pd and Pt. Usually, the binding energies are found to be higher when increasing the electrons on the d orbital until it is half-filled. Hu *et al.* studied the adsorption of 15 different transition metal adatoms – mainly 3d – on graphene using first principles DFT.[361] For metals such as Sc, Ti, V, Fe, Co and Ni the authors found that H_6-sites are the most stable, while Cu, Pd and Pt, which have a filled or near filled d-shell, prefer *B*- and *T*-sites. They also found that the distortion of the graphene layer is quite significant when the adatom occupies *B*- and *T*-sites. This distortion is mainly due to the formation of the covalent bond. Au, Ag, Zn and half-filled transition metal atoms showed small adsorption energies. The intensity of the adsorption energy and the nature of the adsorption site can indicate the nature of the chemical bond. Therefore, Co, Ni, Pt and Pd are considered as chemisorbed while Au, Ag and Cu are physisorbed. Similar results were observed by Tang *et al.*[362] The authors investigated the adsorption energies, stable configurations, electronic structures, and magnetic properties of the graphene with Pt, Ag, and Au, using first principles DFT. They found that the *B*-site is the most stable adsorption site for the Pt adatom; while Au prefers the *T*-site. The Ag adatom can be stabilized almost equally at the *B*- or the *T*-site. Nakada and Ishii studied the adsorption and migration energies of almost all periodic table atoms, from hydrogen to bismuth, including transition metals, on graphene.[337] They used a first principles band calculation technique based on DFT. The authors show that the H_6-sites are the most stable sites for transition metal atoms, except for Cu, Pd and Pt which prefer *B*-sites, and Ag and Au the *T*-sites. The bond energy shows an increasing tendency with an increasing number of d-electrons until the d-orbital is half-filled. For filled or near-filled d-shells the binding energy tends to decrease. Considering that the bond distances (between the adatoms and the graphene layer) are smaller than 2 Å and the migration energies are above 0.5 eV, Ti, V, Cr, Mn, Fe and Co are chemisorbed for the 3d transition metals, Nb, Mo, Tc and Ru for 4*d* transition metals and Ta, W, Re and Os for 5d transition metals. The adsorption of 12 different metal adatoms on graphene was studied using first principles DFT.[363] For the adatoms studied from groups I–III of the Periodic Table, the results are consistent with ionic bonding, and the adsorption is characterized by minimal change in the graphene electronic states and large charge transfer. For transition, noble, and group IV metals, the calculations are consistent with covalent bonding, and the adsorption is characterized by strong hybridization between the electronic states of the adatom and graphene.

Liu *et al.* compared the adsorption of several metal adatoms on graphene by *ab initio* calculations.[338] Most of the metal adatoms favor the H_6-site, while Cu and Au adatoms prefer the *T*-site and Cr, Pd, Pt, and Ag adatoms take the *B*-site. All d-metals studied are chemisorbed with the exception of Ag, Cr and Mn. The diffusion barriers are very small for Cr, Mn, Cu, Ag, and Au adatoms on graphene. This study also points out that metals with near-filled or filled d-orbitals show a weak interaction with graphene. The binding energy of Au, Cr, and Al atoms on the armchair and zigzag edge binding sites of monolayer graphene, and at the high-symmetry adsorption sites of single layer, bilayer, and trilayer graphene was calculated.[364] The contribution to the total binding energy from graphene sub-layers was predicted to be very significant, although the edge binding energies were found to be substantially higher for all atoms in all cases. The adatom migration activation barriers for the lowest energy migration paths on pristine monolayer, bilayer, and trilayer graphene were also calculated and found to be smaller than or within an order of magnitude of k_BT at room temperature, implying very high mobility for all adatoms studied. This suggests that metal atoms evaporated onto graphene samples quickly migrate across the lattice and bind to the energetically favorable edge sites. For many transition metal species, it was reported that they have rather strong binding energy, but diffusion barrier energies are small, because the difference of the value at *T*, *B* and H_6 sites are small.[365] However, other elements such as, V, Cr, Mn, Fe, Co, Mo and particularly Ru have strong adsorption energy and strong migration barrier energy, and could be anticipated as potentially interesting to coat uniformly both inside and outside of the wall of CNTs.

A DFT study of transition metal atoms (from Sc to Zn, Pt, and Au) embedded in single and double vacancies (SV and DV) in a graphene sheet showed that for most metals, the bonding is strong and the metal–vacancy complexes exhibit interesting magnetic behavior.[366] In particular, an Fe atom on a SV is not magnetic, while the Fe@DV complex has a high magnetic moment (Figure 3.19). Surprisingly, Au and Cu atoms at SV are magnetic. Both bond strengths and magnetic moments can be understood within a simple local-orbital picture, involving carbon sp^2 hybrids and the metal spd orbitals.

Most of the transition metal adatoms are chemisorbed on the CNT surface, as observed for graphene, and yield the strongest binding on the η^6-*H_6-site*. Durgun *et al.*[367,368] studied the adsorption of 27 different atoms on (8,0) zigzag SWCNT[365,366] and four atoms on a (6,6) armchair SWCNT.[360,361] The authors calculated the atomic structure, binding geometry, binding energy, and resulting electronic structure of an individual-atom adsorbed on SWCNT. The results indicate that the properties of SWCNTs can be modified by the adsorbed metallic atoms. Most of the adatoms studied yield the strongest binding at the H_6-site. Ni, Pd, Pt, Cu, Ag, and Au, however, seem to prefer

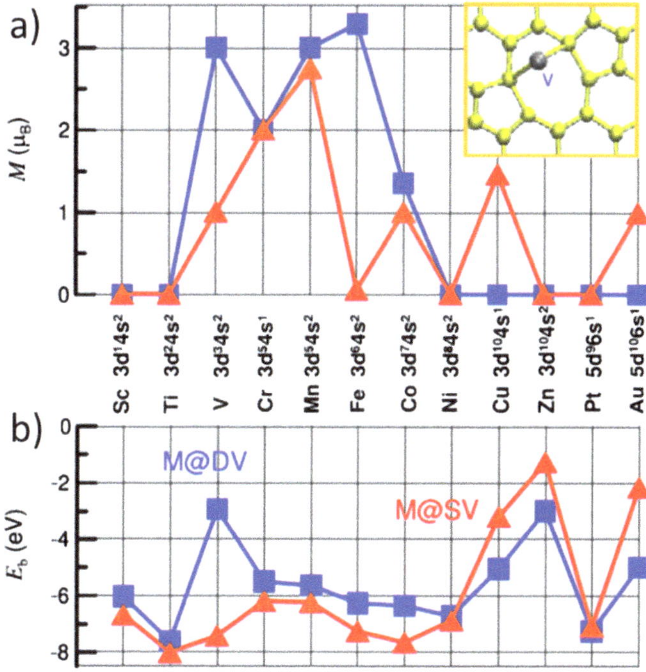

Figure 3.19 Magnetic moments, M (a) and binding energies, E_b (b) of the graphene sheet with TM atoms adsorbed on SVs and DVs (red and blue curves, respectively). The inset shows the configuration for V atom on DV which is different from those for all other atoms. Adapted from ref. 366.

the *B* site. The analyses of the binding energies showed that the transition metals with a few d electrons form strong bonds with the CNT – Sc, Co, Ti, Nb and Ta – with binding energies ranging from 2.4 eV to 1.8 eV. Metals with a near-filled or filled d-orbital showed relatively weak binding, for example Cu, Ag, Au and Zn with binding energies ranging from 0.8 to 0.05 eV.

A systematic DFT study of the 3d transition metal interactions with planar graphene and curved (8,0) SWCNT graphitic surfaces has been reported.[369] The authors show that the adsorption of the adatoms is rather similar for graphene and CNTs. Both graphene and (8,0) SWCNT surfaces presented η^6 hollow adsorption sites (η^6-H_6-*site*) with metals that mainly bound covalently, and ionicity degrees that varied as expected with the metal electronegativity (charge transfer from the metal to the surface). The calculations show that 3d transition metal atoms were chemisorbed onto graphene in H_6-site, with the exception of Cr, Mn and Cu. These atoms were physisorbed and presented no energetically favored adsorption sites. As observed in previous examples, the binding energy increased as one moved in either direction away from the Cr/Mn couple of the 3*d* series. The $sp^{2+\eta}$ hybridization of (8,0) SWCNT surface, interact more strongly with the d orbitals than observed for graphene. Therefore, the magnetic moments were one unit less than the magnetic moments in graphene for systems containing Sc, Ti, or V. However, due to low migration barriers and the strong metal–metal attraction, aggregation of metal atoms in the form of a metallic layer or clusters on typical SWCNTs (diameter of 1–2 nm) seems inevitable.[370]

The curvature of the CNT affects the binding energies as well as the binding geometry. In general, the smaller the diameter of the CNT is, the stronger binding is. Chen *et al.* reported the interaction of Pt adatom on CNT with several radii.[371] The authors observed a decrease of the binding energies for larger CNTs and also a change on the binding geometry. Similar observation were reported by Menon *et al.*, who studied the interaction of Ni on graphite, and (5,5) and (10,10) CNTs.[372] The interaction of the adatom is stronger on the CNT surface. The binding geometry is also affected, while the H_6-site is the most stable for graphite it becomes unstable for CNT, being the *T*-site the most stable.

The interaction of transition metal adatoms with CNTs may also depend on whether the adatoms are adsorbed on the inside or outside surface of the CNT. Gao and Zhao used first principles calculations to investigate the binding energies, geometric structures, and electronic properties of 4d transition metal atoms adsorbed outside/inside SWCNTs.[373] The most energetically favorable site is the same for interior and exterior adsorptions. The 4*d* elements from Y to Ru prefer the η^6-H_6-site, the Rh atom prefer an off-center η^5-H_6-*site* and finally Pd a *B* site. The binding mode is related to the valence electron configurations of the adatoms: for metals with filled or near-filled d-orbital, the sp–d hybridation with carbon becomes more difficult and they prefer to be in contact with less carbon atoms. The theoretical binding energy for Ru adsorption on nanotube or graphene surfaces is around 3.2 eV, which does not depend sensitively on tube diameter and chirality. In general, the binding energy of the Ru atom outside the CNT is higher than that inside the CNT. Such difference becomes smaller as the tube diameter increases; for example, from 0.31 eV for (6,6) tube to 0.06 eV for (12,12) tube. Due to the curvature effect, the distribution of π electrons inside and outside the nanotube wall is uneven, resulting in higher electron density on the external surface of the CNT. When a Ru atom is adsorbed outside a CNT, the enriched π electrons due to the curvature effect enhance the hybridization between the π electrons from carbon and the *d* electrons from Ru, leading to a higher binding energy for exohedral adsorption.

The defects on CNTs also play an important role on the adsorption of transition metals. In a work of Shi *et al.*,[374] first principles methods were applied to study the interaction of Stone–Wales defects in a SWCNT with eight different metal atoms (Fe, Ni, Co, Ti, Cu, Al, Mg, and Mo), frequently used as catalysts in the growth of CNTs. They found that among the eight metals studied only Mo and Ti always exhibited a positive cohesive energy, which indicates that there was no bonding of these metal atoms to CNT. Cohesive energies for Fe, Co, and Ni were relatively low, evidencing a strong attractive interaction with CNTs. In fact, these metals are regularly used as catalysts for the catalytic growth of CNTs. Several works have evidenced that interaction of Ni atoms with intrinsic defects (vacancies and Stone–Wales) in SWCNTs enhances the Ni binding energies.[52,375,376] It was found that the orientation of the SW-defects is closely related to the nature of the Ni adsorption, rather than to tube chirality. The binding affinity trend of metal species toward the pristine and defective SWCNTs is in the order of $M^+ > M^- > M$.[377] This implies that the transfer of electron density between metal species and the nanotubes, the electrostatic attraction, and the Pauli repulsion play an important role in the M-SCWNT system.

In conclusion, single transition metal atoms are generally chemisorbed on graphene and CNT surfaces, except for the filled or near-filled d-orbital metals. Metals such as Cu, Ag or Au show low energy binding, which is between physisorption and chemisorption. The binding geometry is usually on the H_6-site, but also depends on the valence electronic configuration of the metal atoms. Finally, the adsorption configuration and binding energy of transition metals is similar in graphene and CNT; however, when the CNT curvature increases (smaller radius) the interaction between the atom and the CNT increases. Similarly, the adsorption of the adatom outside of small diameter CNT is stronger than inside of the CNT. The presence of defects also increases the binding of transition metals to the carbon support.

Clusters. CNTs and graphene interacting with metal nanoparticles are gaining considerable interest as sensing materials and catalysts. The metal–CNT interface may constitute the central part of these applications. Platinum, palladium and gold clusters are the most studied systems due to their applications in catalysis and are discussed in the following three subsections, followed by a brief section covering other clusters.

Platinum Clusters

Okamoto used DFT calculations to examine the interface between graphene and a icosahedral Pt_{13} cluster.[378] The calculations showed that the Pt_{13} cluster prefers to adsorb in a top site with the face of the icosahedron in contact with the graphene, with adsorption energy of 2.08 eV. These results indicate that the interface of the metal clusters and graphene are stabilized when the metal atoms are just above the carbon atoms, as can be seen in Figure 3.20.

First-principle calculations based on DFT have been applied to investigate the interactions between Pt_n (n = 1–4) clusters and a graphene sheet.[379] For one platinum adatom (Pt_1) the adsorption energy is −2.17 eV on a *B*-site. $Pt_{2\text{-}4}$ clusters also prefer to adsorb near *B*-sites. In the non-defective graphene, the interfacial interaction between Pt_n and graphene becomes weaker as the number of Pt atoms in clusters increases; 2.17 eV per atom for Pt_1, 0.63 eVper atom for Pt_2, and 0.15 eV per atom for a triangular Pt_3. However, the tetrahedral Pt_4 cluster strongly interacts with the non-defective graphene sheet in comparison with the triangular Pt_3 cluster.

Similarly, Dai *et al.* studied, by first principles total energy calculations, the formation and structures of Pt clusters on graphene.[380] The most stable adsorption site of a single Pt atom is the *B*-site. The stable atomic configuration of the Pt_2 is on the *B*-sites as well. As for Pt_3, which prefers a linear configuration, the Pt atoms are located on the *B*-sites and E_{form} is 3.622 eV per atom. For Pt_4 clusters, the tetrahedral configuration is the most stable among the three considered. The adsorption of a tetrahedral cluster is more stable than that of the quadrilateral and linear Pt_4. This is because a tetrahedral Pt_4 consists of four Pt triangles and the interatomic interaction is strong. With an increasing Pt coverage, the formation energy also increases and the absorbed Pt atoms have strong interatomic interactions and preferentially form clusters on graphene. With an increasing numbers of adatoms, the adsorbed Pt can form dimers, linear atom chains and 3D tetrahedral clusters. The 3D tetrahedral cluster is more stable than the planar clusters (dimer and linear chain) on graphene.

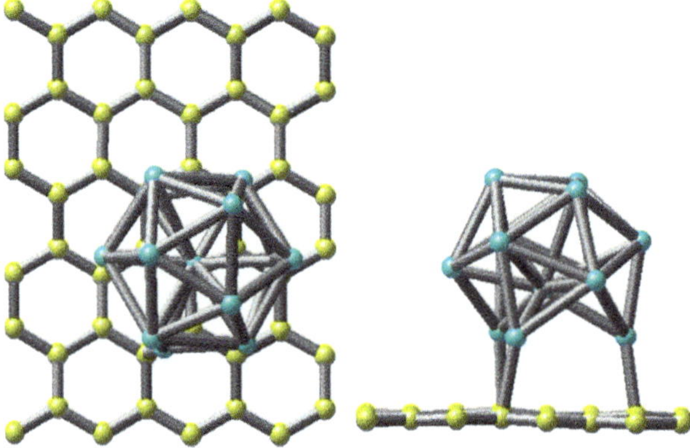

Figure 3.20 Optimized geometry of a Pt_{13} cluster on flat, defect-free graphene sheet: top view (left) and side view (right). Reprinted with permission from ref. 378. Copyright 2006 Elsevier B.V.

Okazaki-Maeda *et al.* studied the stable atomic configurations and the electronic structures of several Pt clusters, from 1 to 13 atoms, by first principles calculations based on DFT.[381] The results for small clusters $Pt_{1\text{-}5}$ show that the interfacial interaction energy determines the stability of the Pt/graphene system, resulting in the stabilization of the vertical adsorption of the planar cluster. However, for the bigger clusters $Pt_{>10}$ it is the cohesive energy of the free Pt cluster that determines the stability of the Pt/graphene system, resulting in the stabilization of the 3D-cluster adsorption.

The defective sites on graphene sheets, the presence of five- and seven-member rings or the presence of doping atoms on graphene have an important role on the stabilization of the Pt nanoparticles. Their presence increased the adsorption energy of the Pt clusters. Okamoto calculated a significant increase of the adsorption energy between a graphene vacancy and an icosahedral Pt_{13} cluster.[378] The adsorption energy of the Pt_{13} cluster on a defect-free graphene was calculated to be 2.08 eV, while on a defective graphene sheet it was 4.71 eV. The presence of five- or seven-member rings on the graphene also raised the adsorption energy to 3.38 and 3.30 eV, respectively. A first principles investigation aiming at controlling the stabilization and catalytic activity of metal nanoclusters supported on graphene by tuning the mechanical strain in graphene was reported.[382] It was shown that a relatively modest tensile strain (10%) applied to graphene greatly increases the adsorption energies of various kinds of metal clusters (Pt_4, Ag_7, Pd_9, Al_{13}, and Au_{16}) by at least 100%, suggesting the greatly strain-enhanced stabilization of these metal clusters on graphene, which is highly desired for graphene-based catalysis.

Fampiou and Ramasubramaniam performed a systematic study of the adsorption energetics, structural features, and electronic structure of platinum nanoclusters $Pt_{1\text{-}4}$ and Pt_{13} supported on both defective and defect-free graphene using a combination of DFT and bond-order potential calculations.[383,384] The Pt clusters bind strongly to the defective graphene. The stronger bonds are at the unreconstructed divacancy, followed by the 5-8-5 defect (see Table 3.6 for adsorption energies).

Table 3.6 Optimal adsorption energies for Pt adatom, Pt dimer, Pt trimer, Pt tetramers and Pt_{13} clusters on pristine graphene, single vacancy, unreconstructed divacancy, 5-8-5 reconstructed divacancy, and 555-777 reconstructed divacancy for both structural relaxation of high-symmetry clusters and annealed and relaxed clusters. EP results are in parentheses. Adapted from ref. 384.

Substrate	Pt adatom E_{ads}/eV	Pt dimer E_{ads}/eV	Pt trimer E_{ads}/eV	Pt tetramer E_{ads}/eV	Pt_{13} cluster E_{ads}/eV
Pristine	−1.57 (−1.99)	−0.78 (−1.94)	−1.35 (−2.54)	−1.13 (−1.63)	−0.25 (−3.11)
Vacancy	−7.45 (−6.68)	−7.26 (−6.93)	−7.61 (−6.52)	−7.27 (−5.51)	−4.67 (−5.86)
Divacancy	−6.97 (−8.60)	−6.40 (−8.66)	−6.33 (−7.95)	−7.06 (−9.26)	−6.59 (−9.58)
5–8–5	−6.12 (−6.94)	−4.14 (−7.34)	−5.54 (−7.50)	−6.01 (−7.4)	−6.28 (−7.33)
555–777	−2.38 (−2.27)	−2.43 (−2.21)	−5.15 (−2.05)	−5.33 (−3.35)	−2.07 (−4.66)

In general the clusters lose their high-symmetry when they bind to the graphene. For Pt_{13} clusters there is no resemblance to their high-symmetry icosahedral or cuboctahedral configurations. Instead, these are low-symmetry, open shapes that are more strongly adsorbed (by several eV in some cases) and are thermodynamically more stable than the structures obtained from relaxation alone (Figure 3.21). The formation of strong bonds with the carbon substrate also increases the Pt–Pt bond length in a cluster. Finally, the electronic structure studies reveal a clear tendency for charge transfer from Pt clusters to the graphene substrate directly related to the strength of binding to the defect.

The effect of the presence of doping atoms on graphene on the adsorption of Pt clusters on the carbon surface was studied by Dai *et al.*[385] The authors studied the adsorption of Pt clusters on graphene and doped graphene (nitrogen-, boron- and silicon-doped) using DFT. All doping atoms enhance the interaction between Pt and graphene. The adsorption energy of Pt clusters on Si-doped graphene is much higher than those on the N- and B-doped graphene. For instance, the adsorption energy on pristine and on Si-doped graphene for a Pt_2 cluster is 2.42 eV and 4.54 eV, respectively; and for a Pt_4 cluster 4.26 eV and 4.63 eV, respectively. Similar

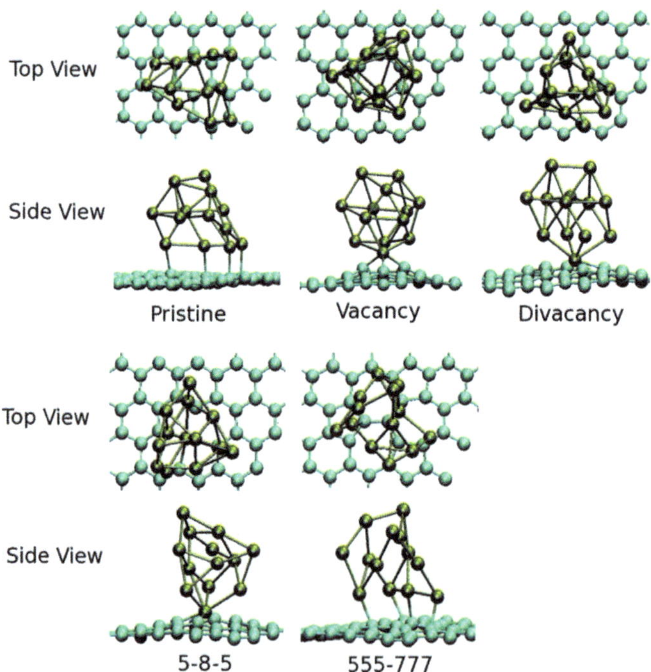

Figure 3.21 Selected low-energy configurations obtained by DFT structural relaxation of EP-based MD annealing of Pt_{13} clusters on graphene. Cyan and gold spheres represent C and Pt atoms, respectively. Reprinted with permission from ref. 384. Copyright 2012, American Chemical Society.

results were reported by Jhi *et al.* for adsorption of a Pt_{13} cluster. Most of the Pt atoms in Pt_{13} that bond to graphene sit at the carbon bridge sites except in pyridine-like N-doped graphene. Due to a mismatch of the bond length between Pt atoms and the carbon bridge, Pt clusters change their morphology on top of boron or pyridine-like nitrogen defects. The binding energy of Pt_{13} on B or N defects is about three times larger than on pristine graphene. The calculated E_b of Pt_{13} was −1.5 eV on the pristine graphene, −2.69 eV on the armchair edge, −4.33 eV on B-doped graphene, and −4.90 eV on N-doped graphene. Due to the mismatch of bond length of Pt atoms to C–C distance, the morphology of Pt_{13} on boron and nitrogen defects differs from the highly symmetric shape. The strong binding of Pt to defective graphene prevents particle migration, which is a dominant process in degradation of Pt catalysts.

For Pt_3Ni, Pt_2Ni_2, and $PtNi_3$ clusters adsorbed on graphene, DFT calculations have predicted binding energy between −0.02 and −0.95 eV.[386] The binding energy and relative stability of the supported clusters depend not just upon the composition but also upon how the cluster is adsorbed onto graphene, whether in the face-on configuration or the edge-on configuration. A PtNi cluster was placed on non-defective graphene and nitrogen-doped graphene surfaces.[387] The PtNi cluster was obtained by replacing the two inner Pt layers from the octahedral Pt_{44} nanoparticle by Ni atoms, resulting in a 1 : 1.2 Pt : Ni cluster. The PtNi cluster deposited on non-defective graphene exhibits adsorption energy equal to −0.77 eV. On the other hand, the cluster deposited on N-doped graphene exhibits a larger adsorption energy of −0.86 eV. This is related to an improved chemical reactivity of N-doped nanocarbons, rationalizing the formation of several Pt–C bonds, where mainly the C atoms around the N_3-islands participate in the chemical bonding to the PtNi cluster due to strong electron localization on pyridine-like nitrogen. From Figure 3.22a it can be observed that the graphene substrate of the PtNi/graphene system is slightly deformed after cluster adsorption, but no chemical bonds are found between the cluster and the substrate, where the cluster is located at ~3.5 Å above the graphene. In contrast, the PtNi/N-doped graphene system exhibits new chemical bonds, and the local geometry at the binding site is modified for both components. As observed in Figure 3.22b, the cluster deposited on the doped substrate exhibits one Pt–N and seven Pt–C bonds with an average bond length of 2.33 and 2.36 Å, respectively. It is also noteworthy that, in this case, the graphene does not lose its planar structure.

DFT calculations have predicted that Pt and PtRu clusters have the highest and lowest stability on boron-doped graphene and pristine graphene, respectively.[388] Even the Pt_{32} metal cluster, which is close to the equivalent of a 1 nm particle, is more stable on the boron-doped graphene when compared to the pristine graphene surface. The stability of the metal clusters on nitrogen-doped and boron–nitrogen co-doped graphene was in between that of pristine and boron-doped graphene. Overall, the stability of the PtRu clusters was greater than the Pt metal clusters on all of the surfaces.

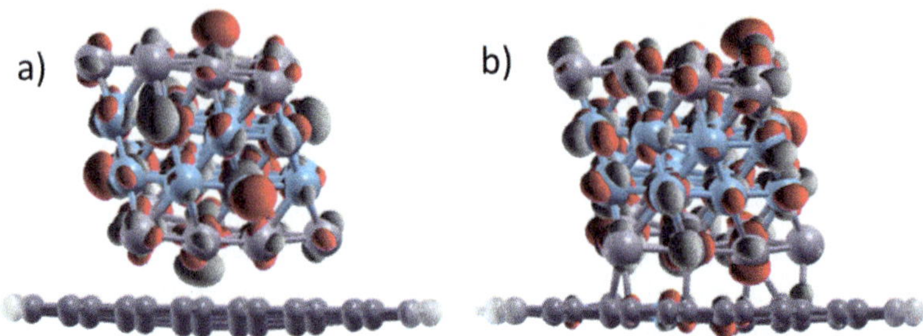

Figure 3.22 Isosurface wave functions of the HOMO of (a) PtNi/graphene, and (b) PtNi–N-doped graphene; the isosurfaces value is ±0.03 e Å^{-3}. Reprinted with permission from ref. 387. Copyright 2014, American Chemical Society.

The adsorption of Pt clusters on CNT has also been intensively studied. Adsorption geometries and binding affinities of metal NPs onto CNTs have been investigated through DFT calculations.[389] Clusters of 13 metal atoms were used as models for metal NPs. Platinum particles strongly chemisorb onto the CNT surface. For monolayer or multilayers of metal atoms on CNTs, most of the metal binding arises from metal–metal interaction rather than metal–graphene interaction.[225] This is due to high cohesive energies of the metals in the bulk crystalline state. If only the metal–graphene part of the interaction is considered, the binding for Pt falls rapidly with layer thickness, indicating perhaps a possible instability of the Pt films beyond a certain thickness. A critical cluster size of a few tens of metal atoms has been determined, below which the metal should wet a CNT surface uniformly, and above which non-uniform clustering is likely. Using a simple model the authors have shown that the critical cluster size for Pt is ~23, which is smaller than that for Pd (~30), implying a higher propensity of Pt to form a non-uniform coating, unless it is deposited in the form of ultrafine nanoparticles.

A first principles electronic structure calculation was performed to explore stability and electronic structure of Pt_{13} cluster adsorbed on CNTs.[390,391] The bridge adsorption sites on the outer wall of CNTs are found to be favorable for Pt atoms. The stability of the Pt_{13} cluster on CNT (*ca.* 4.0 eV) is enhanced significantly compared with that on graphene. The increase of adsorption energy of the Pt_{13} cluster on CNTs can be explained in term of geometric factors. Due to the curvature-induced pyramidalization and misalignment of π-orbitals of carbon atoms in SWCNTs, carbon atoms on the outside wall of the SWCNTs have more sp^3 character than those on a flat graphene sheet. In addition, the nature of Pt–C bonding is hybridization between d-states of Pt atoms and p-states of adjacent C atoms. As a result, Pt atoms can strongly interact with carbon atoms in the case of CNTs, leading to an increase in the adsorption energy. CNTs with high curvature promotes the overlap between p and d wave functions, thus the adsorption energy of a single Pt atom on SWCNTs increases linearly with the curvature.[392] In the case of Pt_{13} cluster, three Pt atoms prefer to adsorb on bridge sites resulting in a deformation of Pt cluster. High curvature makes Pt–C bonding stronger and the larger deformation results in a smaller binding energy of the Pt cluster. Hence, it is possible to predict an optimal curvature of SWCNTs for the adsorption of Pt_{13} cluster on SWCNTs. For a (10,0) semi-conducting SWCNTs, the highest adsorption energy (4.1 eV) is at around 0.06 ($Å^{-1}$) curvature, which corresponds to a semi-conducting SWCNT with a 16.66 Å diameter. It was also reported that charge redistribution among Pt atoms takes place. The electronic structure of carbon supported Pt cluster exhibits metallic characteristics, similar to a Pt metal surface. It was also found that the electronic band structures of Pt/CNT systems are very sensitive to the small changes in the geometries of Pt clusters and chains.[393] In some cases, metallic (5,5)-SWCNT becomes a small-gap semiconducting nanotube with adsorbed Pt clusters and chains. A study investigated the dynamical behavior of different size Pt clusters on different CNTs and finds that the clusters show different movement behaviors on (5,5) and (9,0) CNTs.[394] At temperatures lower than 300 K, Pt cluster movement is not significant. At temperatures higher than 300 K, an obvious increase in movement is noticed. In addition, the Pt_4 and Pt_6 clusters can easily diffuse along the tangent direction of the zigzag CNT. With an increase in cluster size, it is difficult for the Pt cluster to diffuse on the CNT. This suggests that Pt_{13} is the critical cluster size for the stable adsorption of a Pt cluster on the CNT. Molecular dynamic simulations have been used to investigate the effect of the magnitude of the metal–carbon interaction on the structure of supported metal NPs.[395] The morphology of Pt NPs of 130, 249 and 498 atoms supported on graphite and various bundles of CNTs was studied. For the larger NP it was found that, although the details of platinum–carbon interactions are important for correctly capturing the morphological details, the morphology of the support is the primary factor that determines such features. Platinum–carbon interactions affect more significantly the results obtained for metal NPs supported by CNT bundles. In this case, it was found that the deviations become significant for small supported NPs, as well as for NPs of any size supported on CNTs of small diameter.

A Monte Carlo method has been performed to simulate the thermal evolution of an icosahedral Pt_{55} cluster encapsulated in (15, 15) and (20, 20) SWCNTs.[396] A melting-like structural transformation is found for the icosahedral clusters encapsulated in SWCNTs. The melting-like transformation temperatures of the icosahedral clusters encapsulated in (15,15) and (20,20) SWCNTs are estimated from the fluctuations of the total potential energy, and are 280 and 320 K, respectively. The simulations indicate that the melting-like transformation temperature for the encapsulated icosahedral clusters increases with the pore size of SWCNTs. At higher temperatures, a stacked structure in layers is found for the encapsulated icosahedral Pt_{55} clusters. The adhesion of various sizes of Pt clusters on metallic (5,5) CNTs with and without point defects has been investigated by means of DFT.[397] The calculations show that the binding energies of Pt_n (n = 1–6) clusters on the defect free CNTs are more than 2.0 eV. However, the binding energies are increased more than three times on the point defective CNTs. The stronger orbital hybridization between the Pt atom and the carbon atom explains larger charge transfers on the defective CNTs than on the defect free CNTs, which allows strong interactions between Pt clusters and CNTs.

Molecular dynamics simulations have been performed to investigate the structure of a Pt_{100} cluster adsorbed on the three distinct sides of a carbon platelet (a mimic of CNFs).[398] In the simulations, carbon platelet edges both with and without hydrogen terminations have been studied. It is found that the initial mismatch between the atomic structure of the platelet edge and the adsorbed face of the Pt_{100} cluster leads to desorption of a few Pt atoms from the cluster and the subsequent restructuring of the cluster. Consequently, the average Pt–Pt bond length is enlarged in agreement with experimental results. This change in the bond length could play a significant role in the enhancement of the catalytic activity, which is demonstrated by studying the changes in the bond order of the platinum atoms. The effect of the curvature of the carbon support material has been investigated by molecular dynamics simulations of a Pt_{100} and a Ni_{100} cluster adsorbed on fishbone-like carbon nanofibres.[399] Carbon nanocones both with and without hydrogen termination have been considered. Without hydrogen termination, significant differences are found between adsorbed Pt and Ni clusters for the dependence of the bond strain on the curvature. For instance, the support curvature does not seem to have an appreciable effect on the interatomic distances in adsorbed Pt clusters, while increased bond strain is observed for Ni clusters with increasing curvature. Since the bond length is related to a d-band shift as well as to a change in the bond order, it is concluded that the catalytic performance of Ni clusters can be enhanced by optimizing the curvature of the support material. In general, hydrogen termination attenuates the degree of metal–metal bond strain. Molecular dynamics simulations were also performed to examine the effects of the variable morphologies of herringbone-type carbon nanofibers (h-CNFs) on the microstructures of supported Pt_{100} clusters.[400] Four h-CNF cone-helix models with different basal-to-edge surface area ratios and edge plane terminations were employed. Calculated results indicate that, upon adsorption of Pt_{100} clusters, a fraction of Pt atoms migrates from the metal particles onto the h-CNFs either to accumulate at the metal–support interface or to attain a single atom adsorption on the supports. With decreasing apex angle or introduction of H termination, the Pt atoms are more likely to be coordinated to the basal planes and the binding energies of the Pt_{100} clusters to the h-CNFs are lowered, accompanied by a lower degree of the cluster reconstruction. On the contrary, if more h-CNF edge planes are exposed, a higher Pt dispersion, lower surface first-shell Pt–Pt coordination numbers, and longer Pt–Pt surface bonds are attained.

When adsorbing Pt_n (n = 1, 2, 5, 13) clusters on C_{60}, it was found that the systems gain energy when the platinum atoms aggregate on the fullerene surface, forming clusters of different sizes and symmetries.[401] Notable structural variations around the adsorption sites were obtained, consisting in expansions and contractions of the C–C, Pt–Pt, and Pt–C bond lengths as large as 7%. The adsorption energies vary in the range of 1.5–3.1 eV, and there is a notable Pt → C charge transfer (0.15e) that leads to the formation of robust Pt–C bonds. The most important features to mention are that, upon Pt_{13} attachment, global structural transformations on the platinum cluster geometry are observed and that these highly deformed structures are energetically preferred (by values as large as 1.23 eV) when compared with the more symmetric Pt_{13} structure shown in Figure 3.23c. The configurations shown in Figure 3.23a and b are very interesting from the point of view of the chemical reactivity, due to the existence of low-coordinated Pt atoms having both positive as well as negative charges, and a notable reduction in the HOMO–LUMO gap to 0.17 eV, which are two facts that define these fullerene compounds as efficient molecular attractors.

In conclusion, Pt clusters chemisorb on graphene and CNT surfaces. As observed for single adatoms, the presence of defects or doping atoms such as nitrogen, boron or silicon have an important role on the stabilization of the Pt nanoparticles, increasing the binding energies. Also the curvature increases the bonding energies making the Pt_{13} cluster on CNT more stable than on graphene. Finally, Pt does not form stable metallic films, it prefers to form clusters due to the strong interatomic interactions of the absorbed Pt atoms.

Gold clusters

Density functional calculations were performed to study the interfaces of graphene with an icosahedral cluster of Au_{13}.[378] The gold cluster binds weaker than a Pt_{13} cluster on the graphene surface (E_{ad} = 0.90 eV for gold and E_{ad} = 2.08 eV

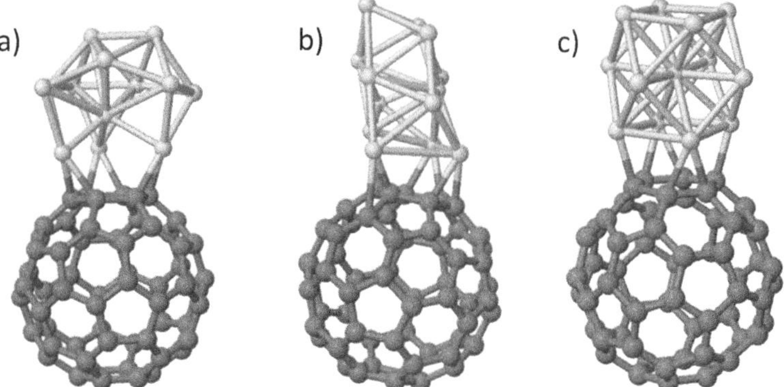

Figure 3.23 Calculated low-energy atomic configurations for various isomers of the $C_{60}Pt_{13}$ fullerene compound: (a) Pt_{13} adsorbed on a hexagonal carbon ring through the triangular facet; (b) adsorption of Pt_{13} through one of its square facets but being initially placed on top of a C–C bond of the fullerene; and (c) Pt_{13} adsorbed on a hexagonal carbon ring through the square facet.

for platinum). The results revealed that the introduction of a carbon vacancy or the presence of defects (five- and seven-member rings) into the graphene increased the adsorption energy of the cluster onto the graphene surface. A similar adsorption energy was found for a cluster of Au_{10} on graphene.[402]

The adsorption of $Au_{1\text{-}4}$ clusters on graphene, as well as their mobility and clustering from pre-adsorbed fragments have been studied by DFT.[403] The adsorption energies on graphene of all studied clusters range from −0.1 to −0.59 eV, and substantially exceed their diffusion barriers, which are 4–36 meV, only. All clusters diffuse along the C–C bonds, since none of them binds to the hollow sites in the centers of the carbon hexagons. The analysis of the densities of states shows which states contribute to the strong intra-cluster bonds, predominantly the 5d states, and which to the chemical bond between gold and carbon, *i.e.* mostly s-d_{z2} cluster states hybridizing with p_z states of carbon. The formation of the strong Au–Au bonds is the driving force behind the tendency to form bigger clusters from smaller pre-adsorbed fragments.

The adsorption of Au_n (n = 1–5) on perfect and defective graphene has been studied by DFT calculations.[404] All the clusters are bonded to graphene through an anchor atom as observed by Martin *et al.*[403] and the geometries are similar to the corresponding free-standing clusters. Charge transfer between clusters and graphene shows strong size dependency, and the amount is larger in the presence of the single vacancy on the graphene than on a pristine sheet. The interactions between Au clusters and graphene sheets with and without B- and N-doping have been studied systematically.[405] The B-substituted graphene was found to be a good support for depositing Au clusters for their stronger combining interaction compared to pristine and N-substituted graphene. This is because of the charge transfer from the graphene sheet to its supported Au clusters, but also due to its stronger interaction at the interface resulting from the doped B atom, which will facilitate more electrons being transferred to the upper surface of the Au cluster.

It has also been shown that the presence of Au_nPt_n clusters on graphene changes the electronic properties in an important way.[406] In almost all cases moderate adsorption energies corresponding to physisorption were found. It was observed that graphene can be metallic or semiconducting depending on number of Au and Pt atoms in the cluster and the charge transfer between cluster and the graphene.

A theoretical study was performed to understand the structures and properties of C_{60}-gold nanocontacts.[407] In this investigation, C_{60} was sandwiched between gold clusters. In the studied clusters, the number of Au atoms varied from 2 to 8 on each side of C_{60}. Geometries of all complexes were optimized under C_{2h} symmetry except for the C_{60}–10Au complex, for which C_2 symmetry was assumed. The complexes in which gold clusters are bonded at the top of the center of fused six-membered rings ($\eta^{2(6)}$-type coordination) are significantly more stable than those bonded at the top of the center of fused six- and five-membered rings ($\eta^{2(5)}$-type coordination). A significant amount of charge transfer from C_{60} to Au clusters was revealed, suggesting an electrostatic type of interaction between the components of the clusters. However, at the microscopic level near the nanojunction, electronic charge is transferred from the gold atom to the interacting carbon atoms of C_{60}.

Palladium clusters

DFT calculations of Pd absorbed on graphene have been performed to study the first stages of Pd coating and/or Pd cluster formation on the graphene surface.[408] It was found that palladium atoms deposited on graphene have a strong tendency to form clusters. Three-dimensional clusters are more stable than planar clusters and the transition from planar to three-dimensional Pd clusters adsorbed on graphene occurs very early as a function of cluster size, as a consequence of the strong Pd–Pd interaction. DFT calculations have been performed on a Pd_9 cluster adsorbed on two different carbonaceous supports, namely two stacked polycircumcoronene units mimicking a double layer of graphite and a portion of an armchair (6,6) CNT.[409] Calculations reveal a major geometrical distortion occurring in the Pd cluster supported on both graphite and nanotubes, which is caused by strong Pd–C interactions. The natural curvature of an armchair (6,6) carbon nanotube seems to be more suitable for supporting small palladium clusters obtained from *fcc* lattice truncation than the flat surface of graphite. To give rise to stronger interactions, the cluster bends the graphite surface so that the process of supporting palladium particles on graphite could result in a local "surface-atoms" extraction process.

The structure, binding energies, and nature of bonding of coronene•••X_2 and coronene•••X_4 (X = Pd, Ag, Au) complexes were investigated at the DFT levels.[410] The calculations demonstrated that the bonds formed by palladium complexes with the surface are considerably stronger than those of gold, which in turn are stronger than silver complexes. The silver and gold clusters bind to carbon surfaces through dispersion and charge-transfer interactions, whereas the palladium clusters are bound by dative bonds. Calculations on coronene•••X complexes indicated that the binding energies of Pd, Ag, and Au clusters increase linearly with the number of metal atoms. The same trend was observed for graphene•••X complexes. The calculated binding energy of a Pd_{13} nanoparticle on a single vacancy graphene is as high as −6.10 eV, owing to the hybridization between the *dsp* states of the Pd particles with the sp^2 dangling bonds at the defect sites.[411] The strong interaction results in the averaged *d*-band center of the deposited Pd NPs shifted away from the Fermi level from −1.02 to −1.45 eV. Doping the single vacancy graphene with B or N will further tune the average d-band center.

DFT calculations on exohedral metallofullerenes Pd_n/C_{60} show that the palladium–fullerene bond energy remains essentially constant for $n = 1$–6.[412] A novel $Pd_2(\eta^2\text{–}C_{60})$ structure with the two metal atoms bridging over a six-membered ring has been identified as the most stable arrangement of two palladium atoms on the surface of C_{60}, although entropy considerations suggest that both isolated atoms and weakly bonded metal aggregates may exist in equilibrium. Both metal atoms benefit from η^2 coordination at (6-6′) junctions as well as some metal-metal interaction. Binding of Pd atoms to the fullerene is preferred over palladium dimerization.

Geometric and electronic structures of Pd_n clusters supported on AC were analyzed using semi-empirical quantum chemical modeling for n = 1–22.[413] Qualitative reliability of the results was checked by DFT calculations for n = 1–6. Supported Pd atoms and clusters are shown to be strongly bound to unsaturated and defect surface sites. In such positions, interaction of Pd atoms with the support was much stronger than that with each other. That provides the driving force for atomic dispersity of Pd/AC catalysts. Geometry of small clusters was determined by morphology of an adhesion position. Nanosized particles form compact three-dimensional structures with close-packed triangular surfaces. AC support causes notable excitations in the electronic structure of metal atoms directly bound to the support, resulting in the direct nucleation of *fcc*-like structures. These excitations are quickly extinguished when moving far from the support surface. The combination of experimental measurements and DFT calculations of binding energies suggests that the Pd nanoclusters are much more effectively anchored on functionalized SWCNT than on the pristine one.[414] The modification of the electronic structure caused by the presence of oxygen atoms is responsible for the enhanced adsorption energy, and leads to the formation of Pd–O bonds on the functionalized SWCNT.

Other metallic clusters

The interaction of graphene with other transition metal clusters with interesting applications in catalysis has also been investigated in theoretical studies. Clusters of Co,[415] Ru,[416] and Fe[404,417,418] show strong interactions with graphene, which increase with the presence of vacancy defects. Large charge redistribution of Fe_n compared to Ni_n lead to stronger interactions in Fe_n@graphene.[418] Small clusters, Fe_{13} and Ni_{13}, prefer to stay on the triangular surface, and large clusters Fe_n and Ni_n (n = 38, 55) are favored on the hexagonal surface parallel to the pristine and strained graphene, while for defective graphene (monovacancy and divacancy), the most stable structure is the vertex atom trapping in the vacancy to form 666 and 5566 structures. In the case of Ru_{13} clusters, the presence of B doping atoms further increases the binding energy between the graphene and the cluster. The possible anchoring modes of Ru nanoparticles on CNT sidewalls were investigated by DFT calculations, by using a simple model made of a Ru_{13} cluster and a graphene monolayer.[419] The DFT calculations showed that the Ru_{13} cluster binds more strongly to a graphene double vacancy (Gr-DV) than to the pristine graphene (Gr). The adsorption energy of the icosahedral Ru_{13} cluster followed the trend: Ru/Gr-DV-$(COOH)_2$ > Ru/Gr-DV > Ru/Gr, where Gr-DV-$(COOH)_2$ is a double vacancy decorated with two carboxylic groups. On defect-free CNTs, the binding affinity with a Cu_{13} is very weak.[420] When various defects such as vacancies, substitutional nickel defects, and nickel adatoms are introduced in CNTs to increase the binding strength, the binding energies of the metal nanoclusters increase substantially irrespective of types of defects.

The adsorption of Fe, Co, and Ni trimers and tetramers on graphene was computed. It was found that for trimers the highest binding energy occurs for configurations that are perpendicularly bound and have the largest charge transfer to graphene.[421] For tetramers, the binding energy is highest for the compact configurations with the largest charge transfer to graphene. Binding is generally strongest at the hole site of the graphene lattice. The charge transfer to graphene is mainly from the base atoms. Thus, the binding energy of mixed trimers and tetramers to graphene is determined largely by the elemental identity of the base atoms. It was also shown that graphene-adsorbed mixed clusters $FeCo_2$ and $FeCo_3$ with a top Fe atom are strongly bound and have large top-atom magnetic moments.

The composition and stability of oxidized cobalt sub-nanometre clusters composed of four metal atoms supported on ultra-nanocrystalline diamond (UNCD) and alumina surfaces were studied using a combination of grazing-incidence X-ray absorption near-edge spectroscopy, grazing incidence small-angle X-ray scattering, and density functional calculations.[422] The calculations indicate that the stability of the cobalt oxide clusters on UNCD is the result of electrostatic and dispersive interactions for the pristine hydrogen-terminated surfaces and covalent bonding between the cluster and defect sites on the surfaces.

In general, the strong interatomic interactions between d-metal atoms prevent the formation of atomic coating, thus forming metal clusters. The d-metal clusters strongly interact with graphene or CNTs. The presence of defects or doping atoms increases the binding energies as observed for single adatoms.

3.4.4.4 *Adsorption of Other Metals, Metalloids and Non-Metals*

The adsorption of other metals, metalloids and non-metals (Groups III to VI) on carbon nanomaterials has been much less studied than the adsorption of alkali, alkaline-earth and d-metals. In the systematic study of Nakada *et al.* about the adsorption of several adatoms on graphene by DFT calculations the authors show that atoms of groups III to VI strongly bond to graphene.[337] Carbon shows the highest bond energy (3.43 eV) followed by sulfur (2.34 eV) and phosphorus (2.20 eV). The selected adsorption behavior of sulfur atoms on SWCNTs was studied by DFT in terms of adsorption energy and charge transfer.[423] It was found that the adsorption on the tips was energetically preferred in perfect SWCNTs; the adsorption energy and charge transfer on five-member rings with one dangling bond defects were significantly larger with respect to perfect sites on SWCNTs, suggesting selective adsorption of sulfur on these defects. Importantly, adsorbed sulfur atoms can effectively stabilize these defects through the formation of S–C bonds, which is not found in perfect SWCNTs. The rest of the elements show bond energies in between 1.07 eV and 1.86 eV. Group III elements prefer to bond on the H_6-site, except for boron which prefers the bridge sites. Group IV to VI prefers to bind on the bridge site.

The adsorption of Al, In and Pb on graphene has been studied by Liu *et al.* who show that Al and In chemisorb on the surface on the hollow site.[338] However, Pb shows a weaker bond energy (0.3 eV) and is physisorbed.

Doped bilayer-graphene is also prone to the attachment of heteroatoms, as reflected by the large adsorption energies for Al (3.91 eV per atom), Si (3.93 eV per atom), P (2.64 eV per atom), and Si (2.23 eV per atom).[424] For Al- and P-doping it was found that the structure with a short interlayer distance and without a covalent bond between the heteroatoms is more stable than that which has a covalent bond and longer interlayer distance, but if the number of P–P bonds per surface area is increased, the structure covalently bonded and with longer inter-sheet distance is expected to become more stable. Yet, for silicon the linked structure is more stable.

The structural and electronic properties of small silicon clusters (Si_n with n = 2–7) adsorbed on graphite (0001) and diamond (100) substrates were studied by DFT within periodic boundary conditions.[425] A three-layer graphene slab with 60 carbon atoms in each layer was used to represent the graphite substrate. The diamond substrate was described by an eight-layer slab involving a p(8 × 4) surface cell. Side adsorption geometry of Si_n on the substrates was considered. Maximum stability was encountered for particle site adsorption of the Si_n clusters on the graphite surface and for bridge site adsorption on diamond. Weak covalent interaction prevails between the cluster and the graphite substrate (0.38 eV for Si_7), while much stronger bonding effects are observed for adsorption on diamond (3.25 eV for Si_7), leading to considerable structural deformation of the cluster.

The adsorption of Al, C, Si, Pb and S on SWCNTs has been investigated using the first principles pseudo-potential plane wave method within DFT.[368] Carbon and silicon form rather strong bonds with the SWCNTs. The calculated binding energies are high (*i.e.*, E_b = 4.2 and 2.5 eV, respectively for carbon and silicon). The bridge site is energetically favorable for both C and Si as observed for graphene. Sulfur also strongly bonds on the *B*-site (2.8 eV) of SWCNTs. Al and Pb prefer to bind on the H_6-site with weaker binding energies (1.3 and 1.6 eV, respectively).

Shi *et al.* studied the adsorption of 10 different atoms (H, B, C, N, O, F, Si, P, Li and Na) over Stone–Wales defects in SWCNTs.[426] They observed that only the adsorption capacity of B, N, F, and Si among the 10 foreign atoms studied would benefit from this defect introduction (the presence of a heptagon reduced the binding energy by about 0.5 eV). Other atoms have a positive or near zero cohesive energy, which implied a weak bonding to the carbon nanotubes.

Interestingly, silicon doping does not destroy the structural and stability of the fullerenes while changing their electronic properties. Si adatom and dimer adsorption on C_{60} has been computed.[427] For Si adatoms, in the most stable configuration (Figure 3.24) the Si atom is placed between two six member rings of fullerene. The bond length between Si and each of two C atoms (which are between two rings) is 1.93 Å and bond length of the adjoining C–C is 1.57 Å, with both bond lengths comparable to Si–C and C–C bonds. In this configuration the Si atom carries +0.6 natural charge. The minimum-energy structures obtained for $C_{60}Si_2$ is shown in Figure 3.24. After optimization, Si–Si and adjoining C–C bond lengths are

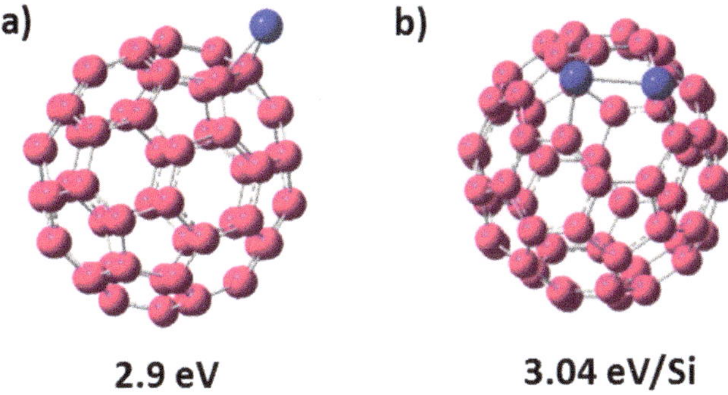

Figure 3.24 Optimized (a) SiC_{60} and (b) Si_2C_{60} configurations.

2.37 Å and 2.71 Å, respectively. In this configuration two Si atoms carry +0.47 natural charge. These bond lengths and this charge indicate that two Si atoms form bonds, and adjoining C–C bonds are broken (E_B per Si-atom = 3.04 eV).

In the case of rare earth metals on graphene, the adsorption energy is also generally high (above 1.56 eV) except for the half-filled and fully-filled f-shell elements Eu and Yb, which have adsorption energies of 0.87 and 0.35 eV, respectively.[224] The diffusion barrier of Nd, Sm, and Gd adatoms are generally higher (0.3 eV) compared to that of Eu, Dy, and Yb adatoms (0.15 eV). The nature (*i.e.*, ionic or covalent) and the strength of the interaction can be qualitatively understood by the analysis of bonding electron distributions The bonding charge distribution (expect for Yb) is intermediate between the alkali and 3*d*-transition metal. The calculations show that the hollow site of graphene is the energetically favorable adsorption site for rare earth adatoms adsorption.[428] The formation of flat Eu islands on graphene can be attributed to its low diffusion barrier and relatively larger ratio of adsorption energy to its bulk cohesive energy. The interactions between the Nd and Gd adatoms and graphene cause noticeable in-plane lattice distortions in the graphene layer. A single Ce atom prefers to adsorb strongly at hollow site of SWCNTs, being similar to the case of Eu doped SWCNTs, where the SWCNTs are simulated under periodic boundary conditions.[429] The Ce–C distance is 2.47 Å with the binding strength E_b = 3.13 eV.

3.5 Concluding Remarks and Future Perspectives

Adsorption phenomena play a pivotal role in catalysis science, from catalyst preparation to reactivity. For catalyst preparation, the adsorption of catalyst precursor and also of the metallic phase on the support should be considered since they will directly impact catalyst dispersion and also stability. Although the literature on adsorption of metals (adatoms and clusters) on nanostructured carbon materials is relatively well documented (particularly for sp^2 and $sp^{2+\delta}$ carbon nanomaterials), the studies on adsorption of metallic salts/complexes is relatively limited, and should be encouraged. Some work should also be devoted to the characterization and modelling of carbon onions and nanodiamonds, so that theoretical adsorption studies could be performed. Adsorption of reactants is also crucial during catalysis. This is particularly true if the carbon material is directly used as catalyst, but also if it is used as a support, since the local reactant concentration around the active phase can be modified upon adsorption on the support. For CNTs, more studies should be devoted to exohedral/endohedral adsorption. Whatever the nature of the nanostructured carbon materials, the adsorption is most of the time strengthened if defects are introduced on their surface. Except for fullerenes, all the synthesized nanostructured carbon materials are not perfect structures and present many defects. In addition, in a single class of material the nature and concentration of defects will be different, due to (often subtle) differences in their synthesis. These considerations should be taken into account if comparison of adsorption on different materials is the aim. Foreign atoms such as N, B, S, P can also be incorporated into nanostructured carbon materials as *substitutional* impurities. In this case, the impurity atom replaces one or two carbon atoms. These doped-carbon materials present specific adsorption properties that make them particularly attractive as catalysts or catalyst supports.

References

1. S.-J. Park and S.-Y. Lee, in *Novel Carbon Adsorbents*, ed. J. M. D. Tascón, Elsevier, Oxford, 2012, pp. 435–467.
2. Y. Ding, X.-b. Yang and J. Ni, *Front. Phys. China*, 2006, **1**, 317–322.
3. R. R. Bacsa, C. Laurent, A. Peigney, W. S. Bacsa, T. Vaugien and A. Rousset, *Chem. Phys. Lett.*, 2000, **323**, 566–571.
4. L. R. Radovic and B. Bockrath, *J. Am. Chem. Soc* 2005, **127**, 5917–5927.
5. M. Acik and Y. J. Chabal, *Jpn. J. Appl. Phys.*, 2011, **50**, 070101.
6. F. Banhart, J. Kotakoski and A. V. Krasheninnikov, *ACS Nano*, 2010, **5**, 26–41.
7. P. Ayala, R. Arenal, M. Rümmeli, A. Rubio and T. Pichler, *Carbon*, 2010, **48**, 575–586.
8. H. Wang, T. Maiyalagan and X. Wang, *ACS Catal.*, 2012, **2**, 781–794.
9. K. F. Mak, J. Shan and T. F. Heinz, *Phys. Rev. Lett.*, 2010, **104**, 176404.
10. M. F. Craciun, S. Russo, M. Yamamoto and S. Tarucha, *Nano Today*, 2011, **6**, 42–60.
11. J. P. Olivier and M. Winter, *J. Power Sources*, 2001, **97–98**, 151–155.
12. T. Placke, V. Siozios, R. Schmitz, S. F. Lux, P. Bieker, C. Colle, H. W. Meyer, S. Passerini and M. Winter, *J. Power Sources*, 2012, **200**, 83–91.
13. M. De La Pierre, M. Bruno, C. Manfredotti, F. Nestola, M. Prencipe and C. Manfredotti, *Mol. Phys.*, 2013, 1–10.

14. A. Dominic, D. David and U. Sergio, *J. Phys.: Condens. Matter*, 1996, **8**, 641.
15. A. A. Stekolnikov, J. Furthmüller and F. Bechstedt, *Phys. Rev. B*, 2002, **65**, 115318.
16. J. Ristein, *Appl. Phys. A*, 2006, **82**, 377–384.
17. M. Tachiki, Y. Kaibara, Y. Sumikawa, M. Shigeno, H. Kanazawa, T. Banno, K. Soup Song, H. Umezawa and H. Kawarada, *Surf. Sci.*, 2005, **581**, 207–212.
18. I. K. Snook, S. P. Russo and A. S. Barnard, *Surf. Rev. Lett.*, 2003, **10**, 233–239.
19. F. Maier, M. Riedel, B. Mantel, J. Ristein and L. Ley, *Phys. Rev. Lett.*, 2000, **85**, 3472–3475.
20. J. van der Weide, Z. Zhang, P. K. Baumann, M. G. Wensell, J. Bernholc and R. J. Nemanich, *Phys. Rev. B*, 1994, **50**, 5803–5806.
21. L.-s Li and X. Zhao, *J. Chem. Phys.*, 2011, **134**, 044711.
22. O. Echt, A. Kaiser, S. Zöttl, A. Mauracher, S. Denifl and P. Scheier, *ChemPlusChem*, 2013, **78**, 910–920.
23. F. Suárez-García, A. Martínez-Alonso and J. M. D. Tascón, in *Adsorption by Carbons*, eds. E. J. Bottani and J. M. D. Tascón, Elsevier, Amsterdam, 2008, pp. 329–367.
24. H. Cheng, A. C. Cooper, G. P. Pez, M. K. Kostov, P. Piotrowski and S. J. Stuart, *J. Phys. Chem. B*, 2005, **109**, 3780–3786.
25. S. Gotovac, H. Honda, Y. Hattori, K. Takahashi, H. Kanoh and K. Kaneko, *Nano Lett.*, 2007, **7**, 583–587.
26. R. Wesołowski, S. Furmaniak, A. Terzyk and P. Gauden, *Adsorption*, 2011, **17**, 1–4.
27. L. Ji, Z. X. Zhang, L. M. Peng, Z. Q. Xue and J. L. Wu, *J. Phys. D: Appl. Phys.*, 2003, **36**, 3034.
28. Y. Kishimoto and K. Hata, *Surf. Interface Anal.*, 2008, **40**, 1669–1672.
29. X. M. Min, D. X. Lan and F. Cheng, *Nanoelectronics Conference*, 2008. INEC 2008. 2nd IEEE International, 2008.
30. Z. Chen, W. Thiel and A. Hirsch, *ChemPhysChem*, 2003, **4**, 93–97.
31. X. Wang, K. Wang, Q. Meng and D. Wang, *Comput. Theor. Chem*, 2014, **1027**, 160–164.
32. H. Mauser, A. Hirsch, N. J. R. E. Hommes and T. Clark, *J. Mol. Model.*, 1997, **3**, 415–422.
33. D. Srivastava, D. W. Brenner, J. D. Schall, K. D. Ausman, M. Yu and R. S. Ruoff, *J. Phys. Chem. B*, 1999, **103**, 4330–4337.
34. D. Ding, J. Wang, Z. Cao and J. Dai, *Carbon*, 2003, **41**, 579–582.
35. N. Chakrapani, Y. M. Zhang, S. K. Nayak, J. A. Moore, D. L. Carroll, Y. Y. Choi and P. M. Ajayan, *J. Phys. Chem. B*, 2003, **107**, 9308–9311.
36. J. M. Hilding and E. A. Grulke, *J. Phys. Chem. B*, 2004, **108**, 13688–13695.
37. T. Humberto, L. Ruitao, T. Mauricio and S. D. Mildred, *Rep. Prog. Phys.*, 2012, **75**, 062501.
38. R. Martinazzo, S. Casolo and G. F. Tantardini, in *GraphITA 2011*, eds. L. Ottaviano and V. Morandi, Springer Berlin Heidelberg, 2012, pp. 137–145.

39. V. N. Mochalin, O. Shenderova, D. Ho and Y. Gogotsi, *Nat. Nano.*, 2012, 7, 11–23.
40. J. Rossato, R. J. Baierle, A. Fazzio and R. Mota, *Nano Lett.*, 2004, **5**, 197–200.
41. I. Gerber, M. Oubenali, R. Bacsa, J. Durand, A. Gonçalves, M. F. R. Pereira, F. Jolibois, L. Perrin, R. Poteau and P. Serp, *Chem. Eur. J.*, 2011, **17**, 11467–11477.
42. S. Berber and A. Oshiyama, *Phys. B*, 2006, **376–377**, 272–275.
43. J. Yuan and K. M. Liew, *Carbon*, 2009, **47**, 1526–1533.
44. W. Orellana and P. Fuentealba, *Surf. Sci.*, 2006, **600**, 4305–4309.
45. F. P. Larkins, *J. Phys. C: Solid State Phys.*, 1973, **6**, L345.
46. L. Z.-H. Wang Kai-Yue, T. Yu-Ming, Z. Yu-Mei, Z. Yuan-Yuan and C. Yue-Sheng, *Acta Phys. Sin.*, 2013, **62**, 67802–067802.
47. A. S. Barnard and M. Sternberg, *J. Comput. Theor. Nanosci.*, 2008, **5**, 2089–2095.
48. M. W. Doherty, N. B. Manson, P. Delaney, F. Jelezko, J. Wrachtrup and L. C. L. Hollenberg, *Phys. Rep.*, 2013, **528**, 1–45.
49. A. J. Stone and D. J. Wales, *Chem. Phys. Lett.*, 1986, **128**, 501–503.
50. H. F. Bettinger, *J. Phys. Chem. B*, 2005, **109**, 6922–6924.
51. S. Picozzi, S. Santucci, L. Lozzi, L. Valentini and B. Delley, *J. Chem. Phys.*, 2004, **120**, 7147–7152.
52. S. H. Yang, W. H. Shin and J. K. Kang, *J. Chem. Phys.*, 2006, **125**, 084705.
53. J. L. Rivera, J. L. Rico and F. W. Starr, *J.Phys. Chem. C*, 2007, **111**, 18899–18905.
54. C. Ataca, E. Akturk, H. Sahin and S. Ciraci, *J. Appl. Phys.*, 2011, **109**, 013704.
55. M. Tsuda, M. Hata and S. Oikawa, *Appl. Surf. Sci.*, 1996, **107**, 116–121.
56. F. Bechstedt, W. G. Schmidt and A. Scholze, *EPL (Europhysics Letters)*, 1996, **35**, 585.
57. K. Alexander, *Int. J. Electrochem. Sci.*, 2007, **2**, 355–385.
58. X.-H. Li and M. Antonietti, *Chem. Soc. Rev.*, 2013, **42**, 6593–6604.
59. J. P. Paraknowitsch and A. Thomas, *Energy Environ. Sci.*, 2013, **6**, 2839–2855.
60. J. Xu, M. C. Granger, Q. Chen, J. W. Strojek, T. E. Lister and G. M. Swain, *Anal. Chem.*, 1997, **69**, 591A–597A.
61. A. S. Barnard and M. Sternberg, *J. Phys. Chem. B*, 2005, **109**, 17107–17112.
62. S. Koizumi, T. Teraji and H. Kanda, *Diamond Relat. Mater.*, 2000, **9**, 935–940.
63. I. Sakaguchi, M. N.-Gamo, Y. Kikuchi, E. Yasu, H. Haneda, T. Suzuki and T. Ando, *Phys. Rev. B*, 1999, **60**, R2139–R2141.
64. A. S. Barnard and M. Sternberg, *Diamond Relat. Mater.*, 2007, **16**, 2078–2082.
65. A. H. Nevidomskyy, G. Csányi and M. C. Payne, *Phys. Rev. Lett.*, 2003, **91**, 105502.
66. B. R. Puri, ed. P. L. Walker, Marcel Dekker, New York, 1970, vol. 6, p. 191.

67. R. E. Smalley, in *Fullerenes., American Chemical Society*, 1992, vol. 481, pp. 141–159.
68. F. Gao, G.-L. Zhao, S. Yang and J. J. Spivey, *J. Am. Chem. Soc.*, 2012, **135**, 3315–3318.
69. A. W. Robertson, A. Bachmatiuk, Y. A. Wu, F. Schäffel, B. Rellinghaus, B. Büchner, M. H. Rümmeli and J. H. Warner, *ACS Nano*, 2011, **5**, 6610–6618.
70. A. T. Blumenau, M. I. Heggie, C. J. Fall, R. Jones and T. Frauenheim, *Phys. Rev. B*, 2002, **65**, 205205.
71. A. T. Blumenau, R. Jones, T. Frauenheim, B. Willems, O. I. Lebedev, G. Van Tendeloo, D. Fisher and P. M. Martineau, *Phys. Rev. B*, 2003, **68**, 014115.
72. C. Pantea, J. Gubicza, T. Ungár, G. A. Voronin, N. H. Nam and T. W. Zerda, *Diamond Relat. Mater.*, 2004, **13**, 1753–1756.
73. Y. Wang and J. T. W. Yeow, *J. Sens.*, 2009, **2009**, 493904.
74. Y. Liu, X. Dong and P. Chen, *Chem. Soc. Rev.*, 2012, **41**, 2283–2307.
75. G. Zollo and F. Gala, *J. Nanomater.*, 2012, **2012**, 152489.
76. C.-W. Chen, M.-H. Lee and S. J. Clark, *Diamond Relat. Mater.*, 2004, **13**, 1306–1313.
77. L. Ji, Z. X. Zhang, L. M. Peng, Z. Q. Xue and J. L. Wu, *J. Phys. D: Appl. Phys.*, 2003, **36**, 3034.
78. E.-C. Lee, Y. S. Kim, Y. G. Jin and K. J. Chang, *Phys. Rev. B*, 2002, **66**, 073415.
79. J. A. Alonso, J. S. Arellano, L. M. Molina, A. Rubio and M. J. Lopez, *IEEE Trans. Nanotechnol.*, 2004, **3**, 304–310.
80. Y. Miura, H. Kasai, W. Dino, H. Nakanishi and T. Sugimoto, *J. Appl. Phys.*, 2003, **93**, 3395–3400.
81. S. Patchkovskii, J. S. Tse, S. N. Yurchenko, L. Zhechkov, T. Heine and G. Seifert, *Proc. Natl. Acad. Sci. U. S. A*, 2005, **102**, 10439–10444.
82. V. Tozzini and V. Pellegrini, *J. Phys. Chem. C*, 2011, **115**, 25523–25528.
83. V. Barone, J. Heyd and G. E. Scuseria, *J. Chem. Phys.*, 2004, **120**, 7169–7173.
84. S. Casolo, O. M. Lovvik, R. Martinazzo and G. F. Tantardini, *J. Chem. Phys.*, 2009, **130**, 054704.
85. D. Stojkovic, P. Zhang, P. E. Lammert and V. H. Crespi, *Phys. Rev. B*, 2003, **68**, 195406.
86. P. Ruffieux, O. Gröning, M. Bielmann, P. Mauron, L. Schlapbach and P. Gröning, *Phys. Rev. B*, 2002, **66**, 245416.
87. M. Yoon, S. Yang, E. Wang and Z. Zhang, *Nano Lett.*, 2007, **7**, 2578–2583.
88. M. D. Ganji, M. Asghary and A. A. Najafi, *Commun. Theor. Phys.*, 2010, **53**, 987.
89. E. Daykova, S. Pisov and A. Proykova, in *Carbon Nanotubes*, eds. V. Popov and P. Lambin, Springer Netherlands, 2006, vol. 222, pp. 209–210.
90. A. Kaiser, S. Zöttl, P. Bartl, C. Leidlmair, A. Mauracher, M. Probst, S. Denifl, O. Echt and P. Scheier, *ChemSusChem*, 2013, **6**, 1235–1244.

91. K. E. Whitener, R. J. Cross, M. Saunders, S.-i. Iwamatsu, S. Murata, N. Mizorogi and S. Nagase, *J. Am. Chem. Soc.*, 2009, **131**, 6338–6339.
92. L. Firlej and B. Kuchta, *Colloids Surf., A*, 2004, **241**, 149–154.
93. D.-L. Chen, W. Al-Saidi and K. Johnson, *APS March Meeting 2011*, Dallas, Texas, 2011.
94. S. Jalili and R. Majidi, *Phys. E.*, 2007, **39**, 166–170.
95. V. V. Simonyan, J. K. Johnson, A. Kuznetsova and J. J. T. Yates, *J. Chem. Phys.*, 2001, **114**, 4180–4185.
96. L. Sheng, Y. Ono and T. Taketsugu, *J. Phys. Chem. C*, 2010, **114**, 3544–3548.
97. Z. Jijun, B. Alper, H. Jie and L. Jian Ping, *Nanotechnology*, 2002, **13**, 195.
98. O. Leenaerts, B. Partoens and F. M. Peeters, *Phys. Rev. B*, 2008, **77**, 125416.
99. Z. B. Nojini and S. Samiee, *J. Phys. Chem. C*, 2011, **115**, 12054–12063.
100. D. A. Britz and A. N. Khlobystov, *Chem. Soc. Rev.*, 2006, **35**, 637–659.
101. D. Quiñonero, A. Frontera and P. M. Deyà, *J. Phys. Chem. C*, 2012, **116**, 21083–21092.
102. I. D. Mackie and G. A. DiLabio, *Phys. Chem. Chem. Phys.*, 2011, **13**, 2780–2787.
103. A. J. Du, C. H. Sun, Z. H. Zhu, G. Q. Lu, V. Rudolph and C. S. Sean, *Nanotechnology*, 2009, **20**, 375701.
104. W.-L. Yim, O. Byl, J. T. Yates and J. K. Johnson, *J. Chem. Phys.*, 2004, **120**, 5377–5386.
105. Q.-H. Yang, P.-X. Hou, S. Bai, M.-Z. Wang and H.-M. Cheng, *Chem. Phys. Lett.*, 2001, **345**, 18–24.
106. K. Masenelli-Varlot, E. McRae and N. Dupont-Pavlovsky, *Appl. Surf. Sci.*, 2002, **196**, 209–215.
107. F. Ashrafi, A. S. Ghasemi, S. A. Babanejad and M. Rahimof, *Res. J. Appl. Sci., Eng. Technol.*, 2010, **2**, 547–551.
108. S. Furmaniak, A. P. Terzyk, P. A. Gauden and G. Rychlicki, *J. Colloid Interface Sci.*, 2006, **295**, 310–317.
109. S. Furmaniak, A. P. Terzyk, P. A. Gauden, P. J. F. Harris, M. Wiśniewski and P. Kowalczyk, *Adsorption*, 2010, **16**, 197–213.
110. C. H. Sun, F. Li, H. M. Cheng and G. Q. Lu, *Appl. Phys. Lett.*, 2005, **87**, 243109.
111. Y.-C. Chiang and C.-Y. Lee, *J. Mater. Sci.*, 2009, **44**, 2780–2791.
112. M. Muris, N. Dupont-Pavlovsky, M. Bienfait and P. Zeppenfeld, *Surf. Sci.*, 2001, **492**, 67–74.
113. R. B. Hallock and Y. H. Kahng, *J. Low Temp. Phys.*, 2004, **134**, 21–30.
114. Z. Jie, W. Yao, L. Wenjun, W. Fei and Y. Yangxin, *Nanotechnology*, 2007, **18**, 095707.
115. L. Heroux, V. Krungleviciute, M. M. Calbi and A. D. Migone, *J. Phys. Chem. B*, 2006, **110**, 12597–12602.
116. A. Fujiwara, K. Ishii, H. Suematsu, H. Kataura, Y. Maniwa, S. Suzuki and Y. Achiba, *Chem. Phys. Lett.*, 2001, **336**, 205–211.

117. C.-M. Yang, H. Kanoh, K. Kaneko, M. Yudasaka and S. Iijima, *J. Phys. Chem. B*, 2002, **106**, 8994–8999.
118. S. Iwata, Y. Sato, K. Nakai, S. Ogura, T. Okano, M. Namura, A. Kasuya, K. Tohji and K. Fukutani, *J. Phys. Chem. C*, 2007, **111**, 14937–14941.
119. D.-H. Yoo, G.-H. Rue, M. H. W. Chan, Y.-H. Hwang and H.-K. Kim, *J. Phys. Chem. B*, 2003, **107**, 1540–1542.
120. C. Matranga and B. Bockrath, *J. Phys. Chem. B*, 2005, **109**, 4853–4864.
121. M. D. Ellison, M. J. Crotty, D. Koh, R. L. Spray and K. E. Tate, *J. Phys. Chem. B*, 2004, **108**, 7938–7943.
122. A. G. Albesa, E. A. Fertitta and J. L. Vicente, *Langmuir*, 2009, **26**, 786–795.
123. M. Bienfait, P. Zeppenfeld, N. Dupont-Pavlovsky, M. Muris, M. R. Johnson, T. Wilson, M. DePies and O. E. Vilches, *Phys. Rev. B*, 2004, **70**, 035410.
124. W.-L. Yim, O. Byl, J. J. T. Yates and J. K. Johnson, *J. Chem. Phys.*, 2004, **120**, 5377–5386.
125. D. S. Rawat, L. Heroux, V. Krungleviciute and A. D. Migone, *Langmuir*, 2005, **22**, 234–238.
126. H. Ulbricht, G. Moos and T. Hertel, *Phys. Rev. B*, 2002, **66**, 075404.
127. H. Ulbricht, J. Kriebel, G. Moos and T. Hertel, *Chem. Phys. Lett.*, 2002, **363**, 252–260.
128. J. Hilding, E. A. Grulke, S. B. Sinnott, D. Qian, R. Andrews and M. Jagtoyen, *Langmuir*, 2001, **17**, 7540–7544.
129. P.-X. Hou, S.-T. Xu, Z. Ying, Q.-H. Yang, C. Liu and H.-M. Cheng, *Carbon*, 2003, **41**, 2471–2476.
130. P. Gauden, A. Terzyk, S. Furmaniak, M. Wiśniewski, P. Kowalczyk, A. Bielicka and W. Zieliński, *Adsorption*, 2013, **19**, 785–793.
131. J. Rouquerol, F. Rouquerol and K. S. W. Sing, *Adsorption by Powders and Porous Solids*, Elsevier Science, 1998.
132. N. Kanellopoulos, *Nanoporous Materials: Advanced Techniques for Characterization, Modeling, and Processing*, Taylor & Francis, 2011.
133. A. Dąbrowski, *Adv. Colloid Interface Sci.*, 2001, **93**, 135–224.
134. K. Yang, L. Zhu and B. Xing, *Environ. Sci. Technol.*, 2006, **40**, 1855–1861.
135. I. Johnson, R. Denoyel, J. Rouquerol and D. H. Everett, *Colloids Surf.*, 1990, **49**, 133–148.
136. C. Moreno-Castilla, *Carbon*, 2004, **42**, 83–94.
137. J. Ma, A. Michaelides, D. Alfè, L. Schimka, G. Kresse and E. Wang, *Phys. Rev. B*, 2011, **84**, 033402.
138. P. Kim, Y. Zheng and S. Agnihotri, *Ind. Eng. Chem. Res.*, 2008, **47**, 3170–3178.
139. S. Agnihotri, P. Kim, Y. Zheng, J. P. B. Mota and L. Yang, *Langmuir*, 2008, **24**, 5746–5754.
140. M. D. Ellison, A. P. Good, C. S. Kinnaman and N. E. Padgett, *J. Phys. Chem. B*, 2005, **109**, 10640–10646.
141. O. Byl, J.-C. Liu, Y. Wang, W.-L. Yim, J. K. Johnson and J. T. Yates, *J. Am. Chem. Soc.*, 2006, **128**, 12090–12097.

142. A. Striolo, A. A. Chialvo, K. E. Gubbins and P. T. Cummings, *J. Chem. Phys.*, 2005, **122**, 234712.
143. J. Marti and M. C. Gordillo, *J. Chem. Phys.*, 2001, **114**, 10486–10492.
144. Y. Maniwa, H. Kataura, M. Abe, S. Suzuki, Y. Achiba, H. Kira and K. Matsuda, *J. Phys. Soc. Jpn.*, 2002, **71**, 2863.
145. D. Mattia and Y. Gogotsi, *Microfluid. Nanofluid*, 2008, **5**, 289–305.
146. K. S. Novoselov, A. K. Geim, S. V. Morozov, D. Jiang, Y. Zhang, S. V. Dubonos, I. V. Grigorieva and A. A. Firsov, *Science*, 2004, **306**, 666–669.
147. G. R. Jenness and K. D. Jordan, *J. Phys. Chem. C*, 2009, **113**, 10242–10248.
148. R. R. Q. Freitas, R. Rivelino, F. d. B. Mota and C. M. C. de Castilho, *J. Phys. Chem. A*, 2011, **115**, 12348–12356.
149. T. O. Wehling, A. I. Lichtenstein and M. I. Katsnelson, *Appl. Phys. Lett.*, 2008, **93**, 202110–202113.
150. M. C. Gordillo and J. Martí, *J. Phys.: Condens. Matter*, 2010, **22**, 284111.
151. A. I. Kolesnikov, J.-M. Zanotti, C.-K. Loong, P. Thiyagarajan, A. P. Moravsky, R. O. Loutfy and C. J. Burnham, *Phys. Rev. Lett.*, 2004, **93**, 035503.
152. J. Rafiee, M. A. Rafiee, Z.-Z. Yu and N. Koratkar, *Adv. Mater.*, 2010, **22**, 2151–2154.
153. A. Politano, A. R. Marino, V. Formoso and G. Chiarello, *AIP Adv.*, 2011, **1**, 042130.
154. G. A. Kimmel, J. Matthiesen, M. Baer, C. J. Mundy, N. G. Petrik, R. S. Smith, Z. Dohnaálek and B. D. Kay, *J. Am. Chem. Soc.*, 2009, **131**, 12838–12844.
155. G. R. Birkett and D. D. Do, *J. Phys. Chem. C*, 2007, **111**, 5735–5742.
156. V. T. Nguyen, D. D. Do and D. Nicholson, *Carbon*, 2014, **66**, 629–636.
157. S. Ji, T. Jiang, K. Xu and S. Li, *Appl. Surf. Sci.*, 1998, **133**, 231–238.
158. S. A. Denisov, G. A. Sokolina, G. P. Bogatyreva, T. Y. Grankina, O. K. Krasil'nikova, E. V. Plotnikova and B. V. Spitsyn, *Prot. Met. Phys. Chem. Surf.*, 2013, **49**, 286–291.
159. M. Tunckol, J. Durand and P. Serp, *Carbon*, 2012, **50**, 4303–4334.
160. B. Wu, D. Hu, Y. Kuang, B. Liu, X. Zhang and J. Chen, *Angew. Chem., Int. Ed.*, 2009, **48**, 4751–4754.
161. Y. Liu, L. Yu, S. Zhang, J. Yuan, L. Shi and L. Zheng, *Colloids Surf. A*, 2010, **359**, 66–70.
162. B. Dong, Y. Su, Y. Liu, J. Yuan, J. Xu and L. Zheng, *J. Colloid Interface Sci.*, 2011, **356**, 190–195.
163. L. Zhao, Y. Li, Z. Liu and H. Shimizu, *Chem. Mater.*, 2010, **22**, 5949–5956.
164. A. I. Frolov, K. Kirchner, T. Kirchner and M. V. Fedorov, *Faraday Discuss.*, 2012, **154**, 235–247.
165. M. H. Ghatee and F. Moosavi, *J. Phys. Chem. C*, 2011, **115**, 5626–5636.
166. Y. Hernandez, V. Nicolosi, M. Lotya, F. M. Blighe, Z. Sun, S. De, M. T. B. Holland, M. Byrne, Y. K. Gun'Ko, J. J. Boland, P. Niraj, G. Duesberg, S. Krishnamurthy, R. Goodhue, J. Hutchison, V. Scardaci, A. C. Ferrari and J. N. Coleman, *Nat. Nanotechnol.*, 2008, **3**, 563–568.
167. X. Wang, P. F. Fulvio, G. A. Baker, G. M. Veith, R. R. Unocic, S. M. Mahurin, M. Chi and S. Dai, *Chem. Commun.*, 2010, **46**, 4487–4489.

168. J. Lee and T. Aida, *Chem. Commun.*, 2011, **47**, 6757–6762.
169. X. Zhou, T. Wu, K. Ding, B. Hu, M. Hou and B. Han, *Chem. Commun.*, 2010, **46**, 386–388.
170. T. T. Tung, T. Y. Kim, J. P. Shim, W. S. Yang, H. Kim and K. S. Suh, *Org. Electron.*, 2011, **12**, 2215–2224.
171. G. P. Rao, C. Lu and F. Su, *Sep. Purif. Technol.*, 2007, **58**, 224–231.
172. H. C. Choi, M. Shim, S. Bangsaruntip and H. Dai, *J. Am. Chem. Soc.*, 2002, **124**, 9058–9059.
173. C. Carrillo-Carrion, R. Lucena, S. Cardenas and M. Valcarcel, *Analyst*, 2007, **132**, 551–559.
174. X. Wang, J. Lu and B. Xing, *Environ. Sci. Technol.*, 2008, **42**, 3207–3212.
175. B. Pan and B. Xing, *Environ. Sci. Technol.*, 2008, **42**, 9005–9013.
176. F. Tournus and J. C. Charlier, *Phys. Rev. B*, 2005, **71**, 165421.
177. M. Yudasaka, J. Fan, J. Miyawaki and S. Iijima, *J. Phys. Chem. B*, 2005, **109**, 8909–8913.
178. S. Agnihotri, M. J. Rood and M. Rostam-Abadi, *Carbon*, 2005, **43**, 2379–2388.
179. S. Y. Bhide and S. Yashonath, *J. Phys. Chem. B*, 2000, **104**, 11977–11986.
180. D. L. Irving, S. B. Sinnott and A. S. Lindner, *Chem. Phys. Lett.*, 2004, **389**, 96–100.
181. J.-W. Lee, H.-C. Kang, W.-G. Shim, C. Kim and H. Moon, *J. Chem. Eng. Data*, 2006, **51**, 963–967.
182. M. Sander, Y. Lu and J. J. Pignatello, *J. Environ. Qual.*, 2005, **34**, 1063–1072.
183. A. Z. AlZahrani, *Appl. Surf. Sci.*, 2010, **257**, 807–810.
184. F. Tournus, S. Latil, M. I. Heggie and J. C. Charlier, *Phys. Rev. B*, 2005, **72**, 075431.
185. T. Kar, H. F. Bettinger, S. Scheiner and A. K. Roy, *J. Phys. Chem. C*, 2008, **112**, 20070–20075.
186. J. Lu, S. Nagase, X. Zhang, D. Wang, M. Ni, Y. Maeda, T. Wakahara, T. Nakahodo, T. Tsuchiya, T. Akasaka, Z. Gao, D. Yu, H. Ye, W. N. Mei and Y. Zhou, *J. Am. Chem. Soc.*, 2006, **128**, 5114–5118.
187. J. Björk, F. Hanke, C.-A. Palma, P. Samori, M. Cecchini and M. Persson, *J. Phys. Chem. Lett.*, 2010, **1**, 3407–3412.
188. O. V. Ershova, T. C. Lillestolen and E. Bichoutskaia, *Phys. Chem. Chem. Phys.*, 2010, **12**, 6483–6491.
189. L. M. Woods, Ș. C. Bădescu and T. L. Reinecke, *Phys. Rev. B*, 2007, **75**, 155415.
190. A. Rochefort and J. D. Wuest, *Langmuir*, 2008, **25**, 210–215.
191. S. Zhang, T. Shao, H. S. Kose and T. Karanfil, *Environ. Sci. Technol.*, 2010, **44**, 6377–6383.
192. V. A. Basiuk, *J. Phys. Chem. B*, 2004, **108**, 19990–19994.
193. C. Rajesh, C. Majumder, H. Mizuseki and Y. Kawazoe, *J. Chem. Phys.*, 2009, **130**, 124911.
194. L. Wang, D. Zhu, L. Duan and W. Chen, *Carbon*, 2010, **48**, 3906–3915.
195. J. D. Wuest and A. Rochefort, *Chem. Commun.*, 2010, **46**, 2923–2925.
196. P. A. Denis and F. Iribarne, *J. Mol. Struct.: THEOCHEM.*, 2010, **957**, 114–119.

197. E. N. Voloshina, D. Mollenhauer, L. Chiappisi and B. Paulus, *Chem. Phys. Lett.*, 2011, **510**, 220–223.
198. J. Beheshtian, A. A. Peyghan and Z. Bagheri, *Appl. Surf. Sci.*, 2012, **258**, 8980–8984.
199. M. Mezgebe, L.-H. Jiang, Q. Shen, C. Du and H.-R. Yu, *Colloids Surf. A*, 2012, **415**, 86–90.
200. C. Lu and F. Su, *Sep. Purif. Technol.*, 2007, **58**, 113–121.
201. S. Zhang, T. Shao, S. S. K. Bekaroglu and T. Karanfil, *Water Res.*, 2010, **44**, 2067–2074.
202. J. J. Davis, M. L. H. Green, H. Allen, O. Hill, Y. C. Leung, P. J. Sadler, J. Sloan, A. V. Xavier and S. Chi Tsang, *Inorg. Chim. Acta.*, 1998, **272**, 261–266.
203. C. Richard, F. Balavoine, P. Schultz, T. W. Ebbesen and C. Mioskowski, *Science*, 2003, **300**, 775–778.
204. S. S. Karajanagi, A. A. Vertegel, R. S. Kane and J. S. Dordick, *Langmuir*, 2004, **20**, 11594–11599.
205. L. Ji, W. Chen, L. Duan and D. Zhu, *Environ. Sci. Technol.*, 2009, **43**, 2322–2327.
206. C. E. Junkermeier, D. Solenov and T. L. Reinecke, *J. Phys. Chem. C*, 2013, **117**, 2793–2798.
207. A. Ferre-Vilaplana, *J. Phys. Chem. C*, 2008, **112**, 3998–4004.
208. T. Wu, X. Cai, S. Tan, H. Li, J. Liu and W. Yang, *Chem. Eng. J.*, 2011, **173**, 144–149.
209. J. Xu, L. Wang and Y. Zhu, *Langmuir*, 2012, **28**, 8418–8425.
210. P. Ganesan, R. Kamaraj and S. Vasudevan, *J. Taiwan Inst. Chem. Eng*, DOI: 10.1016/j.jtice.2013.1001.1029.
211. L. Wang, R. T. Yang and C.-L. Sun, *AIChE J.*, 2013, **59**, 29–32.
212. Y.-Q. Xiong, H.-L. Zhang, Y. Peng, C.-H. Liu, K.-F. Xie, K.-G. Zhou and Y.-H. Zhang, *Int. J. Nanosci.*, 2009, **08**, 5–8.
213. A. H. Norzilah, A. Fakhru'l-Razi, T. S. Y. Choong and A. L. Chuah, *J. Nanomater.*, 2011, **2011**, 495676.
214. S.-C. Hsu and C. L u, *J. Air Waste Manage. Assoc.*, 2009, **59**, 990–997.
215. Q. Liao, J. Sun and L. Gao, *Colloids Surf. A*, 2008, **312**, 160–165.
216. Y. Li, Q. Du, T. Liu, X. Peng, J. Wang, J. Sun, Y. Wang, S. Wu, Z. Wang, Y. Xia and L. Xia, *Chem. Eng. Res. Des.*, 2013, **91**, 361–368.
217. S. Sarkar, M. L. Moser, X. Tian, X. Zhang, Y. F. Al-Hadeethi and R. C. Haddon, *Chem. Mater*, 2013, **26**, 184–195.
218. R. R. Bacsa and P. Serp, in *Carbon Meta-nanotubes*, John Wiley & Sons, Ltd, 2011, pp. 163–221.
219. G. G. Wildgoose, C. E. Banks and R. G. Compton, *Small*, 2006, **2**, 182–193.
220. B. F. Machado and P. Serp, *Catal. Sci. Technol.*, 2012, **2**, 54–75.
221. K. Nakada and A. Ishii, in *Graphene Simulation*, ed. J. R. Gong, InTech, 2011.
222. S. Fullam, D. Cottell, H. Rensmo and D. Fitzmaurice, *Adv. Mater.*, 2000, **12**, 1430–1432.

223. L. Han, W. Wu, F. L. Kirk, J. Luo, M. M. Maye, N. N. Kariuki, Y. Lin, C. Wang and C.-J. Zhong, *Langmuir*, 2004, **20**, 6019–6025.
224. X. Liu, C.-Z. Wang, M. Hupalo, H.-Q. Lin, K.-M. Ho and M. Tringides, *Crystals*, 2013, **3**, 79–111.
225. A. Maiti and A. Ricca, *Chem. Phys. Lett.*, 2004, **395**, 7–11.
226. Y. Sasaki, R. Kitaura, Y. Yamamoto, S. Arai, S. Suzuki, Y. Miyata and H. Shinohara, *Appl. Phys. Express*, 2012, **5**, 065103.
227. D. He, K. Cheng, H. Li, T. Peng, F. Xu, S. Mu and M. Pan, *Langmuir*, 2012, **28**, 3979–3986.
228. J. C. Claussen, A. Kumar, D. B. Jaroch, M. H. Khawaja, A. B. Hibbard, D. M. Porterfield and T. S. Fisher, *Adv. Funct. Mater.*, 2012, **22**, 3399–3405.
229. G. Williams, B. Seger and P. V. Kamat, *ACS Nano*, 2008, **2**, 1487–1491.
230. C. Nethravathi, J. T. Rajamathi, N. Ravishankar, C. Shivakumara and M. Rajamathi, *Langmuir*, 2008, **24**, 8240–8244.
231. S.-M. Paek, E. Yoo and I. Honma, *Nano Lett.*, 2008, **9**, 72–75.
232. S. Guo and S. Sun, *J. Am. Chem. Soc.*, 2012, **134**, 2492–2495.
233. K. Jiang, A. Eitan, L. S. Schadler, P. M. Ajayan, R. W. Siegel, N. Grobert, M. Mayne, M. Reyes-Reyes, H. Terrones and M. Terrones, *Nano Lett.*, 2003, **3**, 275–277.
234. L.-S. Zhang, X.-Q. Liang, W.-G. Song and Z.-Y. Wu, *Phys. Chem. Chem. Phys.*, 2010, **12**, 12055–12059.
235. M. N. Groves, A. S. W. Chan, C. Malardier-Jugroot and M. Jugroot, *Chem. Phys. Lett.*, 2009, **481**, 214–219.
236. A. A. Dzhurakhalov and F. M. Peeters, *Carbon*, 2011, **49**, 3258–3266.
237. D. W. Boukhvalov, M. I. Katsnelson and A. I. Lichtenstein, *Phys. Rev. B*, 2008, **77**, 035427.
238. V. V. Ivanovskaya, A. Zobelli, D. Teillet-Billy, N. Rougeau, V. Sidis and P. R. Briddon, *Eur. Phys. J. B*, 2010, **76**, 481–486.
239. R. Balog, B. Jørgensen, J. Wells, E. Lægsgaard, P. Hofmann, F. Besenbacher and L. Hornekær, *J. Am. Chem. Soc.*, 2009, **131**, 8744–8745.
240. S. Ryu, M. Y. Han, J. Maultzsch, T. F. Heinz, P. Kim, M. L. Steigerwald and L. E. Brus, *Nano Lett.*, 2008, **8**, 4597–4602.
241. Z. Luo, T. Yu, K.-j. Kim, Z. Ni, Y. You, S. Lim, Z. Shen, S. Wang and J. Lin, *ACS Nano*, 2009, **3**, 1781–1788.
242. M. Yoon, S. Yang and Z. Zhang, *J. Chem. Phys.*, 2009, **131**, 064707.
243. M. Sankaran and B. Viswanathan, *Carbon*, 2006, **44**, 2816–2821.
244. I. López-Corral, J. de Celis, A. Juan and B. Irigoyen, *Int. J. Hydrogen Energy*, 2012, **37**, 10156–10164.
245. V. Gayathri and R. Geetha, *Adsorption*, 2007, **13**, 53–59.
246. W. Grochala and P. P. Edwards, *Chem. Rev.*, 2004, **104**, 1283–1316.
247. U. Zimmermann, N. Malinowski, U. Näher, S. Frank and T. P. Martin, *Phys. Rev. Lett.*, 1994, **72**, 3542–3545.
248. R. Loutfy and E. Wexler, in *Perspectives of Fullerene Nanotechnology*, ed. E. Ōsawa, Springer Netherlands, 2002, pp. 281–287.
249. A. Kaiser, C. Leidlmair, P. Bartl, S. Zöttl, S. Denifl, A. Mauracher, M. Probst, P. Scheier and O. Echt, *J. Chem. Phys.*, 2013, **138**, 074311.

250. S. M. Lee, Y. H. Lee, Y. G. Hwang, J. R. Hahn and H. Kang, *Phys. Rev. Lett.*, 1999, **82**, 217–220.
251. G. E. Froudakis, M. Schnell, M. Mühlhäuser, S. D. Peyerimhoff, A. N. Andriotis, M. Menon and R. M. Sheetz, *Phys. Rev. B*, 2003, **68**, 115435.
252. A. Farmany, M. Hatami and R. Sahraei, *Fullerenes, Nanotubes, Carbon Nanostruct.*, 2013.
253. M. Grujicic, G. Cao and R. Singh, *Appl. Surf. Sci.*, 2003, **211**, 166–183.
254. P. Giannozzi, R. Car and G. Scoles, *J. Chem. Phys.*, 2003, **118**, 1003–1006.
255. F. Mehmood, R. Pachter, W. Lu and J. J. Boeckl, *J. Phys. Chem. C*, 2013, **117**, 10366–10374.
256. J. Nakamura, J. Ito and A. Natori, *J. Phys.: Conf. Ser.*, 2008, **100**, 052019.
257. S. Michaelson, R. Akhvlediani, L. Tkach and A. Hoffman, *Phys. Status Solidi A*, 2012, **209**, 1683–1689.
258. H. Guang, M. Aoki, S. Tanaka and M. Kohyama, *Solid State Commun.*, 2013, **174**, 10–15.
259. D. W. Boukhvalov, S. Moehlecke, R. R. da Silva and Y. Kopelevich, *Phys. Rev. B*, 2011, **83**, 233408.
260. B. Akdim, T. Kar, X. Duan and R. Pachter, *Chem. Phys. Lett.*, 2007, **445**, 281–287.
261. X. Lu, Z. Chen and P. v. R. Schleyer, *J. Am. Chem. Soc.*, 2004, **127**, 20–21.
262. G. Lee, B. Lee, J. Kim and K. Cho, *J. Phys. Chem. C*, 2009, **113**, 14225–14229.
263. F. Cataldo, *J. Nanosci. Nanotechnol.*, 2007, **7**, 1439–1445.
264. F. Cataldo, *J. Nanosci. Nanotechnol.*, 2007, **7**, 1446–1454.
265. H. Chang, J. D. Lee, S. M. Lee and Y. H. Lee, *Appl. Phys. Lett.*, 2001, **79**, 3863–3865.
266. S. Tang and Z. Cao, *J. Chem. Phys.*, 2009, **131**, 114706.
267. M. A. Turabekova, T. C. Dinadayalane, D. Leszczynska and J. Leszczynski, *J. Phys. Chem. C*, 2012, **116**, 6012–6021.
268. Y.-H. Zhang, K.-G. Zhou, K.-F. Xie, X.-C. Gou, J. Zeng, H.-L. Zhang and Y. Peng, *J. Nanosci. Nanotechnol.*, 2010, **10**, 7347–7350.
269. S. Tang and Z. Cao, *J. Chem. Phys.*, 2011, **134**, 044710.
270. M. T. Baei, *Heteroat. Chem.*, 2013, **24**, 516–523.
271. M. D. Ganji, M. Mohseni and O. Goli, *J. Mol. Struct.: THEOCHEM.*, 2009, **913**, 54–57.
272. C. J. Tighe, M. V. Twigg, A. N. Hayhurst and J. S. Dennis, *Ind. Eng. Chem. Res.*, 2011, **50**, 10480–10492.
273. M. Toscano and N. Russo, in *Studies in Surface Science and Catalysis*, eds. A. Z. Claudio Morterra and C. Giacomo, Elsevier, 1989, vol. 48, pp. 893–902.
274. B. Sanyal, O. Eriksson, U. Jansson and H. Grennberg, *Phys. Rev. B*, 2009, **79**, 113409.
275. Y.-H. Zhang, L.-F. Han, Y.-H. Xiao, D.-Z. Jia, Z.-H. Guo and F. Li, *Comput. Mater. Sci.*, 2013, **69**, 222–228.
276. D. Borisova, V. Antonov and A. Proykova, *Int. J. Quantum Chem.*, 2013, **113**, 786–791.
277. X. Lin, J. Ni and C. Fang, *J. Appl. Phys.*, 2013, **113**, 034306.

278. B. C. Wood, S. Y. Bhide, D. Dutta, V. S. Kandagal, A. D. Pathak, S. N. Punnathanam, K. G. Ayappa and S. Narasimhan, *J. Chem. Phys.*, 2012, **137**, 054702.
279. J. H. Cho and C. R. Park, *Catal. Today*, 2007, **120**, 407–412.
280. A. A. Al-Ghamdi, E. Shalaan, F. S. Al-Hazmi, A. S. Faidah, S. Al-Heniti and M. Husain, *J. Nanomater.*, 2012, **2012**, 484692.
281. G. E. Froudakis, *Nano Lett.*, 2001, **1**, 531–533.
282. K. R. S. Chandrakumar and S. K. Ghosh, *Nano Lett.*, 2007, **8**, 13–19.
283. Z. Zhou, X. Gao, J. Yan and D. Song, *Carbon*, 2006, **44**, 939–947.
284. F. Yoshitaka and S. Susumu, *J. Phys. Conf. Ser.*, 2011, **302**, 012006.
285. I. Loápez-Corral, E. a Germaán, A. Juan, M. a. A. Volpe and G. P. Brizuela, *J. Phys. Chem. C*, 2011, **115**, 4315–4323.
286. Z. W. Zhang, J. C. Li and Q. Jiang, *J. Phys. Chem. C*, 2010, **114**, 7733–7737.
287. Z. M. Ao, Q. Jiang, R. Q. Zhang, T. T. Tan and S. Li, *J. Appl. Phys.*, 2009, **105**, 074307.
288. D. Kim, S. Lee, S. Jo and Y.-C. Chung, *Phys. Chem. Chem. Phys.*, 2013, **15**, 12757–12761.
289. D. Li, Y. Ouyang, J. Li, Y. Sun and L. Chen, *Solid State Commun.*, 2012, **152**, 422–425.
290. Y. G. Zhou, X. T. Zu, F. Gao, J. L. Nie and H. Y. Xiao, *J. Appl. Phys.*, 2009, **105**, 014309.
291. R. H. Miwa, T. B. Martins and A. Fazzio, *Nanotechnology*, 2008, **19**, 155708.
292. Y. Zhang, D. Zhang and C. Liu, *J. Phys. Chem. B*, 2006, **110**, 4671–4674.
293. R. Wang, D. Zhang, Y. Zhang and C. Liu, *J. Phys. Chem. B*, 2006, **110**, 18267–18271.
294. Y. Sun, L. Chen, F. Zhang, D. Li, H. Pan and J. Ye, *Solid State Commun.*, 2010, **150**, 1906–1910.
295. Z. M. Ao, J. Yang, S. Li and Q. Jiang, *Chem. Phys. Lett.*, 2008, **461**, 276–279.
296. S. Peng and K. Cho, *Nano Lett.*, 2003, **3**, 513–517.
297. L. Bai and Z. Zhou, *Carbon*, 2007, **45**, 2105–2110.
298. R. Wang, D. Zhang, W. Sun, Z. Han and C. Liu, *J. Mol. Struct.: THEOCHEM.*, 2007, **806**, 93–97.
299. J. Dai, J. Yuan and P. Giannozzi, *Appl. Phys. Lett.*, 2009, **95**, 232105.
300. D. J. Mowbray, J. M. García-Lastra, I. L. Arocena, Á. Rubio, K. S. Thygesen and K. W. Jacobsen, in *Chemical Sensors: Simulation and Modeling*, ed. G.Korotcenkov, Momentum Press, 2012.
301. R. Kodi Pandyan, S. Seenithurai and M. Mahendran, *Indian J. Phys.*, 2012, **86**, 677–680.
302. C. Tabtimsai, S. Keawwangchai, B. Wanno and V. Ruangpornvisuti, *J. Mol. Model.*, 2012, **18**, 351–358.
303. C. Tabtimsai, B. Wanno and V. Ruangpornvisuti, *Mater. Chem. Phys.*, 2013, **138**, 709–715.
304. P. Pannopard, P. Khongpracha, M. Probst and J. Limtrakul, *J. Mol. Graphics Modell.*, 2009, **28**, 62–69.

305. C. S. Yeung, L. V. Liu and Y. A. Wang, *J. Phys. Chem. C*, 2008, **112**, 7401–7411.
306. X. Zhou, W. Q. Tian and X.-L. Wang, *Sens. Actuators., B*, 2010, **151**, 56–64.
307. Y. Li, Z. Zhou, G. Yu, W. Chen and Z. Chen, *J. Phys. Chem. C*, 2010, **114**, 6250–6254.
308. Y.-H. Lu, M. Zhou, C. Zhang and Y.-P. Feng, *J. Phys. Chem. C*, 2009, **113**, 20156–20160.
309. E. H. Song, Z. Wen and Q. Jiang, *J. Phys. Chem. C*, 2011, **115**, 3678–3683.
310. Z. Miao, L. Yun-Hao, C. Yong-Qing, Z. Chun and F. Yuan-Ping, *Nanotechnology*, 2011, **22**, 385502.
311. M. R. Sonawane, B. J. Nagare, D. Habale and R. K. Shivade, *Adv. Mater. Res.*, 2013, **678**, 179–184.
312. M. R. Sonawane, D. Habale, B. J. Nagare and R. Gharde, *Int. J. Appl. Phys. Math.*, 2011, **1**, 138–143.
313. E. Cruz-Silva, F. Lopez-Urias, E. Munoz-Sandoval, B. G. Sumpter, H. Terrones, J.-C. Charlier, V. Meunier and M. Terrones, *Nanoscale*, 2011, **3**, 1008–1013.
314. D. Jiayu and Y. Jianmin, *J. Phys.: Condens. Matter*, 2010, **22**, 225501.
315. Z. Yong-Hui, C. Ya-Bin, Z. Kai-Ge, L. Cai-Hong, Z. Jing, Z. Hao-Li and P. Yong, *Nanotechnology*, 2009, **20**, 185504.
316. S. M. Seyed-Talebi, J. Beheshtian and M. Neek-amal, *J. Appl. Phys.*, 2013, **114**.
317. P. Cabrera-Sanfelix, *J. Phys. Chem. A*, 2008, **113**, 493–498.
318. K.-J. K. Lee, *Bull. Korean Chem. Soc.*, 2013, **34**, 3022–3026.
319. P. A. Gauden and M. Wiśniewski, *Appl. Surf. Sci.*, 2007, **253**, 5726–5731.
320. B. C. Wood, S. Y. Bhide, D. Dutta, V. S. Kandagal, A. D. Pathak, S. N. Punnathanam, K. G. Ayappa and S. Narasimhan, *J. Chem. Phys.*, 2012, **137**, 054702.
321. X. Hu, Y. Wu, H. Li and Z. Zhang, *J. Phys. Chem. C*, 2010, **114**, 9603–9607.
322. S. Yang, G.-L. Zhao and E. Khosravi, *J. Phys. Chem. C*, 2010, **114**, 3371–3375.
323. B. Shan and K. Cho, *Chem. Phys. Lett.*, 2010, **492**, 131–136.
324. H. J. Yan, B. Xu, S. Q. Shi and C. Y. Ouyang, *J. Appl. Phys.*, 2012, **112**, 104316.
325. J. Dai and J. Yuan, *Phys. Rev. B*, 2010, **81**, 165414.
326. G. R. Jenness, O. Karalti and K. D. Jordan, *Phys. Chem. Chem. Phys.*, 2010, **12**, 6375–6381.
327. E. Voloshina, D. Usvyat, M. Schutz, Y. Dedkov and B. Paulus, *Phys. Chem. Chem. Phys.*, 2011, **13**, 12041–12047.
328. P. Cabrera-Sanfelix and G. R. Darling, *J. Phys. Chem. C*, 2007, **111**, 18258–18263.
329. M. Oubal, S. Picaud, M.-T. Rayez and J.-C. Rayez, *Surf. Sci.*, 2010, **604**, 1666–1673.
330. M. D. Ellison, S. T. Morris, M. R. Sender, J. Brigham and N. E. Padgett, *J. Phys. Chem. C*, 2007, **111**, 18127–18134.
331. P. Ren, A. Zheng, X. Pan, X. Han and X. Bao, *J. Phys. Chem. C*, 2013, **117**, 23418–23424.

332. E. Schröder, *J. Nanomater.*, 2013, 871706.
333. Z.-R. Tang, *Phys. B*, 2010, **405**, 770–773.
334. P. Lazar, F. Karlický, P. Jurečka, M. Kocman, E. Otyepková, K. Šafářová and M. Otyepka, *J. Am. Chem. Soc.*, 2013, **135**, 6372–6377.
335. R. A. Guirado-López, M. Sánchez and M. E. Rincón, *J. Phys. Chem. C*, 2006, **111**, 57–65.
336. K. Rytkönen, J. Akola and M. Manninen, *Phys. Rev. B*, 2007, **75**, 075401.
337. K. Nakada and A. Ishii, *Solid State Commun.*, 2011, **151**, 13–16.
338. X. Liu, C. Z. Wang, M. Hupalo, W. C. Lu, M. C. Tringides, Y. X. Yao and K. M. Ho, *Phys. Chem. Chem. Phys.*, 2012, **14**, 9157–9166.
339. M. Khantha, N. A. Cordero, L. M. Molina, J. A. Alonso and L. A. Girifalco, *Phys. Rev. B*, 2004, **70**, 125422.
340. A. M. Garay-Tapia, A. H. Romero and V. Barone, *J. Chem. Theory Comput.*, 2012, **8**, 1064–1071.
341. P. V. C. Medeiros, F. d. B. Mota, A. J. S. Mascarenhas and C. M. C. de Castilho, *Solid State Commun.*, 2011, **151**, 529–531.
342. E. Rangel, J. M. Ramirez-de-Arellano and L. F. Magana, *Phys. Status Solidi B*, 2011, **248**, 1420–1424.
343. M. Senami, Y. Ikeda, A. Fukushima and A. Tachibana, *AIP Adv.*, 2011, **1**, 042106.
344. F. Rabilloud, *Comput. Theor. Chem.*, 2011, **964**, 213–217.
345. F. Rabilloud, *J. Phys. Chem. A*, 2010, **114**, 7241–7247.
346. W. Koh, J. I. Choi, S. G. Lee, W. R. Lee and S. S. Jang, *Carbon*, 2011, **49**, 286–293.
347. K. M. O'Donnell, T. L. Martin, N. A. Fox and D. Cherns, *Phys. Rev. B*, 2010, **82**, 115303.
348. A. Lugo-Solis and I. Vasiliev, *Phys. Rev. B*, 2007, **76**, 235431.
349. D. Umadevi and G. N. Sastry, *Chem. Phys. Lett.*, 2012, **549**, 39–43.
350. H. Lee, J. Ihm, M. L. Cohen and S. G. Louie, *Phys. Rev. B*, 2009, **80**, 115412.
351. M. Yoon, S. Yang, C. Hicke, E. Wang, D. Geohegan and Z. Zhang, *Phys. Rev. Lett.*, 2008, **100**, 206806.
352. X. Yang, R. Q. Zhang and J. Ni, *Phys. Rev. B*, 2009, **79**, 075431.
353. C. Ataca, E. Akturk and S. Ciraci, *Phys. Rev. B*, 2009, **79**, 041406.
354. M. C. Nguyen, M.-H. Cha, J. Bae, Y. Kim, M. Kim and J. Ihm, *ChemPhysChem*, 2011, **12**, 777–780.
355. C. Cazorla, S. A. Shevlin and Z. X. Guo, *Phys. Rev. B*, 2010, **82**, 155454.
356. W. Liu, S. Xu, C. Li and G. Yuan, *Diamond Relat. Mater.*, 2012, **29**, 59–62.
357. W. Liu, S. Xu, G. Yuan and Y. Xu, *Phys. B*, 2013, **408**, 46–50.
358. P. A. Denis and F. Iribarne, *Chem. Phys.*, 2014, **430**, 1–6.
359. V. Wang, H. Mizuseki, H. P. He, G. Chen, S. L. Zhang and Y. Kawazoe, *Comput. Mater. Sci.*, 2012, **55**, 180–185.
360. K. Koichi, I. Tetsuji and T. Hiroto, *Jpn. J. Appl. Phys.*, 2014, **53**, 02BD02.
361. L. Hu, X. Hu, X. Wu, C. Du, Y. Dai and J. Deng, *Phys. B*, 2010, **405**, 3337–3341.
362. Y. Tang, Z. Yang and X. Dai, *J. Magn. Magn. Mater.*, 2011, **323**, 2441–2447.
363. K. T. Chan, J. B. Neaton and M. L. Cohen, *Phys. Rev. B*, 2008, **77**, 235430.

364. T. P. Hardcastle, C. R. Seabourne, R. Zan, R. M. D. Brydson, U. Bangert, Q. M. Ramasse, K. S. Novoselov and A. J. Scott, *Phys. Rev. B*, 2013, **87**, 195430.
365. A. Ishii, M. Yamamoto, H. Asano and K. Fujiwara, *J. Phys. Conf. Ser.*, 2008, **100**, 052087.
366. A. V. Krasheninnikov, P. O. Lehtinen, A. S. Foster, P. Pyykkö and R. M. Nieminen, *Phys. Rev. Lett.*, 2009, **102**, 126807.
367. E. Durgun, S. Dag, V. M. K. Bagci, O. Gulseren, T. Yildirim and S. Ciraci, *Phys. Rev. B*, 2003, **67**, 201401.
368. E. Durgun, S. Dag, S. Ciraci and O. Gulseren, *J. Phys. Chem. B*, 2004, **108**, 575–582.
369. H. Valencia, A. Gil and G. Frapper, *J. Phys. Chem. C*, 2010, **114**, 14141–14153.
370. P. O. Krasnov, F. Ding, A. K. Singh and B. I. Yakobson, *J. Phys. Chem. C*, 2007, **111**, 17977–17980.
371. G. Chen and Y. Kawazoe, *Phys. Rev. B*, 2006, **73**, 125410.
372. M. Menon, A. N. Andriotis and G. E. Froudakis, *Chem. Phys. Lett.*, 2000, **320**, 425–434.
373. H. L. Gao and J. J. Zhao, *J. Chem. Phys.*, 2010, **132**, 234704.
374. F. Y. Meng, L. G. Zhou, S.-Q. Shi and R. Yang, *Carbon*, 2003, **41**, 2023–2025.
375. Y. Fan, B. R. Goldsmith and P. G. Collins, *Nat. Mater.*, 2005, **4**, 906–911.
376. S. H. Yang, W. H. Shin, J. W. Lee, S. Y. Kim, S. I. Woo and J. K. Kang, *J. Phys. Chem. B*, 2006, **110**, 13941–13946.
377. C. Inntam and J. Limtrakul, *J. Phys. Chem. C*, 2010, **114**, 21327–21337.
378. Y. Okamoto, *Chem. Phys. Lett.*, 2006, **420**, 382–386.
379. K. Okazaki-Maeda, S. Yamakawa, Y. Morikawa, T. Akita, S. Tanaka, S. Hyodo and M. Kohyama, *J. Phys. Conf. Ser.*, 2008, **100**, 072044.
380. D. Xian-Qi, T. Ya-Nan, Z. Jian-Hua and D. Ya-Wei, *J. Phys.: Condens. Matter*, 2010, **22**, 316005.
381. K. Okazaki-Maeda, Y. Morikawa, S. Tanaka and M. Kohyama, *Surf. Sci.*, 2010, **604**, 144–154.
382. M. Zhou, A. Zhang, Z. Dai, Y. P. Feng and C. Zhang, *J. Phys. Chem. C*, 2010, **114**, 16541–16546.
383. I. Fampiou and A. Ramasubramaniam, *J. Phys. Chem. C*, 2013, **117**, 19927–19933.
384. I. Fampiou and A. Ramasubramaniam, *J. Phys. Chem. C*, 2012, **116**, 6543–6555.
385. X.-Q. Dai, Y.-N. Tang, Y.-W. Dai, Y.-H. Li, J.-H. Zhao, B. Zhao and Z.-X. Yang, *Phys. Rev. B*, 2011, **20**, 056801.
386. J. Wu, S. W. Ong, H. C. Kang and E. S. Tok, *J. Phys. Chem. C*, 2010, **114**, 21252–21261.
387. E. Gracia-Espino, X. Jia and T. Wågberg, *J. Phys. Chem. C*, 2014, **118**, 2804–2811.
388. C. K. Acharya, D. I. Sullivan and C. H. Turner, *J. Phys. Chem. C*, 2008, **112**, 13607–13622.

389. N. Park, D. Sung, S. Lim, S. Moon and S. Hong, *Appl. Phys. Lett.*, 2009, **94**, 073105.
390. N. T. Cuong, D. H. Chi, Y.-T. Kim and T. Mitani, *Phys. Status Solidi B*, 2006, **243**, 3472–3475.
391. N. T. Cuong, A. Fujiwara, T. Mitani and D. H. Chi, *Comput. Mater. Sci.*, 2008, **44**, 163–166.
392. D. H. Chi, N. T. Cuong, N. A. Tuan, Y.-T. Kim, H. T. Bao, T. Mitani, T. Ozaki and H. Nagao, *Chem. Phys. Lett.*, 2006, **432**, 213–217.
393. K. E. Hayes and H.-S. Lee, *Chem. Phys.*, 2012, **393**, 96–106.
394. Y.-C. Wang, H.-L. Chen, S.-P. Ju, J.-Y. Hsieh and C.-Y. Tai, *Int. J. Energy Res.*, 2013, **38**, 1053–1059.
395. B. H. Morrow and A. Striolo, *Mol. Simul.*, 2009, **35**, 795–803.
396. D. Cheng, W. Wang and S. Huang, *J. Phys. Chem. C*, 2007, **111**, 1631–1637.
397. J.-g Wang, Y.-a Lv, X.-n Li and M. Dong, *J. Phys. Chem. C*, 2008, **113**, 890–893.
398. C. F. Sanz-Navarro, P.-O. Astrand, D. Chen, M. Ronning, A. C. T. van Duin, T. Jacob and W. A. Goddard, *J. Phys. Chem. A*, 2008, **112**, 1392–1402.
399. C. F. Sanz-Navarro, P.-O. Åstrand, D. Chen, M. Rønning, A. C. T. van Duin and W. A. Goddard, *J. Phys. Chem. C*, 2010, **114**, 3522–3530.
400. H.-Y. Cheng, Y.-A. Zhu, P.-O. Åstrand, D. Chen, P. Li and X.-G. Zhou, *J. Phys. Chem. C*, 2013, **117**, 14261–14271.
401. R. Méndez-Camacho and R. A. Guirado-López, *J. Phys. Chem. C*, 2013, **117**, 10059–10069.
402. K. Okazaki-Maeda, T. Akita, S. Tanaka and M. Kohyama, *Mater. Trans.*, 2008, **49**, 2441–2444.
403. A. Martin, S. Biplab, E. Olle and V. S. Natalia, *J. Phys.: Condens. Matter*, 2011, **23**, 205301.
404. M. K. Srivastava, Y. Wang, A. F. Kemper and H.-P. Cheng, *Phys. Rev. B*, 2012, **85**, 165444.
405. X. Kong, Z. Sun and Q. Chen, *Phys. Chem. Chem. Phys.*, 2012, **14**, 13564–13568.
406. O. Üzengi Aktürk and M. Tomak, *Phys. Rev. B*, 2009, **80**, 085417.
407. M. K. Shukla, M. Dubey and J. Leszczynski, *ACS Nano*, 2008, **2**, 227–234.
408. I. Cabria, M. J. López and J. A. Alonso, *Phys. Rev. B*, 2010, **81**, 035403.
409. D. Duca, F. Ferrante and G. La Manna, *J. Phys. Chem. C*, 2007, **111**, 5402–5408.
410. J. Granatier, P. Lazar, R. Prucek, K. Šafářová, R. Zbořil, M. Otyepka and P. Hobza, *J. Phys. Chem. C*, 2012, **116**, 14151–14162.
411. X. Liu, L. Li, C. Meng and Y. Han, *J. Phys. Chem. C*, 2011, **116**, 2710–2719.
412. O. Loboda, V. R. Jensen and K. J. B⊘rve, *Fullerenes, Nanotubes, Carbon Nanostruct.*, 2006, **14**, 365–371.
413. I. Efremenko and M. Sheintuch, *J. Catal.*, 2003, **214**, 53–67.
414. T. Prasomsri, D. Shi and D. E. Resasco, *Chem. Phys. Lett.*, 2010, **497**, 103–107.
415. L. Liu, Y. Su, J. Gao and J. Zhao, *Phys. E.*, 2012, **46**, 6–11.

416. X. Liu, K. X. Yao, C. G. Meng and Y. Han, *Dalton Trans.*, 2012, **41**, 1289–1296.
417. D.-H. Lim, A. S. Negreira and J. Wilcox, *J. Phys. Chem. C*, 2011, **115**, 8961–8970.
418. W. Song, M. Jiao, K. Li, Y. Wang and Z. Wu, *Chem. Phys. Lett.*, 2013, **588**, 203–207.
419. B. F. Machado, M. Oubenali, M. Rosa Axet, T. Trang Nguyen, M. Tunckol, M. Girleanu, O. Ersen, I. C. Gerber and P. Serp, *J. Catal.*, 2014, **309**, 185–198.
420. S. Dongchul, P. Noejung, K. Gunn and H. Suklyun, *Nanotechnology*, 2012, **23**, 205204.
421. H. Johll, J. Wu, S. W. Ong, H. C. Kang and E. S. Tok, *Phys. Rev. B*, 2011, **83**, 205408.
422. G. A. Ferguson, C. Yin, G. Kwon, E. C. Tyo, S. Lee, J. P. Greeley, P. Zapol, B. Lee, S. Seifert, R. E. Winans, S. Vajda and L. A. Curtiss, *J. Phys. Chem. C*, 2012, **116**, 24027–24034.
423. X. Lu, C. Sun, F. Li and H.-M. Cheng, *Chem. Phys. Lett.*, 2008, **454**, 305–309.
424. P. A. Denis, *Chem. Phys. Lett.*, 2011, **508**, 95–101.
425. J. H. Wu and F. Hagelberg, *Phys. Rev. B*, 2007, **76**, 155409.
426. L. G. Zhou and S. Q. Shi, *Carbon*, 2003, **41**, 613–615.
427. N. Naghshineh and M. Hashemianzadeh, *Int. J. Hydrogen Energy*, 2009, **34**, 2319–2324.
428. X. Liu, C. Z. Wang, M. Hupalo, Y. X. Yao, M. C. Tringides, W. C. Lu and K. M. Ho, *Phys. Rev. B*, 2010, **82**, 245408.
429. Z. W. Zhang, W. T. Zheng and Q. Jiang, *Phys. Chem. Chem. Phys.*, 2011, **13**, 9483–9489.

CHAPTER 4

Surface Chemistry of Nanostructured Carbon Materials and Preparation of Nanocarbon Supported Catalysts

4.1 Introduction

The adsorption sites of nanostructured carbon materials, described in Chapter 3, can be decorated with a wide variety of elements giving rise to a new class of materials with specific properties. The nanostructure of the surface serves as a template to anchor or organize particles or molecules in a highly dispersed state. Due to their interaction with the carbon surface, the deposited metals, semiconductors, oxides, organic molecules and biomolecules often present unique structure–property relations. For example, it has been observed that the confinement of metal nanoparticles in CNTs gives rise to new properties.[1] As a result, these composite materials find a number of applications in the fields of catalysis, environmental pollution control and energy.[2–8]

The carbon surface can also be suitably modified to attach specific functional groups thanks to the presence of inherent structural defects. For applications in catalysis where carbon supports coated with transition metals have been investigated, modifications can be effected on specific sites so as to obtain high catalyst dispersion resulting in an efficient utilization of the entire catalyst loading. Also, the possibility of macroscopic shaping

RSC Catalysis Series No. 23
Nanostructured Carbon Materials for Catalysis
By Philippe Serp and Bruno Machado

Published by the Royal Society of Chemistry, www.rsc.org

of materials with high specific surface area and full accessibility makes these materials attractive as catalyst supports in large-scale applications.[9,10] Indeed, resistance to abrasion, thermal and dimensional stabilities, and specific adsorption properties are important factors that influence the activity and reproducibility of the catalytic system. In particular, nanostructured carbon materials could replace activated carbon supports in liquid-phase reactions since the surface properties of the current supports cannot be easily controlled, and their increased microporosity has often been detrimental to mass transport. As well as transition metals, including noble metals, a number of inorganic materials have also been coated onto the carbon surface. Understandably, the volume of scientific research on these materials has been increasing each year and review articles are now available that describe the ongoing research in this field.[11–14]

When carbon is used as support, the surface chemistry governs the catalyst dispersion, its loading, as well as the catalytic activity, selectivity and stability. The presence of, often confined, surface functional groups can influence surface diffusion, desorption or provide suitable environments for gradient concentrations that can affect both catalyst activity and selectivity. Knowledge of the surface chemistry of carbon materials is also of great importance as the physico-chemical properties of carbons are strongly influenced by the presence, even in small amounts, of chemical species on the surface. Hence, many of their applications are conditioned by their chemical characteristics. In addition, these properties are known to change during storage of carbons. Thus, the surface functional groups determine the self-organization, the chemical stability and the reactivity in adsorptive and catalytic processes.

In this chapter, our aim is to present a concise overall view of the main synthesis method to produce decorated nanostructured carbon materials.

4.2 Surface Chemistry of Nanostructured Carbon Materials

The surface chemistry of sp^2 and $sp^{2+\delta}$ carbon nanostructured materials is governed by basal and edge carbon atoms, as well as by the presence of defects (see Chapter 3, Section 3.2.4). These imperfections and defects along the edges of graphene layers are the most active sites owing to high densities of unpaired electrons. There, heteroatoms, such as oxygen,[15] hydrogen, nitrogen,[16–18] boron, sulfur[19] and phosphorous, can be chemisorbed leading to stable surface compounds, resulting in a complex surface chemistry, if compared to oxides such as silica or alumina. Oxygen-containing surface groups confer hydrophilic and cation exchange properties. Nitrogen-containing carbons show enhanced anion exchange properties and catalytic activity in redox reactions. Phosphorus-containing carbons show a number of specific characteristics that range from acid surface groups and cation exchange properties to enhanced oxidation stability. The sulfur-containing carbon catalysts exhibit superior sintering-resistant behavior at high temperature. Multiple doping is a versatile synthetic approach for new carbon materials and takes the tuning of carbon properties one step further in comparison to one-type-only heteroatom doping. In particular, nitrogen is reckoned as a peerless carbon dopant, thus N-doping is by far the most ubiquitous when compared with other heteroatoms. Complementing nitrogen as a doping agent, sulfur, boron or phosphorous are receiving increasing attention in carbon materials research. Binary and ternary doping of nitrogen, boron, and phosphorus into carbon can be performed at the carbon growth step.[20] The B-doping reinforces the sp^2-structure of graphite, whereas P-doping enhances the charge delocalization of the carbon atoms and produces carbon structures with many edge sites. A nitrogen doped carbon can also be enriched with oxygen surface groups, resulting in new properties.[21] The various schemes for synthesis and characterization of $B_xC_yN_z$ nano-composites has also been recently reviewed.[22]

Although a direct analogy to the functional groups classified in organic chemistry exists, given the chemical complexity of the carbon surface and the fact that the heteroatoms are often located in a confined space, one cannot fully predict the behavior of those groups based on well-known organic chemistry reaction mechanisms. The concentration and distribution of the surface groups present on the carbon depends on the carbon type and the pre-treatment applied. Although adsorption onto carbons is mainly of the dispersive interaction type, surface chemistry plays an important role when specific interactions are considered. The surface chemistry of carbons determines their moisture content, catalytic properties, acid-base character, and adsorptive properties. It is related to the presence of heteroatoms other than carbon within the carbon matrix, and to the presence of carbene and carbyne structures at the edges of the graphene layer in sp^2 hybridized carbon materials.[23] The most common heteroatoms are oxygen, nitrogen, boron, phosphorus, hydrogen, chlorine, and sulfur. They are bound to the edges of the graphite-like layers and form organic functional groups such as carboxylic acids, lactones, phenols, carbonyls, aldehydes, ethers, amines, nitro compounds, and phosphates. These functional groups can be acidic, basic, or neutral in character. Acidic behavior is often associated with surface complexes or oxygen functionalities such as carboxyls, lactones and phenols. On the other hand, functionalities like amines, pyrones, chromenes, ethers and carbonyls are responsible for basic properties of the carbon surfaces. Basic properties associated with Lewis sites located at the π electron-rich regions within the basal planes of graphitic micro-crystals, away from the edges, have also been proposed. We will concentrate our analysis on oxygen-, nitrogen-, phosphorous-, and sulfur-containing surface groups, since they are the most common and relevant for catalysis. It is however worth mentioning that with the discovery of fullerenes, CNTs and graphene a rich carbon functionalization chemistry as recently emerged.[24–28] Figures 4.1 to 4.4 represent the different types of oxygen-, nitrogen-, phosphorous-, and sulfur-containing functionalities generally observed on nanostructured carbon materials.

Carbon surface oxidation is the most popular way of modifying carbon surfaces.[29–31] It can be done either from a gaseous (oxygen, ozone, air or nitric oxides) or liquid phase (nitric acid, hydrogen peroxide, potassium permanganate, sulfuric acid, sodium peroxydisulfide).[32] Although conditions of gas phase oxidation vary, usually it is done in an oven at elevated temperatures (473–623 K) with a continuous flow of the oxidant. Air oxidation is considered weak, and as a result, various oxygen-containing groups are formed with the predominant population of weakly acidic groups such as phenols. Oxidation in the liquid phase is much more complex and results in more severe changes to the carbon surface chemistry. The oxidations are usually carried out in open vessels with oxidants in a wide range of concentrations, depending on the desired effects. Another important factor in this type of oxidation is temperature. The higher the temperature and stronger the oxidant, the more oxidized the carbon surface becomes. In some cases, strong oxidation, as that obtained with nitric acid at its boiling point, can totally destroy the carbon structure.[33] It is generally accepted that oxidation with strong oxidants such as nitric acid leads to carbons with a high concentration of surface carboxylic groups, whereas treatment with hydrogen peroxide increases mainly the population of phenols. Nitric acid oxidation can also results in incorporation of small amounts of nitrogen as nitro groups, likely attached to the carbons at the edges of graphene planes.[34]

The oxygen-containing functionalities are considered as acidic or basic. The acidic surface groups (*i.e.*, carboxylic acid, lactone, and phenol groups) are formed when the carbon surface (including nanodiamond)[4] is exposed to oxygen *via* reactions with oxidizing agents from solutions or gas phase, either at room or high temperatures. These surface chemical groups disrupt the π-conjugation and introduce surface dipole moments, leading to higher work functions (up to 5.1 eV).[35] Pyrone-like functionalities at the edge of graphene layers are predicted to exhibit a broad range of pK_a values (4–13) depending on the relative position of the ketone and etheric rings.[36] Basic groups are formed when an oxidized surface is reduced by heating in inert atmosphere at high temperatures. The decomposition of acidic groups results in active sites at the edges of the graphene layers, which upon cooling under an inert atmosphere and re-exposure to air, attract oxygen forming basic functional groups such as chromene or pyrone.[37]

Although there is a general agreement about the type of surface functionalities that determines the acidic character of a carbon material, the nature of carbon basic surfaces remains controversial and open for investigation. Generally speaking, oxygen-containing functionalities (*i.e.*, chromene, pyrone, quinone) and non-heteroatomic Lewis base sites, characterized by regions of π electron density on the carbon basal planes, govern carbon basicity.[36]

Introduction of nitrogen into carbon materials can be performed both in the liquid or gas phase using nitrogen-containing precursors. Ammonia is usually used as a gaseous precursor of nitrogen at temperatures between 673 and 1273 K. When the modifications are carried out on the carbon samples, either pre-oxidized or not, in the liquid phase, compounds such as carbazole, nitrogen enriched polymers, acridine, melamine or urea are used. The carbons are impregnated with water or alcoholic solutions of the nitrogen containing compounds and then exposed to heat treatment at temperatures between 673 and 1293 K. Usually, the nitrogen content in carbon materials is very small, unless it is present in the carbon precursor, as for instance acetonitrile, carbazole, nitrogen enriched polymers, acridine, and melamine. So far, it has been demonstrated that the presence of nitrogen can be a key parameter for the performance of carbon materials as catalyst supports and for the catalytic activity.[78]

The type of nitrogen functionalities present on the carbon surface is a function of the treatment applied.[38] This includes the type of nitrogen containing precursor, the chemical activity of the carbon surface and, the most important, the temperature of heat treatment. The latter determines the type of chemistry owing to the fact that some nitrogen containing species are unstable at high temperatures. According to the studies performed, lactam and

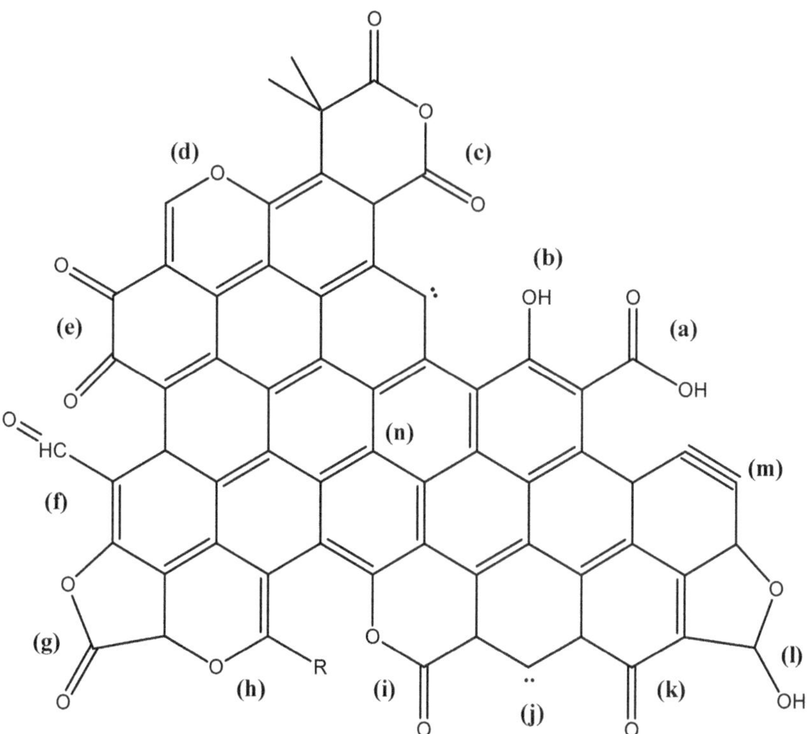

Figure 4.1 Oxygen-containing species, and specific sites generally present on carbon surface and their decomposition by TPD: (a) carboxylic acid, (b) phenol, (c) carboxylic anhydride, (d) ether, (e) quinone, (f) aldehyde, (g) lactone, (h) chromene, (i) pyrone, (j) carbene like species, (k) carbonyl, (l) lactol, (m) carbyne like species, and (n) π electron density on carbon basal plane.

imide structures are mainly formed by ammination, and amides are formed upon ammoxidation;[39] the former is transformed to pyrrole and pyridine by heat treatments.[40] As for oxygen, nitrogen containing functionalities determine the acidic or basic character of carbons and thus its surface chemical reactivity or catalytic activity. Treatments of carbonaceous surfaces with nitrogen-containing reagents at low temperatures (less than 800 K) lead to the formation of lactams, imides, and amines, which are slightly acidic in their nature. On the other hand, treatments at high temperatures result in an increase in the N quaternary (N atoms incorporated in the graphitic layer in substitution of C-atoms), pyridinic and pyrrole-type structures. They are responsible for an increase in the surface polarity of the carbon, as well as basicity since both pyridine and pyrrole-type structures are considered as basic.

The nitrogen functionalities that provide basic properties can enhance the interactions between the carbon surface and acid molecules, such as dipole–dipole, hydrogen bonding, covalent bonding, and so on. The possible structures of the nitrogen functionalities include amides, imides, lactames, pyrrolic and pyridinic groups, they are shown in Figure 4.2.

The introduction of phosphorous in the nanostructured carbon matrix can result from phosphoric acid activation.[41] There is controversy in the literature as to the chemical state of P in phosphorus-containing carbons.[41] Some authors, based on both experimental data and theoretical calculations, proposed that C–O–P bonding is more stable; the others argued that this bond would not be strong enough to survive at high temperature and proposed an

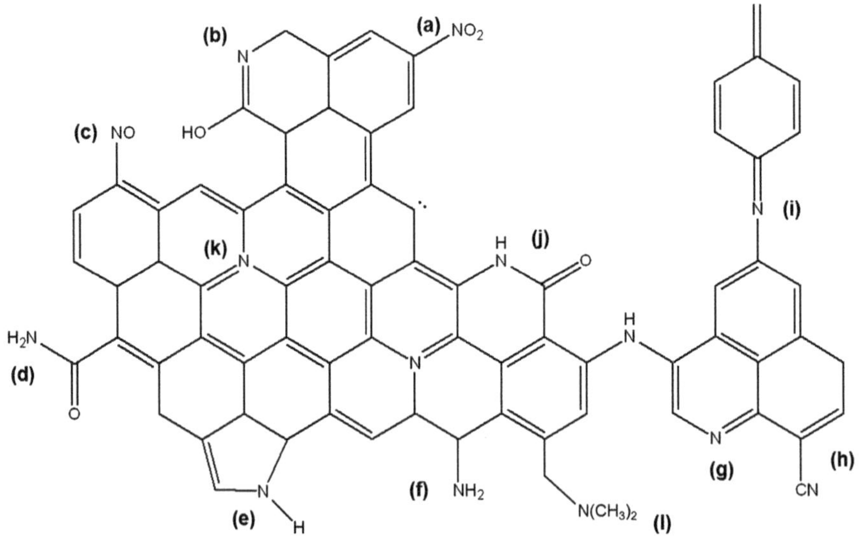

Figure 4.2 Nitrogen-containing species, and specific sites generally present on carbon surface: (a) nitroso group, (b) α-pyridone, (c) nitro group, (d) amide, (e) pyrrole type nitrogen, (f) amine, (g) pyridine like group, (h) nitrile, (i) imine, (j) lactam, (k) quaternary amine, and (l) tertiary amine.

alternative structure with C–P–O bonding. A recent XPS and ^{31}P-NMR study revealed that the most abundant phosphorus species in carbons obtained at 673–1273 K by phosphoric acid activation is a phosphate-like structure with pentavalent tetra-coordinated phosphorus bound to four oxygen atoms.[41] Possible phosphorous-containing surface groups are presented in Figure 4.3.

The introduction of sulfur functionalities onto the carbon surface is usually done either by heating carbons in the presence of elemental sulfur or other sulfur containing compounds, such as hydrogen sulfide or fuming H_2SO_4.[19,40,42] In this way, up to 10% of sulfur can be introduced in the carbon matrix. The temperature of heat treatment varies from 473 to 1273 K. Sulfur is present in carbonaceous materials as elemental sulfur, inorganic species and organo-sulfur compounds. The surface complexes and their possible configurations are presented in Figure 4.4. Its content varies between 0 and 5 wt%. Carbon–sulfur complexes are extremely stable and they are not completely removed from the carbon matrix even at high temperatures reaching 1373 K. Only heating in a reducing hydrogen atmosphere is able to remove sulfur completely form the carbon matrix.

Because of its small size, the boron atom is easily incorporated to the carbon lattice, and can be substituted in both sp^2 and sp^3 configurations. A standard CVD process with BCl_3 is used to produce highly boron-doped carbons, which have a hexagonal, graphite-like structure, and contain up to 17% boron. For CNTs and graphene, B_2H_6 has been generally used as the boron source.[43] Substituted boron atoms in the carbon lattice can accelerate the graphitization and suppress the oxidation of carbon materials.

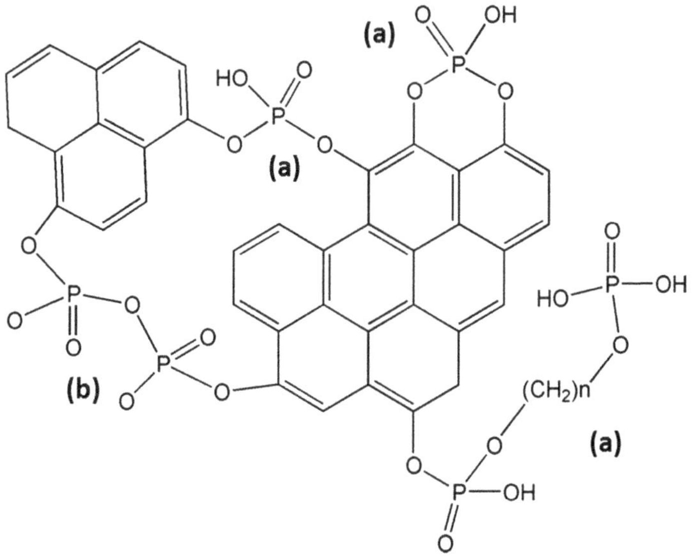

Figure 4.3 Phosphate–carbon complexes as described in the literature: (a) phosphocarbonaceous esters, and (b) pyrophosphate species.

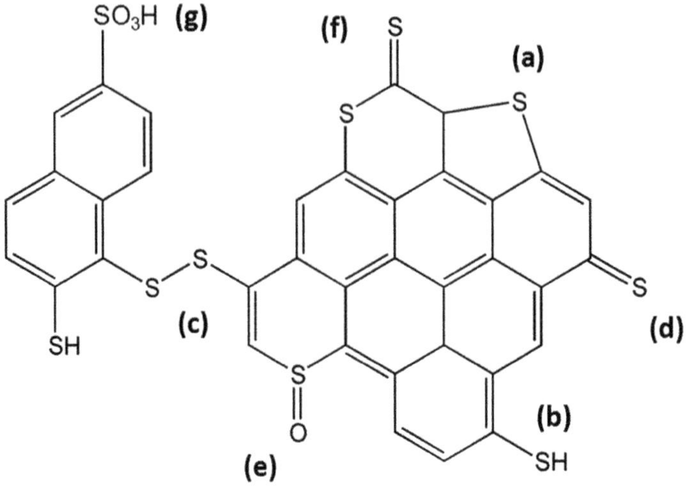

Figure 4.4 Carbon–sulfur surface compounds as described in the literature: (a) sulfide, (b) thiophenol, (c) disulfide, (d) thioquinone, (e) sulfoxide, (f) thiolactone, and (g) sulfonic.

Boron-containing CNTs are predicted to behave as semiconductors over a large range of diameters and chirality.

Very marked effects of carbon surface chemistry have been reported on the adsorption of various species such as aromatics, dyes, heavy metals, pharmaceuticals, polar species, such as alcohols, acids or aldehydes, and even small molecule gases. The species present on the carbon surface can enhance the specific interactions or even alter the porosity *via* blocking of pore entrances for molecules to be adsorbed. Specific interactions include hydrogen bonding, acid/base, complexation, *etc.* These specific interactions are also very important in catalysis.

4.2.1 Characterization of Nanostructured Carbon Material Surface Groups

A complete characterization of carbon surface chemistry necessitates the use of a broad battery of analytical techniques.[15,44–49] The nature of the chemical surface groups is currently determined by Fourier-transformed infrared spectroscopy (FTIR, DRIFT) and X-ray photoelectron spectroscopy (XPS). Their quantitative determination can also be carried out by selective titrations in aqueous solutions at room temperature. The results obtained by selective titrations can be contrasted by other experimental methods.[44] Among them, thermogravimetry (TG), thermal programmed desorption (TPD), and, very frequently, calorimetric measurements are used.

Generally speaking, the methods used to characterize the surface of carbonaceous materials are referred to as "wet" and "dry" techniques. The "wet" technique involves Boehm[30] and potentiometric titrations.[50] In general, the Boehm titration, which has been recently standardized,[51,52] proceeds as follows: carbon is added to three different bases ($NaHCO_3$, Na_2CO_3 and NaOH), the $NaHCO_3$ and Na_2CO_3 aliquots are then back-titrated with NaOH, and the NaOH aliquots are titrated directly with HCl. As the strongest base, NaOH is assumed to neutralize all Brønsted acids, while Na_2CO_3 neutralizes carboxylic and lactonic groups and $NaHCO_3$ neutralizes carboxylic acids. From this knowledge, the types of oxygen surface groups can be determined. Acid–base potentiometric titration can also be used for the determination of pH at the point of zero charge (pH_{PZC}),[53] *i.e.* the pH above which the total surface of the carbon particles is negatively charged. This determination is crucial for determining the adsorption of metallic ions during catalyst preparation.[54] Generally speaking, the combination of these titration methods, together with temperature-programmed desorption is necessary for an in-depth characterization.

"Dry" methods include diffuse reflectance FTIR, X-ray photoelectron spectroscopy, thermal analysis, and thermal programmed desorption. TPD has been reported as one of the most commonly used techniques to identify the oxygenated surface functional groups on carbon (nanostructured) materials,[55] which decompose at different temperatures releasing CO, CO_2, and water. There are some inconsistencies in literature with respect to the assignment of the TPD peaks to specific surface groups, as the peak temperature may be shifted due to the differences in sample mass, heating rate, gas-flow rate, and the reactor dimensions of the experimental set-ups.[56] However, a specific temperature range can be defined for particular functional groups in general. The attributions presented below are mainly based on ref. 48. The CO_2 spectra are generally decomposed into four contributions, corresponding to two types of carboxylic groups, which can be assigned to strongly acidic (513–573 K) and less acidic carboxylic groups (653–703 K), carboxylic anhydrides (803–953 K), and lactones (923–1093 K).[57,58] The carboxylic anhydrides decompose by releasing one CO and one CO_2 molecule. Thus, the CO peak has the same shape and equal magnitude of the CO_2 peak. The generation of anhydride groups by condensation of two adjacent carboxylic groups when the sample is heated has also been proposed; this process is confirmed by the presence of water.[59] In addition to the carboxylic anhydrides, the CO spectrum includes contributions from phenols (943–993 K) and carbonyl/quinones (1093–1153 K). The CO peak that appears at low temperatures (503–703 K) probably comes from the decomposition of carbonyl groups in R-substituted ketones and aldehydes.[48] Basic structures like pyrone and chromene decompose thermally at temperatures above 1273 K leading to CO.[60]

XPS has proved to be a useful analytical tool for monitoring the processing steps by providing information on the relative amounts of different elements with respect to carbon and their valence states.[61] Since other analytical techniques cannot distinguish between the sp^2 and sp^3 carbons very well, XPS can be useful in the semi-quantitative analysis of carbon species in carbon materials. Some common binding energy peak assignments for the carbon 1*s*, oxygen 1*s*, the nitrogen 1*s*, and the S *2p* peaks are seen in Tables 4.1–4.4.[62] Results of the deconvolution of the C *1s*, O *1s*, N *1s* or S *2p* core level spectra may result in some shift in the binding energy from one study to the other.[63]

Table 4.1 Common binding energy assignments for the carbon 1s peak. The observed binding energies will depend on the specific environment where the functional groups are located. Most ranges are ± 0.2 eV, but some can be larger.

Functional group	Binding energy/eV
C sp^2	284.2
Hydrocarbon C–H, C–C	285.0
Amine C–N	286.0
Alcohol, ether C–O–H, C–O–C	286.5
C sp^3 (nanodiamond)	285.3[a]
Carbonyl C=O	288.0
Amide N–C=O	288.2
Acid, ester O–C=O	289.0
Urea N–CO–N	289.2
Carbamate O–C–N	289.9
Carbonate O–C–O	290.3
π–π^* shake-up satellite	291.0
Plasmon	292.0

[a]After correcting of the charge effect.[64]

Table 4.2 Common binding energy assignments for the oxygen 1s peak. The observed binding energies will depend on the specific environment where the functional groups are located. Most ranges are ± 0.2 eV.

Functional group	Binding energy/eV
Carbonyl C=O, O—C=O	532.2
Alcohol, ether C–O–H, C–O–C	532.8
Anhydride, carboxylic acid, lactone, ester C–O–C=O	533.7
Chemisorbed water or O_2	535–536

When applied to porous carbons for the determination of the oxygen surface groups, XPS has the following drawbacks: (i) the external surface area is only a small fraction of the total surface area and it is not representative of all the material; (ii) the roughness of the surface can affect the final results because the surface is not flat; (iii) the analysis is made in high-vacuum, that is, under conditions quite different from those usually used in the applications of the carbon catalyst, and a rearrangement of the surface can occur; and (iv) deconvolution of the N 1s, O 1s and C 1s peaks is not straightforward, and is still a matter of discussion.

Infrared (IR) spectroscopy has been widely used to characterize the surface groups of different carbon nanomaterials.[65–68] However, IR spectra of carbon materials are difficult to obtain because of problems in sample preparation, poor transmission, uneven light scattering related to large particle size, and so forth. Moreover, the electronic structure of carbon materials results in a complete absorption band through the visible region to the infrared. Fortunately, some of these problems can be overcome by improving sample preparation (*e.g.*, carbon films) as well as by using the

Table 4.3 Common binding energy assignments for the nitrogen 1s peak. The observed binding energies will depend on the specific environment where the functional groups are located.

Functional group	Binding energy/eV
Pyridinic nitrogen, Ar–N–Ar	398.6 + 0.3
-N–H	399.4 + 0.3
Pyrrolic nitrogen	
Pyridine nitrogen	
Nitrile	399.8 + 0.3
-O–C=N Pyridine pyrrole	400.2 + 0.1
N, quaternary	401.3 + 0.2
Pyridine-N-oxide	402.5–403.8
NO_x, oxidized, N–O–C	404.5 + 0.4

Table 4.4 Common binding energy assignments for the sulfur 2p peak. The observed binding energies will depend on the specific environment where the functional groups are located.

Functional group	Binding energy/eV
Thiol or carbon bisulphide; PhSH, CS_2	163.5 + 0.3
Sulphides; C–S–C	164.5 + 0.3
Thioethers; R–S–S–OR	
Sulphoxides, sulphite; $R_2S{=}O$, SO_3^{2-}	167.5 + 0.3
Sulphone; $R{-}SO_2{-}R$	168.0
Sulphate, sulphite, sulphonic acids; SO_4^{2-}, SO_3^{2-}, $RO_2{-}S{-}S{-}R$, $R{-}SO_3H$	169 + 0.7

most recently developed IR techniques such as diffuse reflectance Fourier transform IR spectroscopy. The adsorption band and peaks for O-, N-, S-, P-, and B-surface species on carbon materials are given in Table 4.5. Besides technical difficulties in obtaining IR spectra of carbon nanomaterials, their interpretation is often an additional problem because not all of the observed absorption bands may be assigned unequivocally to specific functional groups, most likely owing to the overlap of several bands. In other cases, it is not uncommon that some band assignments differ substantially among the recent IR studies on carbon materials. This is the case for the so-called "quinone band" that has been assigned to the 1660–1670[69] or 1550–1680 cm^{-1} interval.[70] Another controversial assignment for O-surface species corresponds to the band at 1600 cm^{-1}, which is a prominent feature in the IR spectra of carbon materials. The 1600 cm^{-1} band has been attributed to either oxygen surface compounds or ring vibrations of the basal plane. Intriguingly, the presence and intensity of this band is strictly related to the concentration of surface oxides, but IR studies using ^{18}O labelled carbon do not support an assignment to carbonyl-type species.[71] Thus, a widely accepted hypothesis assigns the 1600 cm^{-1} band to the C=C bonds (Table 4.5), whose IR intensity would be reinforced

Table 4.5 Infrared adsorption bands on carbon surfaces and their corresponding assignments to O-, N-, S- and B-containing functionalities.

	Assignment/cm^{-1}		
Group/functionality	500–1500	1500–2050	2050–3700
C–O stretch of ethers	1000–1300		
Ether bridge between rings	1230–1250		
Cyclic ethers containing COCOC groups	1025–1141		
Alcohols	1049–1276		3200–3640
Phenolic groups:			
C–O stretch	1000–1220		
O–H bend/stretch	1160–1200		2500–3620
Carbonates; carboxyl-carbonates	1000–1500	1590–1600	
Aromatic C=C stretching		1585–1600	
Quinones		1550–1680	
Carboxylic acids	1120–1200	1665–1760	2500–3300
Lactones	1160–1370	1675–1790	
Anhydrides	980–1300	1740–1880	
Ketenes (C=C=O)			2080–2200
C–H stretch			2600–3000
C–N stretch	1100–1300		
C–N bending	1070		
–N–CH_3	1370		
C=N		1590–1610	
C≡N			2080, 2180, 2350
NH and/or NH_2			3220–3420
CONH stretch of C=O in cyclic amides	1460	1546,1680	
C–S stretch	780, 920		
SO sulphoxides	1070		
SO_3H assymmetrical stretch of SO_3	1250		
HSO_4, SO, Stretch of S=O and S–O	580–590, 670–675		
H-bonded P=O groups (phosphates or polyphosphates); O–C stretching in P–O–C(aromatic); P=OOH	1160		
P^+–O^- in acid phosphate esters; sym. vibration in polyphosphate P–O–P	1065–1070		
B–C stretching	1020		
B–O stretching	1325		

by chemisorbed oxygen. However, the relationship between the nature of carbon surface oxides and the intensity of the 1600 cm^{-1} band is not yet clear.

Quantum chemical methods have been recently employed to get a more detailed assignment of the IR absorption bands of carbon materials.[71] It was found that the frequencies of the C=O bonds present in acid functional groups were systematically lowered when phenolic groups were close enough to establish hydrogen bonds. Concerning the origin of the 1600 cm^{-1} band of carbons, it was found that, in the case of acid carbons, this band can be

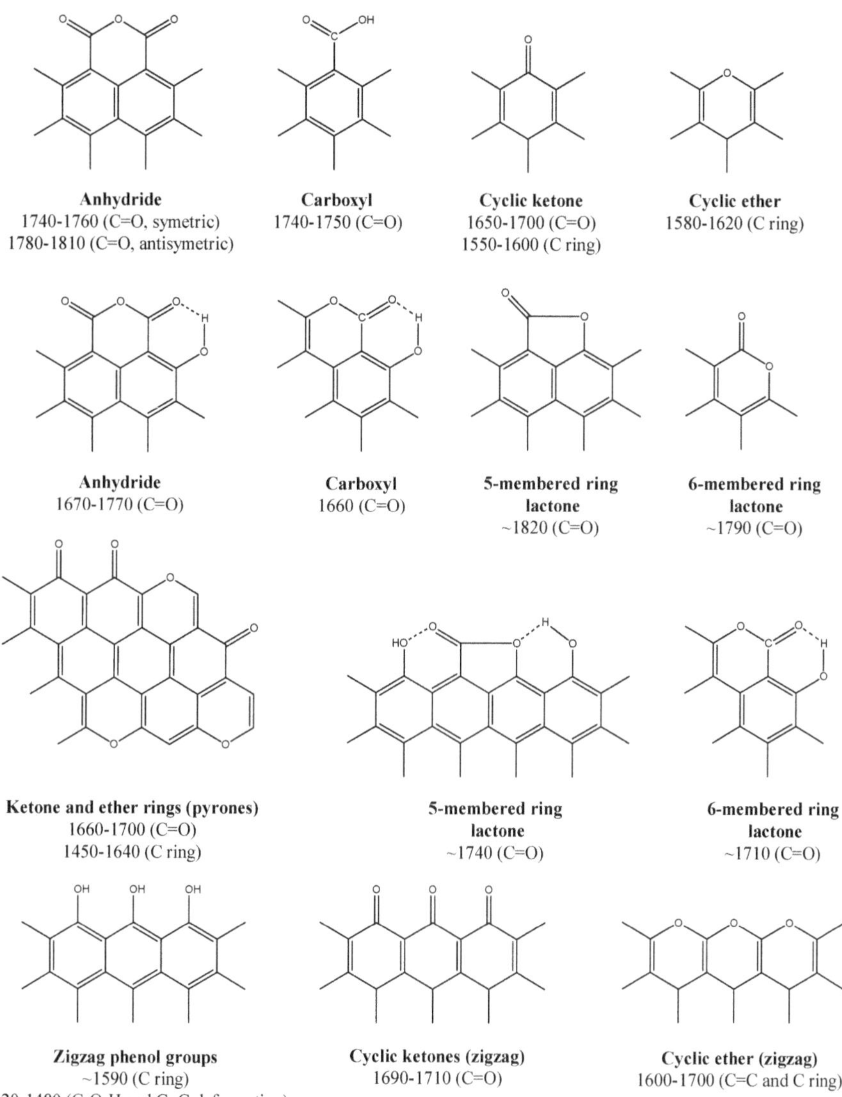

Figure 4.5 Structure of carbon surface oxides and their corresponding IR assignments in the 1400–1900 cm^{-1} region according to B3LYP/6-31G* calculations on polyaromatic systems. Adapted from ref. 71.

assigned to C=C stretching of carbon rings decorated mainly with phenolic groups. Cyclic ethers in basic carbons would also promote absorption in the 1600 cm^{-1} region of the IR spectrum. The calculate (DFT) vibrational frequencies in the 1400–1900 cm^{-1} range of the surface oxygen species are given in Figure 4.5.

Although XPS and DRIFTS provide excellent qualitative information about the carbon surface, a quantitative insight is not straightforward and requires special mathematical treatment using many approximations.

On the other hand, Boehm and potentiometric titrations provide qualitative and quantitative information on the carbon surface. However, the information on acidic groups is limited to compounds such as phenols, lactones and carboxylic acids, neglecting any other groups present.[34] Potentiometric titration results are useful to describe the material behavior in aqueous solutions, where the activity of the surface groups depends not only on their concentrations, but also on their environment.[72] Wet-chemically oxidized carbon nanostructures generally exhibit both covalently-bound acidic functional groups on the surface and surface-adsorbed acidic substances, *i.e.* carbonaceous fragments from the oxidation procedure.[73] Direct potentiometric titration of oxidizable high surface area materials with dynamically desorbing acidic fragments is slow and inaccurate. Adsorbed acidic fragments are deprotonated by sodium hydroxide and form anions in solution, which is not the case for covalently bound acidic groups on the nanocarbon, so filtration after NaOH treatment separates desorbable acidic substances from non-desorbable or covalently bound groups.

Temperature programmed desorption allows one to study functional groups that decompose below 1250 K, but it does not provide direct qualitative results on the carbon chemistry.[74] The quantities of CO and CO_2 released during the TPD experiments correspond to the total amount of surface oxygen groups. The decomposition temperature is related to the bond strength of specific oxygen-containing groups. Thus, the position of the peak maximum at a defined temperature corresponds to a specific oxygen complex at the surface. Deconvolution of the TPD profiles gives quantitative information about surface oxygen groups. The results of TPD experiments and the acid-base titrations indicate that outgassing at gradually rising temperatures with subsequent storage in air not only reduces the total number of acidic groups but also changes their distribution.[75] The decomposed surface groups are single carboxylic groups. Some of the carboxylic groups are converted into cyclic structures (anhydrides, lactones or lactols). Other structures are also formed.

A combination of different NMR methods was used to explore the pore structure and the interaction of adsorbates with the pore walls,[76] and also to determine the nature of surface groups.[77–81] The main feature of ^{13}C NMR spectra in oriented carbon materials is the occurrence of a large shift between the resonance lines achieved for the magnetic field parallel or perpendicular to the *c* axis of the graphite-like structure.[82] This shift is due to the existence of a highly anisotropic magnetic susceptibility, and it leads to a large anisotropy in the ^{13}C NMR powder pattern of graphite and related materials. In disordered materials and heat-treated carbon materials, the bulk magnetic susceptibility effect is not so large and the main cause of broadening of the resonance lines is the occurrence of a high chemical shift anisotropy

Table 4.6 ^{13}C, ^{31}P and ^{15}N NMR data for carbon (nano)materials.

Functional group	Chemical shift (ppm)
^{13}C NMR	
sp^2 ^{13}C in condensed aromatic rings	125–143
C_{60}	143.6
^{13}C–OH	55–75
Epoxide	60
Carboxyl carbon	175–190
Ketones and aldehydes	200–220
Aliphatic carbon	1–60
sp^3 ^{13}C in nanodiamonds	32–38
^{31}P NMR	
Phosphate-like structure (PO_4) and pyrophosphate	-5–0
Phosphonates (C–P bonding)	7–31
^{15}N NMR	
$-NH_2$	−320/−350
sp^3 N–C bonds, amide or pyrrole	−200/−300
sp^2 N–C bonds, pyridine or pyrazine	0/−150
NO_x, oxidized, N–O–C	404.5 + 0.4

associated with the aromatic planes. Due to these facts, the ^{13}C NMR spectra of graphite, and well-carbonized carbon materials are usually broad and present poor resolution. High-resolution techniques, primarily magic-angle spinning and high-power proton decoupling are usually employed for the achievement of informative spectra in the solid state. In the case of CNTs, it has been shown that the line position can be exploited to measure the average diameter of the CNTs (eqn (4.1)), while the line width provides information on the number of walls and/or homogeneity of the samples.[83]

$$\delta = 18.3\,\mathrm{D} + 102.5\,(\text{ppm relative to TMS}) \tag{4.1}$$

Table 4.6 summarises the chemical shift assignments for O-, P- and N-containing groups and different C atoms for carbon (nano)materials.

4.3 Active Phase Deposition

The choice of the coating method depends largely on the application of the coated materials and the scale of operation. Since metallic nanoparticles/nanostructured carbon composites have found a large range of applications, a broad choice of preparation procedures exists, some of them being not so familiar to the catalysis community.[2,11–14] Electrochemical coating methods[84] are obvious choices in electrocatalytic applications. The wet-impregnation method, the most widely used technique for catalyst preparation, is a multistep process that consists of the deposition of a precursor of the coated material followed by calcination and reduction. Self-assembly techniques,[85] though efficient and precise, are difficult to upscale due to the high cost of the reagents. In this section, the different experimental methods developed for coating elements such as metals, oxides or semimetals have been described. Each method is illustrated by representative examples and the list of references is not exhaustive. In all cases it has been found that an efficient coating is possible when the carbon surfaces are suitably functionalized or modified.

4.3.1 Deposition from Solution

4.3.1.1 Wet and Incipient Wetness Impregnation

Due to its simplicity, impregnation is one of the most commonly used techniques for depositing metals on the carbonaceous support. Nanostructured carbon materials can be impregnated with catalyst precursors by mixing the two in an aqueous solution. This is followed by a reduction step, which is required to reduce the catalyst precursor to its metallic state. As reduction occurs after the impregnation step, the nature of the support plays a crucial role in controlling particle size. The reduction step can be chemical or electrochemical. Drawbacks of the impregnation technique are related to using liquid solutions as the processing medium. The easy agglomeration of particles in solution has been observed.

Two different techniques of impregnation are practiced, *i.e.*, wet impregnation (WI) and incipient wetness impregnation (IW or IWI). In WI the support is brought into contact with a large excess of solution containing the metal-precursors. After solvent evaporation, an optional calcination, and reduction the final catalyst is obtained. In general WI results in large particle size since the majority of the metal precursor is present in the solution outside the pore system of the support. Additionally, during solvent evaporation, the metal-precursors often deposit on the outer surface of the support particles resulting in large crystals. When an excess of solution is used but the solvent evaporation step mentioned above is replaced by filtering of the excess of solution, only the adsorbed species form the active (precursor) phase. This method is called ion adsorption (IA).

To make effective use of the pore structure of the support, IWI (also called capillary impregnation or dry impregnation) can be performed. Typically, the active metal precursor is dissolved in an aqueous or organic solution. Then the metal-containing solution is added to a catalyst support containing the same pore volume as the volume of the solution that was added. Capillary action draws the solution into the pores. Solution added in excess of the support pore volume causes the solution transport to change from a capillary action process to a diffusion process, which is much slower. The catalyst can then be dried and calcined to drive off the volatile components within the solution, depositing the metal on the catalyst surface. The maximum loading is limited by the solubility of the precursor in the solution. The concentration profile of the impregnated compound depends on the mass transfer conditions within the pores during impregnation and drying. Compared to WI, a closer contact between metal-precursor and support is achieved for IWI, which in general results in smaller particles, provided the drying, calcination and/or reduction are executed carefully.

All theory on impregnation and ion adsorption is based on the coulombic interaction between the support and active-phase precursor.[86] This is crucial for obtaining a good distribution of the precursor over the support. Basically, a negatively charged surface, *i.e.*, using a pH of the impregnating solution above the point of zero charge (PZC), attracts positively charged precursors, while positively charged surfaces, *i.e.*, below the PZC, attract negatively charged precursors. The PZC determination can be said to be the first step in carbon supported catalyst preparation. It is a very important parameter used to determine the pH range at which the impregnation step should be carried out. The PZC experiment is carried out by contacting the different carbons with pH adjusted, deinoized water.[15] An eleven point PZC measurement is the most common experiment used to obtain the PZC. Figure 4.6 shows an electrostatic interaction between the carbon support and platinum precursor. Mainly, if the carbon surface is positively charge an anionic Pt complex is needed, *i.e.* $[PtCl_6]^{2-}$. If the carbon surface is negatively charge a cationic Pt complex is needed, *i.e.* $[Pt(NH_3)_4]^{2+}$.

The WI method involves the impregnation of the carbon surface with a solution of the metal precursor salt/organometallic complex, followed by solvent evaporation or filtration, calcinations and reduction to give metal particles anchored to the carbon surface. The reduction is carried out either by using H_2 gas or by common reducing agents such as sodium borohydride or ethylene glycol.[11] Indeed, it has been shown that metal-catalyzed etching by H_2 can damage the carbon nanostructure.[87] This method is the most commonly used, mainly due to its simplicity, cost and ease of up-scaling.

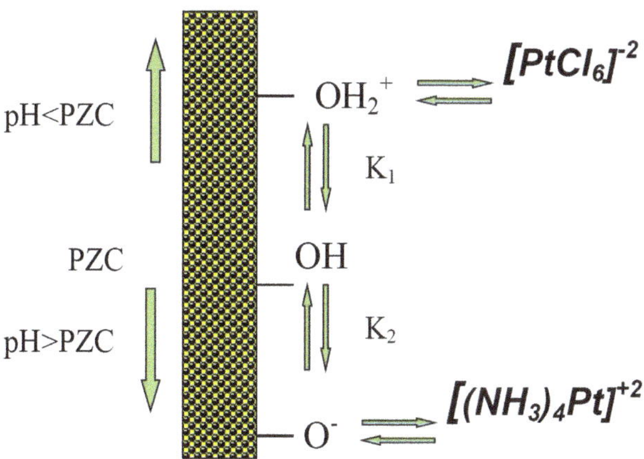

Figure 4.6 Adsorption of platinum precursor on carbon surface according to PZC.

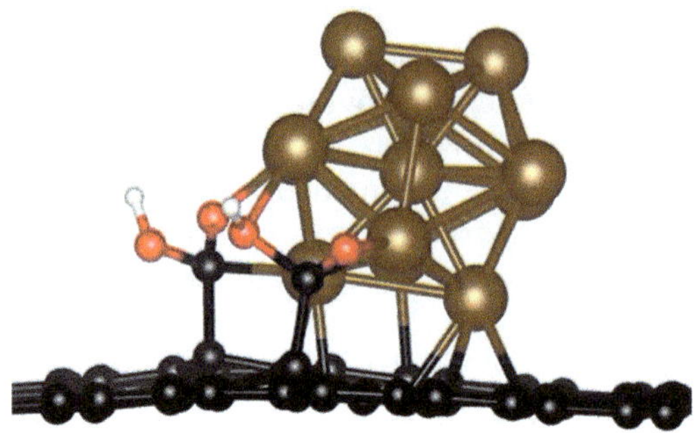

Figure 4.7 Side-views of a Ru_{13} cluster adsorbed on a double vacancy decorated with two carboxylic groups. Adapted from ref. 87.

A first example of such a WI coating was reported as early as 1994 where CNTs were coated by Ru nanoparticles.[88] It has been recently shown that the oxidation of CNTs by nitric acid creating various oxygen surface functional groups on the CNT external surface is a crucial step for Ru grafting from an organometallic Ru(0) complex.[87] In particular, it was demonstrated that carboxylic acid, carboxylic anhydride, and lactone groups act as anchoring centers for the Ru precursor, presumably as surface acetato ligands (Figure 4.7). Lower metal loading was obtained by IA on pristine than on oxidized MWCNTs. The Ru/CNT catalysts prepared on oxidized CNTs are extremely stable, keeping high dispersion and small nanoparticle size even after heat treatment at high temperature in the presence of H_2. This remarkable stability can be attributed to the low edge/basal plane ratio present on the CNT support. Indeed, similar catalysts prepared on CNF or FLG are not so stable, due to catalytic etching. The Ru particle size, its dispersion and its interaction with the support were also clearly influenced by acidic oxygen-containing group concentration on CNF supports.[89]

Uranyl acetate was also used due to its relative ease of observation by electron microscopy even at low concentrations.[90] Arc produced MWCNTs were treated with concentrated HNO_3 that not only dispersed the CNTs in solution[91] but also created active carboxyl groups that were replaced by uranyl groups by simple ion exchange. The raw nanotubes on the other hand were not coated by uranyl acetate, thus leading to the conclusion that oxidizing the nanotubes is an essential preparative step before coating. Following this work, acid treated CNTs were decorated with nanoscale Pt, Au and Ag clusters by reaction in solution of their corresponding oxo acids followed by hydrogen reduction.[92] Once again, the importance of acid functional groups on the CNT surface for effective coating of the nanoparticles was demonstrated. This was confirmed for Fe/CNT,[93,94] Co/CNT,[95] Pt/CNT,[96] Ni/CNT,[97] Au/FLG,[98] Au on various carbon materials,[99] or CuO/FLG.[100] Figure 4.8 shows the results obtained for Ru nanoparticles grafted on pristine (Figure 4.8(a)) and HNO_3 treated FLG (313 K (Figure 4.8(b)) and

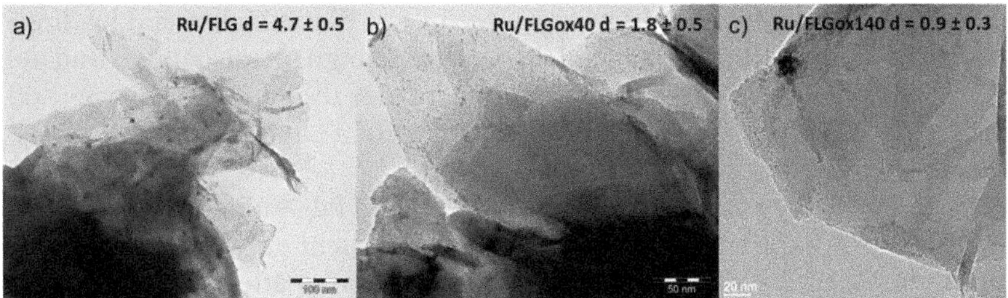

Figure 4.8 Micrographs of a 3% wt. % Ru/FLG catalyst prepared on (a) pristine FLG (scale bar = 100 nm); (b) FLG treated at 313 K with HNO_3 (scale bar = 50 nm); and (c) FLG treated at 393 K with HNO_3 (scale bar = 20 nm).

413 K (Figure 4.8(c))). The mean particle size, measured by TEM was 4.7 ± 1.5, 1.8 ± 0.5, and 0.9 ± 0.3 nm for pristine, HNO_3-313-FLG and HNO_3-413-FLG, respectively.

Carbon nanofibers can also be treated in a similar way,[101] but on this support that presents a high edge/basal plane site ratio, nitric acid oxidation is not a prerequisite to reach high metallic dispersion.[102] Another study has evidenced that the functional groups introduced by oxidation improve significantly the ion-exchange capabilities of CNTs.[103] It was reported that HNO_3-oxidized CNTs are negative net surface charged, which decides that the CNT surface will interact with the cations of the solution. When H_2PtCl_6 is used as precursor, adsorption of $PtCl_6^{2-}$ anions is weak, leading to a very poor dispersion.[104] The introduction of nitrogen functionalities in the carbon matrix also significantly improves the ion-exchange capabilities of carbon materials.[105] Thus high metal dispersions are generally achieved on nitrogen-doped nanostructured carbon materials.[106–112] Molecular dynamics simulations have shown that the nitrogen atoms act as donor-like atoms and not directly as binding sites for the deposited metal active phase.[106] A review article has presented the different surface functionalization approaches that provide efficient avenues for the deposition of metal nanoparticles on CNTs.[113]

4.3.1.2 Homogenous Deposition Precipitation Method (HDP)

The HDP technique has been developed for the preparation of highly loaded and highly dispersed oxide-supported metal catalysts. With HDP, a solvated metal precursor is deposited exclusively onto the surface of a suspended support by deposition precipitation, which can be realized in a number of ways, such as (i) increase of pH; (ii) change of valence of the metal ion; or (iii) removal of a stabilizing ligand of the metal ion.[114] Thus, the slow and homogeneous introduction of a precipitating agent, generally hydroxyl ions, avoids nucleation of a solid precursor compound in the bulk solution. Generally, the hydrolysis of urea at 323–373 K is used to achieve the required slow and homogeneous increase in pH. Among the advantages of HDP over other preparation techniques we can mention (i) its reproducibility, (ii) the high metal loadings that can be achieved, (iii) high metal dispersions at high metal loadings and (iv) a uniform distribution of the active component over the support. This method has been successfully applied to the preparation of a large variety of catalysts, including Pt and Ru/CNF,[115] Co/CNF,[116] Ni/CNF,[117,118] Fe/CNT,[119] Ru/CNF,[120] Co/CNT and Co/Cspheres,[121] Au/CNT,[122] Au/FLG,[123] and Pd–Au/FLG.[124]

While most of the time the metal coating is efficient only on activated carbon nanomaterial surfaces, the technique of impregnation should also be carefully chosen. The group of de Jong has studied the HDP for Pt[115,125] and Ni[117,118] deposition on CNFs, and compared this method to WI. For example, the coating of Ni from nickel nitrate on oxidized CNFs by WI and HDP has been studied.[117] In the first method, nickel nitrate solutions were stirred with acid functionalized CNFs and the composite was dried and reduced in H_2+Ar atmosphere at 773 K to generate nickel particles on the surface of the CNFs. The second method consisted of reacting nickel nitrate and urea in the presence of CNFs so as to precipitate a layer of nickel oxide on the nanofibers that was subsequently reduced in H_2+Ar atmosphere as described above. It was reported that the HDP process yielded a better control of particle size when compared to simple WI. For low Ni loadings (10 wt%), the HDP method gave a particle size distribution of 5–7 nm when compared to 9–11 nm for wet impregnation. For higher Ni loadings (30%) the HDP procedure allowed a much better control of particle size (8–10 nm) compared to wet-impregnation (8–60 nm). Further, the deposition precipitation of 20 wt% Ni onto oxidized CNFs (CNF–O), followed by partial reduction (CNF-OR) or thermal treatment (CNF-OT) has been studied.[118] The CNF–O contains acidic groups of the carboxylic-type. On such supports, extensive adsorption occurs right from the start of the deposition precipitation. At the maximum of the pH curve, already 22% of the nickel present has been deposited onto CNF–O. Nucleation of nickel hydroxide onto adsorbed nickel ion clusters proceeded subsequently. Characterization of the dried Ni/CNF–O samples with TEM and XRD showed well dispersed and thin (5 nm) platelets of nickel hydroxide adhering to the CNFs. After reduction at 773 K in hydrogen the Ni/CNF–O contained metallic nickel particles of 8 nm thickness homogeneously distributed over the fibers. After removal of the surface acidic groups by either chemical reduction (CNF-OR) or thermal treatment (CNF-OT), deposition precipitation takes place separately from the support and large nickel hydroxide platelets (>500 nm) are observed. This study shows clearly that for HDP the presence of carboxylic acid groups is essential to successfully deposit metal on nanostructured carbon materials.

CNF supported Pt catalysts have been prepared using $[Pt(NH_3)_4(NO_3)_2]$ as a precursor by two different ion adsorption techniques, WI and HDP.[115] With both synthesis techniques homogeneously distributed, highly dispersed and thermally stable metal particles were obtained from diluted precursor solutions with an average particle size of 1–2 nm on oxidized CNFs. With the HDP method for the Pt/CNF catalysts, a linear relationship between the number of acidic oxygen-containing groups on the surface of activated

CNF and the metal loading has been found. Furthermore, it has been established that with this procedure higher platinum loadings (*ca.* 4 wt%) can be achieved than with the WI procedure (<2 wt%). It was concluded that the acidic oxygen-containing groups not only increase the wettability but are also the anchoring sites of the metal precursor during the preparation. A study on the nature and role of oxygen-containing surface groups for Pt deposition on CNFs by HDP from $[Pt(NH_3)_4(NO_3)_2]$ has also been reported.[125] Synthesis of Pt/CNF *via* HDP results in metal anchoring on carboxyl as well as phenol oxygen surface groups. It was proposed that during synthesis, both groups become deprotonated and are able to bring about platinum ion adsorption.

4.3.1.3 Microwave Heated Polyol Process

Microwave irradiation through dielectric heating loss is fast, simple, uniform, energy efficient and has been widely used for carbon-based material synthesis.[126–128] The polyol process, in which an ethylene glycol (EG) solution of the metal precursor salt is slowly heated to produce colloidal metal, has recently been extended to produce metal nanoparticles supported on carbon substrates. In this process, a polyol (most commonly EG) solution containing the metal precursor salts is refluxed at 393–443 K, where the polyol decomposes homogeneously to release the reducing agent for metal ion reduction. A support material may be optionally present to capture the depositing metal particles. The size distribution of the metal particles depends largely on the experimental conditions, mainly the polyol/water concentration ratio.

Thus, the microwave assisted polyol process was employed to coat PtRu alloy nanoparticles of uniform size on MWCNTs or CB.[129–133] In this process, the polyol solution containing the metal salt is refluxed at 393–413 K to decompose ethylene glycol that acts as a reducing agent leading to the formation of ultrafine nanoparticles (<3 nm for 20 wt% total metal). It appears that the size of the metal nanoparticles is determined by the rate of reduction of the metal precursor and hence the use of microwave techniques for the decomposition reaction is significant.[131] The dielectric constant (41.4 at 298 K) and the dielectric loss of ethylene glycol are high, and hence rapid heating occurs easily under microwave irradiation. The fast heating by microwave accelerates the reduction of the precursor and results in the instantaneous nucleation of the metal clusters. Additionally, the homogeneous microwave heating of liquid samples reduces the temperature and concentration gradients in the reaction. It was also found that the polyol process produced alloy compositions that are not consistent with the metal ratios in the precursors.[133] EG is only a mild reducing agent requiring a long reaction time for the reduction reactions to go to completion, even for the preparation of Pt/CNT. Its reducing strength, or power, may be strong enough for reduction of Pt ions but a little bit too weak for the reduction of Ru ions. Suitable amounts of sodium hydrogen sulfite ($NaHSO_3$) and calcium hydroxide aqueous solutions were employed as additives to EG forming a modified reducing agent.[129] This new approach was based on the use of more suitable

PtRu deposition environments created by the modified EG near the isoelectric point (IEP) on the acid-oxidized CNT surfaces. Polyoxometallate-stabilized PtRu alloy nanoparticles supported on MWCNTs were also synthesized by a microwave-assisted polyol process.[134] The polyoxometallate formed a self-assembled monolayer on the surface of the PtRu nanoparticles and MWCNTs, which effectively prevented the agglomeration of PtRu nanoparticles, due to the electrostatic repulsive interactions between the negatively charged polyoxometallate monolayers.

For SWCNTs, it is necessary to exfoliate bundles into individual tubes by using a suitable surfactant or a stabilizer molecule to achieve uniform coating.[135] A modified polyol process for the coating of Pt particles on SWCNTs has been reported where polymer wrapping of the nanotubes has been used for the effective dispersion of SWCNTs and to control of the deposition of Pt catalysts. SWCNT bundles were dispersed into individual tubes by dispersing them with polystyrene sulfonate on which Pt was deposited by wet impregnation of H_2PtCl_6 and reduction to Pt was achieved using ethylene glycol at pH = 12.5. The Pt loading could be increased up to 30 wt% and 2–3 nm size Pt particles were deposited in high density. Another approach consists of producing the metallic nanoparticles (Pt or PtRu) with the polyol process and then to deposit them on SWCNTs.[136] In that case, the use of a stabilizer, like a thiol, is mandatory to produce the nanoparticles.

Pt/rGO catalysts were synthesized by employing a fast microwave-assisted polyol process, which facilitated the simultaneous reduction of graphene oxide and formation of Pt nanoparticles.[137,138] It was found that the high concentration of oxygen functional groups on rGO plays a major role on the removal of carbonaceous species on the adjacent Pt sites, underlining a synergetic effect between the oxygen moieties on graphene support and Pt nanoparticles. Detailed analysis of the mechanism of formation of the hybrids indicates a synergistic co-reduction mechanism whereby the presence of the Pt ions leads to a faster reduction of GO and the presence of the defect sites on the reduced GO serves as anchor points for the heterogeneous nucleation of Pt.[137] In the reaction system for the preparation of PtRu/FLG catalysts, EG acts also as solvent and reducing agent for reduction of PtRu nanoparticles from their precursors and reduction of graphene from graphene oxide.[139] The preparation of graphene–metal particle catalysts in a water–ethylene glycol system using graphene oxide as a precursor and metal nanoparticles (Au, Pt and Pd) as building blocks has also been reported.[140] These metal nanoparticles are adsorbed on graphene oxide sheets and play a pivotal role in catalytic reduction of graphene oxide with ethylene glycol, leading to the formation of graphene–metal particle nanocomposites.

Interestingly, it has been reported that the direct solid-state Joule heating of CNTs or graphene with organic metal salts using microwave irradiation is a highly effective and versatile approach for the preparation of a wide variety of metal and metal oxide nanoparticle-decorated substrates.[141] Although reactions using conventional heating may usually take at least an hour to complete, microwave heating of solid metal salt/substrate mixtures resulted in the nearly instantaneous formation of metal or metal oxide NPs at excellent conversion yields. The solvent-free microwave heating method is a much more rapid, and thus higher throughput, alternative to the effective and scal-ableconventional heating method previously reported.

Pt NPs were deposited on bucky diamond (BD) and nanodiamond (ND, average particle size of 5 nm) supports using a microwave-assisted reduction polyol method from $H_2PtCl_6{\cdot}6H_2O$ and ethylene glycol.[142] Surface graphitization of ND led to the formation of core-shell structural bucky diamond with a nanoscale diamond core covered by a graphitic shell. The results showed that BD had a higher affinity with Pt metal than ND, resulting in a higher dispersion of Pt nanoparticles on the BD (3–5 nm on BD *vs.* 3–10 nm on ND). Platinum NPs supported on undoped ND with an average particle size of 50 nm were prepared using a microwave-heating polyol method.[143]

The polyol process was also used to prepare other catalytic systems including RuO_2/MWCNT,[144] Ru/MWCNT,[145] Au/MWCNT,[146] Pd-, Ni- or Sn/MWCNT,[147] PtSn/MWCNT,[148] PtZn/MWCNT,[149] PtCo/MWCNT,[150] PtNi/N-MWCNT,[151] PtAuSn/$C_{nanoparticles}$,[152] Pt/CNF,[153] PtRu/CNF,[154] Pt/CB,[155] PtSn/CB,[156] Au-, Pt- or Pd/FLG,[140] Au/FLG,[157] and Co_3O_4/FLG.[158]

4.3.1.4 Electro-deposition

Electrochemistry is a powerful technique for deposition of various nanoparticles (especially metal NPs), as it enables effective control over nucleation and growth of nanoparticles. Electro-deposition offers many advantages over high-temperature metal deposition for metal NP formation on nanostructured carbon materials. Electro-deposition is the process whereby metal particles are produced during the reduction of a metal salt solution at the electrode (cathode). In conventional electroplating, the anode is made of the material to be coated that dissolves in the electrolyte. The polarization of the cathode results in the metal salts losing their charge and plating out on the surface of the cathode. However, reduction or oxidation can also occur in solution, the anode or cathode merely serving to transport the charges. Another common electrochemical method is electrophoretic deposition wherein colloidal particles of metal suspended in a liquid medium migrate under the influence of an electric field and are deposited onto an electrode that is the substrate to be coated. In contrast to electro-deposition predominantly used for metal coating, all colloidal particles that can be used to form stable suspensions and that can carry a charge can be used in electrophoretic deposition. One of the most significant advantages of electrochemical deposition is the ability to control the size and distribution of NPs by varying potential, time or concentration of the noble metal salts. The parameters controlling NP number density, distribution, and size have been identified, with short deposition times and high driving forces favoring the formation of ultra-small particles at high density. In particular, as shown in detail for Pd, control over the deposition potential and time allows the number density, distribution, and NP size to be controlled.[159] Additionally, novel lead structures ranging from nanowires and mesoparticles with octahedral, decahedral, and icosahedral shapes to porous nanowires, multipods, nanobrushes, and even snowflake-shaped structures were synthesized on HOPG through systematically exploring electro-deposition parameters, including reduction potentials, solution concentration, starting materials, supporting electrolytes, and surfactants.[160] 3D flowerlike Pt NPs were also electrodeposited onto MWCNTs by using a three-step protocol, which is all-electrochemical and involves a key, second step of a potential pulse sequence.[161]

Besides, electrochemical deposition method also has the following advantages: noble metal NPs are produced with very high purity, form rapidly and have good adhesion to the CNT substrate.

Most studies involving metal NP electro-deposition focus on noble metals and their alloys, particularly Pt,[162,163] but also PtRu,[164] PtAu,[165,166] PtPb,[167] Ag, Au, and Pd, with a few exceptions, such as Ni and Cu, primarily due to the need for components for alternative energy sources.[168] CNTs typically do not react with the noble metal salts but act as molecular conducting wires and supports for the deposition of noble metal NPs. As for other preparation methods, oxygen functionalities serve as axial ligands for metal NP precursors to bind to the carbon surface. Therefore, the most common pretreatment methods involve treating them with strong acids or oxidizing agents such as H_2SO_4/HNO_3, H_2SO_4/H_2O_2, HNO_3, O_3, and $KMnO_4$.[168] It has also been shown that anodic activation of the CNTs to introduce carboxyl functionalities to the ends and defects along the sidewalls prior to electro-deposition did not give a preference for the ends or a wider distribution of particle sizes along the side walls.[84]

Platinum NPs were electrochemically deposited at the surface of well-aligned MWCNTs by potential cycling between +0.50 and −0.70 V at a scanning rate of 50 mV s^{-1} in 5 mM Na_2PtCl_6 solution containing 0.1 M NaCl.[169] The diameters of the Pt NPs were around 40–70 nm. Pt NPs tend to deposit primarily on the curved tips and evenly distribute along the sidewalls, which may be due to the different reactivity of carbon atoms, defects and functional groups at MWCNTs. Pt nanowires are formed at MWCNTs with an increase of Pt loading. In another study, it was shown that the Pt nanoparticles (*ca.* 10 nm) that are electrodeposited by applying a high number of repeated potential steps from +0.5 to −0.7 V at the MWCNTs show snowflake-like morphology.[170] Compared to those electrodeposited by applying long step widths, they are larger, have a higher specific surface area and higher roughness factor, *i.e.* higher porosity. Pt NPs electrodeposited at the loosely grown MWCNTs are well dispersed without coalescence at a Pt surface coverage of more than 95%. On SWCNTs, 4–6 nm Pt NPs were electrodeposited from K_2PtCl_4.[171] It was also reported that EG can enhance the dechlorination of the Pt precursor salt ($H_2PtCl_6{\cdot}6H_2O$) and led to the formation of Pt nanoparticles of 4.5 nm on CNTs.[172] In the meantime, EG also acts as a stabilizing surfactant to prevent the particles from agglomeration during the electro-deposition processes. A similar effect was noticed for Pt/CB.[173] Compared with the particles electrodeposited from a solution without EG, those with EG were smaller (4 nm) and well dispersed on the CB surface. A simple method for achieving high dispersion and small Pt NPs, down to only 2–3 nm, on structured carbon supports (CNTs-modified PAN-based carbon fiber and CNTs-modified graphite foil) was also reported.[174] Pulsed electro-deposition of Pt NPs was performed at increased viscosity of the H_2PtCl_6 containing electrolyte by addition of glycerol. The catalyst NP size can be controlled by varying the amount of glycerol added into the aqueous H_2PtCl_6 solution, and adjusting the number of the potential pulses. Finally, the galvanic displacement of an under-potentially deposited Cu monolayer by Pt provides a uniquely convenient way of uniformly depositing a metal monolayer on carbon-supported metal nanoparticles in a surface-limited reaction controllable to a small fraction of a monolayer.[175]

A three-dimensional interconnected graphene monolith was used as an electrode support for pulsed electrochemical deposition of Pt NPs.[176] Pt NPs with well-defined morphology and relatively small size (*ca.* 30 nm) were obtained by controlling electro-deposition potential and time. On multilayered graphene nanosheets, Pt nanoparticles of different size (10–300 nm) and morphology (cauliflower-like, needle-like, rose-like) were electrodeposited by adjusting the current pulses.[177] Highly ordered multi-layered 3D graphene structures decorated with Pd, Pt and Au metal NPs were prepared.[178] The ability to control the morphology, distribution and size of the metal NPs on the 3D graphene support upon changing the electro-deposition conditions was demonstrated. Au nanoflowers and nanoparticle structures (30–130 nm) and Pd nanocubes (30–200 nm) were obtained following electro-deposition onto the 3D graphene support. Large Pt NPs (100 nm) on 3D graphene were spherically shaped with rough surfaces.

Undoped ND powder with average particle sizes of 5 and 100 nm were modified with Pt NPs by electrodeposition from $H_2PtCl_6 \cdot 6H_2O$ solution.[179] The preparation of Pt-modified diamond electrodes by electro-deposition is known to be hampered by poor particle adhesion and a lack of uniformity in the spatial distribution of the deposit over the electrode surface. It was demonstrated that the results can be improved significantly if the electrode is given a simple ultrasonic treatment in the presence of diamond powders prior to electro-deposition.[180] An improvement in spatial distribution and a higher Pt dispersion have been seen and, especially, a greater Pt particle stability is observed. The unsupported electro-deposition of Pt on carbon nano-onions (CNOs) has also been reported.[181] On such supports, extremely small Pt clusters were produced. The electro-deposition process behaves similarly to that for CB as a carbon support. Although both Pt/CNOs and Pt/CB have the same metal loading (11 wt%), the CNO samples exhibited enhanced thermal stability.

For bimetallic NPs, an electrochemical method was developed to deposit Pt and Ni NPs on MWCNTs through an established three-step process but with important improvements.[182] The improvements consist of: (i) generation of oxygen functional groups (quinoid, carbonyl and carboxyl) at the defect sites located at the ends and/or the sidewalls of the CNTs by potential cycling; (ii) oxidation of $PtCl_4^{2-}$ and Ni^{2+} to complexes of Pt(IV) and Ni(III) on the MWCNT surface from $K_2PtCl_4 + NiCl_2 + 0.1$ M K_2SO_4; and (iii) conversion of the surface complexes on the MWCNTs to platinum and nickel NPs through potential cycling from +1.0 V to −0.26 V in 0.1 M H_2SO_4 solutions to the steady state. A homogeneous dispersion of NPs largely spherical in shape and approximately 10 nm in size has been obtained. A facile one-step ultrasonication-assisted electrochemical method to synthesize nanocomposites of graphene and PtNi alloy NPs was reported.[183] It was demonstrated that the obtained nanocomposites exhibit a collection of unique features including well-dispersed NPs with alloy features, high NP loading, and effective reduction of graphene oxide. During the growth stage, the size of loaded metal NPs on the GO surface can be controlled by reaction time in the electro-deposition. These NPs exhibit faceted characteristics and their shapes are mostly spherical, with a diameter of about 80 nm.

4.3.2 Deposition from the Gas Phase

Though not as prevalent as wet-impregnation and electrochemical coating, gas phase coating technologies such as electron beam evaporation, sputtering, atomic layer deposition (ALD) and chemical vapor deposition (CVD) have been employed for the coating of metals and oxides on nanostructured carbon material surfaces. These methods have the advantage of often being a single step process with high coating speeds and can be particularly adapted to the large scale production of coated samples provided a good gas–solid contact can be obtained. In principle, by tuning the evaporation conditions, highly reproducible samples can be obtained if suitable precursors are available that are sufficiently volatile and do not decompose in the homogeneous phase under the experimental conditions. However, sample size is often limited by stringent experimental conditions such as high vacuum in the case of electron beam evaporation and the use of expensive and often toxic precursors in organometallic CVD reactions. Nevertheless, gas phase techniques yield purer products without the formation of insoluble residues that may occur during wet impregnation or electrochemical deposition.

4.3.2.1 Physical Vapor Deposition – Electron Beam Evaporation

In physical vapor deposition, a high kinetic energy beam of electrons is directed at the material for evaporation. Upon impact, the high kinetic energy is converted into thermal energy allowing the evaporation of the target material. This method is mainly used for the preparation of model catalysts.[184] Oxygen and argon plasmas were used to modify MWCNTs to improve their interfacial interaction with subsequently electron beam deposited Pt NPs.[185] In contradiction to what was found in the case of highly oriented pyrolytic graphite, XPS confirms the introduction of chemical functionalizations by oxygen plasma treatment; however, as in the case of HOPG, argon plasma treatment produced physical defects. Transmission electron microscopy provided visual evidence of the interaction of subsequently evaporated Pt with treated CNTs, showing it to have been enhanced by both plasma treatments. XPS and TEM analyses demonstrate that the enhancement is due to similar interactions of Pt NPs with both types of treated CNTs, although not to the same extent: XPS gives no evidence of chemical bonds formed for either plasma treatment. The morphology of the Pt NPs changes with the deposition rate, which may be influenced by the limited availability of the CNT surface. A simple electron beam evaporation process was performed to prepare a composite electrode PtAu/MWCNTs.[186] The diameter of nano-Au/nano-Pt was less than 10 nm. The interaction between evaporated Pd and pristine or oxygen plasma-treated MWCNTs was also investigated.[187] Pd is found to nucleate at defective sites, whether initially present or introduced by oxygen plasma treatment. The plasma treatment induced a uniform dispersion of 3–5 nm Pd NPs at the CNT-surface.

It was observed that atoms deposited by physical vapor deposition onto FLG condense upon annealing to form NPs with an average diameter that is determined by the graphene film thickness.[188] Different metal NPs (Au, Al, Ti, Cr, Pd, and Ni) were decorated by electron beam evaporation on graphene.[189] All studied metals favor sites on the omnipresent hydrocarbon surface contamination rather than on the clean graphene surface and present non-uniform distributions, which never result in continuous films but instead in clusters or nano-crystals, indicating a weak interaction between the metal and graphene. This behavior can be altered to some degree by surface pretreatment (hydrogenation) and high-temperature vacuum annealing. Damaging of graphene has been observed in the presence of Al, Ti, Cr, Pd, and Ni, although their interaction with graphene varies; for example, Al, Cr, and Ti are much more reactive than Pd and Ni. No damaging was observed in the case of gold. Silver metal NPs were decorated by electron beam evaporation on graphene foam.[190] The graphene foam was decorated with Ag NPs that are spherical and agglomerate on the surface. The average particle size was estimated to be 106 nm.

4.3.2.2 Atomic Layer Deposition (ALD)

In this process, an appropriate precursor spray or a reactive gas is alternately pulsed onto the support and the chamber is purged with inert gas in between the cycles. All the processes take place at relatively low temperatures. The precursor coats the active sites on the support in a monolayer making it a self-terminating reaction. The excess of the precursor is removed by the purging gas that helps in controlling the film thickness and also results in the uniform coverage of substrate inhomogeneities such as trenches. The reactant gas is sent in and the precursor molecules react to produce a thin film. The number of pulses determines the thickness of the film. The excellent conformality that can be achieved with ALD also renders it a promising candidate for coating porous structures, *e.g.* for functionalization of large surface area substrates for catalysis, fuel cells, batteries, super-capacitors, *etc.*[191,192]

Conformal deposition of Pt NPs with good dispersion on CNTs was performed by atomic layer deposition.[193] Methylcyclopentadienyl trimethylplatinum [$MeCpPtMe_3$] and air or oxygen have been used as precursors to grow a uniform and thickness-controllable particle-like Pt film on a 3.5 μm-thick layer of acid-treated CNTs. The amount of Pt loading on carbon cloth and the nucleation densities with 100 cycles of ALD increases as the acid-treatment time increases. A similar procedure was used to produce highly dispersed Pt NPs (1.3 nm) on gram quantities of non-functionalized MWCNTs by ALD in a fluidized bed reactor at 573 K.[194] The porous structures of MWCNTs did not change with the deposition of Pt NPs. Platinum NPs were also deposited on oxygen plasma treated CNTs by ALD from [$MeCpPtMe_3$] and oxygen.[195] X-Ray photoelectron spectroscopic analysis indicated that oxygen plasma can graft oxygen-containing functional groups to the CNT surface to act as nucleation sites for growth of Pt NPs. Formation of very uniform and well distributed Pt NPs of a size of 1.6–4.8 nm was achieved. The growth rate of Pt NPs could be controlled by the number of ALD cycles and oxygen plasma treatment time.

Thin tubular TiO_2 coating on MWCNTs were prepared by ALD from tetrakis(dimethylamido)titanium and ozone.[196] The original tube diameter was approximately 10 nm and broadened during ALD to *ca.* 16 nm, meaning that a 3 nm thick TiO_2 layer was deposited onto the MWCNTs, which corresponds to a growth rate of *ca.* 0.03 nm per ALD cycle. ZnO NPs were grown on SWCNTs by ALD using diethylzinc and water.[197] Because of the chemical inertness of nanotubes towards diethylzinc and water molecules, such NPs are not likely to grow on the wall of clean and perfect nanotubes. Rather, the growth of ZnO NPs should be attributed to imperfection of CNTs, such as defects and carbonaceous impurities. Ultra-thin (2–5 nm thick) aluminum oxide layers were grown on non-functionalized individual SWCNT[216] and their bundles by ALD.[198,199] Bundles of SWCNTs were coated continuously, but at the same time, bare individual nanotubes remained uncoated. The successful coating of bundles was explained by the formation of interstitial pores between the individual SWCNTs constituting the bundle, where the precursor molecules can adhere, initiating the layer growth.

The formation of coaxial nanotubes of CNT@Fe_2O_3 *via* ALD using ferrocene and oxygen as precursors was reported.[200] It was found that, of the employed undoped and N-doped CNTs, the undoped ones needed chemical functionalization prior to ALD due to their inert surface nature. Even so, the undoped CNTs functionalized with HNO_3 and/or SDS were only coated with Fe_2O_3 in an arbitrary manner, *i.e.*, the as-deposited Fe_2O_3 was random and non-uniform. In sharp comparison, N-CNTs were more favorable for a uniform and tunable ALD, which was ascribed to their chemically active surface nature induced by intrinsically incorporated N atoms. ALD was also used for the preparation of highly dispersed Ru-decorated Pt NPs onto N-doped MWCNTs at 523 K using [$MeCpPtMe_3$] and bis(ethylcyclopentadienyl)ruthenium [$Ru(EtCp)_2$] and O_2 as the precursors.[201] The average particle size is between 1.5 and 6 nm for ALD cycles between 323 and 423 K.

A practical synthesis for Pt single atoms, sub-nanoclusters, and nanoparticles uniformly deposit on the surface of graphene nanosheets using the ALD technique from [$MeCpPtMe_3$] and oxygen was reported.[202] It was shown that the morphology, size, density and loading of Pt on graphene can be precisely controlled by simply adjusting ALD cycle numbers. An ALD method to synthesize Fe_3O_4/graphene and Ni/graphene composites was reported.[203] The surfaces of graphene are densely covered by narrowly distributed Fe_3O_4 or Ni nanoparticles.

ALD was used to synthesize TiO_2-graphene nanosheet nanocomposites using titanium isopropoxide and water as precursors.[204] The synthesized nanocomposites demonstrated that ALD exhibited many benefits as a controllable method. It was found that the as-deposited TiO_2 was tunable not only in its morphologies but also in its structural phases. As for the former, TiO_2 was transferable from nanoparticles to nanofilms with increased cycles. With regard to the latter, TiO_2 was changeable from the amorphous to crystalline phase. Uniform TiO_2 NP coatings have been also achieved on a porous graphene network using TiO_2 ALD from $TiCl_4$ and water.[205] The three-dimensional open channels in the composites provide exceptional high utilization of the TiO_2 NPs. 323 and 423 K TiO_2 ALD cycles were employed to control the particle size, which varied from 4–10 nm.

Other systems prepared by ALD included V_2O_5/MWCNT,[206,207] SnO_2/N-MWCNT,[208] SnO_2/FLG,[209] and RuO_2/MWCNT.[210]

4.3.2.3 CVD Coating Methods

Chemical vapor deposition (CVD) is a chemical process used to produce high-purity solid materials. In typical CVD, the support is exposed to one or more volatile precursors, which react and/or decompose on the substrate surface to produce the desired deposit.[211] Frequently, volatile by-products are also produced, which are removed by gas flow through the reactor. To ensure homogeneous deposition on powders, the use of fluidized bed reactors is often preferred.[212,213] To date, CVD methods are limited in application mainly due to experimental difficulties encountered in obtaining a uniform coating on nanostructured carbon materials. Generally, organometallic complexes are needed to ensure high volatility of the precursor. Luckily, a variety of organometallic precursors can be efficiently synthesized and are available readily for noble metals and some representative examples are presented.

Controlled-growth of platinum NPs on graphitic substrates such as carbon nanospheres[214] or MWCNTs[215] was achieved by the CVD of [$PtMe_2$(COD)] in a fluidized bed reactor. Under the operating conditions, the platinum germination is controlled by surface chemical reactions, and it was necessary to create oxygen containing groups to deposit Pt.

Fuel-cell electrode catalysts have been prepared by dispersing Pt NPs onto CNT using a CVD method.[216] The [$MeCpPtMe_3$] precursor has been used as a Pt precursor in the CVD process and the CVD conditions have been optimized to disperse small Pt particles onto the CNT. Pt particles synthesized by CVD have a relatively uniform size of approximately 1 nm, which is substantially smaller than in the case of a commercial Pt/CB catalyst ($\leq$4.5 nm) prepared by wet impregnation. Pd/C catalysts were obtained by the atmospheric pressure chemical vapor deposition of [Pd(allyl)Cp] in helium atmosphere at 323 K on a CNF support system in a fixed bed reactor.[217] The functionalization of the CNFs is a necessary step to provide anchoring sites for Pd. Palladium catalysts were prepared from [Pd(allyl)Cp] by a similar procedure on a CNT-CNF composite as support.[218] Oxygen-containing functional groups were found to be essential for the deposition of palladium on the surface of the composite. The application of oxygen plasma proved effective for functionalization of the carbon surfaces. A CVD technique has been successfully applied to produce Pt and Pt–Ru NPs supported on CNT.[219] The CVD technique used resulted in well dispersed NPs with a very narrow size distribution. The CVD method developed produced unique metal clusters/atoms distributed between the graphene sheets of the CNT, and stabilized by the CNT defects. The Pt NPs produced using a WI or CVD method, resulted in *fcc* crystal structures, and the smallest average NPs were observed on CVD samples. Co/CNT catalysts were prepared *via* the CVD method with [$Co_2(CO)_8$] as precursor.[220] The obtained Co/CNTs catalysts feature a narrow size distribution of Co particles centering around 7.5 nm. For the deposition of Cu NPs on MWCNTs, sublimation of [Cu(hfac)(TMVS)] [Cu(hexaflouroacetylacetonate) (trimethylvinylsilyl)] at 393–573 K followed by hydrogen reduction has been used.[221] The influence of different pre-treatment steps such as thermal and plasma activation has been studied in detail. In all cases no copper wires were obtained and the size of the particles depended on the substrate temperature, deposition time and the number and type of functional groups on the CNTs.

RuO_2/CNT nanocomposites were prepared from ruthenium acetyl acetonate by coating an amorphous RuO_x layer on CNTs using CVD followed by post-annealing at a relatively low temperature of 623 K.[222] Arrays of SWCNT bundles organized following different architectures have been coated by a homogeneous deposit of nanocrystalline titania.[223] The TiO_2 deposition was performed at 673 K using the CVD technique and titanium tetraisopropoxide [$Ti(O^iPr)_4$] as a precursor. X-Ray diffraction and Raman spectroscopy evidence the anatase structure of the TiO_2 coatings, formed by grains with an average size of about 55 nm. The structural and compositional characteristics of the TiO_2 deposits are not sensitive to the organization of the nanotube arrays, which maintain their pristine architectures. Considering the relatively large number of works dealing with the preparation of TiO_2 photocatalyst loaded onto activated carbon support using CVD,[224] further studies should be devoted to understanding such systems. A simple and efficient approach for coating MWCNTs with size-controllable SnO_2 NPs by CVD has been developed using tin hydride (SnH_4) gas as the source of SnO_2 at 823 K.[225] Most of the MWCNTs have been covered with many particles with the diameters of 3–15 nm. The size and coverage of SnO_2 NPs can be adjusted by simply controlling the deposition time and the flow rate of the SnH_4/N_2 mixture gas during the CVD procedure. ZnO nanostructures were grown on FLG using catalyst-free metal-organic vapor-phase epitaxy from diethylzinc and oxygen.[226,227] The aspect ratio and density of the nanostructures depended strongly on the growth temperature.[227] At the lower growth temperature of 673 K, the ZnO nanostructures had a mean length of 1.0 ± 0.1 μm and a diameter of 100 ± 10 nm. Furthermore, interesting growth behavior – the formation of aligned ZnO nanoneedles in rows and vertically aligned nanowalls – was observed, presumably resulting from the enhanced nucleation at graphene step edges. On oxygen plasma treated graphene and under similar conditions, ZnO nuclei with uncommon cubic structures and non-epitaxially-grown ZnO clusters were observed in the early stage of growth, followed by evolution to large grains exhibiting an epitaxial relationship with graphene and formation of various defects.[228]

4.3.3 Self-assembly Methods

Recent years have seen an unprecedented development in the self-assembly of NPs to form ordered arrays of functional surfaces.[229–231] These methods take advantage of the electrostatic interaction between the surface of the NP and the chemically functionalized sites present on the support. The electrostatic attraction causes the NPs to anchor on the surface and organize themselves to form interesting ordered arrangements. Needless to say, the nanoscale nature of the support provides an excellent template for a bottom-up organization of a nanocomposite. The advantage of this method is that preformed NPs are used, so that good control of the dispersion is obtained. However, this method often requires carbon nanomaterial organic functionalization, which can modify their properties, and that often involves multi-step preparation procedures.

4.3.3.1 Self-assembly Deposition of Preformed Metal Nanoparticles

Gold NPs were self-assembled onto the surface of solubilized CNTs through an interlinker of a bi-functionalized molecule terminated with a pyrenyl unit at one end and a thiol group at the other end.[232] A generally applicable procedure for the assembly of gold NPs and nanorods on CNTs in aqueous

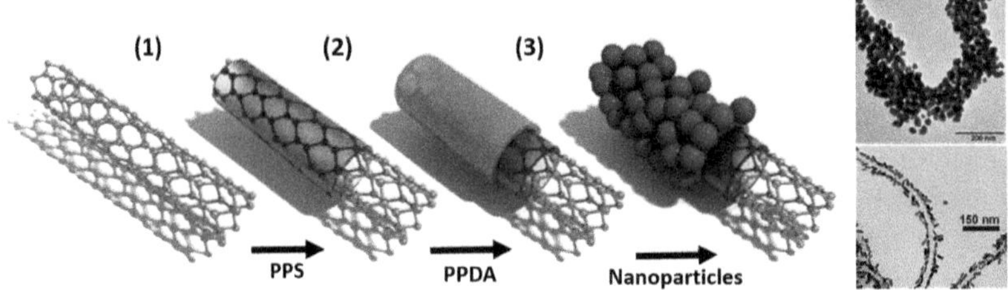

Figure 4.9 Schematic illustration of a non-covalent functionalization of CNTs comprising (1) polymer wrapping using poly(sodium 4-styrenesulfonate) (PSS), (2) self-assembly of poly(diallyldimethylammonium chloride) (PDDA) and (3) Au nanoparticle deposition. TEM images of Au nanoparticles and nanorods. Adapted from ref. 233.

solution has been reported (Figure 4.9).[233] The method makes use of polyelectrolytes for wrapping CNTs and providing them with adsorption sites for electrostatically driven NP deposition. Gold nanowires have been produced by the self-assembly of gold NPs onto MWCNTs. The CNTs were mixed with a suspension of gold NPs, resulting in a binding between the nanotubes and the gold. Subsequent heating in air for 2 min gave rise to continuous nanowires up to 10 μm in length.[234] SWCNTs were functionalized with biotin in a series of chemical reactions.[235] Streptavidin-coated gold NPs were attached to the biotin-modified SWCNTs in solution in a self-assembly process. The biotin–streptavidin interaction is extremely strong, allowing a stable binding of streptavidin-coated gold colloids to the biotin-terminated CNTs simply by adding the NPs to the SWCNT suspension. A robust and effective composite film based on a gold nanoparticles/room temperature ionic liquid/MWCNTs modified glassy carbon electrode was also prepared by a layer-by-layer self-assembly technique.[236]

Au nanoparticles/rGO–ionic liquid hybrids were synthesized by combining supra-molecular self-assembly between rGO sheets and ionic liquids with sonochemistry.[237] Ionic liquid-coated rGOs (ionic liquid was successfully coated on the surface of rGO sheets with average thickness of 1.5 ± 0.2 nm) provide large surface area with abundant nucleation sites for positioning Au NPs (7.2 ± 0.2 nm). As-obtained rGO–ionic liquids were mixed with an Au precursor solution in a sonochemical system, yielding uniform size of Au NPs deposited on the rGO–ionic liquid surface. The use of cationic polyelectrolyte poly(diallyldimethyl ammonium chloride) (PDDA) functionalized FLG as the building block in the self-assembly of FLG/Au nanoparticle heterostructures was reported.[238] Citrate-capped Au NPs were adsorbed onto the graphene by electrostatic interaction, which was the driving force for the self-assembly process. The high-loading Au NPs distributed uniformly on the surface of functionalized FLG. Positively charged gold NPs with diameters of 2–6 nm were self-assembled onto the surfaces of 1-pyrene butyric acid functionalized graphene sheets simply by mixing their aqueous dispersions.[239] The amount of gold NPs assembled on functionalized graphene can be easily modulated by controlling the feeding weight ratio of both components. PtAu NPs with controlled PtAu molar ratios and PtAu NP loadings were also successfully self-assembled onto PDDA-functionalized graphene.[240,241] A PdAu/FLG nanostructure was prepared by self-assembly of alloyed PdAu NPs on ionic liquid-grafted graphene sheets by electrostatic interaction to form graphene–metal hybrid nanomaterials under mild conditions.[242]

A hybrid three-dimensional nanocomposite was prepared by alternatively assembling graphene nanosheets modified by ionic liquid and Pt NPs.[243] In this strategy, an imidazolium salt-based ionic liquid (IS-IL)-functionalized graphene was synthesized by covalently binding 1-(3-aminopropyl)-3-methylimidazolium bromide onto graphene nanosheets. The introduction of IS-IL on the surface of graphene nanosheets results in dispersed graphene nanosheets with a positive charge. Then, the positively charged IS-IL-functionalized graphene nanosheets are strong enough to drive the formation of the 3D nanomaterials with negatively charged citrate-stabilized Pt NPs through electrostatic interaction. FePt NPs were also synthesized and assembled on un-modified FLG by a solution-phase self-assembly method.[244]

A general approach to make CNTs-based nanocomposites *via* self-assembly was reported (Figure 4.10).[85] The method allows for the preparation of binary composites as well as complex systems such as ternary or even quaternary composites where nanoparticles of active phases (*e.g.*, metals and metal oxides) are used as primary building blocks. Six different kinds of binary, ternary, and quaternary nanocomposites, TiO_2/CNTs, Co_3O_4/CNTs, Au/CNTs, Au/TiO_2/CNTs, TiO_2/Co_3O_4/CNTs, and Co/CoO/Co_3O_4/CNTs, have been prepared on nitric acid oxidized CNTs. On the basis of this work, it was shown that highly complex inorganic–organic nanohybrids can be fabricated from preformed units with good control of particle shape, size, and distribution. Anionic sulfate surfactants were used to assist the stabilization of graphene in aqueous solutions and facilitate the self-assembly of *in situ* grown nanocrystalline TiO_2, rutile and anatase, with graphene.[245] TiO_2 nanorods were self-assembled on the GO sheets at the water/toluene interface.[246] The self-assembled GO-TiO_2 nanorod composites can be dispersed in water. The effective anchoring of TiO_2 nanorods on the whole GO sheets was confirmed by TEM, XRD, FTIR, and TGA. Magnetic MWCNTs were prepared by the electrostatic self-assembly approach.[247] Poly(2-diethylaminoethyl methacrylate) (PDEAEMA) was covalently grafted onto the surfaces of MWCNTs by MWCNT-initiated *in situ* atom transfer radical polymerization of 2-DEAEMA. The PDEAEMA-grafted MWCNTs were further quaternized with methyl iodide, resulting in cationic polyelectrolyte-grafted MWCNTs. Magnetic Fe_3O_4 NPs were loaded onto the MWCNT surface by electrostatic self-assembling. Magnetic 3D graphene/Fe_3O_4 architectures have also been prepared *via* an *in situ* self-assembly of graphene obtained by a mild chemical reduction of graphene oxide in water in the presence

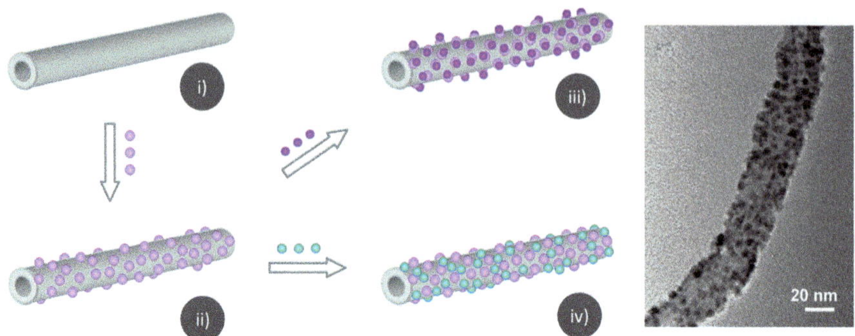

Figure 4.10 Schematic illustration of the self-assembly schemes for preparation of nanocomposites: (i) CNTs, primary support, (ii) a second phase of nanoparticles (in pink) are anchored onto CNTs, (iii) a third phase of nanoparticles (in purple) are attached to the surface of the second phase nanoparticles, and (iv) a third phase of nanoparticles (in blue) are also attached onto the CNTs. Further addition of phases can also be carried out with the same assembling strategy. TEM image of the as-prepared Co_3O_4/CNTs nanocomposites. Adapted from ref. 118.

of Fe_3O_4 nanoparticles.[248] The obtained gel is super-paramagnetic, porous, and light.

MWCNTs were used as a template for hydroxyapatite NP deposition by self-assembly.[249] The assembled apatite was analyzed by TEM and SEM. Defects that were analogous to edge dislocations along the CNT surfaces were the nucleation sites for hydroxyapatite nanoparticles after these defects had been functionalized principally into carboxylic groups.

4.3.3.2 *Self-assembly Deposition from Metallic Precursors*

CNTs were non-covalently functionalized with poly(allylamine hydrochloride) (PAH) and then employed as the support of Pt nanoparticles.[250,251] The negatively charged Pt precursors are adsorbed on positively charged PAH-wrapped CNT surfaces *via* electrostatic self-assembly and then *in situ* reduced in ethylene glycol. X-Ray diffraction and TEM images revealed that Pt NPs with an average size of *ca.* 2.6 nm are uniformly dispersed on CNT surface. An *in situ* synthetic method was reported for preparing and decorating metal NPs at sidewalls of sodium dodecyl sulfate micelle functionalized SWCNTs, MWCNTs and CNFs (C_{NM}/SDS).[252] Typically, H_2PtCl_6 was added into the C_{NM}/SDS solution, and was slowly reduced by 1-dodecanol to zero-valence atoms in the micellar core, forming Pt NPs. The amount and morphology of Pt NPs depend on the type of carbon nanomaterials. The results demonstrate the attachment of metal NPs to the CNT surface *via* non-destructive surfactant modification.

4.3.4 Grafting of Molecular Complexes

There are several approaches to immobilizing transition metal complexes on carbon supports.[253] One way to discuss them is to group them according to the type of interaction between the support and the molecular species to be grafted. In this context, three different groups of support–molecule interactions can be considered: (i) covalent bonding, (ii) non-covalent interactions (including physical adsorption and electrostatic interactions) and (iii) encapsulation. Covalent bonding of metallic complexes onto the support is by far the most used strategy to immobilize transition metal complexes with catalytic properties. It can be effected either directly by reaction of the metal complex with the support surface groups or mediated through a spacer previously grafted to the support or reacted with the complex. The other two methodologies for immobilization of metal complexes, physical adsorption and electrostatic interaction, belong to the group of non-covalent interactions between the support and the metal complex.[254–258] The former case includes π–π interactions, van der Waals, hydrogen bonds and hydrophobic–hydrophilic interactions between the support and the complex. In the second method, there is an electrostatic interaction between the support and the complex, and consequently there must be charges of opposite signs between the support and the complex. Encapsulation implies the physical entrapment of the metal complex within the pores of the support and as a starting-point it is supposed that no other interactions between the support and the metal complex exist besides the physical confinement.

4.3.4.1 Covalent Bonding

The organometallic η^6-complexation of Cr with graphene, graphite and CNTs was studied.[259] All of these systems exhibit some degree of reactivity toward [$Cr(CO)_6$] and [$Cr(CO)_3(\eta^6\text{-}C_6H_6)$], and the formation of (η^6-arene)$Cr(CO)_3$ or (η^6-arene)$_2$Cr, where arene = CNT, or FLG was demonstrated. It was found that the SWCNTs are the least reactive, presumably as a result of the effect of curvature on the formation of the η^6-bond.

Just as the in the case of the immobilization of metal nanoparticles, the degree of surface oxidation of the support is of utmost importance for the grafting of molecular catalysts. Pristine CNTs form only (weak) η^2 coordinates with metal complexes such as [$IrCl(CO)(PPh_3)_2$],[260] which are of limited catalytic value (Figure 4.11(a)). Oxidized SWCNTs formed a hexacoordinated Ir(III) complex after oxidative addition to surface-bound hydroxyl moieties (Figure 4.11(b)).[261] The oxidation of sp^2-hybridized carbons creates mainly alcohols within the plain and predominantly carboxylic acids at the edges and defects. Hence, a different coordination mode was proposed for the impregnation of HNO_3-treated MWCNTs with [$HRh(CO)(PPh_3)_3$] (Figure 4.11(c)).[262] The surface-bound carboxyl groups can be subjected to very simple manipulations to alleviate the formation of organometallic compounds on the CNTs. The [$RuHCl(CO)(PPh_3)_3$] complex was successfully supported on 1,2-ethanedithiol functionalized MWCNTs and characterized

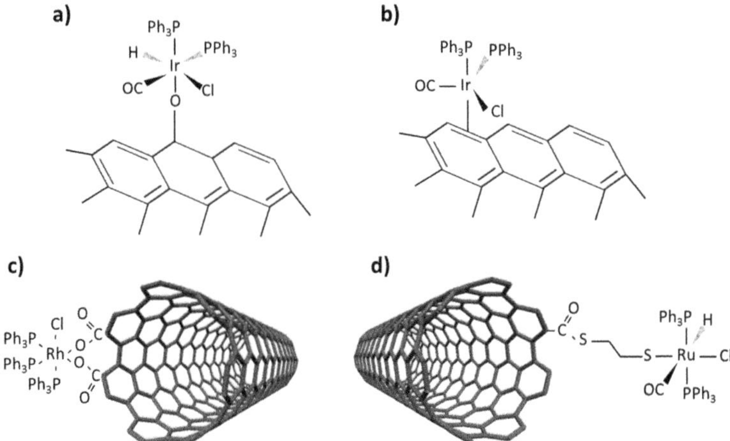

Figure 4.11 (a) η^2-coordination of $[IrCl(CO)(PPh_3)_2]$ on pristine SWCNTs; (b) Hydroxyl-coordinated Ir(III) complex on $KMnO_4$-oxidized SWCNTs; (c) Wilkinson's catalyst coordinated to terminal carboxy groups on HNO_3-oxidized MWCNTs; and (d) Thioester linkage for the grafting of $[RuHCl(CO)(PPh_3)_3]$ on MWCNTs.

(Figure 4.11(d)).[263] The catalyst was characterized by diffuse reflectance UV-vis and FT-IR spectroscopic techniques, SEM, ICP and elemental analysis methods.

A Rh complex, derived from the Wilkinson's catalyst $[RhCl(PPh_3)_3]$, was also immobilized on CNTs through a linear organic molecule (6-amino-1-hexanol), covalently bonded to the support.[264] Giordano *et al.* treated them with sodium carbonate to facilitate the coordination of $[RhCl(CO)_2]_2$,[265] whereas Koningsberger *et al.* treated carboxylated CNFs with thionylchloride and reacted them with anthranilic acid as a ligand for rhodium.[266] The ruthenium complex $[RuCl_2(PPh_3)_3]$ was successfully grafted onto GO surface through coordination interactions with aminosilane ligand spacers.[267]

The Rh diamine complex $[Rh(COD)NH_2CH_2CH_2NH(CH_2)_3Si(OCH_3)_3]^+BF_4^-$ was immobilized on/in CNTs and CNFs.[268–272] The procedure for anchorage is based on the creation of a covalent bond between the trimethoxysilane functionality in the metal complex and phenol type groups created on the surface of the carbon materials. Chiral rhodium hybrid nanocatalysts have also been prepared by covalent anchorage of pyrrolidine-based diphosphine ligands onto functionalized CNTs.[273]

The hydrotris(pyrazol-1-yl)methane iron(II) complex $[FeCl_2(\eta^3\text{-}HC(pz)_3)]$ (pz = pyrazol-1-yl) was immobilized on three different carbon materials (AC, carbon xerogel and MWCNTs) with three different surface treatments (original, treated with nitric acid, and treated with nitric acid followed by sodium hydroxide).[274] The heterogenization process was more efficient for carbon nanotubes treated with nitric acid and sodium hydroxide.

Various Schiff-base have been covalently grafted on CNTs. A styryl functionalized vanadyl Schiff-base has been covalently anchored on mercapto-modified SWCNTs through a radical chain mechanism.[275] SWCNT is more suitable as

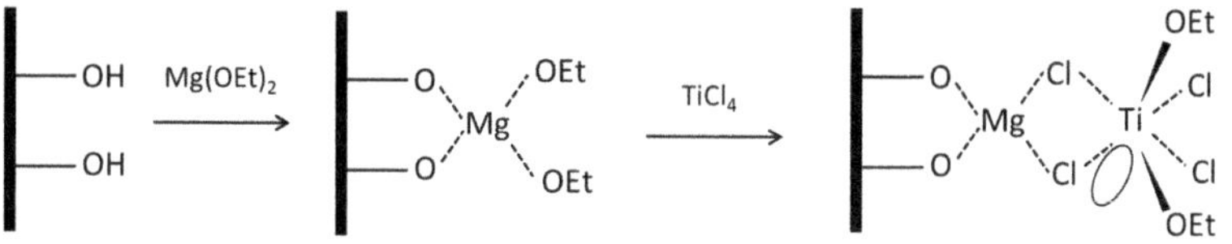

Figure 4.12 Schematic of the mechanism of the supporting Ziegler–Natta catalyst on the MWCNTs in the presence of $Mg(OEt)_2$. Adapted from ref. 279.

support for the vanadyl salen complex than high-surface-area activated carbon because the latter exhibits some adventitious activity. Hydroxyl functionalized copper(II) Schiff-base, *N*,*N*′-bis(4-hydroxysalicylidene)-ethylene-1,2-diamine-copper(II), [Cu((OH)$_2$-salen)], has been also covalently anchored on modified MWCNTs.[276] A hydroxyl functionalized manganese(II) Schiff-base has also been covalently anchored on modified MWCNTs.[277] Immobilization of a chiral vanadyl salen complex having a terminal carbon–carbon double bond onto SWCNTs has been performed.[278] The covalent linkage has been achieved by radical initiated addition of mercapto groups to the C=C double bond. Novel graphene oxide (GO) tethered Cu(II) and Co(II) salen complexes [M-Salen-GO (M = Cu, Co)] were synthesized *via* a stepwise covalent attachment procedure.[279] Basically, amino-functionalized GO was synthesized first by reaction with 3-aminopropyltriethoxysilane. Then, a nucleophilic reaction between chloromethyl-modified salen complexes and the aminopropyl-functionalized GO produced the grafted complexes. Another way to provide the salen ligands involves the covalent modification of GO with an aminosilane, followed by condensation with salicylaldehyde.[280] The copper (salen) complex was subsequently synthesized and simultaneously immobilized onto the GO surface with a designed tetrahedral chelate structure. MoO_2-salen was also successfully tethered onto amino-functionalized graphene oxide.[281] CNTs can be also efficiently functionalized with Ru(II) terpyridine complexes using the simple and mild methodology of direct covalent amidation.[282]

The preparation of single-site chromium (dichloro-η^5-[3,4,5-trimethyl-1-(8-quinolyl)-2-trimethylsilyl-cyclopentadienyl]chromium(III)) catalysts supported on functionalized GO was described.[283] Owing to the presence of surface hydroxyl groups on the FLG, FLG/methylaluminoxane (MAO) nanosheets can be effectively dispersed in *n*-heptane, thus enabling immobilization of the MAO-activated chromium single-site catalyst on FLG. In contrast to nanometre-scale CB and MWCNTs, which failed to form stable dispersions, FLG/MAO/Cr afforded the highest catalyst activities for polyethylene polymerization and excellent morphological control. Hydroxyl-functionalized MWCNTs were used to anchor a Ziegler–Natta catalyst, as shown in Figure 4.12.[284]

Reactions of transition metal complexes with fullerenes have been much more studied, and will not be developed in this chapter. Briefly, the great majority of fullerene organometallic compounds belong to two groups; those with η^2 and those with η^5 hapticities, however some examples of this type of compound manifest the other four possible hapticities (Figure 4.13).

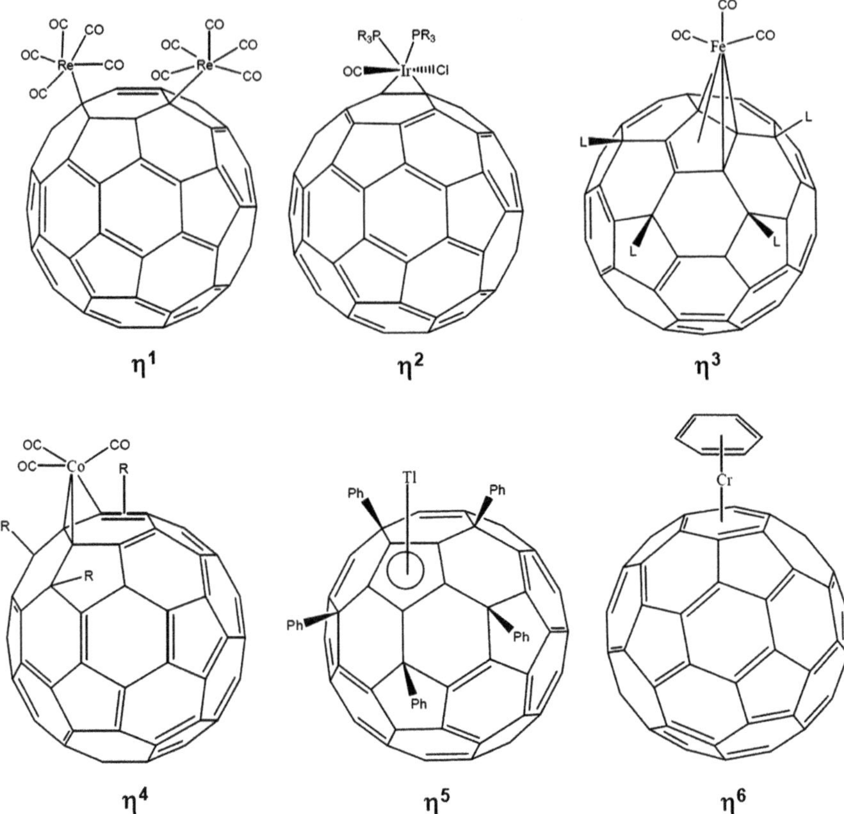

Figure 4.13 All possible hapticities for fullerene C60. η^1 hapticity with the metal atom directly above any carbon atom (an σ bond is expected); η^2 hapticity; this may be (6,6) or (6,5) type, although the second of these is not common; η^3 hapticity; the metal atom linked to three carbon atoms, theoretically this could be on either a six or five-member ring; η^4 hapticity, the metal atom linked to four carbon atoms; η^5 hapticity, metal atom linked directly above the center of a five-member ring; and η^6 hapticity, metal atom linked directly above the center of a six-member ring.

Additionally, a large variety of ligands with different binding modes have been attached to C_{60}, as schematically depicted in Figure 4.14. We invite readers who are interested in that subject to look at the following comprehensive papers that should open interesting perspectives for the development of new catalysts on other nanostructured carbon supports.[285–291]

In conclusion, although the oxidation route is undoubtedly the most common and straightforward method to anchor molecular catalysts on nanostructured carbon materials, it is basically a degradation process that increases the number of defect sites or even creates detached amorphous fragments after severe treatments. It is surprising that the rich chemistry for covalent functionalization of nanostructured carbon materials has had comparatively little impact on the development of catalytic systems, although rather mild conditions can be found among them.

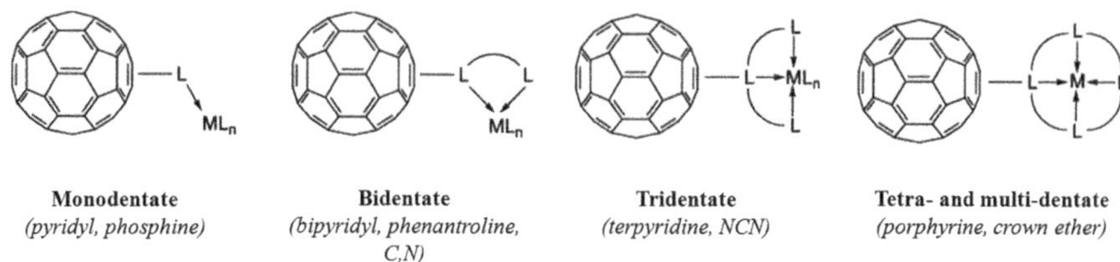

Figure 4.14 Binding modes of metalated bucky-ligands.

4.3.4.2 Non-covalent Interactions

A very elegant grafting approach involves the non-covalent immobilization of complexes *via* π-stacking interactions. The benefits are obvious: pristine supports can be used without any pretreatment or functionalization, and the grafting succeeds simply by mixing catalyst and support in a suitable solvent. The simple adsorption *via* π-stacking interactions of various catalysts, mainly involved in polymerization, have been reported, including: Cp_2ZrCl_2/CNT,[292] $CpTiCl_3$ or Cp^*TiCl_3/CNTs,[293] Cp_2ZrCl_2 or Cp_2TiCl_2/rGO (or on N-doped FLG),[294] aryliminopyridylnickel chloride/CNT (or FLG),[295] pyrene-tagged Fe[296] and Ru[297] complexes/CNT, pyrene-functionalized Ni complexes/CNT,[298] anthracene- and pyrene-tagged bis(oxazoline) Cu complexes/CNT,[299] pyrene-tagged gold(I) complex/CNT,[300] and Ru polypyridyl complexes/CNT.[301]

Different ways of forming hybrid materials between SWCNTs or MWCNTs and biomimetic compounds such as metalloporphyrins, metallophthalocyanines and other MN_4 complexes have been proposed.[302] The non-covalent functionalization, or complexation, of SWCNTs with metalloporphyrins was reported. It was found that metalloporphyrins donate electron density exclusively into semiconducting SWCNTs.[303] Fe- and Co-phthalocyanines and two different Co-porphyrins were immobilized by π–π stacking interactions on various types of CNTs (SWCNTs, DWCNTs, and oxidized or non-oxidized MWCNTs).[304] For C_{60}, an acyclic dimer of a dendritic zinc porphyrin bearing six carboxylic acid functionalities can interact with C_{60} to form "supramolecular peapods", composed of a hydrogen-bonded zinc porphyrin nanotube and fullerenes.[305]

Supported liquid phase catalysts have been also recently prepared on CNTs, being ionic liquids the liquid phase.[306,307] A rhodium catalyst has been prepared by immobilization of a Rh complex in an ionic liquid film on CNTs functionalized with imidazolium-based ionic moieties.

4.3.4.3 Encapsulation

Cobaltocene and ethyl-cobaltocene,[308] ferrocene,[309] and a zinc phenylporphyrin[310] have been encapsulated in CNTs. The nature of the $[CoCp_2]$–CNT interaction appeared to be more complex than pure van der Waals forces. The presence of $[CoCp_2]$ and $[Co(CpEt)_2]$ in SWCNTs was probed by UV-vis absorption spectroscopy, revealing that the oxidation state of Co was changed from +2 to +3 upon their insertion into SWCNTs. It was proposed that, in addition to the Van der Waals components, the metallocenes are expected to have a major electrostatic contribution to their interactions with SWCNTs. Theoretical calculations demonstrate that MCp_2@SWCNT composites are stabilized by weak π-stacking and CH···π interactions, and in the case of the $CoCp_2$@SWCNT composites there is an additional electrostatic contribution as a result of charge transfer from $CoCp_2$ to the nanotube.[311] There is also a Rh complex, [HRh(CO) $(PPh_3)_3$], immobilized into ends-opened and ends-unopened CNTs, in which the authors claimed, on the basis of catalytic activity of supported complexes, that the Rh complex was located in the hollow channels of CNTs.[262]

Endohedral metallofullerenes, a new class of hybrid molecules formed by encapsulation of metallic species inside fullerene cages have also received an important attention.[312] In particular, the interplay between the encapsulated metallic species and the fullerene cage has been well investigated.[313] On one hand, the position and motion of the encapsulated metals can be effectively controlled by exohedral modification. On the other hand, the cage structures, the chemical behaviors of cage carbons and thus the chemical reactivity of the whole molecule are also apparently influenced by the electronic configuration and geometrical conformation of the internal metals *via* strong metal–cage interactions. Multiple addition, multi-step reactions, and catalytic reactions of endohedral metallofullerenes are of importance but represent great challenges.[313]

4.3.5 Confinement of Metallic Nanoparticles in CNTs

The confinement of metal nanoparticles in CNTs has gained an increasing amount of attention due to potential applications in catalysis. Using the spatial restriction effect of CNT channels, a variety of nanomaterials, even of sub-nanometre size, can be synthesized. These materials would be usually intrinsically unstable, particularly under elevated temperature and pressure, or difficult to obtain under mild conditions.[1,2,314–319] Additionally, modified adsorption, diffusion, structural and chemical properties have been reported for a variety of species confined in CNTs compared to their counterparts either in the bulk or deposited on the outer CNT walls.[314,320,321] Stronger adsorption was reported onto the CNT inner surface for CO, H_2, alkanes and alkenes,[322,323] opening the way to selective adsorption/separation processes.[324,325] The diffusion of various molecules, including N_2[326] and H_2O,[327] inside individual CNTs was reported to be faster than the bulk diffusion. γ-Fe NPs with an *fcc* crystal structure, known to be stable between 1185 and 1667 K in the bulk, retained their stability at RT when confined in CNTs.[328] Confined cobalt nanorods showed an *fcc* instead of the stable hexagonal structure.[329] The reduction of iron oxide NPs confined into CNT channels was found to be facilitated with respect to the particles located on the outer walls.[330,331] Due to these confinement effects, the metal NPs inside CNTs usually show better catalytic performances than the ones loaded on the outer surface.[1,314,315,332] Indeed, confinement effects can affect chemical reactions through a host of impacts, such as changes in the thermodynamic state of the system because of interactions with inside walls, selective gas adsorption, and geometrical constraints that affect the reaction mechanism.[321,333] As these effects are expected to be enhanced with reducing CNT inner diameter,[324,331,334–336] preparing NPs selectively localized inside small diameter CNTs is the prerequisite to study this phenomenon.

In order to achieve maximum confinement, wet chemistry appears to be the simplest, most versatile and up-scalable method.[337] However, the capillary effect of CNTs depends on surface functionalization and CNT diameters. CNTs with internal diameters <10 nm are usually filled to a lesser extent, or even remain empty after wet impregnation.[338] Reported methods for filling include: (i) wet impregnation assisted by ultrasound using cut CNTs,[339] (ii) two step biphasic impregnation,[340] (iii) impregnation and selective washing,[341,342] (iv) the use of supercritical CO_2,[343,344] and (iv) molecular recognition.[345] The reported filling yields, which are commonly measured either by conventional 2-D TEM[346] performed with or without sample tilting, or by electron tomography,[347] ranged between 70 and 90%. Among the preparation methods cited, only the first one, originally developed in the group of X. Bao, has been applied for the filling of small diameter (<10 nm) CNTs with Pd and Co,[348] Ru,[348–351] Fe,[330,331,352] Cu,[334] MnO_2,[353] TiO_2[354] and Rh[355] NPs. The first step of this simple procedure involves CNT oxidation with HNO_3 in order to open the CNT tip and to create oxygen surface functional groups.[33] The TEM micrographs shown in Figure 4.15 show bimetallic PtFe nanoparticles and nanowires confined in MWCNTs. Interestingly, it has been shown that air oxidation can result in the motion of metal atoms confined in CNTs.[356] This can be utilized to tailor various hybrid nanostructures, such as core-void-shell nanorods, metal/metal oxide@CNT nanotubes, and porous metal oxide nanotubes.

It is worth noting that this surface oxidation reaction is also used for the preparation of most of the CNT supported catalysts,[2] which often show better performances than their counterparts supported on other supports, including carbonaceous ones.[357] As most of the studies dealing with CNTs for catalysis neglected the possibility of confinement of the active phase in the CNT inner cavity, we may wonder about the possible role of unidentified confinement effects on the catalytic results.

4.4 Concluding Remarks and Future Perspectives

A large variety of active phase/nanostructured carbon material nanohybrids can be synthesized by wet or gas phase processes, most of the time involving surface functionalization of the support. Each of these methods in principle has a common goal, which is to obtain high dispersion and small size of metal nanoparticles on the carbon surface. Tables 4.7 and 4.8 show a collection of data obtained for Pt- and Ni-based catalysts. Obviously, the choice of metal precursor/support couple, the metal loading and the choice of the preparation procedure will be critical to obtain a high dispersion. It is also obvious from this data that it is easier to disperse Pt nanoparticles on

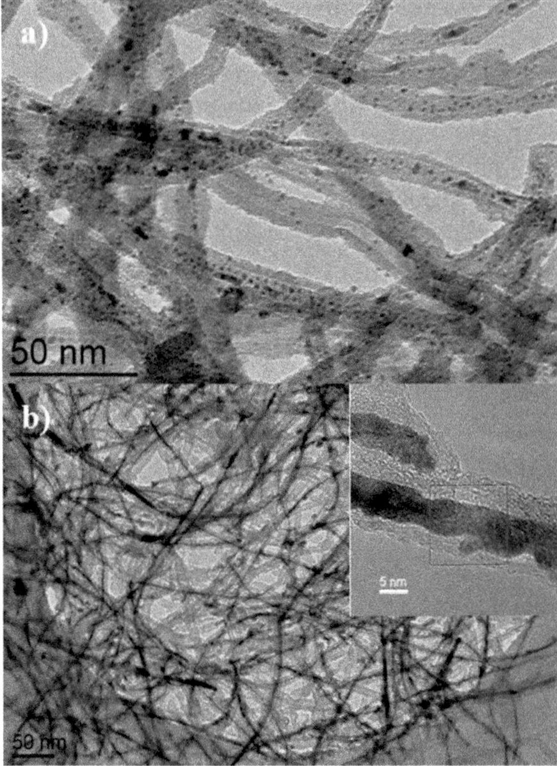

Figure 4.15 TEM micrographs of bimetallic PtFe (a) nanoparticles, and (b) nanowires confined in MWCNTs.

nanostructured carbon materials than Ni nanoparticles, whatever the choice of metal precursor/support couple, the metal loading and the choice of the preparation procedure. As most of the metal–support interaction arises from a direct M–C interface, it is also important to take into consideration the bonding between metal and carbon atoms and the location and diffusivity[358] of metal atoms in graphene layers.

From the binding energies of metal adatoms of M_{13} clusters (M = Pt, Ni, Co) on graphene, there is no clear trend between binding energy and cluster size/density (Table 4.9). The relative rates of diffusion increase with decreasing admetal–carbon bond strengths. Specifically, the larger binding energies and lower rates of migration (Co) did not lead to smaller cluster sizes and higher cluster densities. Although such a correlation was proposed to understand the nucleation and growth of Co, Au, Ni and Pt on TiO_2, the situation seems different on carbon. Obviously, since the binding energies on defects are high whatever the metal, other parameters such as the reactivity of the metallic precursor with the surface should play a significant role on the nucleation step. Further studies are clearly needed to get further information on the

Table 4.7 Nickel catalysts prepared on nanostructured carbon materials.

Catalyst	Preparation	Particle size/nm	References
6% Ni/CNT	WI from Ni nitrate in acetone on MWCNTs (ed = 20 nm; id = 8 nm)	1.2	354
40% Ni/CNT		10	
20% Ni/CNT	IWI from Ni nitrate in acetone on oxidized MWCNTs of 20 $m^2 g^{-1}$	12	97
50% Ni CNT		35	
15% Ni/CNT	IWI from Ni chloride in water on MWCNTs (ed = 30 nm) of 103 $m^2 g^{-1}$	50–70	355
1% Ni/CNT	IWI from Ni nitrate in water on MWCNTs (ed = 90 nm; id = 30 nm; 20 $m^2 g^{-1}$)	3–10	335
20% Ni/CNT	WI from Ni nitrate in water on CNTs (ed = 90 nm; id = 30 nm; 20 $m^2 g^{-1}$)	20–100	356
12% Ni/CNT	IA from Ni nitrate in water on oxidized MWCNTs	10–50	357
Ni/CNT	Electro-deposition	40–200	358
60% NiO/N-CNT	WI from Ni acetate in ethylene glycol/NaOH on N-doped MWCNTs (ed = 50 nm)	3–10	359
7.5 wt% Ni/CNFs	WI from $Ni(NO_3)_2 \cdot 6H_2O$ onto HNO_3-oxidized CNFs	16	360
6% Ni/CNF	DP from Ni acetate Na_2CO_3 aqueous solution on oxidized CNFs of 68 $m^2 g^{-1}$	6	361
14% Ni/CNF			
20% Ni/CNF	DP from Ni nitrate urea aqueous solution on oxidized CNFs of 150 $m^2 g^{-1}$	8	118
65% Ni/rGO	WI from Ni chloride in water on GO + *in situ* reduction	2–4	362
8% Ni/FLG	Pulsed laser ablation	8 ± 3	363

Table 4.8 Platinum catalysts prepared on nanostructured carbon materials.

Catalyst	Preparation	Particle size/nm	References
50% Pt/CNT	WI from H_2PtCl_6 in EG with sodium dodecyl sulphate and $NaBH_4$ on MWCNTs	2.5	364
18% Pt/CNT	WI from H_2PtCl_6 in isopropanol-water on oxidized MWCNTs	9.8 ± 4	365
	IWI from H_2PtCl_6 in isopropanol -water with Na_2CO_3 and KBH_4 on oxidized MWCNTs	4.8 ± 4	
	Colloidal impregnation from H_2PtCl_6 and surfactant in methanol/water on oxidized MWCNTs	2.2 ± 1	
20% Pt/CNT	Electrostatic self-assembly technique from H_2PtCl_6 on poly (allylamine hydrochloride) modified MWCNTs	2.6	251
40% Pt/CNT	Hydrothermal synthesis from H_2PtCl_6 in water/NaOH/H_2CO on oxidized MWCNTs	1.9	366
20% Pt/CNT	Electro-deposition from H_2PtCl_6	20	169
Pt/CNT	Electron beam deposition on plasma treated MWCNTs (Pt surface coverage 70%)	2	185
5%Pt/CNT	ALD on untreated CNT from [$MeCpPtMe_3$]	1.3	194
5–40%Pt/CNT	CVD on oxidized CNT from [$MeCpPtMe_3$]	1	216
5–50%Pt/CNF	WI from H_2PtCl_6 in water	3	367
20% Pt/FLG	Electrochemical reduction from H_2PtCl_6 and GO	10	368
10% Pt/FLG	Chemical reduction of H_2PtCl_6 in EG-water solution on GO	2.7	369
10% Pt/CNT	or oxidized MWCNTs (ed = 10–20 nm)	3.5	
20% Pt/FLG	WI from H_2PtCl_6 in acetone on GO (600 $m^2 g^{-1}$)	2	370
20% Pt/Cspheres	WI from H_2PtCl_6 in formaldehyde on 30 nm C spheres	3	371

Table 4.9 Binding and migration energies for adatoms on graphene, binding energies for clusters on graphene, and mean experimental particle size for Pt, Ni and Co nanoparticles deposited on nanostructured carbon materials.

Metal	$Eb_{M\text{-}C}/eV^{-1}\ at^{-1a}$	Em_M/eV^b	$Eb_{M13\text{-}C}/eV^{-1}\ at^{-1c}$	$Ea_{M13\text{-}CV}/eV\ at^{-1d}$	d/nm^{-1e}
Pt	2.9[5]	0.19[5]	2.08[372]	6.22[372]–7.45[373]	4.5
Ni	3.1[5]	0.40[5]	0.91[374]	6.70[374]	19
Co	3.6[5]	0.77[5]	3.6[375]	—	16[f]

[a] Adsorption energy of the adatom.
[b] Migration energy of the adatom.
[c] Adsorption energy of the M_{13} cluster on pristine graphene per M atom.
[d] Adsorption energy of the M_{13} cluster on a single vacancy of graphene per M atom.
[e] Mean metal particle size calculated from the data of Tables 4.7 and 4.8.
[f] Mean metal particle size of Co/nanocarbon calculated from literature data (10 publications[116,376–390]).

nucleation and growth of metallic nanoparticles on nanostructured carbon supports. Finally, a better understanding is also required regarding the role of inter-particle interactions on the final dispersion and catalyst stability.

Other challenges in catalyst preparation on nanostructured carbon materials include: (i) the balance of conflicts between maintaining the original structure and properties of nanostructured carbon materials and introducing the anchoring groups onto their surface by functionalization; (ii) the size, shape, structure, dispersibility and location of the active phase; and (iii) the stabilization of the active phase on the support.

References

1. P. Serp and E. Castillejos, *ChemCatChem.*, 2010, **2**, 41–47.
2. R. R. Bacsa and P. Serp, in *Carbon Meta-nanotubes*, John Wiley & Sons, Ltd, 2011, pp. 163–221.
3. R. Kumar, R. Kumar Singh, P. Kumar Dubey and I.-K. Oh, *Smart Nanosyst. Eng. Med.*, 2012, **1**, 18–39.
4. V. N. Mochalin, O. Shenderova, D. Ho and Y. Gogotsi, *Nat. Nano.*, 2012, **7**, 11–23.
5. B. F. Machado and P. Serp, *Catal. Sci. Technol.*, 2012, **2**, 54–75.
6. C. Huang, C. Li and G. Shi, *Energy Environ. Sci.*, 2012, **5**, 8848–8868.
7. H. Wang, T. Maiyalagan and X. Wang, *ACS Catal.*, 2012, **2**, 781–794.
8. N. F. Goldshleger, *Fullerene Science and Technology*, 2001, **9**, 255–280.
9. C. Pham-Huu and M.-J. Ledoux, *Top. Catal.*, 2006, **40**, 49–63.
10. Y. Liu, L. D. Nguyen, T. Truong-Huu, Y. Liu, T. Romero, I. Janowska, D. Begin and C. Pham-Huu, *Mater. Lett.*, 2012, **79**, 128–131.
11. V. Georgakilas, D. Gournis, V. Tzitzios, L. Pasquato, D. M. Guldi and M. Prato, *J. Mater. Chem.*, 2007, **17**, 2679–2694.
12. G. G. Wildgoose, C. E. Banks and R. G. Compton, *Small*, 2006, **2**, 182–193.
13. G. Blanita and M. D. Lazar, *Micro and Nanosyst.*, 2013, **5**, 138–146.
14. H. Zhang, X. Huang and F. Boey, *COSMOS*, 2010, **06**, 159–166.

15. H. P. Boehm, *Carbon*, 2002, **40**, 145–149.
16. C. P. Ewels and M. Glerup, *J. Nanosci. Nanotechnol.*, 2005, **5**, 1345–1363.
17. Y. Shao, J. Sui, G. Yin and Y. Gao, *Appl. Catal., B*, 2008, **79**, 89–99.
18. L. Duclaux, *Carbon*, 2002, **40**, 1751–1764.
19. W. Kiciński, M. Szala and M. Bystrzejewski, *Carbon*, 2014, **68**, 1–32.
20. C. H. Choi, S. H. Park and S. I. Woo, *ACS Nano*, 2012, **6**, 7084–7091.
21. Y. Li, Y. Zhao, H. Cheng, Y. Hu, G. Shi, L. Dai and L. Qu, *J. Am. Chem. Soc.*, 2011, **134**, 15–18.
22. D. Jana, C.-L. Sun, L.-C. Chen and K.-H. Chen, *Prog. Mater. Sci.*, 2013, **58**, 565–635.
23. L. R. Radovic and B. Bockrath, *J. Am. Chem. Soc.*, 2005, **127**, 5917–5927.
24. N. Karousis, N. Tagmatarchis and D. Tasis, *Chem. Rev.*, 2010, **110**, 5366–5397.
25. C. Thilgen and F. Diederich, *Chem. Rev.*, 2006, **106**, 5049–5135.
26. K. P. Loh, Q. Bao, P. K. Ang and J. Yang, *J. Mater. Chem.*, 2010, **20**, 2277–2289.
27. D. Tasis, N. Tagmatarchis, A. Bianco and M. Prato, *Chem. Rev.*, 2006, **106**, 1105–1136.
28. D. R. Dreyer, S. Park, C. W. Bielawski and R. S. Ruoff, *Chem. Soc. Rev.*, 2010, **39**, 228–240.
29. J. Lahaye, *Fuel.*, 1998, **77**, 543–547.
30. H. P. Boehm, *Carbon*, 1994, **32**, 759–769.
31. Shen, *Recent Pat. Chem. Eng.*, 2008, **1**, 27–40.
32. A. Stein, Z. Wang and M. A. Fierke, *Adv. Mater.*, 2009, **21**, 265–293.
33. I. Gerber, M. Oubenali, R. Bacsa, J. Durand, A. Gonçalves, M. F. R. Pereira, F. Jolibois, L. Perrin, R. Poteau and P. Serp, *Chem. Eur. J.*, 2011, **17**, 11467–11477.
34. I. I. Salame and T. J. Bandosz, *J. Colloid Interface Sci.*, 2001, **240**, 252–258.
35. H. Ago, T. Kugler, F. Cacialli, W. R. Salaneck, M. S. P. Shaffer, A. H. Windle and R. H. Friend, *J. Phys. Chem. B*, 1999, **103**, 8116–8121.
36. E. Fuente, J. A. Menéndez, D. Suárez and M. A. Montes-Morán, *Langmuir*, 2003, **19**, 3505–3511.
37. J. A. Menéndez, J. Phillips, B. Xia and L. R. Radovic, *Langmuir*, 1996, **12**, 4404–4410.
38. W. Shen and W. Fan, *J. Mater. Chem. A*, 2013, **1**, 999–1013.
39. R. Pietrzak, *Fuel*, 2009, **88**, 1871–1877.
40. T. J. Bandosz, in *Carbon Materials for Catalysis*, John Wiley & Sons, Inc., 2008, pp. 45–92.
41. A. M. Puziy, O. I. Poddubnaya, R. P. Socha, J. Gurgul and M. Wisniewski, *Carbon*, 2008, **46**, 2113–2123.
42. A. P. Terzyk, *Colloids Surf., A*, 2001, **177**, 23–45.
43. L. S. Panchakarla, A. Govindaraj and C. N. R. Rao, *Inorg. Chim. Acta*, 2010, **363**, 4163–4174.
44. M. Domingo-García, F. J. López Garzón and M. J. Pérez-Mendoza, *J. Colloid Interface Sci.*, 2002, **248**, 116–122.
45. M. S. Shafeeyan, W. M. A. W. Daud, A. Houshmand and A. Shamiri, *J. Anal. Appl. Pyrolysis*, 2010, **89**, 143–151.

46. K. Wepasnick, B. Smith, J. Bitter and D. Howard Fairbrother, *Anal. Bioanal. Chem.*, 2010, **396**, 1003–1014.
47. A. M. Kalijadis, M. M. Vukcevic, Z. M. Jovanovic, Z. V. Lausevic and M. D. Lausevic, *J. Serb. Chem. Soc.*, 2011, **76**, 757–768.
48. J. L. Figueiredo, M. F. R. Pereira, M. M. A. Freitas and J. J. M. Órfão, *Ind. Eng. Chem. Res.*, 2007, **46**, 4110–4115.
49. P. Brender, R. Gadiou, J.-C. Rietsch, P. Fioux, J. Dentzer, A. Ponche and C. Vix-Guterl, *Anal. Chem.*, 2012, **84**, 2147–2153.
50. C. Moreno-Castilla, *Carbon*, 2004, **42**, 83–94.
51. S. L. Goertzen, K. D. Thériault, A. M. Oickle, A. C. Tarasuk and H. A. Andreas, *Carbon*, 2010, **48**, 1252–1261.
52. A. M. Oickle, S. L. Goertzen, K. R. Hopper, Y. O. Abdalla and H. A. Andreas, *Carbon*, 2010, **48**, 3313–3322.
53. B. M. Babić, S. K. Milonjić, M. J. Polovina and B. V. Kaludierović, *Carbon*, 1999, **37**, 477–481.
54. J. B. Stelzer, R. Nitzsche and J. Caro, *Chem. Eng. Technol.*, 2005, **28**, 182–186.
55. J. L. Figueiredo, M. F. R. Pereira, M. M. A. Freitas and J. J. M. Órfão, *Carbon*, 1999, **37**, 1379–1389.
56. S. Kundu, Y. Wang, W. Xia and M. Muhler, *J. Phys. Chem. C*, 2008, **112**, 16869–16878.
57. J. L. Figueiredo, M. F. R. Pereira, M. M. A. Freitas and J. J. M. Órfão, *Ind. Eng. Chem. Res.*, 2006, **46**, 4110–4115.
58. N. Mahata, M. F. R. Pereira, F. Suárez-García, A. Martínez-Alonso, J. M. D. Tascón and J. L. Figueiredo, *J. Colloid Interface Sci.*, 2008, **324**, 150–155.
59. Y. Otake and R. G. Jenkins, *Carbon*, 1993, **31**, 109–121.
60. U. Zielke, K. J. Hüttinger and W. P. Hoffman, *Carbon*, 1996, **34**, 983–998.
61. S.-J. Park and K.-S. Kim, in *Microscopy: Science, Technology, Applications and Education eds.* A. Méndez-Vilas and J. Díaz, Formatex, Badajoz, 2010, vol. 3, pp. 1905–1916.
62. B. D. Ratner and D. G. Castner, in *Surface Analysis – The Principal Techniques*, John Wiley & Sons, Ltd, 2009, pp. 47–112.
63. C. Popov, W. Kulisch, S. Boycheva, K. Yamamoto, G. Ceccone and Y. Koga, *Diamond Relat. Mater.*, 2004, **13**, 2071–2075.
64. J. C. Lascovich, R. Giorgi and S. Scaglione, *Appl. Surf. Sci.*, 1991, **47**, 17–21.
65. W. M. Prest W and R. A. Mosher, in Colloids and Surfaces in Reprographic Technology, American Chemical Society, 1982, vol. 200, pp. 225–247.
66. V. Y. Osipov, A. E. Aleksenskiy, A. I. Shames, A. M. Panich, M. S. Shestakov and A. Y. Vul', *Diamond Relat. Mater.*, 2011, **20**, 1234–1238.
67. L.-h. Teng and T.-d. Tang, *J. Zhejiang Univ., Sci., A*, 2008, **9**, 720–726.
68. H. Kuzmany, R. Winkler and T. Pichler, *J. Phys. Condens. Matter*, 1995, 7, 6601.
69. C. Sellitti, J. L. Koenig and H. Ishida, *Carbon*, 1990, **28**, 221–228.
70. P. E. Fanning and M. A. Vannice, *Carbon*, 1993, **31**, 721–730.

71. E. Fuente, J. A. Menéndez, M. A. Díez, D. Suárez and M. A. Montes-Morán, *J. Phys. Chem. B*, 2003, **107**, 6350–6359.
72. H. F. Gorgulho, J. P. Mesquita, F. Gonçalves, M. F. R. Pereira and J. L. Figueiredo, *Carbon*, 2008, **46**, 1544–1555.
73. S. Hanelt, G. Orts-Gil, J. F. Friedrich and A. Meyer-Plath, *Carbon*, 2011, **49**, 2978–2988.
74. M. F. R. Pereira, S. F. Soares, J. J. M. Órfão and J. L. Figueiredo, *Carbon*, 2003, **41**, 811–821.
75. G. S. Szymański, Z. Karpiński, S. Biniak and A. Świątkowski, *Carbon*, 2002, **40**, 2627–2639.
76. M. Krutyeva, F. Grinberg, F. Furtado, P. Galvosas, J. Kärger, A. Silvestre-Albero, A. Sepulveda-Escribano, J. Silvestre-Albero and F. Rodríguez-Reinoso, *Microporous Mesoporous Mater.*, 2009, **120**, 91–97.
77. M. Dubois, K. Guérin, E. Petit, N. Batisse, A. Hamwi, N. Komatsu, J. r. m. Giraudet, P. Pirotte and F. Masin, *J. Phys. Chem. C*, 2009, **113**, 10371–10378.
78. S. Park, Y. Hu, J. O. Hwang, E.-S. Lee, L. B. Casabianca, W. Cai, J. R. Potts, H.-W. Ha, S. Chen, J. Oh, S. O. Kim, Y.-H. Kim, Y. Ishii and R. S. Ruoff, *Nat. Commun.*, 2012, **3**, 638.
79. N. Baccile, G. Laurent, F. Babonneau, F. Fayon, M.-M. Titirici and M. Antonietti, *J. Phys. Chem. C*, 2009, **113**, 9644–9654.
80. W. Cai, R. D. Piner, F. J. Stadermann, S. Park, M. A. Shaibat, Y. Ishii, D. Yang, A. Velamakanni, S. J. An, M. Stoller, J. An, D. Chen and R. S. Ruoff, *Science*, 2008, **321**, 1815–1817.
81. N. Baccile, G. Laurent, C. Coelho, F. Babonneau, L. Zhao and M.-M. Titirici, *J. Phys. Chem. C*, 2011, **115**, 8976–8982.
82. J. C. C. Freitas, F. G. Emmerich, G. R. C. Cernicchiaro, L. C. Sampaio and T. J. Bonagamba, *Solid State Nucl. Magn. Reson.*, 2001, **20**, 61–73.
83. E. Abou-Hamad, M. R. Babaa, M. Bouhrara, Y. Kim, Y. Saih, S. Dennler, F. Mauri, J. M. Basset, C. Goze-Bac and T. Wågberg, *Phys. Rev. B*, 2011, **84**, 165417.
84. B. M. Quinn, C. Dekker and S. G. Lemay, *J. Am. Chem. Soc.*, 2005, **127**, 6146–6147.
85. J. Li, S. Tang, L. Lu and H. C. Zeng, *J. Am. Chem. Soc.*, 2007, **129**, 9401–9409.
86. X. Hao, W. A. Spieker and J. R. Regalbuto, *J. Colloid Interface Sci.*, 2003, **267**, 259–264.
87. B. F. Machado, M. Oubenali, M. Rosa Axet, T. Trang Nguyen, M. Tunckol, M. Girleanu, O. Ersen, I. C. Gerber and P. Serp, *J. Catal.*, 2014, **309**, 185–198.
88. J. M. Planeix, N. Coustel, B. Coq, V. Brotons, P. S. Kumbhar, R. Dutartre, P. Geneste, P. Bernier and P. M. Ajayan, *J. Am. Chem. Soc.*, 1994, **116**, 7935–7936.
89. V. Jiménez, P. Panagiotopoulou, P. Sánchez, J. L. Valverde and A. Romero, *Chem. Eng. J.*, 2011, **168**, 947–954.
90. T. W. Ebbesen, H. Hiura, M. E. Bisher, M. M. J. Treacy, J. L. Shreeve-Keyer and R. C. Haushalter, *Adv. Mater.*, 1996, **8**, 155–157.
91. I. D. Rosca, F. Watari, M. Uo and T. Akasaka, *Carbon*, 2005, **43**, 3124–3131.

92. B. C. Satishkumar, M. V. Erasmus, A. Govindaraj and C. N. R. Rao, *J. Phys. D: Appl. Phys.*, 1996, **29**, 3173.
93. M. A. M. Motchelaho, H. Xiong, M. Moyo, L. L. Jewell and N. J. Coville, *J. Mol. Catal. A: Chem.*, 2011, **335**, 189–198.
94. R. M. Malek Abbaslou, A. Tavasoli and A. K. Dalai, *Appl. Catal., A*, 2009, **355**, 33–41.
95. M. Trépanier, A. Tavasoli, A. K. Dalai and N. Abatzoglou, *Fuel Process. Technol.*, 2009, **90**, 367–374.
96. A. Solhy, B. F. Machado, J. Beausoleil, Y. Kihn, F. Gonçalves, M. F. R. Pereira, J. J. M. Órfão, J. L. Figueiredo, J. L. Faria and P. Serp, *Carbon*, 2008, **46**, 1194–1207.
97. P. Azadi, R. Farnood and E. Meier, *J. Phys. Chem. A*, 2009, **114**, 3962–3968.
98. G. Gonçalves, P. A. A. P. Marques, C. M. Granadeiro, H. I. S. Nogueira, M. K. Singh and J. Graécio, *Chem. Mater.*, 2009, **21**, 4796–4802.
99. L. Prati, A. Villa, A. R. Lupini and G. M. Veith, *Phys. Chem. Chem. Phys.*, 2012, **14**, 2969–2978.
100. T. T. Baby and R. Sundara, *J. Phys. Chem. C*, 2011, **115**, 8527–8533.
101. T. G. Ros, A. J. van Dillen, J. W. Geus and D. C. Koningsberger, *Chem. Eur. J.*, 2002, **8**, 1151–1162.
102. J. Guo, G. Sun, Q. Wang, G. Wang, Z. Zhou, S. Tang, L. Jiang, B. Zhou and Q. Xin, *Carbon*, 2006, **44**, 152–157.
103. A. Stafiej and K. Pyrzynska, *Sep. Purif. Technol.*, 2007, **58**, 49–52.
104. G. Ovejero, J. L. Sotelo, M. D. Romero, A. Rodríguez, M. A. Ocaña, G. Rodríguez and J. García, *Ind. Eng. Chem. Res.*, 2006, **45**, 2206–2212.
105. Y. F. Jia, B. Xiao and K. M. Thomas, *Langmuir*, 2001, **18**, 470–478.
106. L. Mabena, S. Sinha Ray, S. Mhlanga and N. Coville, *Appl. Nanosci.*, 2011, **1**, 67–77.
107. Z. Lei, L. An, L. Dang, M. Zhao, J. Shi, S. Bai and Y. Cao, *Microporous Mesoporous Mater.*, 2009, **119**, 30–38.
108. R. Imran Jafri, N. Rajalakshmi and S. Ramaprabhu, *J. Mater. Chem.*, 2010, **20**, 7114–7117.
109. H. Y. Du, C. H. Wang, H. C. Hsu, S. T. Chang, U. S. Chen, S. C. Yen, L. C. Chen, H. C. Shih and K. H. Chen, *Diamond Relat. Mater.*, 2008, **17**, 535–541.
110. H. Yoon, S. Ko and J. Jang, *Chem. Commun.*, 2007, 1468–1470.
111. K. Chizari, I. Janowska, M. Houllé, I. Florea, O. Ersen, T. Romero, P. Bernhardt, M. J. Ledoux and C. Pham-Huu, *Appl. Catal., A*, 2010, **380**, 72–80.
112. D. C. Higgins, D. Meza and Z. Chen, *J. Phys. Chem. C*, 2010, **114**, 21982–21988.
113. M. S. Saha and A. Kundu, *J. Power Sources*, 2010, **195**, 6255–6261.
114. J. H. Bitter and K. P. de Jong, in *Carbon Materials for Catalysis*, John Wiley & Sons, Inc., 2008, pp. 157–176.
115. M. L. Toebes, M. K. van der Lee, L. M. Tang, M. H. Huis in 't Veld, J. H. Bitter, A. J. van Dillen and K. P. de Jong, *J. Phys. Chem. B*, 2004, **108**, 11611–11619.

116. G. L. Bezemer, P. B. Radstake, V. Koot, A. J. van Dillen, J. W. Geus and K. P. de Jong, *J. Catal.*, 2006, **237**, 291–302.
117. J. H. Bitter, M. K. van der Lee, A. G. T. Slotboom, A. J. van Dillen and K. P. de Jong, *Catal. Lett.*, 2003, **89**, 139–142.
118. M. K. van der Lee, J. van Dillen, J. H. Bitter and K. P. de Jong, *J. Am. Chem. Soc.*, 2005, **127**, 13573–13582.
119. E. van Steen and F. F. Prinsloo, *Catal. Today*, 2002, **71**, 327–334.
120. M. L. Toebes, F. F. Prinsloo, J. H. Bitter, A. J. van Dillen and K. P. de Jong, *J. Catal.*, 2003, **214**, 78–87.
121. H. Xiong, M. A. M. Motchelaho, M. Moyo, L. L. Jewell and N. J. Coville, *J. Catal.*, 2011, **278**, 26–40.
122. E. Castillejos, R. Chico, R. Bacsa, S. Coco, P. Espinet, M. Pérez-Cadenas, A. Guerrero-Ruiz, I. Rodríguez-Ramos and P. Serp, *Eur. J. Inorg. Chem.*, 2010, 5096–5102.
123. K. Szőri, R. Puskás, G. Szőllősi, I. Bertóti, J. Szépvölgyi and M. Bartók, *Catal. Lett.*, 2013, **143**, 539–546.
124. R. Wang, Z. Wu, C. Chen, Z. Qin, H. Zhu, G. Wang, H. Wang, C. Wu, W. Dong, W. Fan and J. Wang, *Chem. Commun.*, 2013, **49**, 8250–8252.
125. A. J. Plomp, D. S. Su, K. P. d. Jong and J. H. Bitter, *J. Phys. Chem. C*, 2009, **113**, 9865–9869.
126. J. Guerra and M. A. Herrero, *Nanoscale*, 2010, **2**, 1390–1400.
127. J. A. Menéndez, A. Arenillas, B. Fidalgo, Y. Fernández, L. Zubizarreta, E. G. Calvo and J. M. Bermúdez, *Fuel Process. Technol.*, 2010, **91**, 1–8.
128. S. C. Motshekga, S. K. Pillai, S. S. Ray, K. Jalama and R. W. M. Krause, *J. Nanomaterials*, 2012, **2012**, 51–51.
129. C.-C. Chien and K.-T. Jeng, *Mater. Chem. Phys.*, 2006, **99**, 80–87.
130. K.-T. Jeng, C.-C. Chien, N.-Y. Hsu, S.-C. Yen, S.-D. Chiou, S.-H. Lin and W.-M. Huang, *J. Power Sources*, 2006, **160**, 97–104.
131. Z. Liu, J. Y. Lee, W. Chen, M. Han and L. M. Gan, *Langmuir*, 2003, **20**, 181–187.
132. W.-X. Chen, J. Y. Lee and Z. Liu, *Mater. Lett.*, 2004, **58**, 3166–3169.
133. L. Li and Y. Xing, *J. Phys. Chem. C*, 2007, **111**, 2803–2808.
134. D. M. Han, Z. P. Guo, Z. W. Zhao, R. Zeng, Y. Z. Meng, D. Shu and H. K. Liu, *J. Power Sources*, 2008, **184**, 361–369.
135. A. Kongkanand, K. Vinodgopal, S. Kuwabata and P. V. Kamat, *J. Phys. Chem. B*, 2006, **110**, 16185–16188.
136. Z. Liu, X. Y. Ling, B. Guo, L. Hong and J. Y. Lee, *J. Power Sources*, 2007, **167**, 272–280.
137. P. Kundu, C. Nethravathi, P. A. Deshpande, M. Rajamathi, G. Madras and N. Ravishankar, *Chem. Mater.*, 2011, **23**, 2772–2780.
138. S. Sharma, A. Ganguly, P. Papakonstantinou, X. Miao, M. Li, J. L. Hutchison, M. Delichatsios and S. Ukleja, *J. Phys. Chem. C*, 2010, **114**, 19459–19466.
139. S. Wang, S. P. Jiang and X. Wang, *Electrochim. Acta*, 2011, **56**, 3338–3344.
140. C. Xu, X. Wang and J. Zhu, *J. Phys. Chem. C*, 2008, **112**, 19841–19845.
141. Y. Lin, D. W. Baggett, J.-W. Kim, E. J. Siochi and J. W. Connell, *ACS Appl. Mater. Interfaces*, 2011, **3**, 1652–1664.

142. J. Zang, Y. Wang, L. Bian, J. Zhang, F. Meng, Y. Zhao, X. Qu and S. Ren, *Int. J. Hydrogen Energy*, 2012, **37**, 6349–6355.
143. L. Y. Bian, Y. H. Wang, J. B. Zang, F. W. Meng and Y. L. Zhao, *Int. J. Hydrogen Energy*, 2012, **37**, 1220–1225.
144. J.-Y. Kim, K.-H. Kim, S.-H. Park and K.-B. Kim, *Electrochim. Acta*, 2010, **55**, 8056–8061.
145. C. Antonetti, M. Oubenali, A. M. Raspolli Galletti, P. Serp and G. Vannucci, *Appl. Catal., A*, 2012, **421–422**, 99–107.
146. M. S. Raghuveer, S. Agrawal, N. Bishop and G. Ramanath, *Chem. Mater.*, 2006, **18**, 1390–1393.
147. T. Ramulifho, K. I. Ozoemena, R. M. Modibedi, C. J. Jafta and M. K. Mathe, *Electrochim. Acta.*, 2012, **59**, 310–320.
148. D. M. Han, Z. P. Guo, R. Zeng, C. J. Kim, Y. Z. Meng and H. K. Liu, *Int. J. Hydrogen Energy*, 2009, **34**, 2426–2434.
149. C.-T. Hsieh, W.-M. Hung, W.-Y. Chen and J.-Y. Lin, *Int. J. Hydrogen Energy*, 2011, **36**, 2765–2772.
150. B. P. Vinayan, R. I. Jafri, R. Nagar, N. Rajalakshmi, K. Sethupathi and S. Ramaprabhu, *Int. J. Hydrogen Energy*, 2012, **37**, 412–421.
151. A. Z. Sadek, C. Zhang, Z. Hu, J. G. Partridge, D. G. McCulloch, W. Wlodarski and K. Kalantar-zadeh, *J. Phys. Chem. C*, 2009, **114**, 238–242.
152. H. Zhu, Y. Liu, L. Shen, Y. Wei, Z. Guo, H. Wang, K. Han and Z. Chang, *Int. J. Hydrogen Energy*, 2010, **35**, 3125–3128.
153. S. L. Knupp, W. Li, O. Paschos, T. M. Murray, J. Snyder and P. Haldar, *Carbon*, 2008, **46**, 1276–1284.
154. M. Tsuji, M. Kubokawa, R. Yano, N. Miyamae, T. Tsuji, M.-S. Jun, S. Hong, S. Lim, S.-H. Yoon and I. Mochida, *Langmuir*, 2006, **23**, 387–390.
155. B. Fang, N. K. Chaudhari, M.-S. Kim, J. H. Kim and J.-S. Yu, *J. Am. Chem. Soc.*, 2009, **131**, 15330–15338.
156. Z. Liu, B. Guo, L. Hong and T. H. Lim, *Electrochem. Commun.*, 2006, **8**, 83–90.
157. K. Jasuja, J. Linn, S. Melton and V. Berry, *J. Phys. Chem. Lett.*, 2010, **1**, 1853–1860.
158. C.-T. Hsieh, J.-S. Lin, Y.-F. Chen and H. Teng, *J. Phys. Chem. C*, 2012, **116**, 15251–15258.
159. T. M. Day, P. R. Unwin and J. V. Macpherson, *Nano Lett.*, 2006, **7**, 51–57.
160. Z.-L. Xiao, C. Y. Han, W.-K. Kwok, H.-H. Wang, U. Welp, J. Wang and G. W. Crabtree, *J. Am. Chem. Soc.*, 2004, **126**, 2316–2317.
161. Y. Zhao, L. Fan, H. Zhong, Y. Li and S. Yang, *Adv. Funct. Mater.*, 2007, **17**, 1537–1541.
162. S. Sharma and B. G. Pollet, *J. Power Sources*, 2012, **208**, 96–119.
163. K. Lee, J. Zhang, H. Wang and D. P. Wilkinson, *J. Appl. Electrochem.*, 2006, **36**, 507–522.
164. Z. He, J. Chen, D. Liu, H. Zhou and Y. Kuang, *Diamond Relat. Mater.*, 2004, **13**, 1764–1770.
165. Y. Hu, H. Zhang, P. Wu, H. Zhang, B. Zhou and C. Cai, *Phys. Chem. Chem. Phys.*, 2011, **13**, 4083–4094.

166. F. Xiao, Z. Mo, F. Zhao and B. Zeng, *Electrochem. Commun.*, 2008, **10**, 1740–1743.
167. C. Hui-Fang, Y. Jian-Shan, L. Xiao, Z. Wei-De and S. Fwu-Shan, *Nanotechnology*, 2006, **17**, 2334.
168. D. Vairavapandian, P. Vichchulada and M. D. Lay, *Anal. Chim. Acta*, 2008, **626**, 119–129.
169. J.-S. Ye, H.-F. Cui, Y. Wen, W. D. Zhang, G. Q. Xu and F.-S. Sheu, *Microchim. Acta*, 2006, **152**, 267–275.
170. H.-F. Cui, J.-S. Ye, W.-D. Zhang, J. Wang and F.-S. Sheu, *J. Electroanal. Chem.*, 2005, **577**, 295–302.
171. D.-J. Guo and H.-L. Li, *J. Electroanal. Chem.*, 2004, **573**, 197–202.
172. M.-C. Tsai, T.-K. Yeh and C.-H. Tsai, *Electrochem. Commun.*, 2006, **8**, 1445–1452.
173. M. Miyake, T. Ueda and T. Hirato, *J. Electrochem. Soc.*, 2011, **158**, D590–D593.
174. X. Chen, N. Li, K. Eckhard, L. Stoica, W. Xia, J. Assmann, M. Muhler and W. Schuhmann, *Electrochem. Commun.*, 2007, **9**, 1348–1354.
175. B. Miomir, Vukmirovic, T. Stoyan, Bliznakov, Kotaro Sasaki, J. X. Wang and R. R. Adzic, *Interface*, 2011, **20**, 33–40.
176. T. Maiyalagan, X. Dong, P. Chen and X. Wang, *J. Mater. Chem.*, 2012, **22**, 5286–5290.
177. J. C. Claussen, A. Kumar, D. B. Jaroch, M. H. Khawaja, A. B. Hibbard, D. M. Porterfield and T. S. Fisher, *Adv. Funct. Mater.*, 2012, **22**, 3399–3405.
178. S. Sattayasamitsathit, Y. Gu, K. Kaufmann, W. Jia, X. Xiao, M. Rodriguez, S. Minteer, J. Cha, D. B. Burckel, C. Wang, R. Polsky and J. Wang, *J. Mater. Chem. A*, 2013, **1**, 1639–1645.
179. L. Y. Bian, Y. H. Wang, J. B. Zang, J. K. Yu and H. Huang, *J. Electroanal. Chem.*, 2010, **644**, 85–88.
180. J. Hu, X. Lu and J. S. Foord, *Electrochem. Commun.*, 2010, **12**, 676–679.
181. D. Santiago, G. G. Rodríguez-Calero, A. Palkar, D. Barraza-Jimenez, D. H. Galvan, G. Casillas, A. Mayoral, M. Jose-Yacamán, L. Echegoyen and C. R. Cabrera, *Langmuir*, 2012, **28**, 17202–17210.
182. Y. Zhao, Y. E, L. Fan, Y. Qiu and S. Yang, *Electrochim. Acta*, 2007, **52**, 5873–5878.
183. H. Gao, F. Xiao, C. B. Ching and H. Duan, *ACS Appl. Mater. Interfaces*, 2011, **3**, 3049–3057.
184. J. Libuda and H. J. Freund, *Surf. Sci. Rep.*, 2005, **57**, 157–298.
185. D.-Q. Yang and E. Sacher, *J. Phys. Chem. C*, 2008, **112**, 4075–4082.
186. T.-F. Hsieh, C.-C. Chuang, Y.-C. Chou and C.-M. Shu, *Mater. Des.*, 2010, **31**, 1684–1687.
187. A. Felten, J. Ghijsen, J. J. Pireaux, W. Drube, R. L. Johnson, D. Liang, M. Hecq, G. Van Tendeloo and C. Bittencourt, *Micron*, 2009, **40**, 74–79.
188. Z. Luo, L. A. Somers, Y. Dan, T. Ly, N. J. Kybert, E. J. Mele and A. T. C. Johnson, *Nano Lett.*, 2010, **10**, 777–781.
189. R. Zan, U. Bangert, Q. Ramasse and K. S. Novoselov, *J. Phys. Chem. Lett.*, 2012, **3**, 953–958.

190. A. Bello, M. Fabiane, D. Dodoo-Arhin, K. I. Ozoemena and N. Manyala, *J. Phys. Chem. Solids*, 2014, **75**, 109–114.
191. M. Knez, K. Nielsch and L. Niinistö, *Adv. Mater.*, 2007, **19**, 3425–3438.
192. C. Detavernier, J. Dendooven, S. Pulinthanathu Sree, K. F. Ludwig and J. A. Martens, *Chem. Soc. Rev.*, 2011, **40**, 5242–5253.
193. C. Liu, C.-C. Wang, C.-C. Kei, Y.-C. Hsueh and T.-P. Perng, *Small*, 2009, **5**, 1535–1538.
194. X. Liang and C. Jiang, *J. Nanopart. Res.*, 2013, **15**, 1–9.
195. H. Yang-Chih, W. Chih-Chieh, L. Chueh, K. Chi-Chung and P. Tsong-Pyng, *Nanotechnology*, 2012, **23**, 405603.
196. S. Deng, S. W. Verbruggen, Z. He, D. J. Cott, P. M. Vereecken, J. A. Martens, S. Bals, S. Lenaerts and C. Detavernier, *RSC Adv.*, 2014, **4**, 11648–11653.
197. Y.-S. Min, E. J. Bae, J. B. Park, U. J. Kim, W. Park, J. Song, C. Seong Hwang and N. Park, *Appl. Phys. Lett.*, 2007, **90**, 263104–263103.
198. D. B. Farmer and R. G. Gordon, *Nano Lett.*, 2006, **6**, 699–703.
199. K. Grigoras, M. Y. Zavodchikova, A. G. Nasibulin, E. I. Kauppinen, V. Ermolov and S. Franssila, *J. Nanosci. Nanotechnol.*, 2011, **11**, 8818–8825.
200. X. Meng, M. Ionescu, M. Banis, Y. Zhong, H. Liu, Y. Zhang, S. Sun, R. Li and X. Sun, *J. Nanopart. Res.*, 2011, **13**, 1207–1218.
201. A. C. Johansson, R. B. Yang, K. B. Haugshøj, J. V. Larsen, L. H. Christensen and E. V. Thomsen, *Int. J. Hydrogen Energy*, 2013, **38**, 11406–11414.
202. S. Sun, G. Zhang, N. Gauquelin, N. Chen, J. Zhou, S. Yang, W. Chen, X. Meng, D. Geng, M. N. Banis, R. Li, S. Ye, S. Knights, G. A. Botton, T.-K. Sham and X. Sun, *Sci. Rep.*, 2013, **3**.
203. G. Wang, Z. Gao, G. Wan, S. Lin, P. Yang and Y. Qin, *Nano Res.*, 2014, **7**, 704–716.
204. M. Xiangbo, G. Dongsheng, L. Jian, L. Ruying and S. Xueliang, *Nanotechnology*, 2011, **22**, 165602.
205. X. Sun, M. Xie, G. Wang, H. Sun, A. S. Cavanagh, J. J. Travis, S. M. George and J. Lian, *J. Electrochem. Soc.*, 2012, **159**, A364–A369.
206. M.-G. Willinger, G. Neri, E. Rauwel, A. Bonavita, G. Micali and N. Pinna, *Nano Lett.*, 2008, **8**, 4201–4204.
207. S. Boukhalfa, K. Evanoff and G. Yushin, *Energy Environ. Sci.*, 2012, **5**, 6872–6879.
208. X. Meng, Y. Zhong, Y. Sun, M. N. Banis, R. Li and X. Sun, *Carbon*, 2011, **49**, 1133–1144.
209. X. Li, X. Meng, J. Liu, D. Geng, Y. Zhang, M. N. Banis, Y. Li, J. Yang, R. Li, X. Sun, M. Cai and M. W. Verbrugge, *Adv. Funct. Mater.*, 2012, **22**, 1647–1654.
210. Y. S. Min, E. J. Bae, K. S. Jeong, Y. J. Cho, J. H. Lee, W. B. Choi and G. S. Park, *Adv. Mater.*, 2003, **15**, 1019–1022.
211. P. Serp, P. Kalck and R. Feurer, *Chem. Rev.*, 2002, **102**, 3085–3128.
212. C. Vahlas, F. Juarez, R. Feurer, P. Serp and B. Caussat, *Chem. Vap. Deposition*, 2002, **8**, 127–144.
213. C. Vahlas, B. Caussat, P. Serp and G. N. Angelopoulos, *Mater. Sci. Eng., R.*, 2006, **53**, 1–72.

214. P. Serp, R. Feurer, Y. Kihn, P. Kalck, J. L. Faria and J. L. Figueiredo, *J. Mater. Chem.*, 2001, **11**, 1980–1981.
215. P. Serp, R. Feurer, J. L. Faria and J. L. Figueiredo, *J. Phys. IV France*, 2002, **12**, 29–36.
216. H. Kim and S. H. Moon, *Carbon*, 2011, **49**, 1491–1501.
217. C. Liang, W. Xia, H. Soltani-Ahmadi, O. Schluter, R. A. Fischer and M. Muhler, *Chem. Commun.*, 2005, 282–284.
218. W. Xia, O. F. K. Schlüter, C. Liang, M. W. E. van den Berg, M. Guraya and M. Muhler, *Catal. Today*, 2005, **102–103**, 34–39.
219. Q.-L. Naidoo, S. Naidoo, L. Petrik, A. Nechaev and P. Ndungu, *Int. J. Hydrogen Energy*, 2012, **37**, 9459–9469.
220. H. Zhang, J. Qiu, C. Liang, Z. Li, X. Wang, Y. Wang, Z. Feng and C. Li, *Catal. Lett.*, 2005, **101**, 211–214.
221. C. Taschner, K. Biedermann, B. B. A. Leonhardt, T. Gemming and K. Wetzig, *Electrochem. Soc., Proc.*, 2005, 396–401.
222. J. D. Kim, B. S. Kang, T. W. Noh, J.-G. Yoon, S. I. Baik and Y.-W. Kim, *J. Electrochem. Soc.*, 2005, **152**, D23–D25.
223. S. Orlanducci, V. Sessa, M. L. Terranova, G. A. Battiston, S. Battiston and R. Gerbasi, *Carbon*, 2006, **44**, 2839–2843.
224. G. Li Puma, A. Bono, D. Krishnaiah and J. G. Collin, *J. Hazard. Mater.*, 2008, **157**, 209–219.
225. Q. Kuang, S.-F. Li, Z.-X. Xie, S.-C. Lin, X.-H. Zhang, S.-Y. Xie, R.-B. Huang and L.-S. Zheng, *Carbon*, 2006, **44**, 1166–1172.
226. J. M. Lee, Y. B. Pyun, J. Yi, J. W. Choung and W. I. Park, *J. Phys. Chem. C*, 2009, **113**, 19134–19138.
227. Y.-J. Kim, J.-H. Lee and G.-C. Yi, *Appl. Phys. Lett.*, 2009, **95**, 213101.
228. J. Jo, H. Yoo, S.-I. Park, J. B. Park, S. Yoon, M. Kim and G.-C. Yi, *Adv. Mater.*, 2014, **26**, 2011–2015.
229. M. Grzelczak, J. Vermant, E. M. Furst and L. M. Liz-Marzán, *ACS Nano*, 2010, **4**, 3591–3605.
230. Z. Nie, A. Petukhova and E. Kumacheva, *Nat. Nano*, 2010, **5**, 15–25.
231. H. Kitching, M. J. Shiers, A. J. Kenyon and I. P. Parkin, *J. Mater. Chem. A*, 2013, **1**, 6985–6999.
232. L. Liu, T. Wang, J. Li, Z.-X. Guo, L. Dai, D. Zhang and D. Zhu, *Chem. Phys. Lett.*, 2003, **367**, 747–752.
233. M. A. Correa-Duarte and L. M. Liz-Marzan, *J. Mater. Chem.*, 2006, **16**, 22–25.
234. S. Fullam, D. Cottell, H. Rensmo and D. Fitzmaurice, *Adv. Mater.*, 2000, **12**, 1430–1432.
235. S. Tobias, B. Udo and P. K. Jörg, *Nanotechnology*, 2005, **16**, 1123.
236. C. Xiang, Y. Zou, L.-X. Sun and F. Xu, *Electrochem. Commun.*, 2008, **10**, 38–41.
237. M. Ho Yang, B. G. Choi, H. Park, T. J. Park, W. H. Hong and S. Y. Lee, *Electroanalysis*, 2011, **23**, 850–857.
238. Y. Fang, S. Guo, C. Zhu, Y. Zhai and E. Wang, *Langmuir*, 2010, **26**, 11277–11282.

239. W. Hong, H. Bai, Y. Xu, Z. Yao, Z. Gu and G. Shi, *J. Phys. Chem. C*, 2010, **114**, 1822–1826.
240. S. Zhang, Y. Shao, H.-g. Liao, J. Liu, I. A. Aksay, G. Yin and Y. Lin, *Chem. Mater.*, 2011, **23**, 1079–1081.
241. S. Wang, X. Wang and S. P. Jiang, *Phys. Chem. Chem. Phys.*, 2011, **13**, 6883–6891.
242. J. Chai, F. Li, Y. Hu, Q. Zhang, D. Han and L. Niu, *J. Mater. Chem.*, 2011, **21**, 17922–17929.
243. C. Zhu, S. Guo, Y. Zhai and S. Dong, *Langmuir*, 2010, **26**, 7614–7618.
244. S. Guo and S. Sun, *J. Am. Chem. Soc.*, 2012, **134**, 2492–2495.
245. D. Wang, D. Choi, J. Li, Z. Yang, Z. Nie, R. Kou, D. Hu, C. Wang, L. V. Saraf, J. Zhang, I. A. Aksay and J. Liu, *ACS Nano*, 2009, **3**, 907–914.
246. J. Liu, H. Bai, Y. Wang, Z. Liu, X. Zhang and D. D. Sun, *Adv. Funct. Mater.*, 2010, **20**, 4175–4181.
247. C. Gao, W. Li, H. Morimoto, Y. Nagaoka and T. Maekawa, *J. Phys. Chem. B*, 2006, **110**, 7213–7220.
248. W. Chen, S. Li, C. Chen and L. Yan, *Adv. Mater.*, 2011, **23**, 5679–5683.
249. S. Liao, G. Xu, W. Wang, F. Watari, F. Cui, S. Ramakrishna and C. K. Chan, *Acta Biomaterialia*, 2007, **3**, 669–675.
250. S. Zhang, Y. Shao, G. Yin and Y. Lin, *Appl. Catal., B*, 2011, **102**, 372–377.
251. S. Zhang, Y. Shao, G. Yin and Y. Lin, *J. Mater. Chem.*, 2010, **20**, 2826–2830.
252. C.-L. Lee, Y.-C. Ju, P.-T. Chou, Y.-C. Huang, L.-C. Kuo and J.-C. Oung, *Electrochem. Commun.*, 2005, **7**, 453–458.
253. C. Freire and A. R. Silva, in *Carbon Materials for Catalysis*, John Wiley & Sons, Inc, 2008, pp. 267–307.
254. J. S. Rafelt and J. H. Clark, *Catal. Today*, 2000, **57**, 33–44.
255. Q.-H. Fan, Y.-M. Li and A. S. C. Chan, *Chem. Rev.*, 2002, **102**, 3385–3466.
256. C. Li, *Catal. Rev.*, 2004, **46**, 419–492.
257. C. E. Song and S.-g. Lee, *Chem. Rev.*, 2002, **102**, 3495–3524.
258. J. M. Notestein and A. Katz, *Chem. Eur. J.*, 2006, **12**, 3954–3965.
259. S. Sarkar, S. Niyogi, E. Bekyarova and R. C. Haddon, *Chem. Sci.*, 2011, **2**, 1326–1333.
260. S. Banerjee and S. S. Wong, *J. Am. Chem. Soc.*, 2002, **124**, 8940–8948.
261. S. Banerjee and S. S. Wong, *Nano Lett.*, 2001, **2**, 49–53.
262. Y. Zhang, H.-B. Zhang, G.-D. Lin, P. Chen, Y.-Z. Yuan and K. R. Tsai, *Appl. Catal., A*, 1999, **187**, 213–224.
263. B. Barati, M. Moghadam, A. Rahmati, V. Mirkhani, S. Tangestaninejad and I. Mohammadpoor-Baltork, *J. Organomet. Chem.*, 2013, **724**, 32–39.
264. M. Pérez-Cadenas, L. J. Lemus-Yegres, M. C. Román-Martínez and C. Salinas-Martínez de Lecea, *Appl. Catal., A*, 2011, **402**, 132–138.
265. R. Giordano, P. Serp, P. Kalck, Y. Kihn, J. Schreiber, C. Marhic and J.-L. Duvail, *Eur. J. Inorg. Chem.*, 2003, **2003**, 610–617.
266. T. G. Ros, A. J. van Dillen, J. W. Geus and D. C. Koningsberger, *Chem. Eur. J.*, 2002, **8**, 2868–2878.
267. Q. Zhao, Y. Li, R. Liu, A. Chen, G. Zhang, F. Zhang and X. Fan, *J. Mater. Chem. A*, 2013, **1**, 15039–15045.

268. L. Lemus-Yegres, I. Such-Basáñez, C. S.-M. de Lecea, P. Serp and M. C. Román-Martínez, *Carbon*, 2006, **44**, 605–608.
269. C. C. Gheorghiu, C. Salinas Martínez de Lecea and M. C. Román Martínez, *ChemCatChem.*, 2013, **5**, 1587–1597.
270. L. J. Lemus-Yegres, M. Pérez-Cadenas, M. C. Román-Martínez and C. S.-M. de Lecea, *Microporous Mesoporous Mater.*, 2011, **139**, 164–172.
271. L. J. Lemus-Yegres, M. C. Román-Martínez, I. Such-Basáñez and C. Salinas-Martínez de Lecea, *Microporous Mesoporous Mater.*, 2008, **109**, 305–316.
272. L. J. Lemus-Yegres, M. C. Román-Martínez and C. S.-M. de Lecea, *J. Nanosci. Nanotechnol.*, 2009, **9**, 6034–6041.
273. C. C. Gheorghiu, B. F. Machado, C. Salinas-Martinez de Lecea, M. Gouygou, M. C. Roman-Martinez and P. Serp, *Dalton Trans.*, 2014, **43**, 7455–7463.
274. L. M. D. R. S. Martins, M. P. de Almeida, S. A. C. Carabineiro, J. L. Figueiredo and A. J. L. Pombeiro, *ChemCatChem.*, 2013, **5**, 3847–3856.
275. C. Baleizão, B. Gigante, H. Garcia and A. Corma, *J. Catal.*, 2004, **221**, 77–84.
276. M. Salavati-Niasari and M. Bazarganipour, *Appl. Surf. Sci.*, 2009, **255**, 7610–7617.
277. M. Salavati-Niasari, S. N. Mirsattari and M. Bazarganipour, *Polyhedron*, 2008, **27**, 3653–3661.
278. C. Baleizão, B. Gigante, H. García and A. Corma, *Tetrahedron*, 2004, **60**, 10461–10468.
279. Z. Li, S. Wu, H. Ding, D. Zheng, J. Hu, X. Wang, Q. Huo, J. Guan and Q. Kan, *New J. Chem.*, 2013, **37**, 1561–1568.
280. Q. Zhao, C. Bai, W. Zhang, Y. Li, G. Zhang, F. Zhang and X. Fan, *Ind. Eng. Chem. Res.*, 2014, **53**, 4232–4238.
281. Z. Li, S. Wu, D. Zheng, J. Liu, H. Liu, H. Lu, Q. Huo, J. Guan and Q. Kan, *Appl. Organomet. Chem.*, 2014, **28**, 317–323.
282. H. Li, J. Wu, Y. Jeilani, C. Ingram and I. Harruna, *J. Nanopart. Res.*, 2012, **14**, 1–14.
283. M. Stürzel, F. Kempe, Y. Thomann, S. Mark, M. Enders and R. Mülhaupt, *Macromolecules*, 2012, **45**, 6878–6887.
284. B. M. Amoli, S. A. A. Ramazani and H. Izadi, *J. Appl. Polym. Sci.*, 2012, **125**, E453–E461.
285. K. Lee, H. Song and J. T. Park, *Acc. Chem. Res.*, 2002, **36**, 78–86.
286. D. V. Konarev, S. S. Khasanov and R. N. Lyubovskaya, *Coord. Chem. Rev.*, 2014, **262**, 16–36.
287. A. L. Balch and C. J. Chancellor, in *Encyclopedia of Inorganic and Bioinorganic Chemistry*, John Wiley & Sons, Ltd, 2011.
288. M. D. Meijer, G. P. M. van Klink and G. van Koten, *Coord. Chem. Rev.*, 2002, **230**, 141–163.
289. A. Hirsch and M. Brettreich, in *Fullerenes*, Wiley-VCH Verlag GmbH & Co. KGaA, 2005, pp. 231–250.
290. A. L. Balch and M. M. Olmstead, *Chem. Rev.*, 1998, **98**, 2123–2166.

291. P. J. Fagan, J. C. Calabrese and B. Malone, *Acc. Chem. Res.*, 1992, **25**, 134–142.
292. S. Park, S. W. Yoon, K.-B. Lee, D. J. Kim, Y. H. Jung, Y. Do, H.-j. Paik and I. S. Choi, *Macromol. Rapid Commun.*, 2006, **27**, 47–50.
293. S. Park and I. S. Choi, *Adv. Mater.*, 2009, **21**, 902–905.
294. B. Choi, J. Lee, S. Lee, J.-H. Ko, K.-S. Lee, J. Oh, J. Han, Y.-H. Kim, I. S. Choi and S. Park, *Macromol. Rapid Commun.*, 2013, **34**, 533–538.
295. L. Zhang, E. Yue, B. Liu, P. Serp, C. Redshaw, W.-H. Sun and J. Durand, *Catal. Commun.*, 2014, **43**, 227–230.
296. L. Zhang, W. Zhang, P. Serp, W.-H. Sun and J. Durand, *ChemCatChem*, 2014, **6**, 1310–1316.
297. S.-N. Ding, D. Shan, S. Cosnier and A. Le Goff, *Chem. Eur. J.*, 2012, **18**, 11564–11568.
298. P. D. Tran, A. Le Goff, J. Heidkamp, B. Jousselme, N. Guillet, S. Palacin, H. Dau, M. Fontecave and V. Artero, *Angew. Chem., Int. Ed.*, 2011, **50**, 1371–1374.
299. D. Didier and E. Schulz, *Tetrahedron: Asymmetry*, 2013, **24**, 769–775.
300. C. Vriamont, M. Devillers, O. Riant and S. Hermans, *Chem. Eur. J.*, 2013, **19**, 12009–12017.
301. D. Jain, A. Saha and A. A. Marti, *Chem. Commun.*, 2011, **47**, 2246–2248.
302. J. H. Zagal, S. Griveau, K. I. Ozoemena, T. Nyokong and F. Bedioui, *J. Nanosci. Nanotechnol.*, 2009, **9**, 2201–2214.
303. D. R. Kauffman, O. Kuzmych and A. Star, *J. Phys. Chem. C*, 2007, **111**, 3539–3543.
304. A. Morozan, S. Campidelli, A. Filoramo, B. Jousselme and S. Palacin, *Carbon*, 2011, **49**, 4839–4847.
305. T. Yamaguchi, N. Ishii, K. Tashiro and T. Aida, *J. Am. Chem. Soc.*, 2003, **125**, 13934–13935.
306. M. Escárcega-Bobadilla, L. Rodríguez-Pérez, E. Teuma, P. Serp, A. Masdeu-Bultó and M. Gómez, *Catal. Lett.*, 2011, **141**, 808–816.
307. L. Rodriguez-Perez, E. Teuma, A. Falqui, M. Gomez and P. Serp, *Chemical Communications*, 2008, 4201–4203.
308. L.-J. Li, A. N. Khlobystov, J. G. Wiltshire, G. A. D. Briggs and R. J. Nicholas, *Nat. Mater.*, 2005, **4**, 481–485.
309. L. Guan, Z. Shi, M. Li and Z. Gu, *Carbon*, 2005, **43**, 2780–2785.
310. H. Kataura, Y. Maniwa, M. Abe, A. Fujiwara, T. Kodama, K. Kikuchi, H. Imahori, Y. Misaki, S. Suzuki and Y. Achiba, *Appl. Phys. A*, 2002, **74**, 349–354.
311. E. L. Sceats and J. C. Green, *J. Chem. Phys.*, 2006, **125**, 154704.
312. X. Lu, L. Feng, T. Akasaka and S. Nagase, *Chem. Soc. Rev.*, 2012, **41**, 7723–7760.
313. X. Lu, T. Akasaka and S. Nagase, *Chem. Commun.*, 2011, **47**, 5942–5957.
314. X. Pan and X. Bao, in *Nanomaterials in Catalysis*, Wiley-VCH Verlag GmbH & Co. KGaA, 2013, pp. 415–441.
315. X. Pan and X. Bao, *Acc. Chem. Res.*, 2011, **44**, 553–562.
316. D. S. Su, S. Perathoner and G. Centi, *Chem. Rev.*, 2013, **113**, 5782–5816.

317. T. W. Chamberlain, T. Zoberbier, J. Biskupek, A. Botos, U. Kaiser and A. N. Khlobystov, *Chem. Sci.*, 2012, **3**, 1919–1924.
318. T. Fujimori and K. Kaneko, *Carbon*, 2014, **69**, 641–642.
319. S. Dargouthi, S. Boughdiri and B. Tangour, *J. Comput. Theor. Nanosci.*, 2014, **11**, 1258–1263.
320. A. La Torre, M. d. C. Giménez-López, M. W. Fay, G. A. Rance, W. A. Solomonsz, T. W. Chamberlain, P. D. Brown and A. N. Khlobystov, *ACS Nano*, 2012, **6**, 2000–2007.
321. N. M. Smith, K. Swaminathan Iyer and B. Corry, *Phys. Chem. Chem. Phys.*, 2014, **16**, 6986–6989.
322. J. Guan, X. Pan, X. Liu and X. Bao, *J. Phys. Chem. C*, 2009, **113**, 21687–21692.
323. P. Kondratyuk, Y. Wang, J. K. Johnson and J. T. Yates, *J. Phys. Chem. B*, 2005, **109**, 20999–21005.
324. W.-H. Zhao, B. Shang, S.-P. Du, L.-F. Yuan, J. Yang and X. C. Zeng, *J. Chem. Phys.*, 2012, **137**, 034501–034508.
325. Y. Jiao, A. Du, M. Hankel and S. C. Smith, *Phys. Chem. Chem. Phys.*, 2013, **15**, 4832–4843.
326. G. Arora, N. J. Wagner and S. I. Sandler, *Langmuir*, 2004, **20**, 6268–6277.
327. A. Alexiadis and S. Kassinos, *Mol. Simul.*, 2008, **34**, 671–678.
328. H. Kim and W. Sigmund, *J. Cryst. Growth*, 2005, **276**, 594–605.
329. S. Liu, J. Zhu, Y. Mastai, I. Felner and A. Gedanken, *Chem. Mater.*, 2000, **12**, 2205–2211.
330. W. Chen, X. Pan, M.-G. Willinger, D. S. Su and X. Bao, *J. Am. Chem. Soc.*, 2006, **128**, 3136–3137.
331. W. Chen, X. Pan and X. Bao, *J. Am. Chem. Soc.*, 2007, **129**, 7421–7426.
332. G. A. Rance, W. A. Solomonsz and A. N. Khlobystov, *Chem. Commun.*, 2013, **49**, 1067–1069.
333. F. Goettmann and C. Sanchez, *J. Mater. Chem.*, 2007, **17**, 24–30.
334. D. Wang, G. Yang, Q. Ma, M. Wu, Y. Tan, Y. Yoneyama and N. Tsubaki, *ACS Catal.*, 2012, **2**, 1958–1966.
335. Q. Shao, L. Huang, J. Zhou, L. Lu, L. Zhang, X. Lu, S. Jiang, K. E. Gubbins, Y. Zhu and W. Shen, *J. Phys. Chem. C*, 2007, **111**, 15677–15685.
336. H. Kyakuno, K. Matsuda, H. Yahiro, Y. Inami, T. Fukuoka, Y. Miyata, K. Yanagi, Y. Maniwa, H. Kataura, T. Saito, M. Yumura and S. Iijima, *J. Chem. Phys.*, 2011, **134**, 244501–244514.
337. D. S. Su, in *Nanomaterials in Catalysis*, Wiley-VCH Verlag GmbH & Co. KGaA, 2013, pp. 331–374.
338. H. Ma, L. Wang, L. Chen, C. Dong, W. Yu, T. Huang and Y. Qian, *Catal. Commun.*, 2007, **8**, 452–456.
339. C. Wang, S. Guo, X. Pan, W. Chen and X. Bao, *J. Mater. Chem.*, 2008, **18**, 5782–5786.
340. J.-P. Tessonnier, O. Ersen, G. Weinberg, C. Pham-Huu, D. S. Su and R. Schlögl, *ACS Nano.*, 2009, **3**, 2081–2089.
341. Q. Fu, W. Gisela and D.-s. Su, *New Carbon Mater.*, 2008, **23**, 17–20.
342. A. La Torre, M. W. Fay, G. A. Rance, M. del Carmen Gimenez-Lopez, W. A. Solomonsz, P. D. Brown and A. N. Khlobystov, *Small*, 2012, **8**, 1222–1228.

343. S. Xu, P. Zhang, H. Li, H. Wei, L. Li, B. Li and X. Wang, *RSC Adv.*, 2014, **4**, 7079–7083.
344. W. A. Solomonsz, G. A. Rance, M. Suyetin, A. La Torre, E. Bichoutskaia and A. N. Khlobystov, *Chem. Eur. J.*, 2012, **18**, 13180–13187.
345. E. Castillejos, P.-J. Debouttière, L. Roiban, A. Solhy, V. Martinez, Y. Kihn, O. Ersen, K. Philippot, B. Chaudret and P. Serp, *Angew. Chem., Int. Ed.*, 2009, **48**, 2529–2533.
346. F. Winter, G. Leendert Bezemer, C. van der Spek, J. D. Meeldijk, A. Jos van Dillen, J. W. Geus and K. P. de Jong, *Carbon*, 2005, **43**, 327–332.
347. H. Friedrich, P. E. de Jongh, A. J. Verkleij and K. P. de Jong, *Chem. Rev.*, 2009, **109**, 1613–1629.
348. T. Trang Nguyen and P. Serp, *ChemCatChem*, 2013, **5**, 3595–3603.
349. S. Guo, X. Pan, H. Gao, Z. Yang, J. Zhao and X. Bao, *Chem. Eur. J.*, 2010, **16**, 5379–5384.
350. H. Friedrich, S. Guo, P. E. de Jongh, X. Pan, X. Bao and K. P. de Jong, *ChemSusChem*, 2011, **4**, 957–963.
351. L. Wang, L. Ge, T. E. Rufford, J. Chen, W. Zhou, Z. Zhu and V. Rudolph, *Carbon*, 2011, **49**, 2022–2032.
352. Z. Yang, S. Guo, X. Pan, J. Wang and X. Bao, *Energy Environ. Sci.*, 2011, **4**, 4500–4503.
353. W. Chen, Z. Fan, L. Gu, X. Bao and C. Wang, *Chem. Commun.*, 2010, **46**, 3905–3907.
354. W. Chen, Z. Fan, B. Zhang, G. Ma, K. Takanabe, X. Zhang and Z. Lai, *J. Am. Chem. Soc.*, 2011, **133**, 14896–14899.
355. X. Pan, Z. Fan, W. Chen, Y. Ding, H. Luo and X. Bao, *Nat. Mater.*, 2007, **6**, 507–511.
356. J. Zhou, H. Song, X. Chen and J. Huo, *J. Am. Chem. Soc.*, 2010, **132**, 11402–11405.
357. P. Serp, in *Carbon Materials for Catalysis*, John Wiley & Sons, Inc., 2008, pp. 309–372.
358. Y. Gan, L. Sun and F. Banhart, *Small*, 2008, **4**, 587–591.
359. H.-S. Kim, H. Lee, K.-S. Han, J.-H. Kim, M.-S. Song, M.-S. Park, J.-Y. Lee and J.-K. Kang, *J. Phys. Chem. B*, 2005, **109**, 8983–8986.
360. T. V. Reshetenko, L. B. Avdeeva, Z. R. Ismagilov and A. L. Chuvilin, *Carbon*, 2004, **42**, 143–148.
361. C. Pham-Huu, N. Keller, V. V. Roddatis, G. Mestl, R. Schlogl and M. J. Ledoux, *Phys. Chem. Chem. Phys.*, 2002, **4**, 514–521.
362. J. Cheng, X. Zhang and Y. Ye, *J. Solid State Chem.*, 2006, **179**, 91–95.
363. Y. I. Golovin, D. Y. Golovin, A. V. Shuklinov, R. A. Stolyarov and V. M. Vasyukov, *Tech. Phys. Lett.*, 2011, **37**, 253–255.
364. W. H. Shin, H. M. Jeong, B. G. Kim, J. K. Kang and J. W. Choi, *Nano Lett.*, 2012, **12**, 2283–2288.
365. S. Van de Vyver, J. Geboers, W. Schutyser, M. Dusselier, P. Eloy, E. Dornez, J. W. Seo, C. M. Courtin, E. M. Gaigneaux, P. A. Jacobs and B. F. Sels, *ChemSusChem.*, 2012, **5**, 1549–1558.
366. C. Wang, J. Qiu, C. Liang, L. Xing and X. Yang, *Catal. Commun.*, 2008, **9**, 1749–1753.

367. Z. Ji, X. Shen, G. Zhu, H. Zhou and A. Yuan, *J. Mater. Chem.*, 2012, **22**, 3471–3477.
368. R. Bajpai, S. Roy, N. kulshrestha, J. Rafiee, N. Koratkar and D. S. Misra, *Nanoscale*, 2012, **4**, 926–930.
369. Y. Wang, X. Xu, Z. Tian, Y. Zong, H. Cheng and C. Lin, *Chem. Eur. J.*, 2006, **12**, 2542–2549.
370. X. Li and I. M. Hsing, *Electrochim. Acta*, 2006, **51**, 5250–5258.
371. M. Wang, K.-D. Woo and D.-K. Kim, *Energy Convers. Manage.*, 2006, **47**, 3235–3240.
372. M. Endo, Y. A. Kim, M. Ezaka, K. Osada, T. Yanagisawa, T. Hayashi, M. Terrones and M. S. Dresselhaus, *Nano Lett.*, 2003, **3**, 723–726.
373. Y.-G. Zhou, J.-J. Chen, F.-b. Wang, Z.-H. Sheng and X.-H. Xia, *Chem. Commun.*, 2010, **46**, 5951–5953.
374. Y. Li, W. Gao, L. Ci, C. Wang and P. M. Ajayan, *Carbon*, 2010, **48**, 1124–1130.
375. R. Kou, Y. Shao, D. Wang, M. H. Engelhard, J. H. Kwak, J. Wang, V. V. Viswanathan, C. Wang, Y. Lin, Y. Wang, I. A. Aksay and J. Liu, *Electrochem. Commun.*, 2009, **11**, 954–957.
376. B. Xu, X. Yang, X. Wang, J. Guo and X. Liu, *J. Power Sources*, 2006, **162**, 160–164.
377. Y. Okamoto, *Chem. Phys. Lett.*, 2006, **420**, 382–386.
378. D.-H. Lim and J. Wilcox, *J. Phys. Chem. C*, 2011, **115**, 22742–22747.
379. W. Song, M. Jiao, K. Li, Y. Wang and Z. Wu, *Chem. Phys. Lett.*, 2013, **588**, 203–207.
380. L. Liu, Y. Su, J. Gao and J. Zhao, *Phys. E.*, 2012, **46**, 6–11.
381. M. Trépanier, A. Tavasoli, A. K. Dalai and N. Abatzoglou, *Appl. Catal., A*, 2009, **353**, 193–202.
382. Z.-J. Liu, Z.-Y. Yuan, W. Zhou, L.-M. Peng and Z. Xu, *Phys. Chem. Chem. Phys.*, 2001, **3**, 2518–2521.
383. A. Tavasoli, R. M. M. Abbaslou, M. Trepanier and A. K. Dalai, *Appl. Catal., A*, 2008, **345**, 134–142.
384. G. L. Bezemer, J. H. Bitter, H. P. C. E. Kuipers, H. Oosterbeek, J. E. Holewijn, X. Xu, F. Kapteijn, A. J. van Dillen and K. P. de Jong, *J. Am. Chem. Soc.*, 2006, **128**, 3956–3964.
385. C.-Y. Lu and M.-Y. Wey, *Fuel.*, 2007, **86**, 1153–1161.
386. L. Guczi, G. Stefler, O. Geszti, Z. Koppány, Z. Kónya, É. Molnár, M. Urbán and I. Kiricsi, *J. Catal.*, 2006, **244**, 24–32.
387. D. G. Larrude, P. Ayala, M. E. H. Maia da Costa and F. L. Freire Jr., *J. Nanomater*, 2012, 695453.
388. G. L. Bezemer, A. van Laak, A. J. van Dillen and K. P. de Jong, *in Studies inSurfaceScienceandCatalysis,* eds.B.XinheandX.Yide,Elsevier,2004,vol. 147, pp. 259–264.
389. S. Yang, G. Cui, S. Pang, Q. Cao, U. Kolb, X. Feng, J. Maier and K. Müllen, *ChemSusChem*, 2010, **3**, 236–239.
390. N. P. Subramanian, S. P. Kumaraguru, H. Colon-Mercado, H. Kim, B. N. Popov, T. Black and D. A. Chen, *J. Power Sources*, 2006, **157**, 56–63.

CHAPTER 5

Nanostructured Carbon Materials as Catalysts

5.1 Introduction

Many reactions involve metals, especially noble metals or metal oxides as catalysts. Although metal-based catalysts have been playing a major role in various industrial processes, they still suffer from multiple competitive disadvantages, including their high cost, susceptibility to poisoning, and detrimental effects on the environment. Owing to their wide availability, environmental acceptability, corrosion resistance, and unique surface properties, certain carbon nanomaterials have recently been demonstrated to be promising metal-free alternatives for low-cost catalytic processes.[1,2] While AC and glassy carbon (GC) have long been used as catalysts for various chemical and electrochemical processes, the recent availability of novel carbon nanomaterials including fullerenes, nanofibers, nanotubes, nanodiamonds and graphene sheets, offer now new opportunities for the development of advanced metal-free materials with improved catalytic performance.[3] The reasons for these enhanced activities are based on several properties characteristic to the carbon matrix: (i) in its sp^2 hybridization state, carbon possesses a number of electrons that can be delocalized; (ii) given that chemical species (*e.g.* H_2, O_2, H_2O, alkanes and olefins) cannot be intercalated within the graphitic layers or form subsurface species, the complexity of the reaction mechanism is only limited to two dimensions; (iii) the edges of the graphitic layers always require saturation of the free bonds with heteroatoms (O and N being the most common), and this can give rise to different acidic/basic functionalities, which in turn are responsible for reactivities in various reactions. Hence, carbon-based catalysts are

RSC Catalysis Series No. 23
Nanostructured Carbon Materials for Catalysis
By Philippe Serp and Bruno Machado

Published by the Royal Society of Chemistry, www.rsc.org

one of the most active research directions in nanomaterials and catalysis.[4] The scope of reactions commonly studied can vary from the gas-phase oxidative dehydrogenation (ODH) of hydrocarbons to liquid-phase oxidation/reduction and acid/base reactions.

5.2 Pristine Nanostructured Carbon Materials as Catalysts

As we have seen in Chapter 2, despite all the different dimensions, curvatures and arrangements, most carbon materials share the same building, *i.e.* a graphitic basal plane. Pristine carbon structures do not have a sufficient number of reactive sites to be a viable catalyst for most reactions, with the only active sites being located at the unsaturated carbons located at the edges or defects of the graphene layers. Nonetheless, the non-covalent binding interactions of graphite, CNTs or fullerenes with the target molecules, along with a fast charge transfer and resonance stabilization of intermediate states by these aromatic platforms, have often been cited to play catalytic role in reactions (Table 5.1).

Natural flake graphite was used to catalyze the cleavage of alkyl ethers using acyl halides (Scheme 5.1a.).[5] The authors observed that tertiary alkyl ethers were selectively cleaved in the presence of secondary ethers, and that primary and secondary alkyl ethers were inert under the experimental reaction conditions tested. A mechanism similar to that proposed for a $Fe^{(III)}$ chloride-acetic anhydride system, where O-acylation of the ether is followed by C–O cleavage to provide a stable carbonium ion, was suggested. Similar reactivities were also observed in the graphite-catalyzed Friedel–Crafts-type substitutions between various acid halides or alkyl halides and electron rich arenes to yield alkyl arenes (Scheme 5.1b.).[6,7] The authors observed that graphite could stabilize electrophilic species as well as cationic intermediates. As a clear evidence of the ability of carbon materials to enhance

Table 5.1 Reactions catalyzed by pristine carbons.

Carbon material	Reaction	Substrate	References
Graphite	Friedel–Crafts	Acyl halides	5
	Friedel–Crafts	Acyl halides	6
	Friedel–Crafts	Acyl halides	7
	Reduction	Nitrobenzenes	8
	Cycloaddition	Citronellal	11
	Cycloaddition	Anthracene	12
	Oxidation	4-Chlorophenol	17
CNTs	Oxidation	Cyclohexane	18
	ODH	Dihydroanthracene	19
Fullerene	Photocatalytic reduction	Nitrobenzenes	15
	Photocatalytic oxidation	Alkenes	20
	Photocatalytic oxidation	Aminodiacetic esters	21
	Thermal reduction	Nitrobenzenes	15,22

reactivity, acylation reactions were found to proceed without the addition of strong Lewis acids.[7]

Van der Waals interaction of the aromatic scaffold in graphite with the target molecules as well as its high electron conductivity is believed to facilitate the reduction of nitroarenes by hydrazine.[8,9] In addition, the high thermal conductivity (19.1 W cm^{-1} K^{-1} at 300 K)[10] of graphite helps to promote thermal cycloadditions[11] and ene reaction.[12] In principle, most of the reactions that can be catalyzed by graphite or CNTs will be expected to work at least equally well with graphene, due to its larger surface area, readily functionalizable surface, and good dispersion properties.

Unfunctionalized carbon materials, such as CNTs and fullerenes, have shown some catalytic activity in the cracking and dehydrogenation of hydrocarbons.[13,14] In addition, fullerenes (both C_{60} and C_{70}) were also reported to be able to catalyze the hydrogenation of nitrobenzene to aniline at room temperature under UV-visible light irradiation.[15] Synergistic effects between neutral and anionic fullerenes (prepared by reacting the neutral fullerene with a nickel–aluminum alloy) were observed, which resulted in improved yields of the aniline product. However, since trace amounts of cobalt, chromium, copper, iron, and silver were detected, speculation exists as to whether the reactivity shown in this work was indeed due to the fullerene or if these metals were the catalytically active species.

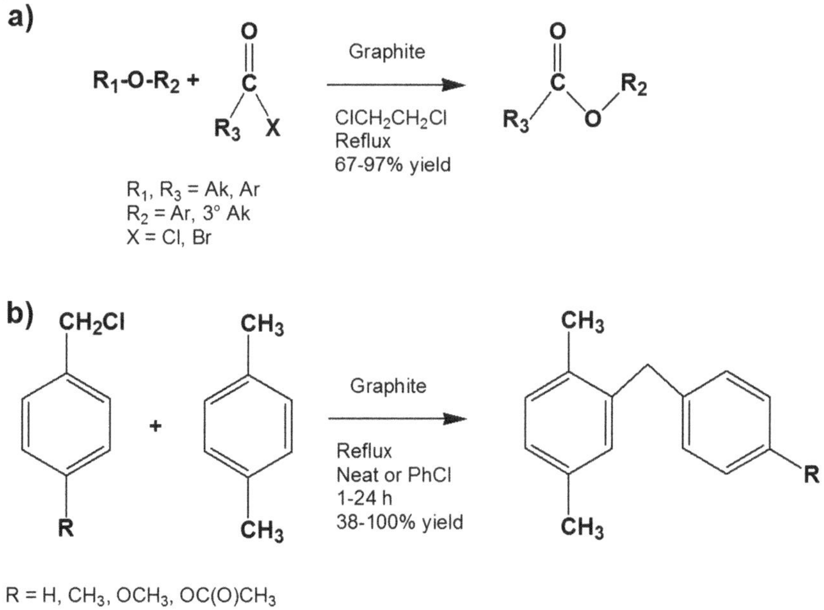

a) R_1-O-R_2 + R_3C(O)X → (Graphite; $ClCH_2CH_2Cl$, Reflux, 67-97% yield) R_3C(O)OR_2

R_1, R_3 = Ak, Ar
R_2 = Ar, 3° Ak
X = Cl, Br

b) CH_2Cl / R-substituted benzene + p-xylene (CH_3, CH_3) → (Graphite; Reflux, Neat or PhCl, 1-24 h, 38-100% yield) product (CH_3, CH_3, R)

R = H, CH_3, OCH_3, OC(O)CH_3

Scheme 5.1 Graphite catalyzed (a) cleavage of ethers, followed by coupling with an acyl halide (adapted from ref. 5); (b) Friedel-Crafts-type substitutions (adapted from ref. 7).

The delocalized π-electron system of pristine graphene layers has also been reported to induce reactivity towards complexation reactions in organometallic catalysis.[16] Haddon *et al.* demonstrated the η^6-complexation reactions of chromium with various forms of graphene, graphite and carbon nanotubes. These extended periodic π-electron systems exhibited varying degrees of reactivity toward the reagents $Cr(CO)_6$ and (η^6-benzene) $Cr(CO)_3$, with the authors reporting the formation of (η^6-arene)$Cr(CO)_3$ or (η^6-arene)$_2$Cr complexes, where arene = SWCNT, exfoliated graphene, epitaxial graphene and highly-oriented pyrolytic graphite.

It must be pointed out that most carbocatalysts (*i.e.*, graphite, CNT, GO, *etc.*) can be contaminated by naturally occurring metals or treatment processes that typically require metal catalysts or metal containing agents. Thus, the role of these metal impurities in catalysis cannot be ruled out unless careful control studies have been carried out.

Despite these few examples of reactivity with pristine carbon materials, the catalytic activity of carbons can be greatly enhanced upon introduction of new active sites either through doping or through functionalization with heteroatoms. In this chapter we are mostly concerned with the reactivity provided by carbon structures containing oxygenated surface groups and defects. For the specific activity of carbons containing other heteroatoms (such as N, B, P and S) please refer to Chapter 6.

5.3 Catalytic Role of Defects and Oxygen Functional Groups

The use of heterogeneous, metal-free carbons instead of metal catalysts in synthetic chemistry has continued to progress over the last decade, in part as a result of diminishing supplies of rare earth metals used in common industrial processes. Table 5.2 provides an overview of the various reactions that can be catalyzed by carbon, together with the type of surface chemistry required or the nature of the active sites, when they have been identified. Some of these results will be summarized over the following sections.

As we have seen in Section 5.2, in the absence of defects, the basal planes are not very reactive; we may therefore expect to find active sites essentially at the edges of the graphene layers, where the unsaturated carbon atoms may chemisorb oxygen or water from exposure to the atmosphere, originating surface groups such as those detailed in Chapter 4. These groups can act as active sites in various reactions, and their type and concentration can be further increased by oxidative treatments, in either gas- or liquid-phase (see Chapter 4). The concentration and type of surface groups plays a very important role on determining the surface chemistry of the carbon phase (Figure 5.1). Hence, among the different types of oxygenated groups commonly found on the surface of carbonaceous materials, acidic groups include carboxylic acids and anhydrides, lactones or lactols, and phenols,

Table 5.2 Overview of several reactions catalyzed by carbons, presenting the surface chemistry and active sites thought to be responsible for the activity. Adapted from ref. 23.

Reaction	Surface chemistry/active sites
Oxidative dehydrogenation	Quinones
Dehydration of alcohols	Carboxylic acids
Dehydrogenation of alcohols	Lewis acids and basic sites
NO_x reduction (selective catalytic reduction with NH_3)	Acidic surface oxides (carboxylic and lactone) + basic sites (carbonyls or N5, N6)
NO oxidation	Basic sites
SO_2 oxidation	Basic sites, pyridinic – N6
H_2S oxidation	Basic sites
Dehydrohalogenation	Pyridinic nitrogen sites
Hydrogen peroxide reactions	Basic sites
Catalytic ozonation	Basic sites
Catalytic wet air oxidation	Basic sites

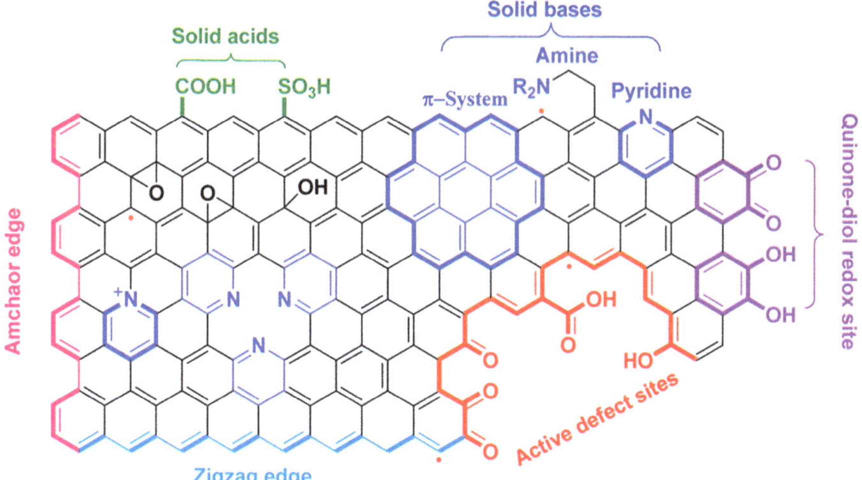

Figure 5.1 Schematic representation showing active catalytic sites in the graphene matrix due to heteroatom incorporation. Reprinted with permission from ref. 24. Copyright 2013 American Chemical Society.

while carbonyls and ethers are considered to be neutral. Basic structures (the nature of which is still open to debate) are attributed to groups such as quinone, chromene, and pyrone. In addition, the π-electron system of the graphitic basal planes also contributes to the carbon basicity.

The amount of surface groups, and thus potential active sites, present on the carbon generally increases with surface area, whereas the concentration depends to a large extent on the microcrystalline structure: small crystallites can expose more edges and therefore more surface groups can be formed, while the role of the basal planes and π electrons will become more important for larger crystallites.

Analysis of the pertinent literature shows that the most relevant conclusions regarding the role of surface chemistry in carbon catalysis have been derived following a common methodology whereby a series of catalysts are prepared from the same carbon material by suitable thermal or chemical treatments while keeping the textural properties essentially unchanged. It is then possible to correlate the catalytic properties with the surface chemistry of the carbons (Table 5.3). The oxydehydrogenation of cyclohexanol over carbon catalysts provides an excellent example of the strong influence of the surface chemistry on product selectivity.[25] Carbon catalysts with surfaces of different chemical nature were investigated. The main products were cyclohexene, cyclohexanone, benzene, and phenol. Dehydration to cyclohexene was found to occur preferentially on the carboxylic acid groups (carbons treated with HNO_3), while the best selectivities to cyclohexanone were obtained on catalysts with high contents in phenol groups (carbons treated with H_2O_2 or oxygen). Carbon treated with N_2O showed high selectivity to phenol, which was explained by the participation of quinone and nitro groups in the oxydehydrogenation mechanism. Unfortunately, the authors were not able to determine precisely the amounts of the different surface groups; thus, no quantitative correlations were established.

Bielawski and co-workers have made a significant contribution to the application of GO as a metal-free catalyst.[38] The use of this material was reported in several catalytic reactions, including the oxidation of alcohols,[29,39] olefins,[29,40] sulfides and thiols,[41] C–H oxidation,[40] alkyne hydrations,[29] Claisen–Schmidt

Table 5.3 Nanostructured carbon materials as carbocatalysts and the types of reactions catalyzed.

Carbon material	Reaction	Substrate	Active site	References
Nanodiamonds	ODH	Ethylbenzene	Quinones	26
	ODH	*n*-Butane	Quinones	27
	DH	Ethyl benzene	Quinones	28
GO	Hydration	Alkynes	Acidic groups	29
	Oxidation	Alcohols	Not clear	29
	Michael-type Friedel–Crafts	Indoles	Acidic groups	30
	Aza-Michael addition	Amines	Acidic groups	31
	Polymerization	Cyclic lactones and lactams	Acidic groups	32
	Photo-oxidation	Tertiary amines	Not clear	33
Chemically modified-GO	Hydrolysis	Esters	Acidic groups	34
	Reduction	Nitrobenzenes	Zig-zag edges	35
	Oxidation	Hydrocarbons	π–π^* or charge transfer interactions	36
	Oxidation	Tertiary amines	Charge transfer interactions	37

condensation,[42] aza Michael addition of amines,[31] polymerization of various olefin monomers,[43] ring opening polymerization of various cyclic lactones and lactams,[32] and dehydration polymerization in the synthesis of carbon reinforced poly(phenylene methylene) composites.[44] Their contributions inspired a great deal of interest among the scientific community, inspiring many other groups to explore the catalytic performance of GO-based materials in other catalytic applications.

5.3.1 Gas-Phase Reactions

Gas-phase reactions, such as ODH, are often carried out at higher temperatures than liquid-phase reactions and thus demand a much stronger structural stability of the carbon framework. Conventional carbons, in particular activated carbons, can undergo unavoidable deactivations due to coking or combustion,[45,46] whereas structured nanostructured carbon materials are much more stable and coke-free.[26,47–49]

5.3.1.1 *Oxidative Dehydrogenation of Alkanes and Aromatics*

Oxidative dehydrogenation of light alkanes (C_2–C_6) in a flow of oxygen for the production of short-chain alkenes is one of the novel routes currently proposed for the exploitation of natural gas as a raw material for highly priced and clean chemicals (eqn (5.1)). The main advantage of ODH is that the reaction is complete (meaning that conversions are not limited by equilibrium), and that it can be carried out at lower temperatures than the conventional dehydrogenation process. In addition, given that the reaction is exothermal ($\Delta H = -116$ kJ mol^{-1}) the amount of heat released and the presence of oxygen in the ODH limits coke deposition, thus preventing a loss of catalytic activity. However, the key point of the ODH is the development of catalysts capable of activating only the C–H bonds of alkanes to alkenes in the presence of oxygen.

$$C_nH_{2n+2} + \frac{1}{2}O_2 \xrightarrow{\text{Catalyst}} C_nH_{2n} + H_2O \tag{5.1}$$

Among others, acid catalysts (such as alumina, zeolites, and metal phosphates) were tested successfully in ODH.[50–52] One of the particular features observed with such systems was the formation of a coke layer, which was found to be the real catalytic surface.[53] These results suggested that carbon materials could be used as catalysts for ODH reactions, being proposed that quinone-type groups on the surface of coke were the active sites for ODH. In order to verify the nature of the active sites, Pereira *et al.* studied the oxidative dehydrogenation of ethylbenzene (ODE) to styrene over ACs with surface properties modified by appropriate thermal or chemical treatments.[54] They found a linear correlation between the activity in the oxidative dehydrogenation of ethylbenzene and concentration of carbonyl/quinone surface groups. Kinetic studies indicated that the ODE over AC followed the

Mars-van-Kreverlan mechanism, in which the catalytic activity is related to the redox ability of the surface oxygen species (quinone surface groups are reduced to hydroquinone by adsorbed ethylbenzene, and re-oxidized back to quinone by oxygen).

With the development and generalization of nanostructured carbon materials, new directions for expanding the activity of metal-free carbon-based ODH catalysts are now more accessible. One interesting improvement is that nanocarbon catalysts do not show loss of activity from coking or combustion, unlike ACs.[3] Schlögl *et al.* carried out the ODE using carbon nanofibers and onion-like carbon materials as catalysts.[26,55] These nanostructured carbon materials showed a higher catalytic activity than highly dispersed graphite, and also had a far superior stability compared to traditional amorphous CB. A series of studies subsequently demonstrated that many nanostructured carbons (carbon nanotubes, nanofibers, onion-like carbons and nanodiamonds) can efficiently catalyze the ODH reaction of ethylbenzene to styrene, and that these catalysts exhibited comparable or even better activities than the traditional iron oxide catalysts.[4,47,49,56,57] This was attributed to the fact that defects of the carbon surface can anchor functional groups as the active sites, while the graphitic structure can tightly hold the active sites to give excellent thermal stability even under an oxidative atmosphere.[58]

Regarding the reaction mechanism, Schlögl *et al.* made a breakthrough on the identification of active species by studying the reaction kinetics of the ODE in combination with *in situ* X-ray photoelectron spectroscopy using different carbon nanotubes and nanodiamonds.[45,59] The reaction pathway and nature of the active sites were shown to be same over both sp^2-nanotube and sp^3-nanodiamond. The π-conjugation effect of the graphite structure did not seem to be essential for the nanocarbon function as a catalyst. Similar to most metal-based catalysts, the reaction catalyzed by carbon materials was found to be based on a dual site Langmuir–Hinshelwood mechanism, with carbonyl/quinone groups on the surface acting as the active sites. The model includes the dissociative adsorption of oxygen molecules and noncompetitive adsorption between ethylbenzene and oxygen molecules, with the overall reaction rate being kinetically limited by the rate of the surface reaction, which mainly involves breaking of two C–H bonds in ethylbenzene at the α and β positions. Carbonyl/quinone groups are suggested to activate the alkane part of ethylbenzene to produce styrene and hydroxyl groups (C–OH) as intermediates. The catalytic cycle is closed by the oxidation of C–OH by gas phase oxygen to C=O and water, which is thermodynamically favorable (Figure 5.2).

To further confirm the active role of quinone groups in the ODE a small molecule mimetic study was carried out by Zhang *et al.*[60] They showed that a phenanthrenequinone macrocyclic trimer could be used as a heterogeneous catalyst for ODE, being, under comparable kinetic conditions, up to 47 times more active than other common solid catalysts (including nanostructured carbon materials, metal phosphates and oxides), confirming the hypothesis that quinone groups do effectively serve as active sites.

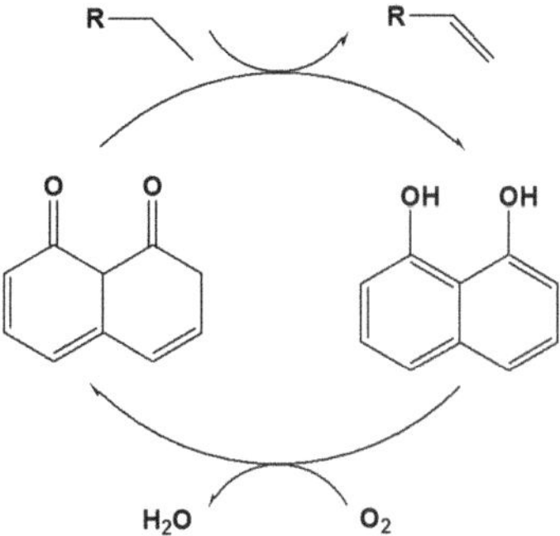

Figure 5.2 Oxidative dehydrogenation of alkanes over nanocarbon catalysts.

Nevertheless, the good performances that some of these carbon nanomaterials are claimed to offer must be put into perspective. Indeed, most of the results have been obtained at temperatures in the range 793–823 K, and the performances of these new catalysts are compared to those of graphite and carbon blacks. On the other hand, activated carbons give high styrene yields at 623 K,[54] a temperature at which graphite and CNTs show no significant activity. Pereira *et al.* compared the performance of CNTs with those of graphite and AC in the ODE at 723 K.[61] AC was found to give the highest conversion and selectivity to styrene, with coke deposition occurring only over the AC and being negligible on both graphite and CNTs.

Even though ethylbenzene is the most frequently studied oxidative dehydrogenation reaction, there are also other substrates that have been reported over different nanostructured carbon materials. Su *et al.* described that the oxidative dehydrogenation of butane to butene and butadiene could be catalyzed by CNTs.[62] When the carbon nanotubes were additionally modified by passivation of the defects with phosphorus oxide, the resulting CNTs gave improved catalytic activity, while the reaction temperature was further decreased by 373–473 K below the industrial catalytic process. The surface oxygen species on the surface of nanocarbon materials can be classified into electrophilic and nucleophilic types. Nucleophilic oxygen species preferentially react with electron-poor saturated bonds in alkanes, leading to the ODH reaction that formed alkenes. Electrophilic oxygen species are electron-deficient and will attack the electron-rich C=C bonds in alkenes, leading to the further oxidation of the reactants and dehydrogenation products and subsequent combustion, thus resulting in a decrease in selectivity. It has also been reported that the doping of a trace amount of boron oxide into the graphitic lattice structure can modify the distribution of the electron cloud and inhibit the activity of the electrophilic oxygen species, leading to improved selectivity to alkene target products.[63] Liang *et al.* found that for oxidative dehydrogenation of isobutane over fullerene-like graphitic carbons, no clear correlation was observed between the catalytic activity and the amount of surface functional groups, while the catalytic activity was related to the openness of the fullerene-like cavities.[64] So far, due to a lack of effective approaches to directly identify and quantify the types and numbers of surface defects, the mechanism and the nature of active sites for most carbon-catalyzed gas phase reactions still remain controversial. Zhang*et al.* achieved a steam- and coke-free dehydrogenation of ethylbenzene using nanodiamonds as catalyst.[28] In the absence of oxygen and steam, the nanodiamonds displayed three times the activity of the commercial iron oxide catalyst at low temperature. After a long reaction time, the surface of the nanodiamond was still free of deposited carbon. The authors found that the nanodiamonds particles are not in a complete sp^3 hybridized state, and the carbon atoms on the surface were affected by the surface curvature and caused graphitization, resulting in a diamond–graphene core-shell structure. The surface graphene layer is highly defective and rich in oxygen-containing groups, which provided an outstanding catalytic performance of nanodiamonds. In a similar work by the same authors, nanodiamonds were also used to catalyze the oxidative dehydrogenation of *n*-butane.[27] The surface of the nanoparticle tended to be transformed into a diamond core with three to ten layers of fullerene shells during reaction. During this phase transformation from the sp^3 to the sp^2 hybridized state, quinoidic carbonyl groups were selectively generated and the formation of electrophilic oxygen species was effectively suppressed, which greatly enhanced product selectivity to the desired butenes.

Besides the oxidative dehydrogenation of ethylbenzene, a similar reaction mechanism to that of Mars-van-Kreverlan has been also proposed for other oxidative dehydrogenation processes, such as 1-butene to butadiene,[65] isobutene to isobutene,[66] propane to propene,[63] and ethane to ethylene.[67]

Given the role of oxygenated surface groups, namely carbonyl/quinones, on the mechanisms that have been presented so far, one could expect that the oxidative dehydrogenation of hydrocarbons could be extremely effective when conducted over the surface of graphene, or more precisely graphene oxide, due to its enhanced surface chemistry. Despite considerable research for GOs, the chemical structure of active sites and catalytic mechanisms of GOs as metal-free catalysts for C–H bond activation is still unclear, both experimentally and theoretically. The oxidative dehydrogenation of propane to propene has been theoretically addressed by Tang *et al.*[68] Using first-principles calculations, the authors observed that the epoxy groups on the GO surface provide active sites for the C–H bond activation. The first C–H bond breaking of propane through H abstraction by epoxide leads to the formation of a propyl radical, which is the rate-determining step for the conversion from propane to propene. The presence of OH groups around the active site can remarkably improve the activity of the epoxy group and facilitate the H abstraction, and the activity enhancement exhibits strong site dependence. The sites of oxygen functional groups on the GO surface can be easily tuned by the diffusion of these groups under an external electric field, which increases the reactivity of GOs towards ODH of propane.

Liang *et al.* reported the utilization of oxygen-functionalized FLG sheets containing variable amounts of oxygen in the heterogeneous catalytic oxidative dehydrogenation reaction of isobutane at 673 K.[66] Although the materials possessed a variation in the oxygen content by more than an order of magnitude, the authors were not able to correlate the activity or selectivity with the total sample oxygen content. Carbonyl groups at the graphene surface were found to be highly stable, and $^{18}O/^{16}O$ switch experiments revealed that lactone and phenol groups were more easily exchanged during the redox cycle of the ODH reaction. Hence, carbon materials with higher sp^3 character and a higher fraction of exposed edges where oxygen active sites can be formed and exchanged are thought to lead to more active catalysts for ODH reactions.

5.3.1.2 Other Reactions

Using carbon molecular sieves (CMSs) in the oxidation of gaseous 1-propanol, 2-propanol, and propanal, their respective aldehyde and ketone products were obtained using air as carrier gas.[69] Interestingly, when the carrier gas was switched to nitrogen, oxidation reaction ceased; upon the re-introduction of air, reactivity rapidly returned. Hence, there was a dependence on an oxygen-rich atmosphere, which was consistent with an alcohol-mediated reduction of the active sites on the surfaces of the CMSs, which were then regenerated by oxygen to complete a catalytic cycle. One possible mechanism that can be used to rationalize the activity involves the reaction between an alcohol and carbocations, which are proposed to exist on the CMS surface. A Lewis acid catalyzed process (similar to that existent with inorganic oxides) should afford the aldehyde product *via* H abstraction followed by deprotonation of the alcohol. Since the CMS may also function as a Brønsted acid, a second possible mechanism can also involve alcohol dehydration.

Schlögl *et al.* demonstrated that several functionalized carbon materials (ACs, CNTs and nanodiamonds) were highly robust and selective catalysts for acrolein oxidation to acrylic acid (eqn (5.2)), as model chemical reaction for the insertion of oxygen into organic molecules.[70]

$$CH_2{=}CHCHO + \frac{1}{2}O_2 \rightarrow CH_2{=}CHCOOH \tag{5.2}$$

The high activity/selectivity of the carbon allotropes was correlated to the exposition of the (0001) basal plane to the outer surface. The authors suggested a mechanism in which a hydrogen is abstracted from acrolein by nucleophilic oxygen species followed by oxygen insertion from the mobile epoxide C–O–O generated by O_2 dissociation on the carbon basal planes (Figure 5.3). A productivity of 26.5 mmol g^{-1} h^{-1}, which is about half of that obtained using an industrial doped MoV mixed metal oxide, was reported.

5.3.2 Liquid-Phase Reactions

5.3.2.1 Oxidations

The use of metal-free heterogeneous carbons as catalysts in liquid-phase synthetic reactions (which Bielawski[29,38] accurately coined as carbocatalysis)[26,35] may be traced as early as the 1920's, when several authors investigated the

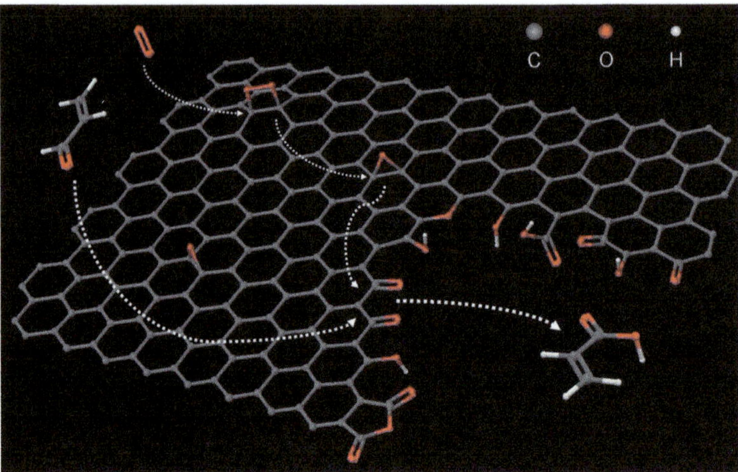

Figure 5.3 Suggested reaction pathway for the oxidation of acrolein at the graphitic carbon surface. The active domain is illustrated as a rectangular section of a planar graphene sheet with a hole defect, which is terminated by arbitrarily positioned oxygen functionalities. O_2 adsorbs dissociatively at the (0001) surface to form mobile epoxy groups, which migrate to the prismatic edge sites. The adsorption of C_3H_4O at the nucleophilic oxygen sites, *i.e.*, the carbonyl/quinone, initiates its oxygenation by epoxy oxygen atoms to form acrylic acid. Reprinted with permission from ref. 70. Copyright 2011 Wiley-VCH Verlag GmbH & Co. KGaA, Weinheim.

ability of charcoals to catalyze aerobic oxidation reactions.[71–74] According to the authors, the acidic surface generated *via* thermal activation of the carbon improved adsorption of gaseous oxygen (compared to unactivated carbons), thus promoting the oxidation reaction. High surface area carbons were particularly useful in carbocatalytic applications because of their high ratio of catalytically active surface area to weight.[75]

Bielawski *et al.* reported a seminal work on the use of GO[29] and graphite oxide[40] as carbocatalysts for aerobic oxidation reactions. The GO catalyst was prepared from graphite using Hummers oxidation and exfoliation. In particular, neat benzyl alcohol was selectively oxidized by ambient oxygen to benzaldehyde in the presence of GO with TON based on a weight of around 10^{-2} mol g^{-1}. In addition, high conversions to the corresponding carbonyl compounds were achieved for diphenylmethanol (>98%), 1,2-diphenylethane-1,2-diol (96%), and cyclohexanol (>98%), while moderate conversions were obtained for 1-phenylethanol (26%) and 2-thienylmethanol (18%). The possibility that traces of transition metals could be responsible for the catalytic activity was ruled out based on the negligible manganese, copper, lead, and nickel content (<36 ppb) in the material. During the course of aerobic oxidation, partial reduction of the used GO material occurred (demonstrated by higher powder conductivity and by the increase of the C/O ratio of the used material with respect to the fresh one). Nevertheless, experiments in the presence or absence of O_2 indicated that GO is acting as a catalyst and not only as an oxidizing agent. The catalyst was reused for up to 10 cycles for oxidation of benzyl alcohol without changes in the catalytic activity when loadings higher than 50 wt% of GO *vs.* substrate were present. However, at lower catalyst loadings (<20 wt%) GO becomes deactivated mainly due to reduction to rGO. In the same work, the authors used GO as catalyst for aerobic oxidation of unsaturated hydrocarbons and observed moderate conversions of *cis*-stilbene (56%), while *trans*-stilbene, 3-hexene, cyclohexane, and β-methylstyrene were found to be unreactive under the studied reaction conditions.[29]

Using the same reaction, Boukhvalow *et al.* performed a computational investigation to better understand the reaction mechanisms for benzyl alcohol oxidation using GO.[76] A GO model with 12.5% of the carbon atoms covered by hydroxyl and epoxy groups was established initially. It was proposed that the reaction occurred *via* transfer of hydrogen atoms from the organic molecule to the GO surface, subsequent ring opening of the epoxide groups present on GO, and final dehydration of the hydroxyl group derived from the epoxides opening after transfer of a second hydrogen atom from the hydroxyl benzyl radical. Thus, in two subsequent steps the α hydrogen atom and the hydrogen bonded to oxygen in the benzyl alcohol have been transferred to the oxygen atom of an epoxide functional group in GO, leading to a locally reduced GO. Importantly, the partially reduced GO can be re-oxidized by molecular oxygen, regenerating the active sites. In contrast, functional group-free carbon materials such as graphite or pristine graphene exhibit higher reaction barriers, rendering them catalytically inactive.

SOx Oxidation. Carbons materials are active catalysts for the oxidation of SO_2 into SO_3 and H_2SO_4 in the presence of O_2 and H_2O. The role of surface chemistry in this process has long been recognized, but the results and interpretations of different authors are often contradictory. A number of mechanisms have been proposed for the catalytic oxidation of SO_2 on activated carbons. Davtyan has indicated that oxygen reversibly adsorbs on AC and oxidizes SO_2 to SO_3.[77] Siedlewski has shown that free radicals on AC act as active centers for the chemisorption of SO_2.[78] Kitagawa *et al.* have found that the capacity for SO_2 removal does not depend on the surface area of the AC but on the number of oxygen atoms in AC.[79] Yamamoto *et al.*[80,81] have also shown that the capacity depends on the preparation conditions and raw materials for activated carbon rather than on the surface area of the AC.[77,78] Kamino *et al.* estimated that basic surface groups were the active sites for the oxidative chemisorption of SO_2.[82,83] Sano *et al.* have shown that certain impurities in the AC, such as surface nitrogen species, seem to have favorable influences on activity for SO_2 removal.[84] Davini has shown that basic oxygen functionalities on AC are the chemisorption sites for SO_2 and that they play an important role to govern the SO_2 adsorption capacity.[85] Mochida *et al.* have reported that polyacrylonitrile-based activated carbon fibers (PAN–ACF) exhibited very high SO_2 adsorption rates and large capacities among the carbon adsorbents in the co-existence of oxygen and water vapor at 373–453 K.[86] Surface functional groups on PAN–ACF, as well as its unique pore distribution, may be responsible for its SO_2 adsorption capacity.

Recovery of H_2SO_4 and regeneration of ACF are performed in two ways as illustrated in Figure 5.4: (1) reduction of SO_3 to SO_2 over the carbon at 673 K, (2) hydration and elution as sulfuric acid.

Reduction of SO_3 leaves one oxygen atom over the carbon surface that produces the oxygen functional group on the surface. Therefore, a half or one carbon atom is lost by the recovery of every SO_2 molecule. Generally, such oxygen functional groups decompose into CO_2 or CO with the heat-treatment. Carbon dioxide is generated at around 673 K, while CO at around 1073 K. The adsorption and oxidizing activities of the carbon surface were found to decrease with the amount of oxygen functionality produced on the surface. Hence, decomposition of the oxygen functionality is required for the regeneration of the adsorption ability. The carboxyl and lactone groups are believed to produce CO_2, while carbonyl, aldehyde, and phenols produce CO. Which carbon on the surface is responsible for such functional groups when heated is not yet known. Activated carbons of higher surface area tend to generate a larger amount of CO. Therefore, the temperature to regenerate adsorption activity depends on the nature of the carbon.

Sulfuric acid produced on the surface must be sufficiently hydrated for its recovery. In this recovery method, there is basically no loss of carbon. In such a system, extraction efficiency by liquid water is low for the hydration of H_2SO_4 in the pores, while the condensation of the steam in the pores is effective for hydration, depending strongly on the relative humidity. Hence, a lower temperature is favorable. ACF is more suitable for the recovery of sulfuric acid compared to granular active carbon because the diffusion of sulfuric acid is much easier in the pores of the fiber.

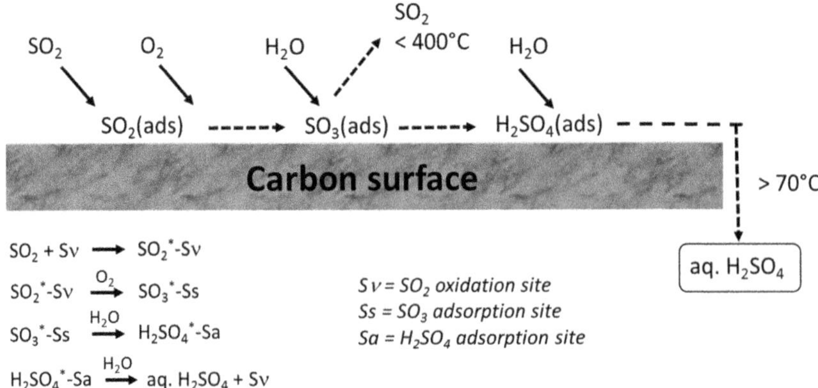

Figure 5.4 De-SO_x mechanisms. Adapted from ref. 87.

The active site for oxidizing SO_2 on the carbon surface was believed to be one of the oxygen functional groups. Among the oxygen functional groups, a basic and/or oxidizing one was considered to be active for the oxidative adsorption of SO_2. However, desufurization reactivity was remarkably enhanced by heat-treatment to remove almost all surface oxygen functional groups from the carbon surface. The surface defects induced by the decomposition of the oxygen functional groups appear to be responsible for SO_2 adsorption and oxygen adsorption and activation. The surface dangling bond may be formed through the decomposition of an oxygen functional group. However, there is no evidence that the free radical detected by ESR is an active site. The hexagonal carbon planes carry two prismatic edges of zig-zag and arm-chair types. So far, there is no evidence to identify which edge is responsible for the activity. The benzene-type bond formed by the elimination of the oxygen functional groups may be preferably located on the zig-zag edge. A larger hexagonal plane may stabilize such sites, suggesting higher activity of pitch-based ACF than PAN-based ACF. PAN–ACF can contain nitrogen atoms in the carbon skeleton. Such sites are believed to form adsorption sites for acidic SO_2 and electronically influence the oxidative active site. In addition to the active site, hydrophobicity of the surface may be another important factor for desulfurization activity because hydration and elution of the sulfuric acid from the active site determines the reaction rate. Both high-temperature treatment of the carbon and a higher degree of graphitization are thus favorable for continuous desulfurization. Such a set of results may provide us a guideline for preparing carbon nanomaterial surfaces of higher activity for SO_2 removal.

Porous GO foams (GOF), prepared from GO suspensions that are subsequently freeze dried, have been reported as catalysts and oxidant reagents for room-temperature transformation of SO_2 to SO_3.[88] It was found that SO_2 oxidation was more effective using aqueous GOF suspensions (TON 6.12 mol SO_2 g_{GOF}^{-1}), where H_2SO_4 was dissolved, than in the gas-phase (TON 0.434 mol SO_2 g_{GOF}^{-1}). Under both conditions GO became reduced during the course of the reaction, rendering it more hydrophobic. Importantly, the oxidation reaction was promoted by O_2, indicating that GO was also acting partly as catalyst besides reactant.

H_2S Oxidation. Hydrogen sulfide (H_2S) is a product of anaerobic digestion and is one of the major toxic and malodorous gases derived from natural gas, coal gasification gas and digester gas.[89–91] Due to the increasing standards of efficiency required by environmental protection agencies, sulfur recovery from the H_2S-containing gases has become more and more important. Besides the conditions of the process, the choice of an adsorbent seems to also be a very important task. The removal of hydrogen sulfide from industrial off-gases is a good example where the adsorptive and catalytic properties of porous carbons can be combined advantageously, particularly when the H_2S concentration is low. Hydrogen sulfide can either be converted into sulfur dioxide or into elemental sulfur (eqn (5.3) and eqn (5.4)).

$$H_2S + \frac{3}{2}O_2 \rightarrow SO_2 + H_2O \tag{5.3}$$

$$H_2S + \frac{1}{2}O_2 \rightarrow S + H_2O \tag{5.4}$$

At low temperatures (<373 K) ACs promote mainly the formation of elemental sulfur. Molten sulfur can be collected by operating above its melting point (392 K). This process has been known since the 1920s. The earlier literature, reviewed by Bansal *et al.* shows that there are different views concerning the reaction mechanisms involved.[92] Accordingly, Puri *et al.*[93] and Cariaso *et al.*[94] proposed a Rideal–Eley mechanism between chemisorbed oxygen and hydrogen sulfide in the gas phase, whilst Steijns *et al.* suggested a mechanism involving dissociatively adsorbed H_2S and chemisorbed oxygen.[95] The elemental sulfur produced was found to promote further oxidation of H_2S. On the other hand, Hedden *et al.* proposed that the reaction occurs *via* intermediate ionic species in a water film formed on the carbon surface, in which both oxygen and hydrogen sulfide are dissolved.[96]

The catalytic oxidation of H_2S requires a wide-pore carbon and a large total pore volume, which is needed for retention of the sulfur formed.[97] Some works suggests that oxygen functional groups on the carbon surface may be involved in the formation of sulfur oxides. Primavera *et al.* investigated the effect of water in the low-temperature catalytic oxidation of H_2S in low concentrations over ACs.[98] The authors showed that the presence of water enhanced the reaction rate and the amount of sulfur that could be loaded before regeneration, and explained their results assuming the formation of a liquid layer inside the pores of the carbon, as proposed by Hedden *et al.*[96] However, these findings contradict the kinetic data reported by Dalai *et al.*, who did not find any effect of water on the H_2S conversion.[89]

Katoh *et al.* showed that H_2S, methanethiol, and dimethylsulfide could be removed simultaneously from gaseous streams at room temperature over wet activated carbon fibers,[99] while Dalai *et al.* reported on the oxidation of methanethiol over an AC.[89] An AC prepared from a cellulosic precursor by CO_2 activation was found to exhibit outstanding performance in the oxidation of H_2S (1000 ppm) in a hydrogen stream at 423 K: namely, 100% conversion for more than 10 hours, and 100% selectivity to sulfur.[100] The selectivity aspects in the oxidation of H_2S to sulfur were addressed in a work by Bashkova *et al.*[101] These authors reported that a high volume of micropores and small mesopores, together with a narrow pore size distribution, is desirable for the retention of SO_2, while a high surface reactivity with a significant amount of basic groups was important for the retention of carbonyl sulfide.

Although nanostructured carbons such as nanodiamonds and graphene have not yet been reported for H_2S oxidation, there have been some attempts to use CNFs and CNTs as a catalyst on this process.[102,103] The highest activities and selectivities (toward sulfur) were obtained on CNFs with the graphene layers oriented perpendicularly to the axis (platelet-CNFs) and on CNTs. However, the carbon materials were contaminated with remains of the metal catalysts used for their synthesis (Ni, Ni–Cu), which may affect the catalytic properties observed.

Peroxide-Based Reactions. Although molecular oxygen as a terminal oxidant should be the preferred oxidizing reagent due to its availability, the use of other tolerable oxidants such as hydrogen peroxide (H_2O_2), *tert*-butylhydroperoxide (TBHP), and peroxymonosulfate (PMS) can also be advantageous in order to achieve higher conversions and selectivities (both in terms of substrate and oxidant agent), envisaging applications in the synthesis of high added value chemicals or for environmental remediation.

Ma *et al.* reported a direct, one-step hydroxylation of benzene to phenol using H_2O_2 as oxidant and rGO as catalyst.[104] The authors observed that the oxidation proceeded with high selectivity and the catalyst was reused seven times without significant loss of catalytic activity. The possibility that the observed catalytic activity could be due to metal traces was ruled out by preparing a rGO containing manganese and observing under these conditions high H_2O_2 decomposition to O_2 (E_a = 32 kJ mol^{-1}) and negligible benzene oxidation. The high activity and selectivity of the catalytic oxidation using rGO was mainly attributed to different factors: (i) the preference of rGO to adsorb benzene with respect to phenol (thus avoiding over-oxidation of phenol to benzoquinone); (ii) the adequate π character of rGO and large surface area (446.5 m^2 g^{-1}) leading to strong adsorption of substrates; and (iii) low H_2O_2 decomposition. The same reaction of benzene hydroxylation to phenol by H_2O_2 using CNTs, provided a much lower catalytic activity (benzene conversion around 6%).[105] These positive results obtained with rGOs were also extended to other substrates including toluene, naphthalene, *p*-xylene, and ethylbenzene, reflecting different π–π interactions between rGO and the aromatics with steric hindrance being the major factor to be considered to rationalize the reactivity.[104]

Hydrogenated graphene (HG) prepared by γ-ray irradiation of GO in aqueous suspension at room temperature was tested as a catalyst for Fenton-like degradation of Orange II and compared against a chemically reduced GO.[106] The authors observed an effective oxidation of Orange II by H_2O_2 promoted by HG, while in the absence of catalyst no oxidation occurred, indicating the potential ability of HG as a metal-free carbocatalyst. In contrast, rGO showed much lower activity than HG, exemplifying the importance of the reduction treatment. HG was reused up to five times with no significant changes in the activity at final time; however, there was some change in the initial reaction rate. The exact nature of the active sites responsible of H_2O_2 activation in HG was not mentioned by the authors, as further studies are necessary to understand the oxidation mechanism.

Other works have focused on the use of rGO-based materials for generation of sulfate radical anions ($SO_4^{-\bullet}$) from peroxymonosulfate (PMS). These radicals have a strong oxidizing power (E_{ox} = 2.5–3.1 V depending on the conditions) and are able to promote the degradation and/or mineralization of different contaminants such as phenol, 2,4-dichlorophenol, and Methylene Blue as probes of organic pollutants in water.[107,108] The catalytic activity of rGO (obtained following a hydrothermal reduction method) to promote pollutant decomposition by PMS was found to be higher than that of CNTs or AC.[108] Furthermore, graphite powder or GO were unable to activate PMS on their own. The activity of rGO was even higher than that of Co_3O_4 NPs, which is a benchmark metal oxide catalyst for this process. In order to explain their results, the authors proposed that electron-rich oxygen-containing groups at the zigzag edges of rGO are the sites promoting formation of sulfate radicals through single-electron reduction to PMS. Unfortunately, the catalytic activity for phenol removal using rGO gradually reduced upon reuse (100–25.5% from first to third cycle). Given that characterization data of the used catalyst did not show significant differences compared to the fresh rGO sample, the authors proposed that strong adsorption of the main reaction intermediates such as 4-hydroxybenzoic acid, *p*-benzoquinone, and 1,2-dihydroxybenzene were responsible for the deactivation of the active sites.

Advanced Oxidation Processes. In addition to selective oxidation reactions, carbon materials have also been reported to be active in advanced oxidation processes, commonly used for pollutant removal from water. Among the most common are catalytic ozonation,[109–111] catalytic wet air oxidation,[112–115] and catalytic wet peroxide oxidation.[116,117] In these processes carbon materials are reported to promote the formation of several highly oxidizing radical species. The first few studies were focused mainly on the application of activated carbons, but with the advent of nanostructured carbon materials, improved removal efficiencies were attained, particularly when using CNTs.[118–123]

Pereira *et al.* studied the effect of oxygen-containing surface groups on CNT for the ozone decomposition and catalytic ozonation of oxalic and oxamic acids.[118] CNTs with different surface chemical properties were prepared by oxidation treatments with nitric acid, peroxide hydrogen and oxygen. A strong correlation was observed between the normalized rate constant for heterogeneous ozone decomposition and pH_{PZC}. More acidic CNTs have a higher content of surface electron withdrawing oxygenated groups (mainly carboxylic acids and anhydrides), which are responsible for the decrease of delocalized π electrons on the surface. Hence, the electron density on the surface is lower than that of less acidic CNTs, disfavoring the adsorption step and the reduction of the ozone molecule, which have electrophilic properties. Consequently, CNTs with lower acidic character presented a higher catalytic activity.

Catalytic wet air oxidation (WAO) is a very attractive process that has been applied around the world to treat highly polluted industrial wastewaters. Oxygen is used to generate active oxygen species, such as hydroxyl (HO$^{\bullet}$) and hydroperoxyl (HOO$^{\bullet}$) radicals, which will oxidize the organic compounds preferentially into carbon dioxide and water, or alternatively into easily biodegradable by-products. Different carbon materials have been applied in CWAO, such as activated carbons,[112,113,124–126] carbon xerogels,[113,114] and CNTs.[119,122,123,127] Mesoporous carbons with high external surface area are usually more efficient for CWAO processes than carbon materials that are essentially microporous. Organic pollutants can diffuse easily to meso- and macropores where the degradation takes place at a faster rate than in the micropores.[125] Therefore, MWCNTs are very attractive materials for CWAO, where the accessibility of the solution and, consequently, of the dissolved pollutants to the carbon surface is expected to be easier than when using activated carbons, which are typically microporous materials.[128]

Figueiredo *et al.* modified CNTs by chemical and thermal treatments to obtain materials with different surface chemical properties.[127] These materials were then directly applied as catalysts in CWAO, without any impregnated metal, using oxalic acid as a model compound at 413 K and 40 bar of total pressure. The authors observed that oxalic acid could be totally degraded in less than 30 min in the presence of CNTs (Figure 5.5a). The rate of oxidation of oxalic acid depended on the chemical properties of CNTs, the apparent initial first-order rate constant being lower for CNTs with a marked acid character (Figure 5.5b). In addition, textural properties of CNTs were stable in cyclic CWAO experiments, but a decrease in basicity led to the reduction of their catalytic activity (nearly total oxalic acid degradation in 45 min and complete mineralization in 120 min). In another work by the same group, Faria *et al.* showed that the adsorption and degradation of the nitrophenolic pollutants on the surface of a carbon xerogel produced a surface modification during the CWAO process. The carboxylic acid groups were found to increase while phenol and quinone functional groups decreased.[114]

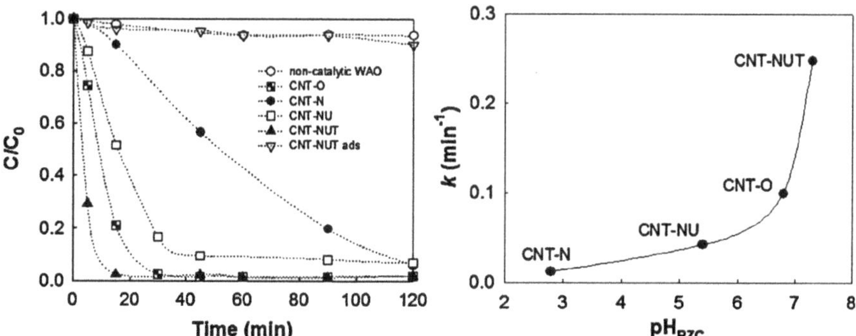

Figure 5.5 (a) Normalized oxalic acid oxidation at 413 K and 40 bar of total pressure in non-catalytic WAO and CWAO experiments, and an adsorption experiment using CNT-NUT under pure nitrogen at the same total pressure (CNT-NUT ads); (b) Apparent first-order initial reaction rate constants (k) *vs.* pH_{PZC} for the original and treated CNTs (CNT-O: original material; CNT-N: nitric acid treated; CNT-NU: CNT-N treated with urea; CNT-NUT: CNT-NU thermally treated). Reprinted with permission from ref. 127. Copyright 2011 Elsevier B.V.

In the case of catalytic wet peroxide oxidation processes, whilst ACs are known as catalysts for the decomposition of hydrogen peroxide,[129–131] only a few studies have been carried out using graphene or CNT materials.[17,132] The reaction mechanisms are not completely established and it seems that free radical species can act as intermediates.[129–131] The formation of these radicals would take place by an electron-transfer reaction, similar to the Fenton mechanism. The recombination of free radical species in the liquid phase and/or onto the AC surface will produce water and oxygen.

Guerrero-Ruiz *et al.* studied the catalytic decolorization of a textile dye (C.I. Reactive Red 241) by wet oxidation with H_2O_2, using different carbon materials as catalysts.[121] Various carbon materials (including activated carbons, a high surface area graphite, carbon nanofibers and carbon nanotubes) differing both in their texture/structure features and in their surface chemistry properties showed that the presence of oxygenated groups on all carbon materials was unfavorable to the dye adsorption. The presence of oxygenated groups led to a reduction in concentration and reactivity at the unsaturated carbon surface, which are required active sites for the H_2O_2 decomposition, where hydroxyl radicals are generated, and thus produce the dye degradation. Hence, it was no surprise that the highest performance in terms of activity for the catalytic wet peroxide oxidation was exhibited by pristine CNTs, achieving a total removal of the dye after 90 min in process. The authors attributed this result to the special morphology, composed of cylindrical mesopores formed with curved graphite sheets and the surface chemistry of the CNT material, promoting the decomposition of H_2O_2 with formation of $HO^{\bullet}$ radicals inside the mesopores.

Enhanced Fuel/Propellant Combustion. An interesting application of graphene-based materials related to aerobic oxidation could be their use as additives to accelerate combustion of future high-speed propellants.[133,134] Thus, it has been reported that graphene dispersed in liquid nitromethane can increase the burning rate of this compound up to 175% over neat nitromethane, outperforming other conventional additives such as silica or alumina NPs. Theoretical studies suggest that the active sites in this process would be defects in the graphene sheet such as carbon vacancies and particularly oxygen functionalities such as hydroxyls, ethers, and carbonyls around sp^2 dangling bonds at the vacancy edges.[133,134]

5.3.2.2 Reductions

While most of the earlier reactions involving carbocatalysts were oxidations (typically involving atmospheric oxygen or hydrogen peroxide as oxidant species), carbon catalyzed reductions were also demonstrated.

Formate and hydrazine hydrate can be used as hydrogen sources to promote hydrogenation. Bao *et al.* reported the use of rGO as catalyst for hydrogenation of nitrobenzene with hydrazine at room temperature.[35] The authors showed the superior activity of rGO with respect to other carbonaceous materials (AC, CB, graphite and GO) when working at room temperature. Importantly, the possible activity derived from the presence of Mn^{2+} coming from the catalyst preparation was excluded for rGO, GO, and CB samples. The presence of unsaturated carbon atoms at the edges of rGO and defects on rGO were proposed as the active sites for nitrobenzene reduction with hydrazine. The stability of rGO was established by performing nine catalytic cycles without observing loss of catalytic activity. It was proposed that the catalytic activity of rGO could be due to the unique electronic structure of zigzag edges in a single layer or a few-layer graphene. DFT calculations also suggest that the carbon atoms at the zigzag edges of graphene can interact with both terminal oxygen atoms of nitrobenzene. Thus, the N–O bonds are weakened and consequently the molecule activated. Furthermore, DFT calculations revealed that nitrobenzene activation is enhanced when the number of layers decreases. These conclusions from theoretical calculations, indicating that single- or few-layer graphene material should exhibit higher catalytic activity than multilayer graphitic materials, are in agreement with the lower and negligible catalytic activity of natural or expanded graphite compared to rGO.

Reduction of 4-nitrophenol to 4-aminophenol is one of the favorite test reactions to evaluate the catalytic activity for reductions of many homogeneous and heterogeneous catalysts. Activation of borohydride is a general reaction that has been considered to require metals as active sites since metal hydrides formed from the BH_4^- reagents are considered intermediates in this type of reduction. For this reason one of the targets in carbocatalysis has to be to find suitable metal-free carbon-containing catalysts that could also be used for this benchmark reaction. In this context, Chen *et al.* recently reported that rGO are suitable carbocatalysts for the reduction of the aromatic nitro group.[135] Based on the combined experimental and theoretical investigations, hydroxyl and alkoxy radicals, as well as holes, were found beneficial to the catalytic performance, while epoxy and carboxyl groups were unfavorable and should be avoided in practical applications. This conclusion was reached because a NaOH-treated rGO sample still containing carbonyl moieties exhibits significantly lower catalytic activity than the rGO.

***NO_x* Reduction.** Carbon materials can be used in de-NO_x processes either as adsorbents, reductants, or catalysts. The impact of pretreatments on the selectivity of carbon materials for NO_x adsorption and reduction in the presence of oxygen was addressed by Xia *et al.*[136] The authors showed that high-temperature (1223 K) hydrogen-treated carbons adsorb NO at room temperature but not oxygen; on the other hand, N_2-treated carbons strongly adsorb both species. Thus, selectivity for NO adsorption can be induced by hydrogen treatment, and this was explained in terms of the different active surface species generated by this treatment. In particular, the authors suggested that the H_2-treated carbon did not contain "dangling carbons" capable of strongly adsorbing oxygen;[137] however, they did contain a high concentration of edge sites (unsaturated basic sites), capable of NO adsorption at 303 K (oxygen only adsorbing there above about 423 K).

A commercially available technology to control the emissions of nitrogen oxides from stationary sources is the selective catalytic reduction with ammonia, generally with oxide catalysts, in the temperature range 573 to 673 K.[138] The reaction can proceed in either the presence or the absence of oxygen, according to eqn (5.5) and eqn (5.6).

$$4NO + 4NH_3 + O_2 \rightarrow 4N_2 + 6H_2O \quad (5.5)$$

$$6NO + 4NH_3 \rightarrow 5N_2 + 6H_2O \quad (5.6)$$

Various reports have shown that this reaction can be efficiently catalyzed by carbon materials at temperatures as low as 373 to 473 K.[139,140] It should be noted that in the presence of carbon catalysts, two different mechanisms may be operative in this process: adsorption of NO_x at low temperatures ($T < 423$ K), and reaction at higher temperatures ($T > 423$ K).

Ahmed *et al.* carried out a detailed study with the objective of identifying the properties of ACs that are important for the selective catalytic reduction of NO.[141] The authors concluded that chemical properties such as surface oxides and mineral matter play a more important role than their physical properties, such as surface area and pore structure. Indeed, they found that the catalyst activity correlated directly with the oxygen content of the carbon samples and inversely with their pH. These results indicate that the NO conversion is favored on more acidic carbons. They also reported that NO reduction by ammonia was negligible in the absence of oxygen. Indeed, it has been shown[142] that oxygen enhances the C–NO reaction through the formation of surface oxygen complexes, which are essential for the C–NO reaction to proceed.[139]

Several authors have also reported an increased catalytic activity after treating the carbon material with sulfuric acid.[140,143,144] This effect was ascribed to the formation of (acidic) oxygen functional groups, which can be the sites for NH_3 adsorption. A linear correlation between NO conversion and the acidity of the ACs was presented by Ku *et al.*[143] However, when such sulfuric acid treatment was applied to PAN–ACFs, a much higher catalytic activity for the selective catalytic reduction of NO was observed in comparison to the activated carbons.[145] Although the authors could not give a precise explanation for this activity increase, they suggested that it might be due to the presence of residual nitrogen in the PAN–ACF. Other authors reported an increased selective catalytic reduction activity by treating the carbon catalyst in NH_3 at high temperatures.[145,146] The activity was found to increase linearly with a concentration of N between 1 and 3%, leveling off at higher concentrations. Although the authors did not provide a clear explanation for the role of the incorporated nitrogen, it became clear that the simultaneous presence of oxygen and nitrogen functionalities on the surface of the carbon materials was key to improving their catalytic activity (this topic is discussed in detail within Chapter 6).

In an attempt to understand the nature of the active site for the NO selective catalytic reduction, Teng *et al.* used ACs with different surface areas as catalysts.[147] In order to achieve this, the authors produced the materials from the same precursor and varied the extent of burn-off in CO_2, thus ensuring that the chemical properties of the carbons were not changed. They found that the NO conversion increased with the surface area of the carbon catalysts. Under those conditions, the number of sites available for the reduction of NO increased with the specific surface area. However, the conversion normalized by the BET area was found to decrease as the burn-off increased; this clearly demonstrated that the surface was not fully accessible to the reactants. Since the effect observed was similar at all temperatures, the authors concluded that spatial limitations for NO adsorption in the required configuration were probably involved. They also observed that the addition of oxygen increased the NO conversion. It was proposed that the process might be controlled by two consecutive steps: the adsorption of NH_3 on active sites, followed by the addition of NO or NO_2 in the neighborhood of the $C(NH_3)$ complexes.

The nature of the active sites was also addressed in a subsequent paper by the same authors, where carbons treated with H_2SO_4 and HNO_3 were used as catalysts.[148] It was observed that the concentration of oxygen-containing surface sites significantly affected the course of the reaction before steady state was reached. In the absence of oxygen, the activity was found to increase as a function of the CO_2 evolved in TPD, while it increased with the CO evolved in the presence of oxygen. Therefore, carboxyl groups were thought to be responsible for the reduction of NO in the absence of oxygen, while hydroxyl and carbonyl groups were involved in the presence of oxygen (Figure 5.6). Thus, the formation of $CO^-(NH_4)^+$ and C(ONO) complexes, and their interactions,

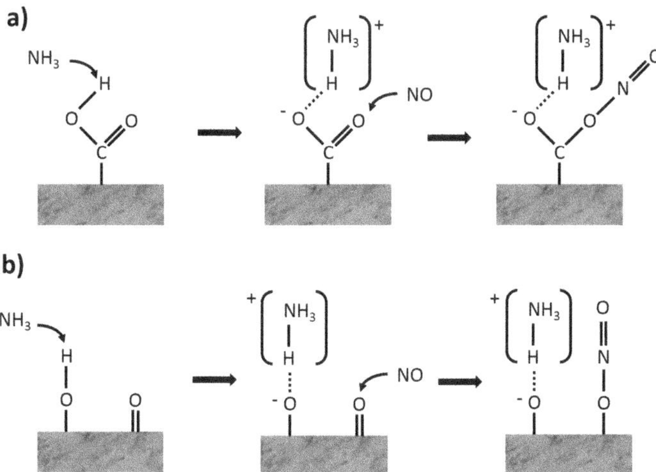

Figure 5.6 Reaction scheme for the formation of surface complexes during NO reduction with NH_3 over carbon catalysts. Adapted from ref. 148.

were identified as important steps in the selective catalytic reduction of NO, which is basically in agreement with the mechanism originally proposed by Mochida *et al.*[146]

5.3.2.3 Acid/Base Reactions

Acid Reactions. Traditional liquid acids are highly efficient acid catalysts; however, they pose serious problems from an environmental point of view and are difficult to recover. Separation of liquid acid catalyst from the reaction mixture is costly, resulting in generation of enormous volumes of liquid wastes if neutralization is needed. In order to overcome these drawbacks of liquid acids, solid acid catalysts such ad amorphous silica–alumina,[149] zeolites,[150] mesoporous materials,[151,152] metal–organic frameworks,[153,154] and carbon materials bearing acid moieties[155,156] have been developed as an alternative to homogeneous catalysts.

Decomposition of alcohols by dehydrogenation or dehydration is a very interesting process, commonly used for studying the acid/basic character of catalyst surfaces. The activity and selectivity of this reaction is governed by textural effects and by the acid/basic and electronic properties of the catalyst.[157] Alcohol dehydrogenation products (aldehydes and ketones) are preferentially formed on basic catalysts, while the dehydration products (olefins and ethers) are favored when acid sites are present. Thus, for catalysts with a large number of acid surface sites, the dehydration follows an E1 (elimination *via* a carbocation intermediate) mechanism whereas the dehydrogenation only occurs to a limited extent *via* an E1cB (elimination *via* a carbanion intermediate) mechanism. When the catalysts have a large number of basic surface sites, both the dehydration and dehydrogenation follow an E1cB mechanism, although the second reaction predominates. Finally, with catalysts with similar acid–base characteristics, the dehydration follows an E2 (bimolecular elimination) concerted mechanism whereas the dehydrogenation occurs by an E1cB mechanism.

The dehydration of methanol to dimethyl ether using oxidized ACs as catalysts was reported by Moreno-Castilla *et al.*[158] By treating the catalyst at increasing temperatures, in order to decompose the most unstable surface groups, the authors confirmed that the dehydration activity was due to the most strong carboxylic acid groups.

Using carbon catalysts oxidized to different extents and subsequently heat treated at different temperatures, it was shown that alcohol dehydration was controlled not only by the number and strength of the carboxylic acid groups, but also by their accessibility.[159,160] At a higher degree of carbon surface oxidation, there is a decrease in catalytic activity during dehydration of secondary aliphatic alcohols.[161,162] It seems that the presence of too many surface groups in the vicinity of catalytically active acid groups – carboxyls – may hinder the interactions between them and the bulky alcohol molecules or even prevent such interactions. It is known that the presence of adjacent polar functional groups leads to the formation of strong intermolecular hydrogen bonds between such groups. The decomposition of some carboxyls, as well as the rearrangement of others, may remove existing steric restrictions to the reaction or facilitate access of the alcohol molecules to the catalytically active acidic groups.

The acidity of rGO, obtained by reduction of GO by $NaBH_4$, was used to promote the condensation of acetone with pyrrole to form dipyrromethane and meso-dimethyl-substituted porphyrins.[163] This condensation requires acid sites of weak or medium strength since the α positions of pyrrole are highly reactive against electrophilic attacks. The selectivity toward the most valuable metal-free porphyrin was the highest using acetone as solvent, reaching a value as high as 25% yield at full pyrrole conversion.

A facile synthesis of sulfated rGO (rGO-SO_3H) was reported by Xiao *et al.* through a hydrothermal sulfation of rGO with fuming sulfuric acid at 453 K.[164] The authors found that rGO-SO_3H exhibited much better catalytic performance (activity and recyclability) than conventional porous solid acids such as SO_3H-functionalized ordered mesoporous carbons (OMC-SO_3H), Amberlyst 15, and SO_3H-functionalized ordered mesoporous silica (SBA-15-SO_3H) for general acid-catalyzed reactions including esterification of acetic acid with cyclohexanol or 1-butanol, Pechmann reaction of resorcinol with ethyl acetoacetate, and hydration of propylene oxide (Figure 5.7). The acidic sites of these catalysts were found to be 1.2, 1.3, 4.7, and 1.26 mmol g^{-1} for rGO-SO_3H, OMC-SO_3H, Amberlyst 15, and SBA-15-SO_3H, respectively. While GO and rGO showed very low activities due to the absence of acidic sites, the best performance of rGO-SO_3H with respect to the other solid acids for cyclohexanol esterification was directly related to the acid strength and 2D morphology of rGO-SO_3H, where most of the sulfate groups are exposed to the reactants without mass transfer limitations. Furthermore, rGO-SO_3H exhibited high recyclability, as a similar conversion (78.3%) for esterification of acetic acid with cyclohexanol than the fresh material even after five cycles was observed. Given the high activity of rGO-SO_3H other practical applications for this material can be envisaged. In particular, biomass hydrolysis has also been reported using similar materials as acid catalysts for production of furfural from aqueous xylose solutions.[165]

Synthesis of xanthenes and benzoxanthenes has been reported using GO and sulfated rGO microsheets as Brönsted acid catalysts.[166] GO was obtained by a modified Hummers method, and rGO-Ar-SO_3H was obtained using an arylsulfonate diazonium salt as intermediate to anchor covalently aryl sulfonic groups on the GO sheet. Excellent xanthene yields were obtained using rGO-Ar-SO_3H (83–97%) compared to GO (63–87%) in water as solvent. Both rGO-Ar-SO_3H and GO were reused for at least five cycles with only a slight decrease of the yield (94% to 81% and 91% to 74%, respectively).

Similarly, GO-SO_3H was prepared by sulfonation of GO with chlorosulfonic acid. GO-SO_3H was found to be a recyclable catalyst for conversion of glucose or fructose (90% conversion) into levulinic acid (76% molar yield), achieving TOFs around 125.[167] In addition, silylation of the –COOH and –OH groups

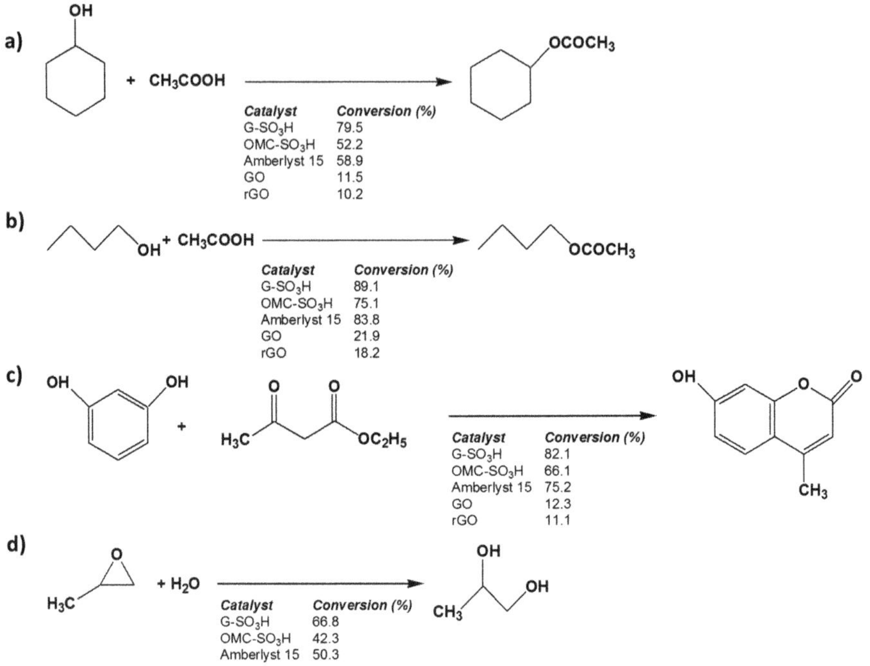

a) Cyclohexanol + CH_3COOH → cyclohexyl acetate ($OCOCH_3$)

Catalyst	Conversion (%)
G-SO_3H	79.5
OMC-SO_3H	52.2
Amberlyst 15	58.9
GO	11.5
rGO	10.2

b) 1-Butanol + CH_3COOH → butyl acetate ($OCOCH_3$)

Catalyst	Conversion (%)
G-SO_3H	89.1
OMC-SO_3H	75.1
Amberlyst 15	83.8
GO	21.9
rGO	18.2

c) Resorcinol + ethyl acetoacetate → 7-hydroxy-4-methylcoumarin

Catalyst	Conversion (%)
G-SO_3H	82.1
OMC-SO_3H	66.1
Amberlyst 15	75.2
GO	12.3
rGO	11.1

d) Propylene oxide + H_2O → propylene glycol

Catalyst	Conversion (%)
G-SO_3H	66.8
OMC-SO_3H	42.3
Amberlyst 15	50.3

Figure 5.7 Reaction catalyzed by GO-SO_3H: (a and b) Esterification of acetic acid with cyclohexanol (a) and 1-butanol (b), (c) Pechmann reaction of resorcinol with ethyl acetoacetate, and (d) Hydration of propylene oxide.

in the GO structure by methoxytrimethylsilane resulted in a decrease of conversion and levulinic yield to 31% and 11%, respectively. Also, use of GO, AC, AC-SO_3H, the ion-exchange resin Amberlyst-36, and the zeolite HY resulted in conversions between 20% and 47% and molar yields toward levulinic acid lower than 15%. Thus, it was proposed that the layered morphology of GO-SO_3H together with the presence of the Brönsted acid –SO_3H sites as well as –COOH and –OH groups could be responsible for the good activity of the catalyst.

Bielawski reported, in a pioneering work, the use of GO for oxidation and hydration reactions.[29] Thus, using GO as solid acid catalyst, high conversions for hydration of phenylacetylene (>98%), 1-butynylbenzene (52%), 1,2-diphenylethyne (41%), 1,4-diethynylbenzene (26%), and 1-decyne (27%) into their corresponding ketones were achieved. Similarly, Garcia and *et al.* reported the use GO as an acid catalyst for the room-temperature ring opening of epoxides using methanol and other primary alcohols as nucleophile and solvent.[168] Importantly, using styrene oxide as reactant, the catalytic reaction could be carried out using only 0.19 wt% of GO while achieving good conversions and selectivity toward the epoxide ring-opening products. This catalyst amount is much lower than that used for benzyl alcohol

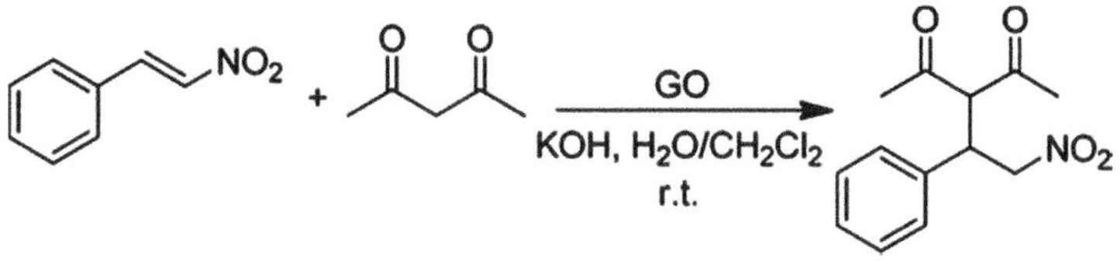

Scheme 5.2 Michael addition catalyzed by GO as phase transfer catalyst.

oxidation or alkynes hydration (200 wt%)[29] or for hydration of propylene oxide (3.4 wt%).[164]

A simple method to perform the room-temperature Michael addition of 2,4-pentanedione to *t*-β-nitrostyrene in dichloromethane using GO as a phase transfer catalyst with potassium hydroxide in water has been reported by Lee *et al.* (Scheme 5.2).[169] The final reaction yield was 83%. The reaction was further extended to various substrates with reasonably good yields. On the basis of the experimental data, it was believed that oxygen functional groups in GO, including carbonyl, carboxylic, lactone, quinone, and especially epoxy and hydroxyl groups, are responsible for interaction with the base cations, increasing the hydroxide ability to react strongly and quickly with reactants. Although it was suggested that the catalyst can be reused, no experimental data have yet been provided. On the other hand, claims that this process is probably green needs to be revised as it uses dichloromethane as one of the solvents.

Similarly, use of GO as an efficient and a recyclable carbocatalyst for aza-Michael addition of amines to activated alkenes leading to β-amino compounds has also been reported by Khatri *et al.* (Scheme 5.3).[31] The authors proposed that oxygen functionalities such as hydroxyl, epoxides, carboxylate, and carbonyl present on the GO were the active sites. This hypothesis was supported using rGO instead of GO under the same reactions conditions, whereby a considerable decrease in the catalytic activity was observed. In general, aliphatic amines were found to be more reactive compared to aromatic amines following the general reactivity trend as nucleophiles. Reuse experiments were carried out by extraction of the aqueous reaction mixture with dichloromethane and using the aqueous phase containing the catalyst in a new cycle. In this way, the carbocatalyst was reused nine times without losing catalytic activity.

GO has also been found to be a promising catalyst for etherification of benzyl alcohol to dibenzyl ether in the absence of added solvent.[170] The dehydrative etherification of alcohols is a reaction that typically is performed by strong Brönsted acids, and the catalytic results suggest that adsorption of benzyl alcohol on the graphene sheet can increase the reaction rate in a comparable way as alternative use of strong acids.

GO has also been reported as an efficient and a mild catalyst for sequential dehydration–hydrothiolation reactions to give unsymmetrical thioethers from a mixture of secondary aryl alcohols and thiols.[171] Under optimized

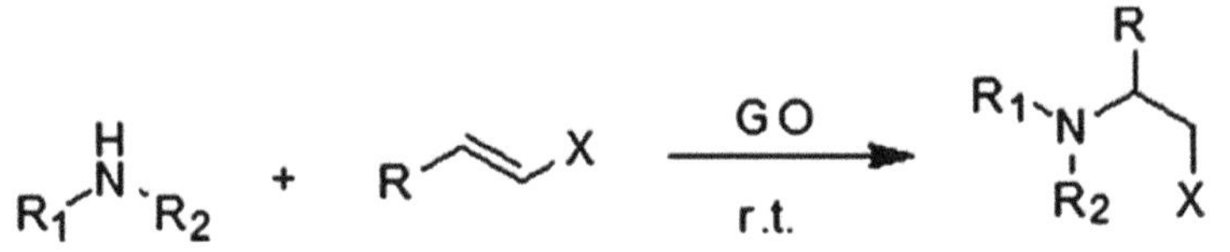

Scheme 5.3 GO-catalyzed aza-Michael addition reaction.

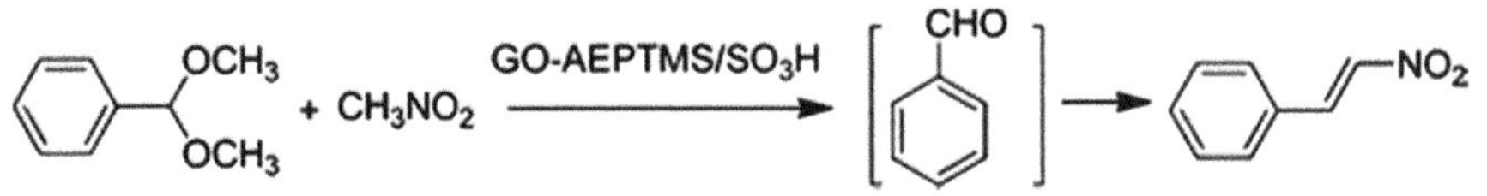

Scheme 5.4 Tandem deacetalization–nitroaldol reaction catalyzed by GO-AEPTMS/SO_3H as catalyst.

reaction conditions, a mixture of 1-(2-naphthyl)ethanol and 4-chlorobenzenethiol in toluene leads to thioether in 79% yield, accompanied by the corresponding disulfide in 8%. The scope of this protocol was expanded to a series of unsymmetrical thioethers in good yields, without the need of a transition metal. It was presumed that GO participates in the overall process, the one-pot dehydration–hydrothiolation reaction, *via* an acid-catalyzed Markovnikov addition to the intermediate alkene. The catalyst was reused five times without noticeable changes in the yields, suggesting its stability. However, the acidity of GO measurements before and after catalysis should give more information on the catalyst stability.

Also, acid–base bifunctional GO hybrids have been prepared by GO modification with aminosilyl and sulfonic acid groups.[172] The obtained bifunctional hybrids displayed high acid and base catalytic activities for the one-pot, acid–base deacetalization–nitroaldol reaction. Amine-functionalized GO (GO-AEPTMS) was prepared by silylation of GO with 3-[2-(2-aminoethylamino)ethylamino]propyl-trimethoxysilane (AEPTMS). Acidic $-SO_3H$ groups, on the other hand, were attached with sulfonic acid-containing aryl radicals that presumably become anchored to isolated C=C bonds of GO. Characterization of the GO-AEPTMS/SO_3H hybrids demonstrated the high density and homogeneous distribution of both amine and sulfonic acid groups on the GO surface. GO-AEPTMS/SO_3H as catalyst promotes complete conversion of benzaldehyde dimethyl acetal to β-nitrostyrene with 95% yield (Scheme 5.4). The catalyst was used for four consecutive cycles without any significant drop in its activity. However, no leaching tests were performed to determine the possible detachment of the active units. In contrast, use of (3-aminopropyl)dimethylmethoxysilane and 3-(2-aminoethyl)-3-aminopropyltrimethoxysilane as amine components in the synthesis of analogous GO hybrid materials in combination with $-SO_3H$ showed 100% conversion of dimethyl acetal, but the yields of β-nitrostyrene were only 44% and 62%, respectively. The superior performance of bifunctional GO-AEPTMS/SO_3H was attributed by the authors to the high density and homogeneous distributions of the acid

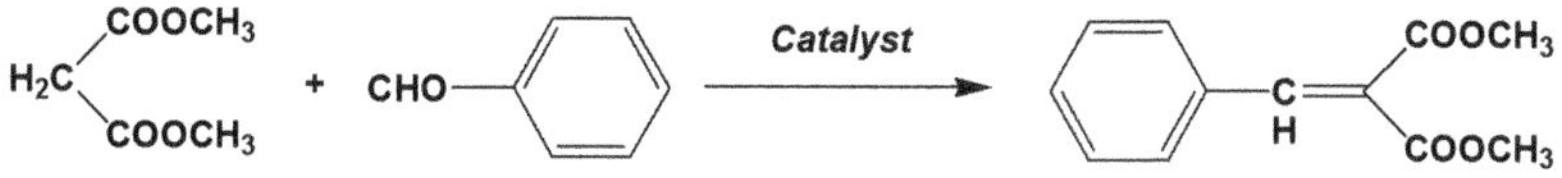

Scheme 5.5 Knoevenagel condensation catalyzed by modified rGO.

and basic functional groups and the unique two-dimensional structure of the graphene supports.

Base Reactions. Triethylamine-modified rGO (rGO-NEt_3) has been synthesized by NEt_3 grafting on rGO and investigated as a solid basic catalyst for hydrolysis of ethyl acetate.[173] The as-prepared rGO-NEt_3 consisted of FLG with wrinkling even more apparent than in rGO. Elemental analysis of rGO-NEt_3 indicated 4.8% nitrogen content after amino grafting. rGO-NEt_3 showed comparable activity to sodium hydroxide (hydrolysis conversion of 67%), used for comparison purposes. Interestingly, rGO exhibited a much lower activity for hydrolysis of around 10%, demonstrating the role of basic sites. rGO-NEt_3 was reused for five cycles and a gradual decrease in the hydrolysis conversion was observed (from 67% to 54%). The authors suggested that the decrease in the nitrogen content from 4.80% to 4.04% in rGO-NEt_3 (due to loss of amino groups during ester hydrolysis), was responsible for catalyst deactivation.

Another application of rGO as a basic catalyst is the Knoevenagel condensation. Liu *et al.* covalently linked rGO to an amino-terminated third-generation poly(amidoamine) (PAMAM) dendrimer and tested it in the Knoevenagel condensation between benzaldehyde and dimethyl malonate (Scheme 5.5).[174] Catalyst preparation involved the reaction between preformed GO and PAMAM dendrimer, with the latter acting both as a reducing and a stabilizing agent of GO. This enabled shorter reducing times, and allowed rGO to be stably dispersed in water and organic solvents, such as DMF and DMSO. The rGO-PAMAM composites led to the condensation product in yields as high as 99%.

A possible mechanism for the reaction was proposed by the authors, and involved the formation of an anion by deprotonation of the dimethyl malonate by the G3-PMR. Its addition to the benzaldehyde creates a metastable adduct that rapidly transforms into the final condensation product by releasing a hydroxyl anion (Scheme 5.6).

5.3.2.4 *Other Applications*

Qu *et al.* have developed a colorimetric method for glucose detection.[175] GO–COOH was prepared by adding NaOH and chloroacetic acid into GO solution to convert the –OH groups to –COOH *via* conjugation of acetic acid moieties. Based on the catalytic performance of GO–COOH samples in the reaction of 3,3,5,5-tetramethylbenzidine oxidation in the presence of H_2O_2, GO–COOH has been demonstrated to possess intrinsic peroxidase-like activity

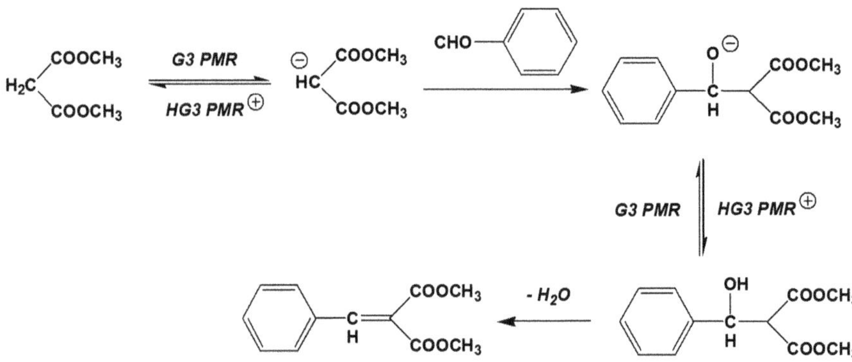

Scheme 5.6 Possible mechanism of Knoevenagel condensation catalyzed by G3-PMR.

to produce a blue color reaction following a ping-pong mechanism, with a maximum reaction rate of $(3.85 \pm 0.22) \times 10^{-8}$ M s^{-1}. This method was simple, cheap and highly selective, and was successfully employed for glucose detection in buffer solution, diluted blood and fruit juice. These findings facilitate GO's utilization in medical diagnostics and biotechnology. Cao *et al.* reported the first convenient metal-free catalytic process for an efficient imine synthesis from various amines under mild and neat conditions with molecular oxygen as the terminal oxidant.[176] Based on the oxidative coupling of amines to imines, Su *et al.* have probed the activity of GO and studied its catalytic mechanism.[177] Through the base and acid treatment of GO, it was found that the carboxylic groups as well as the localized unpaired electrons facilitated synergistic intermolecular arrangements, contributing to the enhanced catalytic activity of oxidative coupling of various primary amines.

5.3.3 Chemical Reactions inside Carbon Nanotubes

The unique tubular morphology of CNTs in the nanometer scale has attracted wide attention for studying chemistry in their channels.[178,179] Molecules frequently exhibit different behavior from that in the bulk when confined within pores. Modified adsorption and diffusion, and structural properties have been reported for a variety of molecules and nanomaterials inserted in the channels compared to their counterparts either on the outer walls of CNTs or in the bulk. The increasing numbers of studies demonstrates that confinement inside CNTs may affect chemical reactions with respect to those occurring on the outer surface of CNTs.[179–181] Such a different behavior originates from both the physical (spatial restriction of the channels without blocking diffusion pathway of molecules) and chemical (interaction of confined species with the CNT walls) situation inside CNTs, and can also be used as metal-free system.

The modified adsorption, diffusion of molecules and electronic structure of confined materials are expected to result in different reactivities. Theoretical studies predict that chemical reactions can be influenced by confinement inside CNT channels due to the significantly reduced volume and the interaction of molecules with CNT walls compared to those in the bulk.[179,182] Santiso *et al.* identified different effects including changes in the thermodynamic state of the system due to interactions with the pore walls, selective adsorption, and geometrical constraints that can affect chemical equilibrium, kinetic rates and reaction mechanism.[179] The relative importance of each factor could be very different for different chemical reactions.

For example, the D and H_2 exchange reaction to form HD and H was predicted to be considerably enhanced inside CNTs compared to the gas phase reaction.[183] Furthermore the reaction probabilities were considerably higher inside smaller CNTs. This enhancement was attributed to be steric in nature, having to do with the alignment of H_2 along the tube axis. Another theoretical study suggested that the isomerization of *n*-butane, 1-butene and 1,3-butadiene could be significantly modified by steric hindrance inside CNTs with a pore size comparable to the molecular dimensions.[184]

Figure 5.8 shows simulated results on ammonia synthesis and bimolecular hydrogen iodide decomposition over models of carbon slit-pores and nanotubes using the Reactive Monte Carlo method.[182] These results indicate that the equilibrium yield and kinetics of chemical reactions could be enhanced by confinement inside pores, and that the confinement effects could be different according to the structure and surface chemistry of the catalyst support material. For example, confinement within slit-shaped pores increased the conversion by almost 40%, and within nanotubes conversion can be increased by an additional 10–20%. The rate was enhanced within narrower pores, as the density was often greater in confined spaces. The enhancement in ammonia synthesis largely resulted from selective adsorption of the ammonia product. If some –COOH groups were introduced on the surface, the reaction could be further shifted toward formation of more NH_3 molecules due to favorable interaction of NH_3 with the surface, and its dipolar character.[182]

Halls *et al.* showed by hybrid DFT that the polarizability of CNTs was responsible of an interaction mechanism for a reaction involving dipolar species.[185] The enthalpy and activation energy of the Menshutkin SN_2 reaction $H_3N + H_3CCl \rightarrow H_3NCH_3^+ + Cl^-$, in which the charged product $[H_3NCH_3]^+[Cl]^-$ was formed, were modified. Dipolar species confined within nanotubes interacted with CNTs *via* an induced image dipole on the tubes. Thus, the produced species were stabilized inside CNTs with a significantly reduced reaction volume relative to the gas phase, making the overall reaction more favorable. For example, the reaction endothermicity was reduced by more than 23 kcal mol^{-1} within the (9, 0) nanotubes. The activation energy was also significantly reduced compared to that in the gas phase. The effect of CNT confinement on reaction enthalpies was proposed to closely resemble that of solvation in low-dielectric solvents.[185]

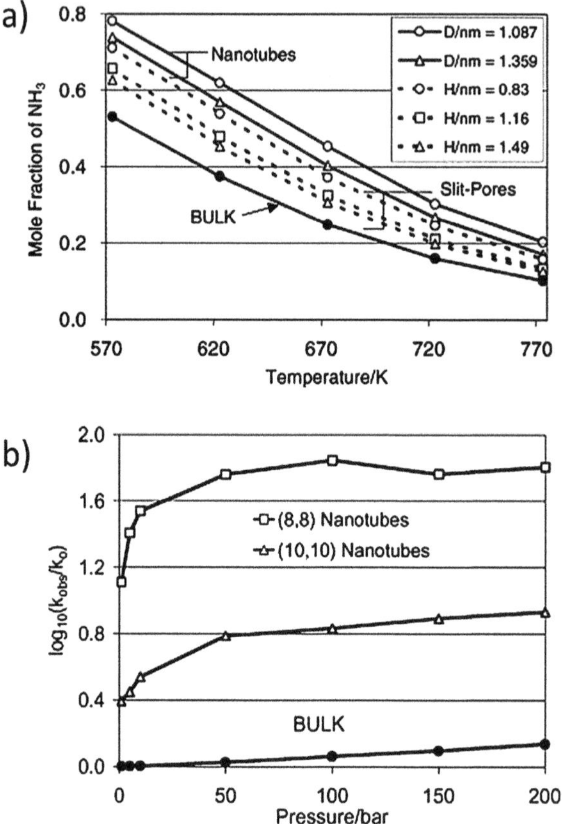

Figure 5.8 Effects of confinement on (a) ammonia synthesis (b) HI decomposition in slit-shaped pores and carbon nanotubes of various sizes. Reprinted with permission from ref. 182. Copyright 2002 Elsevier B.V.

In addition to the above theoretical simulations on metal-catalyst free chemical reactions, which show the effects of confinement inside CNTs, Kondratyuk *et al.* carried out an interesting experimental study on the reaction between 1-heptene and hydrogen atoms.[186] A lower reactivity of physisorbed 1-heptene toward atomic hydrogen was observed inside SWCNTs, compared to 1-heptene molecules adsorbed on external SWCNT sites. This was attributed to the shielding effect of nanotube walls, which prevented the reactive atomic H species from entering and undergoing reaction with the double bond of the confined reactant.[186]

5.3.4 Electrocatalysis

Electrocatalysis is one of the most important catalytic areas and GO possesses excellent activity toward many pivotal transformations. The electro-oxidation of ascorbic acid, uric acid, dopamine and acetaminophen using

GO was reported by Su *et al.*[187] The authors prepared the GO-modified electrode through the *in situ* electrochemical oxidation of graphite. During the positive scans, the van der Waals forces between graphitic sheets were weakened owing to the oxidation process, forming few-layer GO sheets. The obtained low oxidation potential and high oxidation current obtained for ascorbic acid demonstrated very good potential for electrochemical applications, which was explained by the increased surface area and oxygen functional groups of GO compared to graphite. Based on the high sensitivity and apparent peak separations for the different substrates, GO-modified electrodes might find wide applications for biosensors and bioanalysis.

In addition, reduced GO sheet films were also prepared by Li *et al.* as advanced electrode materials, and they have exhibited fast electron-transfer kinetics and possessed excellent electrocatalytic activity toward oxygen reduction.[188,189] ORR was found to be inhibited on the cleavage basal plane of highly ordered pyrolytic graphite due to the lack of the active adsorption sites for molecular O_2 and intermediates, like HO_2^-,[190] whereas the functional groups prevailing on the edge planes enhance the interaction of edge sites with molecular O_2 and produce much larger ORR current.[191,192] Yeager reproduced the Garten and Weiss mechanism, which proposed the participation of surface quinone groups in the reduction of O_2 to peroxide.[191] The ORR activity of quinone on modified GC electrodes was further confirmed by other works.[193–195] Decorating quinone on layer-by-layer assembled CNTs can also increase the ORR activity, *i.e.* smaller onset overpotential and larger cathodic current.[196] Note that the quinone group often leads to the peroxide ORR pathway, differentiating it from nitrogen functionalization (Chapter 6).

It is worthwhile to clarify whether the anchoring position of O-functionalities on nanostructured carbon materials affects the ORR activity. Site-selective oxidation of CNTs at the edge sites (open end of the nanotube) and at the hole defects was conducted to evaluate the ORR activity.[197] The excess O-functionalities on the CNT surface inhibited the electron transfer and occupied the adsorption sites for molecular O_2, which eventually caused additional resistance and adsorption overpotential to the overall ORR. After removing the O-functionalities from the non-defective sites but preserving those on defects the overall overpotential was decreased by 0.19 V. The catalytic effect of the O-groups was confirmed in a blend consisting of graphene oxide and CNTs.[198] In a different study, an improved ORR performance was observed on small graphene sheets which had higher oxygen content and more exposed edges (Figure 5.9).[199] The authors used high energy ball milling of graphene sheets in air to simultaneously reduce the lateral size and increase the oxygen content. Calculation results showed that hydroxyl and epoxy groups sitting on the basal plane of graphene sheets were ORR inactive, which was attributed to the repulsion of the lone pair electrons in O atom from the delocalized π bonding of graphene due to the mismatched orbital orientation. The calculation further

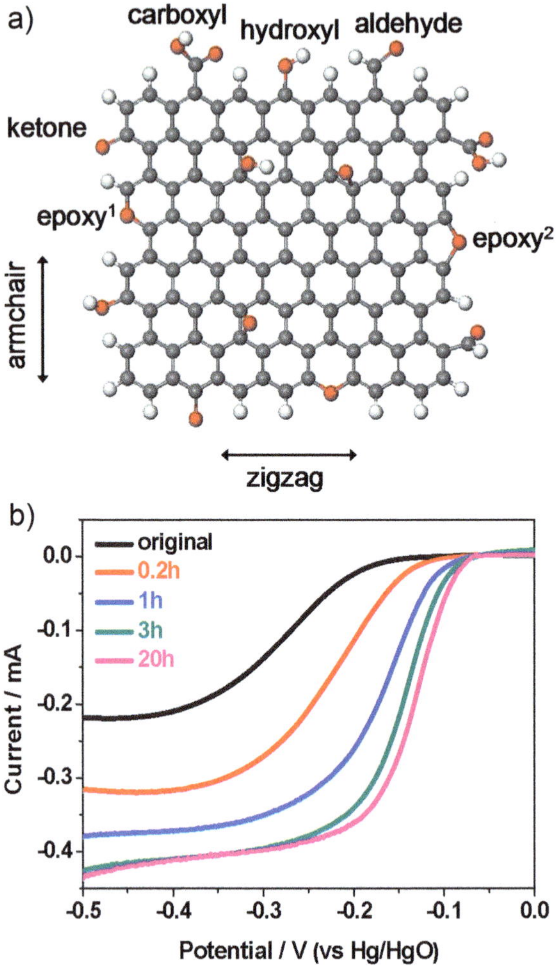

Figure 5.9 (a) A schematic diagram of nanographene showing armchair and zigzag edges with oxygen-containing groups. The grey, red and white balls represent C, O and H atoms, respectively. (b) Oxygen reduction voltammogram of pristine and ball-milled samples resulting from ball milling for different time durations. Reprinted with permission from ref. 199.

demonstrated the selective ORR activity on armchair and zigzag edges of oxygenated graphene sheets, where the former was inactive but the latter was active. The calculations further demonstrated ORR activity on armchair edges of oxygenated graphene sheets were inactive whereas those located over the in zigzag edge were active.

Regarding the electrocatalytic activity of metal-free nanostructured carbon materials, a more detailed discussion can be found on Chapter 6, where the use of heteroatoms other than oxygen is addressed, namely nitrogen, boron, sulfur and phosphorous.

5.3.5 Photocatalysis

Oxygenated species attached to the pristine graphene surface are tightly related to its band structure and the induced band gap opens up possibilities for photocatalytic applications (GO is a semiconductor or an insulator, while pristine graphene has a zero band gap). A GO semiconductor with an apparent band gap of 3.3–4.3 eV for direct transition and 2.4–3.0 eV for indirect transition was synthesized by Teng *et al.*[200] This band gap energy was sufficient to overcome the endothermic energy required for the water splitting reaction (1.23 eV). With the irradiation of both UV and visible light, abundant and continuous hydrogen evolution was obtained for reactions conducted in a 20 vol% aqueous methanol solution. No noticeable degradation of GO was observed, demonstrating its high catalytic performance. Methanol was considered as a sacrificial hole scavenger, which could catch the photo-generated holes in the valence band of GO, thus hindering electron–hole recombination.

Chen *et al.* reported an efficient photocatalytic conversion of CO_2 to methanol using GO as a catalyst (Figure 5.10).[201] Phosphoric acid was used to react with hydroxyl groups forming esters, which could prevent the GO basal plane from further oxidation. GO with a band gap in the range of 3.2–4.4 eV was obtained, which was sufficient to overcome the endothermic characteristics of the CO_2 reduction under solar energy excitation. The O-containing groups would stretch the band gap energy and help the electrons excite from the valence band to the conduction band, leading to photogenerated electrons and holes, which served as oxidizing and reducing radicals, respectively.

Tan *et al.* reported the first example of using GO to facilitate the synthesis of organic compounds under visible light irradiation,[33] GO was employed as a cooperative catalyst to rose Bengal, contributing to the photocatalytic oxidative C–H functionalization of tertiary amines to generate imines, which have great value in the synthesis of many industrially important materials and biologically active compounds.

Given the importance of nanostructured carbon materials in photocatalysis, these materials headline a chapter of their own (Chapter 8), mostly involving composites containing both metal oxides and carbon.

5.4 Concluding Remarks and Future Perspectives

Carbons have found broad use in many different areas. Historically, nanostructured carbon materials have functioned primarily as heterogeneous supports for metal catalysts. The latter being the lead actors in the catalytic

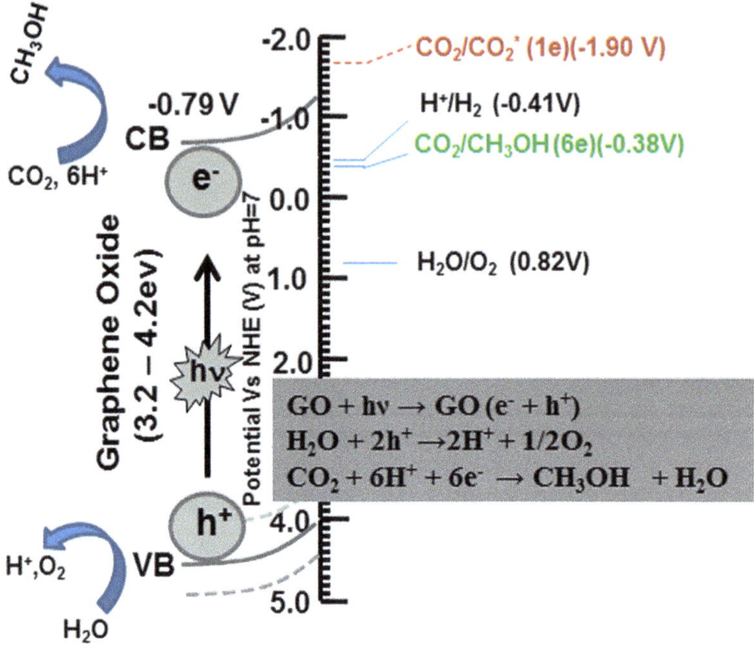

Figure 5.10 Schematic illustration of the photocatalytic CO_2 reduction mechanism on GO. Reprinted with permission from ref. 201.

performance, with carbon material assuming the supporting role. To address multiple competitive disadvantages associated with metal-based catalysts, such as the high cost, susceptibility to gas poisoning, and detrimental effects on the environment, recent research has demonstrated that certain nanostructured carbon materials could be used to replace metal-based catalysts for high-performance and low-cost catalytic processes. Many of the early studies focused on simple redox processes, but the field has progressed to demonstrate that carbons can facilitate more sophisticated reactions, including complex functional group transformations and carbon–carbon or carbon–heteroatom bond formations.

Pristine graphene has a low chemical activity due to its uniform and symmetrical electronic structure. Doped elements, however, can break the perfect π covalent bonds and induce local high electron/spin densities owing to the electronegativity differences, which contribute to its catalytic activity. While oxygen-rich functional group materials (such as GO) exhibit broad reactivity under mild conditions, their structure is not fully understood, which, combined with the heterogeneous nature of the catalysts employed, renders mechanistic elucidation extremely difficult. Considering the complexity of the surface oxygen-containing groups, detailed characterizations are required to understand their distributions.

Among the different catalytic examples on the use of metal-free nanostructured carbon materials discussed in this chapter, one of the most important and promising applications is the gas-phase oxidative dehydrogenation of alkanes. In this case, nanocarbons enabled the reaction to be performed with a high activity at a lower temperature, as well as with no coke formation. Other promising applications for metal-free nanocarbons involve many synthetic transformations such as oxidations, reductions, dehydrogenations, dehydrations, hydrations of various species. It is envisaged that the number of new carbocatalysed transformations will continue to grow and new forms of carbon-related materials, with engineered morphology, electronic properties or functionalities, will emerge as powerful catalysts for mediating synthetic transformations. Given their potential, clear targets for carbocatalysis could involve hydrogenations of alkenes, alkynes, and unsaturated organic compounds in the absence of noble metals, and aerobic oxidations of alkenes to epoxides, benzylic compounds to alcohols and ketones, and aerobic oxidations of hydrocarbons.

References

1. D. S. Su, in *Nanomaterials in Catalysis*, Wiley-VCH Verlag GmbH & Co. KGaA, 2013, pp. 331–374.
2. J. L. Figueiredo and M. F. R. Pereira, in *Carbon Materials for Catalysis*, John Wiley & Sons, Inc., 2008, pp. 177–217.
3. D. S. Su, J. Zhang, B. Frank, A. Thomas, X. Wang, J. Paraknowitsch and R. Schlögl, *ChemSusChem*, 2010, **3**, 169–180.
4. D. Yu, E. Nagelli, F. Du and L. Dai, *J. Phys. Chem. Lett.*, 2010, **1**, 2165–2173.
5. Y. Suzuki, M. Matsushima and M. Kodomari, *Chem. Lett.*, 1998, 319–320.
6. M. Kodomari, Y. Suzuki and K. Yoshida, *Chem. Commun.*, 1997, 1567–1568.
7. G. A. Sereda, V. B. Rajpara and R. L. Slaba, *Tetrahedron*, 2007, **63**, 8351–8357.
8. H. H. Byung, H. S. Dae and Y. C. Sung, *Tetrahedron Lett.*, 1985, **26**, 6233–6234.
9. J. W. Larsen, M. Freund, K. Y. Kim, M. Sidovar and J. L. Stuart, *Carbon*, 2000, **38**, 655–661.
10. P. G. Klemens and D. F. Pedraza, *Carbon*, 1994, **32**, 735–741.
11. B. Garrigues, C. Laporte, R. Laurent, A. Laporterie and J. Dubac, *Liebigs Ann.*, 1996, 739–741.
12. B. Garrigues, R. Laurent, C. Laporte, A. Laporterie and J. Dubac, *Liebigs Ann.*, 1996, 743–744.
13. P. Serp, M. Corrias and P. Kalck, *Appl. Catal., A*, 2003, **253**, 337–358.
14. N. F. Goldshleger, *Fullerene Sci. Technol.*, 2001, **9**, 255–280.
15. B. Li and Z. Xu, *J. Am. Chem. Soc.*, 2009, **131**, 16380–16382.
16. S. Sarkar, S. Niyogi, E. Bekyarova and R. C. Haddon, *Chem. Sci.*, 2011, **2**, 1326–1333.

17. F. Lücking, H. Köser, M. Jank and A. Ritter, *Water Res.*, 1998, **32**, 2607–2614.
18. H. Yu, F. Peng, J. Tan, X. Hu, H. Wang, J. Yang and W. Zheng, *Angew. Chem., Int. Ed.*, 2011, **50**, 3978–3982.
19. D. Bégin, G. Ulrich, J. Amadou, D. S. Su, C. Pham-Huu and R. Ziessel, *J. Mol. Catal. A: Chem.*, 2009, **302**, 119–123.
20. H. Tokuyama and E. Nakamura, *J. Org. Chem.*, 1994, **59**, 1135–1138.
21. Y. Shi, L. Gan, X. Wei, S. Jin, S. Zhang, F. Meng, Z. Wang and C. Yan, *Org. Lett.*, 2000, **2**, 667–669.
22. L. Pacosová, C. Kartusch, P. Kukula and J. A. van Bokhoven, *ChemCatChem*, 2011, **3**, 154–156.
23. J. L. Figueiredo and M. F. R. Pereira, *Catal. Today*, 2010, **150**, 2–7.
24. C. Su and K. P. Loh, *Acc. Chem. Res.*, 2012, **46**, 2275–2285.
25. I. F. Silva, J. Vital, A. M. Ramos, H. Valente, A. M. B. do Rego and M. J. Reis, *Carbon*, 1998, **36**, 1159–1165.
26. N. Keller, N. I. Maksimova, V. V. Roddatis, M. Schur, G. Mestl, Y. V. Butenko, V. L. Kuznetsov and R. Schlögl, *Angew. Chem., Int. Ed.*, 2002, **41**, 1885–1888.
27. X. Liu, B. Frank, W. Zhang, T. P. Cotter, R. Schlögl and D. S. Su, *Angew. Chem., Int. Ed.*, 2011, **50**, 3318–3322.
28. J. Zhang, D. S. Su, R. Blume, R. Schlögl, R. Wang, X. Yang and A. Gajović, *Angew. Chem., Int. Ed.*, 2010, **49**, 8640–8644.
29. D. R. Dreyer, H.-P. Jia and C. W. Bielawski, *Angew. Chem., Int. Ed.*, 2010, **49**, 6813–6816.
30. A. Vijay Kumar and K. Rama Rao, *Tetrahedron Lett.*, 2011, **52**, 5188–5191.
31. S. Verma, H. P. Mungse, N. Kumar, S. Choudhary, S. L. Jain, B. Sain and O. P. Khatri, *Chem. Commun.*, 2011, **47**, 12673–12675.
32. D. R. Dreyer, K. A. Jarvis, P. J. Ferreira and C. W. Bielawski, *Polym. Chem.*, 2012, **3**, 757–766.
33. Y. Pan, S. Wang, C. W. Kee, E. Dubuisson, Y. Yang, K. P. Loh and C.-H. Tan, *Green Chem.*, 2011, **13**, 3341–3344.
34. J. Ji, G. Zhang, H. Chen, S. Wang, G. Zhang, F. Zhang and X. Fan, *Chem. Sci.*, 2011, **2**, 484–487.
35. Y. Gao, D. Ma, C. Wang, J. Guan and X. Bao, *Chem. Commun.*, 2011, **47**, 2432–2434.
36. X.-H. Li, J.-S. Chen, X. Wang, J. Sun and M. Antonietti, *J. Am. Chem. Soc.*, 2011, **133**, 8074–8077.
37. S. Wang, C. T. Nai, X.-F. Jiang, Y. Pan, C.-H. Tan, M. Nesladek, Q.-H. Xu and K. P. Loh, *J. Phys. Chem. Lett.*, 2012, **3**, 2332–2336.
38. D. R. Dreyer and C. W. Bielawski, *Chem. Sci.*, 2011, **2**, 1233–1240.
39. D. R. Dreyer, S. Murali, Y. Zhu, R. S. Ruoff and C. W. Bielawski, *J. Mater. Chem.*, 2011, **21**, 3443–3447.
40. H.-P. Jia, D. R. Dreyer and C. W. Bielawski, *Tetrahedron*, 2011, **67**, 4431–4434.
41. D. R. Dreyer, H.-P. Jia, A. D. Todd, J. Geng and C. W. Bielawski, *Org. Biomol. Chem.*, 2011, **9**, 7292–7295.

42. H.-P. Jia, D. R. Dreyer and C. W. Bielawski, *Adv. Synth. Catal.*, 2011, **353**, 528–532.
43. D. R. Dreyer and C. W. Bielawski, *Adv. Funct. Mater.*, 2012, **22**, 3247–3253.
44. D. R. Dreyer, K. A. Jarvis, P. J. Ferreira and C. W. Bielawski, *Macromolecules*, 2011, **44**, 7659–7667.
45. J. A. Maciá-Agulló, D. Cazorla-Amorós, A. Linares-Solano, U. Wild, D. S. Su and R. Schlögl, *Catal. Today*, 2005, **102–103**, 248–253.
46. M. F. R. Pereira, J. J. M. Órfão and J. L. Figueiredo, *Appl. Catal., A*, 2001, **218**, 307–318.
47. D. S. Su, N. Maksimova, J. J. Delgado, N. Keller, G. Mestl, M. J. Ledoux and R. Schlögl, *Catal. Today*, 2005, **102–103**, 110–114.
48. J. J. Delgado, R. Vieira, G. Rebmann, D. S. Su, N. Keller, M. J. Ledoux and R. Schlögl, *Carbon*, 2006, **44**, 809–812.
49. D. Su, N. I. Maksimova, G. Mestl, V. L. Kuznetsov, V. Keller, R. Schlögl and N. Keller, *Carbon*, 2007, **45**, 2145–2151.
50. F. Cavani and F. Trifirò, in *Basic Principles in Applied Catalysis*, ed. M. Baerns, Springer Berlin Heidelberg, 2004, vol. 75, pp. 19–84.
51. R. Grabowski, *Catal. Rev.*, 2006, **48**, 199–268.
52. M. Sun, J. Zhang, P. Putaj, V. Caps, F. Lefebvre, J. Pelletier and J.-M. Basset, *Chem. Rev.*, 2013, **114**, 981–1019.
53. A. E. Lisovskii and C. Aharoni, *Catal. Rev.*, 1994, **36**, 25–74.
54. M. F. R. Pereira, J. J. M. Órfão and J. L. Figueiredo, *Appl. Catal., A*, 1999, **184**, 153–160.
55. G. Mestl, N. I. Maksimova, N. Keller, V. V. Roddatis and R. Schlögl, *Angew. Chem., Int. Ed.*, 2001, **40**, 2066–2068.
56. T.-J. Zhao, W.-Z. Sun, X.-Y. Gu, M. Rønning, D. Chen, Y.-C. Dai, W.-K. Yuan and A. Holmen, *Appl. Catal., A*, 2007, **323**, 135–146.
57. J. J. Delgado, D. S. Su, G. Rebmann, N. Keller, A. Gajovic and R. Schlögl, *J. Catal.*, 2006, **244**, 126–129.
58. B. Frank, A. Rinaldi, R. Blume, R. Schlögl and D. S. Su, *Chem. Mater.*, 2010, **22**, 4462–4470.
59. J. Zhang, D. Su, A. Zhang, D. Wang, R. Schlögl and C. Hébert, *Angew. Chem., Int. Ed.*, 2007, **46**, 7319–7323.
60. J. Zhang, X. Wang, Q. Su, L. Zhi, A. Thomas, X. Feng, D. S. Su, R. Schlögl and K. Müllen, *J. Am. Chem. Soc.*, 2009, **131**, 11296–11297.
61. M. F. R. Pereira, J. L. Figueiredo, J. J. M. Órfão, P. Serp, P. Kalck and Y. Kihn, *Carbon*, 2004, **42**, 2807–2813.
62. J. Zhang, X. Liu, R. Blume, A. Zhang, R. Schlögl and D. S. Su, *Science*, 2008, **322**, 73–77.
63. B. Frank, J. Zhang, R. Blume, R. Schlögl and D. S. Su, *Angew. Chem., Int. Ed.*, 2009, **48**, 6913–6917.
64. H. Xie, Z. Wu, S. H. Overbury, C. Liang and V. Schwartz, *J. Catal.*, 2009, **267**, 158–166.
65. X. Liu, D. S. Su and R. Schlögl, *Carbon*, 2008, **46**, 547–549.
66. V. Schwartz, W. Fu, Y.-T. Tsai, H. M. Meyer, A. J. Rondinone, J. Chen, Z. Wu, S. H. Overbury and C. Liang, *ChemSusChem.*, 2013, **6**, 840–846.

67. B. Frank, M. Morassutto, R. Schomäcker, R. Schlögl and D. S. Su, *ChemCatChem.*, 2010, **2**, 644–648.
68. S. Tang and Z. Cao, *Phys. Chem. Chem. Phys.*, 2012, **14**, 16558–16565.
69. G. C. Grunewald and R. S. Drago, *J. Am. Chem. Soc.*, 1991, **113**, 1636–1639.
70. B. Frank, R. Blume, A. Rinaldi, A. Trunschke and R. Schlögl, *Angew. Chem., Int. Ed.*, 2011, **50**, 10226–10230.
71. I. M. Kolthoff, *J. Am. Chem. Soc.*, 1932, **54**, 4473–4480.
72. E. K. Rideal and W. M. Wright, *J. Chem. Soc. Trans.*, 1925, **127**, 1347–1357.
73. E. K. Rideal and W. M. Wright, *J. Chem. Soc. (Resumed)*, 1926, **129**, 1813–1821.
74. E. K. Rideal and W. M. Wright, *J. Chem. Soc. (Resumed)*, 1926, **129**, 3182–3190.
75. F. Rodríguez-reinoso, *Carbon*, 1998, **36**, 159–175.
76. D. W. Boukhvalov, D. R. Dreyer, C. W. Bielawski and Y.-W. Son, *ChemCatChem.*, 2012, **4**, 1844–1849.
77. O. K. Davtyan and Y. A. Tkach, *Zh. Fiz. Khim.*, 1961, **35**, 992–998.
78. Y. U. Siedlewski, *Int. Chem. Eng.*, 1965, **5**, 608.
79. H. Kitagawa, N. Yuki, Y. Sanada, S. Watari and H. Honda, *Kogyo Kagaku Zasshi.*, 1969, **72**, 2260–2265.
80. K. Yamamoto and M. Seki, *Kogyo Kagaku Zasshi*, 1971, **74**, 78–83.
81. K. Yamamoto, K. Kaneko and M. Seki, *Kogyo Kagaku Zasshi*, 1971, **74**, 84–88.
82. Y. Kamino, S. Onitsuka and K. Yasuda, *Bull. Jpn. Pet. Inst.*, 1972, **14**, 141–146.
83. Y. Kamino, K. Yasuda, S. Inoue and S. Onitsuka, *Bull. Jpn. Pet. Inst.*, 1972, **14**, 147–152.
84. H. Sano and H. Ogawa, *Sangyo Kogai*, 1974, **10**, 2245–2250.
85. P. Davini, *Carbon*, 1990, **28**, 565–571.
86. I. Mochida, Y. Masumura, T. Hirayama, H. Fujitsu, S. Kawano and K. Goto, *Nippon Kagaku Kaishi*, 1991, **1991**, 269–273.
87. I. Mochida, Y. Korai, M. Shirahama, S. Kawano, T. Hada, Y. Seo, M. Yoshikawa and A. Yasutake, *Carbon*, 2000, **38**, 227–239.
88. Y. Long, C. Zhang, X. Wang, J. Gao, W. Wang and Y. Liu, *J. Mater. Chem.*, 2011, **21**, 13934–13941.
89. A. K. Dalai and E. L. Tollefson, *Can. J. Chem. Eng.*, 1998, **76**, 902–914.
90. T. J. Bandosz and Q. Le, *Carbon*, 1998, **36**, 39–44.
91. M. Seredych and T. J. Bandosz, *Ind. Eng. Chem. Res.*, 2006, **45**, 3658–3665.
92. R. C. Bansal, J.-B. Donnet and F. Stoeckli, eds., *Active Carbon*, Marcel Dekker, New York, 1988.
93. B. R. Puri, B. Kumar and K. C. Kalra, *Indian J. Chem.*, 1971, **9**, 970.
94. O. C. Cariaso and P. L. Walker Jr, *Carbon*, 1975, **13**, 233–239.
95. M. Steijns, F. Derks, A. Verloop and P. Mars, *J. Catal.*, 1976, **42**, 87–95.
96. K. Hedden, L. Humber and B. R. Rao, *Adsorptive reinigung von schwefel was ser stoffhaltigen abgasen*, VDI-Verlag, Düsseldorf, 1976.

97. J. Klein and K.-D. Henning, *Fuel*, 1984, **63**, 1064–1067.
98. A. Primavera, A. Trovarelli, P. Andreussi and G. Dolcetti, *Appl. Catal., A*, 1998, **173**, 185–192.
99. H. Katoh, I. Kuniyoshi, M. Hirai and M. Shoda, *Appl. Catal., B*, 1995, **6**, 255–262.
100. X. Wu, A. K. Kercher, V. Schwartz, S. H. Overbury and T. R. Armstrong, *Carbon*, 2005, **43**, 1087–1090.
101. S. Bashkova, F. S. Baker, X. Wu, T. R. Armstrong and V. Schwartz, *Carbon*, 2007, **45**, 1354–1363.
102. V. V. Shinkarev, V. B. Fenelonov and G. G. Kuvshinov, *Carbon*, 2003, **41**, 295–302.
103. G. G. Kuvshinov, V. V. Shinkarev, A. M. Glushenkov, M. N. Boyko and D. G. Kuvshinov, *China Particuol.*, 2006, **4**, 70–72.
104. J.-H. Yang, G. Sun, Y. Gao, H. Zhao, P. Tang, J. Tan, A.-H. Lu and D. Ma, *Energy Environ. Sci.*, 2013, **6**, 793–798.
105. S. Song, H. Yang, R. Rao, H. Liu and A. Zhang, *Catal. Commun.*, 2010, **11**, 783–787.
106. Y. Zhao, W.-f. Chen, C.-f. Yuan, Z.-y. Zhu and L.-f. Yan, *Chin. J. Chem. Phys.*, 2012, **25**, 335–338.
107. W. Peng, S. Liu, H. Sun, Y. Yao, L. Zhi and S. Wang, *J. Mater. Chem. A*, 2013, **1**, 5854–5859.
108. H. Sun, S. Liu, G. Zhou, H. M. Ang, M. O. Tadé and S. Wang, *ACS Appl. Mater. Interfaces*, 2012, **4**, 5466–5471.
109. F. J. Beltrán, F. J. Rivas, L. A. Fernández, P. M. Álvarez and R. Montero-de-Espinosa, *Ind. Eng. Chem. Res.*, 2002, **41**, 6510–6517.
110. P. C. C. Faria, J. J. M. Órfão and M. F. R. Pereira, *Appl. Catal., B*, 2008, **79**, 237–243.
111. J. Rivera-Utrilla and M. Sánchez-Polo, *Appl. Catal., B*, 2002, **39**, 319–329.
112. M. Santiago, F. Stüber, A. Fortuny, A. Fabregat and J. Font, *Carbon*, 2005, **43**, 2134–2145.
113. H. T. Gomes, B. F. Machado, A. Ribeiro, I. Moreira, M. Rosário, A. M. T. Silva, J. L. Figueiredo and J. L. Faria, *J. Hazard. Mater.*, 2008, **159**, 420–426.
114. Â. C. Apolinário, A. M. T. Silva, B. F. Machado, H. T. Gomes, P. P. Araújo, J. L. Figueiredo and J. L. Faria, *Appl. Catal., B*, 2008, **84**, 75–86.
115. F. Stüber, J. Font, A. Fortuny, C. Bengoa, A. Eftaxias and A. Fabregat, *Top. Catal.*, 2005, **33**, 3–50.
116. L. C. A. Oliveira, C. N. Silva, M. I. Yoshida and R. M. Lago, *Carbon*, 2004, **42**, 2279–2284.
117. V. P. Santos, M. F. R. Pereira, P. C. C. Faria and J. J. M. Órfão, *J. Hazard. Mater.*, 2009, **162**, 736–742.
118. A. G. Gonçalves, J. L. Figueiredo, J. J. M. Órfão and M. F. R. Pereira, *Carbon*, 2010, **48**, 4369–4381.
119. S. Yang, X. Li, W. Zhu, J. Wang and C. Descorme, *Carbon*, 2008, **46**, 445–452.

120. Z.-Q. Liu, J. Ma, Y.-H. Cui and B.-P. Zhang, *Appl. Catal., B*, 2009, **92**, 301–306.
121. M. Soria-Sánchez, E. Castillejos-López, A. Maroto-Valiente, M. F. R. Pereira, J. J. M. Órfão and A. Guerrero-Ruiz, *Appl. Catal., B*, 2012, **121–122**, 182–189.
122. S. Yang, W. Zhu, X. Li, J. Wang and Y. Zhou, *Catal. Commun.*, 2007, **8**, 2059–2063.
123. S. Yang, X. Wang, H. Yang, Y. Sun and Y. Liu, *J. Hazard. Mater.*, 2012, **233–234**, 18–24.
124. C. Aguilar, R. García, G. Soto-Garrido and R. Arriagada, *Appl. Catal., B*, 2003, **46**, 229–237.
125. S. Morales-Torres, A. M. T. Silva, A. F. Pérez-Cadenas, J. L. Faria, F. J. Maldonado-Hódar, J. L. Figueiredo and F. Carrasco-Marín, *Appl. Catal., B*, 2010, **100**, 310–317.
126. T. Cordero, J. Rodríguez-Mirasol, J. Bedia, S. Gomis, P. Yustos, F. García-Ochoa and A. Santos, *Appl. Catal., B*, 2008, **81**, 122–131.
127. R. P. Rocha, J. P. S. Sousa, A. M. T. Silva, M. F. R. Pereira and J. L. Figueiredo, *Appl. Catal., B*, 2011, **104**, 330–336.
128. G. Ovejero, J. L. Sotelo, M. D. Romero, A. Rodríguez, M. A. Ocaña, G. Rodríguez and J. García, *Ind. Eng. Chem. Res.*, 2006, **45**, 2206–2212.
129. A. Georgi and F.-D. Kopinke, *Appl. Catal., B*, 2005, **58**, 9–18.
130. L. B. Khalil, B. S. Girgis and T. A. M. Tawfik, *J. Chem. Technol. Biotechnol.*, 2001, **76**, 1132–1140.
131. H.-H. Huang, M.-C. Lu, J.-N. Chen and C.-T. Lee, *Chemosphere.*, 2003, **51**, 935–943.
132. X. Hu, B. Liu, Y. Deng, H. Chen, S. Luo, C. Sun, P. Yang and S. Yang, *Appl. Catal., B*, 2011, **107**, 274–283.
133. J. L. Sabourin, D. M. Dabbs, R. A. Yetter, F. L. Dryer and I. A. Aksay, *ACS Nano.*, 2009, **3**, 3945–3954.
134. L.-M. Liu, R. Car, A. Selloni, D. M. Dabbs, I. A. Aksay and R. A. Yetter, *J. Am. Chem. Soc.*, 2012, **134**, 19011–19016.
135. X.-k. Kong, Q.-w. Chen and Z.-y. Lun, *J. Mater. Chem. A*, 2014, **2**, 610–613.
136. B. Xia, J. Phillips, C.-K. Chen, L. R. Radovic, I. F. Silva and J. A. Menéndez, *Energy Fuels*, 1999, **13**, 903–906.
137. J. A. Menéndez, J. Phillips, B. Xia and L. R. Radovic, *Langmuir*, 1996, **12**, 4404–4410.
138. J. N. Armor, *Appl. Catal., B*, 1992, **1**, 221–256.
139. K. Knoblauch, E. Richter and H. Jüntgen, *Fuel*, 1981, **60**, 832–838.
140. I. Mochida, M. Ogaki, H. Fujitsu, Y. Komatsubara and S. Ida, *Fuel*, 1983, **62**, 867–868.
141. S. N. Ahmed, J. M. Stencel, F. J. Derbyshire and R. M. Baldwin, *Fuel Process. Technol.*, 1993, **34**, 123–136.
142. T. Suzuki, T. Kyotani and A. Tomita, *Ind. Eng. Chem. Res.*, 1994, **33**, 2840–2845.

143. B. J. Ku, J. K. Lee, D. Park and H.-K. Rhee, *Ind. Eng. Chem. Res.*, 1994, **33**, 2868–2874.
144. S. N. Ahmed, R. Baldwin, F. Derbyshire, B. McEnaney and J. Stencel, *Fuel*, 1993, **72**, 287–292.
145. Y. Komatsubara, S. Ida, H. Fujitsu and I. mochida, *Fuel*, 1984, **63**, 1738–1742.
146. I. Mochida, M. Ogaki, H. Fujitsu, Y. Komatsubara and S. Ida, *Fuel*, 1985, **64**, 1054–1057.
147. H. Teng, Y.-F. Hsu and Y.-T. Tu, *Appl. Catal., B*, 1999, **20**, 145–154.
148. H. Teng, Y.-T. Tu, Y.-C. Lai and C.-C. Lin, *Carbon*, 2001, **39**, 575–582.
149. A. Corma and H. Garcia, *Adv. Synth. Catal.*, 2006, **348**, 1391–1412.
150. A. Corma and H. García, *Chem. Rev.*, 2003, **103**, 4307–4366.
151. D. E. De Vos, M. Dams, B. F. Sels and P. A. Jacobs, *Chem. Rev.*, 2002, **102**, 3615–3640.
152. J. A. Melero, R. van Grieken and G. Morales, *Chem. Rev.*, 2006, **106**, 3790–3812.
153. J. Lee, O. K. Farha, J. Roberts, K. A. Scheidt, S. T. Nguyen and J. T. Hupp, *Chem. Soc. Rev.*, 2009, **38**, 1450–1459.
154. A. Dhakshinamoorthy, M. Alvaro and H. Garcia, *Chem. Commun.*, 2012, **48**, 11275–11288.
155. K. Nakajima and M. Hara, *ACS Catal.*, 2012, **2**, 1296–1304.
156. J. Li, J. Feng, M. Li, Q. Wang, Y. Su and Z. Jia, *Solid State Sci.*, 2013, **21**, 1–5.
157. T. Yashima, H. Suzuki and N. Hara, *J. Catal.*, 1974, **33**, 486–492.
158. C. Moreno-Castilla, F. Carrasco-Marín, C. Parejo-Pérez and M. V. López Ramón, *Carbon*, 2001, **39**, 869–875.
159. F. Carrasco-Marín, A. Mueden and C. Moreno-Castilla, *J. Phys. Chem. B*, 1998, **102**, 9239–9244.
160. G. S. Szymański, Z. Karpiński, S. Biniak and A. Świątkowski, *Carbon*, 2002, **40**, 2627–2639.
161. G. S. Szymański and G. Rychlicki, *Carbon*, 1991, **29**, 489–498.
162. G. S. Szymański and G. Rychlicki, *Carbon*, 1993, **31**, 247–257.
163. S. M. Singh Chauhan and S. Mishra, *Molecules*, 2011, **16**, 7256–7266.
164. F. Liu, J. Sun, L. Zhu, X. Meng, C. Qi and F.-S. Xiao, *J. Mater. Chem.*, 2012, **22**, 5495–5502.
165. E. Lam, J. H. Chong, E. Majid, Y. Liu, S. Hrapovic, A. C. W. Leung and J. H. T. Luong, *Carbon*, 2012, **50**, 1033–1043.
166. A. Shaabani, M. Mahyari and F. Hajishaabanha, *Res. Chem. Intermed.*, 2013, 1–12.
167. P. P. Upare, J.-W. Yoon, M. Y. Kim, H.-Y. Kang, D. W. Hwang, Y. K. Hwang, H. H. Kung and J.-S. Chang, *Green Chem.*, 2013, **15**, 2935–2943.
168. A. Dhakshinamoorthy, M. Alvaro, P. Concepcion, V. Fornes and H. Garcia, *Chem. Commun.*, 2012, **48**, 5443–5445.
169. Y. Kim, S. Some and H. Lee, *Chem. Commun.*, 2013, **49**, 5702–5704.
170. H. Yu, X. Wang, Y. Zhu, G. Zhuang, X. Zhong and J.-g. Wang, *Chem. Phys. Lett.*, 2013, **583**, 146–150.

171. B. Basu, S. Kundu and D. Sengupta, *RSC Adv.*, 2013, **3**, 22130–22134.
172. Y. Li, Q. Zhao, J. Ji, G. Zhang, F. Zhang and X. Fan, *RSC Adv.*, 2013, **3**, 13655–13658.
173. C. Yuan, W. Chen and L. Yan, *J. Mater. Chem.*, 2012, **22**, 7456–7460.
174. T. Wu, X. Wang, H. Qiu, J. Gao, W. Wang and Y. Liu, *J. Mater. Chem.*, 2012, **22**, 4772–4779.
175. Y. Song, K. Qu, C. Zhao, J. Ren and X. Qu, *Adv. Mater.*, 2010, **22**, 2206–2210.
176. H. Huang, J. Huang, Y.-M. Liu, H.-Y. He, Y. Cao and K.-N. Fan, *Green Chem.*, 2012, **14**, 930–934.
177. C. Su, M. Acik, K. Takai, J. Lu, S.-j. Hao, Y. Zheng, P. Wu, Q. Bao, T. Enoki, Y. J. Chabal and K. Ping Loh, *Nat. Commun.*, 2012, **3**, 1298.
178. D. Ugarte, A. Châtelain and W. A. de Heer, *Science*, 1996, **274**, 1897–1899.
179. E. E. Santiso, A. M. George, C. H. Turner, M. K. Kostov, K. E. Gubbins, M. Buongiorno-Nardelli and M. Sliwinska-Bartkowiak, *Appl. Surf. Sci.*, 2005, **252**, 766–777.
180. P. Serp and E. Castillejos, *ChemCatChem.*, 2010, **2**, 41–47.
181. X. Pan and X. Bao, *Acc. Chem. Res.*, 2011, **44**, 553–562.
182. C. H. Turner, J. K. Brennan, J. Pikunic and K. E. Gubbins, *Appl. Surf. Sci.*, 2002, **196**, 366–374.
183. T. Lu, E. M. Goldfield and S. K. Gray, *J. Phys. Chem. C*, 2008, **112**, 2654–2659.
184. E. E. Santiso, M. Buongiorno Nardelli and K. E. Gubbins, *J. Chem. Phys.*, 2008, **128**, 034704.
185. M. D. Halls and H. B. Schlegel, *J. Phys. Chem. B*, 2002, **106**, 1921–1925.
186. P. Kondratyuk and J. T. Yates, *J. Am. Chem. Soc.*, 2007, **129**, 8736–8739.
187. F. Zeng, Z. Sun, X. Sang, D. Diamond, K. T. Lau, X. Liu and D. S. Su, *ChemSusChem.*, 2011, **4**, 1587–1591.
188. D. Chen, H. Feng and J. Li, *Chem. Rev.*, 2012, **112**, 6027–6053.
189. L. Tang, Y. Wang, Y. Li, H. Feng, J. Lu and J. Li, *Adv. Funct. Mater.*, 2009, **19**, 2782–2789.
190. I. Morcos and E. Yeager, *Electrochim. Acta.*, 1970, **15**, 953–975.
191. E. Yeager, *J. Mol. Catal.*, 1986, **38**, 5–25.
192. C. Paliteiro, A. Hamnett and J. B. Goodenough, *J. Electroanal. Chem. Interfacial Electrochem.*, 1987, **233**, 147–159.
193. M. S. Hossain, D. Tryk and E. Yeager, *Electrochim. Acta.*, 1989, **34**, 1733–1737.
194. A. Sarapuu, K. Vaik, D. J. Schiffrin and K. Tammeveski, *J. Electroanal. Chem.*, 2003, **541**, 23–29.
195. K. Vaik, A. Sarapuu, K. Tammeveski, F. Mirkhalaf and D. J. Schiffrin, *J. Electroanal. Chem.*, 2004, **564**, 159–166.
196. M. Zhang, Y. Yan, K. Gong, L. Mao, Z. Guo and Y. Chen, *Langmuir*, 2004, **20**, 8781–8785.
197. K. Matsubara and K. Waki, *Electrochem. Solid-State Lett.*, 2010, **13**, F7–F9.

198. S. Wang, S. Dong, J. Wang, L. Zhang, P. Han, C. Zhang, X. Wang, K. Zhang, Z. Lan and G. Cui, *J. Mater. Chem.*, 2012, **22**, 21051–21056.
199. D. Deng, L. Yu, X. Pan, S. Wang, X. Chen, P. Hu, L. Sun and X. Bao, *Chem. Commun.*, 2011, **47**, 10016–10018.
200. T.-F. Yeh, J.-M. Syu, C. Cheng, T.-H. Chang and H. Teng, *Adv. Funct. Mater.*, 2010, **20**, 2255–2262.
201. H.-C. Hsu, I. Shown, H.-Y. Wei, Y.-C. Chang, H.-Y. Du, Y.-G. Lin, C.-A. Tseng, C.-H. Wang, L.-C. Chen, Y.-C. Lin and K.-H. Chen, *Nanoscale*, 2013, **5**, 262–268.

CHAPTER 6

Doped Nanostructured Carbon Materials as Catalysts

6.1 Introduction

Metallic nanoparticles are undoubtedly the most vibrant catalysts due to their good performance in a series of important chemical reactions in modern chemistry. However, their use on an industrial scale is restricted by limited reserves, high cost and low stability. In moving towards green and sustainable chemistry, carbon nanostructured materials without metal elements have been explored and studied extensively in catalysis.

Heteroatom doped carbon materials represent one of the most prominent families of materials that are used in energy related applications, such as fuel cells, batteries, hydrogen storage or super-capacitors.[1–3] Doping substantially modifies the atomic scale structures, surface energy, chemical reactivity and mechanical properties of carbon nanomaterials.[4,5] While doping carbons with nitrogen atoms has seen great progress throughout the past decades and yielded promising material concepts, other doping candidates such as boron, phosphorus or sulfur have gathered increasing interest over the last few years. Boron is already widely studied, and as its electronic situation is contrary to the one of nitrogen, co-doping carbons with both heteroatoms can probably create synergistic effects. Sulfur and phosphorus have just recently entered the world of carbon synthesis, but already several studies published prove their potential, especially as electrocatalysts in the cathodic compartment of fuel cells. Given that their size and electronegativity are lower than those of carbon, structural distortions and changes of the charge densities are induced in the carbon materials.

RSC Catalysis Series No. 23
Nanostructured Carbon Materials for Catalysis
By Philippe Serp and Bruno Machado

Published by the Royal Society of Chemistry, www.rsc.org

Heteroatoms can be incorporated in the graphitic lattice either during synthesis or with post-synthetic treatments. Typical synthetic protocols for the heteroatom doping of carbon nanomaterials are summarized in the next section. Nevertheless, the objective is not to provide a detailed view of all the techniques used, but instead give an insight into the most common techniques currently used to introduce different heteroatoms in the carbon structure.

6.2 Synthesis of Doped Nanostructured Carbon Materials

Several strategies, including *in situ* doping and post-treatment have been proposed to introduce heteroatoms into the carbon framework.[5–7] In the *in situ* doping route there is a direct carbonization of heteroatom-containing species (*e.g.* polymers or ionic liquids), and includes chemical vapor deposition, segregation growth, solvothermal, and arc-discharge approaches. The heteroatom source for the *in situ* doping should be carefully selected by considering its dissociation temperature, which is usually desired to be close to the carbon growth temperature.[8] During the doping process, there is a competition between C–C and C–heteroatom bonding. This competition is strongly influenced by the synthesis temperature. A high temperature generally favors C–C bonds.[8] Amongst several N-doping sources, NH_3 is the most common choice for N-doping nanostructured carbon materials,[9–12] whereas diborane,[13,14] triphenyl phosphine,[15,16] and elemental sulfur/thiophene[17,18] have been found to be good candidates for B-, P- and S-doping, respectively. The *in situ* heteroatom doping of CNTs can induce dramatic changes in the tubular structure, with N-doping leading to bamboo-like features,[19] S-doping causing branching,[20] and B-doping giving rise to knee-like bending.[21]

In the post-treatment procedure, the carbon matrix is subjected to certain chemical agents (*i.e.* H_2S, NH_3, *etc.*) at a relatively high temperature, comprising thermal and plasma methodologies. Given that doping of chemically stable pristine graphitic plane is generally difficult, more-reactive oxidized nanostructured carbon materials are frequently employed as the starting material for effective doping.[22–26] The high-temperature annealing restores the graphitic structure by removing oxygen functional groups and healing the structural disorder.[27] Furthermore, if organic substances containing heteroatoms are used simultaneously as carbon precursors, not only is the surface functionalized, but also bulk-doped carbons are obtained without the need for additional steps.

The different types of functionalities (containing N, B, S and P) present on the carbon surface, as well as the most commonly used characterization techniques for their identification and quantification, are discussed in detail within Chapter 4.

6.2.1 Nitrogen-Doped Nanostructured Carbon Materials

In order to control the type and amount of different heteroatoms present, different methods using different precursors have been developed.[28] Nitrogen-containing functional groups can be introduced through either reaction with nitrogen-containing reagents (such as NH_3, nitric acid and amines) or carbonization/activation of nitrogen-rich carbon precursors, such as polyacrylonitrile, melamine, quinoline pitch, urea-polymer or ionic liquids.[5,29–32]

Nitrogen-containing nanostructured carbon materials are commonly obtained through impregnation of nitrogen-compounds or reaction with NH_3 at high temperature. Substitution of carbon by nitrogen in MWCNTs was accomplished as early as 1997.[33] This was done by the pyrolysis of aza-aromatic compounds such as pyridine and triazine over cobalt nanoparticles in an Ar atmosphere. The nitrogen content decreased with increasing temperature of pyrolysis, the estimated nitrogen content being around 5, 3.5, and 3 at% in nanotubes prepared by the pyrolysis of pyridine at 973, 1123 and 1273 K, respectively. However, the low content and the instability of nitrogen-containing species can pose a problem. This may be due to either an insufficient number of defects on the carbon matrix, or the high annealing temperature, which will break the C–N bonds. In addition, the nitrogenation of porous carbon affects both its surface chemical nature and porous structure at high temperatures.

Another common approach to N-doping is the pyrolysis or chemical vapor deposition of nitrogen and carbon containing precursors, such as heterocycles, melamine or aminated sugars, by which a direct incorporation of the nitrogen atoms into the forming carbon backbone becomes possible.[34,35] The nitrogen incorporated into the carbon matrix endows some special properties to the porous carbons such as semiconductor characteristics, pseudocapacitance and oxidation/reduction activity. Unfortunately, the pore structure is difficult to control during the physical and/or chemical activation. Although high nitrogen contents (*ca.* 16 at%) have been reported,[9] the amounts obtained with CVD are normally around 4–9 at%. The bonding configuration of nitrogen within *e.g.*, N-graphene, varies with different studies. By using Cu as the catalyst and CH_4/NH_3 (1 : 1) as the precursor, the nitrogen type in N-graphene is mainly quaternary-N;[11] however, when Ni and CH_4/NH_3 (5 : 1) are used, the obtained N-graphene consists of mainly pyridinic N and pyrrolic N.[36] If C_2H_4/NH_3 is used as precursor while keeping Cu as catalyst, then pyridinic N becomes the predominant type.[9]

Other less conventional routes consist of hydrothermal carbonization for deriving carbonaceous materials from carbohydrate rich biomass.[37,38] Using nitrogen-containing biomass as precursor and treating it hydrothermally yields nitrogen-containing carbonaceous materials that offer different possibilities for further treatments and applications. Another approach to prepare N-doped carbons is based on the thermal stability of ionic liquids. Given that some ionic liquids do not decompose completely to volatile products under an inert gas, these ionic liquids can be used as a nitrogen-doped carbon sources.[39–42] In addition, some supra-molecular types of ionic liquids can also form systems suitable for N-doped carbon synthesis.[43]

When carbon materials are placed in a nitrogen plasma atmosphere, carbon atoms will be partly replaced by nitrogen atoms. Hence, this method can be used to prepare N-doped carbon materials, such as CNTs or graphene.[22,44–46] The nitrogen content, which can be controlled by the plasma strength and exposure time, can vary from 3 to 8.5 at%.

Finally, a simple way to produce polymer-based carbons with specific nitrogen functions in controlled concentrations is to prepare phenol–formaldehyde resins in which part of the phenol is substituted by aniline (for amine-type nitrogen) or 3-hydroxypyridine (for pyridine-type nitrogen) or tyrosine (for both types of nitrogen).[47] The main advantage of using polymers as carbon precursors is that the added nitrogen is normally homogeneously distributed. Melamine-formaldehyde resins are very rich in nitrogen, and chars and carbons produced by pyrolysis retain a high nitrogen content at high temperatures. With commercial melamine foam, 13% N was found after decomposition at 1273 K, and about 5% at 1473 K.[48]

6.2.2 Boron-Doped Carbons

Boron is an element with unique properties within the periodic table and it is a highly interesting candidate for carbon materials doping, modifying the properties of pristine carbons. Chemical substitution brings about significant changes in carbon materials as the presence of boron or nitrogen in carbon nanostructures renders them p- or n-type, respectively. Early works on boron doped carbons were inspired by the fascination of stoichiometric boron nitride compounds that can form hexagonal patterns enabling sp^2-carbon-related structures, such as stacked sheets or nanotubes.[49,50] SWCNTs of up to 15 at% boron doping can be obtained by a substitution reaction, in which a mixture of B_2O_3 and SWCNTs is heated in an NH_3 atmosphere to achieve high doping.[51] B-concentration affects both the nature and growth rates of nanotube structures grown by the laser ablation of Co/Ni/B-doped carbon targets.[52] Targets with low B-concentration (below 3 at%) produce ropes of SWCNTs with no detectable boron. At higher concentration (above 3 at%), B poisons the Co–Ni catalyst particles and prevents the growth of SWCNTs. Furthermore, high B-concentrations lead to increased disorder in the graphene sheets, and a small proportion of DWCNTs can be observed.

Boron modification, and also synthesis of B- and N-co-doped carbons, is usually performed by the heat treatment of carbon precursors with boron[53] or boric acid,[54] arc discharge approaches, or by direct chemical vapor deposition of gas-phase carbon and boron sources.[13,55–58]

6.2.3 Phosphorus-Doped Carbons

Phosphorus has the same number of valence electrons as nitrogen and often shows similar chemical properties. Nevertheless, it has a larger atomic radius and higher electron-donating ability, which make it an excellent selection as dopant for carbon materials. Research related to phosphorus doping has focused mainly on the preparation of P-doped carbon materials with diamond sp^3-like binding motifs.[59–62] Regarding carbon materials with sp^2-hybridised binding motifs, examples of P doping involve mainly thermal annealing with triphenylphosphine as the P source.[63–66] As an example, phosphorus-doped ordered mesoporous carbons (POMCs) can be prepared using a metal-free nanocasting method with SBA-15 mesoporous silica and triphenylphosphine and phenol as phosphorus and carbon sources, respectively (Figure 6.1a).[67] Three differently sized POMCs were prepared by altering the SBA-15 templates with different rod lengths. Uniform distributions of nearly spherical carbon microstructures (Figure 6.1b), with the formation of highly ordered uniform pore distribution (Figure 6.1c) were obtained. The resulting P-doped mesoporous carbon had a small amount of P (less than 1.5 at%).

Computational studies predict that P insertion will influence the band-gap of graphene to a greater extent than that calculated for sulfur, while also being energetically more favorable.[68,69] Furthermore, calculations have also shown that P doping improves the electron-donor properties of a carbon material.[70] Apart from carbon materials doped exclusively with phosphorus, some effort has also been put on the design of materials using phosphorus as a co-dopant, mostly in addition to nitrogen, through pyrolysis mixtures of dicyandiamide and phosphoric acid.[71,72]

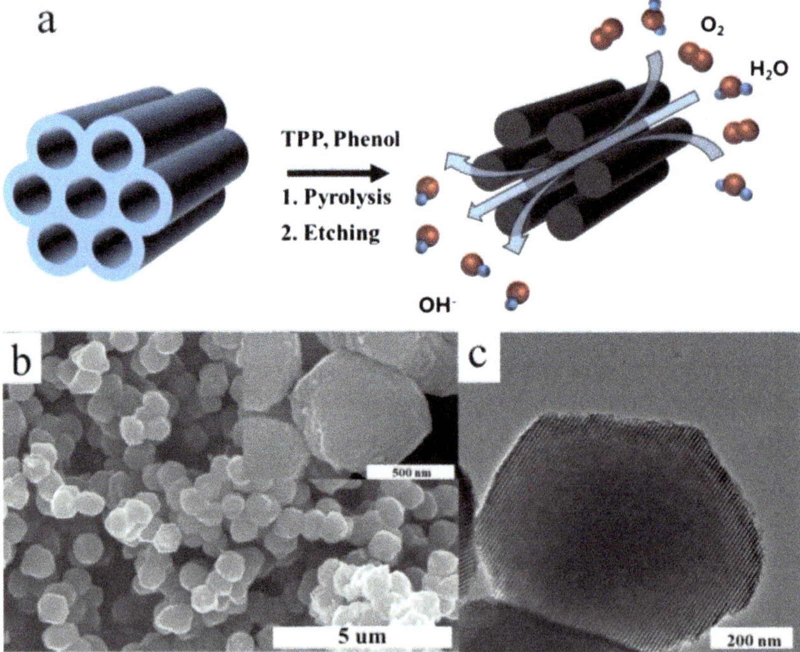

Figure 6.1 (a) Schematic illustration of the preparation of P-doped ordered mesoporous carbons; (b) FESEM image with an inset showing magnified image and (c) TEM image of 0.7 μm length P-OMC. Reprinted with permission from ref. 67. Copyright 2012 American Chemical Society.

6.2.4 Sulfur-Doped Carbons

In comparison to boron, sulfur doping in carbon materials is still quite rare and represents an emerging field within carbon material research.[73] While today the potential of these materials for energy applications is continuously revealed and exploited, only a few years ago, little was known about such sulfur-doped carbonaceous species. Carbon bulk doping with S alters not only the surface chemistry, and thus adsorption, electrochemical and catalytic properties, but also modifies all the fundamental physicochemical properties, *i.e.* induces semi-conductivity, unusual magnetic behavior or increased catalytic photoactivity.[4] Sulfur-doped carbons are usually based on sulfur-rich precursors (hydrogen sulfide, benzyl disulfide) and can be produced *via* pyrolysis.[25,74,75] In fact, thermolysis of heteroatom rich precursors (usually under the flow of inert gas) has been routinely performed to prepare doped carbonaceous materials, *e.g.* using ionic liquids.[76] Liquid salts with dicyanamide anions are an especially versatile precursor for N-doped carbons, and for co-doping with S.[7,77,78]

A facile synthesis process for the preparation of S and N co-doped carbon catalysts with uniform nanospherical morphologies, using polyacrylonitrile nanospheres and sulfur as precursors is presented on Figure 6.2.[79] Samples with sulfur contents between 4.3 and 9.5 wt% were obtained, with sulfur existing both as thiophene-S and oxidized-S.

The sol-gel approach has also been used to synthesize sulfur-doped carbon xerogels *via* polymerization of resorcinol and 2-thiophenecarboxaldehyde.[80] This methodology allows for a conveniently controlled texture and morphology by a simple process modification of the sol-gel conditions, while introducing significant amounts of sulfur (as high as 19.6 wt%, dropping to 14% after calcination at 873 K).

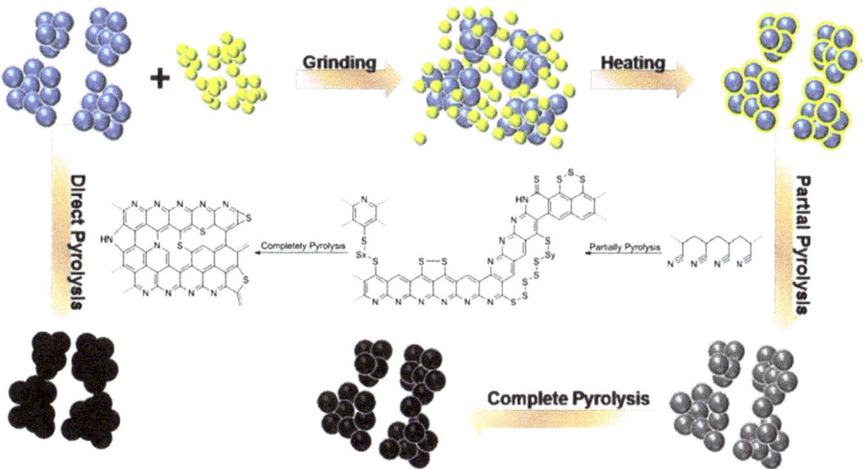

Figure 6.2 Schematic representation of the fabrication process for S and N co-doped carbon nanospheres. Reprinted with permission from ref. 79. Copyright 2013 Elsevier Science Ltd.

6.3 Catalytic Role of Nanostructured Carbon Dopants

Most pristine nanostructured carbon materials are chemically inert and functionalization or surface modification is crucial to render them chemically active. Doping the nanostructured carbon materials with foreign atoms, which are able to create superficial defects and break the chemical inertness of pure nanostructured carbon materials, represents a feasible path to reactivity for a large number of applications.[29,36,81,82] As dopant atoms on a graphitic lattice act as a Brönsted–Lowry acid or base, they become charged moieties when protonated or deprotonated. Typically, for pristine graphitic carbons, the pH_{PZC} value is generally near or below 7, which is dominated by the weak acidic oxygen functionalities imparted from the exposure to ambient conditions.[83] Although, there is a general agreement about the type of surface functionalities that determine the acidic character of a carbon material, the nature of carbon basic surfaces remains controversial and open to investigation.[84] Generally speaking, oxygen-containing functionalities (*i.e.*, chromene, pyrone, ketones, quinones) and non-heteroatomic Lewis base sites, characterized by regions of electron density on the carbon basal planes, govern carbon basicity.[85,86] In addition, air oxidation yields carbons possessing a less acidic surface and consequently, more easily distinguishable amphoteric properties, than those of carbons oxidized with nitric acid. It has been demonstrated that oxidation of activated carbon in the gas phase increases mainly the concentration of hydroxyl and carbonyl surface groups.[86] Oxidation of carbons with hot air results in a greater proportion of relatively weak acidic surface functional groups (*i.e.*, phenolic).[87]

In contrast, 4.0% N-doping induces an alkaline pH_{PZC} value close to 9 due to the prevalence of positively charged N-dopants. This variation in the surface properties of heteroatom doped graphitic carbons can be exploited for enhanced chemical reactivity. The relative abundance and chemical environment of N-containing groups (pyridinic, pyrrolic, quaternary and nitrogen oxides) is expected to determine the carbons' acid/base properties. Therefore, it is imperative to understand the electron donor/acceptor properties of these groups. Of the most common, the sp^2 hybridized quaternary-N, substituting a carbon atom of the graphitic matrix, is under-coordinated and contributes with one electron to the conjugated π system of the aromatic carbon matrix. This results in a delocalization of the electrons of the nitrogen atoms. Subsequently, removing an electron from this nitrogen, *i.e.* the uptake of H^+ by the nitrogen atom, will decrease the aromaticity of the system, which is energetically unfavorable. These nitrogen groups are not expected to display any basic behavior, and the quaternary nitrogen does not contribute to the basicity of the N-CNTs. The sp^3 hybridized nitrogen atom in pyrrole is also part of an aromatic network. Protonation of the nitrogen atom requires electron donation to the H^+ ion, and results in a loss of aromaticity of the five-fold ring, which is energetically not favorable. Therefore, pyrrole will have an acidic character, as was also experimentally established.[88] The N–O species, essentially being an oxidized form of pyridinic N, have also been reported to present an acidic character.[89] pK_a values below 3 were found for pyridine N-oxides and other comparable compounds, determined from protonation experiments. With its electron pair, pyridinic nitrogen can act as a Lewis and/or a Brønsted base, and thus can interact with a proton. Therefore, of all the main N-type groups, only the pyridinic ones are expected to present a basic character. Nevertheless, the presence of amine species cannot be ruled out. Several authors reported on amine species found in nitrogen incorporated in carbon by XPS analysis with a binding energy around 399.4 eV.[90] With XPS however, the amine peak is difficult to distinguish from the adjacent overlapping peaks, *viz.* pyridinic N and pyrrolic N.

Substitution in nanostructured carbon materials has drawn substantial interest due the increased reactivity of hetero-nanotubes to other molecules.[91] The most important and, at the same time, the most abundant heteroatom that affects the use of nanostructured carbon materials is oxygen. It is usually either chemisorbed on the surface or arranged in the form of functional groups analogous to those existing in organic chemistry (see Chapter 4). Since those groups increase the reactivity of the otherwise hydrophobic carbon surface, oxidation often opens the door to further modifications *via* interactions of specific chemicals with the groups and their reactions on the surface. An example is the introduction of nitrogen to the carbon matrix, which is more efficient when the carbon materials are pre-oxidized.[70,92,93] The nitrogen doping can be seen as regular defects, which change the chemical behavior of the material. As an example, the reactivity of doped nanotubes can be estimated to be higher than non-substituted carbon nanotubes of same diameters,[91] with theoretical calculations predicting a localization of the unpaired electrons around the nitrogen-defect in the semiconducting hetero-nanotubes. Boron and nitrogen are the natural choices for doping since they differ only by one valence electron compared to carbon atoms. The boron and nitrogen serve as acceptors and donors of electrons, respectively, since boron has one electron less, and nitrogen has one electron more than the carbon atom. Since the atomic radii are similar to the carbon atom, they only create a small perturbation in the structure in comparison to a perfect one. Other heteroatoms such as hydrogen, phosphorus and sulfur can also be chemisorbed over a large number of imperfections and defects leading to stable surface compounds.[94]

For the sake of clarity, for carbon tubular nanostructures three main categories of doping are now well established, *i.e.*, endohedral, exohedral and in-plane doping. Endohedral doping implies encapsulation, exohedral doping stands for intercalation, and substitutional doping is an in-plane replacement of carbon in the graphene sheet (Figure 6.3).[95] Some atoms or molecules can also be encapsulated in the hollow core of the tubes (endohedral doping) or trapped within bundles intercalated between the outer shells of the tubes (exohedral doping). Substitutional doping is mainly related to the removal of C atoms from the crystalline structure of a pristine CNT. This implies either a direct substitution of one C atom by a heteroatom, or a multiple removal of C and incorporating heteroatoms in their place. This second possibility

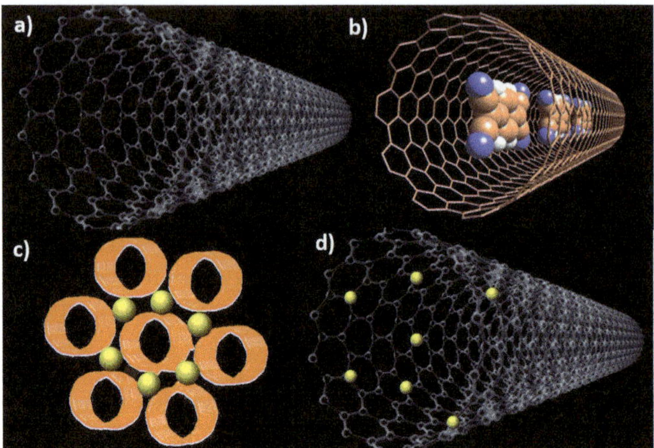

Figure 6.3 Different ways to modify the electronic properties of pristine nanotubes (a): endohedral doping filling with crystals or fullerenes (so-called peapods) (b), intercalation within SWCNTs in a bundle (c), and formation of hetero-nanotubes (with substitutional atoms) (d).

involves the elimination of C atoms from the CNT walls inducing structural defects affecting the CNT properties in distinctive manners.

The incorporation of N atoms into the carbon matrix can lead to three main types of N formats, including graphitic-N with direct substitution structure, and pyridinic-N and pyrrolic-N structures.[96] Graphitic-N means the doping N atom is combined into a hexagonal ring. Pyridinic- and pyrrolic-N donate one and two p electrons to the π system, forming sp^2 and sp^3 hybridized bonds, respectively.[97] More interestingly, it was confirmed that N-doped carbon materials can improve their biocompatibility significantly.[98] Taking N doping of CNTs an example, the three-coordinated N atom within the sp^2-hybridized network (graphitic-N) will induce sharp localized states above the Fermi level due to the presence of additional electrons. These doped nanotubes exhibit n-type conduction, and are expected to strongly react with electron acceptor molecules. N-pyridinic groups induce the presence of localized states below and above the Fermi level. Therefore, substitutional N doping in SWCNTs should result in n-type conducting behavior, whereas pyridine-type N may produce either a p- or n-type conductor, depending on the level of doping, the number of N atoms and the number of removed C atoms within the hexagonal sheet.

6.4 Reactions Catalyzed by Heteroatom-Doped Nanostructured Carbon Materials

This section does not aim to give a complete list of all the reactions tested, but only selected examples, mainly with reference to industrially important reactions, in order to highlight how this class of doped materials represents a potential breakthrough. Doping of carbon nanomaterials have been shown to be a promising approach to the development of metal-free, carbon-based catalysts with a higher electrocatalytic activity and better long-term operation stability than that of commercially available platinum-based electrodes for oxygen reduction in fuel cells.

6.4.1 Electrocatalytic Oxygen Reduction

The oxygen reduction reaction (ORR) takes place at the cathode of a fuel cell and plays a key role in determining cell performance, cost and durability.[3,99] Currently, different metals are regarded as active and efficient for the ORR (see Chapter 9 for metal-based electrocatalysts). However, many issues, including the high cost, CO poisoning and sintering, tend to severely hinder its commercialization. Hence, much effort has been devoted to developing cheaper and more efficient ORR electrocatalysts, mainly with respect to three aspects: (i) increasing Pt efficiency;[100–102] (ii) developing non-precious-metal electrocatalysts;[103–106] and (iii) developing metal-free ORR electrocatalysts.[107–110] The latter has drawn great attention recently and has become a fast-growing branch of ORR electrocatalysis.[109,110]

The electrocatalytic ORR process involves various steps and multiple adsorbed intermediates. The reaction proceeds *via* an inner-sphere electron transfer mechanism, which means the activity and the overall ORR pathways are closely related to the catalyst surfaces.[111,112] The ORR taking place on carbon electrocatalysts without doping, functionalization or lattice substitution usually exhibits a two-plateau peroxide pathway (eqn (6.1)) rather than the one-plateau $4e^-$ pathway (eqn (6.2)) typically observed on Pt/C.[99] In order to produce higher voltage and higher power performances in alkaline fuel cells or metal-air batteries, the ORR catalysts are usually required to conduct the $4e^-$ pathway so that they can deliver large current densities and high onset potentials (the two-plateau peroxide pathway of ORR is often associated with a small current and low onset potential). Therefore, one major goal of the heterogeneous carbon ORR research is to explore efficient carbon catalysts favorable for the $4e^-$ ORR behavior. Properties of the nanocarbon surface, such as functional groups, heteroatoms, basicity/acidity, and hydrophilicity/hydrophobicity, are major parameters influencing the ORR behavior.[99]

$$O_2 + 2H_2O + 2e^- \rightarrow HO_2^- + OH^-$$
$$HO_2^- + H_2O + 2e^- \rightarrow 3OH^- \quad (6.1)$$
$$O_2 + 2H_2O + 4e^- \rightarrow 4OH^- \quad (6.2)$$

Compared with commercial Pt/C catalysts, carbon-based metal-free ORR electrocatalysts show high catalytic activities and perform better in terms of longevity, abundance of raw materials and CO tolerance. As is well known,

sp^2 carbon materials have abundant free-flowing p electrons, with the potential to act as catalysts for reactions such as the ORR. However, these p electrons need activation before they can be used. The most direct method is to insert a higher number of electrons into the carbon p orbitals by, for example, doping sp^2 carbon materials with electron-rich N.[108] The extra electrons donated by N increase the electron density and raise the highest occupied molecular orbital (HOMO) energy level of the sp^2 carbons, thus enabling the ORR. Interestingly, this delocalization of p electrons can also be achieved by conjugation with vacant orbitals, *i.e.*, doping with electron-deficient elements, such as B.[58] In this case, the vacant $2p_z$ orbital of B conjugates with the carbon π system to extract electrons, which then become active due to the low electronegativity of B. Moreover, the typically high specific surface areas of carbon nanomaterials also benefit the ORR performance by increasing the reduction current density.[113] Hence, despite the existence of various doping choices (N, B, O, P, S *etc.*), the idea to use heterogeneous carbon materials for ORR is based on breaking the electroneutrality of nanostructured carbon materials to create charged sites that are favorable for O_2 adsorption, regardless of whether the dopants are electron-rich or electron-deficient. In addition, there should be also an effective utilization of carbon π electrons for the O_2 reduction. The specific synergistic effect of different heteroatoms present in the nanocarbon matrix (N, O, B, P, S) on ORR is discussed over the next few sections.

6.4.1.1 *Nitrogen-Doped Carbon Nanostructures*

Nitrogen functionalization changes not only the carbon surface chemistry, such as basicity and hydrophilicity/hydrophobicity, but also the electronic structure by donating a lone pair of electrons and shifting the neutralized charge distribution on the undoped carbon lattice.[114] Strelko *et al.* suggested that N atoms can lower the band gap and increase the charge mobility of the graphitic lattice, and showed from their quantum chemical calculations that the pyrrole groups at the edge sites and those in combination with valley N atoms (graphitic-N) have the highest charge mobility.[70] These changes in the carbon band structure would eventually lower the electron work function at the interface of carbon with a liquid or gas compared to pure carbon surfaces. It was hypothesized that the altered distribution of charge density and spin density of carbon atoms and heteroatoms played a critical role in the chemisorption of oxygen molecules and the first electron transfer (rate determining step in ORR).[3,99]

The study of N-doped metal-free ORR electrocatalysts originates from studies based on TM-N-C (TM = Fe, Co) ORR catalysts. The general agreement was that TM-N species embedded in the carbon matrix (C–N–M, M = Fe, Co or Ni) were responsible for the ORR activity.[103,115,116] However, some researchers then suggested that the N-C moiety itself could catalyze the ORR. In order to confirm that TM species were in fact unnecessary, Stevenson *et al.* observed that N-doped CNFs showed (while still weak) promising ORR activity at neutral to basic pH through a two-electron process (the activity being attributed to the presence of edge plane defects and nitrogen functionalities within the CNF structure),[82] whereas Ozkan *et al.* found that N-containing carbon deposited on pure alumina with less than 1 ppm metal contamination was also ORR active in acidic conditions.[117] However, the interest in this topic really took off in 2009, when Dai *et al.* reported excellent ORR performance by vertically aligned N-doped CNT (VA-NCNTs) arrays (Figure 6.4) in alkaline media.[108] This metal-free catalyst was shown to catalyze a four-electron ORR process free from CO poisoning, with a much higher electrocatalytic activity and better long-term operation stability than that of commercially available Pt-based (20 wt%) electrodes in alkaline electrolytes. The high surface area, good electrical and mechanical properties, and superb thermal stability of aligned CNTs provided additional benefits for the nanotube electrode to be used in fuel cells under both ambient and harsh conditions (*e.g.*, for high-temperature use). In order to explain their results, the authors performed quantum mechanics calculations using B3LYP hybrid density functional theory. The improved catalytic performance was attributed to the electron-accepting ability of the nitrogen atoms, which create a net positive charge on adjacent carbon atoms in the nanotube carbon plane of VA-NCNTs to readily attract electrons from the anode to facilitate the ORR. The nitrogen-induced charge

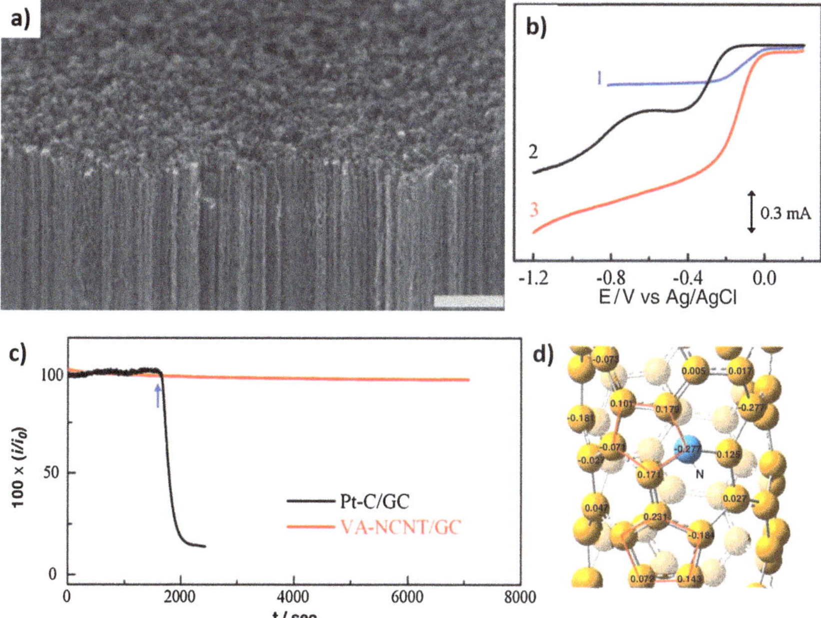

Figure 6.4 (a) SEM image of VA-NCNTs. Scale bars, 2 μm. (b) Rotating ring-disk electrode voltammogram for oxygen reduction in air-saturated 0.1 M KOH at the Pt–C/GC (curve 1), VA-CCNT/GC (curve 2), and VA-NCNT (curve 3) electrodes. (c) CO-poison effect on *i*–*t* chronoamperometric response for the Pt–C/GC and VA-NCNT/GC electrodes. The arrow indicates the addition of CO gas into the O_2 flow; (d) Calculated charge density distribution for the NCNTs. Adapted from ref. 108.

delocalization could also change the chemisorption mode of O_2 from the usual end-on adsorption (Pauling model) at the nitrogen-free CNT (CCNT) surface to a side-on adsorption (Yeager model) onto the NCNT electrodes. The N-induced charge-transfer from adjacent carbon atoms could lower the ORR potential, while the parallel diatomic adsorption could effectively weaken the O–O bonding, facilitating ORR at the VA-NCNT electrodes. Since then, a wide range of N-doped carbon nanostructures with different morphologies and structural characteristics have been synthesized, evidencing striking electrocatalytic performance for ORR.[36,118–123]

The first report on the application of N-doped graphene for ORR as a metal-free catalyst was published by Dai *et al.* in 2010.[36] Graphene films were prepared under the same CVD conditions, with N-doping being achieved with the introduction of NH_3 gas; the authors compared the electrocatalytic activity of these materials with that of a commercially available Pt-loaded carbon (Vulcan XC-72R) supported by a glassy carbon electrode (Pt/C). With an N/C ratio of *ca.* 4 at% for N-graphene, the authors revealed a superb metal-free electrode for ORR in alkaline fuel cells. In a similar manner to that already observed with undoped CNTs, the pristine graphene electrode also exhibited a two electron process for oxygen reduction, with onset potentials of about −0.45 and −0.7 V. As for N-doped CNTs, N-graphene electrode also displayed a one step, four electron pathway, which is more efficient for ORR. The steady catalytic current density of N-graphene was found to be nearly 3 times higher than that of the commercial Pt/C electrode over a large potential range (Figure 6.5). In addition, N-graphene was found to be insensitive to CO, thus resolving the severe drawback of Pt nanoparticle poisoning, showed high

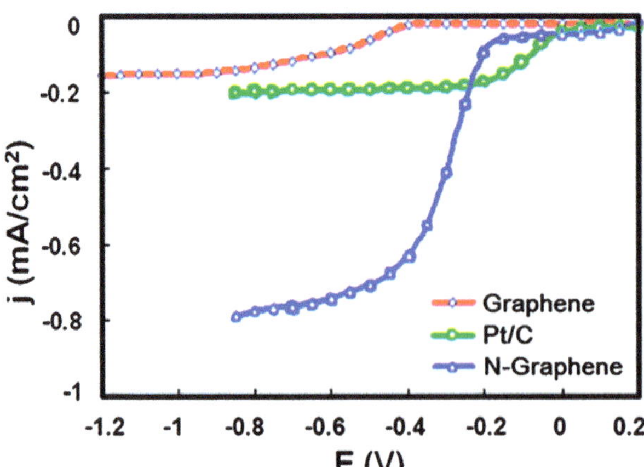

Figure 6.5 Rotating ring disk electrode voltammograms (1000 rpm and 0.01 V s^{-1} using 7.5 μg of material) for the ORR in air-saturated 0.1 M KOH at the pristine graphene electrode (red line), Pt/C electrode (green line), and N-graphene electrode (blue line). Reprinted with permission from ref. 36. Copyright 2010 American Chemical Society.

selectivity, long-term operation stability and fine tolerance to the crossover effect.

Colloidal graphene quantum dots are a new addition to zero-dimensional quantum-confined systems, and have some unique properties due to the dimensionality of graphene.[124] Using a solution chemistry approach, Li *et al.* prepared different graphene quantum dots doped with nitrogen atoms.[125] N-groups still played a vital role in reducing the adsorption barrier and increasing the ORR current, but the authors clearly showed that the size of the sp^2 lattice in the N-functionalized nanostructured carbon materials also influenced the ORR behavior. It was proposed that by increasing the lateral size of graphene quantum dots, the HOMO level was elevated and the N-functionalized graphene quantum dots have greater affinity to oxidation.

Another potentially interesting carbon structure that has yielded excellent results in ORR is N-doped carbon nanocages.[113] These materials were prepared using an *in situ* MgO template method with pyridine as the precursor, and possessed a high N-doping (10 wt%) and large specific surface area (*ca.* 1400 $m^2 g^{-1}$) without any TM residues, meaning that they are an excellent example to demonstrate the N-doping and morphology effects on ORR performance (Figure 6.6). The ORR activities observed were comparable to those of commercial Pt/C electrocatalysts, with superior stability towards methanol crossover and CO poisoning. Electrochemical tests also demonstrated that the onset potential of the N-doped nanocages was almost the same as that of N-doped nanotubes, but with a much higher current density, in alkaline media. The reason for this behavior was attributed to the greater surface area of the nanocages (885 $m^2 g^{-1}$) compared to the nanotubes (*ca.* 200 $m^2 g^{-1}$). Given that there is no interference by ORR-sensitive metal impurities, these experimental results confirm that the ORR activity does originate from N-doped carbon species alone, even though the exact mechanism is still unclear.

Nitrogen content is also related to the ORR activity of N-doped nanostructured carbon materials. Generally speaking, doped carbon materials with high nitrogen content usually provide more active reaction sites than other N-carbon materials and seem to be more feasible metal-free ORR electrocatalysts. Some hydrocarbons with high N content such as melamines, have become remarkable materials for the fabrication of N-doped carbon nanomaterials with graphitic structures.[126,127] Furthermore, graphitic-carbon nitride (g-C_3N_4) has also generated tremendous interest recently among material researchers because of its high nitrogen content and facile synthesis procedure. Shi *et al.* immobilized g-C_3N_4 onto graphene sheets to form gC_3N_4/graphene composites in a liquid-phase solution.[128] These composites exhibited enhanced electrocatalytic activity for ORR and CO tolerance comparable to that of 23 wt% Pt nanoparticles supported on graphene sheets alone.

One of the most controversial aspects of using metal-free N-doped nanostructured carbon materials in ORR is the determination of which of the N configurations is responsible for the ORR activity. Pyridinic-N refers to N atoms at the edges of graphene planes, where each N atom is bonded to two carbon atoms and donates one p-electron to the aromatic π system.

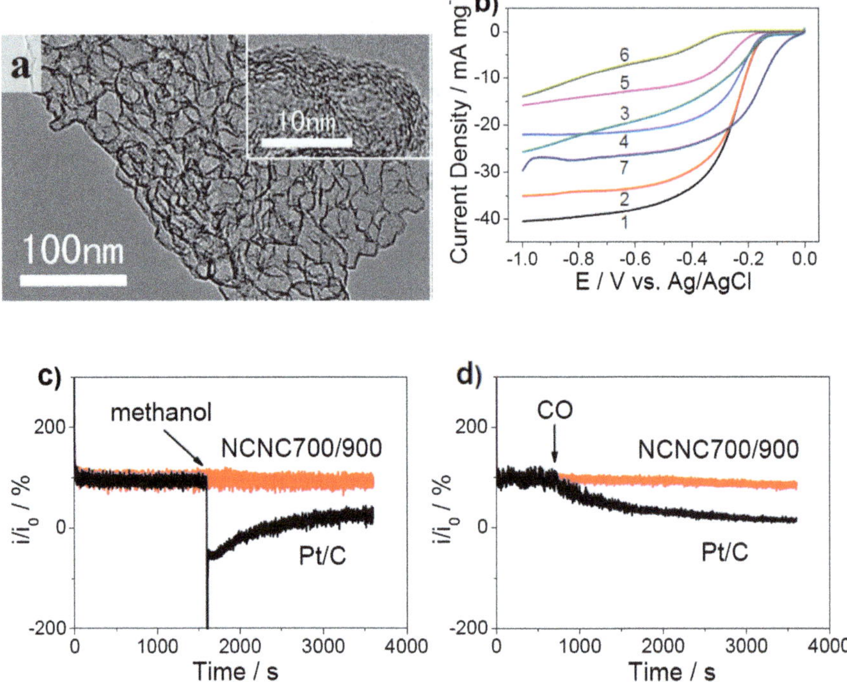

Figure 6.6 (a) Tem image of N-doped carbon nanocages (inset: high-resolution TEM). (b) Rotating disk electrode voltammogram (rotation rate: 2500 rpm, scan rate: 10 mV s^{-1}); Chronoamperometric responses of N-doped carbon nanocages to the presence of (c) methanol and (d) CO in O_2-saturated KOH solution (0.1 M). Reprinted with permission from ref. 113. Copyright 2012 WILEY-VCH Verlag GmbH & Co. KGaA.

Pyrrolic-N atoms are incorporated into five member heterocyclic rings, where each N atom is bonded to two carbon atoms and contributes two p electrons to the π system. Graphitic-N atoms are incorporated into the graphene layer and substitute carbon atoms within the graphene plane (see Chapter 4, Figure 4.2). Among these nitrogen types, pyridinic-N and quaternary-N are sp^2 hybridized and pyrrolic-N is sp^3 hybridized.

Whereas some authors have attributed the enhanced ORR activity to the graphitic-N atoms (substitutional doping),[122,129–133] others, on the other hand, attributed the improvement in electrocatalytic activity to pyridinic- and pyrrolic-N atoms.[36,130,134–136]

Lai *et al.* successful produced N-doped graphenes with well-defined N types by annealing graphene oxide under ammonia and by annealing reduced graphene oxide/N-containing polymer (polyaniline or polypyrrole) composites.[130] The authors found that the electrocatalytic activity of the N-doped graphene catalyst mainly depended on the graphitic N content, which determined the limiting current density, while the pyridinic N species converted the ORR reaction mechanism from a $2e^-$ dominated process to a 4 e^- dominated process and improved the ORR onset potential. Qu *et al.* synthesized N-doped graphene with a limiting current density three times larger than that of 20 wt% Pt/C by chemical vapor deposition of methane in the presence of ammonia. They proposed that its improved ORR activity was due to the pyridinic N-abundant structure.[36] Matter *et al.* also suggested that more active carbon catalysts might contain a greater amount of pyridinic nitrogen with a higher proportion of edge planes.[134]

These conflicting results may be attributed to the variation in experimental conditions. In addition, it is also apparent that pyridinic-N and pyrrolic-N are always located at edge sites, as a marker of edge plane exposure. Thus, it is uncertain whether the pyridinic- or pyrrolic-N provides actual active sites for promoting ORR, or whether the reaction is simply more favorable due to the abundance of highly active edges and/or defect sites. In addition, it is a challenge to determine the exact location of nitrogen atoms in the nanocarbon structures, the chemical nature of the catalytic sites, and the electrochemical kinetics of the N-doped nanocarbon electrodes.

In order to understand the catalytic mechanism, a series of theoretical studies have been conducted to predict the adsorption and reduction of oxygen molecules on the N-functionalized nanostructured carbon materials.[108,112,137–140] These support the hypothesis that the presence of quaternary-type nitrogen favors O_2 adsorption at carbon sites on the zig-zag edges located adjacent to quaternary-type nitrogen. The incorporation of electron-accepting nitrogen atoms creates a relatively high positive charge density on adjacent carbon atoms; nitrogen becomes more electronegative and creates a net positive charge on the adjacent carbon atoms. As a result, the adsorption of O_2 on nitrogen doped carbons becomes easier, leading to an improved electrocatalytic activity of these materials towards ORR in alkaline media (higher than that of Pt/C catalyst). Regarding the nature of the active site, Kim *et al.* reported an inter-conversion between graphitic and pyridinic sites at the edge of a graphene nanoribbon within a catalytic cycle.[129] The authors showed that both the edge structure and N-doping near the edge enhanced the oxygen adsorption, the first electron transfer, and also the selectivity toward the four-electron, rather than the two-electron, reduction pathway. In addition, the graphitic-N edge-site was more active than the in-plane graphitic-N and pyridinic-N ones. The proposed catalytic cycle around the graphitic-N site involves a ring-opening of the cyclic C–N bond at the edge of graphene which results in the pyridinic nitrogen (IV in Scheme 6.1); when a hydrogen atom is attached to the oxygen adatom in step 4, the broken C–N bond zips back and the nitrogen becomes again a graphitic-N (V in Scheme 6.1). This new type of N-doped active site that inter-converts between pyridinic and graphitic-types could potentially reconcile the experimental controversy over whether the pyridinic, graphitic or both types of nitrogen are the ORR active sites for N-doped nanocarbon materials.

In an attempt to overcome the fact that simulations are often carried out considering the separate contribution of graphitic and pyridinic N-doping in the ORR process, Oshima *et al.* investigated the valence electronic states of

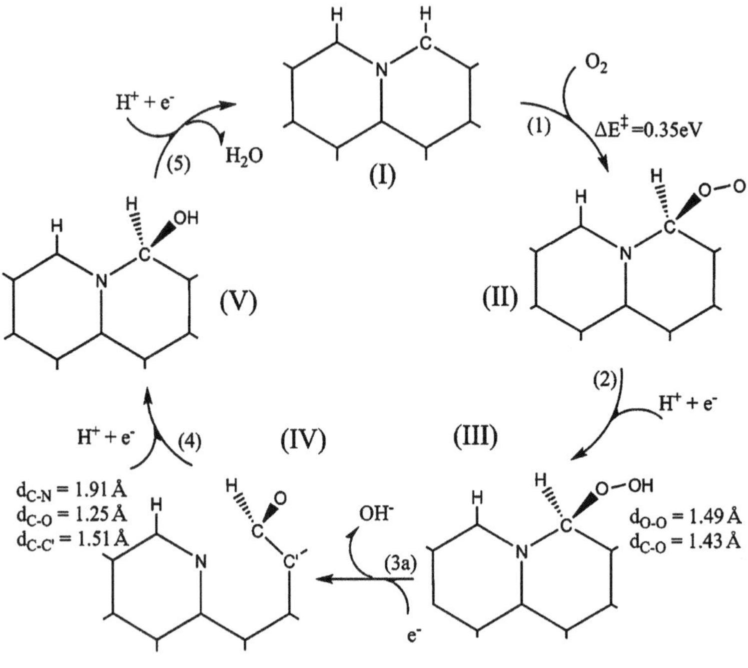

Scheme 6.1 Proposed ORR catalytic cycle at the graphene nanoribbon edge including the inter-conversion of the graphitic-N structure into a pyridinic-N one, and back to graphitic-N.[129]

nitrogen in carbon materials by soft X-ray absorption spectroscopy (XAS) to clarify the role of nitrogen in cathode catalyst activity.[132] The authors showed that catalysts with a relatively larger amount of graphitic N exhibit higher ORR activity than those with a relatively larger amount of pyridinic N atoms.

Xia *et al.* used DFT calculations to study the mechanism of ORR on N-graphene cathodes of fuel cells in acidic environments.[141] The simulation results on the electron transformation process showed that the ORR followed a four-electron pathway on N-graphene but pure graphene did not have such catalytic activities. Interestingly, the catalytic active sites on the N-graphene were found to depend more on spin density distribution and atomic charge distribution. The substituting nitrogen atom introduced no-pair electrons to the graphene and changed the atomic charge distribution on it. Generally, the carbon atoms that possess the highest spin density are the electrocatalytic active catalytic sites. If the negative value of spin density is small, the carbon atoms with large positive atomic charge density may act as the active sites.

Because N-doped carbon materials usually consist of different structures and the content of different nitrogen types usually varies with nitrogen doping level, a higher amount of N incorporated in the carbon matrix alone does not always lead to better performance, which also crucially depends on structural and morphological properties.[134] Development of N-doped carbon materials with high surface area and a well-defined porous structure, which facilitates reactant transport, would be an ideal solution. Yang *et al.* prepared mesoporous nitrogen-doped carbon materials (BET areas up to 1500 m^2 g^{-1}) by the carbonization of nucleobases dissolved in an all-organic ionic liquid *via* hard templating with silica nanoparticles.[118] The authors demonstrated that the resulting nitrogen-doped carbons had very high catalytic activity for ORR, with the materials exhibiting a low onset voltage for ORR in alkaline medium and high methanol tolerance compared with those of commercial 20 wt% Pt/C catalyst.

One important aspect that raises some concerns regarding the mechanism of a metal-free process is related to the presence of metallic impurities, resulting from the material synthesis. The development of truly metal-free N-doped carbon materials with excellent ORR catalytic activity is therefore of both scientific and practical importance.[120,122,142–145] Liu *et al.* reported the preparation of nitrogen-doped ordered mesoporous graphitic arrays by a metal-free nanocasting technique using nitrogen-containing aromatic dyes as the carbon precursors.[122] Huang *et al.* has successfully synthesized metal-free N-doped carbon spheres by directly pyrolyzing a nebulized solution of xylene and ethylenediamine *via* spray pyrolysis.[142] Xia *et al.* have also proposed a facile, catalyst-free thermal annealing approach for large scale synthesis of N-doped graphenes using the low-cost industrial material melamine as the nitrogen source.[143] In all of these cases an ORR electrocatalytic performance better than that of platinum was observed, thus confirming the activity is attributed exclusively to the incorporation of nitrogen on the carbon matrix, and the C–N structure is expected to play a decisive role in the observed ORR activity enhancement.

The data relevant to the ORR performance for both acidic and alkaline media is tabulated in Table 6.1. Unfortunately, kinetic current density values are all obtained at different onset potentials, thus hindering a direct comparison between all the different doped nanostructures gathered here. Nevertheless, a promising indication is given by nitrogen-doped few-layer graphene, evidencing the highest current density value. Co-doping of N-graphene with S also has a very positive impact on the current density value.

6.4.1.2 Boron-Doped Carbon Nanostructures

Research on B-doped carbon nanomaterials for the ORR is rare compared with that on the N-doped counterparts, because the idea that electron-deficient B-doped sp^2 carbons can catalyze the ORR, which requires electrons, is counterintuitive. However, boron atoms are also able to break the uniform charge density of the sp^2 carbon lattice and accept electrons because of their three valence electrons, which shift the Fermi level to the conducting band. These changes in the band structure and electronic states are known to improve ORR activity. Nevertheless, doping of sp^2 carbon materials is more difficult with B than with N, with the highest reported B-doping level (2.24 wt%)[58] being far less than that in the case of N (12%).[118] Furthermore,

Table 6.1 Heteroatom-doped carbon materials on ORR performance.

Catalyst	% Nitrogen	ORR performance	References
Acidic media			
N-SWCNT	3.6	3.54 mA cm^{-2} @ −0.10 V	146
N-CNT	4.5	0.12 mA cm^{-2} @ −0.40 V	147
N-CNF	n.d.	10.9 mA cm^{-2} @ −0.30 V	148
Alkaline media			
N-OMC	1.9	9.2 mA cm^{-2} @ −0.35 V	122
N-CNT	4.0–6.0	4.1 mA cm^{-2} @ −0.22 V	108
N-CNT	3.27	4.71 mA @ −0.40 V	149
N-CNT	2.7	4.49 mA cm^{-2} @ −0.33 V	150
N-CNT	2.91	4.12 mA cm^{-2} @ −0.46 V	150
N-CNT	4.7	5.1 mA cm^{-2} @ −0.20 V	151
N-CNF	4	0.9 mA cm^{-2} @ −0.67 V	152
N-Nanocages	7.1	1.7 mA cm^{-2} @ −0.13 V	113
N-graphitic carboncages	1.9	3.27 mA cm^{-2} @ −0.40 V	153
N-carbon spheres	6.2	5.21 mA cm^{-2} @ −0.29 V	142
N-Hollow mesoporous carbon spheres	3.8	0.87 mA @ 0.14 V	154
N-graphene	7.8	4.2 mA cm^{-2} @ −0.50 V	155
N-graphene	n.d.	6.26 mA cm^{-2} @ −1.00 V	156
N-Graphene sheets	3.7	1.09 mA cm^{-2} @ 0.03 V	157
N-FLG	2.4	11.6 mA cm^{-2} @ −0.40 V	158
CNF@NG aerogels	9.8	5.0 mA cm^{-2} @ −0.40 V	159
Other heteroatoms			
B-N-CNT	10 (4% B)	10.1 mA cm^{-2} @ −0.30 V	160
B-N-graphene	17 (14% B)	5.0 mA cm^{-2} @ −0.60 V	161
P-OMC	— (0.9% P)	4.41 mA cm^{-2} @ −0.50 V	162
P-CX	— (1.6% P)	6.01 mA cm^{-2} @ −0.13 V	163
P-graphite	— (0.3% P)	2.81 mA cm^{-2} @ −0.44 V	66
N-S-aerogels	4.96 (0.74% S)	1.14 mA cm^{-2} @ −0.20 V 1.82 mA cm^{-2} @ −0.90 V	164
N-S-graphene	4.5 (2% S)	15 mA cm^{-2} @ −0.40 V	165
S-graphite	— (1.3% S)	9.3 mA cm^{-2} @ −0.30 V	74
S-graphene	— (1.8% S)	6.99 mA cm^{-2} @ −1.00 V	156

B-doping is often accompanied by a high O content because of the strong affinity between O and B. In fact, the electron accepting nature of boron makes it more reactive to oxygen than to carbon. According to XPS analysis, boron itself can be oxidized when the B-containing nanostructured carbon materials are exposed to air.[58] The chemisorbed O can occupy the active sites on these B-containing carbons prior to the reduction of oxygen in an electrochemical environment. A higher concentration of boron in carbon was obtained by direct pyrolysis of tetraphenylboron sodium as a single precursor for both carbon and boron (3.8 and 5.3 at%).[166] The ORR activity of this B-containing carbon was found to be dependent on the pyrolysis temperature. By increasing the temperature, the kinetic current was increased greatly, which was attributed to the more abundant sites of BC_3 and B_4C that were assumed to be ORR active. The onset potential

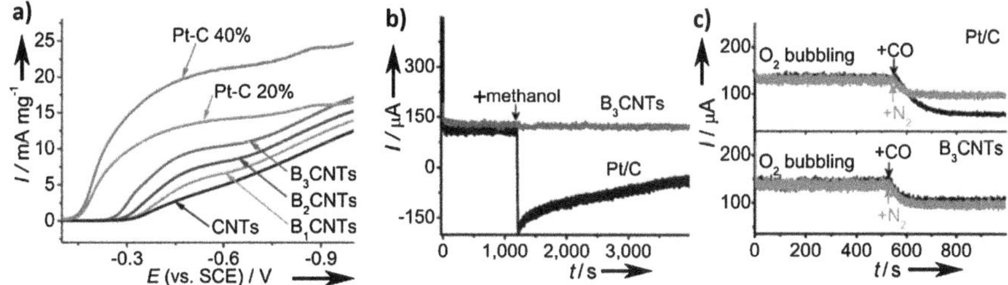

Figure 6.7 (a) RDE voltammograms of catalysts for ORR in O_2-saturated 1M NaOH electrolyte (scan rate 50 mV s^{-1} and rotation speed of 2500 rpm). (b) Methanol crossover tests (1.5 mL of methanol at 1200 s). (c) CO poisoning tests (CO, at the same flow rate as O_2, into the electrolyte at 520 s). Reprinted with permission from ref. 58. Copyright 2011 WILEY-VCH Verlag GmbH & Co. KGaA.

only exhibited a small positive shift by 0.012 V and was 0.155 V lower than that of Pt/C. The onset potentials of the two different B-containing carbons were close however, which indicated a similar conductivity and adsorption barrier for oxygen molecules. B-doped CNTs were prepared by CVD using benzene and triphenylborane as precursors. Depending on the level of B doping achieved (0–2.13 wt%), the morphology of the nanotubes can evolve from straight to bamboo-like, and eventually twisted. This is associated with the B–C bond being about 0.5% longer than the C–C one.[13] Electrochemical tests demonstrated that the ORR catalytic performance in alkaline solution progressively improved with increasing B content (increased reduction current and the positively shifted onset and peak potentials in Figure 6.7a). Just like their N-doped counterparts, B-doped CNTs are also not affected by both methanol crossover and CO poisoning (Figure 6.7b and c). Sheng *et al.* reported similar findings using graphene, *i.e.*, B doping was found to improve the electrocatalytic performance.[167] Theoretical calculations indicated that the electrocatalytic activity of B-doped CNTs for the ORR originates from the conjugation between the vacant $2p_z$ orbital of B and π* electrons of the conjugated system (p orbital of carbon atoms); transfer then readily occurs to the chemisorbed O_2 molecules with boron as a bridge. The transferred charge weakens the O–O bonds and facilitates the ORR on the B-doped CNTs.

6.4.1.3 *Phosphorus-Doped Carbon Nanostructures*

Besides B and N, other non-metallic elements such as P, have also been used to produce doped carbon materials and enhance ORR catalytic activity.[67,74] Peng *et al.* prepared a P-doped graphite layer catalyst without any metal residue, by pyrolysis of toluene and triphenylphosphine, that exhibited high electrocatalytic activity, long-term stability, and excellent tolerance to methanol cross-over effects for ORR in alkaline medium.[66] The better activity was attributed to the positive charge of the P atom (0.652) in the P–C bond in which the C atom had a negative charge (−0.298). However the ORR pathway on these P-containing nanostructured carbon materials was a two-plateau pathway with an evident production of HO_2^-. Usually the adsorption of oxygen is believed to be the rate determining step and the adsorption of HO_2^- can promote its reduction to OH^-. Phosphorus-doped MWCNTs prepared by thermolysis of toluene and triphenylphosphine using $FeMo/Al_2O_3$ as catalyst were also reported to show much higher ORR activity than commercial Pt/C in alkaline fuel cells.[65] The highest ORR current was obtained for CNTs containing only 0.13 at% P, revealing that the ORR performance of the P-doped CNTs is not only related to the P content but also to their morphology.

Given that ordered mesoporous carbons have been a popular choice as electrocatalyst supports for fuel cells (due to their high surface area, tunable pore size and large pore volume with narrow pore size distributions), Yu *et al.* studied the electrocatalytic response of phosphorus-doped ordered mesoporous carbons using a metal-free nanocasting method of SBA-15 mesoporous silica and triphenylphosphine and phenol as phosphorus and carbon sources, respectively.[67] The resulting P-doped mesoporous carbon had a small amount of P (less than 1.5 at%) and exhibited outstanding electrocatalytic activity, long-term stability, and excellent resistance to alcohol crossover effects for ORR in alkaline media. The P-doping was found to induce defects in the carbon framework and increased the electron delocalization due to good the electron-donating properties of P, promoting active sites for ORR. The authors also studied the effect of channel length on catalytic ORR performance, and showed an increase in activity with a decrease in the channel length, probably due to increased surface area and decreased resistance of shorter channels.

6.4.1.4 Sulfur-Doped Carbon Nanostructures

After carbon materials doped with atoms having larger (N) or smaller (P, B) electronegativity than carbon had been confirmed to improve the electrocatalytic performance of ORR, interest turned to the development of carbon materials doped with an element of similar electronegativity to carbon (2.58 *vs.* 2.55). Huang *et al.* successfully prepared sulfur-doped graphene by direct annealing of graphene oxide and benzyl disulfide in argon.[74] The resulting electrocatalytic performances indicated that the S-graphene could exhibit excellent catalytic activity, long-term stability, and high methanol tolerance in alkaline media for ORRs. As a result of the large atomic radius of S and the long S–C bond, S is not likely to participate directly in π conjugated systems but rather create defects and distortions that perturb the delocalized π electrons, and thus improve ORR performance. Yang *et al.* also confirmed that S-doped graphene synthesized *via* thermal reaction between graphene oxide and H_2S gas could act as a metal-free electrocatalyst for oxygen reduction reactions, showing comparable electrocatalytic activity to that of commercially available Pt/C.[168]

Research concerning the origin of ORR activity in N-, B-, or P-doped carbon materials indicates that breaking the electroneutrality of graphitic materials by doping with elements, which have larger (N) or smaller (P, B) electronegativity than carbon, might be an important factor for promoting the ORR activity. Studies have established that the presence of dopants in the carbon framework creates positively charged sites, which are favorable for the side-on O_2 adsorption. This parallel diatomic adsorption could effectively weaken the O–O bonding and facilitate the direct reduction of oxygen to OH^- (H_2O in acidic electrolytes) *via* a four-electron ($4e^-$) process, hence enhancing ORR activity. However, unlike nitrogen (or boron) dopants, sulfur has a similar electronegativity to carbon. Obviously, the previous mechanism could not explain the results obtained. It was then proposed that breaking the electroneutrality of graphitic materials may not be necessary factor for the ORR activity enhancement, while the unique electron structure derived from the conjugation between the sulfur lone-pair electrons and the graphene π system may play a vital role for the high ORR activity of S-graphene. According to this new understanding, it is suggested that tailoring the π electronic system of graphene may be a preferred factor for producing significantly improved materials for ORRs, regardless of the nature of the dopant heteroatoms.

In summary, sulfur in the thiophenic (–C–S–C–) configuration is presumably the active site for promoting ORR.[74,156,169] The presence of thiophenic sulfur improves the overall electrocatalytic activity of the doped carbon in both basic and acidic media. Enhancement of the ORR efficiency could be caused by lone pairs of S, which may contribute to interaction with O_2. Sulfur also induces strain and defects in the carbon matrix, which facilitates charge localization for favorable O_2 chemisorption.

6.4.1.5 Multi-Heteroatom-Doped Carbon Nanostructures

Multiple doping is a versatile synthetic approach for new carbon materials and takes the tuning of carbon properties one step further in comparison to one-type-only heteroatom doping. Because mono-doping can induce ORR activity of sp^2 nanostructured carbon materials, intuitively, co-doping with two (or more) heteroatoms represents an interesting route to optimize the activities of carbon-based metal-free ORR electrocatalysts.

Ozaki *et al.* tried co-doping by carbonization of poly(furfuryl alcohol) containing both a BF_3-CH_3OH complex and melamine.[170] The obtained B-N co-doped carbons give higher reduction currents than their B and N mono-doped counterparts in acidic media. Recently, Dai *et al.* reported that B-N co-doped CNTs[160] and graphene[171] are excellent ORR electrocatalysts in alkaline solution (Figure 6.8). The improved activity was attributed to a synergetic effect arising from co-doping of the CNTs with boron and nitrogen. It seemed that N-groups were more advantageous for increasing the onset potential and the limiting current, while the B-groups appeared only active for increasing the limiting current rather than the onset potential. Despite

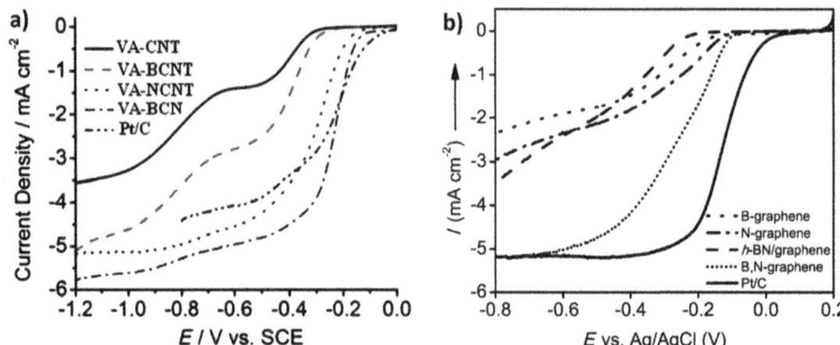

Figure 6.8 Linear-sweep voltammetry curves for (a) vertically-aligned B-, N- and B-N CNTs, and (b) B-N graphene with different compositions in oxygen-saturated 0.1 M KOH electrolyte solution at 10 mV s^{-1} compared with a commercial Pt/C electrocatalyst. Reprinted with permission from ref. 160 and 161. Copyright 2011 and 2013 WILEY-VCH Verlag GmbH & Co. KGaA.

these promising results, the irregular variations in ORR activities with respect to the B/N ratios and contents are problematic and the exact cause is unknown. For example, the ORR performance of a certain B-N co-doped graphene is better than those of commercial Pt/C catalysts, but the performance of some other materials can be much worse, despite their higher B and N contents.[171]

There is a fundamental issue with B-N co-doped carbon materials, *i.e.*, whether the B and N dopants are bonded together or located separately. Due to the compensation effect between p- and n-type dopants, these two cases correspond to totally different electronic structures, with different conjugation effects in the delocalized π system, which can eventually lead to distinct ORR activities. This issue has been systematically studied by Hu *et al.*[172] To obtain co-doped CNTs containing separated B and N, the authors first prepared B-doped CNTs (triphenylborane as B precursor) and then doped it with N using a NH_3 post-treatment. The bonded B-N co-doped CNTs were prepared by a simultaneous co-doping route, *i.e.*, the B and N sources were mixed together in the precursor during the CVD growth. Electrochemical tests showed that the bonded material could not, but the separated one could, convert inert CNTs into ORR electrocatalysts in alkaline solution (Figure 6.9). Theoretical studies indicated that the bonded and separated cases corresponded to un-activated and activated p electrons, respectively, which provided sound evidence for a correlation of the ORR activities of carbon materials with p electron activation. An identical conclusion regarding B-N co-doped graphene was also reached by Qiao *et al.*, who also detected a synergistic coupling effect due to the sequential incorporation of N and B into the graphene domain.[161] The calculation of O_2 adsorption on a B–N co-functionalized zigzag edge demonstrated that the boron atoms in the C–N–B–C and the C–N–B–N–C were the most energetically favorable oxygen adsorption sites.[173] However, when the B–N unit was

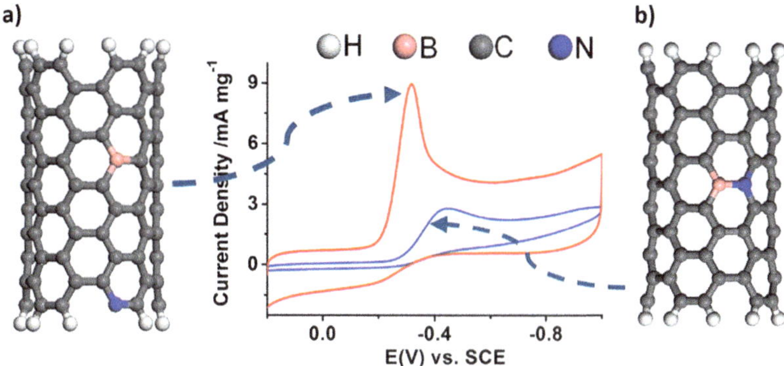

Figure 6.9 Schematic diagram of B-N co-doped CNTs: (a) separated B-N with high ORR activity; (b) bonded B-N with low ORR activity. Reprinted with permission from ref. 172. Copyright 2013 American Chemical Society.

placed in the basal plane, the oxygen adsorption became unfavorable. Once the O_2 was chemisorbed on B, the –B–O–O adduct formed. The authors speculated that the hydrophobicity of the –CH group on the substrate inhibited the solvation of the middle O in the –B–O–O adduct, and thus the terminal O in -B-O-O was the one to be reduced. Interestingly, once the terminal O is removed, the remaining O atom can be further reduced in a barrierless way if on a zigzag-edged surface, or with a small barrier (3 kcal mol^{-1}) if on a zigzag-shaped step edge.

Doping of N-doped carbons with other heteroatoms such as P or S is also an effective strategy in the development of enhanced ORR electrocatalytic activity. Woo *et al.* reported binary P,N-doped carbon prepared *via* pyrolysis of dicyandiamide and phosphoric acid, as sources of C–N and P, respectively.[71] They observed that additional P-doping promoted the catalytic activity of nitrogen-doped carbon, and an optimized phosphorous–nitrogen-doped carbon exhibited a more than four-fold increase in ORR activity compared with that of nitrogen-doping alone. The P–N co-modification was found to generate more edges and wrinkled surfaces, that in turn could disturb the distribution of uniform charge density and spin density and thus improve the ORR performance. A superior ORR activity was also observed when the N and P heteroatoms were simultaneously incorporated in CNTs using pyridine and triphenylphosphine as precursors for injection-assisted CVD.[174] Binary P-N-CNT arrays were shown to exhibit a high ORR electrocatalytic activity, superb long-term durability, and good tolerance to methanol and carbon monoxide. The co-doped CNTs significantly outperformed their counterparts doped with P or N alone, and was comparable to the commercially available Pt/C catalyst (45 wt% Pt on Vulcan XC-72R; E-TEK) due to a demonstrated synergetic effect arising from the co-doping with both P and N.

Co-doped carbon materials comprising nitrogen and sulfur are an alternative to the preparation of binary doped carbon materials. In this case, the

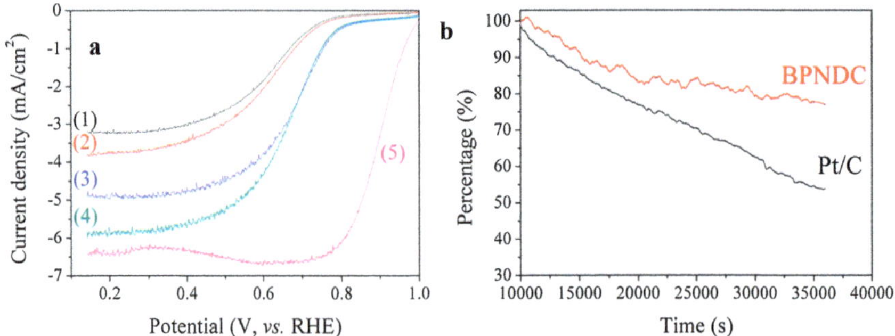

Figure 6.10 (a) ORR results in 1MHClO$_4$ for (1) N, (2) B-N, (3) P-N, (4) B-P-N, and (5) Pt/C (40 wt. %). (b) Current-time chronoamperometric response (0.6 V *vs.* RHE) of B-P-N and Pt/C for 10 h in 1 M HClO$_4$ with continuous oxygen bubbling. Reprinted with permission from ref. 72. Copyright 2012 American Chemical Society.

same authors used five different amino acids (alanine, cysteine, glycine, niacine, and valine) as heteroatom precursors *via* pyrolysis and tested the electrocatalytic activity of the resulting material in ORR.[175] The authors verified that the dual-doped carbon catalysts showed the highest onset potential and electrochemical activity in acidic media, about 43% of that of commercial Pt/C (40 wt%). They concluded that nitrogen doping coupled with sulfur doping of carbon played an essential role in the improvement of the electrocatalytic activity. Wohlgemuth *et al.* reported a simple, metal-free route toward nitrogen- and sulfur-doped, high surface area carbon aerogels based on a glucose–ovalbumin system.[164] By comparing solely nitrogen-doped with nitrogen- and sulfur co-doped carbon aerogels, the authors observed that the presence of sulfur improved the overall electrocatalytic activity of the carbon material in both basic and acidic media. Liao *et al.* demonstrated a facile synthesis process for the preparation of S and N co-doped carbon catalysts with uniform nanospherical morphologies, using polyacrylonitrile nanospheres and sulfur as precursors.[79] The obtained catalysts showed high ORR activity, outstanding long-term stability, and excellent methanol tolerance in an alkaline medium. In addition, sulfur was found to be able to enhance the catalyst's ORR performance significantly by retaining the nanospherical morphology, constructing porous structures and forming active sites.

The first ternary (N,B,P)-doped carbon to be adapted as an ORR catalyst was reported by Woo *et al.*, and demonstrated that additional B and/or P doping modifies carbon characteristics and improves the ORR activity of N-doped carbon catalysts alone.[72] The ORR activities of the prepared catalysts were in the order B-P-N > P-N > B-N > N, with the catalyst showing the highest activity being B-P-N (Figure 6.10a). B-P-N doped carbon also demonstrated a good stability (Figure 6.10b), even after 10 h. The authors proposed that B-doping increased the portion of pyridinic-N sites for various types of N-doping and magnified the amount of sp^2-carbon structures, while P-doping was responsible for the charge delocalization of the carbon atoms and constructed a morphology with many open edge sites that were split and wrinkled. These results indicated that binary or ternary doping of B and P with N into carbon induces remarkable performance enhancements, and that charge delocalization of the carbon atoms and number of edge sites are significant factors in deciding oxygen reduction activity in carbon-based catalysts.

6.4.2 Oxidation Reactions

C–H bond activation is an important reaction traditionally catalyzed by transition metals and organometallic complexes.[176] Cheap metal-free catalysts have emerged as promising candidates for this transformation, with oxidative reactions over heteroatom-doped carbons receiving increasing attention, since they are able to activate oxygen molecules without the assistance of any metals.[177–181]

Oxidative dehydrogenation (ODH) of light alkanes offers a potentially attractive route to alkenes, since the reaction is exothermic and avoids the thermodynamic constraints of non-oxidative routes by forming water as a byproduct. In addition, carbon deposition (coke formation) during ODH is eliminated, leading to stable catalytic activity. Compared with the conventional steam-cracking method of dehydrogenating alkanes to olefins and current catalytic dehydrogenation processes, ODH could reduce costs, lower greenhouse gas emissions, and save energy. One extremely interesting application of this reaction (especially from the industrial point of view) is the preparation of propene (propylene) from propane, given its importance in the petrochemical industry for producing polypropylene, propylene oxide and plastics.

Despite the high yields obtained using V/Mo catalysts,[182] the use of doped carbon materials as metal-free catalysts in this same reaction has also been successfully reported.[183–185] Analysis of the reaction mechanism points out that both the reactivity of absorbed propane/propene and oxygen are key factors for controlling activity and selectivity. A higher reactivity of adsorbed oxygen will result in a high selectivity towards CO and CO_2 instead of the desired propene. Adsorbed oxygen has different reactivities at different edge sites,[186] indicating that the activity and selectivity of ODH might depend on the termination of edges of the graphite sheets in the CNFs, or more precisely on the ratio of armchair and zigzag sites. Furthermore, studies have revealed that π electron interaction between alkenes and graphite plays an important role for alkene chemisorption,[187] resulting in preferred chemisorption of propene than propane. This leads to a challenging task in improving the selectivity towards propene during the ODH of propane.

Dai *et al.* tested both B and P doping to improve the selectivity of CNFs in the oxidative dehydrogenation of propane.[184] The CNF samples were pretreated in concentrated HNO_3 for 7 days. The protection from oxidation of CNFs by B and P doping was clearly evidenced by temperature programmed oxidation, where the temperature at the maximum oxidation rate followed the order: P-CNF > B-CNF > CNF, thus revealing that the resistance to oxidation of doped CNF was higher than that shown by pristine CNFs. The starting temperature of the oxidation of propene to CO and CO_2 was much higher on the doped samples. However, it increased rather rapidly with increasing temperature, indicating a higher activation energy for oxidation of propene to CO and CO_2 on doped samples. The effect of B- and P-doping of CNFs on propane oxidative dehydrogenation showed that the selectivity to CO_2 was lower and selectivity to propene was higher on the P-CNFs than on B-CNFs, indicating that secondary reaction of propene to CO_2 was somewhat suppressed. The fact that the selective blockage of the surface armchair sites by P-containing groups resulted in suppression of the CO_2 formation might indicate that the oxygen situated on the front of armchair sites or between the armchair sites are preferred sites for CO_2 formation from propene.

Recently Su *et al.* reported that doping of CNTs with nitrogen could efficiently enhance the catalytic performance of the ODP.[185] The graphitic nitrogen was suggested to play a determining role in enhancing the catalytic activity by decreasing the overall activation energy and speeding up the activation of oxygen. More interestingly, N-CNTs increased propene selectivity at the same conversion level compared to CNTs. A DFT study concerning the adsorption and activation of O_2 on N-CNTs revealed that the incorporated N atoms in CNTs could reduce the gap between the HOMO and the LUMO, which would make the electron transfer from the CNT to the adsorbed oxygen species easier.[188] The three-coordinated substitutional N atoms induces sharp localized states above the Fermi level due to the presence of additional electrons, since nitrogen has one electron more than C. Therefore, substitutional N doping in CNTs should result in n-type conducting behavior, and make adsorbed surface oxygen species more nucleophilic. As a result, it can selectively enhance C–H bond activation leading to propene production.

These results clearly illustrate that selective blockage of armchair sites by P doping of CNFs, and electron rich N-doping CNTs or CNFs, are powerful methods to improve the selectivity to propene and improve the catalyst stability by means of suppressing the gasification of CNFs even at temperatures as high as 823 K.

The selective oxidation of cyclohexane (C_6H_{12}) is extremely important in the chemical industry because it produces a series of useful chemicals, including cyclohexanone, cyclohexanol, and organic acids such as adipic acid. The biggest challenge of commercial processes is the difficulty in controlling the selectivity to target products. Peng *et al.* demonstrated the use of N-doped CNTs as a metal-free catalyst for the liquid-phase aerobic oxidation of cyclohexane.[180] Using N-doped CNT (4.5 at% N) the whole reaction could be altered to one-step production of adipic acid to give selectivities as high as 60% at a conversion higher than 40% (reaction rate 2–10 times higher than those of Au/ZSM-5 and FeAlPO catalysts).[189,190] Furthermore, no difference in both activity and selectivity was observed after five cycling tests, evidencing an excellent recyclability. In a more recent work by the same group,[178] Yu *et al.* synthesized N-, P- and B-CNTs with controllable heteroatom functionalities. The authors found that N-CNTs and P-CNTs were active for the liquid-phase aerobic oxidation of cyclohexane, although P-CNTs suffered from a low specific surface area. It was revealed that the electron transfer between graphene sheets and reactive radicals played a major role, which could be enhanced by n-type dopants, such as N or P functionalities. B acts in an opposite way due to its electron-deficiency, and was found not to improve the activity of CNTs.

N-doped graphene was reported to exhibit high activity and selectivity for the oxidation of various arylalkanes in aqueous phase.[191] The N-doped sp^2 hybridized carbon was prepared through a CVD process with acetonitrile vapor as the N source (*ca.* 9% N content). Control experiments with different carbon materials suggested the N doping greatly enhanced catalytic activity as well as selectivity, and the activity was not related to the surface area of the catalysts. DFT simulation was carried out to study the catalytic mechanism with graphitic N as the doping format according to the XPS result. Interestingly, the catalytic active site was the *ortho* carbon site, instead of the doped graphitic-N. Based on electronic partial density calculations, it was suggested that the density of states near the Fermi level for *ortho* carbon are higher than those of other nearby C atoms, conferring it a metal-like d-band electronic structure and a more metal-like catalytic performance.

Peng *et al.* demonstrated that CNTs and N-CNTs were efficient metal-free catalysts for the aerobic allylic oxidation of cyclohexene toward the production of the corresponding unsaturated ketone and alcohol.[192] Pristine CNTs produced cyclohexenyl hydroperoxide as the major product with 60–66% selectivity at about 20% conversion (Figure 6.11). Nitrogen doping significantly enhanced the specific activity by up to 6 times. The selectivity pattern of allylic products was also altered by nitrogen doping, evidenced by the considerably improved ketone/alcohol ratio. A systematic investigation on the effects of nitrogen content, residual metals, defects, and oxygen groups indicated that the electronic interaction between radicals and carbon catalysts played a central role in this reaction.

The oxidation of benzyl alcohol has been regarded as one of the most fundamental organic transformations in synthetic chemistry, because benzaldehyde has tremendous application in the perfumery, pharmaceutical, dyestuff and agrochemical industries. Nitrogen-doped carbon nanotubes, prepared by CVD were tested in selective oxidation of benzyl alcohol to benzaldehyde with molecular oxygen as the terminal oxidant under mild reaction conditions.[193] The results showed that the N-CNTs exhibited much higher activity than the undoped CNTs, and the improved catalytic activity was attributed to the introduction of electron-rich nitrogen atoms in the graphitic domains enhancing electron transfer. Moreover, N-CNTs displayed excellent stability without any obvious loss in activity and selectivity for benzyl alcohol oxidation even after eight cycling reactions.

Wang *et al.* have employed a heat treatment procedure in NH_3 to synthesize N-doped graphene and reported its efficient catalytic activity toward the

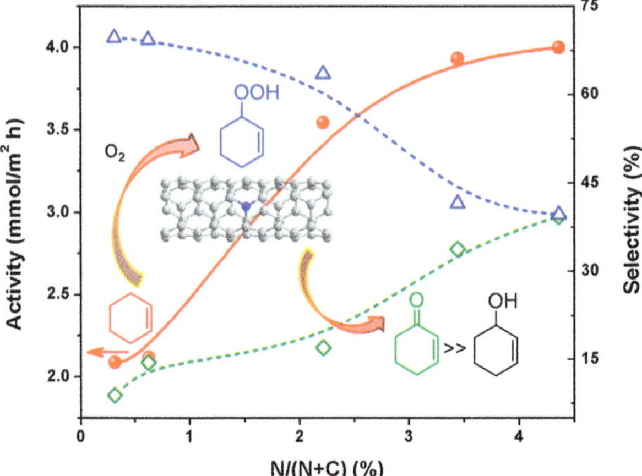

Figure 6.11 Variation of activity and selectivities of allylic products with the nitrogen content in N-CNTs. The selectivities were obtained at similar cyclohexene conversion of about 20%. Reprinted with permission from ref. 192. Copyright 2014 American Chemical Society.

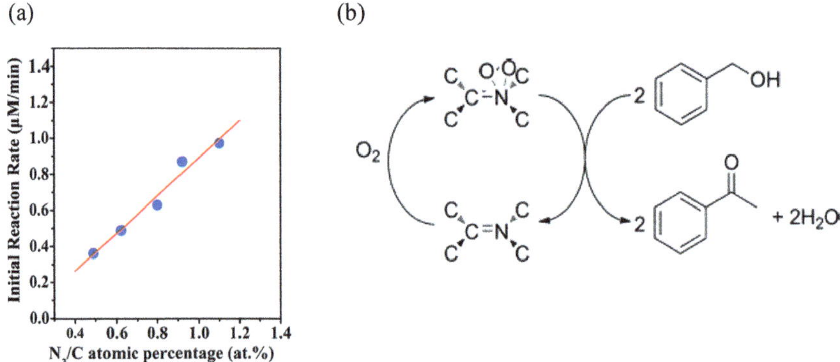

Figure 6.12 (a) Correlation between initial reaction rate and graphitic-N species; (b) Proposed reaction pathway for aerobic alcohol oxidation over N-doped graphene nanosheets. Reprinted with permission from ref. 194. Copyright 2012 American Chemical Society.

aerobic selective oxidation of different benzylic alcohols.[194] They found that the nitridation temperature greatly influenced the N doping concentrations and formats. Graphitic-N was found to be the active site with good linear correlation with activities (Figure 6.12a). The reaction process was proposed based on electron paramagnetic resonance (EPR) and it was suggested that the formation of a sp^2 N–O_2 adduct transition state, could oxidize alcohols directly to aldehydes without any byproduct, including H_2O_2 and carboxylic acids (Figure 6.12b).

6.4.3 Reduction Reactions

Nitro compounds are widely generated as by-products in different industries like agrochemicals, coloring agents and pharmaceutical products. Among various nitrocompounds, 4-nitrophenol is one of the most frequently occurring by-products, presenting a high toxicity to the environment. In pharmaceutical industries, 4-aminophenol (4-AP) is derived from 4-nitrophenol (4-NP) by reduction, and is a vital precursor for the production of medicines such as paracetamol, phenacetin and acetanilide. The reduction of 4-NP to 4-AP was first reported by Pal *et al.* using sodium borohydride ($NaBH_4$).[195] Since then it has become one of the most often used reactions to test the catalytic activity of catalysts in aqueous solution.[196] Almost all of the existing investigations on this reaction are based on metallic materials under mild conditions, and the first report of metal-free catalytic reduction of 4-NP was obtained using visible irradiation by Gazi *et al.*[197] Recently, Chen *et al.* reported the synthesis of N-graphene by hydrothermal treatment of GO and ammonia, and assessed its catalytic behavior in the reduction of 4-NP.[198] N-doped graphene was found to follow pseudo-zero-order kinetics, which was completely different from traditional pseudo-first-order reactions catalyzed by metallic nanoparticles, and this was assigned to the presence of smaller active sites. The authors also performed DFT calculations to simulate the adsorption configuration of 4-NP ions on this metal-free catalyst and proposed a reaction mechanism. Only the carbon atom near the doped N atom could be activated to catch 4-NP species (Figure 6.13) owing to its weaker conjugation compared to other carbon atoms, so the number of active sites on the graphene surface was much smaller than those of metallic nanoparticles. Therefore, the adsorption process of 4-NP on the surface of N-graphene was more pivotal than in the cases of metals, changing its reaction kinetics. Furthermore, the active sites of N-graphene were positively charged owing to the large electronegativity of doped N atoms, so 4-NP preferred to combine with the graphene sheet *via* the atom of hydroxyl group, as based on the Mulliken analysis it had a charge of −0.450 electrons, more negatively charged than the −0.246 electrons of the nitro group.

Based on the outstanding electrocatalytic properties of nitrogen-doped graphene in ORR, Xia *et al.* studied a series of nitroaromatic compounds as probe molecules to investigate the electrocatalytic property of N-graphene.[199] They reported an enhanced electrocatalytic activity toward the reduction of nitroaromatic compounds with a decreased overpotential compared with thermally reduced graphene, which could mean the development of an improved sensing platform for the detection of explosive nitroaromatic compounds. The prominent electrocatalytic activity was ascribed to the pyridine-like nitrogen atoms existing at the edge plane of graphene sheets.

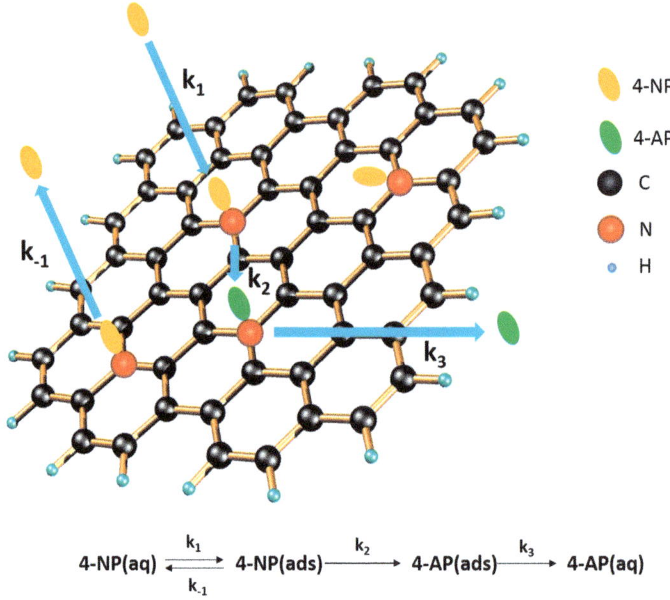

Figure 6.13 Catalytic process of the reduction of 4-nitrophenol on the surface of N-doped graphene. Reprinted with permission from ref. 198.

6.4.4 Other Applications

Hydrogen peroxide is one of most important chemicals in the world. The most elegant and efficient reaction pathway is obviously the direct conversion of elemental hydrogen and oxygen. However, such a process poses a serious safety risks with the threat of explosion. Electrochemical devices, such as polymer electrolyte membrane fuel cells (PEMFC), represent an attractive and much safer alternative by separating oxygen reduction and hydrogen oxidation together with cogeneration of energy. Fellinger *et al.* studied the application of a mesoporous nitrogen-doped carbon (derived from the ionic liquid *N*-butyl-3-methylpyridinium dicyanamide) as a metal-free catalyst for the electrochemical synthesis of hydrogen peroxide.[200] The authors produced H_2O_2 in a three electrode setup at 1600 rpm rotation speed and *ca.* 325 $\mu g\ cm_{geo}^{-2}$ catalyst loading. The electrolyte (0.1 M $HClO_4$) was continuously gas-flushed with O_2, and the working voltage was set constant at $E = 0.1$ V (*vs.* reversible hydrogen electrode, RHE), which yielded a H_2O_2 concentration of *ca.* 20 mg L^{-1}.

Alkene epoxidation is an important industrial process because the products of the reactions are versatile intermediates in fine chemical synthesis, where they serve as building blocks for the fabrication of plasticizers, perfumes, and pharmaceutical products. It was only very recently that the first report on the epoxidation of *trans*-stilbene using non-metal, graphene-based catalytic systems was reported.[201] Nitrogen-doped graphene materials synthesized

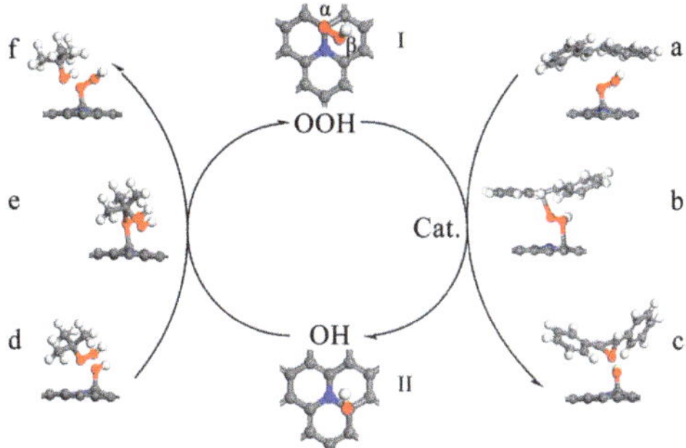

Figure 6.14 The reaction mechanism for the epoxidation of *trans*-stilbene on nitrogen-doped graphene catalyst. Reprinted with permission from ref. 201. Copyright 2014 American Chemical Society.

through a simple ammonia high-temperature treatment method was demonstrated to be very active for the C=C bond epoxidation reaction. They provided a recyclable catalytic activity for the epoxidation of *trans*-stilbene of 95.8% conversion and 94.4% selectivity to *trans*-stilbene epoxide. The authors attributed the outstanding catalytic performance to the unique properties of the nitrogen dopants at the quaternary sites (graphitic-N). DFT calculations were used by the authors in an attempt to better understand the mechanism of the nitrogen-doped sp^2 carbon catalyst and the oxidation of the carbon–carbon double bonds (Figure 6.14). The *ortho*-carbon in nitrogen-doped graphene was the most stable adsorption site of peroxide-like species (OOH). The α-oxygen (that did not connect with hydrogen) in the pristine reactive oxygen species stayed straight above one of the *ortho*-carbons, and the β-oxygen atom was pointing toward another *ortho*-carbon around the graphitic nitrogen (Figure 6.14-I). Because graphene has a large π system, reactants with π orbitals such as stilbene or styrene were able to adsorb on its surface *via* a strong π–π interaction (Figure 6.14a). With the benzene rings stabilized by the graphene domain, the locked internal C=C bond lays on top of reactive oxygen species with a favorable adsorption energy of −0.43 eV. In the transition state (Figure 6.14b), the C=C bond was attacked by the α-oxygen of the reactive oxygen species. Because of the existence of three equivalent *ortho*-carbons, the reaction between the α-oxygen and the double bond, the O–O bonds were elongated while the terminal hydroxyls containing the β-oxygen were hopping toward another adjacent *ortho*-carbon, which were positively charged (average 0.21 eV) and has delocalized spin density. As a result, the epoxide was formed (Figure 6.14c), and the terminal hydroxyls jumped, being accommodated on the adjacent *ortho*-carbon (Figure 6.14c, II). The reaction was exothermic (−1.85 eV) with a moderate barrier of 0.61 eV. Significantly, the surface reactive oxygen species could be recovered by the reaction between TBHP and the surface hydroxyl groups (with a barrier of 0.5 eV, Figure 6.14d–f), which completed the reaction cycle.

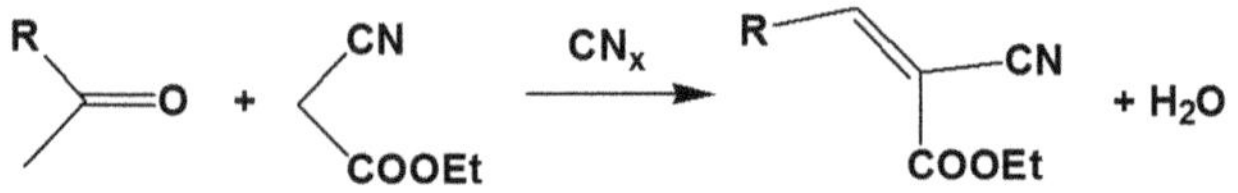

Scheme 6.2 Knoevenagel condensation of aldehydes with ethyl cyanoacetate catalyzed by N-CNT catalysts.

One potentially interesting application of doped carbon materials is as an effective solid-base catalyst for the Knoevenagel condensation (Scheme 6.2).[202,203] The Knoevenagel condensation reaction is one of the most basic routes for the synthesis of α,β-unsaturated carbonyl compounds by condensation of aldehydes or ketones with a C–H acidic methylene group-containing compound. Bitter *et al.* prepared several N-CNT materials, with varying N/C composition, by CVD using acetonitrile and pyridine as N-containing precursors.[202] The catalytic activity of the samples was tested in the Knoevenagel condensation of benzaldehyde with ethyl cyanoacetate to form ethyl-α-cyanocinnamate. According to the authors, the catalytic activity of the samples seemed to be determined by the concentration of pyridinic nitrogen present in the catalyst. Wang *et al.* also used N-CNTs with variable basicity in the Knoevenagel condensation of benzaldehyde and observed that the materials with the highest N content showed effective aldehyde conversion with 100% selectivity.[203] In addition, the N-CNT catalyst could be reused at least four times without significant loss of catalytic activity.

Other applications, such as transesterification, aldol condensations and dehydrogenations reactions that make use of the basic properties of doped carbon materials have also been studied and presented a promising degree of success, using nitrogen-doped nanotubes, graphene and nanoporous carbons.[204–208]

6.5 Heteroatom-Doped Nanostructured Carbon Materials as Catalyst Support

We have discussed over the last few sections the direct effect that metal-free nanostructured carbon materials doped with heteroatoms can have on several catalytic processes. Nevertheless, these materials can also find use as a metal catalyst supports for a number of other catalytic applications. Despite the fact that some advantages inherent to the metal-free catalysts are reduced or even lost when a metallic phase is added, some other advantages from heteroatom doping can render these materials quite interesting as catalyst support. Carbon doping can impact size, composition and stability of supported nanoparticles.[28,209] Furthermore, support doping can induce electronic effects that can transfer to the active metal phase and help modulate activity and selectivity. This enhanced activity can ultimately lead to a reduction of the amount of metal necessary for a given catalytic process, especially in the case of expensive elements such as noble metals.

The functionalization of carbon nanostructures with oxygen groups using oxidants such as nitric or sulfuric acid is a standard procedure for the preparation of supported catalysts.[210,211] This oxidation stage is almost mandatory for the preparation of metals supported on nanotubes/nanofibers and graphene supports, given that initially the materials do not possess enough anchoring sites to attain a high metal dispersion and prevent sintering of the metal active phase. The effect of oxygen present throughout the surface of carbonaceous materials on metal anchoring has already been addressed in the literature.[212,213]

Metals supported on N-containing carbons have shown enhanced performance with respect to their non-doped counterparts for several catalytic processes.[5,214] The major benefit of N-doped carbon supports is attributed to the enhancement of specific interactions between nitrogen present on the carbon matrix and the active metal phase. Besides preventing metal sintering by keeping particles well dispersed, this interaction is often responsible for an improvement on activity/selectivity, due to electronic or steric effects induced by the doped support. Metals like Pt, Ru, Pd or Au, have been used in a myriad of applications such as selective oxidations,[215,216] hydrogenations,[217,218] and electrocatalysis.[219–221]

Given the preference for boron and phosphorous doping for the edges of graphitic planes, where dangling bonds are located, there are important implications regarding metal deposition. It has been well established that phosphorous species will preferentially bond to armchair (1120) faces, whilst boron will tend to favor the attachment to the zigzag (1010) faces.[222,223] Hence, metal deposition can only occur on B or P-doping over zigzag or armchair sites, respectively.

Over the next few chapters we will be addressing the catalytic application of carbon materials (both doped and undoped) as supports for other active phases. We will focus our attention on common catalytic processes such as oxidation/hydrogenation reactions, coupling reactions (C–C, C–O and C–N), among others, but also on very important fields such as photocatalysis and electrocatalysis.

6.6 Concluding Remarks and Future Perspectives

Heteroatom doped carbon nanomaterials represent today one of the most prominent families of materials. With the selection of appropriate precursors, different metal-free heteroatoms can be doped into the carbon matrix, modifying its geometric and electronic structures and inducing interesting active sites, which contribute to their different catalytic performances. For nanocarbon heteroatom-doping, there are two main preparation methods: the first is the synchronous doping method, which depends on CVD technology, with the addition of an appropriate carbon source and heteroatom precursor, and the foreign element can be introduced into the graphene sheets during its growing process. The other method is the post treatment process, in which nanostructured carbon materials were prepared first, and then treated or annealed at high temperatures with the introduction of heteroatom containing molecules, affording different doped graphene catalysts. Despite the improvements already made, it is still very difficult to control the doping concentration, bonding formats, distributional uniformity as well as some other features. This is especially valid for the co-doping cases. All of these features directly or indirectly determine the properties of the doped carbons in catalytic reactions.

A number of rather interesting applications for metal-free nanostructured carbon materials have been thoroughly investigated in the last decade. Some examples discussed in this chapter involve their use as advanced electrodes for fuel cells, but also in oxidation, reduction and condensation reactions, the most important of which is ORR. In light of the fast growing knowledge on metal-free nanocarbon catalysts, it is clear that all types of heteroatoms (N, B, P, S) can to some extent change the catalytic reactivity of parent carbon surfaces for the ORR. However, the exact origin for ORR activity enhancement is still unclear and several mechanisms have been proposed. According to Hu *et al.*, breaking the electroneutrality of graphitic materials creates positive charged sites favorable for side-on O_2 surface adsorption.[58] Using DFT calculations, Xia *et al.* reported that the spin density was likely more important than atomic charge density in determining the catalytic active sites.[141] Huang *et al.* suggested that tailoring the π electronic system of graphene may be a preferred factor for producing significantly improved materials for ORRs, regardless of the existence of dopants.[74] Other authors have also attributed activity enhancements to structural strain in doped carbon materials, as introduction of atoms with a larger atomic size than carbon into carbon materials may cause higher strain at their edges and thereby facilitate charge localization and associated chemisorption of oxygen. Interestingly, the addition of B, S or P to the already N-modified nanostructured carbon materials often induces notable improvements to the ORR activity, by either reducing the onset overpotential (although to a small extent) or increasing the limiting current (sometimes quite obviously). In this case, the underlying synergistic effect between N and other heteroatoms might be related to the strengthened disturbance to the homogeneous electronic structures of clean nanocarbon surfaces.

However, an exact correlation between the nature of the active site and the reactivity of the nanocarbon catalysts is still limited to a large extent. For instance, the ORR active N-containing groups have not been completely identified. The difficulty lies in the complex properties (electron donator and acceptor, basicity and acidity, hydrophilicity and hydrophobicity, polarity and nonpolarity, among others) of heterogeneous carbon surfaces, which are closely dependent on the synthesis chemistry and post-treatment. Furthermore, the reactivity of these active sites is also largely related to the edge/basal planes ratio, topological defects and synergistic interactions among them. Hence, a deeper insight into the identification of active sites exposed in the electrochemical environment is needed in order to prepare specifically designed metal-free doped nanostructured carbon materials that can provide competitive performances regarding those exhibited by metal-based catalysts.

Although the industrial application of heterogeneous nanocarbon is uncertain, these new metal-free catalysts have already exhibited excellent properties in many catalytic reactions, some of which are extremely promising, and we believe a bright future lies ahead for the research and development of this new cost-effective alternative material.

References

1. P. Trogadas, T. F. Fuller and P. Strasser, *Carbon*, 2014, **75**, 5–42.
2. W. Y. Wong, W. R. W. Daud, A. B. Mohamad, A. A. H. Kadhum, K. S. Loh and E. H. Majlan, *Int. J. Hydrogen Energy*, 2013, **38**, 9370–9386.
3. Z. Yang, H. Nie, X. a Chen, X. Chen and S. Huang, *J. Power Sources*, 2013, **236**, 238–249.
4. U. N. Maiti, W. J. Lee, J. M. Lee, Y. Oh, J. Y. Kim, J. E. Kim, J. Shim, T. H. Han and S. O. Kim, *Adv. Mater.*, 2014, **26**, 40–67.
5. W. Shen and W. Fan, *J. Mater. Chem. A*, 2013, **1**, 999–1013.
6. A. Stein, Z. Wang and M. A. Fierke, *Adv. Mater.*, 2009, **21**, 265–293.
7. J. P. Paraknowitsch and A. Thomas, *Macromol. Chem. Phys.*, 2012, **213**, 1132–1145.
8. Y. Xue, B. Wu, L. Jiang, Y. Guo, L. Huang, J. Chen, J. Tan, D. Geng, B. Luo, W. Hu, G. Yu and Y. Liu, *J. Am. Chem. Soc.*, 2012, **134**, 11060–11063.
9. Z. Luo, S. Lim, Z. Tian, J. Shang, L. Lai, B. MacDonald, C. Fu, Z. Shen, T. Yu and J. Lin, *J. Mater. Chem.*, 2011, **21**, 8038–8044.
10. X. Wang, X. Li, L. Zhang, Y. Yoon, P. K. Weber, H. Wang, J. Guo and H. Dai, *Science*, 2009, **324**, 768–771.
11. D. Wei, Y. Liu, Y. Wang, H. Zhang, L. Huang and G. Yu, *Nano Lett.*, 2009, **9**, 1752–1758.
12. D. H. Lee, W. J. Lee and S. O. Kim, *Nano Lett.*, 2009, **9**, 1427–1432.
13. L. S. Panchakarla, K. S. Subrahmanyam, S. K. Saha, A. Govindaraj, H. R. Krishnamurthy, U. V. Waghmare and C. N. R. Rao, *Adv. Mater.*, 2009, **21**, 4726–4730.
14. Z. Wang, C. H. Yu, D. C. Ba and J. Liang, *Vacuum*, 2007, **81**, 579–582.
15. E. Cruz-Silva, D. A. Cullen, L. Gu, J. M. Romo-Herrera, E. Muñoz-Sandoval, F. López-Urías, B. G. Sumpter, V. Meunier, J.-C. Charlier, D. J. Smith, H. Terrones and M. Terrones, *ACS Nano*, 2008, **2**, 441–448.
16. S. Some, J. Kim, K. Lee, A. Kulkarni, Y. Yoon, S. Lee, T. Kim and H. Lee, *Adv. Mater.*, 2012, **24**, 5481–5486.
17. G. Hui, L. Zheng, S. Li, G. Wenhua, G. Wei, C. Lijie, R. Amrita, Q. Weijin, V. Robert and M. A. Pulickel, *Nanotechnology*, 2012, **23**, 275605.
18. J. Xu, G. Dong, C. Jin, M. Huang and L. Guan, *ChemSusChem*, 2013, **6**, 493–499.
19. B. G. Sumpter, V. Meunier, J. M. Romo-Herrera, E. Cruz-Silva, D. A. Cullen, H. Terrones, D. J. Smith and M. Terrones, *ACS Nano*, 2007, **1**, 369–375.

20. J. M. Romo-Herrera, B. G. Sumpter, D. A. Cullen, H. Terrones, E. Cruz-Silva, D. J. Smith, V. Meunier and M. Terrones, *Angew. Chem., Int. Ed.*, 2008, **47**, 2948–2953.
21. D. P. Hashim, N. T. Narayanan, J. M. Romo-Herrera, D. A. Cullen, M. G. Hahm, P. Lezzi, J. R. Suttle, D. Kelkhoff, E. Muñoz-Sandoval, S. Ganguli, A. K. Roy, D. J. Smith, R. Vajtai, B. G. Sumpter, V. Meunier, H. Terrones, M. Terrones and P. M. Ajayan, *Sci. Rep.*, 2012, **2**.
22. H. M. Jeong, J. W. Lee, W. H. Shin, Y. J. Choi, H. J. Shin, J. K. Kang and J. W. Choi, *Nano Lett.*, 2011, **11**, 2472–2477.
23. S. K. Hwang, J. M. Lee, S. Kim, J. S. Park, H. I. Park, C. W. Ahn, K. J. Lee, T. Lee and S. O. Kim, *Nano Lett.*, 2012, **12**, 2217–2221.
24. X. Li, H. Wang, J. T. Robinson, H. Sanchez, G. Diankov and H. Dai, *J. Am. Chem. Soc.*, 2009, **131**, 15939–15944.
25. S. Yang, L. Zhi, K. Tang, X. Feng, J. Maier and K. Müllen, *Adv. Funct. Mater.*, 2012, **22**, 3634–3640.
26. S. Park, Y. Hu, J. O. Hwang, E.-S. Lee, L. B. Casabianca, W. Cai, J. R. Potts, H.-W. Ha, S. Chen, J. Oh, S. O. Kim, Y.-H. Kim, Y. Ishii and R. S. Ruoff, *Nat. Commun.*, 2012, **3**, 638.
27. X. Feng, C. Matranga, R. Vidic and E. Borguet, *J. Phys. Chem. B*, 2004, **108**, 19949–19954.
28. H.-P. Boehm, in *Carbon Materials for Catalysis*, John Wiley & Sons, Inc., 2008, pp. 219–265.
29. H. Wang, T. Maiyalagan and X. Wang, *ACS Catal.*, 2012, **2**, 781–794.
30. P. Ayala, R. Arenal, M. Rümmeli, A. Rubio and T. Pichler, *Carbon*, 2010, **48**, 575–586.
31. C. P. Ewels and M. Glerup, *J. Nanosci. Nanotechnol.*, 2005, **5**, 1345–1363.
32. J. P. Paraknowitsch and A. Thomas, *Energy Environ. Sci.*, 2013, **6**, 2839–2855.
33. R. Sen, B. C. Satishkumar, A. Govindaraj, K. R. Harikumar, M. K. Renganathan and C. N. R. Rao, *J. Mater. Chem.*, 1997, **7**, 2335–2337.
34. Y. Wu, S. Fang and Y. Jiang, *J. Mater. Chem.*, 1998, **8**, 2223–2227.
35. D. P. Kim, C. L. Lin, T. Mihalisin, P. Heiney and M. M. Labes, *Chem. Mater.*, 1991, **3**, 686–692.
36. L. Qu, Y. Liu, J.-B. Baek and L. Dai, *ACS Nano*, 2010, **4**, 1321–1326.
37. B. Hu, K. Wang, L. Wu, S.-H. Yu, M. Antonietti and M.-M. Titirici, *Adv. Mater.*, 2010, **22**, 813–828.
38. M.-M. Titirici and M. Antonietti, *Chem. Soc. Rev.*, 2010, **39**, 103–116.
39. J. P. Paraknowitsch, A. Thomas and M. Antonietti, *J. Mater. Chem.*, 2010, **20**, 6746–6758.
40. J. P. Paraknowitsch, J. Zhang, D. Su, A. Thomas and M. Antonietti, *Adv. Mater.*, 2010, **22**, 87–92.
41. X. Wang and S. Dai, *Angew. Chem., Int. Ed.*, 2010, **49**, 6664–6668.
42. J. S. Lee, X. Wang, H. Luo and S. Dai, *Adv. Mater.*, 2010, **22**, 1004–1007.
43. M. C. Gutierrez, D. Carriazo, C. O. Ania, J. B. Parra, M. L. Ferrer and F. del Monte, *Energy Environ. Sci.*, 2011, **4**, 3535–3544.

44. D. Golberg, Y. Bando, L. Bourgeois, K. Kurashima and T. Sato, *Carbon*, 2000, **38**, 2017–2027.
45. R. Imran Jafri, N. Rajalakshmi and S. Ramaprabhu, *J. Mater. Chem.*, 2010, **20**, 7114–7117.
46. Y. Wang, Y. Shao, D. W. Matson, J. Li and Y. Lin, *ACS Nano*, 2010, **4**, 1790–1798.
47. F. Kapteijn, J. A. Moulijn, S. Matzner and H. P. Boehm, *Carbon*, 1999, **37**, 1143–1150.
48. M. Kodama, J. Yamashita, Y. Soneda, H. Hatori and K. Kamegawa, *Carbon*, 2007, **45**, 1105–1107.
49. D. Golberg, Y. Bando, Y. Huang, T. Terao, M. Mitome, C. Tang and C. Zhi, *ACS Nano*, 2010, **4**, 2979–2993.
50. R. Ma, D. Golberg, Y. Bando and T. Sasaki, *Philos. Trans. R. Soc. London Ser. A*, 2004, **362**, 2161–2186.
51. G. G. Fuentes, E. Borowiak-Palen, M. Knupfer, T. Pichler, J. Fink, L. Wirtz and A. Rubio, *Phys. Rev. B*, 2004, **69**, 245403.
52. P. L. Gai, O. Stephan, K. McGuire, A. M. Rao, M. S. Dresselhaus, G. Dresselhaus and C. Colliex, *J. Mater. Chem.*, 2004, **14**, 669–675.
53. L. R. Radovic, M. Karra, K. Skokova and P. A. Thrower, *Carbon*, 1998, **36**, 1841–1854.
54. D. H. Zhong, H. Sano, Y. Uchiyama and K. Kobayashi, *Carbon*, 2000, **38**, 1199–1206.
55. W. Cermignani, T. E. Paulson, C. Onneby and C. G. Pantano, *Carbon*, 1995, **33**, 367–374.
56. T. Shirasaki, A. Derré, M. Ménétrier, A. Tressaud and S. Flandrois, *Carbon*, 2000, **38**, 1461–1467.
57. E. Kim, I. Oh and J. Kwak, *Electrochem. Commun.*, 2001, **3**, 608–612.
58. L. Yang, S. Jiang, Y. Zhao, L. Zhu, S. Chen, X. Wang, Q. Wu, J. Ma, Y. Ma and Z. Hu, *Angew. Chem., Int. Ed.*, 2011, **50**, 7132–7135.
59. C. A. Davis, Y. Yin, D. R. McKenzie, L. E. Hall, E. Kravtchinskaia, V. Keast, G. A. J. Amaratunga and V. S. Veerasamy, *J. Non-Cryst. Solids*, 1994, **170**, 46–50.
60. S. Bohr, R. Haubner and B. Lux, *Diamond Relat. Mater.*, 1995, **4**, 133–144.
61. R. Haubner, S. Bohr and B. Lux, *Diamond Relat. Mater.*, 1999, **8**, 171–178.
62. M. T. Kuo, P. W. May, A. Gunn, M. N. R. Ashfold and R. K. Wild, *Diamond Relat. Mater.*, 2000, **9**, 1222–1227.
63. C. Zhang, N. Mahmood, H. Yin, F. Liu and Y. Hou, *Adv. Mater.*, 2013, **25**, 4932–4937.
64. F. Niu, L.-M. Tao, Y.-C. Deng, Q.-H. Wang and W.-G. Song, *New J. Chem.*, 2014, **38**, 2269–2272.
65. Z. Liu, F. Peng, H. Wang, H. Yu, J. Tan and L. Zhu, *Catal. Commun.*, 2011, **16**, 35–38.
66. Z.-W. Liu, F. Peng, H.-J. Wang, H. Yu, W.-X. Zheng and J. Yang, *Angew. Chem., Int. Ed.*, 2011, **50**, 3257–3261.
67. D.-S. Yang, D. Bhattacharjya, S. Inamdar, J. Park and J.-S. Yu, *J. Am. Chem. Soc.*, 2012, **134**, 16127–16130.

68. P. A. Denis, *Chem. Phys. Lett.*, 2010, **492**, 251–257.
69. P. A. Denis, *Comput. Mater. Sci.*, 2013, **67**, 203–206.
70. V. V. Strelko, V. S. Kuts and P. A. Thrower, *Carbon*, 2000, **38**, 1499–1503.
71. C. H. Choi, S. H. Park and S. I. Woo, *J. Mater. Chem.*, 2012, **22**, 12107–12115.
72. C. H. Choi, S. H. Park and S. I. Woo, *ACS Nano*, 2012, **6**, 7084–7091.
73. W. Kiciński, M. Szala and M. Bystrzejewski, *Carbon*, 2014, **68**, 1–32.
74. Z. Yang, Z. Yao, G. Li, G. Fang, H. Nie, Z. Liu, X. Zhou, X. a Chen and S. Huang, *ACS Nano*, 2011, **6**, 205–211.
75. H. L. Poh, P. Šimek, Z. Sofer and M. Pumera, *ACS Nano*, 2013, **7**, 5262–5272.
76. T.-P. Fellinger, A. Thomas, J. Yuan and M. Antonietti, *Adv. Mater.*, 2013, **25**, 5838–5855.
77. J. P. Paraknowitsch, B. Wienert, Y. Zhang and A. Thomas, *Chem. - Eur. J.*, 2012, **18**, 15416–15423.
78. N. Fechler, T.-P. Fellinger and M. Antonietti, *J. Mater. Chem. A*, 2013, **1**, 14097–14102.
79. C. You, S. Liao, H. Li, S. Hou, H. Peng, X. Zeng, F. Liu, R. Zheng, Z. Fu and Y. Li, *Carbon*, 2014, **69**, 294–301.
80. W. Kiciński and A. Dziura, *Carbon*, 2014, **75**, 56–67.
81. Y. Shao, S. Zhang, M. H. Engelhard, G. Li, G. Shao, Y. Wang, J. Liu, I. A. Aksay and Y. Lin, *J. Mater. Chem.*, 2010, **20**, 7491–7496.
82. S. Maldonado and K. J. Stevenson, *J. Phys. Chem. B*, 2005, **109**, 4707–4716.
83. S. Maldonado, S. Morin and K. J. Stevenson, *Carbon*, 2006, **44**, 1429–1437.
84. E. Fuente, J. A. Menéndez, D. Suárez and M. A. Montes-Morán, *Langmuir*, 2003, **19**, 3505–3511.
85. M. A. Montes-Morán, D. Suárez, J. A. Menéndez and E. Fuente, *Carbon*, 2004, **42**, 1219–1225.
86. M. S. Shafeeyan, W. M. A. W. Daud, A. Houshmand and A. Shamiri, *J. Anal. Appl. Pyrolysis*, 2010, **89**, 143–151.
87. V. Strelko Jr, D. J. Malik and M. Streat, *Carbon*, 2002, **40**, 95–104.
88. I. Shimoyama, G. Wu, T. Sekiguchi and Y. Baba, *J. Electron Spectrosc. Relat. Phenom.*, 2001, **114–116**, 841–848.
89. M. Balón, M. C. Carmona, M. A. Muñoz and J. Hidalgo, *Tetrahedron*, 1989, **45**, 7501–7504.
90. S. Kundu, W. Xia, W. Busser, M. Becker, D. A. Schmidt, M. Havenith and M. Muhler, *Phys. Chem. Chem. Phys.*, 2010, **12**, 4351–4359.
91. A. H. Nevidomskyy, G. Csányi and M. C. Payne, *Phys. Rev. Lett.*, 2003, **91**, 105502.
92. Y. El-Sayed and T. J. Bandosz, *Langmuir*, 2005, **21**, 1282–1289.
93. A. Bagreev, J. A. Menendez, I. Dukhno, Y. Tarasenko and T. J. Bandosz, *Carbon*, 2005, **43**, 208–210.
94. B. R. Puri, in *Chemistry and Physics of Carbon*, ed. P. L. Walker, Marcel Dekker, New York, 1970, vol. 6, pp. 191–282.
95. M. Terrones, A. S. Filho and A. Rao, in *Carbon Nanotubes*, ed. A. Jorio, G. Dresselhaus and M. Dresselhaus, Springer Berlin Heidelberg, 2008, vol. 111, pp. 531–566.

96. D. Usachov, O. Vilkov, A. Grüneis, D. Haberer, A. Fedorov, V. K. Adamchuk, A. B. Preobrajenski, P. Dudin, A. Barinov, M. Oehzelt, C. Laubschat and D. V. Vyalikh, *Nano Lett.*, 2011, **11**, 5401–5407.
97. J. Bai, Q. Zhu, Z. Lv, H. Dong, J. Yu and L. Dong, *Int. J. Hydrogen Energy*, 2013, **38**, 1413–1418.
98. J. C. Carrero-Sánchez, A. L. Elías, R. Mancilla, G. Arrellín, H. Terrones, J. P. Laclette and M. Terrones, *Nano Lett.*, 2006, **6**, 1609–1616.
99. D.-W. Wang and D. Su, *Energy Environ. Sci.*, 2014, **7**, 576–591.
100. V. R. Stamenkovic, B. S. Mun, M. Arenz, K. J. J. Mayrhofer, C. A. Lucas, G. Wang, P. N. Ross and N. M. Markovic, *Nat. Mater.*, 2007, **6**, 241–247.
101. S. Jiang, Y. Ma, G. Jian, H. Tao, X. Wang, Y. Fan, Y. Lu, Z. Hu and Y. Chen, *Adv. Mater.*, 2009, **21**, 4953–4956.
102. D. Wang, H. L. Xin, R. Hovden, H. Wang, Y. Yu, D. A. Muller, F. J. DiSalvo and H. D. Abruña, *Nat. Mater.*, 2013, **12**, 81–87.
103. G. Wu, K. L. More, C. M. Johnston and P. Zelenay, *Science*, 2011, **332**, 443–447.
104. R. B. Levy and M. Boudart, *Science*, 1973, **181**, 547–549.
105. F. Jaouen, E. Proietti, M. Lefevre, R. Chenitz, J.-P. Dodelet, G. Wu, H. T. Chung, C. M. Johnston and P. Zelenay, *Energy Environ. Sci.*, 2011, **4**, 114–130.
106. Z. Chen, D. Higgins, A. Yu, L. Zhang and J. Zhang, *Energy Environ. Sci.*, 2011, **4**, 3167–3192.
107. B. Winther-Jensen, O. Winther-Jensen, M. Forsyth and D. R. MacFarlane, *Science*, 2008, **321**, 671–674.
108. K. Gong, F. Du, Z. Xia, M. Durstock and L. Dai, *Science*, 2009, **323**, 760–764.
109. Y. Zheng, Y. Jiao, M. Jaroniec, Y. Jin and S. Z. Qiao, *Small*, 2012, **8**, 3550–3566.
110. D. Yu, E. Nagelli, F. Du and L. Dai, *J. Phys. Chem. Lett.*, 2010, **1**, 2165–2173.
111. A. J. Bard, *J. Am. Chem. Soc.*, 2010, **132**, 7559–7567.
112. L. Yu, X. Pan, X. Cao, P. Hu and X. Bao, *J. Catal.*, 2011, **282**, 183–190.
113. S. Chen, J. Bi, Y. Zhao, L. Yang, C. Zhang, Y. Ma, Q. Wu, X. Wang and Z. Hu, *Adv. Mater.*, 2012, **24**, 5593–5597.
114. R. Arrigo, M. Hävecker, S. Wrabetz, R. Blume, M. Lerch, J. McGregor, E. P. J. Parrott, J. A. Zeitler, L. F. Gladden, A. Knop-Gericke, R. Schlögl and D. S. Su, *J. Am. Chem. Soc.*, 2010, **132**, 9616–9630.
115. M. Lefèvre, E. Proietti, F. Jaouen and J.-P. Dodelet, *Science*, 2009, **324**, 71–74.
116. J. Yang, D.-J. Liu, N. N. Kariuki and L. X. Chen, *Chem. Commun.*, 2008, 329–331.
117. P. Matter and U. Ozkan, *Catal. Lett.*, 2006, **109**, 115–123.
118. W. Yang, T.-P. Fellinger and M. Antonietti, *J. Am. Chem. Soc.*, 2010, **133**, 206–209.
119. Y. Tang, B. L. Allen, D. R. Kauffman and A. Star, *J. Am. Chem. Soc.*, 2009, **131**, 13200–13201.
120. W. Xiong, F. Du, Y. Liu, A. Perez, M. Supp, T. S. Ramakrishnan, L. Dai and L. Jiang, *J. Am. Chem. Soc.*, 2010, **132**, 15839–15841.

121. T. C. Nagaiah, S. Kundu, M. Bron, M. Muhler and W. Schuhmann, *Electrochem. Commun.*, 2010, **12**, 338–341.
122. R. Liu, D. Wu, X. Feng and K. Müllen, *Angew. Chem., Int. Ed.*, 2010, **49**, 2565–2569.
123. D. Yu and L. Dai, *J. Phys. Chem. Lett.*, 2009, **1**, 467–470.
124. M. Bacon, S. J. Bradley and T. Nann, *Part. Part. Syst. Charact.*, 2014, **31**, 415–428.
125. Q. Li, S. Zhang, L. Dai and L.-s Li, *J. Am. Chem. Soc.*, 2012, **134**, 18932–18935.
126. M. Terrones, H. Terrones, N. Grobert, W. K. Hsu, Y. Q. Zhu, J. P. Hare, H. W. Kroto, D. R. M. Walton, P. Kohler-Redlich, M. Rühle, J. P. Zhang and A. K. Cheetham, *Appl. Phys. Lett.*, 1999, **75**, 3932–3934.
127. Z. Sun, Z. Yan, J. Yao, E. Beitler, Y. Zhu and J. M. Tour, *Nature*, 2011, **471**, 124–124.
128. Y. Sun, C. Li, Y. Xu, H. Bai, Z. Yao and G. Shi, *Chem. Commun.*, 2010, **46**, 4740–4742.
129. H. Kim, K. Lee, S. I. Woo and Y. Jung, *Phys. Chem. Chem. Phys.*, 2011, **13**, 17505–17510.
130. L. Lai, J. R. Potts, D. Zhan, L. Wang, C. K. Poh, C. Tang, H. Gong, Z. Shen, J. Lin and R. S. Ruoff, *Energy Environ. Sci.*, 2012, **5**, 7936–7942.
131. D. Geng, Y. Chen, Y. Chen, Y. Li, R. Li, X. Sun, S. Ye and S. Knights, *Energy Environ. Sci.*, 2011, **4**, 760–764.
132. H. Niwa, K. Horiba, Y. Harada, M. Oshima, T. Ikeda, K. Terakura, J.-i. Ozaki and S. Miyata, *J. Power Sources*, 2009, **187**, 93–97.
133. Z. Lin, G. Waller, Y. Liu, M. Liu and C.-P. Wong, *Adv. Energy Mater.*, 2012, **2**, 884–888.
134. P. H. Matter, L. Zhang and U. S. Ozkan, *J. Catal.*, 2006, **239**, 83–96.
135. S. Kundu, T. C. Nagaiah, W. Xia, Y. Wang, S. V. Dommele, J. H. Bitter, M. Santa, G. Grundmeier, M. Bron, W. Schuhmann and M. Muhler, *J. Phys. Chem. C*, 2009, **113**, 14302–14310.
136. Y. Sun, C. Li and G. Shi, *J. Mater. Chem.*, 2012, **22**, 12810–12816.
137. K. A. Kurak and A. B. Anderson, *J. Phys. Chem. C*, 2009, **113**, 6730–6734.
138. T. Ikeda, M. Boero, S.-F. Huang, K. Terakura, M. Oshima and J.-i Ozaki, *J. Phys. Chem. C*, 2008, **112**, 14706–14709.
139. Y. Okamoto, *Appl. Surf. Sci.*, 2009, **256**, 335–341.
140. D. W. Boukhvalov and Y.-W. Son, *Nanoscale*, 2012, **4**, 417–420.
141. L. Zhang and Z. Xia, *J. Phys. Chem. C*, 2011, **115**, 11170–11176.
142. X. Zhou, Z. Yang, H. Nie, Z. Yao, L. Zhang and S. Huang, *J. Power Sources*, 2011, **196**, 9970–9974.
143. Z.-H. Sheng, L. Shao, J.-J. Chen, W.-J. Bao, F.-B. Wang and X.-H. Xia, *ACS Nano*, 2011, **5**, 4350–4358.
144. C. Jin, T. C. Nagaiah, W. Xia, B. Spliethoff, S. Wang, M. Bron, W. Schuhmann and M. Muhler, *Nanoscale*, 2010, **2**, 981–987.
145. X. Wang, J. S. Lee, Q. Zhu, J. Liu, Y. Wang and S. Dai, *Chem. Mater.*, 2010, **22**, 2178–2180.

146. D. Yu, Q. Zhang and L. Dai, *J. Am. Chem. Soc.*, 2010, **132**, 15127–15129.
147. J.-i Ozaki, S.-i Tanifuji, N. Kimura, A. Furuichi and A. Oya, *Carbon*, 2006, **44**, 1324–1326.
148. J. Yin, Y. Qiu and J. Yu, *J. Electroanal. Chem.*, 2013, **702**, 56–59.
149. H. Li, H. Liu, Z. Jong, W. Qu, D. Geng, X. Sun and H. Wang, *Int. J. Hydrogen Energy*, 2011, **36**, 2258–2265.
150. Z. Chen, D. Higgins and Z. Chen, *Electrochim. Acta*, 2010, **55**, 4799–4804.
151. D. Higgins, Z. Chen and Z. Chen, *Electrochim. Acta*, 2011, **56**, 1570–1575.
152. P. H. Matter, E. Wang, M. Arias, E. J. Biddinger and U. S. Ozkan, *J. Mol. Catal. A: Chem.*, 2007, **264**, 73–81.
153. J. Yan, H. Meng, W. Yu, X. Yuan, W. Lin, W. Ouyang and D. Yuan, *Electrochim. Acta*, 2014, **129**, 196–202.
154. J. Yan, H. Meng, F. Xie, X. Yuan, W. Yu, W. Lin, W. Ouyang and D. Yuan, *J. Power Sources*, 2014, **245**, 772–778.
155. Y. Zhang, K. Fugane, T. Mori, L. Niu and J. Ye, *J. Mater. Chem.*, 2012, **22**, 6575–6580.
156. J.-e Park, Y. J. Jang, Y. J. Kim, M.-S. Song, S. Yoon, D. H. Kim and S.-J. Kim, *Phys. Chem. Chem. Phys.*, 2014, **16**, 103–109.
157. S.-Y. Yang, K.-H. Chang, Y.-L. Huang, Y.-F. Lee, H.-W. Tien, S.-M. Li, Y.-H. Lee, C.-H. Liu, C.-C. M. Ma and C.-C. Hu, *Electrochem. Commun.*, 2012, **14**, 39–42.
158. Z. Lin, G. H. Waller, Y. Liu, M. Liu and C.-p Wong, *Carbon*, 2013, **53**, 130–136.
159. T.-N. Ye, L.-B. Lv, X.-H. Li, M. Xu and J.-S. Chen, *Angew. Chem., Int. Ed.*, 2014, **53**, 6905–6909.
160. S. Wang, E. Iyyamperumal, A. Roy, Y. Xue, D. Yu and L. Dai, *Angew. Chem., Int. Ed.*, 2011, **50**, 11756–11760.
161. Y. Zheng, Y. Jiao, L. Ge, M. Jaroniec and S. Z. Qiao, *Angew. Chem., Int. Ed.*, 2013, **52**, 3110–3116.
162. D.-S. Yang, D. Bhattacharjya, M. Y. Song and J.-S. Yu, *Carbon*, 2014, **67**, 736–743.
163. J. Wu, Z. Yang, Q. Sun, X. Li, P. Strasser and R. Yang, *Electrochim. Acta*, 2014, **127**, 53–60.
164. S.-A. Wohlgemuth, R. J. White, M.-G. Willinger, M.-M. Titirici and M. Antonietti, *Green Chem.*, 2012, **14**, 1515–1523.
165. J. Liang, Y. Jiao, M. Jaroniec and S. Z. Qiao, *Angew. Chem., Int. Ed.*, 2012, **51**, 11496–11500.
166. G. Jo and S. Shanmugam, *Electrochem. Commun.*, 2012, **25**, 101–104.
167. Z.-H. Sheng, H.-L. Gao, W.-J. Bao, F.-B. Wang and X.-H. Xia, *J. Mater. Chem.*, 2012, **22**, 390–395.
168. Y. Tan, C. Xu, G. Chen, X. Fang, N. Zheng and Q. Xie, *Adv. Funct. Mater.*, 2012, **22**, 4584–4591.
169. S. Inamdar, H.-S. Choi, P. Wang, M. Y. Song and J.-S. Yu, *Electrochem. Commun.*, 2013, **30**, 9–12.
170. J.-i Ozaki, N. Kimura, T. Anahara and A. Oya, *Carbon*, 2007, **45**, 1847–1853.

171. S. Wang, L. Zhang, Z. Xia, A. Roy, D. W. Chang, J.-B. Baek and L. Dai, *Angew. Chem., Int. Ed.*, 2012, **51**, 4209–4212.
172. Y. Zhao, L. Yang, S. Chen, X. Wang, Y. Ma, Q. Wu, Y. Jiang, W. Qian and Z. Hu, *J. Am. Chem. Soc.*, 2013, **135**, 1201–1204.
173. T. Ikeda, M. Boero, S.-F. Huang, K. Terakura, M. Oshima, J.-i. Ozaki and S. Miyata, *J. Phys. Chem. C*, 2010, **114**, 8933–8937.
174. D. Yu, Y. Xue and L. Dai, *J. Phys. Chem. Lett.*, 2012, **3**, 2863–2870.
175. C. H. Choi, S. H. Park and S. I. Woo, *Green Chem.*, 2011, **13**, 406–412.
176. J. A. Labinger and J. E. Bercaw, *Nature*, 2002, **417**, 507–514.
177. Y. Cao, X. Luo, H. Yu, F. Peng, H. Wang and G. Ning, *Catal. Sci. Technol.*, 2013, **3**, 2654–2660.
178. Y. Cao, H. Yu, J. Tan, F. Peng, H. Wang, J. Li, W. Zheng and N.-B. Wong, *Carbon*, 2013, **57**, 433–442.
179. X. Yang, H. Wang, J. Li, W. Zheng, R. Xiang, Z. Tang, H. Yu and F. Peng, *Chem. – Eur. J.*, 2013, **19**, 9818–9824.
180. H. Yu, F. Peng, J. Tan, X. Hu, H. Wang, J. Yang and W. Zheng, *Angew. Chem., Int. Ed.*, 2011, **50**, 3978–3982.
181. J. Luo, F. Peng, H. Yu, H. Wang and W. Zheng, *ChemCatChem*, 2013, **5**, 1578–1586.
182. F. Cavani, N. Ballarini and A. Cericola, *Catal. Today*, 2007, **127**, 113–131.
183. B. Frank, J. Zhang, R. Blume, R. Schlögl and D. S. Su, *Angew. Chem., Int. Ed.*, 2009, **48**, 6913–6917.
184. Z.-j. Sui, J.-h Zhou, Y.-c Dai and W.-k Yuan, *Catal. Today*, 2005, **106**, 90–94.
185. C. Chen, J. Zhang, B. Zhang, C. Yu, F. Peng and D. Su, *Chem. Commun.*, 2013, **49**, 8151–8153.
186. P. J. Hart, F. J. Vastola and P. L. Walker Jr, *Carbon*, 1967, **5**, 363–371.
187. W. P. Hoffman, F. J. Vastola and P. L. Walker Jr, *Carbon*, 1984, **22**, 585–594.
188. X. Hu, Y. Wu, H. Li and Z. Zhang, *J. Phys. Chem. C*, 2010, **114**, 9603–9607.
189. R. Zhao, D. Ji, G. Lv, G. Qian, L. Yan, X. Wang and J. Suo, *Chem. Commun.*, 2004, 904–905.
190. M. Dugal, G. Sankar, R. Raja and J. M. Thomas, *Angew. Chem., Int. Ed.*, 2000, **39**, 2310–2313.
191. Y. Gao, G. Hu, J. Zhong, Z. Shi, Y. Zhu, D. S. Su, J. Wang, X. Bao and D. Ma, *Angew. Chem., Int. Ed.*, 2013, **52**, 2109–2113.
192. Y. Cao, H. Yu, F. Peng and H. Wang, *ACS Catal.*, 2014, **4**, 1617–1625.
193. J. Luo, F. Peng, H. Wang and H. Yu, *Catal. Commun.*, 2013, **39**, 44–49.
194. J. Long, X. Xie, J. Xu, Q. Gu, L. Chen and X. Wang, *ACS Catal.*, 2012, **2**, 622–631.
195. N. Pradhan, A. Pal and T. Pal, *Colloids Surf., A*, 2002, **196**, 247–257.
196. P. Herves, M. Perez-Lorenzo, L. M. Liz-Marzan, J. Dzubiella, Y. Lu and M. Ballauff, *Chem. Soc. Rev.*, 2012, **41**, 5577–5587.
197. S. Gazi and R. Ananthakrishnan, *Appl. Catal., B*, 2011, **105**, 317–325.
198. X.-k Kong, Z.-y Sun, M. Chen, C.-l Chen and Q.-w Chen, *Energy Environ. Sci.*, 2013, **6**, 3260–3266.
199. T.-W. Chen, J.-Y. Xu, Z.-H. Sheng, K. Wang, F.-B. Wang, T.-M. Liang and X.-H. Xia, *Electrochem. Commun.*, 2012, **16**, 30–33.

200. T.-P. Fellinger, F. Hasché, P. Strasser and M. Antonietti, *J. Am. Chem. Soc.*, 2012, **134**, 4072–4075.
201. W. Li, Y. Gao, W. Chen, P. Tang, W. Li, Z. Shi, D. Su, J. Wang and D. Ma, *ACS Catal.*, 2014, **4**, 1261–1266.
202. S. van Dommele, K. P. de Jong and J. H. Bitter, *Chem. Commun.*, 2006, 4859–4861.
203. L. Wang, L. Wang, H. Jin and N. Bing, *Catal. Commun.*, 2011, **15**, 78–81.
204. X. Yuan, M. Zhang, X. Chen, N. An, G. Liu, Y. Liu, W. Zhang, W. Yan and M. Jia, *Appl. Catal., A*, 2012, **439–440**, 149–155.
205. A. Villa, J.-P. Tessonnier, O. Majoulet, D. S. Su and R. Schlögl, *ChemSusChem*, 2010, **3**, 241–245.
206. N. Kan-nari, S. Okamura, S.-i. Fujita, J.-i. Ozaki and M. Arai, *Adv. Synth. Catal.*, 2010, **352**, 1476–1484.
207. E. Asedegbega-Nieto, M. Perez-Cadenas, M. V. Morales, B. Bachiller-Baeza, E. Gallegos-Suarez, I. Rodriguez-Ramos and A. Guerrero-Ruiz, *Diamond Relat. Mater.*, 2014, **44**, 26–32.
208. L. Faba, Y. A. Criado, E. Gallegos-Suárez, M. Pérez-Cadenas, E. Díaz, I. Rodríguez-Ramos, A. Guerrero-Ruiz and S. Ordóñez, *Appl. Catal., A*, 2013, **458**, 155–161.
209. Y. Marco, L. Roldán, S. Armenise and E. García-Bordejé, *ChemCatChem*, 2013, **5**, 3829–3834.
210. T. G. Ros, A. J. van Dillen, J. W. Geus and D. C. Koningsberger, *Chem. – Eur. J.*, 2002, **8**, 1151–1162.
211. T. G. Ros, D. E. Keller, A. J. van Dillen, J. W. Geus and D. C. Koningsberger, *J. Catal.*, 2002, **211**, 85–102.
212. P. Serp, M. Corrias and P. Kalck, *Appl. Catal., A*, 2003, **253**, 337–358.
213. J. H. Bitter, *J. Mater. Chem.*, 2010, **20**, 7312–7321.
214. W. J. Lee, U. N. Maiti, J. M. Lee, J. Lim, T. H. Han and S. O. Kim, *Chem. Commun.*, 2014, **50**, 6818–6830.
215. E. Castillejos, R. Chico, R. Bacsa, S. Coco, P. Espinet, M. Pérez-Cadenas, A. Guerrero-Ruiz, I. Rodríguez-Ramos and P. Serp, *Eur. J. Inorg. Chem.*, 2010, **2010**, 5096–5102.
216. A. Villa, D. Wang, P. Spontoni, R. Arrigo, D. Su and L. Prati, *Catal. Today*, 2010, **157**, 89–93.
217. K. Chizari, I. Janowska, M. Houllé, I. Florea, O. Ersen, T. Romero, P. Bernhardt, M. J. Ledoux and C. Pham-Huu, *Appl. Catal., A*, 2010, **380**, 72–80.
218. Y. Motoyama, Y. Lee, K. Tsuji, S.-H. Yoon, I. Mochida and H. Nagashima, *ChemCatChem*, 2011, **3**, 1578–1581.
219. C.-H. Hsu, H.-M. Wu and P.-L. Kuo, *Chem. Commun.*, 2010, **46**, 7628–7630.
220. P. L. Kuo, C. H. Hsu, H. M. Wu, W. S. Hsu and D. Kuo, *Fuel Cells*, 2012, **12**, 649–655.
221. H. Huang and X. Wang, *J. Mater. Chem. A*, 2014, **2**, 6266–6291.
222. D. W. McKee, C. L. Spiro and E. J. Lamby, *Carbon*, 1984, **22**, 507–511.
223. D. J. Allardice and P. L. Walker Jr, *Carbon*, 1970, **8**, 375–385.

CHAPTER 7

Heterogeneous Catalysis on Nanostructured Carbon Material Supported Catalysts

7.1 Introduction

For almost twenty years there has been an increasing interest in the use of nanostructured carbon materials as catalyst supports.[1–16] Following the nanotechnology wave, a wide variety of carbon nanomaterials has been investigated as catalyst supports, including fullerenes, CNTs, CNFs, carbon onions, nanodiamonds, FLG or GO. CNTs and CNFs, which constitute a new family of support offering a good compromise between the advantages of AC and high surface area graphite have been particularly studied. The main advantages offered by CNT or CNF supports are:[17] (i) the high purity of the material can avoid self-poisoning; (ii) the mesoporous nature of these supports can be of interest for liquid-phase reactions, thus limiting the mass transfer; (iii) the high thermal conductivity that limits hot-spots that can damage the catalyst; (iv) their well-defined and tunable structure; (v) their rich surface chemistry that offers numerous perspectives for adsorption and dispersion of the active phase; and (vi) the specific metal support interactions, due to high electrical conductivity of the support and to the tunable orientation of the graphene layers, that exist and that can directly affect catalytic activity and selectivity. The possibility to perform catalysis in a confined space is also of great interest.[6,18] From a theoretical viewpoint, graphene provides the ultimate two-dimensional model of a catalytic support. The reason is related to its high theoretical surface area (*ca.* 2630 $m^2\ g^{-1}$), high conductivity, unique graphitized basal plane structure and chemical tolerance. FLG,

RSC Catalysis Series No. 23
Nanostructured Carbon Materials for Catalysis
By Philippe Serp and Bruno Machado

Published by the Royal Society of Chemistry, www.rsc.org

although having a lower surface area, presents the advantage of a potentially low manufacturing cost. Additionally, the scaled up and reliable manufacture of graphene derivatives, such as GO and reduced GO, offers a wide range of possibilities to synthesize graphene-based functional materials that can be used either as supports for immobilizing active species or as metal-free catalysts.

Owing to the advantageous properties of nanostructured carbon materials as supports, several studies have been carried out on different catalytic reactions. In particular, much attention has been dedicated to liquid-phase reactions. Liquid-phase reactions are mainly used for the production of fine chemicals, and noble metals supported on carbon are the mostly used catalysts. The processes are conducted in stirred tank batch reactors. About 30% of the catalysts consist of supported palladium, most of them being used for hydrogenation reactions, the synthesis of amines from nitro compounds, or the C–C coupling reactions such as Suzuki or Heck reactions. Activated carbon and carbon black are common supports for precious metal catalysts. One of the shortcomings of AC supported catalysts is the leaching of the active phase during the reaction. Nanostructured carbon materials, and in particular the functionalized ones, provide a strong metal-support interaction and therefore a strong anchoring of metal nanoparticles. This strong metal–support interaction is also the reason why catalysts supported on nanostructured carbon materials exhibit higher selectivity in some liquid-phase reactions, due to the facile electron transfer from the support to catalyst. Another important reason to use nanostructured carbon materials instead of AC as support is that their unique mesoporous structure without micropores permits a facile mass transport and prevents the coking by blocking of micropores by reactants. Catalytic gas phase reactions occur mostly at high temperature, and in many case also under high pressure (for instance, for ammonia synthesis or hydrogenation reactions). Mechanical and thermal stability of the support must be considered when choosing which one ought to be used. In addition, the support should present high external surface area to facilitate mass transport under robust reaction conditions. A suitable metal–support interaction is needed to hinder the sintering of metal particles at high temperature under oxidative or reductive condition. Thus, nanostructured carbon materials have been tested as support for many gas phase reactions.

If nanostructured carbon materials are often considered as model supports, the real driving force for the rapid development of these materials in catalysis has been the increasing demand for new catalysts that should be more efficient, energy-saving, resource-saving and environment-friendly, together with an increasing demand for renewable and alternative energies needed for the sustainable development of our society. The resulting achievements or progress (detailed in the previous chapters) have made the wide application of nanostructured carbon materials in catalysis possible: (i) they can be produced at (semi)industrial scale, becoming commercially available at relative low price; (ii) techniques for their functionalization have been developed and well-established, allowing the addition to their surface of certain groups with heteroatoms as active reaction centers, or as anchoring sites for metal complexes or particles; (iii) the synthetic methods for the preparation of metal supported catalysts became mature and novel methods to selectively localize nanoparticles have been developed.

In this chapter, we will present the catalytic results obtained for each type of reaction both in liquid- and gas-phase and, when possible, we will try to rationalize these results by comparison with other carbonaceous supports.

7.2 Hydrogenation Reactions

One of the most investigated catalytic reactions is hydrogenation, and four types of reactions have been particularly studied: alkene and alkyne hydrogenation, regio- or chemoselective selective hydrogenation, –OH or $-NO_2$ species reduction, and CO hydrogenation. Several studies report that for liquid-phase hydrogenation the use of CNT or CNF supports possessing large mesoporosity gave better catalytic activities than microporous AC where mass transfer limitations are operating.

7.2.1 Alkene and Alkyne Hydrogenation

Baker's research team has conducted several studies on ethylene, but-1-ene and buta-1,3-diene hydrogenation on nickel catalysts supported on different types of CNFs, γ-alumina and AC.[19–21] The authors state that the Ni crystallite activity and selectivity can undergo important modifications by interactions with the support; indeed, it was found that the catalyst supported on CNFs gives higher conversion than those employing γ-alumina and AC, even though the metallic particle size was larger (6.4–8.1 nm on CNFs, 5.5 nm on AC and 1.4 on Al_2O_3). These results point to the fact that catalytic hydrogenation might be extremely sensitive to the nature of the metal-support interaction. HRTEM studies have been performed to get a deeper insight on metallic particle morphology: on CNF supports the deposited crystallites were found to adopt very thin, hexagonal structures, and the relevant growth pathways are generally believed to be followed in situations in which strong metal–support interactions are present to cause the spreading of the metal onto the support surface. In contrast, a globular particle geometry prevails when nickel is supported on γ-Al_2O_3, providing a somewhat weaker metal–support interaction. CNF supported nickel catalysts with high activity and selectivity for liquid-phase hydrogenation of benzene were also reported.[22] It has been demonstrated that the CNF structure has a significant effect on the interactions of Pd nanoparticles.[23] The strength of these interactions follows the order: Pd/p-CNFs > Pd/h-CNFs > Pd/r-CNT.

A larger angle between the graphene sheet and the axis of the CNF yields a stronger interaction compared to a CNF with smaller-angle graphene sheets. Interestingly, monodisperse (4.9 ± 0.4) nm tetrahedral rhodium nanoparticles on AC were compared to (4.8 ± 0.4) nm spherical Rh nanoparticles on AC and commercial Rh/AC as a catalyst for the hydrogenation of anthracene.[24] The former was 5.8- and 109-times more active than the latter two, respectively. It also shows a higher selectivity and excellent activity in the hydrogenation of several other arenes. The state of highly dispersed Pd particles supported on filamentous carbon was studied using HRTEM, XPS, and X-ray diffraction analysis.[25] Three types of filamentous carbon were used, in which the basal planes of graphite were arranged along, across, and at an angle to the nanofiber axis. The structure of the carbon support was found to affect the properties of the active component for the reaction of selective 1,3-butadiene hydrogenation to butenes. Highly dispersed Pd particles exhibited the strongest interaction with a carbon surface formed by the edges of graphite (002) layers. This interaction resulted in electron transfer from the metal to the support and in the stabilization of Pd in the most dispersed state. A change in the properties of Pd particles caused a change in the catalytic properties of Pd/C catalysts in the reaction. The strong interaction of Pd^{2+} with the edges of graphite resulted in the stabilization of palladium in an ionic state. An increase in the fraction of Pd^{2+} in the catalysts was responsible for a decrease in both the overall activity and selectivity of Pd/C catalysts.

Efremenko and Sheintuch performed an orbital analysis of the interactions between carbon materials and Pd atoms, as well as Pd clusters, thus providing fundamental insight into Pd/C bonding, functional-group effects, and cluster growth on carbon surfaces.[26] This study revealed that unsaturated surface carbon atoms formed much stronger bonds with Pd atoms than the basal graphite plane, owing to the formation of several Pd–C bonds and to the stabilization of the π-electronic system of several nearest-neighboring aromatic rings. This orbital analysis showed that the Pd 5s orbital accepted significant electron density from the nearest C atoms, whereas the Pd 4d orbitals donated a large amount of electron density in the opposite direction. Owing to the excellent electronic-transfer properties of graphene, the donated electrons from the Pd 4d orbitals are re-localized in close proximity to the edges of the graphene, which leads to a partial compensation for the deficiency in the aromatic electronic structure of the unsaturated rings. At very low loadings, Pd atoms on the AC surface are positively charged (+1.32 eV), due to the extremely effective delocalized donor–acceptor interactions with the π system of the support. Saturation of the dangling surface bonds suppresses the distant interactions between supported Pd atoms and the aromatic π system. This suppression leads to partial relaxation of the Pd electronic structure and causes weakening of the donor–acceptor Pd–AC interactions. The reason for the high activity of Pd/CNF (platelets and fishbones) catalysts could be that the Pd metal center is more electrophilic and this favors alkene coordination, which is often the rate determining reaction step in alkene hydrogenation. Rhodium nanoparticles supported on a CNFs show high catalytic activity toward arene hydrogenation under mild conditions in high turnover numbers without leaching the Rh species.[27] The reaction was highly tolerant to epoxy groups, which often undergo ring-opening hydrogenation with conventional catalysts.

The processes involved in the butene hydro-isomerization, occurring on a small Pd_9 cluster in the presence of dissociated hydrogen, have been investigated by means of DFT and DFT/MM approaches.[28] This study has been performed both on an unsupported Pd_9 cluster and on the same cluster supported on a portion of a CNT. The main aspects of the parallel reaction steps of the whole hydro-isomerization mechanisms are not strongly affected by the presence of the support, which does, however, modify the energetics involved, likely due to the presence of strong metal support interaction (SMSI) effects. Palladium supported on CNTs catalyzes efficiently hydrogenation reactions.[29] The C≡C bond of tolane, phenylacetylene and 1-heptyne were hydrogenated to give a mixture of intermediate ethylenic and completely reduced hydrocarbons.[30] The composition of hydrogenation products depends greatly on the structure of the starting reagent and reaction conditions. CNT-supported metallic NPs including Pt, Rh, and bimetallic Pd–Rh were reported to be effective catalysts for the hydrogenation of neat benzene at room temperature, which cannot be achieved by carbon-based Pd and Rh catalysts available commercially.[31] The rate of the hydrogenation reaction was zero-order with respect to benzene and first-order with respect to hydrogen and the catalyst. CNT-supported Pd and Au NPs show negligible activity for room-temperature hydrogenation of benzene. The bimetallic Pd–Rh/CNT NP catalyst exhibits a strong synergistic effect relative to the individual single metal NPs for catalytic hydrogenation of benzene, toluene, and 1-phenyl-1-cyclohexene. Rhodium NPs supported on MWCNT showed also a remarkable catalytic activity for arene hydrogenation with enhanced turnover numbers of 10 000.[32] Disubstituted arenes showed selective conversion of thermodynamically less favorable *cis* products (>80%). A series of arenes have been tested using various Rh-based catalysts, and a comparison of the results with that of reported rhodium catalysts shows unique selectivity under mild conditions. Size-tunable Rh NPs were deposited uniformly on MWCNTs.[33] The hydrogenation of neat *o*-, *m*-, and *p*-xylene catalyzed by the MWCNT-supported metallic Rh NP catalysts reveals a negative particle size effect (antipathetic structure sensitivity). The stereoselectivity of this reaction measured by the ratio of *cis*-/*trans*-dimethylcyclohexane increases with decreasing size of Rh NPs. Ru catalysts supported on three types of CNFs (CNF-h, CNF-p, CNF bamboo-like), have been tested for selective hydrogenation of paracetamol.[34] The influence of the support nanostructure on their catalytic performance was studied. An inverse correlation between the heats of CO adsorption and the activity and stereoselectivity to the *trans*-4-acetamidocyclohexanol was found. These results corroborate a pseudo-morphological growth of Ru particles on CNF support that depends on its structure, *i.e.* edges *vs.* basal planes. The different metal-support interactions that can be related to the metallic particle location on the surface carbon microdomains is what ultimately controls the morphology and the electronic features of the Ru nanocrystallites, and consequently, the catalytic properties. The use of Ru/CNF-p allows highly efficient catalytic partial hydrogenation of 1,1′-bi-2-naphthol and -naphthyl-amine derivatives.[35] The reactions proceed in high turnover numbers without racemization of the axial chirality, offering a practical procedure for the production of optically pure 5,5′,6,6′,7,7′,8,8′-octahydro-1,1′-binaphthyls in good to high yields.

Benzene hydrogenation was used as a probe reaction for the study of catalytic activity of a CNT-supported NiP amorphous catalyst (NiP/CNT). In comparison with the NiP amorphous alloy, the benzene conversion on NiP/CNT catalyst was lower, but the specific activity of NiP/CNT was higher, which is attributed to the dispersion produced by the support, an electron-donating effect, and the hydrogen-storage ability of CNT. The NiP/CNT thermal stability was improved because of the dispersion and electronic effects and the good heat-conduction ability of the CNT support. CNT-supported Rh NPs are highly active and reusable catalysts for hydrogenation of benzene and its derivatives at room temperature.[36] Complete ring saturation of polycyclic aromatic hydrocarbons can be achieved under mild hydrogenation conditions using the Rh/MWCNT catalyst that cannot be done by commercially available Rh catalysts. MWCNTs filled with Pd clusters were used as a catalyst for the liquid-phase benzene hydrogenation.[37] The results indicated that the confined Pd clusters exhibited higher activity than other supports such as Y zeolite and AC. A high-performance Pd catalyst for selective olefin hydrogenation was synthesized by supporting Pd NPs on nitrogen-doped CNTs.[38] The Pd NPs were stabilized on N-CNTs with narrower size distribution compared with oxygen-functionalized CNTs. The XPS analysis revealed that the nitrogen functional groups favor the reduction of Pd on CNTs suggesting an electronic promoter effect. The Pd/N-CNT catalyst showed extraordinary catalytic performance in terms of activity, selectivity and stability in the selective hydrogenation of cyclooctadiene, which is related to the structural and electronic promoting effect of the N-CNT support.

The effect of the degree of graphitization of the support on the dispersion, sintering resistance, and catalytic activity of a series of Pt catalysts supported on heat-treated CB has been determined.[39] Benzene hydrogenation was used as a test reaction to assess the possible effect of the support on the catalytic behavior of Pt. Platinum dispersions as high as 99% were achieved, and they can be well correlated with the surface acidity characteristics of the supports. The highest resistance to sintering was found for the catalyst supported on the CB with the highest degree of graphitization, and this was attributed to an interaction between the metal particles and the π sites on the support. However, this interaction is not strong enough to affect the catalytic behavior for benzene hydrogenation. Fixation of Pt on fullerene black through non-conjugate double bonds or hydroxyl groups allows platinum particles of 3–4 nm to be obtained.[40] These compositions catalyze 1-decene and nitrobenzene hydrogenation and exceed in activity the traditional catalyst

Pt/C with platinum particle size of 70–80 nm. The average size of the fixed Pt particles affects the catalytic activity stronger than the specific surface area and the electron conduction of the support. Palladium loaded on detonation nanodiamond was also used as a catalyst for the C=C and C≡C bonds hydrogenation.[41] By the example of diphenyl acetylene hydrogenation, the Pd(0)/nanodiamond was found to greatly surpass in catalytic activity other Pd(0)/nanocarbon catalysts (fullerenes and CNTs). Such difference in the activity may be explained by structural features of these carbon materials. Investigation of the detonation nanodiamond structure showed that the purification by conventional oxidation methods is unsuitable for the entire removal of impurities of fullerene- or graphite-like agglomerates, which form a thin shell on the surface of nanodiamond cores. This could play a positive role if nanodiamond is used as a support for the preparation of the metal-containing catalysts. As would be expected, the sp^2-carbon centers of such surface fragments enable more strong coordination of metal clusters, including palladium clusters, compared to the non-modified surface containing only sp^3-carbon centers. At the same time, the formation of σ-bonds between palladium and the surface sp^3-carbon atoms cannot be ruled out. Nickel catalysts supported on nanodiamonds were also investigated for hydrogenation.[42,43] The ultra-dispersed diamond powder appears very promising for deposition of metallic nickel, active in the catalytic hydrogenation of toluene. Stable Ru or Rh metal NPs were supported on chemically derived graphene surfaces with small and uniform particle sizes (Ru 2.2 ± 0.4 nm and Rh 2.8 ± 0.5 nm).[44] The graphene-supported hybrid NPs were shown to be active and could be re-used at least 10 times as catalysts for the hydrogenation of cyclohexene and benzene under organic-solvent-free conditions with constant activities up to 1570 mol cyclohexane mol $metal^{-1}$ h^{-1} at 4 bar and 348 K. Iron NPs supported on chemically-derived graphene were prepared and identified as an effective catalyst for the hydrogenation of alkenes and alkynes.[45] The catalyst can easily be separated by magnetic decantation.

The impact of carbon substrate–Ru NP interactions on benzene and hydrogen adsorption that is directly related to the performance in catalytic hydrogenation of benzene has been investigated by first-principles based calculations.[46] The stability of Ru_{13} nanoparticles is enhanced by the defective graphene substrate due to the hybridization between the dsp states of the Ru_{13} particle with the sp^2 dangling bonds at the defect sites. The strong interfacial interaction results in the shift of averaged d-band center of the deposited Ru nanoparticle, from −1.41 eV for a freestanding Ru_{13} particle, to −1.17 eV for the Ru/graphene composites (Ru_{13}/SV), and to −1.54 eV on mesocellular carbon foam (Ru_{13}/C). Accordingly, the adsorption energies of benzene are decreased from −2.53 eV for the Ru_{13}/C, to −2.62 eV on freestanding Ru_{13} particles, and to −2.74 eV on Ru_{13}/SV. A similar change in hydrogen adsorption is also observed, and all these can be correlated to the shift of the d-band center of the nanoparticle. The enhanced adsorption of H and benzene suggests a possible higher activity of Ru_{13}/SV in catalytic hydrogenation of

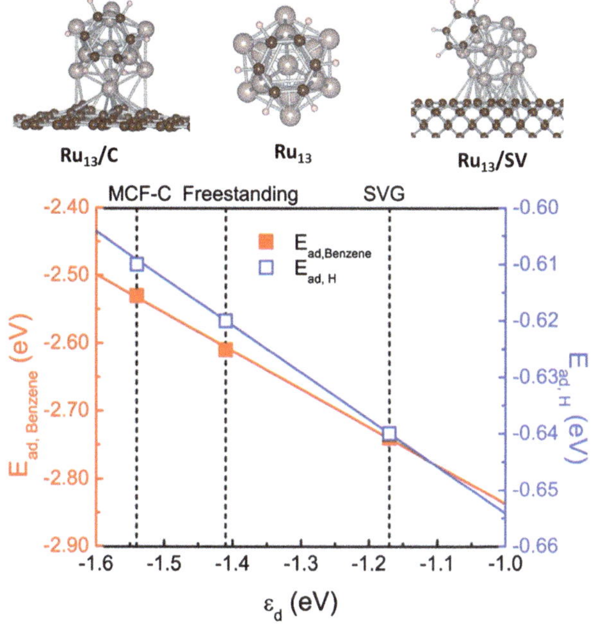

Figure 7.1 The calculated adsorption energy, E_{ads} plotted *versus* ε_d. The calculated adsorption energies correlate well with ε_d. Reprinted with permission from ref. 46.

benzene. And, the enhancement in E_{ads} can be directly correlated with the upshift of the ε_d that originates from the interaction between the Ru nanoparticle and the graphene substrate (Figure 7.1). This thus demonstrates the effectiveness of interfacial interactions in tuning the electronic structure of the deposited TM particles, and the possibility of improving the activity of TM nanocatalysts by careful selection of the substrate.

Supported homogeneous systems have also been investigated for C=C hydrogenation reactions. Oxidized SWCNTs have been reacted with Wilkinson's complex (Figure 7.2(a)).[47] It has been found that the Rh metal coordinates to these nanotubes through the increased number of oxygen atoms, forming a hexacoordinate structure around the Rh atom. The functionalized SWCNT-Wilkinson's complex adducts have been found to catalyze the hydrogenation of cyclohexene to cyclohexane at room temperature. Hybrid catalysts have been prepared by the immobilization of a Rh diamine complex (Figure 7.2(b)) on four MWCNTs of different inner diameters (4–10 nm).[48] The complex has been anchored to the support by the reaction of the $-Si(OCH_3)_3$ functionality in the rhodium complex with phenol type –OH groups on the support surface, that have been created by oxidation with air. The hybrid catalysts have been tested in the hydrogenation of cyclohexene and carvone. In general terms, the obtained hybrid catalysts are more active than the Rh complex in homogeneous phase and they are recyclable. It has been found that the nanotube dimensions have

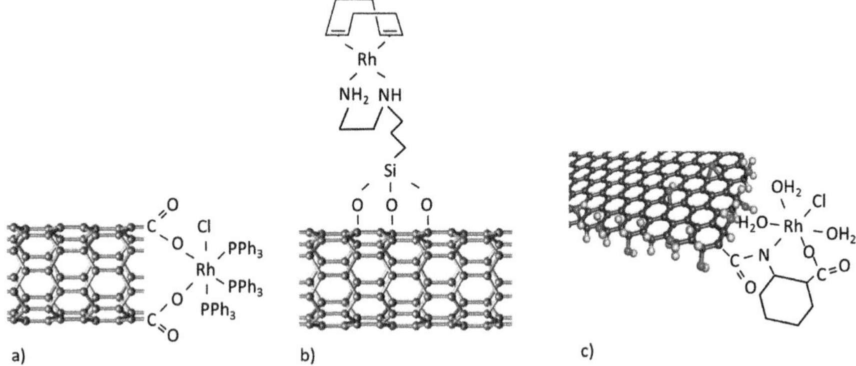

Figure 7.2 Supported homogeneous catalysts on carbon nanostructured materials: (a) Wilkinson catalyst; (b) a Rh diamine complex, and (c) a Rh^{III} anthranilic acid complex.

an influence on the catalytic properties. In the case of smaller nanotubes (inner Ø *ca.* 4 nm) the support has no positive effect, while the catalysts prepared with nanotubes of inner diameter around 7 nm give the best results. Highly active rhodium catalysts for hexane hydrogenation have been prepared by immobilization of an ionic liquid (IL) film on MWCNTs functionalized with imidazolium-based ionic moieties, the rhodium complex being immobilized in the ionic liquid film.[49] The immobilization of a rhodium/anthranilic acid complex onto CNFs-h was executed by means of the following steps: (1) CNF surface oxidation, (2) conversion of the oxygen-containing surface groups into acid chloride groups, (3) attachment of anthranilic acid and (4) complexation of rhodium by the attached anthranilic acid (Figure 7.2(c)).[50] The as-synthesized Rh^{III} complex itself is not active in the liquid-phase hydrogenation of cyclohexene. However, reduction with sodium borohydride yields small particles (d = 1.5–2 nm) of rhodium metal that are highly active. The results indicate that different activation procedures for the immobilized Rh/anthranilic acid system should be applied, such as reduction with a milder reducing agent or direct complexation of the rhodium in the Rh^{I} state. Two complexes containing an *N*-heterocyclic carbene ligand with a pyrene tag were immobilized onto the surface of rGO by π-stacking.[51] The catalytic properties of the parent molecular complexes and hybrid materials have been studied in the palladium-catalyzed hydrogenation of alkenes and the ruthenium-catalyzed alcohol oxidation. The results show that the catalytic properties are improved in the hybrid materials, compared to the catalytic outcomes provided by the homogeneous analogues. Although the palladium-catalyzed reactions may be due to the formation of Pd NPs, the ruthenium-catalyzed ones are facilitated by the supported molecular catalysts. The catalyst stability was analyzed by means of recyclability studies, hot filtration test, and large-scale experiments. Both hybrid materials have been reused up to 10 times without any decrease in activity, affording quantitative yields of the products. The hot filtration experiment reveals that the catalysis is heterogeneous in nature without any detectable leaching or boomerang effect. Finally, rhodium(I) complexes have been immobilized on functionalized graphene oxide through coordination interaction.[52] The obtained catalyst can be readily recycled and shows enhanced activity in the catalytic hydrogenation of cyclohexene.

7.2.2 Other Hydrogenation (〉C=O, –OH, $-NO_2$)

7.2.2.1 Carbonyl Compounds

SWCNTs were used as supporting materials for Pd NPs generated *in situ* in ionic liquid.[53] The Pd catalysts exhibit superior reactivity for hydrogenation of aryl ketones in ionic liquid under mild conditions and can be reused over 10 times without any loss of catalytic activity. Ruthenium oxide nanorods supported on graphene nanoplatelets were found to be an effective and reusable heterogeneous catalyst for the transfer hydrogenation of aromatic aldehydes and ketones in good yield with excellent selectivity.[54,55] A ruthenium supported catalyst has been synthesized by covalently bonding a $[RuCl_2(PPh_3)_3]$ onto the GO surface through coordination interaction with aminosilane ligand spacers.[56] The supported catalyst showed enhanced catalytic performance towards hydrogenation of olefins and ketones compared with the homogeneous analogue, and it could be readily recycled and reused several times without discernible loss of its activity. Ru on C_{60} was active for the liquid-phase hydrogenation of 2-cyclohexenone at atmospheric pressure.[57] Hydrogenation of benzaldehyde is a typical consecutive reaction, since the intermediate benzyl alcohol is suitable for further hydrogenation.[58] It was demonstrated that the selectivity of benzyl alcohol can be tuned *via* functionalization of CNTs, which are used as the support of Pd. With the original CNTs, the selectivity of benzyl alcohol is 88% at a 100% conversion of benzaldehyde. With introduction of oxygen-containing groups onto CNTs, it drops to 27%. In contrast, doping CNTs with N atoms, the selectivity reaches 96% under the same reaction conditions. The kinetic study shows that hydrogenation of benzyl alcohol is significantly suppressed, which can be attributed to weakened adsorption of benzyl alcohol. This is most likely related to the modified electronic structure of Pd species *via* interaction with functionalized CNTs, as shown by XPS characterization.

7.2.2.2 Alcohols

The catalytic action of 10 wt% Pd supported on two forms of CNFs for phenol hydrogenation has been assessed, and compared with the performance of 10 wt% Pd on SiO_2, Ta_2O_5, AC, and graphite.[59] The specific activities exhibited the following sequence of increasing values: Pd/AC < Pd/CNF < Pd/SiO_2 ≈ Pd/graphite < Pd/Ta_2O_5. A diversity of product composition responses to variations in reaction conditions points to the involvement of Pd particle size distribution, Pd particle geometry, and electronic character in determining overall catalytic behaviour. A Rh/CNF catalyst with an average size of 2–3 nm presented a high activity in the hydrogenation of phenol in a medium of supercritical CO_2 ($scCO_2$) at a low temperature of 323 K.[60] The presence of compressed CO_2 retards hydrogenation of cyclohexanone to cyclohexanol under the reaction conditions used, and this is beneficial for the selective formation of cyclohexanone.

CNTs and AC supported Pd and Ni catalysts were prepared for the hydrogenation of phenol to cyclohexanone and cyclohexanol.[61] The hydrophobic/hydrophilic properties of the catalysts were tailored by pretreating the carbonaceous support with HNO_3. The catalytic results suggested that Pd and Ni supported on CNTs show significantly higher activity than that supported on ACs. Pretreating the CNTs with HNO_3 increases the local hydrophilicity of the active phase (by introducing oxygenated groups), which result in an increase of the cyclohexanone selectivity and strongly decrease the phenol conversion. DFT calculations suggested that the adsorption/desorption behaviors of phenol, methanol, H_2O, and cyclohexanone on the catalysts might be highly influenced by the hydrophobic/hydrophilic properties. The hydrophilic catalysts show high selectivity in cyclohexanone by lower conversion in phenol and *vice versa*. A catalyst made of Pd NPs supported on polyaniline-functionalized CNTs, Pd-PANI/CNT, was shown to be highly active towards the direct hydrogenation of phenol to cyclohexanone.[62] Phenol conversion exceeding 99% was achieved with a cyclohexanone selectivity of >99% under atmospheric pressure of hydrogen in aqueous media. The generality of the catalyst for this reaction was demonstrated by selective hydrogenation of other hydroxylated aromatic compounds with similar performance, again under green and mild conditions. The results also indicate that phenol conversion is presumably related to the conductive property of PANI/CNT, whereas cyclohexanone selectivity is attributed to the nitrogen-containing nature of PANI/CNT.

7.2.2.3 Nitro Compounds

CNT-supported Pt catalysts have been tested for nitrobenzene hydrogenation under atmospheric pressure and ambient temperature.[63] The results show that the catalysts, both of low and higher Pt loading, show high activity for nitrobenzene, directly hydrogenating to aniline under mild conditions. The highly dispersed Pt and mesoporosity structure of acid-oxidized CNT-supported Pt catalyst are responsible for the extraordinary activity. A facile and efficient route to deposit ultrafine Pt particles onto MWCNTs with the aid of tip sonication was reported.[64] The loading of Pt on the MWCNTs could attain the very high level of 50 wt% and the size of the Pt particles could be controllably tuned in the range 1.9–3.5 nm with narrow size distributions. The resultant nanocomposites were applied to catalyze the hydrogenation of nitrobenzene under solvent-free conditions. It was demonstrated that the Pt/MWCNT catalysts showed excellent activity with a high turnover frequency (*e.g.*, 69 900 h^{-1}) as well as superior selectivity to aniline (*e.g.*, >99%) in this reaction. Granulated Pt/CNTs were found to have a much better catalytic activity in the liquid-phase hydrogenation of nitrobenzene than Pt/AC.[65] The granulated CNTs had much larger pores than the AC particles, which gave a faster mass transfer rate of H_2 that helped to produce aniline with high selectivity. The effect of two types of catalysts on the activity of the catalytic hydrogenation of nitrobenzene was studied.[66] The influence of the supports' (AC and a mixture of AC and MWCNTs) surface area was investigated. Despite having a similar Pd particle size (4–5 nm), the catalyst prepared on a mixture of AC and nanotubes was significantly less active than the catalyst prepared on pure AC; the rate of this reaction was approximately 30% lower than the initial reaction rate. This feature could be attributed to the lower specific surface area of the Pd/AC/CNT (531 $m^2\ g^{-1}$) in comparison with the Pd/AC (692 $m^2\ g^{-1}$).

Reduced graphene oxide (rGO) was used as a catalyst for the reduction of nitrobenzene at room temperature.[67] High catalytic activity and stability were exhibited in circular experiments, presumably due to metal impurities. Reduced graphene oxide (rGO)-supported Pt catalyst was used for the hydrogenation of nitroarenes.[68] The yield of aniline over the Pt/rGO catalyst reached 70.2 mol aniline $mol_{Pt}^{-1}\ min^{-1}$ at 273 K, which is 12.5 and 19.5 times higher than that of MWCNT- and AC-supported Pt catalysts, respectively. When the reaction temperature was increased to 293 K, the catalytic activity of Pt/rGO jumped to 1138 mol aniline $mol_{Pt}^{-1}\ min^{-1}$, and it was also extremely active for the hydrogenation of a series of nitroarenes. The unique catalytic activity of Pt/rGO is not only related to the well dispersed Pt clusters on the rGO sheets but also to the high dispersion of Pt/rGO in the reaction mixture. Ni/graphene nanocomposites were used for the hydrogenation of *p*-nitrophenol to *p*-aminophenol.[69] The heterostructures with a large 2D surface enable the superior catalytic activity and selectivity toward hydrogenation reaction for *p*-nitrophenol. Complete conversion of *p*-nitrophenol was achieved with selectivity to *p*-aminophenol as high as 90% at room temperature. This catalyst can be efficiently recycled with long lifetime and stability over 10 successive cycles. A Au/graphene hydrogel exhibited excellent catalytic performance towards the reduction of 4-nitrophenol to 4-aminophenol, which is about 90 times larger than previously reported values for spongy Au nanoparticles and 14 times more than the highest value among the polymer supported Au nanoparticle catalysts.[70] The high catalytic activity arises from the synergistic effect of graphene: (i) the high adsorption ability of graphene towards 4-nitrophenol, providing a high concentration of 4-nitrophenol near the Au nanoparticles on graphene; and (ii) electron transfer from graphene to Au nanoparticles, facilitating the uptake of electrons by 4-nitrophenol molecules. Carbon supported nickel (1 wt%) catalysts have been prepared by deposition-precipitation with urea.[71] Two structured, *i.e.* CNF (129 $m^2\ g^{-1}$) and nanospheres (CNS, 15 $m^2\ g^{-1}$), and one unstructured, *i.e.* AC (686 $m^2\ g^{-1}$), supports were used to deposit Ni.[71] The three supports were treated with $HNO_3 + H_2SO_4$ to generate oxygen-containing surface groups that served as anchoring sites for Ni introduction. The three catalysts were tested in the gas-phase hydrogenation of nitrobenzene to aniline. The reaction generated aniline as the sole product where the specific rate constant increased in the order Ni/CNF < Ni/CNS < Ni/AC. Hydrogenation activity was insensitive to Ni particle size and exhibited a proportional increase with increasing surface acidity (mainly carboxylic anhydrides). This response was attributed to an enhanced activation of nitrobenzene at surface acid sites.

Fullerenes (C_{60} and C_{70}) were recently shown to be active in the hydrogenation of nitrobenzene, after activation to form the fullerene anion.[72] However, the catalyst in this reaction is shown to contain residual nickel from the activation of the catalyst, which, as indicated by a series of control experiments, is most likely responsible for the catalytic conversion.[73]

The kinetics of nitrobenzene hydrogenation on a palladium triphenylphosphine catalyst supported on nanodiamonds was studied.[74] It was found that the reaction is of first-order with respect to the catalyst and hydrogen and of zero-order with respect to nitrobenzene. The apparent constant and activation energy of the reaction were calculated. A probable reaction mechanism was proposed. The effects of the triphenylphosphine-to-palladium ratio and the nature of the solvent and an aromatic nitro compound on the activity of the test catalyst were demonstrated. The catalytic activity of Pt- and Pd-containing nanodiamonds has been investigated in liquid-phase nitrobenzene, allyl alcohol, and cyclohexene hydrogenation and propanal hydroamination with 4-aminobenzoic acid as model reactions.[75] These catalysts were significantly more active than commercial Pd/C. The catalysts with a low metal loading were the most effective in liquid-phase catalytic hydrogenation.

7.2.3 Selective Hydrogenation

7.2.3.1 Selective Hydrogenation of Acetylene

The main industrial method for ethylene manufacture is hydrocarbon pyrolysis (ethane–ethylene fraction, propane–propylene fraction, and butane fraction). It is necessary to remove acetylene (usually 0.4–2.2%) from the ethane–ethylene fraction of pyrolysis gases produced at ethylene plants. Alkynes can be selectively hydrogenated into alkenes on solid palladium catalysts. There is great interest to use carbon nanostructured materials in highly selective catalysts for hydrogenation of diene and acetylene hydrocarbons to olefins. One of the most important reasons to use them as palladium supports is the absence of acid sites that can promote undesired acetylene polymerization to so called "green oil" contamination. The selective hydrogenation requires a strong modification of the near-surface region of palladium, in which carbon (from fragmented feed molecules) occupies interstitial lattice sites.[76] *In situ* X-ray photoelectron spectroscopic measurements under reaction conditions indicated that much less carbon was dissolved in palladium during unselective, total hydrogenation. Additional studies of hydrogen content indicated that unselective hydrogenation proceeds on hydrogen-saturated β-hydride, whereas selective hydrogenation was only possible after decoupling bulk properties from the surface events. Thus, the population of subsurface sites of palladium, by either hydrogen or carbon, governs the hydrogenation events on the surface.

Monodispersed Pd nanoparticles (8, 11, and 13 nm in diameter) were deposited on CNFs and used to study the effect of the support nature on the selective acetylene hydrogenation.[77] Antipathetic size dependence of TOF disappeared at particle sizes bigger than 11 nm. TOF was found to increase from 15 s^{-1} up to 24 s^{-1} with the Pd particle size increasing in the range of 8–13 nm. Initial selectivity to ethylene was found to be size-independent. The deactivation due to coke deposition was faster for smaller particles. The structure-sensitivity relations for the catalysts investigated were discussed in terms of the "geometric" and "electronic nature" of the size effect and rationalized regarding Pd-C_x phase formation, which is size-dependent. For the smaller particles, the C/Pd ratio is higher, and carbonaceous deposits block a major part of the active Pd surface. The initial TOF was found to decrease with the increasing acidity of the CNF surface, while the selectivity to ethylene slightly increased. The by-product (ethane and oligomers) distribution was found to be shifted toward the ethane formation on "acidic" supports. The observed effects were attributed to the lower electronic density of the activated CNF, which affects the strength and mode of acetylene adsorption. After 6 h on-stream, coke formation on the catalyst causes a decrease in activity and selectivity to ethylene. This decrease was less pronounced for CNF-based supports with higher acidity obtained by oxidative pretreatment. Supported ionic liquid-phase Pd NPs on CNF were also tested for the selective hydrogenation of acetylene to ethylene and showed excellent long-term stability.[78] The IL cation–anion network surrounding the NPs suppressed the formation of active-site ensembles, known to catalyze the oligomerization of acetylene, responsible for the catalyst deactivation. The reaction rate was controlled by the internal diffusion of the reactants through the IL phase.

The state of palladium deposited on CNFs with stacked structure in 0.04–0.5 wt% loading was studied by XRD, electron microscopy and EXAFS; and the results were correlated to catalytic performances in acetylene selective hydrogenation (Figure 7.3).[79] Palladium was found to exist as single atoms attached to CNF in the samples with Pd loading of 0.2 wt% or less. In concentrations lower than 0.04 wt% Pd penetrates deeply into the bulk of CNF and is unavailable to gas-phase molecules. The location of these palladium atoms in the graphite lattice was analyzed using quantum-chemical calculations. At the Pd location the separation between the carbon layers increases from 3.35 Å in graphite to 3.75 Å. The carbon–carbon bond length also slightly increases from 1.42 to 1.44–1.45 Å. At 0.04–0.2 wt% loading range Pd atoms populate the CNF surface. This kind of Pd atom is responsible for selective catalytic olefin hydrogenation. It can be concluded from Pd/CNF catalyst performance results that Pd clusters appear starting from 0.1 wt% Pd loading. It is the reason why starting from 0.1 wt% Pd the selectivity is slightly decreased. These clusters are capable of catalyzing ethylene hydrogenation.

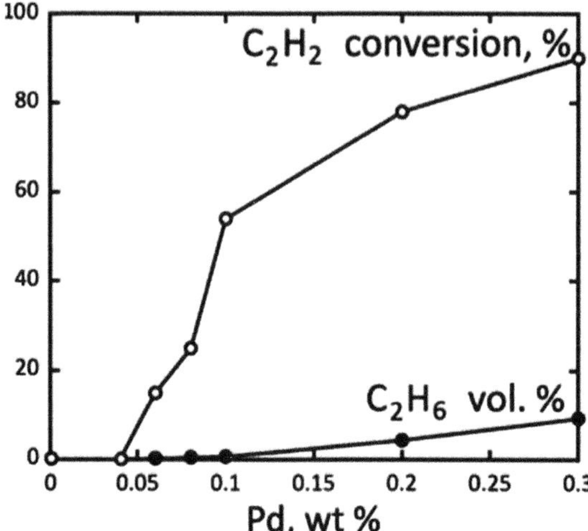

Figure 7.3 Observed acetylene conversion (circles) and ethane concentration (filled circles) in the outlet flow gas mixture of the reactor using Pd-catalyst supported on stacked-type CNFs. The reaction temperature is 363 K. Reprinted with permission from ref. 79. Copyright 2012, Elsevier Science Ltd.

Palladium on a MWCNT support was applied to selective hydrogenation of acetylene in an ethylene-rich flow stream.[80] This material displayed a very promising selectivity toward ethylene production with increasing temperature, and also suppressed oligomer formation during acetylene hydrogenation. New operating conditions for selective hydrogenation of acetylene in an ethylene-rich flow were introduced. This catalyst gave a considerably higher yield, as high as 93%, than that previously obtained for ethylene production. It was postulated that the governing mechanism for acetylene hydrogenation over 0.5 wt% Pd/MWCNT was hydrogen transfer. It was also revealed that the catalytic performance at high temperatures did not depend upon the Pd particle size or heterogeneous energetic sites formed during functionalizing of the MWCNT material.[81] Catalytic measurement for Pd_2Ga/CNT was conducted for acetylene hydrogenation and results were compared with the commonly used $Pd_{20}Ag_{80}$ catalyst.[82] Oxidized vacancies and localized double bonds on CNTs inhibit sintering and loss of the Pd_2Ga NPs during reactions. Nanocrystalline intermetallics have abundant low coordination sites (edge, stepped, and kink) on surfaces, and thereby lead to a high activity. The surface and structure of obtained Pd_2Ga NPs are thermally stable under reaction conditions. Establishing covalent interactions within nanocrystalline intermetallics forms a high barrier for subsurface chemistry and reduces large active ensembles, which can be reflected in the improved selectivity.

Ni–B and Ni–B/CNTs amorphous alloy catalysts were prepared by chemical reduction and impregnation–chemical reduction methods, and their catalytic activities were evaluated in acetylene selective hydrogenation reaction.[83]

Based on characterization results, the effects of carbon nanotubes on Ni–B amorphous alloy were attributed to: (i) a structural effect, dispersing Ni–B particles, leading to higher surface area of active nickel and enhancing the thermal stability; and (ii) an electronic effect, resulting in electron-rich nickel centers.

According to the TEM and HT data, the size of Pd particles in Pd/C catalysts prepared *via* interaction of the $[Pd(\eta^1\text{-}C_3H_5)(\eta^5\text{-}C_5H_5)]$ complex with the surface of carbon supports and reduced with hydrogen at 573 K ranges from 1 to 2 nm.[84] After increasing the reduction temperature up to 873 K, dispersion of Pd/Altunit (ultra-dispersed diamond) remains practically the same, while the size of metal particles in Pd/Sibunit (graphite-like carbon) increases up to 10 nm. This difference is due to the peculiarities of the interaction between Pd particles and the surface of graphite and diamond in hydrogen atmosphere. In the reactions of acetylene and vinylacetylene hydrogenation the specific activity of Pd/C is comparable to that of Pd/SiO_2, but is 1–2 orders of magnitude lower in reactions of ethane hydrogenolysis and CO hydrogenation.

A comparative study was performed of the catalytic activity of nickel NPs deposited on detonation synthesis nanodiamonds, coal and crystalline quartz in the hydrogenation of acetylene.[85] It was shown that Ni/nanodiamond is an active catalyst of acetylene hydrogenation, considerably surpassing Ni/quartz and Ni/C.

7.2.3.2 Selective Hydrogenation of α,β-Unsaturated Substrates

The hydrogenation of α,β-unsaturated substrates on nanostructured carbon supported catalysts has been the subject of several studies. Although, the C=C bond is easier to hydrogenate than the carbonyl group, the unsaturated alcohol is often the desired product. For selective hydrogenation, the relative inertness of the carbon surface is of importance since the catalytic systems are usually constituted by more than one metal. The carbon inertness facilitates the interaction between the metals and/or between the metals and the promoters, yielding more active and selective catalysts than those supported on other common supports. Various carbon and graphite materials have been used for selective hydrogenation of α,β-unsaturated aldehydes, including CNFs, CNTs, graphite, AC, CBs, and fullerenes.

Hydrogenation of Cinnamaldehyde. One of the first reported applications of CNTs in heterogeneous catalysis was their use as supports for 3–7 nm Ru nanoparticles in the hydrogenation of cinnamaldehyde (CAL).[86] CAL contains both a C=C and a C=O bond in an α,β-unsaturated arrangement. Depending on which bond is activated, hydrocinnamaldehyde (HCAL), cinnamyl alcohol (COL) or phenyl propanol (HCOL) can be obtained *via* the hydrogenation reaction. The possible reaction pathways are shown in Scheme 7.1. In fact, the selective hydrogenation of the carbonyl group of the α,β-unsaturated aldehyde yielding the unsaturated alcohol remains a challenging task. This is of particular interest because of the importance of such alcohols in the fine chemical industry.[87] In general, the catalytic systems obtained by deposition

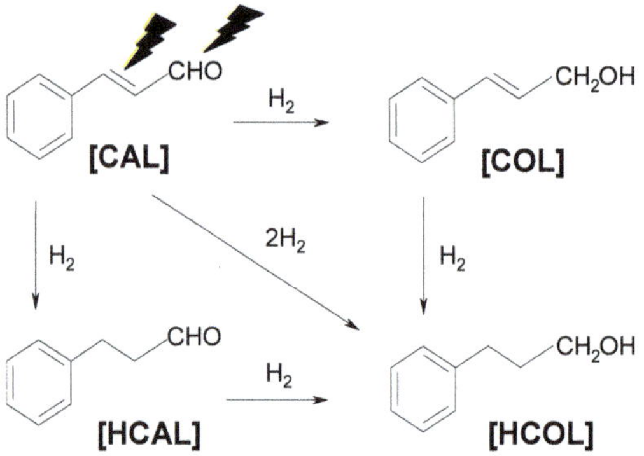

Scheme 7.1 Reaction pathways for the hydrogenation of cinnamaldehyde.

on CNTs or CNFs are more active than their counterparts on AC or oxides. Nanostructured carbon catalysts present a dynamic mesoporous structure that should limit clogging and enhance diffusion phenomena. Catalysts based on graphitic (nano)materials are also more selective towards COL than AC. It is possible to interpret this in terms of an electronic ligand effect. Since metal particles are preferentially located on steps and edges of graphitic planes, the π-electrons of the graphitic planes can be easily extended to the metal particles, thus increasing the charge density of the metal. In fact, the graphitic support acts like a macro electron donating ligand. The increased charge density on the nanoparticles decreases the probability of adsorption *via* the C=C bond, so that the selectivity towards the COL increases.

Pt, Pd, Rh, Ru metal nanoparticles and Pt–Co, Pt–Ru bimetallic nanoparticles supported on nanostructured carbon materials show a high catalytic activity in this reaction. Pd and Rh are highly selective for HCAL, but not active for COL formation. For example, when using a 1 wt% Rh/CNT for the production of HCAL, a selectivity of 100% was obtained.[88] The catalytic activity of Rh/CNT was three times higher than that of a 1 wt% Rh/C, despite the fact that the latter has a surface area of 700 $m^2 g^{-1}$, three times higher than that of CNTs (180 $m^2 g^{-1}$).

Aligned and strongly attached MWCNTs on the wall of a silica reactor were efficiently used as catalyst support for palladium in the selective hydrogenation of cinnamaldehyde.[89] The selectivity towards the HCAL remained high and constant regardless of the level of conversion, *i.e.* selectivity of 90% at a conversion of about 80%. Pham-Huu *et al.* reported the interesting confinement effect of Pd NP catalysts deposited on the inner walls of CNTs for the selective hydrogenation of CAL to HCAL,[90] explained by the unusual interaction between Pd NPs and the inner walls of the CNTs coupled with the relative lack of oxygenated surface groups on them. The nanotube-based catalyst exhibits higher catalytic activity and an extremely high selectivity towards the C=C bond hydrogenation when compared to a commercial catalyst supported on a high surface area AC. A peculiar metal-support interaction and the absence of micropores and of oxygenated surface groups on the CNT support were sought to explain these results. Nitrogen doped CNTs were also tested as catalyst support in the liquid-phase hydrogenation of cinnamaldehyde using palladium as the active phase, and compared with catalysts based on palladium supported on CNT and AC.[91] Nitrogen atom incorporation led to a significant improvement of the hydrogenation activity compared to that observed on the N-free carbon nanotube catalysts. The activity improvement was attributed to possible electronic or morphologic modifications of the active phase leading to a higher turnover frequency of the catalytic site. The results obtained also highlighted the fact that the selectivity towards C=C bond hydrogenation was strongly influenced by the active phase particle size. The N-doped support exhibits a relatively high selectivity towards the C=C bond hydrogenation compared to the one obtained on samples without nitrogen incorporation regardless the slight difference in terms of particle size. The C=C bond hydrogenation selectivity observed in this work on the N-doped catalyst is among the highest that has ever been reported up to now in the literature. Another study by the same authors showed that the type of nitrogen species incorporated in the CNT structure can also influence the catalytic activity.[92] Recycling tests confirmed the high stability of the catalyst as neither palladium leaching nor deactivation were observed.

MWCNTs having different outer diameters but similar inner diameters were loaded with 3.0 wt% Pt and tested for hydrogenation of cinnamaldehyde.[93] All the catalysts showed reasonably high catalytic activity but significantly different product selectivity. The highly selective hydrogenation of C=C bonds into HCAL occurred over Pt-supported on MWCNT with the largest outer diameters (>50 nm). In the case of the other three catalysts, however, C=O bonds were more selectively hydrogenated (60–80% CAL) depending on the catalyst and CAL conversions. Irrespective of the catalyst, the Pt particles have an average size of about 11 nm. Besides other factors, the electronic effects induced by the significantly varied tube diameters of the MWCNTs seems to play a major role in determining the selective behaviors of different catalysts. Two kinds of CNTs with different inner diameter (less than 10 nm: CNTs-1, and between 60 and 100 nm: CNTs-2) were used as catalyst supports for platinum particles.[94] The Pt particles were deposited on the outside surface (CNTs-1) and inside (CNTs-2). The catalysts exhibit high activities in the selective hydrogenation of cinnamaldehyde. But the selectivity of these two catalysts was quite different under the same reaction conditions. The high selective hydrogenation of the C=O bond was observed over catalyst 3% Pt/CNTs-2, while the completely hydrogenation of both C=C and C=O bonds was found over catalyst 3% Pt/CNTs-1.

Bimetallic Pt-based systems enable a better control of selectivity. PtNi/MWCNT (135 $m^2\ g^{-1}$, particles 2–5 nm) are more selective towards hydrocinnamaldehyde formation than Pt/MWCNTs, with values of 25 and 88%, respectively.[95] The use of PtCo/MWCNT[96] or PdRu/SWCNT[97] directs selectivity towards cinnamyl alcohol formation to *ca.* 90% and 57%, respectively.

An Fe promoted Pt catalyst also exhibited high activity (6 340 h^{-1} in turnover frequency) and selectivity (89.1% for COL) in the hydrogenation of cinnamaldehyde, which was attributed to the uniform distribution of Fe promoters on Pt surfaces.[98] In a study on the selective hydrogenation of CAL to COL using monometallic (Pt, Ru) and bimetallic (Pt–Ru) catalysts supported on various carbon supports (MWCNT, SWCNT, CNF, AC), Serp *et al.* found that nanocarbon supported catalysts present higher CAL conversions than those obtained with activated carbon as support, but poor selectivity toward COL (<32%).[99] After a heat treatment under nitrogen at 973 K, Pt/CNT and especially PtRu/CNT catalysts exhibit an increase in the conversion rate and the selectivity toward COL (>66%). Heat treatment did not influence the conversion, and led to a decrease in selectivity towards COL, on AC supported catalysts. Such activity change is related to the influence of carbon surface chemistry on the adsorption of different aromatic compounds.[100,101] On heat-treated CNT samples, a π–π interaction between the CAL π electron ring and the basic π sites of the CNT surface may increase the CAL adsorption capacity. By removal of the electronegative oxygen atoms from the carbon surface, an electron transfer from the CNT to the metal is enhanced leading to the increase in COL selectivity. The electron transfer from the nanotube support to metal particles has already been reported.[102] For bimetallic PtRu/CNT catalysts a particle size and a structural effect that permit an increase in the selectivity have also been evidenced, and WAXS and EXAFS point to the formation of alloyed PtRu nanoparticles after heat treatment.[103]

Interestingly, in the case of Ru/CNF supported catalysts it was also found that the surface oxygen group concentration had a significant effect on the reaction selectivity.[104] The selectivity for COL decreased from 48 to 8%, owing to an enhanced rate of HCAL production with a decreasing number of surface oxygen groups (which was effected by increasing the treatment temperature). The hydrogenation of C=C bonds was significantly enhanced on CNFs with fewer oxygen groups. De Jong and co-workers also studied the influence of oxygen groups on the CNFs on their performance in the liquid-phase hydrogenation of CAL on CNF-supported Pt catalysts.[105] A linear decrease in hydrogenation activity with an increase in the number of acidic groups on the CNF supports was observed. The nature of the effect of oxygen groups on the properties of the Pt NPs and, hence, the catalytic performance has also been investigated. XPS experiments that were performed at 308 K showed no clear evidence of a change in the electronic structure of the Pt particles induced by the presence of oxygen-containing groups on the CNFs. Therefore, it was suggested that hydrogenation was assisted by the adsorption of cinnamaldehyde onto the carbon support after removal of the oxygen-containing surface groups. Bitter *et al.* have studied the influence of the nanostructure of filamentous carbon, *i.e.* CNF-h, CNT and CNF-p on the reducibility of Pt deposited on these carbons.[106] A retarded reduction for Pt/CNT was related to the higher amount of acidic oxygen surface groups on this support resulting in a strong stabilization of the cationic platinum species. A higher reduction temperature for that sample increased the amount of metallic Pt, however the platinum particle size was larger (2–11 nm) compared to that of Pt/CNF-h and Pt/CNF-p (both 1–3 nm). The orientation of the graphene sheets had a significant influence on the selectivity during cinnamaldehyde hydrogenation: Pt/CNF-p resulted in a higher selectivity towards cinnamyl alcohol compared to Pt/CNF-h. Herringbone and platelet CNFs and AC were also used as support of nickel catalyst for the selective hydrogenation of CAL to HCAL.[107] The catalytic activity and selectivity of the nickel supported on carbons of different structure were evaluated based on the determination of the rate constants of the hydrogenation reaction. The hydrogenation of CAL and its partially hydrogenated products HCAL and COL over carbon supported nickel catalyst was found to be strongly dependent on the metal particle size. Regardless of the type of carbon support, the smaller the nickel particles, the better the performance of the catalyst in the liquid-phase hydrogenation of CAL to HCAL at 433 K under a hydrogen pressure of 30 bar. For a given category of carbon support, the higher surface area enhances the dispersion of the nickel particles. It was found that Ni/CNF-p and Ni/AC catalysts show the highest catalytic activity in the hydrogenation reactions. CNFs grown on anatase wash-coated cordierite monolith were use as support for Pd and tested for cinnamaldehyde hydrogenation.[108] Although there is a relatively slow reaction rate over the as-prepared Pd/CNF/TiO_2/monolith due to its small BET surface area and low Pd dispersion (about 15%), the selectivity to HCAL remained high (about 90%) at 95% CAL conversion, the same as that over powdered Pd/CNF (about 93%), and much higher than that over Pd/FAC (about 45%), Pd/MC, and Pd/CNF/TiO_2/monolith-O (each about 82%).

A Pt-based catalyst supported on a silica grafted with C_{60} was prepared and evaluated in the liquid-phase hydrogenation of cinnamaldehyde.[109] The very high density of C_{60} grafting confers a strong hydrophobic character to the surface. A fairly good selectivity to COL was observed on this catalyst at a cinnamaldehyde conversion above 90%.

Hydrogenation of Crotonaldehyde and Citral. The effect of a previous oxidation treatment of the support on the catalytic behavior of Pt in the gas-phase hydrogenation of crotonaldehyde over Pt/CB catalysts has been analysed.[110] Two heat-treated CBs, one of them subjected to an oxidizing treatment with H_2O_2, have been used as supports for platinum. Both the specific catalytic activity and the selectivity towards the formation of the unsaturated alcohol (crotyl alcohol) have been found to be higher for the catalyst with the oxidized support, and they are enhanced after reduction at increasing temperatures. Pt–Sn/CB were also active and presented higher selectivity towards C=O hydrogenation.[111,112]

The catalytic properties of Fe nanoparticles of size 4–8 nm, deposited on CNTs, in the hydrogenation of crotonaldehyde have been studied.[113] It has been shown that the activity of the catalysts depends in an irregular manner on their dimensions and covers the range $(1.6–5.0) \times 10^{-6}$ mol m^{-2} s^{-1}. It has been established that the selectivity of the hydrogenation products is maximal for supported NPs with 6.8 nm. Gas-phase hydrogenation of crotonaldehyde to crotyl alcohol was conducted on 5 wt% Ni catalysts supported on CNF-p, CNF-r and γ-Al_2O_3.[114] Even though the mean particle size greatly differs according to the support used, from a narrow distribution centered at 1.4 nm for γ-Al_2O_3 to a broad distribution centered at 7 nm for CNFs, the higher activity and selectivity to crotyl alcohol were obtained on the CNF supported catalysts. Possible reasons for these differences are (i) different Ni crystallographic face exposure according to the support; and (ii) the possibility of charge transfer.

The heterogeneously catalyzed selective-hydrogenation of 3,7-dimethyl-2,6-octadienal (citral) is one of the more feasible ways to obtain its appreciated unsaturated-alcohols, nerol and geraniol, which are present in over 250 essential oils. The selective hydrogenation of citral is not easy, because citral is an α,β-unsaturated aldehyde which possesses three double bonds that can be hydrogenated: an isolated C=C bond and the conjugated C=O and C=C bonds (Scheme 7.2). Carbon materials are very interesting supports for this type of hydrogenation reaction due to their unique chemical and textural properties.[115] It should be mentioned that the amount of work discovered for citral hydrogenation is smaller than that of other α,β-unsaturated aldehydes. Thus, we consider that it is still possible to improve the catalyst development for this reaction.

MWCNTs decorated with nickel NPs were tested for citral hydrogenation.[116] The experimental results showed that the citronellal, an important raw material for flavoring and perfumery industries, is the favorable product with a percentage as high as 86.9%, which is 7 times higher than that obtained with a Ni/AC catalyst.

Crystalline Pt NPs were electrolessly deposited on poly(acrylic acid) (PAA) grafted MWCNTs.[117,118] The catalytic activity and selectivity of hydrogenation of citral on Pt/CNTs with different morphologies was studied. The irregularly shaped polycrystalline Pt NPs showed the highest conversion of citral, 92.2%, in comparison to 24.7% from Pt/AC, and 45.0%, and 40.4% from polyhedrons and polypods Pt single crystals on CNTs, respectively. The unusually high conversion on Pt/CNTs suggests a synergistic effect between the PAA grafted CNTs and the Pt NPs. In the study of selectivity, tetra- and octahedron Pt NPs showed the highest ratio of geraniol/nerol (5.2), 3 times higher than that of Pt/AC, whereas those of irregularly shaped polycrystalline and tetra- and octapods of single crystalline Pt on the CNTs were 1.9 and 2.2, respectively. Pt-based catalysts supported on MWCNTs were developed to study the influence of the support surface composition on the catalytic performance in the selective hydrogenation of citral to the corresponding unsaturated alcohols.[119] Supports with different oxidative and thermal treatments were characterized and used to prepare Pt and PtFe catalysts. Three factors seem to be important to obtain an adequate catalytic phase: (i) the size of the Pt metallic particles, (ii) the interaction of the Pt particles with the support, and (iii) the presence of a second metal as promoter. The results indicated that the Fe content in these catalysts and a low amount of oxygenated groups on the support surface, which produce an optimized Pt particle size, have a

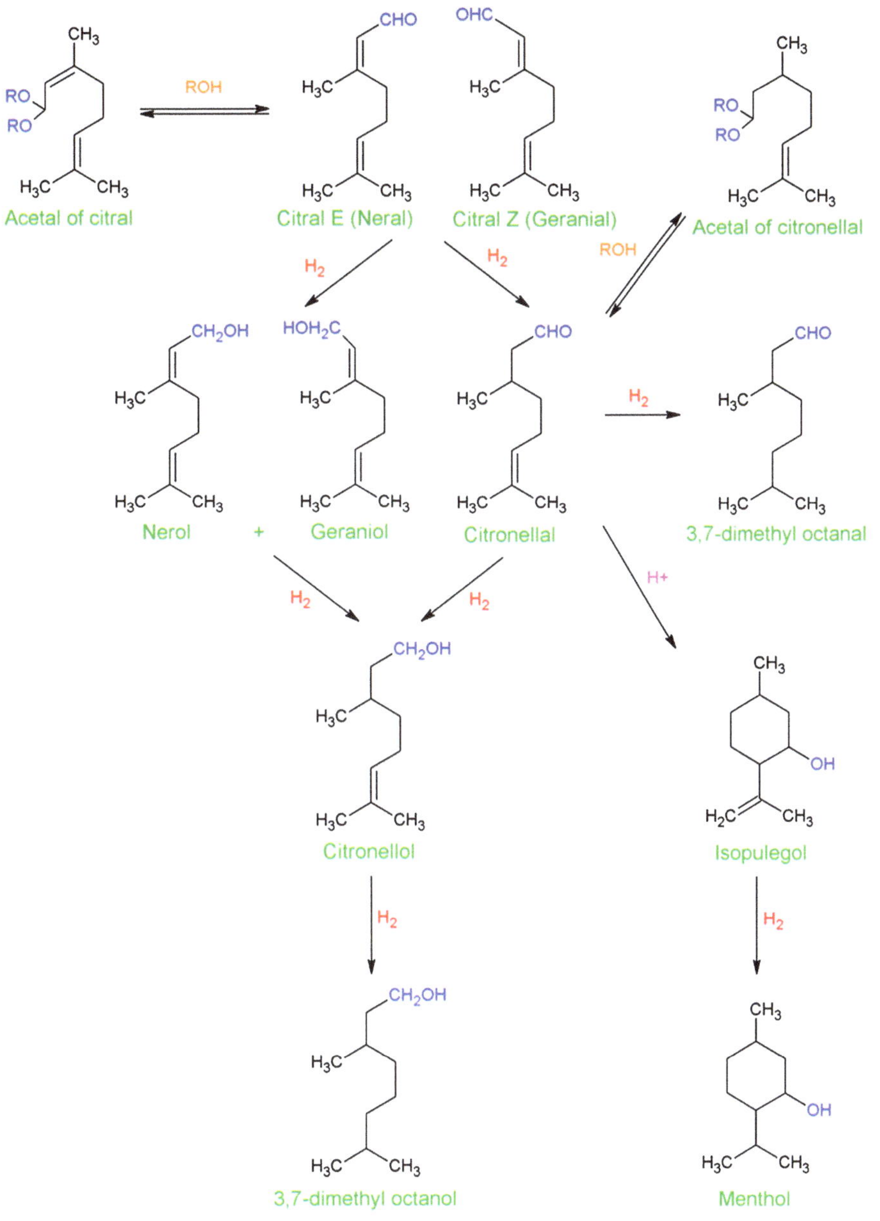

Scheme 7.2 Reaction scheme of the citral hydrogenation.[115]

fundamental role over the selectivity of the reaction to unsaturated alcohols. In fact, the combination of these effects gives a very high selectivity (about 96%) to unsaturated alcohols.

The addition of a second metal has an important role in the preparation of the catalysts for citral selective hydrogenation. The selective hydrogenation of citral to unsaturated alcohols by using two series of PtSn and PtFe catalysts with different metallic loadings and supported on CNTs and CB was reported.[120] The second metal modifies the structure of the Pt particles in a different way according to the support and the promoter used to prepare the bimetallic catalysts. For PtFe catalysts supported on both supports, Fe ionic species prevail on the bimetallic surfaces producing mainly a geometric effect. As it was expected from the correlation of the characteristics of these bimetallic structures with the performance of the catalysts in the citral hydrogenation, the higher the modification of the metallic phase, the higher the selectivity to unsaturated alcohols (up to 90–95%). A lower activity was observed in catalysts supported on CB compared to those supported on CNTs, this being inverted after thermal treatment with N_2 at high temperature, so an effect of the support takes place. For PtSn catalysts, reduced Sn species prevail on the bimetallic surfaces producing mainly an electronic effect. Surprising results were obtained with these bimetallic phases in the citral selective hydrogenation. The highest selectivities to unsaturated alcohols were nearby 80% for the PtSn/CB and 90% for the PtSn/CNT catalysts. In the case of the PtSn/CNT catalyst the thermal treatment with N_2 significantly enhances the selectivity to unsaturated alcohol toward values of 98% at 95% citral conversion, keeping a good activity in relation to that of the parent catalyst without any treatment. The latter effect would be related to the formation of a more adequate bimetallic phase, after sinterization, on the CNT support than on the CB one.

Gold nanoparticles of 10–24 and 5–8 nm in size were obtained on acid-treated MWCNTs and on ZnO/MWCNT composites.[121] The catalytic activity of the composites was studied for the selective hydrogenation of citral. The Au/ZnO/MWCNT composite favors the formation of unsaturated alcohols (selectivity = 50% at a citral conversion of 20%) due to the presence of single-crystalline, hexagonal gold particles, whereas saturated aldehyde formation is favored in the case of the Au/MWCNT nanocomposite that contains spherical gold particles.

7.2.3.3 Selective Hydrogenation of Nitroarene Compounds

The hydrogenation of aromatic nitro compounds with heterogeneous catalysts is in many cases the method of choice for the production of the corresponding aniline derivatives.[122,123] The hydrogenation of simple aromatic nitroarenes (see section 7.2.2.3) poses few problems and is carried out catalytically on very large scales. When other reducible groups are present in the same molecule, it is often difficult to catalytically reduce the nitro group in a selective manner. The main focus is on catalytic systems capable of reducing nitro groups with very high chemoselectivity in substrates containing carbon–carbon or carbon–nitrogen double or triple bonds, carbonyl or benzyl groups, and multiple Cl, Br, or I substituents. Analysis of the literature published in the last decade reveals that two approaches have been most successful; the modification of classical heterogeneous catalysts, such as Pt, Pd, or Ni, and the application of supported metals, such as Au and Ag, which have not to date been used for this reaction. Homogeneous catalysts and different reducing agents have also been investigated but the potential of these approaches seems to be much lower. A variety of supports have been tested to render Pt, Ni, or Ru more selective but most of the time only chloronitrobenzenes have been used as test substrate.

Ni nanoparticles and filamentous carbon were generated simultaneously by decomposing methane over Raney® nickel catalyst.[124] The nickel NPs were stabilized by filamentous carbon. The Ni-filamentous carbon composite materials were tested as catalysts for the hydrogenation of *ortho-* and *para-*chloronitrobenzene, producing chloroanilines selectively (99% yield). The catalysts exclusively reduce the nitro group in chloronitrobenzenes, leaving the C–Cl bond untouched. CNF-supported nickel catalysts for hydrogenation of *o*-chloronitrobenzene were prepared by deposition precipitation method in ethylene glycol.[125] The Ni/CNFs catalyst prepared at 353 K showed the best catalytic activity. For all of the Ni/CNFs catalysts prepared in ethylene glycol, over 98% conversion of *o*-chloronitrobenzene and 97% selectivity can be obtained at 20 bar and 413 K. The as-synthesized Ni/CNFs catalysts are also active for the hydrogenation of *m*-chloronitrobenzene and *p*-chloronitrobenzene to the corresponding chloroanilines.

Three types of carbon nanofibers (CNF-p, CNF-r, CNF-h) were used as supports for iridium deposition, and the resulting catalysts tested for the chemoselective hydrogenation of functionalized nitroarenes and imines.[126] Among the three Ir/CNF samples, Ir/CNF-r showed an excellent catalytic activity and chemoselectivity towards hydrogenation of functionalized nitroarenes and imines; the corresponding aniline derivatives were obtained with high turnover numbers at ambient temperature under 10 bar of H_2, and the catalyst is reusable. In particular, complete suppression of the reductive dehalogenation was achieved in the reaction of halonitrobenzenes over Ir/CNF-r; 4-chloro-, 4-bromo-, and 4-iodonitrobenzene were converted into the corresponding halogenated anilines in almost quantitative yields without contamination by the dehalogenated aniline. Furthermore, the nitro group was selectively reduced to the amino moiety in the presence of benzyloxy and carbonyl groups; the reduction of the keto group and the hydrogenolytic cleavage of the C=O bond in the benzyloxy group, which are often seen in the Pd/C-catalyzed hydrogenation, were not observed. Since the average size of the Ir particles is independent of the type of CNFs, the authors considered that the catalytic activity can be mainly attributed to steric and/or electronic effects derived from interactions between Ir nanoparticles and the CNF support rather than to the size of iridium particles. An interesting approach to obtain selective catalysts was described by Takasaki *et al.*,[127] who prepared CNFs to support Pt or Pd particles. Indeed, both Pd and Pt catalysts achieved selectivities greater than 99.9% at full conversion for 4-chloronitrobenzene when supported on CNF-p or CNF-h and in presence of *n*-octylamine as a catalyst poison. The Pt catalysts were also very selective for the hydrogenation of 3-chloronitrobenzene (S = 99.9%), 4-bromonitrobenzene (S = 99.7%), 4-benzyloxynitrobenzene (S > 99%), and 4-acetylnitrobenzene. Some more demanding substituents, such as 4-CH=CHCOOEt or 4-iodo gave somewhat lower selectivities of 91–96 %. The same authors were also interested in nitrogen-doped CNF as a support for metal nanoparticles.[128] The combination of carbon nanostructures and the doping of nitrogen atoms were expected to offer an efficient poisoning catalyst for chemoselective catalytic reactions. Platinum nanoparticles were immobilized on the surface of nitrogen-doped CNF-h, and the formed Pt/N–CNF-h compounds were highly efficient reusable catalysts for the nitro-selective hydrogenation of functionalized nitroarenes. The catalytic activity can be controlled by an appropriate choice of the Pt/N ratio. The hydrogenation of substituted nitroarenes over 3 wt% Pt/N–CNF-h afforded the corresponding anilines in high yields without promoting the reduction of other reducible functional groups, whereas reactions with 1 wt% Pt/N–CNF-h gave the intermediate *N*-aryl hydroxylamines with high selectivity.

Carbon nanotubes have been employed for the preparation of supported ruthenium NPs, and these systems were used in the selective hydrogenation of *p*-chloronitrobenzene (*p*-CNB) to *p*-chloroaniline (*p*-CAN).[129] Ru/CNT are efficient systems for the selective reduction of the nitro group in *p*-CNB under mild reaction conditions (333 K and 40 bar of H_2), while the C–Cl bond remains intact, thus allowing the almost complete substrate conversion with total selectivity to the target product. The catalytic activity of these materials for the hydrogenation of *p*-CNB at 35 bar and 333 K is shown to reach values as high as 18 $mol_{p\text{-CNB}}\ g_{Ru}^{-1}\ h^{-1}$, which is one order of magnitude higher than that obtained with a commercial Ru/Al_2O_3 catalyst.[130] CNTs, γ-Al_2O_3 and SiO_2 supported Pt and Pd catalysts were produced by laser vaporization deposition of respective bulk metals, and tested for selective hydrogenation of *p*-chloronitrobenzene.[131] The results show that the catalytic properties are greatly affected by the supports. Pt/CNTs catalyst exhibits the best catalytic performance among the Pt-based catalysts, producing *o*-CAN with 99.6% selectivity at complete conversion. The Pd/CNTs catalyst exhibits the best catalytic performance among the Pd-based catalysts, giving *o*-CAN with 95.2% selectivity at complete conversion. For Pt-based catalysts, geometric effects and the textures and properties of the supports play important roles in catalytic properties. On the other hand, geometric effects, electronic effects and the textures and properties of the supports simultaneously influence the catalytic properties of the Pd-based catalysts. Hydrogenation properties of *p*-CNB have been studied over Pt/CNTs and PtM/CNTs catalysts (M = Mn, Fe, Co, Ni and Cu).[132] The results show that the hydrogenation of *p*-CNB can be carried out over PtM/CNTs catalysts. Both catalytic activities and yields of *p*-CAN are improved. The PtFe/CNTs catalyst exhibits the best catalytic activity (TOF is 0.47 s^{-1}), and the PtMn/CNTs catalyst exhibits the highest yield of *p*-CAN (98.5 mol%). The effect of rare earths (Sm, Pr, Ce, Nd and La) on the hydrogenation properties of *p*-CNB over Pt/CNTs catalyst was studied in ethanol at 303 K and normal pressure.[133] The results showed that the hydrogenation of *p*-CNB could be carried out over $PtMO_x$/CNTs catalysts. Both catalytic activities and yields of *p*-CAN were all improved. The $PtCeO_x$/CNTs catalyst exhibited the best catalytic activity (TOF was 0.47 s^{-1}) and the highest yield of *p*-CAN (97.5 mol%). The $PtCeO_x$/CNTs (1.0 wt%) catalyst exhibited good stability for the hydrogenation of *p*-CNB. A Pt–Sn–B/CNTs catalyst was also evaluated for the liquid-phase hydrogenation of CNB.[134] The results showed that the catalyst had higher catalytic performance than common hydrogenation catalysts. The conversion of CNB could reach 99.9%, and the dechlorination of chloroaniline (CAN) was less than 1.9% when catalyzed by Pt–Sn–B/CNTs and more than 8.0% when catalyzed by common hydrogenation catalysts.

A rGO supported ruthenium catalyst was applied for the selective hydrogenation of *p*-CNB to *p*-CAN, exhibiting a turnover frequency (TOF) of 1800 h^{-1} and a selectivity of 99.6% at complete conversion of *p*-CNB.[135,136] The Ru/rGO catalyst displayed excellent stability and was extremely active for the hydrogenation of a series of nitroarenes, which can be ascribed to the fine dispersion of the Ru nanoparticles on the rGO sheets and their electron-deficient state.

Monodisperse and uniform palladium NPs were deposited on biomass-based carbon nanospheres (CSs) with controllable sizes.[137] Compared with other Pd catalysts, the Pd/CSs showed good activities and selectivities for nitroaromatic hydrogenation in the absence of additives or special pre-treatment.

7.2.4 Asymmetric Hydrogenation

Few reports deal with asymmetric hydrogenation. A recyclable chiral catalyst system was developed by absorption of pyrene-modified Pyrphos rhodium catalyst onto CNTs *via* π–π stacking interaction.[138] This modified catalyst was successfully applied in the asymmetric hydrogenation of α-dehydroamino esters for nine cycles without obvious deterioration of activity and enantioselectivity. Hybrid catalysts have been prepared by the immobilization of a Rh complex on CNTs and CNFs.[139] To anchor the complex, a siloxane type bond has been created by reaction of a trimetoxisilane end of a ligand and –OH phenol-type groups on the support's surface. The hybrid catalysts have been tested in the hydrogenation of three different substrates: cyclohexene, carvone and 2-methyl acetamidoacrylate (to evaluate enantioselectivity, using BINAP as chiral ligand). The obtained results show that the hybrid catalysts are more active than the homogeneous complex. The enhanced activity has been related to a confinement effect, produced as a consequence of the metal complex location inside the tubular structures of the support. The enantioselectivity is opposite for the heterogeneous and the homogeneous complex. Chiral rhodium hybrid catalysts have been prepared by covalent anchorage of pyrrolidine-based diphosphine ligands onto functionalized CNTs.[140] The catalysts have been tested in the asymmetric hydrogenation of two different substrates: methyl 2-acetamidoacrylate and α-acetamidocinnamic acid. The catalysts have shown to be active and enantioselective in the hydrogenation of α-acetamidocinnamic acid. A good recyclability of the catalysts with low leaching and without loss of activity and enantioselectivity was observed. A series of SWCNT-supported Pt nanoparticle catalysts with different controlled Pt loadings, (5–20 wt. %) were prepared.[141] By modification with (−)-cinchonidine, these Pt/SWCNT catalysts were found to be efficient for the asymmetric hydrogenation of ethyl pyruvate, providing (*R*)-ethyl lactate in high activity and moderate enantioselectivity. A kinetic study showed that the Pt/SWCNT-catalyzed asymmetric hydrogenation of ethyl pyruvate was a "ligand-accelerated" reaction. CNT-supported Pd catalysts were active in the enantioselective hydrogenation of α,β-unsaturated carboxylic acids using cinchonidine as a chiral modifier.[142] A highly active and enantioselective heterogeneous asymmetric catalyst was fabricated by confining Pt NPs that are modified with cinchonidine within the channels of MWCNTs.[143] A high turnover frequency and enantioselectivity were achieved when using this catalyst for the asymmetric hydrogenation of α-ketoesters. The application of Pd nanoparticle–FLG (Pd/FLG) catalysts in the asymmetric hydrogenation of aliphatic α,β-unsaturated carboxylic acids using cinchonidine as chiral modifier was reported.[144] The Pd/FLG modified by cinchonidine can act as efficient catalysts in the asymmetric hydrogenation of α,β-unsaturated carboxylic acids for producing optically enriched saturated carboxylic acids.

7.2.5 CO Hydrogenation – Fischer–Tropsch Synthesis

Although ACs are not often employed as supports for these reactions, nanostructured carbon materials have been investigated for Fischer–Tropsch reactions, and methanol or higher alcohol synthesis.

7.2.5.1 Synthesis of Methanol and Higher Alcohols

A new catalyst for methanol synthesis, ZnO-promoted rhodium supported on MWCNTs, was developed.[145] It was found that the Rh–ZnO/CNTs catalyst had a high activity of 411.4 mg CH_3OH g_{cat}^{-1} h^{-1} and selectivity of 96.7% for methanol at 10 bar and 523 K. The activation energy of methanol synthesis on this catalyst is approximately 68.8 kJ mol^{-1}, which is lower than that of Cu–ZnO–Al_2O_3 catalyst for methanol synthesis (95.0 kJ mol^{-1}). It was suggested that the CNTs favored both the couple transfer of the proton and electron over the surface of the catalyst and the uptake of hydrogen, which was favorable to methanol synthesis. These results also indicated that the inner diameter of carbon nanotubes plays an important role in this reaction. MWCNTs have also been used as promoters to improve the catalytic activity of Cu–ZnO–Al_2O_3 catalysts for methanol synthesis using H_2/CO/CO_2 mixtures.[146] The catalyst, containing 10–15% of MWCNTs permits a significant increase in methanol formation rate compared to unpromoted systems. The observed single-pass CO-conversion and methanol-STY over a $Cu_6Zn_3Al_1$-12.5%CNTs catalyst reached 64% and 1210 mg h^{-1} g^{-1}, which was about 68%

and 66% higher than those (38% and 730 mg h^{-1} g^{-1}) over the corresponding CNT-free catalyst and $Cu_6Zn_3Al_1$, respectively.[147] The action of CNTs as promoters could arise from (i) a dispersing role of MWCNTs, allowing a significant increase in Cu specific surface area; and (ii) the efficiency of MWCNTs to act as a hydrogen reservoir, favoring CO/CO_2 hydrogenation reactions.

A striking enhancement of the catalytic activity of Rh particles confined inside CNTs for the conversion of CO and H_2 to ethanol was reported by Bao *et al.*[148] The overall formation rate of ethanol $\left(30.0 \text{ mol mol}_{Rh}^{-1} \text{ h}^{-1}\right)$ inside the CNTs exceeds that on the outside of the nanotubes by more than an order of magnitude, although the latter is much more accessible. Alkali-promoted Ni–Co–Mo catalysts supported on MWCNTs were studied for higher alcohols synthesis from synthesis gas.[149] An alkali-promoted trimetallic catalyst with 3 wt% Ni showed the highest total alcohol yield of 0.284 g g_{cat}^{-1} h^{-1}, ethanol selectivity of 20%, and higher alcohols selectivity of 32% at 603 K and 90 bar using a H_2 to CO molar ratio of 1.25. Alkali-modified trimetallic Co–Rh–Mo sulfided catalysts supported on commercial ACs with different textural characteristics were tested for the synthesis of higher alcohols from synthesis gas and compared with a similar catalyst supported on MWCNTs.[150,151] Addition of metals (Co, Rh, and Mo) to the microporous and mesoporous activated carbons, and the MWCNT supports, increased the mean pore diameter and % mesoporosity of the catalysts. The difference in metal dispersion of the sulfided catalysts on different supports was: MWCNTs < ACs. The higher total alcohol STY and selectivity of 0.296 g g_{cat}^{-1} h^{-1} and 35.6%, respectively, were found to be on the MWCNTs-supported alkali-modified trimetallic C–Rh–Mo catalyst compared to similar catalysts supported on AC. Little or no sintering of metal species was observed on the spent catalyst supported on MWCNTs, whereas the agglomeration of catalytic species is high on the microporous AC support compared to that of a mesoporous AC supported catalysts. A metal cobalt-decorated MWCNT-promoted Co–Mo–K oxide-based catalyst was developed, with excellent performance for the selective formation of C_{2-9}-alcohols from syngas.[152–154] The addition of a minor amount of the Co-decorated MWCNTs into the $Co_1Mo_1K_{0.05}$ host catalyst caused little change in the apparent activation energy for the higher alcohol synthesis, but led to an increase of surface concentration of the two kinds of catalytically active species, $CoO(OH)/Co_3O_4$ and Mo^{4+}, both closely associated with the alcohol generation. An excellent adsorption performance of the Co-decorated MWCNTs as promoter for H_2 would be conducive to generating a surface micro-environment with a high concentration of H-adspecies on the functioning catalyst, thus increasing the rate of surface hydrogenation reactions in the higher alcohol synthesis. In addition, a high concentration of H-adspecies on the catalyst would, through synergistic action with the CO_2 in the feed gas, greatly inhibit the water–gas-shift side-reaction. All these factors contribute to an increase in the yield of alcohols. Synthesis of higher alcohols from syngas on potassium promoted molybdenum sulfide supported on MWCNT catalysts was also reported.[155] Addition of K increased the formation of alcohols and suppressed the formation of hydrocarbons.

A catalyst with 15 wt% Mo and 9 wt% K supported on MWCNT showed the highest yield (0.11 g g_{cat}^{-1} h^{-1}) and selectivity (25.6%) towards alcohols. The optimum conditions for producing the higher alcohols from synthesis gas (mole ratio of H_2 and CO equal to 2) are determined to be 593 K and 96.5 bar. h-CNFs have been used as supports for bimetallic Co–Cu catalysts, and the resulting material affords high selectivity towards the formation of higher alcohols, in particular butanol from H_2/CO/CO_2 mixtures under 50 bar and at 523–583 K.[156,157] Also in this case, it has been proposed that a high concentration of reversibly adsorbed H-species on the CNT-promoted catalyst favors surface hydrogenation reactions. The promotional effects of Rh (0 to 2 wt%) on Mo–K/MWCNT catalysts for higher alcohol synthesis from synthesis gas (molar ratio of H_2 to Co equal to 1) were examined.[158] Metal dispersions were increased from 30% to 42%, with an increase in Rh content from 0 to 2 wt%. The maximum total alcohol yield of 0.211 g g_{cat}^{-1} h^{-1}, ethanol selectivity of 16%, and higher alcohols selectivity of 25% were observed on the catalyst with 1.5 wt% Rh at 593 K and 83 bar.

It is also worth noting that CO_2 direct hydrogenation to methanol and CO/CO_2 hydrogenation to higher alcohols has also been reported for CNT-based catalysts.[159]

7.2.5.2 Fischer–Tropsch Synthesis

Syngas conversion (Fischer–Tropsch synthesis, FTS) is one of the most important reactions to convert coal or biomass to fuel.[160,161] The product distribution (alkane, alkene, alcohol) are usually too broad and unselective, extending from C_1 to C_{80} chains. Various supported catalysts such as Fe, Co, Ru on different supports have been tested for this reaction. Developing FT-catalyst with a controllable selectivity toward the hydrocarbons produced remains a challenge.[162]

Iron Catalysts. CNT supported iron catalysts were prepared by incipient wetness, deposition/precipitation using K_2CO_3, and deposition/precipitation using urea.[163] The incipient wetness method and the deposition/precipitation technique using urea yielded highly dispersed Fe^{3+} on the carbon nanotubes support. The deposition/precipitation technique using K_2CO_3 also yielded larger Fe_2O_3 crystallites. After reduction the three catalysts had similar metal surface areas. Nevertheless, the activity of these catalysts in the FTS differed significantly, with the catalyst prepared by IWI being the most active one. It is speculated that the differences in the performance of the catalysts might be attributed to the different crystallite size distributions, which would result in a variation in the amount of the different phases present in the catalyst under reaction conditions. The selectivity in the Fischer–Tropsch synthesis over the three catalysts seems to be independent of the method of preparation. Iron-based catalysts supported on CNTs for use in the FTS were prepared either by incipient wetness or a deposition precipitation method using urea and then promoted with potassium and/or copper.[164] The FTS was carried out in a fixed-bed micro reactor (548 K, 8 bar, $CO/H_2 = 2$). The effect of Cu and promoters on CO conversion, product selectivity and FTS activity were investigated and compared with data reported on other carbon-based supports. The K promoted catalysts gave higher yields of CO_2 and C_2 olefins and the lowest methane selectivity when compared to the unpromoted catalysts. Copper, while enhancing catalyst activity, did not have an effect on the FT product selectivity.

The effects of acid treatment on the activity, product selectivity and life span of iron FT catalysts supported on CNTs was investigated.[165] Two different types of CNTs with low surface area (*ca.* 25 $m^2 g^{-1}$) and high surface area (*ca.* 170 $m^2 g^{-1}$) were prepared and treated with 35 wt% HNO_3 at 298 and 383 K for 16 h. 10 wt% Fe/CNTs were prepared using incipient wetness impregnation. The acid treatments at 298 and 383 K increased the BET surface area by 18% and 31%. The acid treatment on both families of CNTs resulted in a decrease in metal particle sizes. The FTS was carried out in a fixed-bed microreactor (548 K, 20 bar, $CO/H_2 = 2$) for 120 h. Among the catalysts studied, the Fe catalyst supported on pre-treated CNTs at 383 K was stable and active, while the other catalysts experienced rapid deactivations. The Fe catalyst supported on CNTs with low surface area and larger diameter showed much lower CH_4 and higher C_5^+ selectivity. The effects of pore diameter and structure of iron catalysts supported on CNTs on the reaction rates and product selectivity were also studied.[166] Two types of CNTs with different average pore sizes (12 and 63 nm), but with similar surface areas were used. A vast majority (*ca.* 80%) of the iron oxide particles were deposited inside the nanotubes' pores. The iron oxide particles on the wider pore catalyst, Fe/wp-CNT, (17 nm) were larger than those on narrow pore catalysts, Fe/np-CNT sample, (11 nm). The extent of reduction of the Fe/np-CNT catalyst was 17% higher compared to that of the Fe/wp-CNT catalyst. The catalytic performances of both catalysts were evaluated in a fixed-bed reactor for FT reactions at 20 bar and 548 K. The activity of the np-CNT catalyst (CO conversion of 30%) was 2.5 times that of the wp-CNT catalyst (CO conversion of 12%). In addition, the Fe/wp-CNT was more selective toward lighter hydrocarbons with a methane selectivity of 41% compared to that of the np-CNT catalyst with methane selectivity of 14.5%. Deposition of metal particles on the CNT with narrow pore size resulted in a more active and selective catalyst due to higher degree of reduction and higher metal dispersion.

Following previous findings that confinement within CNTs can modify the redox properties of encapsulated iron oxides,[167] Bao *et al.* demonstrated how this can affect the catalytic reactivity of iron catalysts in FTS.[168] The investigation reveals that the distribution of iron carbide and oxide phases is modulated in the CNT-confined system. The iron species encapsulated inside CNTs prefer to exist in a more reduced state, tending to form more iron carbides under the reaction conditions, which have been recognized to be essential to obtain high FTS activity. This causes a remarkable modification of the catalytic performance. The yield of C_{5+} hydrocarbons over the encapsulated iron catalyst is twice that over the iron catalyst outside CNTs and more than

6 times that over an AC-supported iron catalyst. The catalytic activity enhancement is attributed to the effect of confining the iron catalyst within the CNT channels. As demonstrated by temperature-programmed reduction in H_2 and in CO atmospheres, the reducibility of the iron species is significantly improved when they are confined. In order to study the effects of the catalytic site position on FT reactions, a method was developed to control the position of the catalytic sites either on the inner or outer surface of CNTs.[169] More than 70–80% of iron oxide particles can be controlled to be positioned on the inner or outer surface of the nanotubes. Based on H_2-TPR analysis, deposition of iron oxide inside the nanotube pores resulted in easier reduction of the oxide at a lower temperature (from 691 to 654 K). Catalytic performances of the catalysts in terms of FT experiments were tested in a fixed-bed reactor. According to the results, both catalysts showed similar initial CO conversion (*ca.* 90%). However, the catalyst with catalytic sites inside the pores exhibited higher selectivity to heavier hydrocarbons. In addition, deposition of catalytic sites on the interior surface of the CNTs resulted in a more stable catalyst, while its counterpart experienced deactivation within a period of 125 h due to sintering of catalytic sites.

The influence of the iron carbide particle size of promoted and unpromoted CNF supported catalysts on the conversion of synthesis gas has been investigated at 613–623 K, H_2/CO = 1, and pressures of 1 and 20 bar.[170] The surface-specific activity (apparent TOF) based on the initial activity of unpromoted catalysts at 1 bar increased 6–8-fold when the average iron carbide size decreased from 7 to 2 nm, while methane and lower olefins selectivity were not affected (Figure 7.4). The same decrease in particle size for catalysts promoted by Na plus S at 20 bar resulted in a 2-fold increase of the apparent TOF based on initial activity, which was mainly caused by a higher yield of methane for the smallest particles. Presumably, methane formation takes place at highly active low coordination sites residing at corners and edges, which are more abundant on small iron carbide particles. Lower olefins are

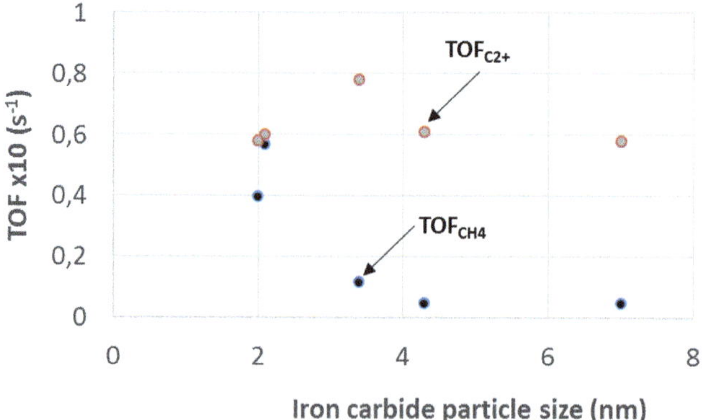

Figure 7.4 Dependence of TOF on the mean size of iron carbide particles for FTS on Fe/CNF catalysts.

produced at promoted (stepped) terrace sites that are available and active, quite independent of size. These results demonstrate that the iron carbide particle size plays a crucial role in the design of active and selective FTS catalysts. De Jong *et al.* have also reported on the conversion of synthesis gas to C_2 through C_4 olefins with selectivity up to 60 wt%, using iron nanoparticles (promoted by sulfur plus sodium) homogeneously dispersed on CNF supports.[171]

By means of the anchoring effect and the intrinsic basicity of nitrogen-doped CNTs, iron nanoparticles were conveniently immobilized on N-CNTs without surface premodification.[172]

High-temperature FTS for the production of short-chain olefins over iron catalysts supported on MWCNTs was investigated to elucidate the influence of nitrogen and oxygen functionalization of the CNTs on the activity, selectivity, and long-term stability.[173] After reduction in pure H_2 at 653 K, the Fe/N-CNT and Fe/O-CNT catalysts were applied in FTS, in which they showed comparable initial conversion values with an excellent olefin selectivity [$S(C_3–C_6) > 85\%$] and low chain growth probability ($\alpha \le 0.5$). TEM analysis of the used catalysts detected particle sizes of 23 and 26 nm on O-CNTs and N-CNTs, respectively, and Fe_5C_2 was identified as the major phase by using XRD, with only traces of Fe_3O_4. After 50 h on stream under steady-state conditions, an almost two-fold higher activity compared to the Fe/O-CNT catalysts had been maintained by the Fe/N-CNT catalysts, which are considered excellent FT catalysts for the production of short-chain olefins owing to their high activity, high selectivity to olefins, low chain growth probability, and superior long-term stability. The so-constructed Fe/N-CNTs catalyst shown good catalytic performance in FTS with high selectivity for lower olefins of up to 46.7% as well as high activity and stability. The excellent performance was well-correlated with enhanced dissociative CO adsorption, inhibition of secondary hydrogenation of lower olefins, and promoted formation of the active phase of χ-Fe_5C_2. All of these merits result from participation of the nitrogen.

Iron-based FT catalysts supported on high surface area graphene nanosheets shown high activity and selectivity for the production of long-chain hydrocarbons.[174] The Fe–K nanoparticle catalyst supported on graphene exhibits high activity and selectivity toward C_8 and higher hydrocarbons with excellent stability and recyclability. In comparison with other carbon supports, such as CNTs, the graphene support shows a unique tendency for minor formation of the low-value and undesirable products methane and carbon dioxide, respectively. The water–gas shift activity is reduced on the graphene support as compared with CNTs, and as a result, the formation of CO_2 is significantly reduced. Evidence was presented for the formation of the active Fe_5C_2 iron carbide phase during the FTS on the graphene-supported Fe catalysts. The high activity and selectivity of the catalysts supported on graphene are correlated with the presence of defects within the graphene lattice that act as favorable nucleation sites to anchor the metal nanoparticles, thus providing tunable metal-support interactions.

Ruthenium Catalysts. Ruthenium nanoparticles supported on CNTs were reported to be a highly selective FT catalyst for the production of C_{10}–C_{20} hydrocarbons (diesel fuel).[175] The C_{10}–C_{20} hydrocarbon selectivity strongly depends on the mean size of the Ru nanoparticles (Figure 7.5). Nanoparticles with a mean size around 7 nm exhibit the highest C_{10}–C_{20} selectivity (*ca.* 65%) and a relatively higher turnover frequency for CO conversion (34%) compared to Ru/AC (11%) or Ru/graphite (20%). Moreover, Ru/CNT exhibits the highest selectivity to diesel fuel.

This study revealed that the surface chemistry of CNTs determines the product selectivity. Indeed, the pretreatment of CNTs with concentrated HNO_3 is necessary to obtain high C_{10}–C_{20} selectivity. The acid groups (–COOH) on the CNT generated by liquid-phase HNO_3 treatment do enhance the hydrocracking of heavier hydrocarbons, thus affording higher diesel fuel selectivity. This is an important advantage when using CNTs as support since the surface of CNTs is tunable by pretreatment. FTS were carried out in a slurry phase over Mn-modified Ru/CNT catalysts.[176] CO conversion and C_{5+} selectivity were dependent on the Ru and Mn concentrations as well as the reaction temperature. The activity of the catalyst containing optimized amounts of Ru and Mn was very similar to that of Ru–Mn/γ-Al_2O_3, although the initial activity of CNT-based catalysts after 3 h was about 10% lower than that of Al_2O_3-based catalyst. FTS was carried out in a water/oil mixture medium, using a Ru catalyst supported on a MWCNT/MgO–Al_2O_3 hybrid as a catalyst support.[177] The nanohybrid particles at the water/oil interface facilitated and stabilized the formation of a water-in-oil emulsion, giving rise to an oil/emulsion/water trilayer liquid structure. FTS occurred in the emulsion phase with much higher conversion rates than those in oil single-phase reactions, yielding products with Anderson–Schulz–Flory distribution. Alkane-enriched hydrocarbons migrate to the top oil phase, while short alcohols remain in the bottom water phase. Thus, this multiphase liquid structure facilitates the separation of products according to their solubility in different phases. Another positive effect of using the biphasic system arises from the enhanced FTS activity observed in the presence of condensed

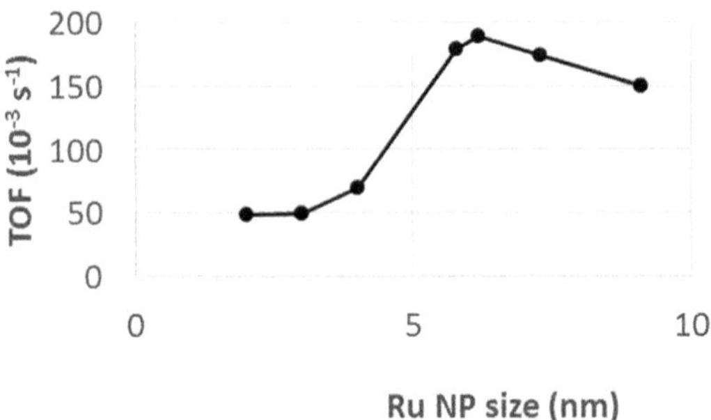

Figure 7.5 Dependence of TOF on the mean size of Ru particles for FTS on Ru/CNT catalysts.

aqueous phase.[178] Finally, the presence of an emulsion seems to improve the C_1/C_{5+} product balance, which can be explained by a dual-site model recently proposed in the literature. The influence of organic capping agents on the performance of Ru nanoparticles in aqueous-phase FTS was investigated.[179] To exclude the effects of particle size, the capping agents were placed onto CNF-supported Ru nanoparticles of size 3.4 nm. Kinetic data suggest that the FT mechanism was largely unaffected by the presence of capping agents; thus, their binding to active centers explains activity trends.

Fe–Ru bimetallic nanoparticles supported on purified CNTs were used for FTS.[180] The Fe–Ru/CNT catalysts were promoted with potassium and/or copper. The effect of Cu and K on CO conversion, product selectivity and FT synthesis activity was investigated. The observed behavior of the small particles (2.1 nm) obtained in this study followed similar trends to what has been observed before for Fe promoted catalysts suggesting that support interactions do not strongly affect the promoter properties of the metals. All the catalysts were found to be stable in the FT reaction (*ca.* 120 h) indicating that the Fe–Ru clusters possess remarkable stability in the FT reaction when supported on CNTs. This resistance to sintering is attributed to the metal support interaction characteristic of CNT supports. CNT-supported Pt or Ru promoted cobalt catalysts were also investigated.[181] Promotion with platinum and ruthenium significantly decreased the cobalt species reduction temperature. The presence of noble metal promoters had no significant effect on the size of cobalt nanoparticles. The Pt and Ru promoted cobalt catalysts exhibited CO conversion of 37.1 and 31.4%, respectively. The Pt promoted cobalt supported on CNTs yielded better catalytic stability than that of the monometallic cobalt catalyst. Bimetallic Co–Ru/CNT catalyst activity and selectivity in FTS have been assessed in a fixed-bed microreactor.[182,183] A very narrow particle size distribution has been obtained by the microemulsion technique and small Co particles (2–7 nm) were mostly confined inside the CNTs. Activity and selectivity were found to be dependent on the catalyst preparation method (microemulsion technique or impregnation). CO conversion increased from 59.1 to 75.1% and the FTS rate increased from 0.291 to 0.372 $g_{HC}\ g_{cat}^{-1}\ h^{-1}$, for the impregnation and microemulsion technique, respectively. C_{5+} liquid hydrocarbons selectivity decreased from 92.4 to 87.6%.

Cobalt Catalysts. CO hydrogenation was investigated over Co or Fe catalysts supported on CNTs.[184] Two types of catalysts were prepared. For the simple impregnation method, metal acetate precursors were deposited onto the surface of CNTs. These catalysts were denoted as "I-samples." The second family of catalysts was prepared by deposition of pre-prepared metal oxide nanoparticles onto the CNT support. These samples were labeled "P-samples." TPR measurements showed easy reducibility of the metal ions in the I–Co and I–Fe samples. TEM images generally pointed to a rather uniform particle size both before and after the reaction. The highest catalytic activity and high selectivity toward C_2–C_4 and C_{5+} fractions, as well as for olefin formation, were found for I–Fe. Catalytic activity was lower for P–Co and P–Fe.

The nature of cobalt species in Co/CNT catalysts and their catalytic performance in FTS were investigated.[185] The catalysts were prepared by IWI using solutions of cobalt nitrate followed by calcination under nitrogen. The characterization techniques uncovered that acid pre-treatment oxidized the CNT surface and removed impurities. Small and irregularly shaped cobalt oxide particles with 8–10 nm were anchored onto the outer CNT surface. The catalysts displayed high cobalt reducibility, which was slightly affected by the pre-treatment with nitric acid and nanotube outer diameter. Cobalt catalysts supported on CNTs exhibited high catalytic activity in FTS. Pre-treatment with nitric acid leads to a 25% increase in hydrocarbon yield, while CNT diameter does not seem to significantly affect the FT performances. The influence of acid treatment on CNT-supported cobalt catalysts has been studied.[186] Cobalt catalysts supported on fresh and acid treated CNTs were prepared by IWI with a cobalt loading of 10 wt%. The TEM analyses of the acid-treated support catalysts showed that the majority of the cobalt particles were homogenously distributed inside the nanotubes. The acid treatments: (a) increased the BET surface area; (b) decreased the cobalt particle size and increased the cobalt dispersion; (c) increased the reducibility of the catalysts and (d) increased the FTS activity and CO conversion. Finally, the product selectivity showed a distinct shift to lower molecular weight hydrocarbons. The influence of cobalt crystallite size on catalyst performance in FTS has been investigated using a functionalized CNT-supported Co catalyst.[187] The catalysts were synthesized by the core reverse micelle reactions with cobalt crystallites of various sizes. Small cobalt crystallites (3–8 nm) were synthesized and they were mostly confined inside the functionalized CNTs. The deposition of cobalt nanoclusters on the functionalized CNTs shift the reduction peaks to a lower temperature, indicating higher reducibility for uniform cobalt crystallites. The catalyst prepared with functionalized CNTs increased the FTS rate from 0.64 to 0.78 g HC g_{cat}^{-1} h^{-1}, C_{5+} selectivity increased 7.4% and CH_4 selectivity decreased 44%, compared to that catalyst prepared on common CNTs. An extensive study of FTS on Co/CNTs with different loadings of cobalt, ruthenium and potassium was reported.[188] Up to 30 wt% of Co, 1 wt% of Ru and 0.0066 wt% of K were added to the catalyst by the co-impregnation method. For the 15 wt% Co/CNT catalyst, most of the metal particles were homogeneously distributed inside the tubes and the rest on the outer surface of the CNTs. Increasing the Co loading to 30 wt% increased the amount of Co on the outer surface of the CNTs, increased the cobalt cluster sizes and decreased the reduction temperature and dispersion. Increasing the Co loading from 15 to 30 wt% increased the CO conversion from 48 to 86% and the C_{5+} selectivity from 70 to 77%. Ruthenium was found to enhance the reducibility of Co_3O_4 to CoO and that of CoO to Co(0), increase the dispersion and decrease the average cobalt cluster size. However, potassium was responsible in shifting the reduction temperatures to higher temperatures. 0.5 wt% Ru increased the FTS rate of 15 wt% Co/CNT catalyst by a factor of 1.4 while addition of 0.0066 wt% K decreased the FTS rate by a factor of 7.5. Both promoters enhanced the selectivity of FTS towards the higher molecular weight hydrocarbons; however, the effect of Ru is less pronounced. Potassium increased the olefin to paraffin ratio from 0.73 to 3.5 and the C_{5+} selectivity from 70 to 87%. 10 and 40 wt% Co/CNT and 10 and 40 wt% Co/SBA-15 catalysts were prepared *via* IWI.[189] 10% Co/CNT and 10% Co/SBA-15 did not perform well in an autoclave slurry system. It was observed that the performance of 40 wt% Co/SBA-15 in a slurry reactor was higher than other catalysts in terms of production of longer chain paraffins. Langmuir–Hinshelwood (LH) and power rate equations were applied to describe the kinetics of the FT reaction on cobalt catalysts and manganese-doped cobalt catalysts supported on CNTs.[190] LH-based kinetics characterize the activity behavior of the unpromoted Co/CNT system satisfactorily, but fail with respect to the Mn-promoted Co/CNT catalyst. An extensive study of FTS on CNT supported and γ-alumina-supported cobalt catalysts with different amounts of cobalt was reported.[191] Up to 40 wt% of cobalt was added to the supports by the impregnation method. The effect of the support on the reducibility of the cobalt oxide species, dispersion of the cobalt, average cobalt clusters size, water–gas shift activity and activity and selectivity of FTS was investigated. Using CNTs as cobalt catalyst support was found to cause the reduction temperature of cobalt oxide species to shift to lower temperatures. The strong metal–support interactions were reduced to a large extent and the reducibility of the catalysts improved significantly. CNT aided the dispersion of metal clusters and the average cobalt cluster size decreased. The hydrocarbon yield obtained by Co/CNT catalyst was surprisingly much larger than that obtained from Co/Al_2O_3. The maximum concentration of active surface Co(0) sites and FTS activity for alumina and CNT supported catalysts were achieved at 34 wt% and 40 wt% cobalt loading, respectively. CNT caused a slight decrease in the FTS product distribution to lower molecular weight hydrocarbons.

The effect of catalyst confinement and pore size on FTS over Co/CNT was investigated.[192] Three types of CNTs with average pore sizes of 5, 11, and 17 nm were used as support. The catalysts were prepared by selectively impregnating cobalt nanoparticles either inside or outside CNTs. As in the case of iron catalysts,[168] the catalyst with Co particles inside the CNTs was easier to be reduced than those with Co outside the CNTs, and the reducibility of cobalt oxide particles inside the CNTs decreased with increasing cobalt oxide particle size. The activity of the catalyst with Co inside the CNTs was higher than that of catalysts with Co particles outside the CNTs. Smaller CNT pore size also appears to enhance the catalyst reduction and FTS activity due to the little interaction between cobalt oxide with carbon and the enhanced electron shift on the non-planar CNT surface. A correlation between the location, pre-treatment, and surface chemistry of the cobalt nanoparticles and the catalytic selectivity in FTS was built.[193] It was found that selectivity towards C_{5+} molecules through FTS on Co/CNT depends on activation temperatures and surface chemistry of the cobalt nanoparticles. A pre-treatment at 573 K in H_2 flow results in a different surface chemistry for Co-in-CNT than for Co-on-CNT, which leads to a difference in selectivity to C_{5+} molecules. Pre-treatment at a relatively high temperature, 673 K, in H_2 flow produces completely reduced Co NPs in Co-in-CNT and Co-on-CNT. There was no significant difference in catalytic selectivity between the two catalysts upon pre-treatment at 673 K. The absence of a significant difference in catalytic selectivity of metallic Co-on-CNT and metallic Co-in-CNT suggests that the electronic effect of the CNT support does not significantly affect the C_{5+} selectivity of cobalt catalysts in FTS. A 10% Co-in-CNT catalyst showed better CO conversion compared to a 10% Co-on-CNT catalyst at various reaction temperatures.[194] An improvement on the catalyst performance was demonstrated in the case of particle deposition inside the pores of the nanotubes with higher C_{5+} and C_2–C_4 selectivity. TEM showed that 480 h continuous FTS increased the average particles size of the particles located inside the pores from 7 to 7.4 nm, while the average particles size of the particles located outside of the tubes increased from 11.5 to 25 nm.[195] XRD analysis of the used catalyst confirmed cobalt re-oxidation and interaction between cobalt and CNTs and creation of carbide phases. The pore-confined Co-in-CNT catalyst was found to reduce easily and showed a higher reducibility in comparison to the Co-out-CNT catalyst.[196] The enhancement in reducibility is ascribed to a pore confinement effect. Simultaneously, Co-in-CNT showed a higher FTS catalytic activity and this is proposed to result from the higher reducibility and dispersion. Finally, the C_{5+} hydrocarbon selectivity showed no observable change between pore-confined Co-in-CNT and Co-out-CNT catalysts for cobalt particles.

A catalyst based on aligned MWCNT arrays was tested for the FTS reaction in a microchannel reactor.[197] The microstuctural catalyst possessed superior thermal conductivity inherent from CNTs, which allows efficient heat removal from catalytic active sites during exothermic FTS reaction. The concept was tested and demonstrated in a micro-channel fixed-bed FTS reactor. FTS turn-over activity was found to enhance by a factor of four owing to potential improvement in mass transfer in the unique microstructure. Cobalt supported on CNT-covered alumina has been recently developed and successfully utilized as a catalyst in FTS.[198] Regular γ- and nano-structured alumina as well as CNTs-covered regular γ- and nano-structured alumina supports were impregnated by cobalt nitrate solution to make new cobalt-based catalysts, which were also promoted by Ru. Catalyst evaluations indicated that nano-structured Al_2O_3 was superior to regular γ-Al_2O_3 and that CNT-covered alumina supports were favored over non-covered ones in terms of activity and heavy hydrocarbon selectivity. CNT-covered catalysts also showed higher wax selectivity and better resistance to deactivation. Furthermore, the cobalt aluminate phase, which is responsible for the permanent deactivation of alumina supported Co-based catalysts, did not form on alumina supported Co-based catalysts covered with CNTs due to weaker interactions between cobalt and alumina.

CNT-supported nano-size monometallic and noble metal (Pt and Ru) promoted cobalt catalysts were prepared by IWI using a solution of cobalt nitrate.[181] Promotion with platinum and ruthenium significantly decreased the cobalt species reduction temperature. The presence of noble metal promoters had no significant effect on the size of cobalt particles. Promotion with small amounts of Pt and Ru resulted in a slight increase in FT cobalt time yield. The Pt and Ru promoted cobalt catalysts exhibited CO conversions of 37.1 and 31.4%, respectively. C_{5+} hydrocarbon selectivity was attained at 80%. The Pt promoted Co/CNT yielded better catalytic stability than that of the monometallic cobalt catalyst.

Several nanostructured carbon materials (CNFs, CNTs and carbon micro-coils) were used as supports for Co.[196] It was found that the different carbon supports affected the catalyst reducibility (Table 7.1) and FTS performance. For cobalt supported on different carbon materials, the Co/CNT showed the highest FTS activity. This was ascribed to a higher reducibility and an optimal dispersion for this catalyst, resulting from the interaction between cobalt and surface groups of carbon.

Co/CNF are also active FT catalysts.[199] The influence of cobalt particle size in the range of 2.6–27 nm on the performance in FTS has been investigated using well-defined catalysts based on an inert CNF support material (Figure 7.6).[200] Cobalt was metallic, even for small particle sizes, after an

Table 7.1 Results obtained from H_2 chemisorption and O_2 titration on the cobalt catalysts.

Catalyst	H_2 chemisorption		O_2 titration	
	Amount of H_2 chemisorbed/ mmol g_{cat}^{-1}	Co dispersion (%)	O_2 consumption/ mmol g_{cat}^{-1}	Extent of reduction (%)
Co/CNF	0.103	8.1	0.8	63.1
Co/CNT	0.056	4.4	1.06	83.7
Co/CMC	0.037	2.94	0.87	68.1
Co-in-CNT	0.13	10.3	1.12	87.6
Co-out-CNT	0.07	5.5	1.07	84.5

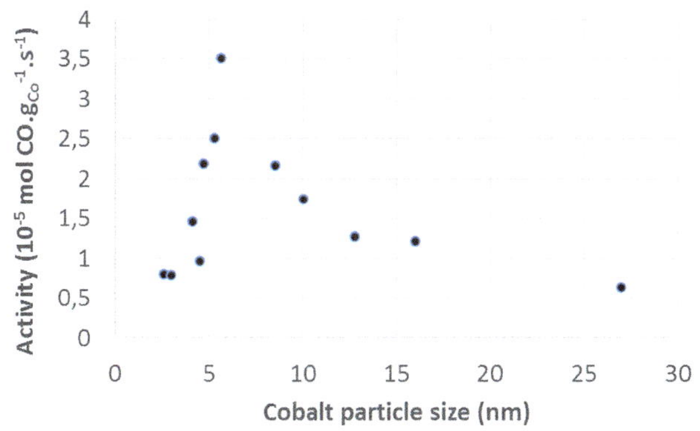

Figure 7.6 Dependence of activity on the mean size of Co particles for FTS on Co/CNF catalysts.

in situ reduction treatment, which is a prerequisite for catalytic operation and is difficult to achieve using traditional oxidic supports.

The turnover frequency for CO hydrogenation was independent of cobalt particle size for catalysts with sizes larger than 6 nm (1 bar) or 8 nm (35 bar), while both the selectivity and the activity changed for catalysts with smaller particles. At 35 bar, the TOF decreased from 23×10^{-3} to 1.4×10^{-3} s^{-1}, while the C_{5+} selectivity decreased from 85 to 51 wt% when the cobalt particle size was reduced from 16 to 2.6 nm. This demonstrates that the minimal required cobalt particle size for FT catalysis is larger (6–8 nm) than can be explained by classical structure sensitivity. Other explanations raised in the literature, such as formation of CoO or Co carbide species on small particles during catalytic testing, were not substantiated by experimental evidence from X-ray absorption spectroscopy. Interestingly, it was found with EXAFS a decrease of the cobalt coordination number under reaction conditions, which points to reconstruction of the cobalt particles. It was argued that the cobalt particle size effects can be attributed to non-classical structure sensitivity in combination with CO-induced surface reconstruction. MnO_x was reported to be a promoter element for cobalt-based FT. De Jong *et al.* have shown that the study of promoter effects can benefit a lot from the use of inert supports like CNF, as even after reduction the promoter is not found separately on the CNF surface.[201] Thus, the addition of very small amounts of Mn (0.15 wt%) brought about a 25% increase of C_{5+} selectivity.

In the FTS, the catalytic behavior of Co-loaded powdered oxidized diamond (O-Dia) catalysts, having a surface area of 24 m^2 g^{-1} was examined.[202] The catalytic activity of 9 wt% Co-loaded oxidized diamond catalyst exhibited a high CO conversion of 44.5% at 523 K with a total pressure of 10 bar containing 1:2 CO and H_2 mixed gas, and a space velocity of 4500 h^{-1}. This conversion is much higher than that of Co-loaded SiO_2 catalysts of two different surface areas, where CO conversions of 38.4 and 12.2% were obtained for SiO_2 having a surface area of 190 and 12 m^2 g^{-1}, respectively. The effect of the cobalt precursor salt was examined, and an increased CO conversion was obtained with the use of cobalt acetate as a precursor for Co. A much higher dispersion of Co with the use of cobalt acetate was observed by temperature-programmed reduction, CO adsorption, and TEM observations. A weaker interaction between the O-Dia surface and cobalt oxide seems to contribute to higher catalytic activity for the FT synthesis.

7.2.6 Water–Gas Shift Reaction

The water-gas shift reaction (WGSR) is an essential step in the purification process for on-board fuel reforming and hydrogen production. It is of particular importance for low-temperature polymer electrolyte membrane fuel cell applications, where CO levels must be decreased to <1 ppm at relatively low temperatures to prevent poisoning of the anode catalysts. The current industrial Cu–ZnO-based low-temperature WGSR catalysts are not well-suited for this process because of their pyrophoricity and susceptibility to deactivation under ambient conditions. Relatively few studies deal with the WGSR on carbon supported catalysts.

Nanocomposite catalysts containing CNF and Cu–Ce–Zr mixed metal oxide (MMO) have been prepared by homogeneous co-precipitation with urea.[203] The WGSR has been used as test reaction. The CNF-containing nanocomposite catalysts exhibit similar overall catalytic activity and stability as the corresponding CNF-free catalyst. 13 wt% of the MMO could be replaced by CNF without decreasing the overall activity and stability of the catalyst. The specific activity of the nanocomposites based on the total metal oxide content was similar or higher than the activity of the CNF-free material, depending on the CNF content. Tavasoli *et al.* have shown that using CNT as cobalt catalyst support instead of γ-alumina dramatically enhances the WGSR rates.[191] The reduction temperature of cobalt oxide species shifted to lower temperatures and the reducibility of the catalyst improved significantly. CNT provided good dispersion of metal clusters, and the average cobalt cluster size decreased. The direct promotional effect of sodium on the WGS activity of Pt/CNT was reported.[204] Whereas the Na-free Pt catalysts were shown to be completely inactive, the addition of sodium was found to improve the WGS activity to levels comparable to those obtained with highly active Pt catalysts on metal oxide supports. Atomically dispersed platinum species were stabilized by the addition of sodium. Oxidized platinum Pt–OH_x contributions are higher in the presence of sodium, providing evidence for a previously reported active-site structure of the form Pt–Na_x–O_y–$(OH)_z$. Pt remained oxidized, even when a H_2-rich gas mixture was used. A strong inhibitory effect of hydrogen was observed on the reaction kinetics, effectively raising the apparent activation energy from 70 ± 5 kJ mol^{-1} (in product-free gas) to 105 ± 7 kJ mol^{-1} (in full reformate gas). Increased hydrogen uptake was observed on these materials when both Pt and Na were present on the catalyst, suggesting that hydrogen desorption might limit the WGSR rate under such conditions. Removal of the oxygen groups and partial removal of sodium by annealing the Na-modified CNTs to 1073 K in inert atmosphere creates a surface on which Pt is initially very active, but of lower stability due to the absence of a sufficient number of surface sodium anchoring sites.[205]

Interestingly, the WGS process was separated into two half-cell electrochemical reactions (H^+ reduction and CO oxidation), catalyzed by enzymes attached to conducting FLG particles.[206] In this study the H^+ reduction reaction was catalyzed by a hydrogenase, Hyd-2, from *Escherichia coli*, and CO oxidation was catalyzed by a carbon monoxide dehydrogenase (CODH) from *Carboxydothermus hydrogenoformans*. This results in a highly efficient heterogeneous catalyst with a turnover frequency, at 303 K, of at least 2.5 s^{-1} per minimum functional unit (a CODH/Hyd-2 pair), which is comparable to conventional high-temperature catalysts.

7.2.7 Hydroformylation

Different types of CNFs were used as support for rhodium in the ethylene hydroformylation reaction at temperatures over the range 453–573 K.[207] The performance of these systems was compared with that of a catalyst where the same metal loading was dispersed on silica. It was found that in general, while the activity of all the catalyst systems was similar, the CNF supported rhodium catalysts exhibited a higher selectivity for the formation of propionaldehyde than the corresponding silica supported system. Furthermore, among the various Rh/CNF catalysts, the ribbon type nanofibers appeared to give the highest selectivity to the desired product. It was proposed that the morphological characteristics acquired by rhodium when dispersed on the CNF edges is a critical factor, rather than the size of the individual crystallites.

Rhodium catalysts supported on CNTs, SiO_2, carbon molecular sieves, AC, and a copolymer of styrene with divinylbenzene, were prepared, and their catalytic behaviors for propene hydroformylation were investigated and compared.[208] The results showed that, over the CNT-supported Rh-catalyst, C_3H_6 conversion and regioselectivity were pronouncedly improved: the average turnover frequency for the catalytic hydroformylation of propene was 0.079 s^{-1} at 393 K, which was 2.1 times faster than that over Rh/SiO_2, and the *n*/*i* ratio of the aldehyde products reached 11.6, which was 1.9 times higher than that over Rh/SiO_2. Propene hydroformylation catalyzed by a Rh-phosphine complex catalyst supported on CNTs and CNFs was investigated,[209] and compared to that catalyzed by a Rh-phosphine complex catalyst supported on SiO_2, a carbon molecular sieve, AC, and a polymer carrier.[210] Activity assays of the catalysts showed that the CNT-supported Rh-phosphine complex catalysts displayed not only high activity of propene conversion but also excellent regioselectivity to butyl-aldehyde. The results obtained, together with the results of comparative studies of the Rh-phosphine complex catalysts supported by several other supports, strongly implied that the tubular channels with an inner diameter of *ca.* 3 nm in the carbon nanostructures and its hydrophobic surface consisting of six-membered C-rings played an important role in enhancing the activity of propene hydroformylation, especially the regioselectivity of butyl-aldehyde.

The excellent catalytic performances of Rh/CNT catalysts for 1-hexene hydroformylation,[88] and that of Co/CNF[211] and Co/CNT[212] for the 1-octene hydroformylation were also reported. The $[RhH(CO)(C_{60})(PPh_3)_2]$ complex, has been synthesized by reacting the hydroformylation catalyst, $[RhH(CO)(PPh_3)_3]$ with an equimolar amount of C_{60}.[213] This compound is an efficient catalyst for the hydroformylation of alkenes to aldehydes at 363 K.

7.3 Ammonia Synthesis and Decomposition

7.3.1 Ammonia Synthesis

The use of ruthenium/carbon catalysts for ammonia synthesis could constitute an alternative method to replace conventional iron-based systems that operate under high pressure and at high temperature. However, extensive studies have shown that Ru/AC deactivation occurs upon prolonged reaction owing to metal sintering, metal leaching and methanation of the support. Hence, the use of nanostructured carbon materials as CNFs or CNTs, intrinsically more stable than AC has been investigated.[214]

A novel ammonia synthesis catalyst, alkali-promoted ruthenium supported on CNTs has been developed.[215] Various alkali-promoters and carbon-based supports were compared. The effects of the contents of Ru and K, the treatment of CNT, and the reaction temperature on ammonia synthesis activity were investigated. It was found that, as a support, the CNT is much better than other carbon-based supports. The yield of NH_3 was 47.423 mL NH_3 h^{-1} g_{cat}^{-1} at 693 K, at atmospheric pressure and $N_2/3H_2$ flowrate of 1800 mL h^{-1} for the catalyst with K/Ru/CNT = 4/4/100 (wt%). Small and uniform Ru nanoparticles (1.3–2.0 nm) supported on the mixture of MgO and CNTs, Ru/MgO-CNT and K–Ru/MgO-CNT exhibited high activity for ammonia synthesis.[216] The highest activity was obtained for the K–Ru/MgO-CNT catalyst under 2 bar at 683 K.

CNTs have been shown to modify some properties of nanomaterials and to modify chemical reactions confined inside their channels. Bao *et al.* have studied ammonia synthesis over Ru as a probe reaction to understand the effect of the electron structure of CNTs on the confined metal particles and their catalytic activity.[217] The catalyst with Ru nanoparticles dispersed almost exclusively on the exterior nanotube surface exhibits a higher activity than the CNT-confined Ru, although both have a similar metal particle size. Several analyses suggest that the outside Ru exhibits a higher electron density than the inside Ru. As a result, the dissociative nitrogen adsorption, which is an electrophilic process and the rate-determining step of ammonia synthesis, is more facile over the outside Ru than that over the inside one.

Nitric acid treatment of the CNTs-cordierite monolith changes the amount of Mg, Si, Al and oxygen-containing functional groups, thereby influencing the surface area and pore size distribution of composite materials.[218] Appropriate treatment of CNT-cordierite with HNO_3 increases the surface area and the amount of micropores slightly, but significantly improves the activity for ammonia synthesis, which might be a consequence of the variation of the amount of Mg, Si, Al and oxygen-containing functional groups. The ammonia synthesis activity of Ba–Ru/CNT-cordierite increases by more than 30% if the support material is treated at 303 K for 4 h with nitric acid.

CNF-supported ruthenium catalysts were used for ammonia synthesis in a fixed bed microreactor.[219] The TEM micrographs of the Ru/CNF and Ru–Ba/CNF catalysts indicated that the Ru particles are in the range of 2–4 nm. The activity of Ru–Ba/CNF catalysts was higher than that of Ru–Ba/AC by about 25%. The methanation reaction on the Ru/CNF catalyst was remarkably inhibited compared with a Ru/AC catalyst. High graphitization of CNF is likely to be the reason for the high resistance to the methanation reaction. Catalysts with the promoters Ba, K and Cs showed large differences in activity for ammonia synthesis. The catalyst promoted with Ba (Ba/Ru = 0.2 molar ratio) was found to be the most active, whereas that with a K promoter was the least active.

7.3.2 Hydrogen Production *via* Ammonia or Hydrazine Decomposition and Ammonia-Borane Hydrolysis

The catalytic decomposition of ammonia or hydrazine to generate CO-free hydrogen for fuel cells is receiving an increasing attention since this process is more economical than using methanol as a H_2 source. Also in this case, ruthenium is the metal of choice, and the possibility to use nanostructured carbon materials to support the metal has been investigated.

7.3.2.1 Ammonia Decomposition

The effects of the active component (Ru, Rh, Pt, Pd, Ni, Fe) and support (CNTs, AC, Al_2O_3, MgO, ZrO_2, TiO_2) on the catalysis of ammonia decomposition were studied for the generation of CO_x-free hydrogen.[220] It was shown that the Ru catalyst using CNTs as support exhibits the highest conversion of NH_3. The performance can be further improved by modifying CNTs with KOH. Ru dispersion was the highest on CNTs. Being in the range of 2–5 nm, the particle size of Ru on CNTs was the smallest among the supported Ru catalysts; the Ru particles on the metal oxides were in the 3–16 nm range. It seems that larger Ru particles are more active for NH_3 decomposition in terms of TOF. Further investigation on the relationship between support basicity and catalytic activity disclosed that a support material of strong basicity is essential for high catalytic performance. In the N_2-TPD studies of supported Ru catalysts, desorption was promoted over catalysts of strong basicity, suggesting that N_2 desorption is the rate-determining step in ammonia decomposition. Very small Ru nanoparticles on CNTs were prepared by means of impregnation and H_2-reduction.[221] The CNT-supported Ru nanoparticles are highly active and stable in catalysing the decomposition of ammonia for the generation of CO_x-free hydrogen. The modification of Ru/CNTs with potassium ions leads to a remarkable improvement in the activity. The excellent catalytic performances of Ru/CNTs and K–Ru/CNTs are related to the high dispersion of Ru, and to the high graphitization and purity of CNTs. The use of MgO-CNT nanocomposites as support yielded more efficient Ru catalysts for the generation of CO_x-free hydrogen from NH_3 decomposition.[222] Indeed, magnesia–carbon nanotubes nanocomposites were thermally more stable than CNTs in a H_2 flow.

A fundamental study on the structural effects of CNTs, when used as supports for Ru nanoparticles, and on the localization of Ru nanoparticles, on ammonia decomposition was reported.[223] The observed trend is counter-intuitive to the general notion about a beneficial effect of reduced particle size for better activity in a nonselective reaction. The fact that the support offers different sites for Ru-*in* and Ru-*out* masks the possible beneficial effect of the enhanced electron density of thermally annealed CNTs. The size of Ru particles can be controlled effectively by a combination of modifying the site density of defects on the support and localization of the active metal inside or outside the cavity of CNTs. At the inside of the CNTs it was not possible to sufficiently enhance the electron density of the support to strongly activate the rough and small metal particles. Figure 7.7 summarizes the catalytic performances correlated with structural descriptors. The catalytic activity of Ru nanoparticles scales with their size; larger particles exhibit more active sites per particle than small particles, suggesting that no relationship between exposed surface and active site density exists for this reaction. Extrapolating from ammonia synthesis, being the reverse reaction, this finding is in line with the known fact that the reaction is totally dominated by surface steps.

Ru catalysts supported on different carbons were used for catalytic ammonia decomposition.[224] The influence of the porous and graphitic structures of carbon supports on the activities of the catalysts were examined. The catalytic activity over supported Ru catalysts is ranked as Ru/graphitic carbon > Ru/CNT > Ru/CB > Ru/CMK ≈ Ru/AC. The optimum range of Ru particle sizes is around 3–4 nm. On the support side, the graphitic structure of the carbons is critical to the activity of the supported Ru catalyst, while the surface area of carbons is less important. In comparison to Ru catalysts supported on AC, MgO, Al_2O_3, and TiO_2 under similar reaction conditions, Ru/CNT was found to show high NH_3 conversion for the generation of hydrogen.[225] The modification of the catalyst with KOH leads to a significant improvement in activity, either in term of NH_3 conversion or TOF. The excellent catalytic activity of Ru/CNTs could be ascribed to the high dispersion of Ru particles, and to the high graphitization and high purity of the CNT material. The residual Cl originated from the $RuCl_3$ precursor was found to be a strong inhibitor for the decomposition reaction.

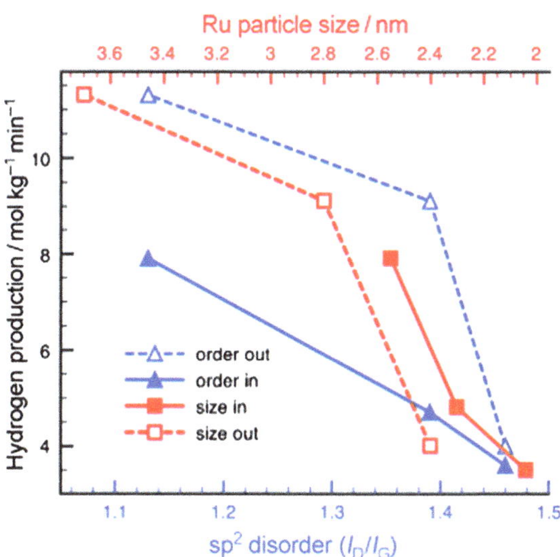

Figure 7.7 Correlations of steady catalytic performance (Ru/CNT catalyst) with structural descriptors for the support and for the active mass.[223]

Ru catalysts supported on two different carbon materials, MWCNTs and bamboo-like CNTs doped with nitrogen were investigated.[226] The Ru catalysts were tested in the catalytic ammonia decomposition reaction. High yields towards hydrogen production were achieved. Carbon nanotubes were heated in an inert atmosphere at temperatures up to 1773 K in order to study the effects of such support treatments on the ammonia decomposition reaction. The elimination of acidic groups from the surfaces, prior to catalyst preparation, and/or the surface graphitization of the materials produced a higher catalytic activity during the reaction. The catalytic activity of Ru particles was significantly improved when supported on carbon nanotubes doped with nitrogen. N-CNTs were also employed to investigate the effects of nitrogen doping on the structure of CNTs and the state of Ru particles supported on CNTs.[227] Two types of nitrogen species, pyridinic and quaternary nitrogen, were found on the surface of N-CNTs. Pyridinic nitrogen atoms may have a strong interaction with Ru. The average Ru particle size decreases with the increase of pyridinic nitrogen content. The activity of Ru/CNTs catalysts in ammonia decomposition is dependent on the dispersion of Ru particles and remaining nitrogen on the surface of CNTs.

Fe–Co alloy NPs inside the tubular channel of CNTs were used to demonstrate an unusual synergism in alloy catalysis.[228] The alloy nanoparticles with widely varying Co/Fe ratio are kept as active as Co catalysts for the H_2 production from NH_3 decomposition. The stability of Co was significantly improved by alloying with Fe. Confinement inside CNTs was found to improve the thermal stability rather than the turnover rate of the nanoparticles.

Interestingly, fresh commercial CNTs containing residual Co or Fe nanoparticle catalysts were found to be highly active for NH_3 decomposition while the microstructure of CNTs remains unchanged.[229] Fe NPs on the tip of CNFs were used as catalysts for ammonia decomposition.[230] The size and shape of Fe particles on the tip of CNFs depended on the Fe particle reconstruction and CNF morphology. The catalyst is highly active and stable because the Fe particles are highly dispersed and physically isolated by CNFs, and the surfaces are largely exposed to the reactants. Similarly, Ni catalysts at the tips of CNFs showed high activity for ammonia decomposition.[231]

Ru nanoparticles were supported on monoliths that were coated with various functionalized CNFs, that is, undoped CNFs, CNFs that had been post-treated with H_2O_2, and CNFs that had been doped with nitrogen during their growth.[232] The Ru uptake onto N-doped CNFs was larger compared to that on their undoped and O-doped counterparts. The functionalization of the CNF support did not play a significant role in determining the size of the deposited Ru nanoparticles, but it substantially impacted on the sintering under the reaction conditions and on the electron density of the reduced metal. Among the catalysts on the different CNF supports, Ru on N–CNF exhibited the highest H_2 productivity from ammonia decomposition, which pointed to electronic effects that were induced by functionalization of the support. CNFs with fishbone graphene alignment and CNTs were used to support ruthenium for ammonia decomposition.[233] The Ru nanoparticles on the CNF supports were more active than those on CNT supports. The Ru particle size was adjusted by changing the Ru loading or by introducing oxygen containing groups onto the CNF surface. The site activity increases when the Ru crystal size also increases. The oxygen groups on the CNFs have a remarkable effect on ammonia decomposition over the Ru NPs. On identically sized Ru crystals, oxygen on the CNFs clearly enhances ammonia decomposition over the Ru/CNFs.

7.3.2.2 Hydrazine Decomposition

Hydrazine has a wide application as a fuel in thrusters for space missions. The hydrazine decomposition reaction produces ammonia, nitrogen and hydrogen. The standard commercial catalyst used for thrusters and gas generators is iridium on Al_2O_3, chosen because it combines reasonable activity and high thermal stability. There are several problems with deactivation of such catalysts in long exposures to hydrazine. High-temperature reaction conditions may induce sintering of the small metal particles, and strongly bound nitrogen species left on the catalyst surface may poison the catalyst.

Graphite felt supporting 40 nm diameter CNFs was synthesized and successfully used as a support for a high loaded iridium catalyst (30 wt%) in the decomposition of hydrazine.[234–236] A strong mechanical resistance and a high thermal conductivity led to a very efficient and stable catalyst as compared to that used industrially: iridium supported on a high surface area alumina.

Nearly mono-dispersed (2–5 nm) nanostructured tungsten carbide particles on CNTs have successfully been synthesized.[237] The resulting materials are active catalysts for hydrazine decomposition and exhibit high selectivity to hydrogen.

Well-dispersed RhNi nanoparticles grown on graphene exert 100% selectively and exceedingly high activity to complete the decomposition reaction of hydrous hydrazine at room temperature.[238] This excellent catalytic performance might be due to the synergistic effect of the graphene support and the RhNi nanoparticles and a promotion effect of NaOH.

Fe–B/MWCNT were developed for hydrolysis of hydrous hydrazine.[239] A high H_2 generation rate of 34.2 L h^{-1} g_{cat}^{-1} for a 9.86 wt% Fe–B/MWCNTs was achieved with 97% H_2 selectivity at room temperature. The Fe–B/MWCNTs are stable catalysts for N_2H_4 hydrolytic dehydrogenation as they provide 114 480 total turnovers in 30 h.

7.3.2.3 Ammonia-Borane Hydrolysis

Ammonia-borane is one of the most promising hydrogen storage materials due to its high hydrogen content and outstanding environmental stability and non-toxicity, which has stimulated substantial effort devoted to developing highly efficient dehydrogenation catalysts.

Ruthenium(0) NPs supported on MWCNT were formed *in situ* during the hydrolysis of ammonia-borane.[240] The results reveal that ruthenium(0) nanoparticles in the range 1.4–3.0 nm are well-dispersed on MWCNTs. They were found to be highly active catalysts in hydrogen generation from the hydrolysis of ammonia-borane with a turnover frequency value of 329 min^{-1}. Chemically derived graphene was used as a support for palladium NPs.[241] The resulting materials are highly active and stable catalysts for the dehydrogenation and hydrolysis of ammonia-borane. In addition to their high activity and stability, these Pd NPs are also a reusable catalyst in both dehydrogenation and hydrolysis of ammonia-borane, preserving 85% and 95% of initial activity after the 5th and 10th runs, respectively. Ru/FLG exhibit superior catalytic activity towards the hydrolytic dehydrogenation of ammonia-borane.[242] Ru/graphene NPs exhibit higher catalytic activity than its graphene free counterparts, and retain 72% of their initial catalytic activity after 4 reaction cycles. The activation energy for the hydrolysis of ammonia-borane in the presence of Ru/graphene NPs has been measured to be 11.7 kJ mol^{-1}, which is the lowest value ever reported for the catalytic hydrolytic dehydrogenation of ammonia-borane. Graphene supported Pd@Co core-shell nanocatalysts with magnetically recyclability showed satisfactory catalytic activity (916 L mol^{-1} min^{-1}) and recycling stability for hydrolytic dehydrogenation of ammonia-borane.[243] A weak, branched polyelectrolyte, polyethyleneimine (PEI), was utilized to assist the deposition of Fe–Ni NPs on graphene oxide.[244] Compared to the Fe–Ni NPs directly deposited on GO, these NPs on PEI-decorated GO reveal a dehydrogenation rate of 0.9 L mL^{-1} min^{-1} at 293 K for the hydrolysis of ammonia-borane, which is 18 times faster than that of the former and nearly comparable to that of the platinum catalyst deposited on carbon under the same conditions. Well-dispersed magnetically recyclable core-shell Ag@M (M = Co, Ni, Fe) nanoparticles supported on graphene have been synthesized and their catalytic activity toward hydrolysis of ammonia-borane was studied.[245] Although the Ag@Fe/graphene NPs are almost inactive, the as-prepared Ag@Co/graphene NPs are the most reactive catalysts, followed by Ag@Ni/graphene NPs. Additionally, the Ag@Co NPs supported on graphene exhibit higher catalytic activity than the catalysts with other conventional supports, such as SiO_2, CB, and γ-Al_2O_3. The as-synthesized Ag@Co/graphene NPs demonstrate satisfactory catalytic activity, with a TOF value of 102.4 mol H_2 min^{-1} ${mol_{Ag}}^{-1}$, and an activation energy value of 20.03 kJ mol^{-1}. Furthermore, the as-synthesized Ag@Co/graphene NPs show good recyclability and magnetic reusability. Ag@CoNi nanoparticles supported on graphene were also employed.[246] The as-synthesized NPs exhibit much higher catalytic activities for hydrolytic dehydrogenation of ammonia-borane than the monometallic, bimetallic, trimetallic alloy (AgCoNi/graphene), and graphene free (Ag@CoNi) counterparts. Kinetic studies indicate that the catalytic hydrolysis by the Ag@CoNi/graphene catalyst is first order, with the activation energy measured to be 36.15 kJ mol^{-1}.

7.4 Hydrotreatment

Sulfide transition metal catalysts have been widely used for petroleum refining and hydroprocessing, such as hydrodesulfurization (HDS) and hydrodenitrogenation (HDN). The traditional commercial hydrotreating catalysts usually contain cobalt or nickel and molybdenum or tungsten commonly deposited on γ-alumina supports. Alumina's desirable properties include its strong mechanical and textural properties as well as the great dispersion it provides for active transition metals. That being said, a negative aspect of γ-Al_2O_3 is the strong chemical interaction that can exist between the support and the transition metal catalysts in their oxide phase. This can make the sulfidation of the metal oxides a difficult procedure and can prevent the catalyst from reaching its maximum potential activity. The opportunity presents itself to apply a catalyst support that can offer great dispersion of supported metals while providing limited interaction with the metals while in the intermediate oxide phase. At the beginning of any hydrotreating process, the loaded catalyst undergoes an initial period of catalyst deactivation. The main cause of this catalyst deactivation is the deposition of carbonaceous species (*i.e.* coke) on the catalyst surface. After initially being deposited very rapidly, the coke eventually reaches a steady-state level within the reactor. A large loss in accessible surface area is also observed along with the rapid decline in catalyst activity. This can be attributed to the pores of the catalyst support being plugged by coke deposits, preventing reactants from accessing active sites within the pore volume. Based on this theory, it can be concluded that applying a catalyst support with a large average pore diameter and a maintained high surface area could help minimize the detrimental effects of the precoking phase. Defunctionalization reactions such as dehydration, decarboxylation/decarbonylation, and hydrodeoxygenation (HDO) play a central role in the conversion of over-functionalized carbohydrates to chemicals. Ideally, the HDO catalyst should favor C–O over C–C bond cleavage and preferentially produce unsaturated compounds to minimize H_2 consumption. Noble metals (Pt) and transition metal carbides (Mo_2C, WC, W_2C) catalyze HDO reactions.

7.4.1 Hydrodesulfurization

The results of several investigations indicate that carbon-supported catalysts have a potential advantage over HDS catalysts in current use, both with respect to high desulfurization activity and low coking propensity.[247]

Co–Mo/CNT catalysts for selective HDS of fluid catalytic cracking (FCC) gasoline were studied, using di-isobutylene, cyclohexene, 1-octene and thiophene as model compounds to simulate FCC gasoline.[248] The results show that the Co–Mo/CNT has very high HDS activity and HDS/hydrogenation selectivity comparing with the Co–Mo/Al_2O_3 and Co–Mo/AC catalysts. The Co/Mo atomic ratio was found to be one of the most important key factors in influencing the hydrogenation selectivity and HDS activity, and the most suitable Co/Mo atomic ratio was 0.4. Co/CNT and Mo/CNT monometallic catalysts showed lower HDS activity and selectivity than the Co–Mo/CNT bimetallic ones. Co–Mo/CNT catalysts were prepared by an IWI technique and their catalytic performances for HDS of thiophene were studied and compared with the reference systems of Co–Mo/Al_2O_3 and Co–Mo/AC.[249] Under the reaction conditions of $(C_4H_4S)/(H_2) = 2.3/97.7$, GHSV = 2200 mL g^{-1} h^{-1}, $P = 1$ bar and $T = 623$ K, the specific reaction rate of thiophene HDS over 7.2% Co–Mo/CNT catalyst reached 0.64×10^{-3} s^{-1}, which was 1.68 and 2.28 times higher than that over 9.7% Co–Mo/Al_2O_3 and 16.9% Co–Mo/AC, respectively. Using the CNT in place of γ-Al_2O_3 or AC as the catalyst support caused little change in the apparent activation energy for thiophene HDS reaction, but led to a significant increase in concentration of active Mo species (Mo^{4+}) on the surface of the functioning catalyst. In addition, the catalyst Co–Mo/CNT could reversibly adsorb a greater amount of hydrogen under atmospheric pressure at temperatures from room temperature to 673 K. This unique feature should help to generate microenvironments with higher stationary-state concentration of active hydrogen-adspecies on the surface of the functioning catalyst. Both factors were favorable to increase the rate of the thiophene HDS reaction. The effect of the catalytic support and sulfiding method on the chemical state of supported Co–Mo catalysts was studied by XPS.[250] After sulfidation, the majority of molybdenum in CNT-supported CoMo catalyst is transferred to species with a formal chemical state Mo^{4+} in the MoS_2 phase, and the rest to Mo^{5+}, which consists of Mo coordinated both to O and S, such as $MoO_2S_2^{2-}$ and $MoO_3S_2^{-}$. In the case of a sulfided CoMo/γ-Al_2O_3 catalyst, a fraction of molybdenum is transferred to MoS_2, but there is still an amount of unreduced Mo^{6+} phase, which is resistant to sulfidation. XPS analyses results suggest that the CNT support facilitates the reduction and sulfidation of active species to a large extent, and that the alumina support strongly interacts with active species, producing a fraction of the phase which resists complete sulfiding. Catalytic measurements of catalysts in the HDS of dibenzothiophene show that CoMo/CNT catalysts have higher HDS activity and selectivity than CoMo/γ-A_2O_3 catalyst, which is in good agreement with the sulfiding behavior of the corresponding catalyst. TPR studies also revealed that active species in oxide state CoMo/CNT catalysts were more easily reduced at relatively lower temperatures in comparison to those in CoMo/γ-Al_2O_3, indicating that the CNT support promoted the reduction of the active species.[251] This shows that the Co/Mo atomic ratio has a great effect on the reducibility of active species on CNTs and their HDS activities and that the incorporation of cobalt improved the dispersion of molybdenum species on CNTs. It was also found that re-dispersion could occur during the sulfiding process. The HDS of dibenzothiophene showed that the CoMo/CNT catalyst was more active than CoMo/γ-Al_2O_3 and the hydrogenolysis/hydrogenation selectivity of the CoMo/CNT catalyst was also much higher than CoMo/γ-Al_2O. To better understand the nature of CoMo/CNTs for selective HDS of FCC gasoline, *in situ* Fourier transform infrared spectroscopy was carried out.[252] The HDS experimental results suggested that the HDS activity and selectivity of

CoMo/CNTs catalysts were affected by the Co/Mo ratio; the optimal Co/Mo atomic ratio was about 0.4, and the optimum reaction temperature was 533 K. On the basis of FT-IR results, it was deduced that the thiophene HDS reaction occurred mainly through a direct hydrogenolysis route, whereas thiophene HDS and diisobutylene hydrogenation reaction over CoMo/CNT catalysts might occur on two different kinds of active sites. A CNT coated cordierite monolith was utilized as the support for the CoMo catalyst.[253] The CNTs were distributed uniformly on the surface of the monolith leading to a high BET surface area and relatively well-adhered mesoporous layer of CNTs. The catalytic activity of the resulting catalyst was determined in an HDS reaction of naphtha. It was concluded that the activity of the CoMo catalyst over the CNT coated monolithic support was higher than that of the CoMo deposited onto the acid-treated monolith and the CoMo/γ-Al_2O_3 conventional catalyst. Ultimately, the HDS reaction over the CoMo/CNT material under optimized operating conditions reduced the sulfur content of naphtha from 2670 to 13 ppm. Ni_2P nanoparticles supported by functionalized CNTs were prepared, and the catalytic activity of the prepared samples was studied in HDS of naphtha.[254] The catalytic study of the prepared samples confirmed that these compounds were stable and active in the process. Removal of sulfur from the desired naphtha in presence of the samples depended on the nickel loadings and increased with temperature in all cases.

Co, Mo, NiMo and CoMo catalysts supported on alumina, and fishbone and platelet CNFs have been prepared.[255] HDS of thiophene was used as a model reaction to compare the activity of different catalysts. The activity tests showed that the alumina supported catalysts exhibited higher activity compared to the corresponding CNF supported catalysts, and the NiMo catalysts were more active than the corresponding CoMo catalysts. The thiophene HDS activity was correlated with the dispersion of the molybdenum species and the reducibility of different catalysts. Interestingly, the CNF supported Co catalysts have higher thiophene HDS activity than the CNF supported Co(Ni)Mo catalysts. The catalytic activity of the isolated Ni–Mo and Co–Mo NPs located on the tips of separate CNFs was tested in thiophene hydrodesulfurization (HDS) at 573 K.[256] The composition of these particles and the state of Ni, Co and Mo were examined by EXAFS spectroscopy. The HDS activities of the CNF catalysts were found to be comparable with those of highly dispersed bimetallic sulfide catalysts supported on Sibunit carbon.

The HDS activity of sulfided Mo (W) catalysts supported on CB composites (CBC) is affected by the kind of functional groups present on the CBC surface.[257] Oxidation of CBC with $(NH_4)_2S_2O_8$ produces functionalities with the highest acid strength and the corresponding catalyst exhibits the highest HDS activity. Sulfided W/CBC is less active in the thiophene HDS than the corresponding Mo counterpart. The rate of thiophene HDS over Mo/"basic" CBC does not depend on the method of Mo deposition. Different CBC supported molybdenum carbides were synthesized.[258] The support was functionalized with nitric acid at different pH of the impregnation solution (pH 5.2 and pH 0), in order to improve the active phase dispersion. A kinetic study of the HDS of dibenzothiophene was performed. It was found that the HDS of dibenzothiophene proceeds *via* the two classical parallel routes: the hydrogenation route leading to cyclohexylbenzene and bicyclohexylbenzene and the direct desulfurization route leading to biphenyl. Furthermore, when the CBC support was pre-oxidized with HNO_3, the dispersion of the molybdenum carbide active phase was improved and a higher HDS activity was observed. CBCs have been prepared by pyrolyzing a mixture of a CB with polyfurfuryl alcohol and then pretreated by oxidation with HNO_3, gasification with water steam or ammoxidation.[259] The effects of the chemical character of the support surface, nature of the active metal phase and pH value of the impregnation solution on the catalytic activity towards the HDS of thiophene of the CBC supported Mo (Co) catalysts were determined. It was stated that the catalytic properties of the CBC supported sulfides of Mo or Co and of Mo carbides are affected by the chemical character of the support surface. Generally, catalysts supported over basic surface CBC exhibit higher activity than those supported over CBC possessing acidic surface character. Co catalysts supported on acidic surface show lower activity (per mol of active metal) than Mo-based ones supported on the same support. In the case of catalysts supported on basic CBC, Co exhibits distinctly higher activity than Mo. A commercial CB was subjected to oxidative and/or thermal treatment to modify its surface and structural properties, and used as a support for Mo.[260] The catalytic activity in thiophene HDS was determined. The knowledge of carbon surface chemistry was shown to provide the necessary framework for the understanding of the variations in catalytic activity. It was concluded that two conflicting requirements complicate the preparation of highly active (*i.e.*, highly dispersed) molybdenum species on carbon surfaces. On one hand, the introduction of oxygen functional groups provides anchoring sites for catalyst precursor adsorption and thus, the potential for its high initial dispersion. On the other hand, this also renders the support surface negatively charged over a wide range of pH conditions. At very low pH conditions, below the isoelectric point of the support, when the attractive forces prevail between the Mo anions and the positively charged carbon surface, Mo polymerization is thought to contribute to catalyst agglomeration. Final catalyst dispersion (*i.e.*, catalytic activity) is also influenced by the thermal stability of the oxygen functional groups on the carbon surface. No significant correlation between structural parameters of the support and catalytic activity was found.

Graphene-supported palladium catalysts were used for HDS of carbonyl sulfide (COS) in coal gas.[261] The Pd/FLG catalyst shows higher catalytic efficiency in the COS hydrogenation compared to a traditional Pd/C catalyst. The improved performance of Pd/FLG for COS conversion was attributed to the small size and uniform dispersion of Pd nanoparticles on graphene sheets. A graphene sheet decorated by nanometer scaled MoS_2 with effective HDS activity for carbonyl sulfide conversion at low temperature (<573 K) was also reported.[262] Compared to MoS_2/AC, the MoS_2/FLG catalyst showed excellent performance for the HDS of carbonyl sulfide.

MoS_2-graphene monolithic catalysts were also developed.[263] The catalytic performance of the hybrid monoliths was investigated by evaluating the activity for the HDS of carbonyl sulfide. It was demonstrated that MoS_2-graphene monolithic catalysts showed an excellent activity for HDS of carbonyl sulfide compared with the traditional $MoS_2/\gamma\text{-}Al_2O_3$ catalyst.

7.4.2 Other Hydroprocesses: Hydrodenitrogenation, Hydrodeoxygenation and Hydrodemetallisation

The HDN and HDS activities of sulfide NiMo/CNT using bitumen-derived light gas oil were carried out at different temperatures under industrial conditions.[264] The HDN and HDS activities of the catalysts increased with increasing Ni content up to 3 wt% and Mo content up to 12 wt%. Based on weight of the catalyst, the HDN and HDS activities of NiMo/CNT are significantly higher than those over conventional Al_2O_3-based catalysts. The introduction of 2.5 wt% P to the CNT-based catalyst was found to show a decrease in hydrotreating activity. MWCNTs were applied as supports for NiMo hydroprocessing (HDN and HDS) catalysts.[265] Power law models were best fit with reaction orders of 2.6 and 1.2, and activation energies of 161 and 82.3 kJ mol^{-1}, for the HDS and HDN reactions, respectively. Generalized Langmuir–Hinshelwood models were found to have reaction orders of 3.0 and 1.5, and activation energies of 155 and 42.3 kJ mol^{-1}, for the HDS and HDN reactions, respectively. CNT-supported CoMo catalysts were used for pyrrole HDN reactions, and compared with the reference system supported on AC.[266] Over the CoMo/CNT catalyst, the specific HDN activity was higher than that of the AC-based counterpart. It was experimentally found that using CNTs instead of AC as support of the catalyst caused little change in the apparent activation energy for the pyrrole HDN reaction. Mo_2C/CNT with different loadings were reported as efficient catalysts for one-step hydrodeoxygenation and isomerization of vegetable oils.[267]

A CBC support was synthesized and activated with HNO_3.[268] These samples were impregnated by ammonium heptamolybdate and carburized to prepare the molybdenum carbide active phase. The HNO_3 treatment of the carbon increased the dispersion of the active phase. The kinetic study of the HDN of indole demonstrated that the catalyst activity was dependent on the active phase dispersion, and the global kinetic order was found to vary between zero and one, depending on the dispersion of the active phase on carbon support. Similar results were obtained when the support was activated with H_2O_2.[269] CB- and alumina-supported Fe, Mo, Fe–Mo, and Co–Mo sulfide catalysts were compared for their ability to catalyze hydrotreating reactions such as dibenzothiophene HDS, quinoline HDN, dibenzofuran HDO, and hydrogenation (HYD) of butenes, naphthalene, biphenyl, and coal extract.[270] Carbon-supported sulfide catalysts are much more active than those supported on alumina for HDS, HDO, HDN, and HYD model reactions when a relatively low hydrogen pressure (below 50 bar of H_2) is employed. However, a further H_2 pressure increase was most effective for the alumina catalysts. The activity differences between carbon- and alumina-supported catalysts were ascribed to differences in active phase-support interaction. Generally, iron sulfide was found to be less active than molybdenum sulfide, and iron sulfide also appeared to be less effective than cobalt sulfide in promoting molybdenum sulfide. However, the Fe–Mo sulfide showed the higher selectivity for hydrogenation relative to HDS. Carbon supported iron sulfide (low-cost catalyst) had considerable activity for coal extract hydrogenation. CB- and CNT-supported Pt catalysts are highly active and reusable for the aqueous-phase HDO of phenols as lignin models without adding any acids.[271] It was suggested that Pt/CB facilitates the hydrogenation of phenols and the hydrogenolysis of the resulting cyclohexanols.

Catalysts based on functionalized CNFs coated with Ni-decorated MoS_2 nanosheets were tested in the hydroprocessing of a vacuum residue and the results were compared against a benchmark alumina-supported NiMo catalyst.[272] Higher asphaltene conversions were obtained for the CNF-supported catalysts, which out-performed the Al_2O_3-supported benchmark catalyst. However, the catalytic performance in hydrodesulfurisation and hydrodemetallisation of the CNF-based catalysts was slightly lower than that of the benchmark catalyst. CNF-supported tungsten-based materials were used as catalyst for the deoxygenation of biomass-derived glyceride feeds.[273] Supported tungsten oxides were shown to be selective toward decarboxylation/decarbonylation products, whereas supported tungsten carbide was selective towards hydrodeoxygenation products. In both cases high yields of unsaturated hydrocarbons could be obtained from a glyceride feed with saturated hydrocarbon chains, thus allowing an upgrade of the feed to higher-value products, even in the presence of H_2. HDO studies over CNF-supported W_2C and Mo_2C catalysts were performed on guaiacol, a prototypical substrate to evaluate the potential of a catalyst for valorization of depolymerized lignin streams.[274] Combined selectivity of up to 87 and 69% to phenol and methylated phenolics were obtained at 648 K for W_2C/CNF and Mo_2C/CNF at > 99% conversion, respectively. The molybdenum carbide-based catalyst showed a higher activity than W_2C/CNF and yielded more completely deoxygenated aromatic products, such as benzene and toluene. Catalyst recycling experiments performed with and without regeneration of the carbide phase, showed that the Mo_2C/CNF catalyst was stable during reusability experiments. Mo_2C/CNF with different loadings were prepared and used as catalysts for HDO of vegetable oils.[275] The optimal reaction conditions with model compounds on Mo_2C/CNF had a conversion of 98.03% and yield of 95.26%. The Mo_2C nanoparticles on the outside of the CNFs showed high catalytic activity compared to ones on the inside of the CNFs. The Mo_2C/CNF catalyst was recycled 5 times without any apparent loss of catalytic activity. Catalytic performances of Mo_2C/CNF, Mo_2C/AC and Mo_2C/CNT were examined using methyl palmitate and maize oil. Compared with Mo_2C/AC and Mo_2C/CNT catalysts, no branched hydrocarbons were detected in the liquid product on Mo_2C/CNF. Ru catalysts with a wide range of dispersion on carbon (including CNFs), silica, alumina, and titania supports were synthesized, characterized and evaluated for HDO activity using phenol as a model compound.[276] High dispersion of ruthenium on the supports converts more phenol to products. The majority of catalysts predominantly catalyze the hydrogenation route typical of noble metal catalysts. There are no apparent support, pore size or morphology effects for the hydrogenation pathway. A highly dispersed Ru/TiO_2 catalyst shows unusually high selectivity toward direct deoxygenation and outstanding activity. Low surface-area Inconel monoliths were coated with *in situ*-grown CNFs, which were subsequently impregnated with Pt, Sn, and bimetallic Pt–Sn.[277] These monoliths were tested for the HDO of guaiacol and anisole (products of lignin pyrolysis), two of the most deactivating compounds present in pyrolysis oil. The main products obtained from these feeds on the monolithic catalysts were phenol and benzene. Coating with CNFs provides increased surface area and anchoring sites for the active species (Pt and Sn), thus increasing the yield of desired products. The bimetallic Pt Sn catalysts showed higher activity and stability than monometallic Pt and Sn catalysts.

The influence of support polarity on Pd/CNF for the deoxygenation of fatty acids was studied.[278] Catalysts with a low and a high amount of oxygen containing groups on the support were prepared. The latter were introduced *via* a HNO_3 gas-phase oxidation treatment on Pd loaded supports. The presence of oxygen containing groups was beneficial for the activity of Pd for the deoxygenation of the amphiphilic stearic acid. This is attributed to a favorable mode of adsorption of the reactant *via* the carboxylic acid group on the more polar support in the vicinity of the catalytically active Pd nanoparticles.

Carbon nanohorns were also proposed as supports for NiMo for hydrotreating gas oils.[279] The results obtained with carbon nanomaterial supported metal catalysts for the hydrodechlorination reaction are reported on Table 7.3 in Section 7.8 of this chapter.

7.5 Oxidation Reactions

7.5.1 Alcohol Oxidation Reactions

Metal-catalyzed alcohol oxidation in the liquid-phase using oxygen as oxidant represents a well-known interesting catalytic process, but its industrial exploitation is limited by the strong deactivation of the catalyst.[280] Ru, Pt, Pd and Au have been used as monometallic catalysts. The main requirement for a high selectivity and long durability has not been achieved yet.

Liquid-phase oxidation of benzyl alcohol to benzaldehyde has been investigated to reveal the differences between CNTs and AC as support for metal particles.[281] Catalytic tests have shown that Pd/CNT behaves differently from Pd/AC. Indeed, Pd/CNT exhibits a lower activity, but a higher selectivity toward benzaldehyde than Pd/AC. The Pd/AC catalyst provides a slightly higher selectivity toward toluene. The catalytic performance is also strongly dependent on the solvent used. A maximum of selectivity (92%) was obtained on Pd/CNT for a 50 vol% of cyclohexane; the corresponding value on Pd/AC is 74%. Modification of CNTs with ammonia before catalyst preparation can enhance the activity and increase the TOF of the same reaction significantly.[282] In liquid-phase reactions, the leaching of the active catalytic species remains a serious problem for their application. In the above benzyl alcohol oxidation reaction, Pd on both AC and CNTs dissolves into the solution (from this the question arises whether the reaction is homogeneously or heterogeneously catalyzed). However, Pd leaching can be limited when CNTs are used as support: Pd/CNT exhibits a slightly more stable cycling (50% activity loss after 7 runs) than Pd/AC (70% activity loss after 7 runs), even if a structural change in the catalysts is observed at the end of the reaction. A consistent improvement in the long-term use of such catalysts has been observed by modifying the monometallic Pd with gold. On recycling, the Pd–Au bimetallic catalyst shows a reduced Pd leaching and exhibits stable catalytic performances over 8 runs. Under similar conditions Au–Pd on AC is less stable.[281] Nanoparticles (2–10 nm) of palladium have been deposited on SWCNT.[283] Pd/C is more active than palladium NPs deposited on SWCNT for the catalytic oxidation by molecular oxygen of cinnamyl alcohol to cinnamaldehyde.

Gold NPs were deposited on CNTs to provide access to a nanohybrid structure, which was involved in the aerobic oxidation of alcohols.[284] The reported system is effective without any added O_2 and requires no heating. While alcohols and aldehydes were chemoselectively oxidized to the corresponding acids, under anhydrous conditions, primary alcohols could also be converted to the corresponding aldehydes. The role of water in the oxidation process was elucidated using ^{18}O-labeled water. This CNT-based system compares favorably to other supported gold NPs in terms of recyclability, catalytic activity, selectivity, and mildness of the operating conditions with regard to the oxidation of alcohols.

CNT-supported ruthenium catalysts, assembled at the interfaces of emulsion droplets, show excellent activity, selectivity, and stability for the selective oxidations of benzyl alcohol to benzaldehyde with oxygen or air as oxidant in the presence of water.[285] The as-made Ru/CNTs catalysts are also active for the aerobic oxidation of a variety of alcohols with a sulfur or nitrogen atom or a carbon–carbon double bond in the multiphase reaction system. The catalytic activity of Ru/CNTs for the selective oxidation of benzyl alcohol is greatly enhanced by water, which is due to the formation of emulsion droplets where the CNT-supported catalysts assemble at the interfaces. Moreover, after the reactions, the catalysts can be easily separated and recycled by sedimentation. CNT-TiO_2 nanohybrids with the potential to control the phase of an emulsion were successfully developed.[286] When ruthenium was deposited on these CNT-TiO_2 nanohybrid supports, they demonstrated catalytic potential for the selective oxidation of benzyl alcohol to benzaldehyde. The relationship between the emulsion properties and the catalytic mechanism was also discussed.

Carboxylic-functionalized CNF-supported Ru catalysts were prepared.[287] These Ru catalysts show a good performance in oxidation of benzyl alcohol with molecular oxygen. In contrast, after partial removal of carboxylic groups on the surface of CNF-supported Ru catalysts, the activities were reduced significantly. These results suggest that the carboxylic species on CNF-supported Ru catalysts play an important role for the promotion of catalytic activity.

Palladium catalysts supported on different carbon materials, *i.e.* FLG, CNT and AC, have been prepared by improved wet impregnation and employed in the solvent-free aerobic oxidation of aromatic alcohols using molecular oxygen as oxidant.[288] High catalytic activity as well as high selectivity to corresponding carbonyl compounds can be obtained on Pd/graphene. Typically, an extremely high turnover frequency of 30 137 $mol\,h^{-1}\,mol_{Pd}^{-1}$ is observed in the aerobic oxidation of benzyl alcohol to benzaldehyde using Pd/FLG as catalyst. The Pd/FLG catalyst shows good recyclability and no palladium leaching was observed during reaction. TEM and XPS showed that the catalytically-active palladium species exist in a quite similar state, *i.e.* in terms of particle sizes and electronic structure, to palladium catalysts on different carbon supports. FT-IR analyses revealed that the adsorption of reactant benzyl alcohol is distinctly promoted compared to that on Pd/AC and Pd/CNT, due to the formation of π-electron interactions between the benzene skeleton and graphene. The O_2-TPD profiles reveal that the adsorption of oxygen on Pd/FLG is distinctly promoted due to oxygen spill-over from Pd sites to the adjacent bridge sites of graphene. Therefore, the superior reactivity of Pd/FLG in the aerobic oxidation of aromatic alcohols should be attributed to the promotion role of graphene support in the adsorption of reactant alcohol and oxygen.

Various Au/C catalysts were supported on different carbonaceous supports including reduced rGO, AC and graphite (GC) using the sol-immobilization method.[289] Au/rGO shows a much higher activity than Au/AC and Au/GC in the liquid-phase aerobic oxidation of benzyl alcohol. The superior catalytic performance of Au/rGO may be related to the presence of surface O-containing functional groups and moderate graphite character of rGO supports.

A stable, inexpensive, and reusable Co_3O_4 supported on a nitrogen-doped CB surface catalyst was shown to be active and selective for the direct oxidative esterification of alcohols using 1 bar of molecular oxygen as oxidant.[290] The catalyst was prepared by pyrolysis of amino-ligated cobalt(II) acetate on commercial Vulcan XC72R. It allows for the synthesis of a series of structurally diverse methyl esters as well as other alkyl esters in good to excellent yields. The oxidative self-esterification of both aliphatic and aromatic alcohols was well demonstrated. This process is simple, cost-effective, and environmentally benign.

Oxidation of alcohols (benzyl alcohol, 1-phenylethanol and cinnamyl alcohol) using an oxidized diamond (O-Dia) supported Pd catalyst was studied.[291] The O-Dia supported catalyst exhibited a high catalytic activity among the support materials of Pd-loaded catalysts in the oxidation of benzyl alcohol at 358 K, showing TOFs of 850 h^{-1}. The addition of a small amount of CeO_2 to O-Dia supports further improved catalytic activity in the oxidation of 1-phenylethanol with molecular oxygen. The promotion effect of CeO_2 was the suppression of acidic properties of the Pd/O-Dia, leading to dehydration and hydrogenolysis reactions of 1-phenylethanol. In addition, the Pd/CeO_2/O-Dia catalyst showed a higher catalytic activity in the oxidation of various alcohols than that of Pd/O-Dia. Moreover, the basic metal oxide-promoted catalysts exhibited high catalytic activity in the dehydrogenation of 1-phenylethanol without molecular oxygen at 393 K.

7.5.2 Catalytic Wet Air Oxidation

CNTs and CNFs have been used as supports for Pt,[292–296] Ru,[297,298] Pd,[299] Cu,[300] and iron complexes[301] in the catalytic wet air oxidation (CWAO) of various molecules, including aniline and phenol. Under the employed conditions no stability problem was noticed for the support, but the use of high surface area AC supports resulted in better performances, except for the case of Ru/AC catalysts. Different types of carbon materials, including multi-walled nanotubes (MWCNT), carbon xerogels (CX) and activated carbon (AC), were used as supports to prepare platinum catalysts (1 wt%), which were tested in the treatment of aqueous aniline solutions by CWAO.[292] All catalysts presented a very high activity for the removal of aniline and total organic carbon (TOC). Catalyst activity and selectivity toward CO_2 formation were found to depend on the nature of the support and concentration of oxygen containing functional groups on the surface of the materials. CNTs were functionalized with HNO_3, HNO_3/H_2SO_4, and HNO_3/Na_2CO_3 and used as supports for Pt, Ru and Cu.[294] The catalysts were tested in continuous CWAO of aniline. For the catalysts prepared with metal precursors, activities were found in the order of Pt/CNT > Cu/CNT > Ru/CNT.

CNFs were used to prepare supported Pt, Pd and Ru monometallic catalysts by means of IWI method.[299] Solids were tested in CWAO of phenol aqueous solution (453–513 K and 10 bar of oxygen partial pressure) carried out in a continuous-flow trickle-bed reactor. Trends of phenol and total organic carbon conversion demonstrate that the CNF support and Pt/CNF catalyst did not exhibit constant activity for CWAO of phenol. A decrease of catalyst activity, detection of carbon dioxide in the off-gas stream while examining catalyst stability and observing significant textural changes, provide evidence that under net oxidizing reaction conditions gasification of the CNF support occurs. The effect of nitrogen content in N-doped carbon nanofibers (N-CNFs) on the catalytic activity of Ru/N-CNFs in the wet air oxidation of phenol has been studied.[298] The N-CNFs, irrespective of nitrogen content and Sibunit, are shown to have low activity. In the case of Ru-containing catalysts, nitrogen in N-CNFs was found to be responsible for both the increased activity and stability of the catalysts toward deactivation. XPS data showed the formation of carbon–oxygen structures with hydroxyl (carbonyl) end groups blocking ruthenium on the surface of the catalysts without nitrogen. For the catalysts with nitrogen, ruthenium NPs were not blocked in the course of the reaction and mainly the carboxyl (carbonate) surface groups were formed.

Iron acetylacetonate complexes anchored on oxidized CNFs were prepared, and the functionalized CNFs and the iron complexes anchored on the CNFs were tested as heterogeneous catalysts for the wet oxidation of phenol with pure oxygen.[301] Complete phenol conversion and high mineralization values were achieved with fresh and reused Fe-acac/CNFs catalysts, which demonstrate the improved stability of the catalysts under the phenol degradation reaction conditions. Furthermore these conditions are comparatively mild, typically 413 K of reaction temperature and 20 bar of oxygen pressure.

Platinum and ruthenium catalysts have been supported both on mixed silica-titania and CBC supports.[302] Both supports exhibit small catalytic activity in WAO of phenol. Activated CBC support is more selective in direct total phenol oxidation than the silica-titania one. Both investigated supports are altered by process conditions: in the case of CBC under mild reaction conditions this effect is quite small. It was demonstrated that developed catalysts are selective in CWAO of phenol and their activity is commensurate with that of the catalysts reported in the literature. It seems that the Pt catalysts favor direct oxidation of phenol while oxidation over the Ru catalyst proceeds *via* an intermediate that is, at least to some extent, difficult to oxidize.

7.5.3 Other Oxidation Reactions

Ultrafine ruthenium oxide nanoparticles with an average diameter of 1.3 nm were anchored on FLG.[303] The resultant RuO_2/FLG was used as a heterogeneous catalyst for the N-oxidation of tertiary amines. After the optimization of reaction conditions for N-oxidation of triethylamine, the scope of the reaction was extended to various aliphatic, alicyclic and aromatic tertiary amines. The RuO_2/FLG showed excellent catalytic activity in terms of yields even at a very low amount of Ru catalyst. The RuO_2/FLG was heterogeneous in nature, chemically as well as physically very stable and could be reused up to 5 times.

Catalytic ozone dissociation and hydroxyl radical generation reactions are important for advanced oxidation processes in the wastewater treatment field and were used as probe reactions to investigate by DFT the synergetic effect of CNTs on a CuO/CNTs catalysts.[304] The 2p orbitals of C and the 4d of the Cu electronically interact rather than the C and O, which contributes to electronic deformation of the CuO/CNTs composite. CNTs not only act as a support and stabilizer but also play a direct role in regulating the electronic structure of a CuO/CNTs hybrid. Furthermore, an electron is transferred from the CNTs to CuO during each CuO/CNTs-catalytic reaction, accelerated by ozone decomposition and hydroxyl radical generation. Electronic charge transfer between CuO and CNTs as well as charge transfer between ozone and a CuO/CNTs composite are non-linear relationships, which verify a synergy effect for such host–guest catalyst systems. The same authors also studied a Pt/graphene system.[305] When Pt particles were supported on graphene, an electronic interaction occurred between Pt atoms and the graphene, leading to electron transfer from graphene to Pt atoms. Further electron transfer can be observed during further catalytic reactions on the surface of Pt/graphene composite, including the ozone decomposition reaction, superoxide radical generation reaction, and hydroxyl radical generation reaction. These results demonstrate that graphene not only acts as a support and stabilizer, but also plays a direct role in regulating the electronic structure of the Pt NPs during reactions.

The results obtained with metal supported catalysts and homogeneous supported catalysts for alkene epoxidation reactions are reported on Table 7.3 in Section 7.8 of this chapter.

7.6 Carbon–Carbon, Carbon–Oxygen and Carbon–Nitrogen Coupling Reactions

Homogeneous palladium catalysis has gained enormous relevance in various coupling reactions such as Heck, Stille, Suzuki, Sonogashira, and Buchwald–Hartwig reactions. Many products could be synthesized by this methodology for the first time or in a much more efficient way than before. This type of catalysis provides high reaction rate and high turnover numbers (TON) and often affords high selectivity and yields. However, homogeneous catalysis has a number of drawbacks, in particular, the lack of reuse of the catalyst or at least the problem of recycling the catalyst. This leads to a loss of expensive metal and ligands and to impurities in the products and the need to remove residual metals. These problems have to be overcome for the application of homogeneous Pd-catalyzed coupling reactions in industry and are still a challenge. In order to address these problems, heterogeneous Pd (nano) catalysis is a promising option.[306–308] The use of Pd/C avoids contamination of the product with palladium and expedites recovery of the metal. The reactions are practical and have been applied to cost-effective large-scale production of multifunctional compounds for drug synthesis.[309]

Palladium nanoparticles deposited on SWCNT exhibit higher catalytic activity than Pd/AC for the Heck reaction of styrene and iodobenzene and for the Suzuki coupling of phenylboronic and iodobenzene.[283] This fact was attributed to the influence of the Pd particle size on the activity of the catalyst for C–C bond forming reactions as compared to other reaction types that are less demanding from the point of view of the particle size. A convenient process has been developed to decorate carboxylic acid functionalized palladium SWCNTs with Pd NPs, in which the NPs are attached by carboxylic acid groups onto the SWCNT surface.[310] The Pd/SWCNT catalyst has been tested as an efficient heterogeneous nanocatalytic system for a copper-free acyl Sonogashira reaction. A library of ynones was synthesized in high yields under mild reaction condition. The catalyst was recovered and recycled successfully up to seven times. This simple protocol was further broadened in the synthesis of trimethylsilyl-ynones and explored in the one-pot multicomponent synthesis of 2,4-disubstituted pyrimidines.

Pd/CNT particles showed excellent catalytic activity and selectivity for Suzuki cross-coupling reactions[311,312] and Suzuki–Miyaura coupling of arylbromides under ligandless and additive-free conditions in aqueous media.[313] The catalytic activity did not deteriorate, even after repeated applications, and was supposed to result from the small size, narrow size distribution, and high surface area of Pd nanoparticles. The Pd/CNT particles catalyst could be separated and recovered from reaction solutions easily by simple filtration. Coupling reactions of Suzuki–Miyaura, Heck and Sonogashira were also efficiently catalyzed by Pd/MWCNT, as well as hydrogenation of the triple carbon–carbon bond.[314] The catalyst was re-usable and can work in aqueous media. Palladium, rhodium, and bimetallic Pd/Rh nanoparticles deposited directly on surfaces of functionalized CNTs were active catalysts for hydrogenation of olefins, for C–C bond formation, and for carbon–oxygen bond cleavage reactions.[315] High loadings of Pd NPs (up to 40 wt%) having sizes between 3 and 5 nm were deposited on the surface of MWCNTs and these materials serve as efficient catalysts in C–C coupling (Heck and Suzuki) as well as in hydrogenation reactions, all characterized by high conversion rates using a small amount of catalyst, high turnover frequency values and good recyclability.[316] A Pd/CNT nanocatalyst was prepared by covalent grafting of poly(lactic acid) onto carbon CNTs and subsequent deposition of Pd NPs.[317] These catalysts were found to be effective in the promotion of Heck cross-coupling reaction between aryl halides and *n*-butyl acrylate. The nanocatalyst was regenerated for three cycles of reaction without any significant loss in its activity. Carboxylated polymer-modified CNTs served as efficient platforms for the *in situ* synthesis and massive loading of 3 nm sized Pd NPs.[318] The CNT@PCOOH@Pd heterostructures prepared exhibited an efficient catalytic effect in the C–C Suzuki coupling reaction and were regenerated up to four times without any significant loss of catalytic activity. Polyamidoamine dendrimers were grown onto the surface of functionalized MWCNTs, and used to immobilize Pd NPs on the surface of MWCNTs.[319] The hybrid materials were found to be very active in cross-coupling reactions of aryl iodides, bromides and also chlorides with olefinic compounds in Heck reactions with short reaction time duration and high yields. The catalyst can be recycled several times without loss in activity. Preformed Pd NPs have been immobilized onto functionalized MWCNTs through displacement of the stabilizing ligand, and the composite formed is an active and recyclable catalyst for the Suzuki reaction.[320] CuO NPs deposited on α-Fe_2O_3-CNTs have been prepared and used as magnetically separable catalytic system in Ullmann-type coupling of aryl halides with phenols.[321] The catalyst synthesized showed good activity in C–O cross coupling reactions affording the highest rate of completion. The magnetic feature of the catalyst helped facile separation of it from the reaction medium. The catalyst could also be reused up to six times without any loss of its activity.

5 wt% Pd catalysts supported on *p*-CNFs has been prepared by IWI.[322] These catalysts have been applied to catalyze Heck reactions of various activated and non-activated aryl substrates. The activity increased exponentially with a decrease in Pd particle size. The high surface area, mesoporous structure of CNFs and highly dispersed palladium species on CNFs make

this one of the most active and reusable heterogeneous catalysts for Heck coupling reactions. Pd/p-CNF appears to be an excellent catalyst due to high activity, low sensitivity towards oxygen, almost no or low issues with leaching and high stability in multi-cycles. In order to prepare stable and reusable heterogeneous catalysts, palladium NPs were supported on *p*-CNF, h-CNF, and r-CNF.[323] It was found that the Pd loading and the particles size as well as the amount of Pd leaching during the Heck reaction were strongly related to the potential difference between the CNF surfaces and the palladium colloids. Pd leaching in the Heck reaction can be significantly suppressed by increasing the potential difference between the CNF surfaces and the palladium colloids during catalyst preparation. The Pd activity decreased exponentially with the Pd particle size. Palladium NPs supported on *p*-CNF, h-CNF, and r-CNF have also shown good catalytic activity in the Suzuki–Miyaura cross coupling and in the Mizoroki–Heck reaction.[324]

Pd/graphite oxide was applied successfully to Suzuki–Miyaura couplings of some aryl chlorides and to the Mizoroki–Heck as well as the Sonogashira reaction showing relatively high activities and good selectivities.[325] Beside its straightforward preparation and its stability in air, the system combines the advantages of both homogeneous and heterogeneous catalysis. Pd/graphite oxide also results in high performance catalysts for the Sonogashira cross coupling reaction.[326] TEM images showed that Pd NPs were distributed quite uniformly on the graphene sheet without obvious aggregation, and the mean size of Pd NPs was determined to be less than 2 nm in diameter. The resulting Pd/graphite oxide showed excellent catalytic efficiency in the Sonogashira reaction and offers significant advantages over inorganic supported catalysts, such as simple recovery and recycling. The Pd/graphite oxide also demonstrated excellent catalytic activity for the Suzuki and Heck cross-coupling reactions with a broad range of utility under ligand-free ambient conditions in an environmentally friendly solvent system.[327] It also offers a remarkable turnover frequency (108 000 h^{-1}) in the microwave-assisted Suzuki cross-coupling reactions with easy removal from the reaction mixture, recyclability with no loss of activity, and significantly better performance than the well-known commercial Pd/AC catalyst. The remarkable reactivity toward Suzuki cross-coupling reactions was attributed to the high degree of the dispersion and concentration of Pd(0) NPs supported on graphene sheets with small particle sizes of 7–9 nm. Well-dispersed Pd NPs with small and narrow size distributions were synthesized conveniently on a graphene oxide surface.[328] The catalyst was found to be a highly efficient and recyclable catalyst for the carbonylative cross-coupling reaction between arylboronic acids and aryl and carboranyl iodides, respectively. Benzophenone and a series of carboranylaryl ketones, 1-R-2-[(C=O)Ar]-1,2-$C2B_{10}H_{10}$ (R = H, Me, Ph; Ar = C_6H_5, C_6H_4-4-OMe and C_6H_4-4-F), were synthesized and fully characterized. The catalyst was recyclable at least three times with sustained activity. Pd NPs supported on single layer graphene oxide showed a high catalytic performance (TON and TOF = 237 000) in the Suzuki–Miyaura cross-coupling reaction.[329] The influence of catalyst preparation, carbon functionalization, and catalyst morphology on the Pd/graphite oxide catalytic activity was investigated for the Suzuki–Miyaura coupling reactions.[330] In contrast to the conventional Pd/AC catalyst, graphite oxide and graphene-based catalysts gave much higher activities with turnover frequencies exceeding 39 000 h^{-1}, accompanied by very low palladium leaching (<1 ppm). Pd nanoparticle catalysts supported on partially reduced graphene oxide nanosheets were used for carbon–carbon cross-coupling reactions.[331] The Pd/rGO catalyst generated in water demonstrates excellent catalytic activity for Suzuki, Heck, and Sonogashira cross coupling reactions, with good recyclability for Suzuki coupling with a turnover number of 7800 and a remarkable turnover frequency of 230 000 h^{-1} at 393 K under microwave heating. The results indicate that the defect sites on the rGO nanosheets play a major role in imparting the exceptional catalytic properties to these catalysts. Rh, Au and Rh–Pt nanodendrites as well as Pd NPs can be deposited on FLG *via* rapid co-reduction by Ti^{3+} at room temperature of graphene oxide.[332] These surface clean Pd NPs show high catalytic activity and selectivity in Suzuki and Heck coupling reactions. 3D macrostructures have been assembled from single-layered graphene oxide and noble-metal nanocrystals (Au, Ag, Pd, Ir, Rh, Pt, *etc.*).[333] They show excellent mechanical properties and have been utilized as a fixed-bed catalyst for the Heck reaction, which proceeded with nearly 100% selectivity and conversion.

Pd/FLG can act as an efficient catalyst for the Suzuki reaction under aqueous and aerobic conditions, with the reaction reaching completion in as little as 5 min.[334] The influence of the preparation conditions on the catalytic activities of the hybrids was also investigated. The Suzuki–Miyaura reaction was employed to analyze the catalytic activity of Pd/FLG catalyst.[335] The yield of biphenyl was 97% when the reaction only lasted 10 min. Calf thymus DNA modified Pd/FLG nanoparticle hybrid materials showed high catalytic activity for organic Suzuki reaction in aqueous solution under aerobic conditions without any preactivation.[336] The main advantages of using DNA are not only because the aromatic nucleobases in DNA can interact through π–π stacking with graphene basal surface but also because they can chelate Pd *via* dative bonding in such defined sites along the DNA lattice. Fe_3O_4 and Pd NPs were assembled on sulfonated FLG.[337] The resulting material could be dispersed homogeneously in water or water–ethanol and further used as an excellent semi-heterogeneous catalyst for the Suzuki–Miyaura cross-coupling reaction in an environmentally friendly solvent. The high heterogeneous catalytic activity appears to be due to the small size of Pd NPs and homogeneous distribution of the NPs on the Fe_3O_4/FLG matrix. In addition, the catalytic activity did not deteriorate even after repeated applications, which may be due to the easy and efficient magnetic separation of the catalyst and the high dispersion and stability of the catalyst in an aqueous solution. The introduction of N-doping of the graphene modified nucleation and growth kinetics during Pd@PdO catalyst deposition, which results in smaller Pd@PdO particles and uniform dispersion.[338] The Pd@PdO-NDG catalysts exhibited high yields and well structure stability for the Suzuki–Miyaura C–C cross coupling reaction. Graphene was successfully modified with Au nanoparticles in a facile route by reducing chloroauric acid in the presence of sodium dodecyl sulfate.[339] The Au/FLG catalysts can act as efficient catalysts for the Suzuki reaction in water under aerobic conditions. The catalytic activity of Au/FLG was influenced by the size of the gold NPs. Au/graphene oxide nanocomposites also exhibited unexpected catalytic activity for the Suzuki–Miyaura coupling reaction of chlorobenzene and phenylboronic acid.[340] Ni@Pd core-shell nanoparticles were supported on FLG and used as a catalyst in Suzuki–Miyaura cross-coupling reactions.[341] Among three different kinds of Ni@Pd nanoparticles tested, the Ni@Pd (Ni/Pd = 3/2) particles were found to be the most active catalyst for the Suzuki–Miyaura cross-coupling of arylboronic acids with aryl iodides, bromides and even chlorides in a dimethylformamide/water mixture by using K_2CO_3 as a base at 383 K. The Ni@Pd/FLG was also stable and reusable, providing 98% conversion after the 5th catalytic run without showing any noticeable Ni/Pd composition change. A highly active catalyst based on PdCo alloy NPs supported on polypropylenimine dendrimers, grown on FLG, was synthesized and used for Sonogashira cross-coupling reactions.[342] The results showed that the catalyst could be easily recovered and reused several times without significant loss of activity.

Pd nanoparticles were deposited with high dispersion and stability on nitrogen-doped magnetic carbon nanoparticles (MCNPs) by a simple impregnation method, and their catalytic performance was investigated for Heck, Suzuki, and Sonogashira coupling reactions (Table 7.2).[343] The Pd/N-MCNPs gave high conversion yields for all the three different reactions. Monodisperse spherical carbon NPs allow the easy diffusion of reactants and highly dispersed palladium NPs offer catalytically active sites for reactants, which may be responsible for the remarkable catalytic activities. The Pd/N-MCNPs could be readily reused several times without losing their catalytic activity.

Finally, a cross-dehydrogenative coupling reaction between tertiary amines and nitroalkanes has been realized under an oxygen atmosphere in water by using RuO_2/FLG as the catalyst.[344]

Table 7.2 Catalytic activities of Pd on nitrogen-doped magnetic carbon nanoparticles for Heck, Suzuki and Sonogashira coupling reactions.

Reaction	Reagent1	Reagent2	T/°C	Product	Yield (%)	
					1st cycle	3rd cycle
Heck	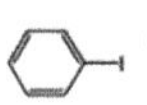 I	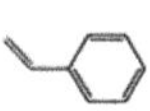	120	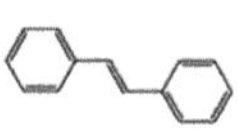	97.6	97.3
Suzuki	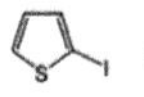 S, I	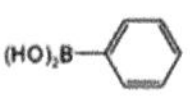 $(HO)_2B$	80	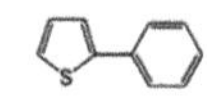 S	94.2	93.7
Sonogashira	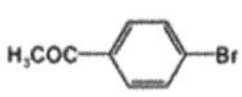 H_3COC, Br	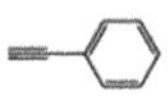	100	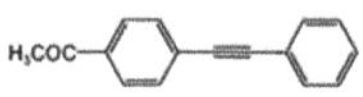 H_3COC	91.2	91.2

It is worth noting that in terms of C–C coupling reactions, relatively few studies compared the performances of nanostructured carbon supports with conventional supports such as AC or oxides, so that it is difficult to reach a conclusion on the superiority of the former. Moreover, to date the real catalytic site where the reaction takes place remains unknown, although there is some evidence for Pd leaching and therefore these highly active homogeneous Pd species could be the real catalyst. In this situation, the Pd nanoparticles would act as a slow source of active Pd species.

7.7 Polymerization

Both SWCNTs and MWCNTs are very robust along their longitudinal axis and possess excellent thermal and electrical conductivities, and hence could be used as fillers in polymer-based advanced composites. However, the homogeneous dispersion of CNTs is a difficult task mainly due to their poor solubility. Three different approaches have been adopted to produce polymer-functionalized CNTs:[345] (i) a non-covalent functionalization method by which polymers are produced by ring-opening metathesis polymerization (ROMP);[346,347] (ii) a covalent functionalization performed by first grafting polymerization initiators onto the tubes through covalent bonds and then exposing these CNT-based macroinitiators to the monomer, the polymer being obtained by anionic polymerization,[348] reversible addition and fragmentation chain transfer (RAFT) polymerization,[349] atom-transfer radical polymerization (ATRP)[350,351] or ROMP[352] and (iii) a functionalization performed by first grafting polymerization catalysts or co-catalysts (olefin polymerization *via* anchored metallocene catalysts[353–356] or ROMP[357,358]). In this section, we will only discuss the latter approach.

7.7.1 Polymerization Catalysis through Covalent Immobilization of the Catalyst/Co-catalyst

Covalent functionalization involves formation of covalent bond(s) between various functional groups and the sidewalls or defect sites on tips of carbon nanomaterials. Purified and oxidized nanocarbons bear functional groups such as hydroxyl and carboxyl on their surface. Those groups can react with some complexes with polar groups and the immobilization catalysts are produced through covalent bonding.

7.7.1.1 Carbon Nanomaterial Supported Catalyst Through an Amide Linkage

The amide bond is a typical way to covalently attach organic compounds to the functional group of nanostructured carbon materials.[359] It is reported in the literature that oxidized carbon materials are the most promising for the coating process. Before reaction of the compound containing the amine group with the oxidized carbon, a necessary step is the

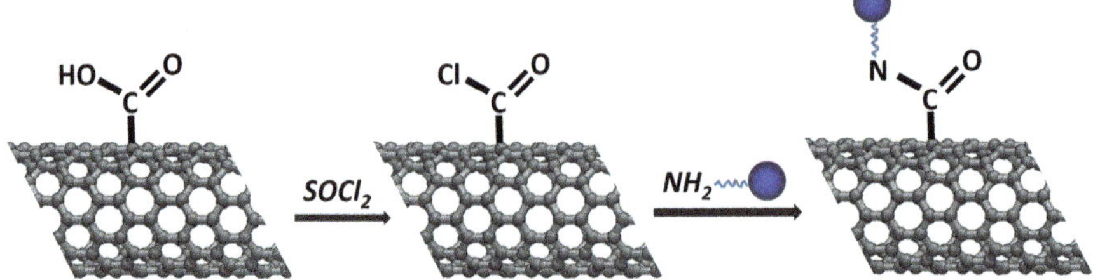

Figure 7.8 Amide functionalization of oxidized carbon nanomaterial.

reaction of the carboxylic function of the nanocarbon with thionyl chloride (see Figure 7.8).

Instead of the amine group, compounds containing an alcohol function can also link to the carbon surface by carboxylic groups.[352] According to the above reaction mechanism, some ROMP catalytic systems were investigated and carbon composites have been produced.[358,360] Covalent functionalization is irreversible and the covalent bond is relatively strong, so that the polymer produced can load well around the surface of the carbon nanomaterial. This *grafting-from* strategy is efficient to graft the catalyst on the sidewall of the nanocarbon. The method is also a useful way to prepare ethylene polymerization heterogeneous catalysts. Due to the covalent bond and the nanocarbon's unique physical and chemical properties, the nanocarbon as a support ligand can have a great influence on the polymerization activity.

7.7.1.2 Carbon Nanomaterial Supported Catalyst through Other Linkages

CNTs were functionalized with a titanium alkoxide catalyst (Figure 7.9(a)) through a Diels–Alder cycloaddition reaction.[361] The catalyst-functionalized CNTs were used for the surface initiated titanium-mediated coordination polymerizations of L-lactide, ε-caprolactone and *n*-hexyl isocyanate. The presence of thick layers of polymers around the CNTs was observed. SWCNTs were oxidatively shortened and functionalized with ruthenium-based olefin metathesis catalysts.[358] These catalyst-functionalized nanotubes were shown to be effective in the ROMP of norbornene, resulting in rapid polymerization from the catalyst sites on the nanotube. The resulting polynorbornene-functionalized nanotubes were found to exhibit solubility in organic solvents, whereas the starting materials and catalyst-functionalized nanotubes were completely insoluble. On MWCNTs, 1st and 2nd generation Grubbs catalysts (Figure 7.9(b)) and Hoveyda–Grubbs retain a very similar activity to bare Grubbs catalysts in the synthesis, *via* ROMP, of polynorbornene.[357,362] SWCNTs were oxidatively shortened and functionalized with ruthenium-based olefin metathesis catalysts.[358] These catalyst-functionalized CNTs were shown to be effective in the ring-opening metathesis polymerization of norbornene, resulting in rapid polymerization from

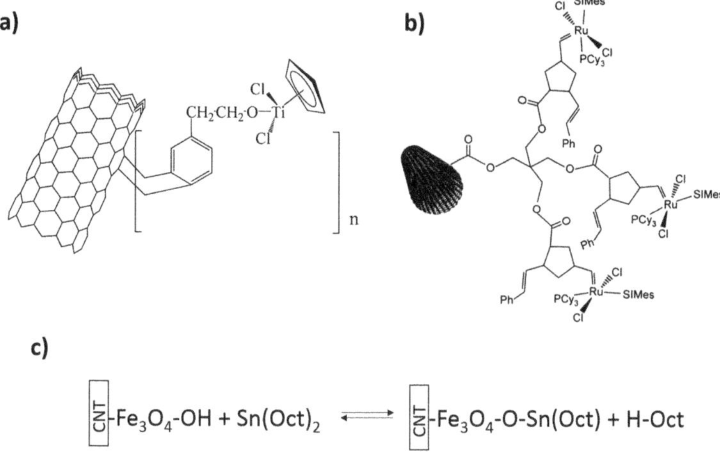

Figure 7.9 CNT supported polymerization catalysts. Adapted from ref. 358, 361 and 364.

the catalyst sites on the nanotube. It was found that high polymer molecular weights could be reached, and the molecular weight increased linearly with polymerization time. The resulting polynorbornene-functionalized nanotubes were found to exhibit solubility in organic solvents, whereas the starting materials and catalyst-functionalized nanotubes were completely insoluble.

SWCNTs-supported nickel(II) carborane complexes have been tested as moderately active catalysts for olefin polymerization in the presence of the co-catalyst MAO.[363]

Later, the inorganic support Fe_3O_4/nOH was used to functionalize the surface of CNTs and support the [$Sn(Oct)_2$] (Oct = octanoate) catalyst for ring-opening polymerization (Figure 7.9(b)).[364] It is well known that the nanocarbons contain many deviations from the pure hexagonal structure, which are potentially reactive and can be attacked by nucleophiles.

7.7.1.3 Carbon Nanomaterial Supported Catalyst through Alkylaluminium

Alkylaluminiums such as methylaluminoxane (MAO) or modified methylaluminoxane (MMAO) were reported by several groups as the medium to produce heterogeneized catalysts. This approach has been successfully applied for producing nanostructured carbon material supported polymerization catalysts.[365–370] Most of the supported catalysts are group IV-based homogeneous catalysts, in particular catalysts containing cyclopentadienyl (Cp) as ligand. The distinct features of this method are: (1) co-catalysts, MAO, for example, reacts with well dispersed CNTs[371] or FLG.[372–375] This leads to the immobilization of MAO molecules on the nanocarbon surface by loose ionic interactions and, to a lesser extent, by virtue of covalent bonding

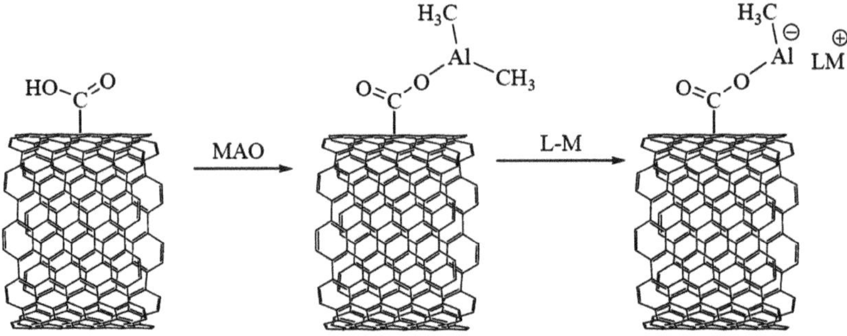

Figure 7.10 Immobilization of the catalyst on nanocarbon through MAO. Reprinted with permission from ref. 359. Copyright 2005 WILEY-VCH Verlag GmbH & Co. KGaA.

(Figure 7.10) to –COOH or –OH groups, which are inherent to partially oxidized nanocarbon, and (2) the formation of catalytic active sites is accomplished by means of heterogenization of the metallocene catalyst precursor on the nanocarbon surface owing to the chemical interaction of metallocene with MAO.[376–378] Then the active sites are formed directly during the initial stage of polymerization at lowered temperature.

MAO molecules chemically grafted to the surface of CNTs form catalytic active species, yielding polymer chains attached directly to nanotubes.[377] This represents an excellent way to improve the metallocene/alkylaluminoxane catalytic system and a new method to cover the carbon nanomaterial by polymer. Polyethylene/CNT nanocomposites were prepared by *in situ* polymerization of ethylene on CNTs whose surface had been previously treated with a metallocene catalytic system.[353] The effects of composition (5–22% CNTs) and structure of the nanotube (SWCNT, DWCNT, and MWCNT) were evaluated, and an excellent nucleating effect on polyethylene matrix was found regardless of the CNT type in comparison to neat high-density polyethylene prepared under identical conditions. The CNTs were found to be more efficient in nucleating the high-density polyethylene than its own crystal fragments, a result obtained by self-nucleation studies. The preparation of isotactic polypropylene nanocomposites filled with crude, purified and oxidized MWCNTs was accomplished by polymerization of propylene with a metallocene/MAO catalyst and *in situ* coating.[369] A homogenous distribution and a good wetting of the nanotubes with polymer could be achieved, especially by using oxidized MWCNTs where a coating phase had been carried out. The coating process and accordingly the covering of the nanotubes with polymer started at the tube edges and led to a good matrix adhesion. Bis(pentamethyl-η^5-cyclopentadienyl)zirconium(IV) dichloride [Cp_2*$ZrCl_2$] was used as a typical polymerization catalyst.[354] Upon reaction with the anchored MAO, [Cp_2*$ZrCl_2$] classically forms the methylated cationic zirconocene species, $[Cp_2{*}ZrMe]^+$, which remains immobilized at the vicinity of the nanotube surface by electrostatic interactions with simultaneously formed MAO counter-anions anchored onto the nanotube surface. Upon ethylene addition, polyethylene is exclusively formed near the CNT surface and, with increasing molecular mass, precipitates onto the nanotubes. Immobilization of methylaluminoxane onto the surface of the CNTs was evidenced by scanning electron microscopy (SEM) and confirmed by XPS and time-of-flight secondary ion mass spectrometry (TOF-SIMS).[355] Such homogeneous polyethylene coating allows for the break-up of the native nanotube bundles. The catalytic system consisting of dichloro[*rac*-ethylene-bis(indenyl)]zirconium(IV) [*rac*-$Et(Ind)_2ZrCl_2$] and MAO was also effective for the *in situ* generation of high density polyethylene/CNT composites.[356]

Owing to the presence of surface hydroxyl groups on functionalized graphene nanosheets (FLG), single FLG/MAO nanosheets can be effectively dispersed in *n*-heptane, thus enabling immobilization of a MAO-activated chromium single-site catalyst on FLG.[379] In contrast to CB, CNT, graphite, and nanoboehmite, which failed to form stable dispersions, FLG/MAO/Cr afforded the highest catalyst activities and excellent morphological control for ethylene polymerization. The unique physical and chemical properties of nanocarbons are hoped to have a positive influence on the activity of the metallocene/alkylaluminoxane catalytic system, due to the covalent bonding between MAO and carbon nanomaterial.

Besides metallocene catalysis, poly(methyl methacrylate)/CNT nanocomposites were prepared *via in situ* polymerization induced by nickel(II) acetylacetonate/MAO ([$Ni(acac)_2$]/MAO) catalyst.[380]

7.7.1.4 Carbon Nanomaterial Supported Ziegler–Natta Catalyst

Similar to the abovementioned MAO, some other inorganic oxides were used to functionalize the surface of the carbon nanomaterial. About ten years ago, it was reported that nanocarbons could support the Ziegler–Natta catalyst though $MgCl_2/n$OH.[381] Even though the mechanism of this supported catalyst was not clear, the polymerization process has successfully been accomplished to produce a polypropylene-CNT nanocomposite by *in situ* polymerization.[382] This strategy has been extended to granular CNTs.[383]

Ultrahigh-molecular-weight polyethylene/MWCNT nanocomposites with different nanotube concentrations (0.5–3.5 wt%) were prepared in the slurry phase *via in situ* polymerization with a bi-supported Ziegler–Natta catalytic system.[384] Magnesium ethoxide [$Mg(OEt)_2$] and surface-functionalized CNTs were used as the support of the catalyst. Titanium tetrachloride and triethylaluminium constituted the Ziegler–Natta catalytic system that interacts with CNT hydroxyl groups. Polyacetylene filled MWCNTs were also prepared by *in situ* polymerization in supercritical carbon dioxide, from CNTs filled with a Ziegler–Natta catalyst.[385] In order to assure controlled polymerization inside the CNTs, the catalyst outside was removed before the reaction.

Polymerized isotactic polypropylene/graphene nanocomposites were prepared *via* an *in situ* slurry polymerization method with graphene-supported

Ziegler–Natta catalyst.[386] Ultrahigh-molecular-weight polyethylene/graphene nanocomposites with molecular weights as high as 3×10^6 g mol^{-1} were prepared *via in situ* polymerization using a bi-supported Ziegler–Natta catalytic system.[387] Polypropylene/graphene oxide (PP/GO) nanocomposites were also prepared *via* an *in situ* Ziegler–Natta polymerization.[388] A Mg/Ti catalyst species was incorporated into GO *via* surface functional groups including –OH and –COOH, yielding a supported catalyst system primarily structured at the nanoscale, predominantly single GO sheet. Subsequent propylene polymerization led to the *in situ* formation of a PP matrix, which was accompanied by the nanoscale exfoliation of GO, as well as its gradual dispersion. A Grignard reagent, *n*-BuMgCl, was found to be able to reduce graphite oxide to form loosely aggregated graphene sheets immobilized with a few Mg–Cl species.[389] Further complexing with $TiCl_4$ leads to a graphene-based supported catalyst that was active for *in situ* olefin polymerization to access electrically conductive polyolefin/graphene nanocomposites. PP/graphene nanocomposites thus prepared possess a rather low electrical percolation threshold (approximately 0.2 vol%) and high conductivities (*e.g.*, 3.92 S m^{-1} at 1.2 vol%, 28.5 S m^{-1} at 4.1 vol%, and 163.1 S m^{-1} at 10.2 vol%).

7.7.2 Polymerization Catalysis through Non-covalent Immobilization of the Catalyst

Non-covalent functionalization is based on attraction between the hydrophobic end of an adsorbed molecule and the surface of nanostructured materials *via* van der Waals forces or π–π interactions. Non-covalent functionalization can also be performed using cation–π electrostatic interactions. This type of functionalization does not interfere with the electronic structure of nanocarbons, as it does not involve covalent bonds. Therefore, to functionalize without damaging the π-electronic structure of nanocarbons, non-covalent attachments seem to be more promising than the covalent ones.[390–392] Among the available non-covalent functionalization approaches, the simplest method involves direct π–π stacking of compounds containing aromatic groups. This method has been successfully reported in several publications about the π–π interaction between the surface of the nanocarbons and the aromatic moieties of various metallocenes.[366,367,393–401] Theoretical calculations have shown that MCp_2@SWCNT composites were stabilized by weak π-stacking and CH~π interactions, and in the case of the $CoCp_2$@SWCNT composites there was an additional electrostatic contribution as a result of charge transfer from $CoCp_2$ to the nanotube.[402] However, compared to a covalent bond, the π–π interaction is much weaker. Therefore the incorporation procedure will be greatly influenced by the reaction conditions, like the solvent used and the reaction temperature. Electronic interactions and binding modes between zirconocenes and CNTs were investigated by theoretical and experimental methods, and both methods indicated that the sidewalls of CNTs could form a coordination bond with the positively charged active species and act as a ligand with strong electron donation.[401]

MWCNT/ultrahigh-molecular-weight polyethylene composites, where pristine MWCNTs were well dispersed in the matrices, were produced using MWCNT/half-titanocene hybrids as catalysts.[395] The diameter of the polyethylene-coated MWCNT strands produced is about 30–70 nm, while the diameter of the pristine MWCNTs used is 10–15 nm. Composites with a molecular weight greater than 2×10^6 were produced.

Graphene plays a role not only as a support for adsorption but also as an additional ligand to $[(n\text{-BuCp})_2ZrCl_2]$ during ethylene polymerization.[403] In the π–π stacking between graphene and $[(n\text{-BuCp})_2ZrCl_2]$, the rich electron environment and bulky surface of graphene increase the electron density of $[(n\text{-BuCp})_2ZrCl_2]$ and inhibit the β-hydrogen elimination, resulting in a higher molecular weight and polydispersion of polyethylene than those of homogenous $[(n\text{-BuCp})_2ZrCl_2]$. Carbon-based nanomaterials such as graphene nanoplatelets with and without N-doping and CNTs were used as a ligand for ethylene polymerization.[404] Zirconocenes or titanocenes were immobilized on such carbon nanomaterials. Polyethylenes produced by such hybrids showed a great increase in molecular weight relative to those produced by the free catalysts. In particular, ultra-high-molecular-weight polyethylenes were produced from the polymerizations at low temperature using the hybrid with N-doped graphene nanoplatelets.

New composite materials based on isotactic polypropylene and fullerene were prepared *via* an *in situ* polymerization with the use of an isospecific metallocene catalytic system.[405,406]

Aqueous high molecular weight linear polyethylene nanocrystal/graphene composite dispersions were produced *in situ* by ethylene insertion polymerization in water or alternatively *via* a post-polymerization technique.[407] Graphene was used without any prior compatibilizing surface modification. Remarkably, the Ni(II) salicylaldiminato catalyst employed is polymerization active in the presence of the functional groups of graphene in the absence of any scavengers or activating agents in the aqueous system. A series of 1-aryliminoethylpyridine ligands was synthesized by condensation of 2-acetylpyridine with 1-aminonaphthalene, 2-aminoanthracene or 1-aminopyrene.[408] Reaction with nickel dichloride afforded the corresponding nickel(II) chloride complexes. Upon activation with MAO, all nickel complexes exhibit high activities for ethylene polymerization, producing waxes of low molecular weight and narrow polydispersion. The presence of MWCNTs or FLG in the catalytic medium can lead to an increase in productivity associated to a modification of the polymer structure.

Metal-CNT-graft-polymer (MCNT-g-P) nanocomposites were synthesized from acid-treated CNTs filled by silver nanoparticles.[409] The CNT containing metal nanoparticles were used as a macroinitiators for ring opening polymerization of ε-caprolactone. Length of grafted polymer arms onto the MWCNT was controlled using MWCNT/ε-caprolactone ratio.

7.8 Other Reactions

The peculiarities of nanostructured carbon-based catalysts have prompted scientists to explore their possible use as supports in a large variety of reactions (Table 7.3), on which their properties could exert important impacts compared to traditional carbon or oxide supports.

7.9 CNTs for Enzyme Immobilization

The use of nanomaterials for enzyme immobilization may offer some advantages regarding enzyme stabilization, control of the pore size to protein molecule dimensions, multiple sites for interaction, and reduced mass-transfer limitations.[481,482] These robust nano-scaffolds have an inherently large surface area which leads to high enzyme loading and consequently high volumetric enzyme activity. Although nanostructured carbon materials have been mainly used for biosensors,[483,484] where their electrical and electrochemical properties are of particular interest, recent studies have evaluated the possibility to use such materials as support for biocatalysts.[485,486] Proteins can interact with nanostructured carbon materials with multiple types of interactions. Non-covalent (physisorption) and covalent (chemical bond formation) conjugations have been reported for the immobilization of various enzymes on nanostructured carbon materials.[487–489] Non-covalent attachment preserves the unique properties of both enzymes and nanostructured carbon materials, but the immobilized protein can be gradually lost during the use of the carbon–enzyme complex. Covalent conjugation provides durable attachment, but the enzyme structure may be more disrupted.[490] Functionalization of nanostructured carbon materials with organic, polymeric, and biological molecules can provide biocompatible carbon composites with specific groups on their surface. No matter what method is used, the main challenge is promoting the stable attachment of enzymes while maintaining their activity and function as closely as possible to their native state. The performance of the carbon–enzyme complexes are affected by a combination of the nanostructured carbon material chemistry and immobilization method.

Enzymes have been immobilized either *via* non-covalent[491–497] (through charge complementary, π–π stacking or hydrophobic interactions) or covalent[498–504] surface functionalization of nanostructured carbon materials.

It is worth noting that even physical adsorption of enzymes onto carbon surfaces may induce secondary structural perturbations as a result of protein interaction with the carbon surface,[495] and that substantial perturbation can induce either nearly complete loss[505] or significant increases[506,507] in catalytic activity. Structural and biochemical characterization have revealed that the curvature of the carbon nanomaterial affects the immobilization yield, the catalytic behavior and the secondary structure of enzymes.[508] The catalytic behavior of the immobilized enzyme also depends on the nature of the terminal group of the carbon surface, the concentration of the enzyme and the immobilization method employed.[509] Furthermore, the strong interaction between the

Table 7.3 Catalytic results obtained with nanostructured carbon materials based catalysts for various transformations.

Reaction	Catalyst	Comments	References
Skeletal *n*-hexane reaction	Pt/CNF	5%Pt/CNF (D = 20%) is more active at lower temperature than a commercial 6%Pt/SiO_2 (EUROPT-1, D = 60%). Both activity and isomer formation are better on Pt/h-CNF than on Pt/p-CNF	410,411
C_6 olefin skeletal isomerization	WO_3/MWCNT	17%WO_3/MWCNT catalyst is significantly more active and much more stable than a commercial 25% WO_3/ZrO_2 catalyst	412
Ethylbenzene	Ni/CNT and Co/CNT	Co/CNTs showed the highest styrene selectivity (92.3%) at 723 K. The catalyst also showed a high thermal stability.	413
	Pd/CNT	The percentage conversion of EB over Pd/CNTs (45.7%) was lower than that over metal-free CNTs (64%)	414
Cyclohexanol	Co/MWCNT	15%Co/MWCNT presents higher activity and selectivity to cyclohexanone than a 15%Co/AC, and HNO_3 treated MWCNTs permit higher conversions. The effect of K addition, attributed to electronic promotion, was also stronger on Co–K/CNT than on Co–K/AC.	415,416
Isopropanol	Cu/CNF	Cu/CNF show similar activity to Cu/AC	417
	Pd/CNT and Ru/CNT	Pd catalysts were 100% selective towards acetone for a conversion of 100%, whereas the Ru catalysts led to dehydration and dehydrogenation products.	418
Dehydrogenation	Pt/MWCNT	Low loading Pt/MWCNT present higher TOFs than their counterparts on activated carbon or graphite flakes	419
Cyclohexane and *CH_3–C_6H_{11}*	Pt/CNT	Pt/hollow carbon particles > Pt/AC > Pt/MWCNTs > Pt/CB	420–422
	Pt/MWCNT	0.25%Pt/h-CNF presents same activity and selectivity as commercial 1%Pt/Al_2O_3	423
	Co/ND	Co/ND > Co/Al_2O_3, ZnO, and MgO	424
Hydroxymatairesinol	Pd/CNF	Both activity and selectivity increase with the concentration of acid sites on CNFs	425
Propane	V_2O_5/CNT	Oxidative dehydrogenation of propane	426
Chlorobenzene	Pd/CNF	Initial activity: Pd/CNF ≈ Pd/AC > Pd/graphite. Catalyst stability: Pd/CNF ≈ Pd/graphite > Pd/AC. Bulk Pd outperformed carbon supported Pd but was less active than Pd on the oxide supports.	427,428
	Pd/CNF	The surface acid groups inhibited activity, a response most pronounced for Pd/AC, less so for Pd/graphite, while the effect was slight for Pd/CNF	429

(continued)

Table 7.3 *(continued)*

Reaction		Catalyst	Comments	References
Hydrodechlorination		Pd/N–CNF	The presence of pyridine-type species helps to remove the byproduct HCl from the Pd catalyst, thus preventing the inhibition effect of HCl.	430
		Pd/ND Ni/SiO_2–CNF	The activity of Pd/ND was found to be several orders of magnitude higher than that of Pd/ACActivity of Ni/SiO_2–CNF was maintained by long reaction times on stream, to ultimately exceed the level of dechlorination achieved with Ni/SiO_2	431–434,435
MIBK synthesis	*Tetrachloroethylene*	Pd/CNF	The surface chemistry of the support markedly affects the catalysts performance	436
		Pd/CNT	The high catalytic performance of Pd/CNTs was attributed to the highly dispersed Pd nanoparticles	437
	4-chlorophenol	Rh/r-GO	The high catalytic performance of Rh/r-GO was attributed to the electron-deficient state of the highly dispersed Rh nanoparticles	438
		Hydrotalcites/CNF	Specific activity of supported hydrotalcites is found to be 4 times higher than that of unsupported catalysts	439,440
Dimethyl oxalate synthesis		Pd/CNF	Significant improvement of catalytic activity compared to Pd/Al_2O_3	441
Dimethyl carbonate synthesis		CuNi/CNT	The high catalytic activity of CuNi/CNTs can be attributed to the synergetic effects of metal Cu, Ni and Cu–Ni alloy and the interaction between the metal particles and the support.	442
Hydrosilylation		Rh/CNF and RhPt/CNF	The effects of confinement in CNF on the regioselectivity of addition to triple carbon–carbon bonds is reported	443,444
Cyanosilylation of aldehydes		[V]/SWCNT	The asymmetric version was also performed using a chiral vanadyl complex obtaining 66% of enantiomeric excess	445,446
Ammonium decomposition	Perchlorate	Cu/MWCNT	Catalytic performances of supported catalyst are found to be better than those of unsupported one	447
		Nd_2O_3/CNT	$Nd_2O_3/CNT > Nd_2O_3$	448
		Co/CNT	The catalytic activity of Co/CNT is higher than that of pure Co NPs	449
		Co_3O_4/GO	Unusual catalytic effect due to the concerted effects of GO and Co_3O_4	450,451
		Ni/FLG	High catalytic activity	452
		Mn_3O_4/FLG	Synergistic effect of Mn_3O_4 and FLG	452,453
		Fe_2O_3/FLG	FLG improved the catalytic activity of Fe_2O_3 due to its high specific area	454

	MnOOH/FLG	Unusual catalytic performance	455
Self-condensation of acetone	La_2O_3/CNF	La_2O_3/CNF ≫ La_2O_3	456
Alkene epoxidation	[Mn]/MWCNT	High reusability, the catalyst was reused several times without significant loss of its catalytic activity	457,458
	[Mn]/MWCNT	Epoxidation of alkenes with sodium periodate under mild conditions	459,460
	[Mo]/MWCNT	High stability and reusability without loss of its catalytic activity	461,462
	[W]/MWCNT	High stability and reusability without loss of its catalytic activity	463
	[Co]/MWCNT	Stable for the epoxidation of cyclohexene	464
	[V]/MWCNT	Liquid-phase oxidation of cyclooctene with molecular oxygen to a mixture of cyclooctene-1-one, cyclooctene-1-ol and cyclooctene epoxide	465
		Vanadium-containing polyphosphomolybdate catalyst	466
	[PVMo]/MWCNT	Oxidized Au nanoparticles are active in the epoxidation of styrene with air	467
	Au/CNT	The Au/CNTs catalysts with smaller gold particle size were related to higher epoxide yield, indicating a size effect of gold nanoparticles on the catalytic performance	468
	Au/CNT		
		Styrene conversion over 99% and epoxide selectivity of 93%	
	Co-in/CNT	Enhance catalytic activity in the presence of SWCNT	469
	[Mo]/SWCNT	Good recoverability without significant loss in activity and selectivity within successive run	470,471
	[Cu]/FLG, [Co]/FLG	Superior activity and stability for the epoxidation of styrene compared to bulk Co_3O_4 and N-free graphene supported Co_3O_4	472
	Co_3O_4/N-FLG	A Theoretical study on propene epoxidation with $H_2/H_2O/O_2$ mixtures	473
	Au/FLG		474
Selective catalytic reduction of NO	V_2O_5/CNT	Good catalytic activity, the larger the CNT diameter the higher the activity	475
	Ce_2O/CNT	Ce_2O/CNT > Ce_2O > CNT	476
	Ce_2O/CNT	Highly dispersed CeO_2 on the CNTs, strong CeO_2-CNTs interaction and the large amounts of strong acid sites explained the excellent performances	477
	MnO_x/CNT	High concentration of MnO_2 and low concentration of MnO lead to high SCR activity	478
	MnO_x–CeO_x/CNT	The catalysts presented a large acid amount and strong acid strength	479,480

non-covalently anchored enzyme and the carbon surface may result in leaching-stable biocatalytic materials.[510] A simple, novel, and efficient approach was described for improving the structures and activities of enzymes bound to graphene oxide such that bound enzymes are nearly as active as those of the corresponding unbound enzymes.[511] The strategy was to pre-adsorb highly cationized bovine serum albumin (cBSA) to passivate GO, and cBSA/GO (bGO) served as an excellent platform for enzyme binding. The binding of met-hemoglobin, glucose oxidase, horseradish peroxidase, BSA, catalase, lysozyme, and cytochrome c indicated improved binding, structure retention, and activities. Nearly 100% of native-like structures of all the seven proteins/enzymes were noted at near monolayer formation of cBSA on GO (400 wt%), and all bound enzymes indicated 100% retention of their activities.

Covalent functionalization of MWCNTs by lipase gives a biocomposite that exhibits remarkable solubility in organic solvents.[501] Water-soluble carbon nanotube-enzyme conjugates were also produced by covalent immobilization.[512]

Multifunctional materials have also been reported. For example, bifunctional graphene/γ-Fe_2O_3 hybrid aerogels with quite low density (30–65 mg cm^{-3}), large specific surface area (270–414 m^2 g^{-1}), high electrical conductivity (0.5–5×10^{-2} S m^{-1}), and superior saturation magnetization (23–54 emu g^{-1}) have been produced.[513] A pH tunable, temperature sensitive magnetoresponsive graphene-based nano-bio carrier for cellulase immobilization was also reported.[514] A graphene oxide–ferroferric oxide–hemoglobin (GO–Fe_3O_4–Hb) showed improved enzyme thermal stability and wider active pH range compared with Fe_3O_4–Hb and soluble Hb.[515] Furthermore, the GO–Fe_3O_4–Hb composite can be easily separated under a magnetic field from the reaction solution and reused. Association of nanostructured carbon materials with metallic nanoparticles and enzymes have also been reported.[516,517] The possibility of integrating biologically derived silica with SWCNTs to create a conductive matrix for immobilization of enzymes was also reported.[518] Some attempts to prepare anchored chemzymes that can not only simply duplicate and imitate the properties of natural enzymes, but also introduce additional new features for practical applications, have also been reported, such as for a peroxidase mimic.[519] Interestingly, a powerful enzymatic mimetic has been fabricated by employing GO nanocolloids to disperse conductive carbon supports of hydrophobic CNTs before and after the loading of Pt nanoparticles.[520] The resulting GO@CNT-Pt nanocomposites present improved aqueous dispersion stability and Pt spatial distribution. They showed greatly enhanced peroxidase-like catalysis and electrocatalytic activities in water.

Carbon nanomaterials-based biocatalysts have been successfully employed for *p*-cresol oxidation with hydrogen peroxide,[505] NADH oxidation,[521] (trans) esterification[508,522–524] and hydrolysis.[507] Recent works have also emphasized biocatalysis with carbon nanomaterials for enzyme-based biofuel cells,[525] and encouraging results have been reported on the use of enzyme–nanocarbon composites for electrocatalytic H_2 oxidation,[526,527] oxygen reduction[528], glucose oxidation,[529,530] or ethanol oxidation.[531]

In order to produce effective and recyclable catalysts for enantioselective trans-esterification in industrial applications, alkaline lipase from *Pseudomonas fluorescens* was non-covalently immobilized on pristine and oxidized MWCNTs.[524] The differences in performance of CNTs and oxidized-CNTs as solid supports were found to be based on the geometry of the pores, dominating hydrophobic interactions and the absence/presence of surface polar groups. An excellent activity and reusability was obtained with the oxidized nanotube-lipase catalysts. Lipase-MWCNTs catalysts were applied for synthesis of the flavor ester 'pentyl valerate' in cyclohexane and effects of solvent, temperature and agitation on ester synthesis were studied.[504] Upon subject to reusability studies for 50 cycles, the biocatalysts were found to be highly sturdy and exhibited *ca.* 79% activity. Lipase attachment onto MWCNTs has significantly affected the performance of the enzyme in terms of temperature dependence and resolution efficiency for the resolution of a model compound (*R*,*S*)-1-phenyl ethanol in *n*-heptane.[532] The activity of the MWCNT-lipase was less temperature-dependent compared with that of the native lipase. The resolution efficiency was much improved with MWCNT-lipase. MWCNT-lipase retained the selectivity of the native lipase for (*R*)-1-phenyl ethanol after consecutive uses. *Yarrowia lipolytica* lipase was covalently immobilized on iron oxide nanoparticles-loaded CNTs.[533] Magnetic CNT-lipase was utilized for the resolution of (*R*,*S*)-1-phenyl ethanol in heptane. Compared to the native lipase, the lipase immobilized on CNTs has significantly improved its enzymatic activity and can be easily recovered after catalysis. Carboxyl-functionalized graphene oxide was utilized to immobilize lipase.[534] For catalyzing the enantioselective reaction in heptane, the catalysis efficiency of the immobilized lipase was 1.6 times that of native lipase, and the immobilized lipase retains the selectivity of the native lipase. Nanoparticles, nanofibers, nanotubes, nanopores, nanosheets and nanocomposites were used in the context of lipase-mediated nanobiocatalysis for biodiesel production.[535]

Hydrolases exhibit higher esterification activity (up to 20-fold) when immobilized on CNTs compared to GO.[508] The covalently immobilized enzymes exhibited comparable or even higher activity compared to the physically adsorbed ones, while they presented higher operational stability. The enzyme–carbon nanomaterial interactions significantly affect the catalytic behavior of enzymes, resulting in an increase of up to 60% of the catalytic efficiency of lipases and a decrease of up to 30% of the esterase.[536] The enhanced catalytic behavior observed for most of the hydrolases covalently immobilized on amine-functionalized CNTs indicate that these functionalized nanomaterials are suitable for the development of efficient nanobiocatalytic systems.

Horseradish peroxidase (HRP) and soybean peroxidase adsorbed onto SWCNTs endure a notable loss in secondary structure and catalytic activity.[537] This structurally and functionally deformed enzyme-SWCNT when confined in CTAB reverse micelles showed 7- to 9-fold enhancement in activity compared to that in water, and also importantly 1500–3500 times higher activity than that of the enzymes in aqueous–organic biphasic mixtures. The activation observed for this nanobiocomposite was due to the (i) possible localization of the enzyme–SWCNT hybrid at the micellar interface; (ii) facile transport of substrates across the microscopic interface of reverse micelles; and (iii) greater local concentration of substrates at the augmented interfacial space in the presence of SWCNT. The physical properties and catalytic activity of GO immobilized HRP and its application in phenolic compound removal were described.[538] HRP loading on GO was found to be much higher than that on reported substrates. The GO immobilized HRP showed improved thermal stability and a wide active pH range, attractive for practical applications. The removal of phenolic compounds from aqueous solution using the GO immobilized HRP was explored with seven phenolic compounds as model substrates. The GO immobilized HRP exhibited overall a high removal efficiency to several phenolic compounds in comparison to soluble HRP.

Compared with native and graphene oxide bound enzymes, it was found that the glucose oxidase (GOD) or glucoamylase (GA) immobilized on rGO exhibited significantly higher enzymatic activity, due to the positive effect of the rGO support.[539] The obtained artificial biocatalyst was characterized by UV/Vis, FTIR, AFM, TEM and SEM. This multi-enzyme microsystem was employed as a biocatalyst to accomplish the starch-to-gluconic acid reaction in one pot, and the yield of gluconic acid could reach 82% within 2 hours. It was also proved that the stability of the multi-enzyme biocatalyst immobilized on rGO was dramatically enhanced compared with the GO microsystem. About 85% of the activity of the biocatalyst could be preserved after four cycles.

A 3D-structured graphene-Rh-complex hydrogel worked as a robust catalyst for electroenzymatic reactions.[540] When α-ketoglutarate was electroenzymatically converted to L-glutamate in the presence of graphene-Rh-complex hydrogel, the L-glutamate yield increased more than 10 times compared to that for a free Rh complex.

Finally, as CNTs can be directed to aqueous–organic interfaces with the aid of surfactants, this phenomenon can also be used to transport enzymes to the interface to effect biphasic biotransformations.[541] Consequently, CNT-enzyme conjugates enhance the rate of catalysis by up to 3 orders of magnitude relative to the rates obtained with native enzymes in similar biphasic systems.

To conclude, enzyme immobilization on nanostructured carbon materials allows researchers to take advantage of the fascinating properties of nanostructured carbon materials. The activity of immobilized enzymes is influenced by the methods and procedures of immobilization. Non-covalent approaches have been more studied than covalent ones. Using non-covalent approaches, enzymes can be less denatured upon immobilization and the intrinsic electronic structure and properties of nanostructured carbon materials are preserved. Recently, more attention has been paid to the controlled immobilization of enzymes on nanostructured carbon materials. To that end, specific groups are introduced onto nanostructured carbon materials by functionalization with organic, polymeric, or biological molecules. Through the functional groups, enzymes can be specifically and precisely bound onto nanostructured carbon materials. It is also necessary to study how the linking molecules interact with enzymes and affect the enzyme structure and the arrangement of enzymes on the carbon surface. These results could significantly contribute to preparing a carbon–enzyme complex with a specific functionality.

7.10 Concluding Remarks and Future Perspectives

The use of nanostructured carbon materials in catalysis has been a cutting-edge research area of chemistry and materials science. In particular, the use of CNTs, CNFs and FLG as catalyst supports brings new insight into nanocatalysis and provides a better understanding of the chemistry and physics of carbon materials. A number of different aspects dealing with the use of nanostructured carbon materials as catalyst support should be highlighted: (i) enhanced characteristics as a support for catalytic functionalities; (ii) stabilization of small catalytic particles with enhanced catalytic behavior, particularly with N-doped nanocarbons[542] or at edge sites of CNFs;[16] (iii) the direct catalytic role of nanocarbon functional groups, for example for oxidative dehydrogenation;[543] (iv) the indirect role of nanocarbon functional groups (hydrophobic/hydrophilic effect) and the edge sites/basal plane ratio on the catalytic activity, for example with aromatic substrates; (v) the confinement effects in CNT cavity;[6,18] (v) the electron-transfer that induces changes in the properties of supported nanoparticles; and (vi) the defect-related catalytic reactivity. With their unique features and specific characteristics, nanostructured carbon materials should be used to manipulate the catalytic properties of nanocatalysts by tuning the properties of the carbon, including the orientation and curvature of the graphene sheets, crystallinity, conductivity, *etc.*

However, the surface chemistry and dynamics of catalytic reactions on nanostructured carbon material supported catalysts is still far from well understood. The number and type of defects on carbon nanomaterial surface, which act as anchoring sites for nanoparticles, cannot be efficiently controlled during their production. Systematic studies on the correlation between the nature of the nanocarbons and the catalytic reactivity of supported catalysts are still limited. This is a direction in which research should be intensified, particularly with characterization and modelling studies, which are clearly the key enabling elements for a rational design of nanocarbons in advanced catalytic applications. The next step should be to take advantage of the unique possibility in nanostructured carbon materials for the tuning of their characteristics, largely related to the flexibility of their nanostructure to account for different types of defects, heteroatoms, and functional groups. It is thus necessary to focus on the design of nanostructured carbon materials and their nano-engineering to develop more active and selective catalysts.

References

1. K. P. De Jong and J. W. Geus, *Catal. Rev.*, 2000, **42**, 481–510.
2. P. Serp, M. Corrias and P. Kalck, *Appl. Catal., A*, 2003, **253**, 337–358.
3. N. F. Goldshleger, *Fullerene Sci. Technol.*, 2001, **9**, 255–280.
4. N. M. D. A. Coelho, J. L. B. Furtado, C. Pham-Huu and R. Vieira, *Mater. Res.*, 2008, **11**, 353–357.
5. V. A. Likholobov, V. F. Surovikin, G. V. Plaksin, M. S. Tsekhanovich, Y. V. Surovikin and O. N. Baklanova, *Catal. Ind.*, 2009, **1**, 11–16.
6. P. Serp and E. Castillejos, *ChemCatChem*, 2010, **2**, 41–47.
7. J. H. Bitter, *J. Mater. Chem.*, 2010, **20**, 7312–7321.
8. H. Gao, L. Song, W. Guo, L. Huang, D. Yang, F. Wang, Y. Zuo, X. Fan, Z. Liu, W. Gao, R. Vajtai, K. Hackenberg and P. M. Ajayan, *Carbon*, 2012, **50**, 4476–4482.
9. B. F. Machado and P. Serp, *Catal. Sci. Technol.*, 2012, **2**, 54–75.
10. C. Huang, C. Li and G. Shi, *Energy Environ. Sci.*, 2012, **5**, 8848–8868.
11. B. Garg and Y.-C. Ling, *Green Mater.*, 2013, **1**, 47–61.
12. X. Zhou, J. Qiao, L. Yang and J. Zhang, *Adv. Energy Mater.*, 2014, **4**, 1301925.
13. B. Moosa, K. Fhayli, S. Li, K. Julfakyan, A. Ezzeddine and N. M. Khashab, *J. Nanosci. Nanotechnol.*, 2014, **14**, 332–343.
14. B. Coq, J. Marc Planeix and V. Brotons, *Appl. Catal., A*, 1998, **173**, 175–183.
15. D. S. Su, S. Perathoner and G. Centi, *Chem. Rev.*, 2013, **113**, 5782–5816.
16. J. Zhu, A. Holmen and D. Chen, *ChemCatChem*, 2013, **5**, 378–401.
17. P. Serp, in *Carbon Materials for Catalysis*, John Wiley & Sons, Inc., 2008, pp. 309–372.
18. X. Pan and X. Bao, *Acc. Chem. Res.*, 2011, **44**, 553–562.
19. C. Park and R. T. K. Baker, *J. Phys. Chem. B*, 1998, **102**, 5168–5177.
20. A. Chambers, T. Nemes, N. M. Rodriguez and R. T. K. Baker, *J. Phys. Chem. B*, 1998, **102**, 2251–2258.
21. C. Park and R. T. K. Baker, *J. Phys. Chem. B*, 1999, **103**, 2453–2459.
22. P. Liu, H. Xie, S. Tan, K. You, N. Wang and H. a Luo, *React. Kinet. Catal. Lett.*, 2009, **97**, 101–108.
23. T.-J. Zhao, D. Chen, Y.-C. Dai, W.-K. Yuan and A. Holmen, *Top. Catal.*, 2007, **45**, 87–91.
24. K. H. Park, K. Jang, H. J. Kim and S. U. Son, *Angew. Chem., Int. Ed.*, 2007, **46**, 1152–1155.
25. V. V. Chesnokov, I. P. Prosvirin, N. A. Zaitseva, V. I. Zaikovskii and V. V. Molchanov, *Kinet. Catal.*, 2002, **43**, 838–846.
26. I. Efremenko and M. Sheintuch, *J. Catal.*, 2003, **214**, 53–67.
27. Y. Motoyama, M. Takasaki, S.-H. Yoon, I. Mochida and H. Nagashima, *Org. Lett.*, 2009, **11**, 5042–5045.
28. V. D'Anna, D. Duca, F. Ferrante and G. La Manna, *Phys. Chem. Chem. Phys.*, 2010, **12**, 1323–1330.
29. R. S. Oosthuizen and V. O. Nyamori, *Platinum Met. Rev.*, 2011, **55**, 154–169.
30. E. V. Starodubtseva, M. G. Vinogradov, O. V. Turova, N. A. Bumagin, E. G. Rakov and V. I. Sokolov, *Catal. Commun.*, 2009, **10**, 1441–1442.

31. B. Yoon, H.-B. Pan and C. M. Wai, *J. Phys. Chem. C*, 2009, **113**, 1520–1525.
32. B. A. Kakade, S. Sahoo, S. B. Halligudi and V. K. Pillai, *J. Phys. Chem. C*, 2008, **112**, 13317–13319.
33. H.-B. Pan and C. M. Wai, *J. Phys. Chem. C*, 2010, **114**, 11364–11369.
34. E. Asedegbega-Nieto, B. Bachiller-Baeza, D. G. Kuvshinov, F. R. García-García, E. Chukanov, G. G. Kuvshinov, A. Guerrero-Ruiz and I. Rodríguez-Ramos, *Carbon*, 2008, **46**, 1046–1052.
35. M. Takasaki, Y. Motoyama, S.-H. Yoon, I. Mochida and H. Nagashima, *J. Org. Chem.*, 2007, **72**, 10291–10293.
36. H.-B. Pan and C. M. Wai, *J. Phys. Chem. C*, 2009, **113**, 19782–19788.
37. A. M. Zhang, J. L. Dong, Q. H. Xu, H. K. Rhee and X. L. Li, *Catal. Today*, 2004, **93–95**, 347–352.
38. P. Chen, L. M. Chew, A. Kostka, M. Muhler and W. Xia, *Catal. Sci. Technol.*, 2013, **3**, 1964–1971.
39. F. Coloma, A. Sepulvedaescribano and F. Rodriguezreinoso, *J. Catal.*, 1995, **154**, 299–305.
40. S. D. Kushch, N. S. Kuyunko and B. P. Tarasov, *Russ. J. Gen. Chem.*, 2009, **79**, 1106–1112.
41. O. V. Turova, E. V. Starodubtseva, M. G. Vinogradov, V. I. Sokolov, N. V. Abramova, A. Y. Vul and A. E. Alexenskiy, *Catal. Commun.*, 2011, **12**, 577–579.
42. V. Mavrodinova, M. Popova, I. Kolev, S. Stavrev and C. Minchev, *Appl. Surf. Sci.*, 2007, **253**, 7115–7123.
43. V. Mavrodinova, M. Popova, D. Mitev, S. Stavrev, S. Vassilev and C. Minchev, *Catal. Commun.*, 2007, **8**, 1502–1506.
44. D. Marquardt, C. Vollmer, R. Thomann, P. Steurer, R. Mülhaupt, E. Redel and C. Janiak, *Carbon*, 2011, **49**, 1326–1332.
45. M. Stein, J. Wieland, P. Steurer, F. Tölle, R. Mülhaupt and B. Breit, *Adv. Synth. Catal.*, 2011, **353**, 523–527.
46. X. Liu, C. Meng and Y. Han, *Nanoscale*, 2012, **4**, 2288–2295.
47. S. Banerjee and S. S. Wong, *J. Am. Chem. Soc.*, 2002, **124**, 8940–8948.
48. L. J. Lemus-Yegres, M. Pérez-Cadenas, M. C. Román-Martínez and C. S.-M. de Lecea, *Microporous Mesoporous Mater.*, 2011, **139**, 164–172.
49. L. Rodriguez-Perez, E. Teuma, A. Falqui, M. Gomez and P. Serp, *Chem. Commun.*, 2008, 4201–4203.
50. T. G. Ros, A. J. van Dillen, J. W. Geus and D. C. Koningsberger, *Chem. – Eur. J.*, 2002, **8**, 2868–2878.
51. S. Sabater, J. A. Mata and E. Peris, *ACS Catal.*, 2014, **4**, 2038–2047.
52. Q. Zhao, D. Chen, Y. Li, G. Zhang, F. Zhang and X. Fan, *Nanoscale*, 2013, **5**, 882–885.
53. J. K. Lee and M.-J. Kim, *Tetrahedron Lett.*, 2011, **52**, 499–501.
54. M. Gopiraman, S. G. Babu, Z. Khatri, K. Wei, M. Endo, R. Karvembu and I. S. Kim, *Catal. Sci. Technol.*, 2013, **3**, 1485–1489.
55. M. Gopiraman, S. Ganesh Babu, Z. Khatri, W. Kai, Y. A. Kim, M. Endo, R. Karvembu and I. S. Kim, *J. Phys. Chem. C*, 2013, **117**, 23582–23596.

56. Q. Zhao, Y. Li, R. Liu, A. Chen, G. Zhang, F. Zhang and X. Fan, *J. Mater. Chem. A*, 2013, **1**, 15039–15045.
57. T. Braun, M. Wohlers, T. Belz and R. Schlögl, *Catal. Lett.*, 1997, **43**, 175–180.
58. Y. Zhou, J. Liu, X. Li, X. Pan and X. Bao, *J. Nat. Gas Chem.*, 2012, **21**, 241–245.
59. C. Park and M. A. Keane, *J. Colloid Interface Sci.*, 2003, **266**, 183–194.
60. H. Wang, F. Zhao, S.-i. Fujita and M. Arai, *Catal. Commun.*, 2008, **9**, 362–368.
61. Y. Xiang, L. Kong, P. Xie, T. Xu, J. Wang and X. Li, *Ind. Eng. Chem. Res.*, 2014, **53**, 2197–2203.
62. J. Chen, W. Zhang, L. Chen, L. Ma, H. Gao and T. Wang, *ChemPlusChem*, 2013, **78**, 142–148.
63. C.-H. Li, Z.-X. Yu, K.-F. Yao, S.-f. Ji and J. Liang, *J. Mol. Catal. A: Chem.*, 2005, **226**, 101–105.
64. Z. Sun, Y. Zhao, Y. Xie, R. Tao, H. Zhang, C. Huang and Z. Liu, *Green Chem.*, 2010, **12**, 1007–1011.
65. S. Jin, W. Qian, Y. Liu, F. Wei, D. Wang and J. Zhang, *Aust. J. Chem.*, 2010, **63**, 131–134.
66. Ľ. Pikna, M. Heželová, S. Demčáková, M. Smrčová, B. Plešingerová, M. Štefanko, M. Turáková, M. Králik, P. Puliš and P. Lehocký, *Chem. Pap.*, 2014, **68**, 591–598.
67. Y. Gao, D. Ma, C. Wang, J. Guan and X. Bao, *Chem. Commun.*, 2011, **47**, 2432–2434.
68. R. Nie, J. Wang, L. Wang, Y. Qin, P. Chen and Z. Hou, *Carbon*, 2012, **50**, 586–596.
69. Y.-g Wu, M. Wen, Q.-s Wu and H. Fang, *J. Phys. Chem. C*, 2014, **118**, 6307–6313.
70. J. Li, C.-y. Liu and Y. Liu, *J. Mater. Chem.*, 2012, **22**, 8426–8430.
71. A. Nieto-Márquez, S. Gil, A. Romero, J. L. Valverde, S. Gómez-Quero and M. A. Keane, *Appl. Catal., A*, 2009, **363**, 188–198.
72. B. Li and Z. Xu, *J. Am. Chem. Soc.*, 2009, **131**, 16380–16382.
73. L. Pacosová, C. Kartusch, P. Kukula and J. A. van Bokhoven, *ChemCatChem*, 2011, **3**, 154–156.
74. I. I. Obraztsova, N. K. Eremenko and Y. N. Velyakina, *Kinet. Catal.*, 2008, **49**, 401–406.
75. N. A. Magdalinova, M. V. Klyuev, N. N. Vershinin and O. N. Efimov, *Kinet. Catal.*, 2012, **53**, 482–485.
76. D. Teschner, J. Borsodi, A. Wootsch, Z. Révay, M. Hävecker, A. Knop-Gericke, S. D. Jackson and R. Schlögl, *Science*, 2008, **320**, 86–89.
77. M. Ruta, N. Semagina and L. Kiwi-Minsker, *J. Phys. Chem. C*, 2008, **112**, 13635–13641.
78. M. Ruta, G. Laurenczy, P. J. Dyson and L. Kiwi-Minsker, *J. Phys. Chem. C*, 2008, **112**, 17814–17819.

79. D. I. Kochubey, V. V. Chesnokov and S. E. Malykhin, *Carbon*, 2012, **50**, 2782–2787.
80. H. Bazzazzadegan, M. Kazemeini and A. M. Rashidi, *Appl. Catal., A*, 2011, **399**, 184–190.
81. H. Bazzazzadegan, M. Kazemeini and A. M. Rashidi, *Res. Chem. Intermed.*, 2013, 1–12.
82. L. Shao, W. Zhang, M. Armbrüster, D. Teschner, F. Girgsdies, B. Zhang, O. Timpe, M. Friedrich, R. Schlögl and D. S. Su, *Angew. Chem., Int. Ed.*, 2011, **50**, 10231–10235.
83. C.-Y. Hu, X.-N. Liao, F.-Y. Li, R.-B. Zhang, R.-F. Zhang and T.-Z. Liu, *J. Chin. Chem. Soc.*, 2007, **54**, 1471–1476.
84. O. S. Alekseev, L. V. Nosova and Y. A. Ryndin, in *Studies in Surface Science and Catalysis*, eds. F. S. L. Guczi and T. P., Elsevier, 1993, vol. 75, pp. 837–847.
85. E. A. Tveritinova, I. I. Kulakova, Y. N. Zhitnev, A. N. Kharlanov, A. V. Fionov, W. Chen, I. Buyanova and V. V. Lunin, *Russ. J. Phys. Chem.*, 2013, **87**, 1114–1120.
86. J. M. Planeix, N. Coustel, B. Coq, V. Brotons, P. S. Kumbhar, R. Dutartre, P. Geneste, P. Bernier and P. M. Ajayan, *J. Am. Chem. Soc.*, 1994, **116**, 7935–7936.
87. P. Gallezot and D. Richard, *Catal. Rev.: Sci. Eng.*, 1998, **40**, 81–126.
88. R. Giordano, P. Serp, P. Kalck, Y. Kihn, J. Schreiber, C. Marhic and J.-L. Duvail, *Eur. J. Inorg. Chem.*, 2003, 610–617.
89. I. Janowska, G. Winé, M.-J. Ledoux and C. Pham-Huu, *J. Mol. Catal. A: Chem.*, 2007, **267**, 92–97.
90. J.-P. Tessonnier, L. Pesant, G. Ehret, M. J. Ledoux and C. Pham-Huu, *Appl. Catal., A*, 2005, **288**, 203–210.
91. J. Amadou, K. Chizari, M. Houllé, I. Janowska, O. Ersen, D. Bégin and C. Pham-Huu, *Catal. Today*, 2008, **138**, 62–68.
92. K. Chizari, I. Janowska, M. Houllé, I. Florea, O. Ersen, T. Romero, P. Bernhardt, M. J. Ledoux and C. Pham-Huu, *Appl. Catal., A*, 2010, **380**, 72–80.
93. Z.-T. Liu, C.-X. Wang, Z.-W. Liu and J. Lu, *Appl. Catal., A*, 2008, **344**, 114–123.
94. H. Ma, L. Wang, L. Chen, C. Dong, W. Yu, T. Huang and Y. Qian, *Catal. Commun.*, 2007, **8**, 452–456.
95. Y. Li, R.-X. Zhou and G.-H. Lai, *React. Kinet. Catal. Lett.*, 2006, **88**, 105–110.
96. Y. Li, P.-F. Zhu and R.-X. Zhou, *Appl. Surf. Sci.*, 2008, **254**, 2609–2614.
97. J. Qiu, H. Zhang, X. Wang, H. Han, C. Liang and C. Li, *React. Kinet. Catal. Lett.*, 2006, **88**, 269–276.
98. Z. Guo, C. Zhou, D. Shi, Y. Wang, X. Jia, J. Chang, A. Borgna, C. Wang and Y. Yang, *Appl. Catal., A*, 2012, **435–436**, 131–140.
99. H. Vu, F. Gonçalves, R. Philippe, E. Lamouroux, M. Corrias, Y. Kihn, D. Plee, P. Kalck and P. Serp, *J. Catal.*, 2006, **240**, 18–22.
100. T. García, R. Murillo, D. Cazorla-Amorós, A. M. Mastral and A. Linares-Solano, *Carbon*, 2004, **42**, 1683–1689.

101. D. M. Nevskaia, E. Castillejos-Lopez, A. Guerrero-Ruiz and V. Muñoz, *Carbon*, 2004, **42**, 653–665.
102. H. C. Choi, M. Shim, S. Bangsaruntip and H. Dai, *J. Am. Chem. Soc.*, 2002, **124**, 9058–9059.
103. J. Teddy, A. Falqui, A. Corrias, D. Carta, P. Lecante, I. Gerber and P. Serp, *J. Catal.*, 2011, **278**, 59–70.
104. M. L. Toebes, F. F. Prinsloo, J. H. Bitter, A. J. van Dillen and K. P. de Jong, *J. Catal.*, 2003, **214**, 78–87.
105. M. L. Toebes, Y. Zhang, J. Hájek, T. Alexander Nijhuis, J. H. Bitter, A. Jos van Dillen, D. Y. Murzin, D. C. Koningsberger and K. P. de Jong, *J. Catal.*, 2004, **226**, 215–225.
106. A. Plomp, T. Schubert, U. Storr, K. de Jong and J. Bitter, *Top. Catal.*, 2009, **52**, 424–430.
107. S. Gryglewicz, A. Śliwak, J. Ćwikła and G. Gryglewicz, *Catal. Lett.*, 2014, **144**, 62–69.
108. J. Zhu, Y. Jia, M. Li, M. Lu and J. Zhu, *Ind. Eng. Chem. Res.*, 2012, **52**, 1224–1233.
109. B. Coq, V. Brotons, J. M. Planeix, L. C. de Ménorval and R. Dutartre, *J. Catal.*, 1998, **176**, 358–364.
110. F. Coloma, J. Narciso-Romero, A. Sepúlveda-Escribano and F. Rodríguez-Reinoso, *Carbon*, 1998, **36**, 1011–1019.
111. F. Coloma, A. Sepúlveda-Escribano, J. L. G. Fierro and F. Rodríguez-Reinoso, *Appl. Catal., A*, 1996, **136**, 231–248.
112. F. Coloma, A. Sepúlveda-Escribano, J. L. G. Fierro and F. Rodríguez-Reinoso, *Appl. Catal., A*, 1996, **148**, 63–80.
113. I. B. Bychko, Y. Y. Kalishyn and P. E. Strizhak, *Theor. Exp. Chem.*, 2012, **48**, 194–198.
114. F. Salman, C. Park and R. T. K. Baker, *Catal. Today*, 1999, **53**, 385–394.
115. E. Bailón-García, F. Maldonado-Hódar, A. Pérez-Cadenas and F. Carrasco-Marín, *Catalysts*, 2013, **3**, 853–877.
116. Y. Tang, D. Yang, F. Qin, J. Hu, C. Wang and H. Xu, *J. Solid State Chem.*, 2009, **182**, 2279–2284.
117. G. Guo, F. Qin, D. Yang, C. Wang, H. Xu and S. Yang, *Chem. Mater.*, 2008, **20**, 2291–2297.
118. F. Qin, W. Shen, C. Wang and H. Xu, *Catal. Commun.*, 2008, **9**, 2095–2098.
119. P. D. Zgolicz, J. P. Stassi, M. J. Yañez, O. A. Scelza and S. R. de Miguel, *J. Catal.*, 2012, **290**, 37–54.
120. J. P. Stassi, P. D. Zgolicz, S. R. de Miguel and O. A. Scelza, *J. Catal.*, 2013, **306**, 11–29.
121. J. Khanderi, R. C. Hoffmann, J. Engstler, J. J. Schneider, J. Arras, P. Claus and G. Cherkashinin, *Chem. – Eur. J.*, 2010, **16**, 2300–2308.
122. H.-U. Blaser, H. Steiner and M. Studer, *ChemCatChem*, 2009, **1**, 210–221.
123. M. Pietrowski, *Curr. Org. Synth.*, 2012, **9**, 470–487.
124. N. Mahata, A. F. Cunha, J. J. M. Órfão and J. L. Figueiredo, *Catal. Commun.*, 2009, **10**, 1203–1206.
125. C. Wang, J. Qiu, C. Liang, L. Xing and X. Yang, *Catal. Commun.*, 2008, **9**, 1749–1753.

126. Y. Motoyama, M. Taguchi, N. Desmira, S.-H. Yoon, I. Mochida and H. Nagashima, *Chem. – Asian J.*, 2014, **9**, 71–74.
127. M. Takasaki, Y. Motoyama, K. Higashi, S. H. Yoon, I. Mochida and H. Nagashima, *Org. Lett.*, 2008, **10**, 1601–1604.
128. Y. Motoyama, Y. Lee, K. Tsuji, S.-H. Yoon, I. Mochida and H. Nagashima, *ChemCatChem*, 2011, **3**, 1578–1581.
129. C. Antonetti, M. Oubenali, A. M. Raspolli Galletti, P. Serp and G. Vannucci, *Appl. Catal., A*, 2012, **421–422**, 99–107.
130. M. Oubenali, G. Vanucci, B. Machado, M. Kacimi, M. Ziyad, J. Faria, A. Raspolli-Galetti and P. Serp, *ChemSusChem*, 2011, **4**, 950–956.
131. L. Jiang, H. Gu, X. Xu and X. Yan, *J. Mol. Catal. A: Chem.*, 2009, **310**, 144–149.
132. X. X. Han, Q. Chen and R. X. Zhou, *J. Mol. Catal. A: Chem.*, 2007, **277**, 210–214.
133. X. X. Han, J. R. Li and R. X. Zhou, *Chin. Chem. Lett.*, 2009, **20**, 96–98.
134. X. Yan, J. Sun, Y. Fang, Z. Xu and W. Wang, *Front. Chem. China.*, 2006, **1**, 41–44.
135. G. Fan, W. Huang and C. Wang, *Nanoscale*, 2013, **5**, 6819–6825.
136. G.-Y. Fan and W.-J. Huang, *Chin. Chem. Lett.*, 2014, **25**, 359–363.
137. Y.-M. Lu, H.-Z. Zhu, W.-G. Li, B. Hu and S.-H. Yu, *J. Mater. Chem. A*, 2013, **1**, 3783–3788.
138. L. Xing, J.-H. Xie, Y.-S. Chen, L.-X. Wang and Q.-L. Zhou, *Adv. Synth. Catal.*, 2008, **350**, 1013–1016.
139. L. J. Lemus-Yegres, M. C. Román-Martínez and C. S.-M. de Lecea, *J. Nanosci. Nanotechnol.*, 2009, **9**, 6034–6041.
140. C. C. Gheorghiu, B. F. Machado, C. Salinas-Martinez de Lecea, M. Gouygou, M. C. Roman-Martinez and P. Serp, *Dalton Trans.*, 2014, **43**, 7455–7463.
141. L. Xing, F. Du, J.-J. Liang, Y.-S. Chen and Q.-L. Zhou, *J. Mol. Catal. A: Chem.*, 2007, **276**, 191–196.
142. G. Szőllősi, Z. Németh, K. Hernádi and M. Bartók, *Catal. Lett.*, 2009, **132**, 370–376.
143. Z. Chen, Z. Guan, M. Li, Q. Yang and C. Li, *Angew. Chem., Int. Ed.*, 2011, **50**, 4913–4917.
144. K. Szőri, R. Puskás, G. Szőllősi, I. Bertóti, J. Szépvölgyi and M. Bartók, *Catal. Lett.*, 2013, **143**, 539–546.
145. J. Wang, H. B. Chen, H. He, Y. Cal, J. D. Lin, J. Yi, H. B. Zhang and D. W. Liao, *Chin. Chem. Lett.*, 2002, **13**, 1217–1220.
146. X. Dong, H.-B. Zhang, G.-D. Lin, Y.-Z. Yuan and K. R. Tsai, *Catal. Lett.*, 2003, **85**, 237–246.
147. X. Dong, B. Shen, H. Zhang, G. Lin and Y. Yuan, *J. Nat. Gas Chem.*, 2003, 49–55.
148. X. Pan, Z. Fan, W. Chen, Y. Ding, H. Luo and X. Bao, *Nat. Mater.*, 2007, **6**, 507–511.
149. V. R. Surisetty, J. Kozinski and A. K. Dalai, *J. Catal.*, 2013, 942145.
150. V. R. Surisetty, A. K. Dalai and J. Kozinski, *Appl. Catal., A*, 2011, **393**, 50–58.

151. V. R. Surisetty, A. K. Dalai and J. Kozinski, *Ind. Eng. Chem. Res.*, 2010, **49**, 6956–6963.
152. X.-M. Wu, Y.-Y. Guo, J.-M. Zhou, G.-D. Lin, X. Dong and H.-B. Zhang, *Appl. Catal., A*, 2008, **340**, 87–97.
153. X. Ma, G. Lin and H. Zhang, *Chin. J. Catal.*, 2006, **27**, 1019–1027.
154. X.-M. Ma, G.-D. Lin and H.-B. Zhang, *Catal. Lett.*, 2006, **111**, 141–151.
155. V. R. Surisetty, A. Tavasoli and A. K. Dalai, *Appl. Catal., A*, 2009, **365**, 243–251.
156. H.-B. Zhang, X. Dong, G.-D. Lin, X.-L. Liang and H.-Y. Li, *Chem. Commun.*, 2005, 5094–5096.
157. X. Dong, X.-L. Liang, H.-Y. Li, G.-D. Lin, P. Zhang and H.-B. Zhang, *Catal. Today*, 2009, **147**, 158–165.
158. V. R. Surisetty, A. K. Dalai and J. Kozinski, *Appl. Catal., A*, 2010, **381**, 282–288.
159. H.-B. Zhang, X.-L. Liang, X. Dong, H.-Y. Li and G.-D. Lin, *Catal. Surv. Asia*, 2009, **13**, 41–58.
160. Q. Zhang, J. Kang and Y. Wang, *ChemCatChem*, 2010, **2**, 1030–1058.
161. G. Liu, E. D. Larson, R. H. Williams, T. G. Kreutz and X. Guo, *Energy Fuels*, 2010, **25**, 415–437.
162. H. Jahangiri, J. Bennett, P. Mahjoubi, K. Wilson and S. Gu, *Catal. Sci. Technol.*, 2014, **4**, 2210–2229.
163. E. van Steen and F. F. Prinsloo, *Catal. Today*, 2002, **71**, 327–334.
164. M. C. Bahome, L. L. Jewell, D. Hildebrandt, D. Glasser and N. J. Coville, *Appl. Catal., A*, 2005, **287**, 60–67.
165. R. M. Malek Abbaslou, A. Tavasoli and A. K. Dalai, *Appl. Catal., A*, 2009, **355**, 33–41.
166. R. M. M. Abbaslou, J. Soltan and A. K. Dalai, *Appl. Catal., A*, 2010, **379**, 129–134.
167. W. Chen, X. Pan and X. Bao, *J. Am. Chem. Soc.*, 2007, **129**, 7421–7426.
168. W. Chen, Z. Fan, X. Pan and X. Bao, *J. Am. Chem. Soc.*, 2008, **130**, 9414–9419.
169. R. M. M. Abbaslou, A. Tavassoli, J. Soltan and A. K. Dalai, *Appl. Catal., A*, 2009, **367**, 47–52.
170. H. M. Torres Galvis, J. H. Bitter, T. Davidian, M. Ruitenbeek, A. I. Dugulan and K. P. de Jong, *J. Am. Chem. Soc.*, 2012, **134**, 16207–16215.
171. H. M. Torres Galvis, J. H. Bitter, C. B. Khare, M. Ruitenbeek, A. I. Dugulan and K. P. de Jong, *Science*, 2012, **335**, 835–838.
172. J. Lu, L. Yang, B. Xu, Q. Wu, D. Zhang, S. Yuan, Y. Zhai, X. Wang, Y. Fan and Z. Hu, *ACS Catal.*, 2014, **4**, 613–621.
173. H. J. Schulte, B. Graf, W. Xia and M. Muhler, *ChemCatChem*, 2012, **4**, 350–355.
174. S. O. Moussa, L. S. Panchakarla, M. Q. Ho and M. S. El-Shall, *ACS Catal.*, 2013, **4**, 535–545.
175. J. Kang, S. Zhang, Q. Zhang and Y. Wang, *Angew. Chem., Int. Ed.*, 2009, **48**, 2565–2568.

176. K. Murata, K. Okabe, M. Inaba, I. Takahara and Y. Liu, *J. Jpn. Pet. Inst.*, 2009, **52**, 16–20.
177. D. Shi, J. A. Faria, A. A. Rownaghi, R. L. Huhnke and D. E. Resasco, *Energy Fuels*, 2013, **27**, 6118–6124.
178. D. Shi, J. Faria, T. N. Pham and D. E. Resasco, *ACS Catal.*, 2014, **4**, 1944–1952.
179. X.-Y. Quek, R. Pestman, R. A. van Santen and E. J. M. Hensen, *ChemCatChem*, 2013, **5**, 3148–3155.
180. M. C. Bahome, L. L. Jewell, K. Padayachy, D. Hildebrandt, D. Glasser, A. K. Datye and N. J. Coville, *Appl. Catal., A*, 2007, **328**, 243–251.
181. H. Zhang, W. Chu, C. Zou, Z. Huang, Z. Ye and L. Zhu, *Catal. Lett.*, 2011, **141**, 438–444.
182. A. Tavasoli and S. Taghavi, *J. Energy Chem.*, 2013, **22**, 747–754.
183. A. Tavasoli, S. Taghavi, S. Tabyar and S. Karimi, *Int. J. Ind. Chem.*, 2014, **5**, 1–11.
184. L. Guczi, G. Stefler, O. Geszti, Z. Koppány, Z. Kónya, É. Molnár, M. Urbán and I. Kiricsi, *J. Catal.*, 2006, **244**, 24–32.
185. H. Zhang, C. Lancelot, W. Chu, J. Hong, A. Y. Khodakov, P. A. Chernavskii, J. Zheng and D. Tong, *J. Mater. Chem.*, 2009, **19**, 9241–9249.
186. M. Trépanier, A. Tavasoli, A. K. Dalai and N. Abatzoglou, *Fuel Process. Technol.*, 2009, **90**, 367–374.
187. A. Karimi, B. Nasernejad, A. M. Rashidi, A. Tavasoli and M. Pourkhalil, *Fuel*, 2014, **117**, 1045–1051.
188. M. Trépanier, A. Tavasoli, A. K. Dalai and N. Abatzoglou, *Appl. Catal., A*, 2009, **353**, 193–202.
189. A. Yakubov, M. G. Kutty, S. Pei Lee, M. Sh Shaharun, S. B. Abd Hamid and P. V, *Adv. Mater. Res.*, 2012, **364**, 70–75.
190. A. Rose, J. Thiessen, A. Jess and D. Curulla-Ferré, *Chem. Eng. Technol.*, 2014, **37**, 683–691.
191. A. Tavasoli, K. Sadagiani, F. Khorashe, A. A. Seifkordi, A. A. Rohani and A. Nakhaeipour, *Fuel Process. Technol.*, 2008, **89**, 491–498.
192. W. Xie, Y. Zhang, K. Liew and J. Li, *Sci. China Chem.*, 2012, **55**, 1811–1818.
193. Y. Zhu, Y. Ye, S. Zhang, M. E. Leong and F. Tao, *Langmuir*, 2012, **28**, 8275–8280.
194. V. R. Surisetty, E. Epelde, M. Trépanier, J. Kozinski and A. K. Dalai, *Int. J. Chem. React. Eng.*, 2012, **10**, 1542–6580.
195. M. Trépanier, A. Tavasoli, S. Anahid and K. D. A, *Iran. J. Chem. Chem. Eng.*, 2011, **30**, 37–47.
196. H. Xiong, M. A. M. Motchelaho, M. Moyo, L. L. Jewell and N. J. Coville, *Catal. Today*, 2013, **214**, 50–60.
197. Y.-H. Chin, J. Hu, C. Cao, Y. Gao and Y. Wang, *Catal. Today*, 2005, **110**, 47–52.
198. M. R. Hemmati, M. Kazemeini, F. Khorasheh, J. Zarkesh and A. Rashidi, *J. Nat. Gas Chem.*, 2012, **21**, 713–721.
199. J. A. Díaz, M. Martínez-Fernández, A. Romero and J. L. Valverde, *Fuel*, 2013, **111**, 422–429.

200. G. L. Bezemer, J. H. Bitter, H. P. C. E. Kuipers, H. Oosterbeek, J. E. Holewijn, X. Xu, F. Kapteijn, A. J. van Dillen and K. P. de Jong, *J. Am. Chem. Soc.*, 2006, **128**, 3956–3964.
201. G. L. Bezemer, U. Falke, A. J. van Dillen and K. P. de Jong, *Chem. Commun.*, 2005, 731–733.
202. T.-o Honsho, T. Kitano, T. Miyake and T. Suzuki, *Fuel*, 2012, **94**, 170–177.
203. F. Huber, Z. Yu, J. C. Walmsley, D. Chen, H. J. Venvik and A. Holmen, *Appl. Catal., B*, 2007, **71**, 7–15.
204. B. Zugic, S. Zhang, D. C. Bell, F. F. Tao and M. Flytzani-Stephanopoulos, *J. Am. Chem. Soc.*, 2014, **136**, 3238–3245.
205. B. Zugic, D. C. Bell and M. Flytzani-Stephanopoulos, *Appl. Catal., B*, 2014, **144**, 243–251.
206. O. Lazarus, T. W. Woolerton, A. Parkin, M. J. Lukey, E. Reisner, J. Seravalli, E. Pierce, S. W. Ragsdale, F. Sargent and F. A. Armstrong, *J. Am. Chem. Soc.*, 2009, **131**, 14154–14155.
207. R. Gao, C. D. Tan and R. T. K. Baker, *Catal. Today*, 2001, **65**, 19–29.
208. Y. Zhang, F. Wu, H. Zhang, G. Lin, Y. Yllan and T. K, *Acta Phys.-Chim. Sin.*, 1997, **12**, 000.
209. H. B. Zhang, Y. Zhang, G. D. Lin, Y. Z. Yuan and K. R. Tsai, in *Studies in Surface Science and Catalysis*, eds. F. V. M. S. M. Avelino Corma and G. F. José Luis, Elsevier, 2000, vol. 130, pp. 3885–3890.
210. Y. Zhang, H.-B. Zhang, G.-D. Lin, P. Chen, Y.-Z. Yuan and K. R. Tsai, *Appl. Catal., A*, 1999, **187**, 213–224.
211. J. Qiu, H. Zhang, C. Liang, J. Li and Z. Zhao, *Chem. – Eur. J.*, 2006, **12**, 2147–2151.
212. H. Zhang, J. Qiu, C. Liang, Z. Li, X. Wang, Y. Wang, Z. Feng and C. Li, *Catal. Lett.*, 2005, **101**, 211–214.
213. J. B. Claridge, R. E. Douthwaite, M. L. H. Green, R. M. Lago, S. C. Tsang and A. P. E. York, *J. Mol. Catal.*, 1994, **89**, 113–120.
214. N. Saadatjou, A. Jafari and S. Sahebdelfar, *Chem. Eng. Commun.*, 2014, **202**, 420–448.
215. H.-B. Chen, J.-D. Lin, Y. Cai, X.-Y. Wang, J. Yi, J. Wang, G. Wei, Y.-Z. Lin and D.-W. Liao, *Appl. Surf. Sci.*, 2001, **180**, 328–335.
216. Q.-C. Xu, J.-D. Lin, J. Li, X.-Z. Fu, Y. Liang and D.-W. Liao, *Catal. Commun.*, 2007, **8**, 1881–1885.
217. S. Guo, X. Pan, H. Gao, Z. Yang, J. Zhao and X. Bao, *Chem. – Eur. J.*, 2010, **16**, 5379–5384.
218. X. Yu, B. Lin, B. Gong, J. Lin, R. Wang and K. Wei, *Catal. Lett.*, 2008, **124**, 168–173.
219. Z. Li, C. Liang, Z. Feng, P. Ying, D. Wang and C. Li, *J. Mol. Catal. A: Chem.*, 2004, **211**, 103–109.
220. S.-F. Yin, Q.-H. Zhang, B.-Q. Xu, W.-X. Zhu, C.-F. Ng and C.-T. Au, *J. Catal.*, 2004, **224**, 384–396.
221. S.-F. Yin, B.-Q. Xu, C.-F. Ng and C.-T. Au, *Appl. Catal., B*, 2004, **48**, 237–241.
222. S. F. Yin, B. Q. Xu, S. J. Wang, C. F. Ng and C. T. Au, *Catal. Lett.*, 2004, **96**, 113–116.

223. W. Zheng, J. Zhang, B. Zhu, R. Blume, Y. Zhang, K. Schlichte, R. Schlögl, F. Schüth and D. S. Su, *ChemSusChem*, 2010, **3**, 226–230.
224. L. Li, Z. H. Zhu, Z. F. Yan, G. Q. Lu and L. Rintoul, *Appl. Catal., A*, 2007, **320**, 166–172.
225. S. F. Yin, B. Q. Xu, W. X. Zhu, C. F. Ng, X. P. Zhou and C. T. Au, *Catal. Today*, 2004, **93–95**, 27–38.
226. F. R. García-García, J. Álvarez-Rodríguez, I. Rodríguez-Ramos and A. Guerrero-Ruiz, *Carbon*, 2010, **48**, 267–276.
227. J. Chen, Z. H. Zhu, S. Wang, Q. Ma, V. Rudolph and G. Q. Lu, *Chem. Eng. J.*, 2010, **156**, 404–410.
228. J. Zhang, J.-O. Müller, W. Zheng, D. Wang, D. Su and R. Schlögl, *Nano Lett.*, 2008, **8**, 2738–2743.
229. J. Zhang, M. Comotti, F. Schuth, R. Schlogl and D. S. Su, *Chem. Commun.*, 2007, 1916–1918.
230. X. Duan, G. Qian, X. Zhou, Z. Sui, D. Chen and W. Yuan, *Appl. Catal., B*, 2011, **101**, 189–196.
231. J. Ji, X. Duan, G. Qian, X. Zhou, D. Chen and W. Yuan, *Ind. Eng. Chem. Res.*, 2013, **52**, 1854–1858.
232. Y. Marco, L. Roldán, S. Armenise and E. García-Bordejé, *ChemCatChem*, 2013, **5**, 3829–3834.
233. X. Duan, J. Zhou, G. Qian, P. Li, X. Zhou and D. Chen, *Chinese J. Catal.*, 2010, **31**, 979–986.
234. R. Vieira, C. Pham-Huu, N. Keller and M. J. Ledoux, *Chem. Commun.*, 2002, 954–955.
235. R. Vieira, D. Bastos-Netto, M.-J. Ledoux and C. Pham-Huu, *Appl. Catal., A*, 2005, **279**, 35–40.
236. R. Vieira, P. Bernhardt, M.-J. Ledoux and C. Pham-Huu, *Catal. Lett.*, 2005, **99**, 177–180.
237. C. Liang, L. Ding, A. Wang, Z. Ma, J. Qiu and T. Zhang, *Ind. Eng. Chem. Res.*, 2009, **48**, 3244–3248.
238. J. Wang, X.-B. Zhang, Z.-L. Wang, L.-M. Wang and Y. Zhang, *Energy Environ. Sci.*, 2012, **5**, 6885–6888.
239. D. G. Tong, W. Chu, P. Wu, G. F. Gu and L. Zhang, *J. Mater. Chem. A*, 2013, **1**, 358–366.
240. S. Akbayrak and S. Özkar, *ACS Appl. Mater. Interfaces*, 2012, **4**, 6302–6310.
241. Ö. Metin, E. Kayhan, S. Özkar and J. J. Schneider, *Int. J. Hydrogen Energy*, 2012, **37**, 8161–8169.
242. N. Cao, W. Luo and G. Cheng, *Int. J. Hydrogen Energy*, 2013, **38**, 11964–11972.
243. J. Wang, Y.-L. Qin, X. Liu and X.-B. Zhang, *J. Mater. Chem.*, 2012, **22**, 12468–12470.
244. X. Zhou, Z. Chen, D. Yan and H. Lu, *J. Mater. Chem.*, 2012, **22**, 13506–13516.
245. L. Yang, W. Luo and G. Cheng, *ACS Appl. Mater. Interfaces*, 2013, **5**, 8231–8240.
246. L. Yang, J. Su, X. Meng, W. Luo and G. Cheng, *J. Mater. Chem. A*, 2013, **1**, 10016–10023.

247. G. M. K. Abotsi and A. W. Scaroni, *Fuel Process. Technol.*, 1989, **22**, 107–133.
248. W. Yin, M. Li, H. Shang and F. W, *J. Nat. Gas Chem.*, 2005, **14**, 163–167.
249. K. Dong, X. Wu, G. Lin and Z. H, *Chin. J. Catal.*, 2005, **26**, 550–556.
250. H.-Y. Shang, C.-G. Liu, R.-Y. Zhao, M.-B. Wu and F. Wei, *Chin. J. Chem.*, 2004, **22**, 1250–1256.
251. H. Shang, C. Liu, Y. Xu, J. Qiu and F. Wei, *Fuel Process. Technol.*, 2007, **88**, 117–123.
252. J. Zhang, W. Yin and S. H, *J. Nat. Gas Chem.*, 2008, **17**, 165–170.
253. E. Soghrati, M. Kazemeini, A. M. Rashidi and K. J. Jozani, *J. Taiwan Inst. Chem. Eng.*, 2014, **45**, 887–895.
254. M. Kiani, H. Aghabozorg, K. J. Jozani, A. Rashidi and M. Mohsennia, *Phosphorus, Sulfur Silicon Relat. Elem.*, 2012, **188**, 1254–1261.
255. Z. Yu, L. E. Fareid, K. Moljord, E. A. Blekkan, J. C. Walmsley and D. Chen, *Appl. Catal., B*, 2008, **84**, 482–489.
256. Z. R. Ismagilov, A. E. Shalagina, O. Y. Podyacheva, V. A. Ushakov, V. V. Kriventsov, D. I. Kochubey and S. A. N, *Int. Sci. J. Altern. Energy Ecol. I.*, 2007, **3**, 150–158.
257. P. Gheek, S. Suppan, J. Trawczyński, A. Hynaux, C. Sayag and G. Djega-Mariadssou, *Catal. Today*, 2007, **119**, 19–22.
258. A. Hynaux, C. Sayag, S. Suppan, J. Trawczynski, M. Lewandowski, A. Szymanska-Kolasa and G. Djéga-Mariadassou, *Appl. Catal., B*, 2007, **72**, 62–70.
259. S. Suppan, J. Trawczyński, J. Kaczmarczyk, G. Djéga-Mariadassou, A. Hynaux and C. Sayag, *Appl. Catal., A*, 2005, **280**, 209–214.
260. J. M. Solar, F. J. Derbyshire, V. H. J. de Beer and L. R. Radovic, *J. Catal.*, 1991, **129**, 330–342.
261. W. Xu, X. Wang, Q. Zhou, B. Meng, J. Zhao, J. Qiu and Y. Gogotsi, *J. Mater. Chem.*, 2012, **22**, 14363–14368.
262. N. Liu, X. Wang, W. Xu, H. Hu, J. Liang and J. Qiu, *Fuel*, 2014, **119**, 163–169.
263. X.-z Wang, N. Liu, H. Hu, X.-p Wang and J.-S. Qiu, *Carbon*, 2014, **76**, 471.
264. I. Eswaramoorthi, V. Sundaramurthy, N. Das, A. K. Dalai and J. Adjaye, *Appl. Catal., A*, 2008, **339**, 187–195.
265. S. Sigurdson, A. K. Dalai and J. Adjaye, *Can. J. Chem. Eng.*, 2011, **89**, 562–575.
266. K. Dong, X. Ma, H. Zhang and G. Lin, *J. Nat. Gas Chem.*, 2006, **15**, 28–37.
267. J. Han, J. Duan, P. Chen, H. Lou, X. Zheng and H. Hong, *Green Chem.*, 2011, **13**, 2561–2568.
268. C. Sayag, M. Benkhaleda, S. Suppanb, J. Trawczynskib and G. Djéga-Mariadassoua, *Appl. Catal., A*, 2004, **275**, 15–24.
269. C. Sayag, S. Suppan, J. Trawczyński and G. Djéga-Mariadassou, *Fuel Process. Technol.*, 2002, **77–78**, 261–267.
270. C. K. Groot, V. H. J. De Beer, R. Prins, M. Stolarski and W. S. Niedzwiedz, *Ind. Eng. Chem. Prod. Res. Dev.*, 1986, **25**, 522–530.
271. H. Ohta, H. Kobayashi, K. Hara and A. Fukuoka, *Chem. Commun.*, 2011, **47**, 12209–12211.

272. J. L. Pinilla, H. Purón, D. Torres, S. de Llobet, R. Moliner, I. Suelves and M. Millan, *Appl. Catal., B*, 2014, **148–149**, 357–365.
273. R. W. Gosselink, D. R. Stellwagen and J. H. Bitter, *Angew. Chem., Int. Ed.*, 2013, **52**, 5089–5092.
274. A. L. Jongerius, R. W. Gosselink, J. Dijkstra, J. H. Bitter, P. C. A. Bruijnincx and B. M. Weckhuysen, *ChemCatChem*, 2013, **5**, 2964–2972.
275. Y. Qin, P. Chen, J. Duan, J. Han, H. Lou, X. Zheng and H. Hong, *RSC Adv.*, 2013, **3**, 17485–17491.
276. C. Newman, X. Zhou, B. Goundie, I. T. Ghampson, R. A. Pollock, Z. Ross, M. C. Wheeler, R. W. Meulenberg, R. N. Austin and B. G. Frederick, *Appl. Catal., A*, 2014, **477**, 64–74.
277. M. Á. González-Borja and D. E. Resasco, *Energy Fuels*, 2011, **25**, 4155–4162.
278. R. W. Gosselink, W. Xia, M. Muhler, K. P. de Jong and J. H. Bitter, *ACS Catal.*, 2013, **3**, 2397–2402.
279. E. Aryee, A. Dalai and J. Adjaye, *Top. Catal.*, 2014, **57**, 796–805.
280. T. Mallat and A. Baiker, *Chem. Rev.*, 2004, **104**, 3037–3058.
281. A. Villa, D. Wang, N. Dimitratos, D. Su, V. Trevisan and L. Prati, *Catal. Today*, 2010, **150**, 8–15.
282. A. Villa, D. Wang, P. Spontoni, R. Arrigo, D. Su and L. Prati, *Catal. Today*, 2010, **157**, 89–93.
283. A. Corma, H. Garcia and A. Leyva, *J. Mol. Catal. A: Chem.*, 2005, **230**, 97–105.
284. R. Kumar, E. Gravel, A. Hagege, H. Li, D. V. Jawale, D. Verma, I. N. N. Namboothiri and E. Doris, *Nanoscale*, 2013, **5**, 6491–6497.
285. X. Yang, X. Wang and J. Qiu, *Appl. Catal., A*, 2010, **382**, 131–137.
286. C. Yu, L. Fan, J. Yang, Y. Shan and J. Qiu, *Chem. – Eur. J.*, 2013, **19**, 16192–16195.
287. T. Tang, C. Yin, N. Xiao, M. Guo and F.-S. Xiao, *Catal. Lett.*, 2009, **127**, 400–405.
288. G. Wu, X. Wang, N. Guan and L. Li, *Appl. Catal., B*, 2013, **136–137**, 177–185.
289. X. Yu, Y. Huo, J. Yang, S. Chang, Y. Ma and W. Huang, *Appl. Surf. Sci.*, 2013, **280**, 450–455.
290. R. V. Jagadeesh, H. Junge, M.-M. Pohl, J. Radnik, A. Brückner and M. Beller, *J. Am. Chem. Soc.*, 2013, **135**, 10776–10782.
291. T. Yasu-eda, R. Se-ike, N.-o. Ikenaga, T. Miyake and T. Suzuki, *J. Mol. Catal. A: Chem.*, 2009, **306**, 136–142.
292. H. T. Gomes, P. V. Samant, P. Serp, P. Kalck, J. L. Figueiredo and J. L. Faria, *Appl. Catal., B*, 2004, **54**, 175–182.
293. J. Garcia, H. T. Gomes, P. Serp, P. Kalck, J. L. Figueiredo and J. L. Faria, *Catal. Today*, 2005, **102–103**, 101–109.
294. G. Ovejero, J. L. Sotelo, M. D. Romero, A. Rodríguez, M. A. Ocaña, G. Rodríguez and J. García, *Ind. Eng. Chem. Res.*, 2006, **45**, 2206–2212.
295. G. Ovejero, J. L. Sotelo, A. Rodríguez, C. Díaz, R. Sanz and J. García, *Ind. Eng. Chem. Res.*, 2007, **46**, 6449–6455.

296. G. Ovejero, A. Rodríguez, A. Vallet and J. García, *Color. Technol.*, 2011, **127**, 10–17.
297. J. Garcia, H. T. Gomes, P. Serp, P. Kalck, J. L. Figueiredo and J. L. Faria, *Carbon*, 2006, **44**, 2384–2391.
298. A. B. Ayusheev, O. P. Taran, I. A. Seryak, O. Y. Podyacheva, C. Descorme, M. Besson, L. S. Kibis, A. I. Boronin, A. I. Romanenko, Z. R. Ismagilov and V. Parmon, *Appl. Catal., B*, 2014, **146**, 177–185.
299. C. D. Taboada, J. Batista, A. Pintar and J. Levec, *Appl. Catal., B*, 2009, **89**, 375–382.
300. A. Rodríguez, G. Ovejero, M. D. Romero, C. Díaz, M. Barreiro and J. García, *J. Supercrit. Fluids*, 2008, **46**, 163–172.
301. M. Soria-Sánchez, A. Maroto-Valiente, J. Álvarez-Rodríguez, I. Rodríguez-Ramos and A. Guerrero-Ruíz, *Carbon*, 2009, **47**, 2095–2102.
302. A. Cybulski and J. Trawczyński, *Appl. Catal., B*, 2004, **47**, 1–13.
303. M. Gopiraman, H. Bang, S. G. Babu, K. Wei, R. Karvembu and I. S. Kim, *Catal. Sci. Technol.*, 2014, **4**, 2099–2106.
304. W. Qin, L. Wei, L. Wang, C. Dong, X. Xiao, Z. Zheng and Y. Yang, *Chem. Phys. Lett.*, 2013, **572**, 53–57.
305. W. Qin and X. Li, *J. Phys. Chem. C*, 2010, **114**, 19009–19015.
306. Yin and J. Liebscher, *Chem. Rev.*, 2006, **107**, 133–173.
307. A. Fihri, M. Bouhrara, B. Nekoueishahraki, J.-M. Basset and V. Polshettiwar, *Chem. Soc. Rev.*, 2011, **40**, 5181–5203.
308. J. Guerra and M. A. Herrero, *Nanoscale*, 2010, **2**, 1390–1400.
309. M. Seki, *Synthesis*, 2006, 2975–2992.
310. S. Santra, P. Ranjan, P. Bera, P. Ghosh and S. K. Mandal, *RSC Adv.*, 2012, **2**, 7523–7533.
311. H. B. Pan, C. H. Yen, B. Yoon, M. Sato and C. M. Wai, *Synth. Commun.*, 2006, **36**, 3473–3478.
312. X. Chen, Y. Hou, H. Wang, Y. Cao and J. He, *J. Phys. Chem. C*, 2008, **112**, 8172–8176.
313. P.-P. Zhang, X.-X. Zhang, H.-X. Sun, R.-H. Liu, B. Wang and Y.-H. Lin, *Tetrahedron Lett.*, 2009, **50**, 4455–4458.
314. V. I. Sokolov, E. G. Rakov, N. A. Bumagin and M. G. Vinogradov, *Fullerenes, Nanotubes, Carbon Nanostruct.*, 2010, **18**, 558–563.
315. B. Yoon and C. M. Wai, *J. Am. Chem. Soc.*, 2005, **127**, 17174–17175.
316. M. Cano, A. Benito, W. K. Maser and E. P. Urriolabeitia, *Carbon*, 2011, **49**, 652–658.
317. G. M. Neelgund and A. Oki, *Appl. Catal., A*, 2011, **399**, 154–160.
318. S. Mahouche Chergui, A. Ledebt, F. Mammeri, F. d r. Herbst, B. Carbonnier, H. Ben Romdhane, M. Delamar and M. M. Chehimi, *Langmuir*, 2010, **26**, 16115–16121.
319. M. R. Nabid, Y. Bide and S. J. Tabatabaei Rezaei, *Appl. Catal., A*, 2011, **406**, 124–132.
320. J. A. Sullivan, K. A. Flanagan and H. Hain, *Catal. Today*, 2009, **145**, 108–113.
321. D. Saberi, M. Sheykhan, K. Niknam and A. Heydari, *Catal. Sci. Technol.*, 2013, **3**, 2025–2031.

322. J. Zhu, J. Zhou, T. Zhao, X. Zhou, D. Chen and W. Yuan, *Appl. Catal., A*, 2009, **352**, 243–250.
323. J. Zhu, T. Zhao, I. Kvande, D. Chen, X. Zhou and W. Yuan, *Chin. J. Catal.*, 2008, **29**, 1145–1151.
324. K. Yamamoto and T. Thiemanna, *J. Chem. Res.*, 2011, **35**, 246–250.
325. L. Rumi, G. M. Scheuermann, R. Mülhaupt and W. Bannwarth, *Helv. Chim. Acta.*, 2011, **94**, 966–976.
326. K. H. Lee, S.-W. Han, K.-Y. Kwon and J. B. Park, *J. Colloid Interface Sci.*, 2013, **403**, 127–133.
327. A. R. Siamaki, A. E. R. S. Khder, V. Abdelsayed, M. S. El-Shall and B. F. Gupton, *J. Catal.*, 2011, **279**, 1–11.
328. A. O. Biying, V. R. Vangala, C. S. Chen, L. P. Stubs, N. S. Hosmane and Z. Yinghuai, *Dalton Trans.*, 2014, **43**, 5014–5020.
329. S.-i Yamamoto, H. Kinoshita, H. Hashimoto and Y. Nishina, *Nanoscale*, 2014, **6**, 6501–6505.
330. G. M. Scheuermann, L. Rumi, P. Steurer, W. Bannwarth and R. Mülhaupt, *J. Am. Chem. Soc.*, 2009, **131**, 8262–8270.
331. S. Moussa, A. R. Siamaki, B. F. Gupton and M. S. El-Shall, *ACS Catal.*, 2011, **2**, 145–154.
332. G. Xiang, J. He, T. Li, J. Zhuang and X. Wang, *Nanoscale*, 2011, **3**, 3737–3742.
333. Z. Tang, S. Shen, J. Zhuang and X. Wang, *Angew. Chem., Int. Ed.*, 2010, **49**, 4603–4607.
334. Y. Li, X. Fan, J. Qi, J. Ji, S. Wang, G. Zhang and F. Zhang, *Nano Res.*, 2010, **3**, 429–437.
335. N. Li, Z. Wang, K. Zhao, Z. Shi, S. Xu and Z. Gu, *J. Nanosci. Nanotechnol.*, 2010, **10**, 6748–6751.
336. K. Qu, L. Wu, J. Ren and X. Qu, *ACS Appl. Mater. Interfaces*, 2012, **4**, 5001–5009.
337. J. Hu, Y. Wang, M. Han, Y. Zhou, X. Jiang and P. Sun, *Catal. Sci. Technol.*, 2012, **2**, 2332–2340.
338. B. Jiang, S. Song, J.-Q. Wang, Y. Xie, W. Chu, H. Li, H. Xu, C. Tian and H. Fu, *Nano Res.*, 2014, **7**, 1280–1290.
339. Y. Li, X. Fan, J. Qi, J. Ji, S. Wang, G. Zhang and F. Zhang, *Mater. Res. Bull.*, 2010, **45**, 1413–1418.
340. N. Zhang, H. Qiu, Y. Liu, W. Wang, Y. Li, X. Wang and J. Gao, *J. Mater. Chem.*, 2011, **21**, 11080–11083.
341. Ö. Metin, S. Ho, C. Alp, H. Can, M. Mankin, M. Gültekin, M. Chi and S. Sun, *Nano Res.*, 2013, **6**, 10–18.
342. A. Shaabani and M. Mahyari, *J. Mater. Chem. A*, 2013, **1**, 9303–9311.
343. H. Yoon, S. Ko and J. Jang, *Chem. Commun.*, 2007, 1468–1470.
344. Q.-Y. Meng, Q. Liu, J.-J. Zhong, H.-H. Zhang, Z.-J. Li, B. Chen, C.-H. Tung and L.-Z. Wu, *Org. Lett.*, 2012, **14**, 5992–5995.
345. P. Liu, *Eur. Polym. J.*, 2005, **41**, 2693–2703.
346. W. Jeong and M. R. Kessler, *Chem. Mater.*, 2008, **20**, 7060–7068.
347. F. J. Gomez, R. J. Chen, D. Wang, R. M. Waymouth and H. Dai, *Chem. Commun.*, 2003, 190–191.

348. L. Cui, J. Yu, X. Yu, Y. Lv, G. Li and S. Zhou, *Polym J.*, 2013, **45**, 834–838.
349. X. Pei, J. Hao and W. Liu, *J. Phys. Chem. C*, 2007, **111**, 2947–2952.
350. S. Qin, D. Qin, W. T. Ford, D. E. Resasco and J. E. Herrera, *J. Am. Chem. Soc.*, 2003, **126**, 170–176.
351. Y.-L. Liu and W.-H. Chen, *Macromolecules*, 2007, **40**, 8881–8886.
352. Y. Xu, C. Gao, H. Kong, D. Yan, Y. Z. Jin and P. C. P. Watts, *Macromolecules*, 2004, **37**, 8846–8853.
353. M. Trujillo, M. L. Arnal, A. J. Müller, E. Laredo, S. Bredeau, D. Bonduel and P. Dubois, *Macromolecules*, 2007, **40**, 6268–6276.
354. D. Bonduel, M. Mainil, M. Alexandre, F. Monteverde and P. Dubois, *Chem. Commun.*, 2005, 781–783.
355. D. Bonduel, S. Bredeau, M. Alexandre, F. Monteverde and P. Dubois, *J. Mater. Chem.*, 2007, **17**, 2359–2366.
356. J. Kim, S. M. Hong, S. Kwak and Y. Seo, *Phys. Chem. Chem. Phys.*, 2009, **11**, 10851–10859.
357. C. Costabile, F. Grisi, G. Siniscalchi, P. Longo, M. Sarno, D. Sannino, C. Leone and P. Ciambelli, *J. Nanosci. Nanotechnol.*, 2011, **11**, 10053–10062.
358. Y. Liu and A. Adronov, *Macromolecules*, 2004, **37**, 4755–4760.
359. K. Balasubramanian and M. Burghard, *Small*, 2005, **1**, 180–192.
360. L. Qu, L. M. Veca, Y. Lin, A. Kitaygorodskiy, B. Chen, A. M. McCall, J. W. Connell and Y.-P. Sun, *Macromolecules*, 2005, **38**, 10328–10331.
361. D. Priftis, N. Petzetakis, G. Sakellariou, M. Pitsikalis, D. Baskaran, J. W. Mays and N. Hadjichristidis, *Macromolecules*, 2009, **42**, 3340–3346.
362. C. Costabile, M. Sarno, F. Grisi, N. Latorraca, P. Ciambelli and P. Longo, *J. Polym. Res.*, 2014, **21**, 1–9.
363. Z. Yinghuai, S. L. P. Sia, K. Carpenter, F. Kooli and R. A. Kemp, *J. Phys. Chem. Solids*, 2006, **67**, 1218–1222.
364. J. Feng, W. Cai, J. Sui, Z. Li, J. Wan and A. N. Chakoli, *Polymer*, 2008, **49**, 4989–4994.
365. A. A. Koval'chuk, A. N. Shchegolikhin, V. G. Shevchenko, P. M. Nedorezova, A. N. Klyamkina and A. M. Aladyshev, *Macromolecules*, 2008, **41**, 3149–3156.
366. S. Park, S. W. Yoon, K.-B. Lee, D. J. Kim, Y. H. Jung, Y. Do, H.-j. Paik and I. S. Choi, *Macromol. Rapid Commun.*, 2006, **27**, 47–50.
367. A. Toti, G. Giambastiani, C. Bianchini, A. Meli, S. Bredeau, P. Dubois, D. Bonduel and M. Claes, *Chem. Mater.*, 2008, **20**, 3092–3098.
368. X. Dong, L. Wang, L. Deng, J. Li and J. Huo, *Mater. Lett.*, 2007, **61**, 3111–3115.
369. A. Funck and W. Kaminsky, *Compos. Sci. Technol.*, 2007, **67**, 906–915.
370. X. Dong, L. Wang, T. Sun, J. Zhou and Q. Yang, *J. Mol. Catal. A: Chem.*, 2006, **255**, 10–15.
371. S. Bredeau, S. Peeterbroeck, D. Bonduel, M. Alexandre and P. Dubois, *Polym. Int.*, 2008, **57**, 547–553.
372. M. A. Milani, D. González, R. Quijada, N. R. S. Basso, M. L. Cerrada, D. S. Azambuja and G. B. Galland, *Compos. Sci. Technol.*, 2013, **84**, 1–7.

373. S. V. Polschikov, P. M. Nedorezova, A. N. Klyamkina, A. A. Kovalchuk, A. M. Aladyshev, A. N. Shchegolikhin, V. G. Shevchenko and V. E. Muradyan, *J. Appl. Polym. Sci.*, 2013, **127**, 904–911.
374. F. d. C. Fim, N. R. S. Basso, A. P. Graebin, D. S. Azambuja and G. B. Galland, *J. Appl. Polym. Sci.*, 2013, **128**, 2630–2637.
375. F. d. C. Fim, J. M. Guterres, N. R. S. Basso and G. B. Galland, *J. Polym. Sci., Part A: Polym. Chem.*, 2010, **48**, 692–698.
376. W. Kaminsky and K. Wiemann, *Compos. Interfaces*, 2006, **13**, 365–375.
377. W. Kaminsky, A. Funck and C. Klinke, *Top. Catal.*, 2008, **48**, 84–90.
378. W. Kaminsky and A. Funck, *Macromol. Symp.*, 2007, **260**, 1–8.
379. M. Stürzel, F. Kempe, Y. Thomann, S. Mark, M. Enders and R. Mülhaupt, *Macromolecules*, 2012, **45**, 6878–6887.
380. L. Cui, N. H. Tarte and S. I. Woo, *Macromolecules*, 2009, **42**, 8649–8654.
381. X. Tong, C. Liu, H.-M. Cheng, H. Zhao, F. Yang and X. Zhang, *J. Appl. Polym. Sci.*, 2004, **92**, 3697–3700.
382. R. J. B. de Oliveira, J. d. S. Santos and M. d. F. V. Marques, *Polímeros*, 2014, **24**, 13–19.
383. Y. Qin, N. Wang, Y. Zhou, Y. Huang, H. Niu and J.-Y. Dong, *Macromol. Rapid Commun.*, 2011, **32**, 1052–1059.
384. B. M. Amoli, S. A. A. Ramazani and H. Izadi, *J. Appl. Polym. Sci.*, 2012, **125**, E453–E461.
385. J. Steinmetz, H.-J. Lee, S. Kwon, D.-S. Lee, C. Goze-Bac, E. Abou-Hamad, H. Kim and Y.-W. Park, *Curr. Appl. Phys.*, 2007, **7**, 39–41.
386. S. Zhao, F. Chen, C. Zhao, Y. Huang, J.-Y. Dong and C. C. Han, *Polymer*, 2013, **54**, 3680–3690.
387. M. Shafiee, A. Ramazani and S.A, *Int. J. Polym. Mater. Polym. Biomater*, 2014, **63**, 815–819.
388. Y. Huang, Y. Qin, Y. Zhou, H. Niu, Z.-Z. Yu and J.-Y. Dong, *Chem. Mater.*, 2010, **22**, 4096–4102.
389. Y. Huang, Y. Qin, N. Wang, Y. Zhou, H. Niu, J.-Y. Dong, J. Hu and Y. Wang, *Macromol. Chem. Phys.*, 2012, **213**, 720–728.
390. Y.-L. Zhao and J. F. Stoddart, *Acc. Chem. Res.*, 2009, **42**, 1161–1171.
391. H. Li, S. I. Song, G. Y. Song and I. Kim, *J. Nanosci. Nanotechnol.*, 2014, **14**, 1425–1440.
392. P. Bilalis, D. Katsigiannopoulos, A. Avgeropoulos and G. Sakellariou, *RSC Adv.*, 2014, **4**, 2911–2934.
393. A. Schaetz, M. Zeltner and W. J. Stark, *ACS Catal.*, 2012, **2**, 1267–1284.
394. F. Shehzad, S. P. Thomas and M. A. Al-Harthi, *Thermochim. Acta*, 2014, **589**, 226–234.
395. S. Park and I. S. Choi, *Adv. Mater.*, 2009, **21**, 902–905.
396. J. Kim, S. Kwak, S. M. Hong, J. R. Lee, A. Takahara and Y. Seo, *Macromolecules*, 2010, **43**, 10545–10553.
397. A. Kovalchuk, V. Shevchenko, A. Shchegolikhin, P. Nedorezova, A. Klyamkina and A. Aladyshev, *J. Mater. Sci.*, 2008, **43**, 7132–7140.
398. W. Kaminsky, A. Funck and H. Hahnsen, *Dalton Trans.*, 2009, 8803–8810.

399. W. Kaminsky, A. Funck and K. Wiemann, *Macromol. Symp.*, 2006, **239**, 1–6.
400. S. Li, H. Chen, D. Cui, J. Li, Z. Zhang, Y. Wang and T. Tang, *Polym. Compos.*, 2010, **31**, 507–515.
401. S. Park, S. W. Yoon, H. Choi, J. S. Lee, W. K. Cho, J. Kim, H. J. Park, W. S. Yun, C. H. Choi, Y. Do and I. S. Choi, *Chem. Mater.*, 2008, **20**, 4588–4594.
402. E. L. Sceats and J. C. Green, *J. Chem. Phys.*, 2006, **125**, 154704.
403. J. S. Lee and Y. S. Ko, *Catal. Today*, 2014, **232**, 82–88.
404. B. Choi, J. Lee, S. Lee, J.-H. Ko, K.-S. Lee, J. Oh, J. Han, Y.-H. Kim, I. S. Choi and S. Park, *Macromol. Rapid Commun.*, 2013, **34**, 533–538.
405. S. V. Polschikov, P. M. Nedorezova, T. V. Monakhova, A. N. Klyamkina, A. N. Shchegolikhin, V. G. Krasheninnikov, V. E. Muradyan, A. A. Popov and A. L. Margolin, *Polym. Sci. Ser. B*, 2013, **55**, 286–293.
406. S. V. Polshchikov, P. M. Nedorezova, O. M. Komkova, A. N. Klyamkina, A. N. Shchegolikhin, V. G. Krasheninnikov, A. M. Aladyshev, V. G. Shevchenko and V. E. Muradyan, *Nanotechnol. Russia*, 2014, **9**, 175–183.
407. A. Tchernook, M. Krumova, F. J. Tölle, R. Mülhaupt and S. Mecking, *Macromolecules*, 2014, **47**, 3017–3021.
408. L. Zhang, E. Yue, B. Liu, P. Serp, C. Redshaw, W.-H. Sun and J. Durand, *Catal. Commun.*, 2014, **43**, 227–230.
409. R. Sepahvand, M. Adeli, B. Astinchap and R. Kabiri, *J. Nanopart. Res.*, 2008, **10**, 1309–1318.
410. R. T. K. Baker, K. Laubernds, A. Wootsch and Z. Paál, *J. Catal.*, 2000, **193**, 165–167.
411. R. T. K. Baker, N. Rodriguez, Á. Mastalir, U. Wild, R. Schlögl, A. Wootsch and Z. Paál, *J. Phys. Chem. B*, 2004, **108**, 14348–14355.
412. B. Pietruszka, F. Di Gregorio, N. Keller and V. Keller, *Catal. Today*, 2005, **102–103**, 94–100.
413. X.-F. Guo, J.-H. Kim and G.-J. Kim, *Catal. Today*, 2011, **164**, 336–340.
414. D. S. Su, N. Maksimova, J. J. Delgado, N. Keller, G. Mestl, M. J. Ledoux and R. Schlögl, *Catal. Today*, 2005, **102–103**, 110–114.
415. Z.-J. Liu, Z. Xu, Z.-Y. Yuan, D. Lu, W. Chen and W. Zhou, *Catal. Lett.*, 2001, **72**, 203–206.
416. Z.-J. Liu, Z.-Y. Yuan, W. Zhou, L.-M. Peng and Z. Xu, *Phys. Chem. Chem. Phys.*, 2001, **3**, 2518–2521.
417. I. Kvande, D. Chen, M. Rønning, H. J. Venvik and A. Holmen, *Catal. Today*, 2005, **100**, 391–395.
418. A. Benyounes, M. Kacimi, M. Ziyad and P. Serp, *Chin. J. Catal.*, 2014, **35**, 970–978.
419. K. Niesz, A. Siska, I. Vesselényi, K. Hernadi, D. Méhn, G. Galbács, Z. Kónya and I. Kiricsi, *Catal. Today*, 2002, **76**, 3–10.
420. J. Du, C. Song, J. Zhao and Z. Zhu, *Appl. Surf. Sci.*, 2008, **255**, 2989–2993.
421. J.-p Du, C. Song, J.-l Song, J.-h Zhao and Z.-p Zhu, *J. Fuel Chem. Technol.*, 2009, **37**, 468–472.
422. J. Du, R. Zhao and G. Jiao, *Int. J. Hydrogen Energy*, 2013, **38**, 5789–5795.
423. Y. Wang, N. Shah and G. P. Huffman, *Energy Fuels*, 2004, **18**, 1429–1433.

424. S. Lee, M. Di Vece, B. Lee, S. Seifert, R. E. Winans and S. Vajda, *ChemCatChem*, 2012, **4**, 1632–1637.
425. H. Markus, A. J. Plomp, T. Sandberg, V. Nieminen, J. H. Bitter and D. Y. Murzin, *J. Mol. Catal., A: Chem.*, 2007, **274**, 42–49.
426. M. Fattahi, M. Kazemeini, F. Khorasheh and A. M. Rashidi, *Ind. Eng. Chem. Res.*, 2013, **52**, 16128–16141.
427. C. Amorim, G. Yuan, P. M. Patterson and M. A. Keane, *J. Catal.*, 2005, **234**, 268–281.
428. C. Amorim, X. Wang and M. A. Keane, *Chin. J. Catal.*, 2011, **32**, 746–755.
429. C. Amorim and M. A. Keane, *J. Chem. Technol. Biotechnol.*, 2008, **83**, 662–672.
430. Q. Liu, Z.-M. Cui, Z. Ma, S.-W. Bian and W.-G. Song, *J. Phys. Chem. C*, 2008, **112**, 1199–1203.
431. S. A. Kachevskii, E. V. Golubina, E. S. Lokteva and V. V. Lunin, *Russ. J. Phys. Chem.*, 2007, **81**, 866–873.
432. E. S. Lokteva, T. N. Rostovshchikova, S. A. Kachevskii, E. V. Golubina, V. V. Smirnov, A. Y. Stakheev, N. S. Telegina, S. A. Gurevich, V. M. Kozhevin and D. A. Yavsin, *Kinet. Catal.*, 2008, **49**, 748–755.
433. E. S. Lokteva, E. V. Golubina, S. A. Kachevskii, A. N. Kharlanov, A. V. Erokhin and V. V. Lunin, *Kinet. Catal.*, 2011, **52**, 145–155.
434. E. V. Golubina, E. S. Lokteva, S. A. Kachevsky, A. O. Turakulova and V. V. Lunin, in *Studies in Surface Science and Catalysis*, eds. M. D. S. H. P. A. J. J. A. M. E. M. Gaigneaux and P. Ruiz, Elsevier, 2010, vol. 175, pp. 293–296.
435. M. Keane and P. Patterson, *Catal. Lett.*, 2005, **99**, 33–39.
436. S. Ordóñez, E. Díaz, R. F. Bueres, E. Asedegbega-Nieto and H. Sastre, *J. Catal.*, 2010, **272**, 158–168.
437. H. Deng, G. Fan and Y. Wang, *Synth. React. Inorg., Met.-Org., Nano-Met. Chem.*, 2014, **44**, 1306–1311.
438. Y. Ren, G. Fan and C. Wang, *J. Hazard. Mater.*, 2014, **274**, 32–40.
439. F. Winter, V. Koot, A. J. van Dillen, J. W. Geus and K. P. de Jong, *J. Catal.*, 2005, **236**, 91–100.
440. F. Winter, M. Wolters, A. J. van Dillen and K. P. de Jong, *Appl. Catal., A*, 2006, **307**, 231–238.
441. T.-J. Zhao, D. Chen, Y.-C. Dai, W.-K. Yuan and A. Holmen, *Ind. Eng. Chem. Res.*, 2004, **43**, 4595–4601.
442. J. Bian, M. Xiao, S.-J. Wang, Y.-X. Lu and Y.-Z. Meng, *Appl. Surf. Sci.*, 2009, **255**, 7188–7196.
443. W. A. Solomonsz, G. A. Rance, M. Suyetin, A. La Torre, E. Bichoutskaia and A. N. Khlobystov, *Chem. – Eur. J.*, 2012, **18**, 13180–13187.
444. W. A. Solomonsz, G. A. Rance, B. J. Harris and A. N. Khlobystov, *Nanoscale*, 2013, **5**, 12200–12205.
445. C. Baleizão, B. Gigante, H. Garcia and A. Corma, *J. Catal.*, 2004, **221**, 77–84.
446. C. Baleizão, B. Gigante, H. García and A. Corma, *Tetrahedron*, 2004, **60**, 10461–10468.

447. C. Ping, F. Li, Z. Jian and J. Wei, *Propellants, Explos., Pyrotech.*, 2006, **31**, 452–455.
448. L. Zhao, Z. Wang, D. Han, D. Tao and G. Guo, *Mater. Res. Bull*, 2009, **44**, 984–988.
449. J. Sui, C. Zhang, J. Li, Z. Yu and W. Cai, *Mater. Lett.*, 2012, **75**, 158–160.
450. C. Xu, X. Wang, J. Zhu, X. Yang and L. Lu, *J. Mater. Chem.*, 2008, **18**, 5625–5629.
451. J. Zhao, Z. Liu, Y. Qin and W. Hu, *CrystEngComm*, 2014, **16**, 2001–2008.
452. N. Li, M. Cao, Q. Wu and C. Hu, *CrystEngComm*, 2012, **14**, 428–434.
453. N. Li, Z. Geng, M. Cao, L. Ren, X. Zhao, B. Liu, Y. Tian and C. Hu, *Carbon*, 2013, **54**, 124–132.
454. Y. Yuan, W. Jiang, Y. Wang, P. Shen, F. Li, P. Li, F. Zhao and H. Gao, *Appl. Surf. Sci.*, 2014, **303**, 354–359.
455. S. Chen, J. Zhu, H. Huang, G. Zeng, F. Nie and X. Wang, *J. Solid State Chem.*, 2010, **183**, 2552–2557.
456. A. M. Frey, J. H. Bitter and K. P. de Jong, *ChemCatChem*, 2011, **3**, 1193–1199.
457. M. Moghadam, I. Mohammadpoor-Baltork, S. Tangestaninejad, V. Mirkhani, H. Kargar and N. Zeini-Isfahani, *Polyhedron*, 2009, **28**, 3816–3822.
458. S. Tangestaninejad, M. Moghadam, V. Mirkhani, I. Mohammadpoor-Baltork and M. S. Saeedi, *Appl. Catal., A*, 2010, **381**, 233–241.
459. M. Zakeri, M. Moghadam, I. Mohammadpoor-Baltork, S. Tangestaninejad, V. Mirkhani and A. R. Khosropour, *J. Coord. Chem.*, 2012, **65**, 1144–1157.
460. M. Zakeri, M. Moghadam, I. Mohammadpoor-Baltork, S. Tangestaninejad, V. Mirkhani, A. Khosropour and M. Alizadeh, *Transition Met. Chem.*, 2012, **37**, 45–53.
461. M. Moghadam, S. Tangestaninejad, V. Mirkhani, I. Mohammadpoor-Baltork, A. Mirjafari and N. S. Mirbagheri, *J. Mol. Catal. A: Chem.*, 2010, **329**, 44–49.
462. M. Moghadam, S. Tangestaninejad, V. Mirkhani, I. Mohammadpoor-Baltork and N. S. Mirbagheri, *J. Organomet. Chem.*, 2010, **695**, 2014–2021.
463. M. Nooraeipour, M. Moghadam, S. Tangestaninejad, V. Mirkhani, I. Mohammadpoor-Baltork and N. Iravani, *J. Coord. Chem.*, 2011, **65**, 226–238.
464. M. Salavati-Niasari, E. Esmaeili, H. Seyghalkar and M. Bazarganipour, *Inorg. Chim. Acta*, 2011, **375**, 11–19.
465. M. Salavati-Niasari, A. Badiei and K. Saberyan, *Chem. Eng. J.*, 2011, **173**, 651–658.
466. H. Salavati, S. Tangestaninejad, M. Moghadam, V. Mirkhani and I. Mohammadpoor-Baltork, *Ultrason. Sonochem.*, 2010, **17**, 453–459.
467. L. Alves, B. Ballesteros, M. Boronat, J. R. Cabrero-Antonino, P. Concepción, A. Corma, M. A. Correa-Duarte and E. Mendoza, *J. Am. Chem. Soc.*, 2011, **133**, 10251–10261.
468. B. Li, P. He, G. Yi, H. Lin and Y. Yuan, *Catal. Lett.*, 2009, **133**, 33–40.

469. Z.-Q. Shi, Z.-P. Dong, J. Sun, F.-W. Zhang, H.-L. Yang, J.-H. Zhou, X.-H. Zhu and R. Li, *Chem. Eng. J.*, 2014, **237**, 81–87.
470. M. Salavati-Niasari and M. Bazarganipour, *J. Mol. Catal. A: Chem.*, 2007, **278**, 173–180.
471. M. Salavati-Niasari and M. Bazarganipour, *Transition Met. Chem.*, 2008, **33**, 751–757.
472. Z. Li, S. Wu, H. Ding, D. Zheng, J. Hu, X. Wang, Q. Huo, J. Guan and Q. Kan, *New J. Chem.*, 2013, **37**, 1561–1568.
473. R. Nie, J. Shi, W. Du, W. Ning, Z. Hou and F.-S. Xiao, *J. Mater. Chem., A*, 2013, **1**, 9037–9045.
474. A. Pulido, M. Boronat and A. Corma, *J. Phys. Chem., C*, 2012, **116**, 19355–19362.
475. B. Huang, R. Huang, D. Jin and D. Ye, *Catal. Today*, 2007, **126**, 279–283.
476. X. Chen, S. Gao, H. Wang, Y. Liu and Z. Wu, *Catal. Commun.*, 2011, **14**, 1–5.
477. C. Fang, D. Zhang, L. Shi, R. Gao, H. Li, L. Ye and J. Zhang, *Catal. Sci. Technol.*, 2013, **3**, 803–811.
478. L. Wang, B. Huang, Y. Su, G. Zhou, K. Wang, H. Luo and D. Ye, *Chem. Eng. J.*, 2012, **192**, 232–241.
479. D. Zhang, L. Zhang, L. Shi, C. Fang, H. Li, R. Gao, L. Huang and J. Zhang, *Nanoscale*, 2013, **5**, 1127–1136.
480. X. Fan, F. Qiu, H. Yang, W. Tian, T. Hou and X. Zhang, *Catal. Commun.*, 2011, **12**, 1298–1301.
481. K. Ariga, Q. Ji, T. Mori, M. Naito, Y. Yamauchi, H. Abe and J. P. Hill, *Chem. Soc. Rev.*, 2013, **42**, 6322–6345.
482. J. Kim, J. W. Grate and P. Wang, *Chem. Eng. Sci.*, 2006, **61**, 1017–1026.
483. T. Kuila, S. Bose, P. Khanra, A. K. Mishra, N. H. Kim and J. H. Lee, *Biosens. Bioelectron.*, 2011, **26**, 4637–4648.
484. K. Balasubramanian and M. Burghard, *Anal. Bioanal. Chem.*, 2006, **385**, 452–468.
485. I. V. Pavlidis, M. Patila, U. T. Bornscheuer, D. Gournis and H. Stamatis, *Trends Biotechnol.*, 2014, **32**, 312–320.
486. J. Filip and J. Tkac, *Electrochim. Acta*, 2014, **136**, 340–354.
487. N. Saifuddin, A. Z. Raziah and A. R. Junizah, *J. Chem.*, 2013, **2013**, 18.
488. M. Calvaresi and F. Zerbetto, *Acc. Chem. Res.*, 2013, **46**, 2454–2463.
489. W. Feng and P. Ji, *Biotechnol. Adv.*, 2011, **29**, 889–895.
490. C.-R. Jason Teng and P. Giorgia, *Nanotechnology*, 2009, **20**, 255102.
491. R. J. Chen, Y. Zhang, D. Wang and H. Dai, *J. Am. Chem. Soc.*, 2001, **123**, 3838–3839.
492. R. J. Chen, S. Bangsaruntip, K. A. Drouvalakis, N. W. Kam, M. Shim, Y. Li, W. Kim, P. J. Utz and H. Dai, *Proc. Natl. Acad. Sci. U. S. A*, 2003, **100**, 4984–4989.
493. B. Reuillard, A. Le Goff, M. Holzinger and S. Cosnier, *J. Mater. Chem. B*, 2014, **2**, 2228–2232.
494. M. Pumera and B. Smid, *J. Nanosci. Nanotechnol.*, 2007, **7**, 3590–3595.

495. J. Zhang, F. Zhang, H. Yang, X. Huang, H. Liu, J. Zhang and S. Guo, *Langmuir*, 2010, **26**, 6083–6085.
496. A. I. Gopalan, S. Komathi, G. Sai Anand and K.-P. Lee, *Biosens. Bioelectron.*, 2013, **46**, 136–141.
497. S. C. Tsang, J. J. Davis, M. L. H. Green, H. A. O. Hill, Y. C. Leung and P. J. Sadler, *J. Chem. Soc., Chem. Commun.*, 1995, 1803–1804.
498. L. Wang and R. Jiang, *Methods Mol. Biol. (Clifton., N.J.)*, 2011, **743**, 95–106.
499. B. Zhang, Y. Xing, Z. Li, H. Zhou, Q. Mu and B. Yan, *Nano Lett.*, 2009, **9**, 2280–2284.
500. S. I. Voicu, A. C. Nechifor, O. Gales and G. Nechifor, 2011, Proc. SPIE 8068, Bioelectronics, Biomedical, and Bioinspired Systems V; and Nanotechnology V, 80680Y (May 03, 2011), DOI:10.1117/12.888780.
501. Q. Shi, D. Yang, Y. Su, J. Li, Z. Jiang, Y. Jiang and W. Yuan, *J. Nanopart. Res.*, 2007, **9**, 1205–1210.
502. S. Alwarappan, C. Liu, A. Kumar and C.-Z. Li, *J. Phys. Chem. C*, 2010, **114**, 12920–12924.
503. L. Jin, K. Yang, K. Yao, S. Zhang, H. Tao, S.-T. Lee, Z. Liu and R. Peng, *ACS Nano.*, 2012, **6**, 4864–4875.
504. T. Raghavendra, A. Basak, L. M. Manocha, A. R. Shah and D. Madamwar, *Bioresour. Technol.*, 2013, **140**, 103–110.
505. S. S. Karajanagi, A. A. Vertegel, R. S. Kane and J. S. Dordick, *Langmuir*, 2004, **20**, 11594–11599.
506. S. Shah, K. Solanki and M. N. Gupta, *Chem. Cent. J.*, 2007, **1**, 30.
507. J. M. Gómez, M. D. Romero and T. M. Fernández, *Catal. Lett.*, 2005, **101**, 275–278.
508. I. V. Pavlidis, T. Vorhaben, T. Tsoufis, P. Rudolf, U. T. Bornscheuer, D. Gournis and H. Stamatis, *Bioresour. Technol.*, 2012, **115**, 164–171.
509. I. V. Pavlidis, T. Tsoufis, A. Enotiadis, D. Gournis and H. Stamatis, *Adv. Eng. Mater.*, 2010, **12**, B179–B183.
510. K. Rege, N. R. Raravikar, D.-Y. Kim, L. S. Schadler, P. M. Ajayan and J. S. Dordick, *Nano Lett.*, 2003, **3**, 829–832.
511. A. Pattammattel, M. Puglia, S. Chakraborty, I. K. Deshapriya, P. K. Dutta and C. V. Kumar, *Langmuir*, 2013, **29**, 15643–15654.
512. P. Asuri, S. S. Karajanagi, E. Sellitto, D. Y. Kim, R. S. Kane and J. S. Dordick, *Biotechnol. Bioeng.*, 2006, **95**, 804–811.
513. L. Chen, B. Wei, X. Zhang and C. Li, *Small*, 2013, **9**, 2331–2340.
514. A. A. Gokhale, J. Lu and I. Lee, *J. Mol. Catal. B: Enzym.*, 2013, **90**, 76–86.
515. J. Zhu, M. Xu, X. Meng, K. Shang, H. Fan and S. Ai, *Process Biochem.*, 2012, **47**, 2480–2486.
516. T. T. Baby, S. S. J. Aravind, T. Arockiadoss, R. B. Rakhi and S. Ramaprabhu, *Sens. Actuators, B*, 2010, **145**, 71–77.
517. T. Liu, M. Xu, H. Yin, S. Ai, X. Qu and S. Zong, *Microchim. Acta.*, 2011, **175**, 129–135.
518. D. Ivnitski, K. Artyushkova, R. A. Rincón, P. Atanassov, H. R. Luckarift and G. R. Johnson, *Small*, 2008, **4**, 357–364.

519. C. Xu, C. Zhao, M. Li, L. Wu, J. Ren and X. Qu, *Small*, 2014, **10**, 1841–1847.
520. H. Wang, S. Li, Y. Si, N. Zhang, Z. Sun, H. Wu and Y. Lin, *Nanoscale*, 2014, **6**, 8107–8116.
521. L. Wang, L. Wei, Y. Chen and R. Jiang, *J. Biotechnol.*, 2010, **150**, 57–63.
522. S. Shah, K. Solanki and M. Gupta, *Chem. Cent. J.*, 2007, **1**, 1–6.
523. B. Eker, P. Asuri, S. Murugesan, R. Linhardt and J. Dordick, *Appl. Biochem. Biotechnol.*, 2007, **143**, 153–163.
524. S. Boncel, A. Zniszczoł, K. Szymańska, J. Mrowiec-Białoń, A. Jarzębski and K. Z. Walczak, *Enzyme Microb. Technol.*, 2013, **53**, 263–270.
525. J. Kim, H. Jia and P. Wang, *Biotechnol. Adv.*, 2006, **24**, 296–308.
526. É. Lojou, X. Luo, M. Brugna, N. Candoni, S. Dementin and M. T. Giudici-Orticoni, *J. Biol. Inorg. Chem.*, 2008, **13**, 1157–1167.
527. M. A. Alonso-Lomillo, O. Rüdiger, A. Maroto-Valiente, M. Velez, I. Rodríguez-Ramos, F. J. Muñoz, V. M. Fernández and A. L. De Lacey, *Nano Lett.*, 2007, **7**, 1603–1608.
528. B. Kowalewska, M. Skunik, K. Karnicka, K. Miecznikowski, M. Chojak, G. Ginalska, G. Belcarz, and P. J. Kulesza, *Electrochim. Acta*, 2008, **53**, 2408–2415.
529. A. Vaze, N. Hussain, C. Tang, D. Leech and J. Rusling, *Electrochem. Commun.*, 2009, **11**, 2004–2007.
530. X. Wu, B. Zhao, P. Wu, H. Zhang and C. Cai, *J. Phys. Chem. B*, 2009, **113**, 13365–13373.
531. Y.-M. Yan, O. Yehezkeli and I. Willner, *Chem. – Eur. J.*, 2007, **13**, 10168–10175.
532. P. Ji, H. Tan, X. Xu and W. Feng, *AIChE J.*, 2010, **56**, 3005–3011.
533. H. Tan, W. Feng and P. Ji, *Bioresour. Technol.*, 2012, **115**, 172–176.
534. Q. Li, F. Fan, Y. Wang, W. Feng and P. Ji, *Ind. Eng. Chem. Res.*, 2013, **52**, 6343–6348.
535. M. Verma, C. Barrow and M. Puri, *Appl. Microbiol. Biotechnol.*, 2013, **97**, 23–39.
536. I. Pavlidis, T. Vorhaben, D. Gournis, G. Papadopoulos, U. Bornscheuer and H. Stamatis, *J. Nanopart. Res.*, 2012, **14**, 1–10.
537. D. Das and P. K. Das, *Langmuir*, 2009, **25**, 4421–4428.
538. F. Zhang, B. Zheng, J. Zhang, X. Huang, H. Liu, S. Guo and J. Zhang, *J. Phys. Chem. C*, 2010, **114**, 8469–8473.
539. F. Zhao, H. Li, Y. Jiang, X. Wang and X. Mu, *Green Chem.*, 2014, **16**, 2558–2565.
540. J. S. Lee, S. H. Lee, J. Kim and C. B. Park, *J. Mater. Chem. A*, 2013, **1**, 1040–1044.
541. P. Asuri, S. S. Karajanagi, J. S. Dordick and R. S. Kane, *J. Am. Chem. Soc.*, 2006, **128**, 1046–1047.
542. W. J. Lee, U. N. Maiti, J. M. Lee, J. Lim, T. H. Han and S. O. Kim, *Chem. Commun.*, 2014, **50**, 6818–6830.
543. D. Chen, A. Holmen, Z. Sui and X. Zhou, *Chin. J. Catal.*, 2014, **35**, 824–841.

CHAPTER 8

Photocatalysis on Nanostructured Carbon Supported Catalysts

8.1 Introduction

Heterogeneous photocatalysis is an advanced oxidation process, which has been the subject of a huge amount of studies related to air cleaning and water purification. By definition, photocatalysis has been described as a change in the rate of a chemical reaction or its initiation under the action of ultraviolet, visible or infrared radiation in the presence of a substance – the photocatalyst – that is involved in the combination of light and oxygen. The particularity of a photocatalytic reaction, when compared to conventional catalysis, consists in the way the catalyst is activated, with thermal activation being replaced by photonic activation. It was not until the breakthrough pioneered by Fujishima *et al.* in 1972 that a dramatic focus was brought to this field of research.[1] The authors showed that the rutile phase of TiO_2 could split water and thereby release hydrogen at the cathode and oxygen at the anode. However, this was solely accomplished when (i) the reaction was specifically irradiated with UV light and (ii) there was a bias present (the pH of the electrolyte solution for the working electrode was maintained at elevated levels). The area of photocatalysis basically exploded following this publication, which was primarily because the authors unambiguously showed that when an oxide semiconductor was irradiated with light of wavelengths shorter than its band gap (*ca.* 415 nm for 3 eV band gap), a photocurrent could be generated that flowed from the counter electrode to the TiO_2-irradiated electrode through an external circuit. Hence, the direction of the current revealed that

RSC Catalysis Series No. 23
Nanostructured Carbon Materials for Catalysis
By Philippe Serp and Bruno Machado

Published by the Royal Society of Chemistry, www.rsc.org

the oxidation reaction could be initiated at the TiO_2 electrode and the reduction reaction occurs at the counter electrode (Pt was used by Fujishma *et al.*).

In a photocatalytic reaction, catalysts are invariably semiconductor materials that photo-generate electron/hole pairs, which are formed on the surface (*e.g.* TiO_2, ZnO, CeO_2, *etc.*). The band model (Figure 8.1) is generally used to describe photocatalysis, and the main involved processes are: (i) photon absorption and electron/hole pair generation; (ii) charge separation and migration to surface reaction sites or recombination sites; and (iii) surface chemical reaction at active sites. In order for charge carrier separation to occur, a photon with high enough energy (equal or greater than that of the band-gap of the semi-conductor) is absorbed and leads to the separation; an electron (e^-) is promoted from the valence band (VB, HOMO) of the semiconductor catalyst to the conduction band (CB, LUMO), and a hole (h^+) is generated in the valence band. For a semiconductor photocatalyst to be efficient, the different interfacial electronic processes involving e^- and h^+ must compete effectively with the major deactivation process: electron/hole recombination. This recombination can occur in the bulk, at the surface of the photocatalyst, and also in lattice defects. The rate of recombination is fast (few nanoseconds); the majority of photogenerated electron/hole pairs recombine with dissipation of heat. One possible way that it is thought to reduce the probability of charge carrier recombination is the development of nanocatalysts. This is mainly related to the fact that in nanostructured materials the distance that the photogenerated electrons and holes need to travel to surface reaction sites is reduced. Once these charge carriers reach the surface, they react with adsorbed electron acceptors (for e^-) and donors (for h^+).[2] The migration and surface reactions are slow processes occurring from tens of nanoseconds to milliseconds.

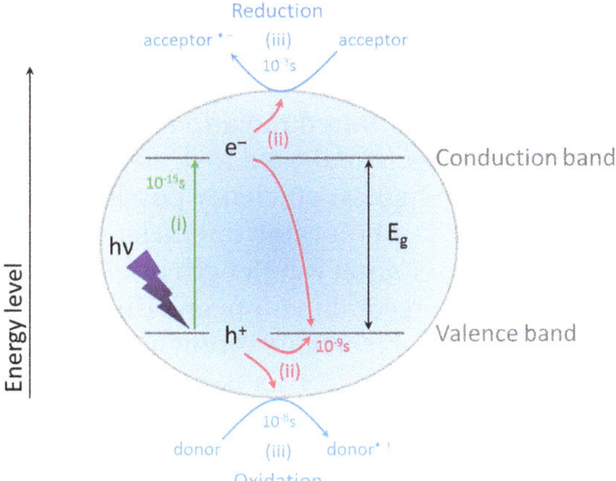

Figure 8.1 Illustration of the basic steps in semiconductor photocatalysis mechanism: (i) photon absorption and electron/hole pair generation; (ii) charge separation and migration; and (iii) surface chemical reaction at active sites.

The desired process consists in the reaction of the activated electron with an oxidant (Ox), yielding a reduced species (Ox^-), and the reaction of the photogenerated hole with a reducing agent (Red) to produce an oxidized species (Red^+), as indicated in eqn (8.1).

$$(Ox)_{ads} + (Red)_{ads} \xrightarrow{SC,\, h\nu \geq E_g} Ox^- + Red^+ \tag{8.1}$$

Under aerobic conditions, molecular oxygen adsorbed on the surface of the semiconductor acts as electron acceptor, while adsorbed water molecules and hydroxyl anions act as electron donors, leading to the formation of very powerful oxidizing $HO^\bullet$ radicals. In the presence of organic molecules adsorbed on the catalyst surface, hydroxyl radicals are the primary oxidizing entities, reacting by adduct formation, followed by structural breakdown into several intermediates until, eventually, total mineralization occurs.

In addition to the degradation of environmental pollutants in aqueous contamination and wastewater treatment,[3–5] photocatalysis applications of wide reaching importance also include water splitting for hydrogen generation,[6,7] carbon dioxide remediation,[8] self-cleaning activity,[9,10] and air purification.[11,12] The photocatalytic activity is usually evaluated by measuring the time dependence of the concentration loss of a degraded compound. The degradation usually follows the Langmuir–Hinshelwood reaction kinetics, which predicts an exponential decay of substrate concentration as a function of time. Alternatively, the progress of a reaction and relative photocatalytic activity may be tracked by measurement of production of a given product.

By far, the most researched photocatalytic material is titanium dioxide, because it has provided the most efficient photocatalytic activity, highest stability, lowest cost and lowest toxicity.[13] Unfortunately, it also suffers from low photonic yield and narrow solar light response range. TiO_2 can only be excited under UV light irradiation at wavelengths <380 nm (the band-gap of TiO_2 is 3.0 and 3.2 eV for rutile and anatase, respectively – Table 8.1). The solar spectrum has been analyzed to contain approximately 7% UV radiation, 46% visible radiation and 47% infrared radiation (range of 5–0.5 eV).[14,15] As a result, various efforts have been dedicated to decreasing the band-gap and harvest the sunlight with the highest efficiency possible. The main areas of activity devoted to improve TiO_2 photocatalytic efficiency involve (i) band-gap tuning and/or extension of excitation wavelength by use of photosensitizers, (ii) minimizing charge carrier recombination, and (iii) promotion of the forward reaction and adsorbance of reactants through provision of an adequate quality and quantity of active sites.[16] This is most often achieved through catalyst modification by doping metal and non-metal ions into the TiO_2 lattice, dye photosensitization, deposition of noble metals, mixing with other semiconductors and addition of inert supports. In order to facilitate the use of TiO_2 as a photocatalyst, dispersion on a suitable support is then highly desirable. High surface area materials have been used to prepare TiO_2 composite catalysts, since they can work as co-adsorbents in the photocatalytic reaction process. Due to their unique pore structure, adsorption capacity,

Table 8.1 Band-gap energy of common photocatalysts.

Photocatalyst	Band-gap energy/eV
MnO_2	0.3
WSe_2	1.2
CuO	1.7
Fe_2O_3	2.2
CdS	2.4
WO_3	2.7
TiO_2 (rutile)	3.0
α-Fe_2O_3	3.1
TiO_2 (anatase)	3.2
ZnO	3.2
$SrTiO_3$	3.4
SnO_2	3.5
ZnS	3.7
SnO	4.2

tunable surface chemistry and electronic properties, carbon-based materials, including carbon nanomaterials of different origins can be used for this purpose.[16] Of course, when irradiation takes place at wavelengths shorter than the absorption edge of the TiO_2, it is impossible to quantify the effect on exciting the carbon phase. This is because below that threshold both phases compete for photon absorption. However, when the light used is in the range $\lambda > 400$ nm, photodegradation can only arise from a photosensitized process.[17] In that case, two major pathways can be considered: (i) the carbon phase acts as a photosensitizer without the contribution of TiO_2, or (ii) the carbon phase is excited and transfers an electron to the conduction band of TiO_2, triggering the conventional semiconductor photocatalysis.[2,18]

Most of the studies combining TiO_2 and carbon materials have evidenced that by introducing materials derived from carbon in a TiO_2 matrix, beneficial effects in the photocatalytic activity of the TiO_2 can be obtained, in addition to the expected increased dispersion. Significant attention is now being directed towards designing and controlling photocatalysts at a more fundamental level, *i.e.* on the nanoscale and below. New opportunities are offered by novel nanostructured carbons such as CNTs, fullerenes, graphene and more.

The advantages or beneficial effects provided by the introduction of the carbon phase are often described in terms of a synergy factor (R), originally defined by Matos *et al.*[19] This factor has been used by several authors, and is defined as the ratio between the apparent first-order rate constants (k_{app}) measured for the composite catalyst containing both phases and just for bare TiO_2 (eqn (8.2)).

$$R = \frac{k_{app}(TiO_2 + C)}{K_{app}(TiO_2)} \tag{8.2}$$

Although it can reasonably be well quantified within a certain set of experiments, this factor is very difficult to use as an independent standard to compare the results in different works of photocatalytic degradation performances (of carbon-modified TiO_2). This is due to the way kinetic parameters are obtained that is inherently linked to experimental details such as the reaction volume, unit weight and area of irradiated photocatalyst.[13]

Over the next sections, we will discuss how different (nano)carbon materials are used to modify TiO_2 with the aim of producing more efficient photocatalysts. A view of the role of each nanocarbon and its interaction with the TiO_2 phase in the composite catalyst concerning the mechanisms involved in heterogeneous photocatalysis is given in each section. Applications of photocatalysis will focus mainly on the liquid-phase photodegradation of organic pollutants. Even though the focus of our discussion will be centered on nanocarbon-TiO_2 composites, the photocatalytic performance of other alternatives to TiO_2, such as ZnO will also be addressed.

8.2 Photocatalysis with Activated Carbons

The term activated carbon defines a group of materials with highly developed internal surface area and porosity (up to 1200 $m^2\ g^{-1}$), and hence a large capacity for adsorption (see Chapter 1).[20] It is necessary that an activated carbon has not only a highly developed internal surface, but accessibility to that surface *via* a network of pores of different diameters. All ACs contain micro-, meso- and macropores within their structures but the relative proportions vary considerably according to the raw material. The typically large surface area and high porosity of ACs favor the dispersion of the active phase over the support and increase its resistance to sintering. Given the need for high-surface area, as a mean to increase photocatalytic activity, researchers have long realized the important benefits that could be obtained from the combination of AC and TiO_2 into a unique composite material. Unfortunately, while AC can enhance the formation of smaller nanosized TiO_2 particles,[21] analysis suggests that the smaller pores are rarely accessed, with the TiO_2 remaining on the outer macropores.[22,23] Hence this potentially leaves much of the nanotexture and nanoscale phenomena in the AC underexplored.

Various methods are commonly used for the preparation of AC-TiO_2 catalysts, such as chemical vapor deposition, precipitation, dip-coating, hydrothermal and sol-gel.[24–26] The selection process of a suitable deposition method depends on the type of catalyst support, the type of pollutant to be degraded and the degradation phase (*e.g.* liquid- or gas-phase). This is because loading the TiO_2 phase onto a support can have a profound and irreversible effect on its photocatalytic properties. One example is the alteration of the TiO_2 chemical and physical structure (including its microcrystalline structure, due to the preparation temperature of the supported titania) in addition to the chemical bond formed between the support and the titania particles that can drastically change the TiO_2 energy band gap, which in turn determines the effectiveness of the catalyst in producing hydroxyl radicals in an aqueous system.

In this respect, activated carbons have proved to be a valuable support in promoting the photocatalytic process[27–30] and in providing a synergistic effect by creating a common interface between both the activated carbon phase and the TiO_2 particle phase.

8.2.1 Mechanism of Photocatalytic Enhancement in AC-composites

One of the main proposed mechanisms for the synergistic enhancement in AC-TiO_2 composites over TiO_2 alone is due to the porosity of the support, progressively allowing an increased quantity of substrate to come in contact with the TiO_2 through means of adsorption. This is important because researchers have established that the oxidizing species ($HO^{\bullet}$) generated by the photocatalyst do not migrate very far from the active centers of the TiO_2 and, therefore, degradation takes place virtually on the catalyst surface (Figure 8.2).[31]

Of potential significance is that a few studies suggest that the AC support itself is capable of a significant level of self-photocatalytic activity.[22,32,33] A possible mechanism underlying the photoactivity of the ACs is illustrated in Figure 8.3.[32] According to the authors, the photons from

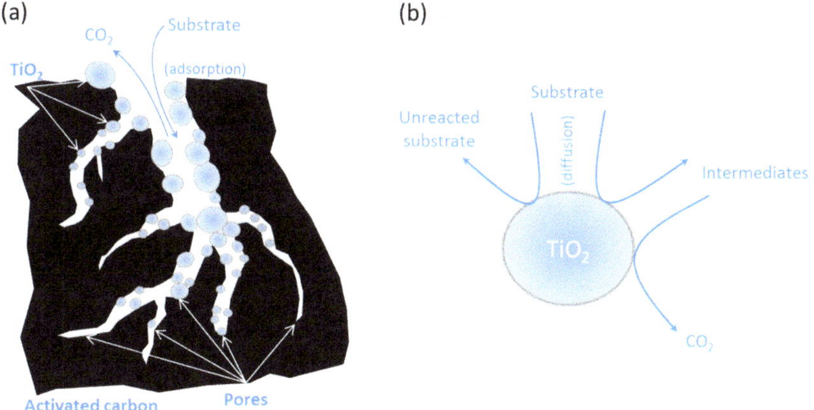

Figure 8.2 Proposed mechanism of synergistic enhancement in AC-TiO_2 composites. (a) Adsorption onto the AC provides a high concentration of reactants near to the TiO_2, which may then be photocatalyzed, possibly passing through intermediate stages. (b) Without a highly adsorbent AC support, reactants must collide with the TiO_2 by chance, and remain in contact for the photocatalysis to proceed. When this is not achieved, the reactants or intermediate products will pass back into solution and can only react further when they collide with TiO_2 again.

UV light would fall on the ACs and generate electron/hole pairs through their irradiation with a sufficient amount of energy to promote electrons from the valence band to the conduction band. The photogenerated electrons would spread throughout the graphene layers and reach molecules of the absorbed sodium diatrizoate and oxygen molecules. The electrons reduce the adsorbed O_2 to form superoxide radicals ($O_2^{\bullet -}$), which can react with the water molecule and trigger the formation of oxidizing radical species,[34] that will interact with the compound, contributing to its degradation (eqn (8.3) to (8.5)). Additionally, the presence of adsorbed oxygen avoids recombination of the electron with the positive hole (eqn (8.3)), allowing interaction between the water molecule and the free hole and increasing the effectiveness of the photocatalytic process. The positive holes are directly responsible for the generation of hydroxyl radicals by

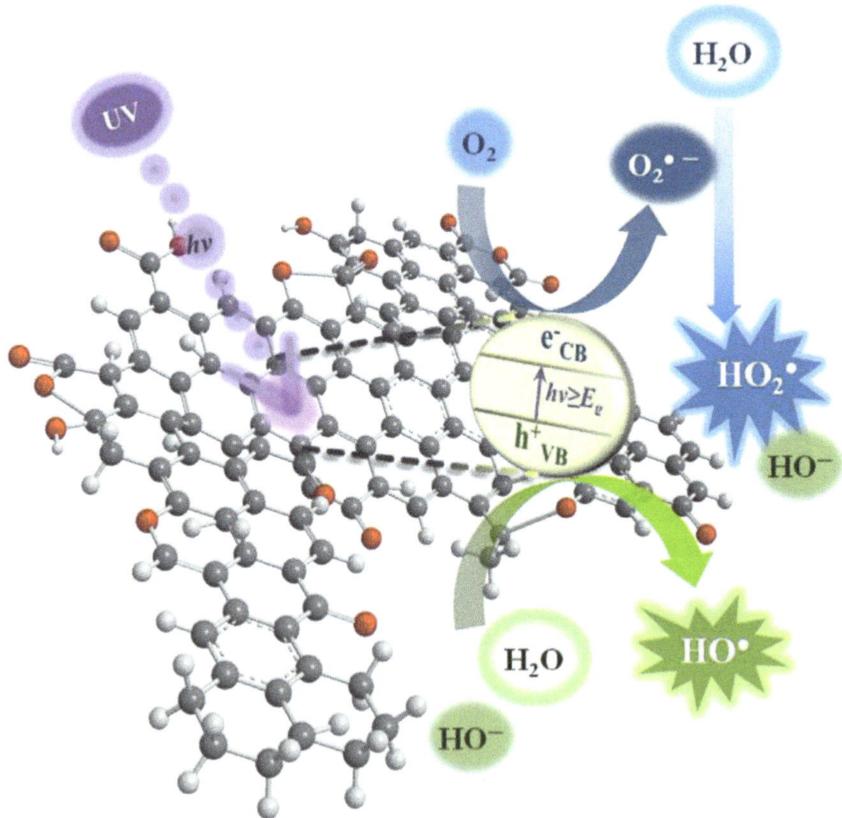

Figure 8.3 Action mechanism of activated carbon as photocatalyst in the presence of UV light. Reprinted with permission from ref. 32. Copyright 2013 Elsevier Science Ltd.

interaction with HO^- groups of the carbon surface and by capture of water molecules (eqn (8.6)).

$$e^-_{CB} + O_2 \rightarrow O_2^{\bullet -} \tag{8.3}$$

$$O_2^{\bullet -} + H_2O \rightarrow HO_2^{\bullet} + HO^- \tag{8.4}$$

$$2HO_2^{\bullet} \rightarrow O_2 + H_2O_2 \rightarrow O_2 + 2HO^{\bullet} \tag{8.5}$$

$$h^+_{VB} + H_2O \rightarrow HO^{\bullet} + H^+ \tag{8.6}$$

8.2.2 Photodegradation of Pollutants with AC-based Materials

Wang *et al.* prepared a high performance composite of TiO_2 deposited on activated carbon by the dip-hydrothermal method.[35] They observed that the AC matrix not only acted as a support for the TiO_2 deposition, but also prevented the growth of TiO_2 particles and the transformation from anatase to rutile phase. The TiO_2-AC composite photocatalysts presented better performances than two physical mixtures containing TiO_2 + AC and a commercial one with TiO_2 P25 (from Evonik Industries) + AC in the photodegradation of methyl orange under UV irradiation. Gao *et al.* studied the effect of AC supported TiO_2 nanoparticles for the photocatalytic degradation of acid red 88 (AR), using visible light irradiation.[36] The TiO_2-AC composites exhibited much higher efficiency for the photocatalytic degradation of the dye in comparison with the unsupported TiO_2, with the highest photocatalytic activity being observed with a 20 wt% AC-containing composite (Figure 8.4).

Tryba *et al.* reported the preparation of AC-TiO_2 composites from tetraorthotitanate solutions.[37] Higher removal of phenol was achieved by combination of adsorption, which occurred in the pores of activated carbon, with decomposition on the anatase particles. A decrease in adsorption was observed after TiO_2 deposition, because TiO_2 particles blocked the entrance to the pores of the AC. However, these TiO_2 particles could decompose the adsorbed organic molecules in the pores of the AC, enhancing the total removal of the pollutant. Matos *et al.*,[19] Araña *et al.*,[38,39] and Liu *et al.*[25] have also observed a synergistic effect for the mixture of TiO_2 with AC using mostly phenol as substrate under UV irradiation. Improvement of the photocatalytic properties of the AC-TiO_2 composite was explained by the high adsorption of the impurities on the surface of AC and their transfer to the TiO_2 surface.

Contrarily to porosity, the effect of the carbon surface chemistry has been scarcely studied. Ocampo-Perez *et al.* have investigated the photocatalytic activity of a series of AC-TiO_2 composites, where the carbon phase was chemically modified by wet oxidation treatments.[40] An improved photocatalytic

Figure 8.4 (a) SEM image of TiO_2-20AC sample; (b) Influence of the content of AC on dye photocatalytic degradation rate. Reprinted with permission from ref. 36. Copyright 2011 Elsevier Science Ltd.

degradation of cytarabine with of AC-TiO_2 composites was observed after oxidation of the ACs. The authors speculated that this behavior was related to interactions between the surface functionalities of the carbon and the radical species generated from the TiO_2 particles. More specifically, the catalytic activity showed a dependence on the concentration of carboxylic acid groups, which would act as catalytic centers capturing the electrons from the photoactivation of TiO_2. This reaction would produce the reduction of carboxylic groups to carbonyls generating H_2O_2, which would be decomposed into $HO^•$

radicals and, subsequently, the generated carbonyls would be transformed into a phenolic superficial group generating additional $HO^{\bullet}$ radicals (eqn (8.7) to (8.9)). The authors support this theory based on the slight increase in the phenolic and lactone groups detected in the carbons after the photocatalytic experiments. In order to further confirm this mechanism, Rivera-Utrilla *et al.* conducted a series of experiments where they observed that the presence of ozonated ACs with a high content of carboxylic acid groups enhanced the photodegradation of 2,4-dichlorophenoxyacetic acid using a UV irradiation system.[41] The improvement was attributed to the additional generation of $HO^{\bullet}$ radicals by interaction of carboxylic groups in the graphene planes of the activated carbon with the electrons produced by the UV/TiO_2.

$$-COOH \xrightarrow{e^-_{aq}} -COH + H_2O_2 \tag{8.7}$$

$$H_2O_2 + h\upsilon(UV) \rightarrow 2HO^{\bullet} \tag{8.8}$$

$$-COH \xrightarrow{2e^-_{aq}} -CH_2OH + 2HO^{\bullet} \tag{8.9}$$

The self-photochemical activity of ACs under UV irradiation has been demonstrated in the absence of semiconductor additives.[22,33] An AC showed an improved photo-oxidation of phenol in aqueous solution compared to both the photolytic reaction and the photodegradation using bare or carbon-immobilized titania. It was suggested that the lower photocatalytic activity in the AC-TiO_2 was due to the drop in porosity of the composite, and blockage of the photoactive centers in the AC support after immobilization of TiO_2. The results showed that beyond the synergistic effect of carbon-TiO_2 composites, the AC alone was capable of a significant level of self-photo-activity under UV irradiation.[22,33] Similar to carbon-TiO_2, when the AC alone was used as a photocatalyst the photodegradation pathway of phenol was modified, through the preferential photo-oxidation of phenol through catechol route.[22] For carbon-TiO_2 composites, this observation had been linked to weak interactions occurring between the carbon material and the semiconductor particles.[22] The fact that the same is observed when carbon alone is used as a photocatalyst confirmed the role of carbon in the photo-oxidation mechanism of the pollutant. The effect of this singular photocatalytic behavior has not yet been fully understood, despite the fact that several hypotheses and scenarios have been postulated based on the carbon physicochemical and structural features.[33] Given that ACs often contain a high amount of metallic impurities, one can also admit the possibility that the metals present in the porous system of the AC might be playing a significant photocatalytic role.

Although TiO_2 was the first effective photocatalyst to be applied, many other semiconductors have been also been investigated for environmental applications. Among them, ZnO (band-gap of 3.17 eV) appears to be a suitable alternative since its photodegradation mechanism is similar to that of TiO_2, and it has been reported to be more active than titanium dioxide in some processes, such as bleaching wastewater, phenol treatment, and the degradation of some dyes.[41,42] Furthermore, the reported optimal pH value for ZnO process is close to the neutral one, whereas the optimum pH for TiO_2 mostly lies in the acidic region. The importance of the pH is related to the fact that under acidic or alkaline conditions the surface of the photocatalyst can be protonated or deprotonated. In addition, the electronic state of the substrate molecule in the reaction media will depend on its acidic properties, namely pK_a. Thus, the effect of the pH on the reaction rate can be interpreted in terms of electrostatic interactions between charged photocatalyst particles and the substrate, which affects the adsorption process. Activated carbon is by far the most widely used carbon material for the immobilization of ZnO, due to the versatility of morphologies and structures.

Melián *et al.* studied the effect of AC on ZnO for the photocatalytic decomposition of aqueous pollutants.[41] Results showed that in addition to a synergistic effect of the AC-ZnO combination, AC content modifies the ZnO particle properties and consequently photocatalytic behavior. For 2,4-dichlorophenol degradation with AC-ZnO catalysts, a negative synergistic effect between AC and ZnO was observed. Due to the high adsorption affinity of dichlorophenol on AC, molecule diffusion to the photoactive centers at ZnO surface was hindered, leading to a slower rate of mineralization and an incomplete degradation. Phenol, on the other hand adsorbs less strongly onto AC and the addition of AC to the ZnO catalyst improved the rate of photocatalytic mineralization of the pollutant and also lowered *in situ* concentrations of the toxic hydroquinone and catechol intermediates.

Sobana *et al.* reported the degradation efficiency of an azo dye direct blue 53 with a AC-ZnO composite photocatalyst using solar light.[43] It was shown that the addition of a commercial AC to ZnO under solar irradiation could induce a substantial synergistic effect by a factor of 4.2 ($k_{App,AC\text{-}ZnO}/k_{App,\ ZnO}$) in the efficiency of the photocatalyst. The synergistic effect observed was ascribed to an extended adsorption of DB53 on activated carbon followed by its transfer to ZnO where it was photocatalytically degraded.

8.3 Photocatalysis with Carbon Black

Carbon blacks (a very pure form of soot) are a group of materials that are characterized by having near spherical carbon particles of colloidal size (see Chapters 1 and 2). The key properties for CB are considered to be fineness (primary particle size), structure (aggregate size/shape), porosity, and surface chemistry, with its surface area being generally considered more accessible than other forms of high surface-area carbon.

Although CB is not in widespread use as a catalyst support (except for applications in fuel cells) there have been a limited number of studies investigating the potential of nanosized carbon black for photocatalytic enhancement of TiO_2. Mao *et al.* compared the photocatalytic activities of composites prepared by a sol-gel technique containing CB and AC.[44] They found an enhanced degradation of methyl orange with CB-TiO_2 compared to AC-TiO_2, highlighting the contributions from the higher surface area and adsorption of reactants, smaller crystallite size, and better dispersion of TiO_2 on CB, coupled with greater retardation of electron/hole pair recombination due to the higher electrical conductivity. Rincon *et al.* also used the sol-gel method to prepare nanometric core–shell CB-TiO_2 composites, and compared their photocatalytic activity to simple physical mixtures using methyl violet as substrate.[45–47] The role of surface area and adsorption sites for effective photocatalytic activity was also highlighted by the authors. In addition, they found a strong red shift in absorption wavelength, and a photosensitizing effect for CB-TiO_2, which was attributed to the formation of mid band-gap carbon states. Yu *et al.* reported a novel baking method for preparation of nanosized TiO_2 films modified by CB, with long term stability.[48] The authors also reported a red shift in absorption wavelength and enhanced degradation of benzamide, although no mechanisms were discussed.

8.4 Photocatalysis with Fullerenes

Fullerenes have attracted extensive attention due to their unique electronic properties. One of the most remarkable properties of C_{60} is its electron-transfer process: they can rapidly photoinduce a charge separation while having a relatively slow charge recombination rate.[49,50] Given that one of the methods to increase photocatalytic activity is by preventing the immediate recombination of photoinduced electron/hole pairs, the combination of photocatalysts with C_{60} may provide an ideal system to achieve an enhanced charge separation by electron transfer. In addition, C_{60} also present a unique three-dimensional structure with delocalized π electrons, which can act as sensitizer when attached to a photocatalyst, thus assisting in the surface photochemical process.

Given the small dimensions of the carbon-phase compared to TiO_2, some authors have chosen to only adsorb the C_{60} on the surface of TiO_2 through electrostatic forces.[51–53] As a result, a close contact between the C_{60} and the TiO_2 matrix cannot be guaranteed and the composites often do not present a synergistic behavior. In order to overcome this limitation, C_{60}–TiO_2 composites have been prepared using mostly hydrothermal,[54,55] but also sol-gel[56] or electrodeposition methods.[57]

The most commonly tested reactions are the photodegradation of organic molecules, mainly dyes[52,55–60] and alcohols.[53,54,61] Other applications of C_{60}-composites also include solar cells[62,63] and antibacterial action.[64]

8.4.1 Mechanism of Photocatalytic Enhancement in C_{60}–TiO_2 Composites

The basis for the enhancement of TiO_2 photocatalysis by C_{60} fullerenes comprises two different types of mechanism, depending on whether fullerene is considered as an electron acceptor or an electron donor. As an electron acceptor, C_{60} provides enhanced photocatalytic activity through the minimization of electron/hole recombination. On the other hand, as an electron donor, C_{60} provides enhanced photocatalytic activity through a sensitization effect of TiO_2.

The mechanism for the enhancement of TiO_2 photocatalysis by C_{60} fullerenes as an electron acceptor was summarized by Apostolopoulou *et al.*[51] C_{60} absorbs moderately in the visible and strongly in the UV range of the electromagnetic spectrum.[65] When irradiated with UV/visible radiation, it undergoes excitation from its ground state to a transient (*ca.* 1.2 ns) singlet excited state ($^1C_{60}^*$), which then goes through rapid intersystem crossing to a longer lasting (>40 μs) lower lying triplet state ($^3C_{60}^*$).[66] Photoexcited fullerenes are also excellent electron acceptors, capable of accepting up to six electrons.[67] $^3C_{60}^*$ has a higher electron-accepting ability than ground state $^1C_{60}$, and electron-donating compounds can reduce $^3C_{60}^*$ to the C_{60} radical anion ($^3C_{60}^-$).[68] Since the potential of the TiO_2 conduction band is −0.5V and the potential of the transformation ($^3C_{60}^*/^3C_{60}^{\bullet-}$) is equal to −0.2 V,[69] electrons may be transmitted from the conduction band of TiO_2 into the deposited $^3C_{60}^*$, resulting in the formation of $^3C_{60}^{\bullet-}$. This radical species may react further with reactants adsorbed at the interface. However, the excited states of C_{60} may also act as electron donors to TiO_2, depending on the environment and experimental conditions. This was demonstrated by Kamat *et al.* using laser excitation to enable a bi-photonic electron ejection mechanism.[70] Furthermore, the donor–acceptor properties may be modified when fullerene is functionalized,[71] which may be necessary to enable adsorption onto TiO_2 and/or water solubility.[58]

Two different types of mechanism, depending on whether the photocatalyst was irradiated with visible or UV light, were proposed by Long *et al.* when they studied the effect of C_{60} on the photocatalytic activity of TiO_2 nanorods for the degradation of Rhodamine B (RhB).[55] When the sample was irradiated with visible light (λ > 420 nm), RhB was excited to its excited state (RhB*), and the electrons of RhB* adsorbed on the surface of TiO_2 nanorods were injected into the conduction band of the TiO_2 nanorods, because of the suitable energy level between the conduction band of TiO_2 nanorods and the RhB*. Meanwhile, RhB* was converted into the cationic radical ($RhB^{\bullet+}$). In the absence of C_{60}, most of the electrons quickly recombined with $RhB^{\bullet+}$. According to the conduction band position of TiO_2 and the potential of $C_{60}/C_{60}^{\bullet-}$, the authors deduced that, when C_{60} was incorporated into the TiO_2 nanorods, the electrons could be transferred from the conduction band of the TiO_2 nanorods to C_{60}, which resulted in charge separation, stabilization and prevented recombination. In turn, the electrons shuttled freely along C_{60} to react with the adsorbed O_2 to yield active oxygen species ($O_2^{\bullet-}$, $HOO^\bullet$, and $HO^\bullet$ radicals). Subsequently, RhB was degraded or mineralized by these active oxygen radicals (Figure 8.5a).

On the other hand, when the sample was irradiated with UV light, the valence electrons of TiO_2 nanorods were excited to the conduction band (Figure 8.5b). Similarly, the photogenerated electrons were scavenged by the

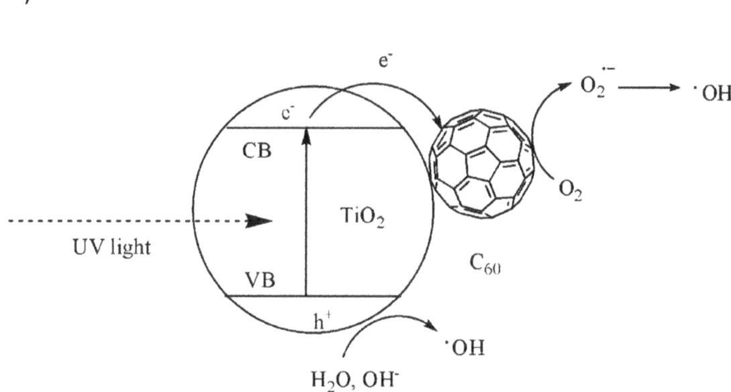

Figure 8.5 Proposed photocatalytic mechanisms for the RhB photodegradation on C_{60}–TiO_2 under: (a) visible irradiation and (b) UV light irradiation. Reprinted with permission from ref. 55. Copyright 2009 American Chemical Society.

absorbed molecular oxygen to produce reactive oxygen radicals. Meanwhile, the holes reacted with the surface HO groups and/or the surface H_2O molecules to form the surface $HO^•$ radicals. Then, RhB was degraded or mineralized by these radicals. In the same way, C_{60} can act as an electron trap so as to enhance the charge separation. However, no $RhB^{•+}$ exists in the reaction system under UV irradiation. Therefore, under visible irradiation, C_{60} showed a stronger effect on the photocatalytic activity of TiO_2 nanorods. In an attempt to explain the photoactivities in different atmospheres (inert and oxidative) the authors observed that C_{60}–TiO_2 under O_2 atmosphere showed higher photocatalytic activity than that under N_2 atmosphere. The predominance indicated that C_{60} could play a role in separation of RhB^+ and e^- both under nitrogen and O_2 atmosphere. It was also confirmed that under O_2 atmosphere C_{60} could accelerate the formation of active oxygen species ($O_2^{•-}$, $HOO^•$, and

HO• radicals) from the adsorbed O_2, and consequently promote the degradation of RhB. This coincided with the author's assumption that the mineralization of RhB was induced by the attack of one of the active oxygen species. However, under nitrogen atmosphere, H_2O adsorbed on the surface of TiO_2 suffered a protonation process and then became HO adsorbed on the surface of TiO_2. After photoelectrons were generated, they were scavenged by the -OH on the surface of TiO_2 to continue the photodegradation process, showing a relatively high activity for C_{60}–TiO_2 in the photodegradation of RhB under nitrogen atmosphere.

8.4.2 Photodegradation of Pollutants with C_{60}

Krishna *et al.* explored the efficiency of water soluble polyhydroxy fullerenes (PHFs) for the enhancement of TiO_2 photocatalysis.[58] The authors attributed enhanced degradation of Procion Red under UV conditions to an improved affinity between PHFs and the titanium dioxide surface, enabling electron scavenging properties of PHF to be exploited without the need for chemical bonding between the PHF molecule and TiO_2. The simplistic nature of the preparation procedure due to the adsorption of PHFs onto the TiO_2 surface was suggested as a particular advantage of C_{60} derivatives over other nanocarbons. Lin *et al.* produced C_{60}-modified TiO_2 nanotube arrays, and found an increased photocatalytic degradation of methylene blue by a factor of 2.3 with respect to TiO_2 nanotube arrays.[57] The high activity was attributed to an effective charge separation due to the presence of C_{60}, which influenced the charge distribution of the electrical double layer and reduced the impedances of the Helmholtz and depletion layers.

Mu *et al.* modified TiO_2 with $C_{60}(CHCOOH)_2$ for the reduction of aqueous Cr(VI) ions.[72] They attributed enhanced photoreduction efficiency compared to TiO_2 under nitrogen atmospheres to enhanced electron/hole separation, as indicated by increased photoluminescence intensity. However, in air P25 outperformed the composite, which was suggested to be due to competition from oxygen reduction on the C_{60} reduction sites. In this work, the authors emphasized the importance of chemical bonding between the modified fullerene and TiO_2 for effective electron transfer, as evidenced by the enhanced photocatalytic activity of the $C_{60}(CHCOOH)_2$–TiO_2 composite compared to simple mechanical mixtures of C_{60} and TiO_2. Long *et al.* also emphasized the need for close interaction between both phases of the composite, while preparing C_{60}–TiO_2 nanorods by a hydrothermal method.[55] The authors found enhanced degradation of Rhodamine blue compared to pure TiO_2 nanorods and P25 (by factors of 3.3 and 2.7, respectively), also proposing the use of C_{60} as an electron scavenging agent.

Oh *et al.* have reported a number of investigations using C_{60}–TiO_2 composites, all involving the degradation of methylene blue (MB).[59,73–75] Whereas C_{60}–TiO_2 and Pt–C_{60}/TiO_2 composites were prepared by an oxidation method, the V–C_{60}/TiO_2 composite was produced *via* a sol-gel process. All composite materials showed enhanced degradation compared to TiO_2 alone. The roles of effective electron transfer from the fullerene surface to TiO_2, and adsorption of MB with subsequent transfer to TiO_2 were emphasized as being key steps in the photocatalytic mechanism, which varied slightly depending on the presence of Pt or V as co-catalysts.

Although it has attracted less attention than titania, ZnO has also been successfully incorporated to fullerenes and improved its photocatalytic properties. Fu *et al.* reported the preparation of hybrid C_{60}–ZnO composites that showed higher photocatalytic activities than pure ZnO in the degradation of methylene blue under UV irradiation.[76] This was achieved even for low amounts of fullerene (*i.e.*, up to 2.5 wt%, the optimum being attained at 1.5 wt%). An additional benefit of the incorporation of C_{60} as additive (besides the efficient charge transfer and decreased recombination) was the inhibition of photocorrosion in the C_{60}–ZnO composite, compared to ZnO alone.

8.5 Photocatalysis with Carbon Nanotubes and Nanofibers

Due to their properties, CNTs show the potential to contribute to all three of the main routes of increasing photocatalytic activity: (i) high-surface area and high quality active sites; (ii) retardation of electron/hole recombination, and (iii) visible light catalysis by modification of band-gap and/or sensitization.[13,77] Specific surface areas typically range between 400 and 900 $m^2\ g^{-1}$ for SWCNTs and between 200 and 400 $m^2\ g^{-1}$ for MWCNTs. This enables the preparation of composite materials with higher surface areas than the initial TiO_2 powders by preventing agglomeration between the TiO_2 particles. Hence, CNTs have the potential to provide reactive surface areas approaching those obtained with activated carbons. Furthermore, CNTs also improve morphology control, with a variety of different structural forms of CNT-TiO_2 photocatalysts being reported.[78]

In addition to morphology control, the surface chemistry of CNTs can also be tailored to promote specificity towards certain adsorbents. This represents an advantage over ACs, which are usually non-selective and therefore have a lower pollutant degradation rate due to the breakdown of both target and neutral species. Modification is usually achieved *via* functionalization with strong acids to form surface groups such as phenols, carbonyls and carboxylic acids, which can be further modified to enhance the adsorption of the target species (see Chapter 4 for an overview of the functionalization process). In addition, CNT functionalization has also been reported to induce stronger TiO_2 anchoring by improving surface contact between the carbon-phase and oxide nanoparticles.[79] This stronger interface interaction also favors electron transfer pathways,[80] thus reducing the probability for electron/hole recombination.

The use of CNTs in photocatalysis is mainly as support for photocatalytically active materials to form composite catalysts. Synthesis routes cover the range of techniques previously outlined in Chapter 4. Most of the TiO_2-CNT composites have been prepared either by sol-gel,[81–84] ultrasonic irradiation,[85] or hydrothermal methods.[84] In the most referenced configuration, TiO_2 nanoparticles bind to the surface of CNTs, TiO_2 nucleates and grows on CNTs dispersed in a liquid phase. Selection of the precursor is especially important because the reactivity of the precursor with the solvent will affect whether the nucleation is homogeneous in solution or heterogeneous on the CNT surface. This in turn, affects the nanoparticle growth rate (which can be adjusted using acids or bases as catalysts), as fast condensation rates usually result in large particle-size distributions.

The most commonly tested reactions are the photodegradation of organic molecules, mainly azo dyes,[79] using CNT-TiO_2 composites as catalysts. Other applications of this composite also include phenol,[81,82] acetone,[85] and NO_x photodegradation,[86,87] photocatalytic H_2 generation from water/alcohol mixtures,[88] and CO_2 reduction with water.[84]

8.5.1 Mechanism of Photocatalytic Enhancement in CNT-TiO_2 Composites

The reaction mechanism of photocatalytic degradation is believed not to be changed by the introduction of the carbon phase,[19,30] thus proceeding by the electron transfer from the water solvent molecules (H_2O and HO^-) to the positive charged holes (h^+) forming the very oxidizing non-selective $HO^•$ radical on the surface of the catalyst. Molecular oxygen, which must be present, is adsorbed on the surface of the catalyst and acts as electron acceptor, forming the superoxide radical anion $O_2^{•-}$. In the presence of an organic molecule, the generated $HO^•$ radical reacts by adduct formation, which then breaks down into several intermediates until, eventually, total mineralization occurs. If the organic reactant is able to compete with adsorbed oxygen and water molecules for the active sites of the photocatalyst, then direct electron transfer to an active hole is also conceivable as the first step of an oxidative degradation producing adsorbed organic radicals. This overall process is commonly accepted and can be described by a set of sequential and concurrent multi-electron transfers.[2,18]

Two mechanisms are being discussed to explain the enhancement of the photocatalytic properties of CNT-TiO_2 composites. One mechanism was proposed by Wang *et al.* whereby the CNTs act as sensitizers and transfer electrons to the TiO_2.[82] The photogenerated electron is injected into the conduction band of the TiO_2, allowing for the formation of superoxide radicals by adsorbed molecular oxygen. Once this occurs, the positively charged CNTs remove an electron from the valence band of the TiO_2 leaving a hole. The now positively charged TiO_2 can then react with adsorbed water to form hydroxyl radicals (Figure 8.6a). The second is a modified mechanism proposed by Hoffmann *et al.* (Figure 8.6b).[2] According to this, a high-energy photon excites an electron from the valence band to the conduction band of anatase TiO_2. Photogenerated electrons formed in the space-charge regions (situated at the Schottky barrier of CNT-TiO_2 interface) are transferred into the CNTs. TiO_2 is an n-type semiconductor, but in the presence of CNTs, photogenerated electrons may move freely towards the CNT surface, which may have a lower Fermi level. Given its properties, such as high electrical conductivity and electron storage capacity (1 electron for every 32 carbon atoms), CNTs can act very efficiently as an electron sink. This leaves an excess of holes in the valence band of the TiO_2, which can migrate to the surface and take part in redox reactions; the TiO_2 therefore effectively behaving as a p-type semiconductor.[89]

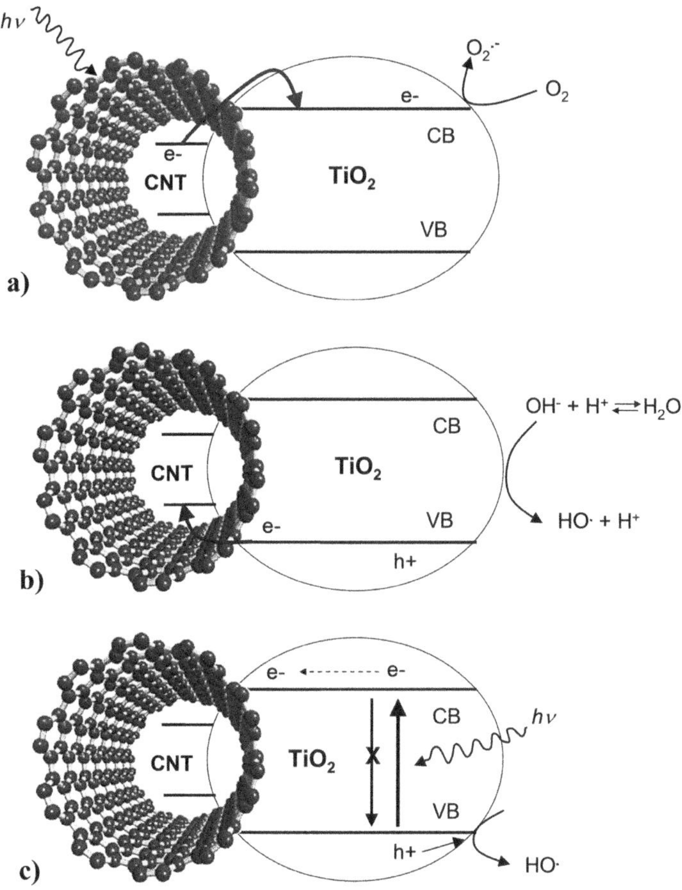

Figure 8.6 Proposed mechanisms of synergistic enhancement in TiO_2-CNT composites: (a) photosensitizing mechanism based on electron/hole pair generation in the CNT. Depending on the relevant positions of the bands, the electron or hole may be injected into the TiO_2, generating O_2^- or $HO^{\cdot}$ species; (b) CNTs inhibit recombination by acting as sinks for photogenerated electrons in TiO_2; and (c) CNTs act as dopants through the Ti–O–C bonds.

However, the CNT-TiO_2 nanocomposite system proves to be more complex. Using CNTs from two common synthesis methods (arc-discharge and CVD, see Chapter 1), Pyrgiotakis *et al.* produced TiO_2-containg composite materials through a sol-gel process.[90] Although both nanocomposites were structurally similar, the photocatalytic dye degradation rate for the arc-discharge CNTs was ten-fold higher than for the CVD-grown CNT nanocomposite. X-Ray photoelectron spectroscopy and Fourier-transform infrared spectroscopy were used to confirm the presence of Ti–O–C bonds in both types of materials (Figure 8.6c), similar to a carbon-doped TiO_2, which extends light absorption to longer wavelengths. However, arc-discharge-synthesized CNTs possessed a higher electrical conductivity and fewer defects. The difference in activity was therefore attributed to the electronic nature of the CNTs, as the electronic-band structure of the CNT was considered a more important factor than the chemical bond between the CNT and TiO_2.

8.5.2 Photodegradation of Pollutants with Carbon Nanotubes and Nanofibers

It is important to note that during photocatalytic degradation studies some degree of CNT oxidation is expected to occur, though this has not been well characterized in the literature.[77] The oxidized portions of the CNTs may initially exhibit opportunities for defect states allowing for enhanced photogeneration of electron-hole pairs; however, the long-term effects, *i.e.* complete degradation of the CNTs, is expected to diminish the photocatalytic ability of the composite system.

This aspect has received little attention in the literature, with the recent notable exception of the suggested self-photocatalytic activity of MWCNTs in the visible range by Luo *et al.*[91] This was observed using highly defective MWCNTs, prepared using a heat treatment technique, which was deliberately intended to introduce defect states. Luo *et al.* proposed that vacancies, local lattice reordering and inter-tube reorientation could initiate defect states in the band-gap, leading to visible light photocatalytic activity in the MWCNTs. Although the observed enhancement in the degradation of hydrogen peroxide was modest, ongoing general interest in the modification of the electronic properties of CNTs by introduction of defects may contribute to progress and growth in attention regarding self-photocatalytic properties.

For composite catalysts, the influence of CNT loadings has been widely investigated. Since the gradual increase in the amount of CNTs in the composite catalysts does not provide a significant increase in their adsorption capacities (measured in terms of the adsorption equilibrium constant), the synergistic effect cannot be merely due to CNTs acting as an adsorbent. In most cases of TiO_2 nanoparticles loaded onto or randomly mixed with CNTs, photocatalytic activity increased up to *ca.* 85 wt% CNT, after which it decreased.[85,88] However, the optimum percentage of CNT appears to be highly dependent on the morphology of the photocatalyst: for mixtures/composites of CNTs loaded onto larger TiO_2 nanoparticles/nanotubes optimal activity has been found at around 20 wt% CNT[92,93] as for AC.[36] Yen *et al.* also found *ca.* 20 wt% CNT to be optimum for TiO_2 nanoparticles loaded on CNTs.[86] In either case, there exists a compromise between an increased synergistic effect from higher CNT loadings and insufficient amounts of TiO_2 and/or blockage of TiO_2 active sites. The apparently contradictory findings could be partially explained based on whether the CNT is acting as a photosensitizer (Figure 8.6a and c) or as an electron sink (Figure 8.6b). If the former mechanism is active, the CNT is the photoactive phase, and optimal activity may be achieved at higher CNT loadings and promotion of the exposed CNT surface area. If the latter mechanism is active, since the TiO_2 is the photoactive phase, it may be beneficial to have higher percentages of TiO_2 and to promote the exposed TiO_2 surface area.

The most tested reactions are the photodegradation of organic molecules, mainly dyes, using CNT-TiO_2 composites as catalysts. It was reported by Wang *et al.* that anchoring TiO_2 onto CNTs apparently induces a synergistic effect on phenol removal, which can result in the complete disappearance of phenol in 4 h, whereas the complete removal of phenol (>95%) is observed only after 6 h for TiO_2 under UV light.[81,83] In the case of visible light irradiation, this synergistic effect is even more remarkable as the complete removal of phenol is accomplished in 5 h, with more than double the phenol removal rate being observed. Bare TiO_2 can only reach 41% of removal within the same reaction time, and the complete elimination of phenol is not attained even after 10 h.[81,83] The enhanced photodegradation of phenol under visible light irradiation was also observed by An *et al.*[94] The MWCNTs modified with TiO_2 nanoparticles exhibited a significant increment of photoactivity in comparison to the pure TiO_2 and the mechanical mixture of TiO_2 and MWCNTs. This phenomenon was explained by a form of band-gap engineering between TiO_2 and MWCNTs, although it is likely due to the loss of crystallinity of the TiO_2 nanoparticles, as evidenced by the red-shift of the composite catalyst in UV-Vis spectra compared to TiO_2 P25 samples.

Regarding the application of CNFs in photocatalysis, there are only a few studies reported. Due to their peculiar structure and morphology, CNFs differ from CNTs since they are composed of graphene layers arranged as stacked cones, cups or plates (see Chapter 1 for more details), instead of concentrically aligned cylinders. The edges of graphene sheets in CNFs make them chemically more active for anchoring functional groups and catalyst particles. Unfortunately, CNFs present a rather high electrical resistivity (See Chapter 2). This structural difference between CNFs and CNTs may have some important consequences on their photocatalytic properties, especially with respect to electrical conductivity and electron storage capacity.

Kim *et al.* studied the photocatalytic activity of TiO_2-embedded CNFs in the oxidation of gaseous acetaldehyde.[95] TiO_2 embedded in the composite fibers was partly reduced during carbonization and then re-oxidized in a post-oxidation process. This post-oxidized TiO_2–CNF composite showed efficient activities in the photocatalytic oxidation of CH_3CHO with the concomitant production of CO_2 under UV irradiation. The oxidized TiO_2 in the CNF composites was considered to be responsible for the photocatalytic oxidation of CH_3CHO under UV irradiation.

Kim *et al.* reported the use of TiO_2–CNF composites containing Ag nanoparticles on the photocatalytic degradation of methylene blue using visible light irradiation.[96] The Ag/TiO_2–CNF photocatalyst degraded methylene blue 17 times faster than TiO_2–CNF with no Ag composites after 3 h. The Ag nanoparticles acted as electron acceptors and trapped the photogenerated electrons immediately, which caused an increase in the photodegradation rate and reduced electron/hole pair recombination. In addition, exposed CNF surfaces with high surface area acted as centers of the physical adsorption process of the dye.

The photocatalytic activities of CNF-containing composites comprising metal oxides other than TiO_2 have also been addressed in the literature. Mu *et al.* fabricated ZnO-CNFs nanocomposites *via* electrospinning and hydrothermal methods.[97] The obtained ZnO-CNFs showed high photocatalytic activity towards the degradation of Rhodamine B dye under UV light irradiation. This was attributed to the structure of the composite material, which improved the separation of photogenerated electrons and holes. Moreover, the ZnO-CNFs could be easily recycled without the decrease in photocatalytic activity due to their one-dimensional nanostructural properties.

Pant *et al.* reported the preparation of CNFs decorated with a binary semiconductor (TiO_2/ZnO).[98] The synthesized nanocomposite exhibited a strong photocatalytic activity for the decomposition of methylene blue under UV irradiation and showed good antibacterial properties as well (Figure 8.7). The enhanced photocatalytic performance of the nanocomposite was attributed to the adsorption characteristic of CNFs and the matched band potentials of TiO_2 and ZnO.

Wang *et al.* reported the synthesis of cuprous oxide-CNF composite films and their application to the photocatalytic degradation of methyl orange

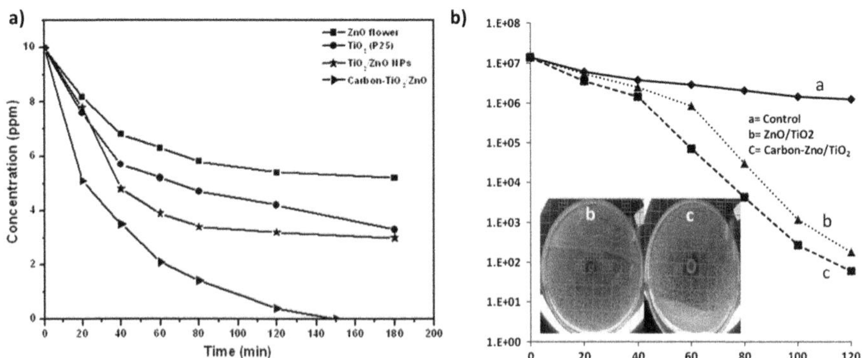

Figure 8.7 (a) Comparison of methylene blue photodegradation in different materials with UV radiation. (b) Antibacterial efficiency of different photocatalysts for gram-negative *E. coli* bacteria under mild UV radiation (insets are the respective zones of inhibition). Reprinted with permission from ref. 98. Copyright 2013 Elsevier Science Ltd.

(MO) as a model organic compound under visible light irradiation at pH 3.0.[99] Results showed that the shape of Cu_2O nanocrystals could be rearranged from irregular nanoparticles to cubic, flower-like particle assembled by Cu_2O nanocubes by simply changing the reaction conditions. After 5 h of irradiation, the photocatalytic degradation ratio of MO reached 99%, whereas the MO solution alone had only 10% degradation.

8.6 Photocatalysis with Graphene and Graphene-related Nanomaterials

Theoretical calculations have shown that the highest surface area possible for a single layer graphene is 2630 $m^2\ g^{-1}$,[100] making graphene highly attractive as a high-surface area 2D photocatalyst support. Furthermore, high quality graphene sheets allow ballistic transport, meaning that electron mobilities can exceed *ca.* 15 000 $m^2\ V^{-1}\ s^{-1}$ without scattering at room temperature,[101–103] making them potentially ideal electron sinks or electron transfer bridges. The one-atom-thick structure also provides high transparency (2.3%),[104] and the surface properties can be adjusted *via* chemical modification, permitting their use as part of a composite material.[105,106]

Graphene sheets decorated with metal oxide nanoparticles combine the outstanding properties of both and might result in some particular properties because of the synergetic effect between them. As we have seen in Chapter 4, there are different preparation methods that can be typically used to prepare composites materials comprising nanocarbons and metal oxides. Among the most frequently used for deposition on graphene are the sol-gel method and hydrothermal/solvothermal process. The sol-gel method is a popular approach for preparation of metal oxide structures and film coatings due to the fact that surface hydroxyl groups of the GO/rGO sheets can act as nucleation sites for the hydrolysis step. On the other hand, the hydrothermal process allows for one-pot preparations that can give rise to nanostructures with high crystallinity without post-synthetic annealing or calcination, while at the same time reducing GO to rGO. Particle intercalation within the graphene layers has the added benefit of preventing aggregation of graphene sheets, thus increasing the available surface area.

Possible applications of graphene-based materials in photocatalysis involve mainly the degradation of pollutants[107–114] and water splitting for hydrogen generation.[115,116]

8.6.1 Mechanism of Photocatalytic Enhancement in Graphene-TiO_2 Composites

Graphene and semiconductor materials have been used together in photocatalysis for a relatively short period, and the exact mechanisms resulting in enhanced photocatalysis by these composite photocatalysts are currently still under investigation. However, a few interpretations arise from the literature:

(i) when graphene is introduced into a photocatalytic system, because the conduction band of most of semiconductors is higher than the Fermi level of graphene, the photogenerated electrons are readily transferred from the semiconductor to graphene by passing through the interface between them. The electrons are then rapidly transferred to target pollutants because the two-dimensional planar π-conjugated structure of graphene facilitates charge transfer. Therefore, the mean free path of photogenerated electrons is prolonged, which promotes reactions such as the formation of free radicals (hydroxyl and superoxide radicals) with high activity, non-selective oxidative degradation of organic pollutants, photocatalytic sterilization, reduction of H^+ to H_2 and reduction of CO_2 to organic fuels. When the conduction band of a semiconductor is lower than the Fermi level of graphene, the photogenerated electrons cannot be transferred from the semiconductor to graphene. If there are sensitizers in the system, they will be sensitized by absorbing photons to generate electrons. Because the Fermi level of graphene is higher than the conduction band of these semiconductors, electrons are transferred from sensitizers to graphene and then to the semiconductor. Also, because of the extremely high conductivity and unique two-dimensional planar structure of graphene, it can increase the rate of transfer of photogenerated charge carriers from graphene to semiconductors and then to the surfaces of reactants. This increases the mean free path of electrons, and reduces the recombination of photogenerated electrons and holes, which enhances the photocatalytic quantum efficiency of semiconductor/graphene composites.[117–119] (ii) When a semiconductor is combined with graphene to form a composite under favorable reaction conditions, a number of M–C or M–O–C (M represents metal) chemical bonds form between the surfaces of the semiconductor and graphene to a certain depth, which is analogous to doping a semiconductor with carbon. As a result, doping energy levels are produced, which narrows the band gap of the semiconductor. This induces a red shift that increases the response of the semiconductor to visible light. (iii) The large number of π electrons in graphene and its unique two-dimensional planar single atomic layer can interact with pollutant molecules *via* π–π conjugation, which enhances the adsorption of pollutants by composite photocatalysts, thereby improving the efficiency of photocatalytic degradation. Compared with other carbon materials, including graphite, CB, AC, CNFs, CNTs and fullerenes, graphene possesses the largest specific surface area because of its single-layer two-dimensional planar structure, which can provide additional reaction space. Meanwhile, graphene also helps to disperse semiconductors by decreasing agglomeration, and promoting the contact between semiconductors and pollutants.

8.6.2 Photodegradation of Pollutants with Graphene

In a demonstration of how graphene can act as an electron-transfer medium, Lightcap *et al.*[120] showed that graphene was able to store and transport electrons through a stepwise electron transfer process: electrons were photogenerated in TiO_2 and then transferred to GO; then, part of these electrons

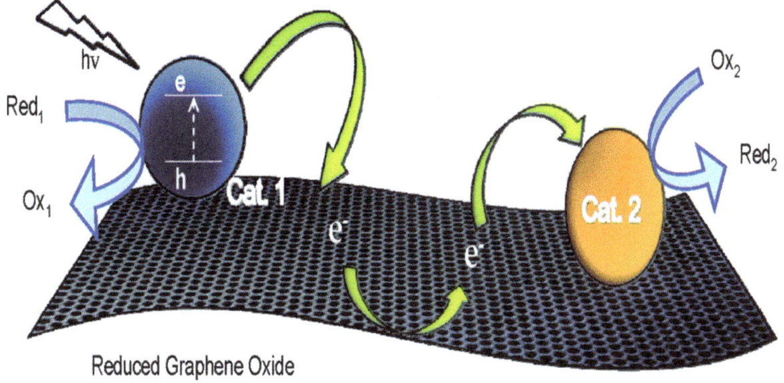

Figure 8.8 Illustration of the three-step electron transfer process involved in making a two-dimensional conducting support. Reprinted with permission from ref. 120. Copyright 2010 American Chemical Society.

were involved in the reduction of GO, whereas the remaining were stored in the rGO sheets; finally, upon introduction of silver nitrate, the stored electrons were used to reduce Ag^+ to Ag^0 (Figure 8.8). Hence, graphene could be regarded as an effective tool to be used in the prevention of electron-hole recombination by accepting and transporting photoelectrons.

Based on these properties, several authors have already reported enhanced photocatalytic activity in the degradation of various molecules using composite materials containing graphene. The advancement is mostly attributed to a high adsorption ability coupled with an improved interaction between both phases, leading to a more effective prevention of the charge recombination process.

Yang *et al.* synthesized a series of TiO_2-graphene, -CNT, and -fullerene (C_{60}) nanocomposite photocatalysts with different carbon contents *via* a sol-gel process along with a hydrothermal post-treatment.[54] The addition of different carbon nanomaterials had no significant influence on the crystal phase, particle size and the morphology of TiO_2, but was able to extend the adsorption edge of all the composites to the visible light region. Using the photocatalytic selective oxidation of benzyl alcohol to benzaldehyde as a model reaction, the authors did not observe much difference between the three different carbon allotropes, as very similar photocatalytic activities for all the composites with low optimum ratios (TiO_2–0.1% GR, TiO_2–0.5% CNT and TiO_2–1.0% C_{60}), along with similar irradiation times, were obtained. In addition, the composites can also promote an efficient separation of the photoexcited electron/hole pairs in a similar way, and the underlying reaction mechanism suggests that TiO_2–nanocarbon photocatalysts follow an analogous oxidation mechanism toward selective oxidation of benzyl alcohol.

Zhang *et al.* used a graphene-P25 TiO_2 catalyst in the photocatalytic degradation of methylene blue dye under both UV and visible lights and compared it against bare P25 TiO_2 and a CNT-P25 TiO_2 composite (Figure 8.9).[108]

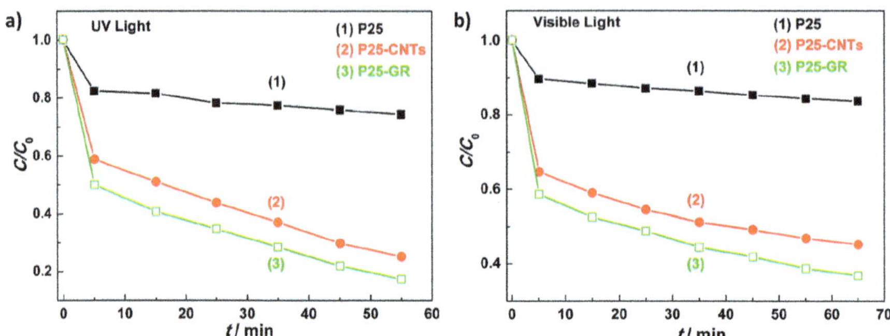

Figure 8.9 Photocatalytic degradation of methylene blue under (a) UV and (b) visible light (λ > 400 nm) with P25 TiO_2, CNT-P25 TiO_2 and grapheme-P25 TiO_2. Reprinted with permission from ref. 108. Copyright 2010 American Chemical Society.

The graphene composite was found to have high dye adsorption capacity, extended light absorption range, and enhanced charge separation and transportation properties. The two-dimensional conjugated structure of graphene assisted in the effective reduction of charge recombination because of improved contact between graphene and TiO_2 nanoparticles.

It is not only TiO_2 nanoparticles that are used in the preparation of graphene-based composite materials.[109,121,122] Yang *et al.* investigated the application of TiO_2 microspheres prepared by ultrasonic spray pyrolysis of an aqueous suspension of GO containing TiO_2 nanoparticles (Aeroxide P25).[121] Graphene–TiO_2 porous microspheres displayed higher photocatalytic activity for the degradation of methylene blue solution than pristine TiO_2 microspheres under the irradiation of a Xe lamp. The microspheres were found to increase the light absorption in the whole visible region and display a red-shifted absorption edge with the increasing of graphene oxide content. The effect of graphene on the photocatalytic activity of porous microspheres was seen in terms of an enhanced charge separation by the graphene–TiO_2 heterojunction, as well as an increased absorption of the visible light.

The preparation of composites containing one dimensional (1D) TiO_2, such as nanotubes or nanorods, has also captured the attention of various research groups. Perera *et al.* studied the hydrothermal synthesis of rGO-TiO_2 nanotube composites on the photodegradation of malachite green.[122] They found that the amount of the graphene in the composite greatly affected the photocatalytic performance. Although a higher graphene content allows for larger quantities of pollutants to be adsorbed, it also blocks the irradiation from reaching the Ti nanotubes surface, leading to lower photocatalytic performances. Thus, composites with 5 and 10 wt% graphene showed enhanced photocatalytic decomposition compared with Ti nanotubes alone, whereas an increase in carbon loading to 15% and 20% induced lower photodegradation rates; the composite with 10% reduced graphene oxide showed the highest photocatalytic activity, with a 3-fold enhancement in photocatalytic efficiency over pure TiO_2 nanotubes in both UV and a broad visible wavelength range.

Liu *et al.* reported the application of a TiO_2 nanorod-graphene composite for the degradation of methylene blue under UV light irradiation.[109] They observed an effective reduction in charge recombination because of improved contact between graphene and TiO_2 nanorods, increasing the photocatalytic activity.

All these results are very promising and could open important perspectives for the improvement of photocatalytic activity of graphene–TiO_2 composites by optimizing the morphology and distribution of TiO_2 nanoparticles on graphene sheets.

In addition to graphene–TiO_2 composites, other materials have also been used as efficient photocatalysts for decomposition of different pollutants in water, namely graphene–ZnO[113,123,124] and graphene–SnO_2.[125]

Lv *et al.* studied the synthesis of ZnO-rGO composites *via* microwave-assisted reduction of GO dispersion with zinc nitrate.[124] The photocatalytic results indicated that the incorporation of rGO enhanced the photocatalytic performance and photo-stability of ZnO in the degradation of methylene blue. A ZnO-rGO composite with 1.1 wt% rGO achieved a degradation efficiency of 88% at pH of 7 and 92% at pH of 9. The increased light absorption and the reduced charge recombination with the introduction of rGO were responsible for the enhanced photocatalytic activity of the ZnO-rGO composite. Using the same methylene blue as a model molecule, Xu *et al.* synthesized a ZnO–graphene composite by reducing GO coated on the surface of ZnO nanoparticles using hydrazine.[113] The degree of photocatalytic activity enhancement strongly depended on the coverage of graphene on the surface of ZnO nanoparticles (Figure 8.10a). The sample with 2 wt% graphene hybridized ZnO showed the highest photocatalytic activity in the degradation of methylene blue, which was about 4 times as that of pristine ZnO (Figure 8.10b).

Li *et al.* studied the photocatalytic properties of a rGO-ZnO composite using Rhodamine B dye.[123] An enhanced photocatalytic activity was obtained for the rGO-ZnO due to the interaction between the graphene sheets and the ZnO nanoparticles. When compared to ZnO alone, the composite exhibited enhanced photosensitized electron injection and slower electron recombination rates.

The effect of dispersing 1D nanostructures on graphene was also addressed using ZnO.[126] Yang *et al.* obtained a hybrid architecture of GO-ZnO nanorods with the rods attached parallel to GO sheets.[126] This hybrid architecture preserved the morphology and crystallinity of the preformed ZnO nanorods and exhibited strong stability and reliability of the heterojunction structure. Despite not yet being tested in photocatalytic applications these materials are thought to possess improved electronic transport properties, given that the effective electron transfer from the excited ZnO to GO sheets was confirmed by luminescence quenching of yellow-green emission of the ZnO nanorods.

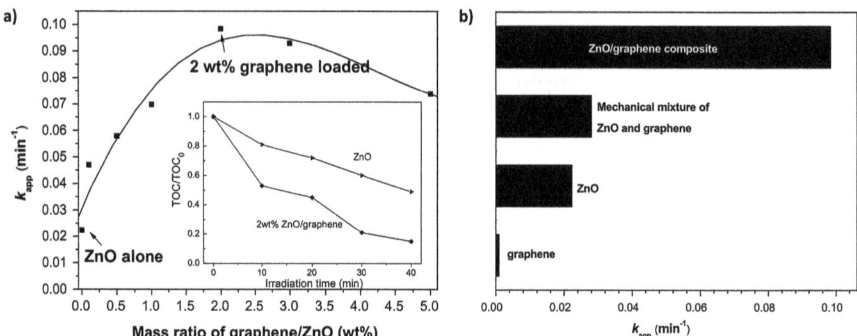

Figure 8.10 (a) The effect of the graphene loading on the apparent rate constant (k) of MB photodegradation (inset: total organic carbon removal on ZnO and ZnO/graphene composite); and (b) the apparent rate constant of MB photodegradation on graphene, ZnO, mechanical mixture of ZnO and graphene (2.0 wt%), and the ZnO/graphene (graphene 2.0 wt%) composite. Reproduced with permission from ref. 113. Copyright 2010 Elsevier Science Ltd.

Despite being used to a smaller extent, other metal oxides (such as SnO_2) have also been coupled with graphene and tested in photodegradation reactions. Zhang *et al.* prepared rGO-SnO_2 and rGO-TiO_2 composites *via* a direct redox reaction.[125] During this reaction, GO was reduced to rGO while Sn^{2+} ($SnCl_2$) and Ti^{3+} ($TiCl_3$) were oxidized to SnO_2 and TiO_2, depositing on the surface of the rGO. The composite materials were found to exhibit excellent photocatalytic performances in the degradation of Rhodamine B under visible light irradiation. The rGO-SnO_2 system exhibited a better photocatalytic activity than P25 TiO_2 and rGO-TiO_2 due to a better electrical conductivity and more effective charge separation due to the presence of rGO.

8.7 Concluding Remarks and Future Perspectives

Despite the huge volume of literature available on carbon–semiconductor composites, there is an evident lack of comparability between studies, making it extremely difficult to compare results obtained in different systems. Hence, comparison of photocatalytic efficiency from one publication to another is currently a non-trivial task. There are innumerous parameters that can dramatically influence the photoefficiency of the system, namely mass of catalyst used, wavelength of irradiation, distance between the reaction container and the light source, intensity (flux) of energy used, length of "dark reaction" (*i.e.*, time the catalyst and pollutant spend while not being irradiated with light), temperature of the reaction while it is being monitored (ideally it should be kept under isothermal conditions), and finally most importantly, the model organic compound must be impervious to the irradiation that is being used to excite the oxide. This last aspect is usually not considered vital by many authors since a large variety of publications recently detailed are evaluating photocatalysts (using visible light) to monitor degradation of dyes (that also absorb in the visible range). As a result, extensive sensitization (electron transfer from dye to oxide) is known to occur.

Novel nanocarbon–TiO_2 combinations have been developed over the last few years, and now offer opportunities for the design of new photocatalytic systems. However, the primary challenge to further exploiting the synergistic effects lies in better understanding the mechanisms of photocatalytic enhancement. When comparing the different mechanisms for carbon–TiO_2 composites, a clear evolution of the understanding is perceived. Synthesis and application of AC-TiO_2 composites to photocatalysis started before the massive application of nanocarbons and its beneficial effects were mostly attributed to the existence of a large surface area, which assisted in both adsorption and dispersion of the TiO_2 phase. With the introduction of nanocarbons and the development of new more elaborate preparation methods, catalyst design evolved and started taking into account other mechanisms for enhanced photocatalytic activity. Hence, in addition to the provision of highly adsorptive active sites (due to their porosity and surface area), the basis for the enhancement of TiO_2 photocatalysis by nanocarbons now comprises two different types of mechanism, depending on whether they are considered as electron acceptors or electron donors. As an electron acceptor, the nanocarbon provides enhanced photocatalytic activity through the minimization of electron/hole recombination, acting as an electron sink. On the other hand, as an electron donor, it provides enhanced photocatalytic activity through a band-gap tuning/photosensitization effect of TiO_2 (for visible light use). In both cases, there is a clear advantage of nanocarbon materials compared with other carbons for the electron storage and rapid charge transfer to target pollutants because of the π-conjugated structure. This means that the mean free path of photogenerated electrons is prolonged by reducing the recombination of generated electrons and holes, thus promoting the formation of free radicals (such as hydroxyl and superoxide) with high activity and enhancing the photocatalytic quantum efficiency of nanocarbon–semiconductor composites. Despite the different range of nanocarbons reported here, there is one crucial aspect to photocatalytic improvement. The formation of a close contact between the nanocarbon and the TiO_2 interface allows for a more efficient charge transfer and/or substrate adsorption. Although some studies suggest that simple mixing is sufficient to achieve a synergistic effect, the presence of an intimate contact is generally shown to further improve charge–reactant transfer. Moreover by achieving chemical bonding between structured nanocarbons and TiO_2, it appears feasible to introduce band-gap tuning by doping-like effects and/or sensitization.

References

1. A. Fujishima and K. Honda, *Nature*, 1972, **238**, 37–38.
2. M. R. Hoffmann, S. T. Martin, W. Choi and D. W. Bahnemann, *Chem. Rev.*, 1995, **95**, 69–96.
3. D. S. Bhatkhande, V. G. Pangarkar and A. A. C. M. Beenackers, *J. Chem. Technol. Biotechnol.*, 2002, **77**, 102–116.
4. K. Pirkanniemi and M. Sillanpää, *Chemosphere*, 2002, **48**, 1047–1060.
5. M. N. Chong, B. Jin, C. W. K. Chow and C. Saint, *Water Res.*, 2010, **44**, 2997–3027.
6. A. Kudo and Y. Miseki, *Chem. Soc. Rev.*, 2009, **38**, 253–278.
7. F. E. Osterloh, *Chem. Mater.*, 2007, **20**, 35–54.
8. S. C. Roy, O. K. Varghese, M. Paulose and C. A. Grimes, *ACS Nano*, 2010, **4**, 1259–1278.
9. I. P. Parkin and R. G. Palgrave, *J. Mater. Chem.*, 2005, **15**, 1689–1695.
10. X. Zhao, Q. Zhao, J. Yu and B. Liu, *J. Non-Cryst. Solids*, 2008, **354**, 1424–1430.
11. J. Zhao and X. Yang, *Build. Environ.*, 2003, **38**, 645–654.
12. A. Fujishima, T. N. Rao and D. A. Tryk, *J. Photochem. Photobiol., C*, 2000, **1**, 1–21.
13. J. L. Faria and W. Wang, in *Carbon Materials for Catalysis*, John Wiley & Sons, Inc., 2008, pp. 481–506.
14. A. L. Linsebigler, G. Lu and J. T. Yates, *Chem. Rev.*, 1995, **95**, 735–758.
15. C. Chen, W. Ma and J. Zhao, *Chem. Soc. Rev.*, 2010, **39**, 4206–4219.
16. R. Leary and A. Westwood, *Carbon*, 2011, **49**, 741–772.
17. C. Lettmann, K. Hildenbrand, H. Kisch, W. Macyk and W. F. Maier, *Appl. Catal., B*, 2001, **32**, 215–227.
18. A. Mills and S. Le Hunte, *J. Photochem. Photobiol., A*, 1997, **108**, 1–35.
19. J. Matos, J. Laine and J. M. Herrmann, *Carbon*, 1999, **37**, 1870–1872.
20. H. Marsh and F. R. Reinoso, *Activated Carbon*, Elsevier Science Ltd, Oxford, 2006.
21. X. Zhang, M. Zhou and L. Lei, *Carbon*, 2006, **44**, 325–333.
22. L. F. Velasco, J. B. Parra and C. O. Ania, *Appl. Surf. Sci.*, 2010, **256**, 5254–5258.
23. D.-K. Lee, S.-C. Kim, S.-J. Kim, I.-S. Chung and S.-W. Kim, *Chem. Eng. J.*, 2004, **102**, 93–98.
24. G. Li Puma, A. Bono, D. Krishnaiah and J. G. Collin, *J. Hazard. Mater.*, 2008, **157**, 209–219.
25. S. X. Liu, X. Y. Chen and X. Chen, *J. Hazard. Mater.*, 2007, **143**, 257–263.
26. Y. Li, S. Zhang, Q. Yu and W. Yin, *Appl. Surf. Sci.*, 2007, **253**, 9254–9258.
27. N. Takeda, T. Torimoto, S. Sampath, S. Kuwabata and H. Yoneyama, *J. Phys. Chem.*, 1995, **99**, 9986–9991.
28. M. Sheintuch and Y. I. Matatov-Meytal, *Catal. Today*, 1999, **53**, 73–80.
29. J.-M. Herrmann, J. Matos, J. Disdier, C. Guillard, J. Laine, S. Malato and J. Blanco, *Catal. Today*, 1999, **54**, 255–265.
30. J. Matos, J. Laine and J.-M. Herrmann, *Appl. Catal., B*, 1998, **18**, 281–291.

31. C. Minero, F. Catozzo and E. Pelizzetti, *Langmuir*, 1992, **8**, 481–486.
32. I. Velo-Gala, J. J. López-Peñalver, M. Sánchez-Polo and J. Rivera-Utrilla, *Appl. Catal., B*, 2013, **142–143**, 694–704.
33. L. F. Velasco, I. M. Fonseca, J. B. Parra, J. C. Lima and C. O. Ania, *Carbon*, 2012, **50**, 249–258.
34. H. P. Boehm, *Carbon*, 2012, **50**, 3154–3157.
35. X. Wang, Z. Hu, Y. Chen, G. Zhao, Y. Liu and Z. Wen, *Appl. Surf. Sci.*, 2009, **255**, 3953–3958.
36. B. Gao, P. S. Yap, T. M. Lim and T.-T. Lim, *Chem. Eng. J.*, 2011, **171**, 1098–1107.
37. B. Tryba, A. W. Morawski and M. Inagaki, *Appl. Catal., B*, 2003, **41**, 427–433.
38. J. Araña, J. M. Doña-Rodríguez, E. Tello Rendón, C. Garriga i Cabo, D. O. González, J. A. Herrera-Melián, J. Pérez-Peña, G. Colón and J. A. Navío, *App. Catal., B*, 2003, **44**, 153–160.
39. J. Araña, J. M. R. Doña, E. Tello Rendón, C. Garriga i Cabo, D. O. González, J. A. Herrera-Melián, J. Pérez-Peña, G. Colón and J. A. Navío, *Appl. Catal., B*, 2003, **44**, 161–172.
40. R. Ocampo-Pérez, M. Sánchez-Polo, J. Rivera-Utrilla and R. Leyva-Ramos, *Appl. Catal., B*, 2011, **104**, 177–184.
41. E. Pulido Melián, O. González Díaz, J. M. Doña Rodríguez, G. Colón, J. Araña, J. Herrera Melián, J. A. Navío and J. P. Peña, *Appl. Catal., A*, 2009, **364**, 174–181.
42. K. M. Joshi and V. S. Shrivastava, *Int. J. ChemTech Res.*, 2010, **2**, 427–435.
43. N. Sobana and M. Swaminathan, *Sol. Energy Mater. Sol. Cells*, 2007, **91**, 727–734.
44. C.-C. Mao and H.-S. Weng, *Chem. Eng. J.*, 2009, **155**, 744–749.
45. M. E. Rincón, M. E. Trujillo-Camacho and A. K. Cuentas-Gallegos, *Catal. Today*, 2005, **107–108**, 606–611.
46. M. E. Rincón, M. E. Trujillo-Camacho, A. K. Cuentas-Gallegos and N. Casillas, *Appl. Catal., B*, 2006, **69**, 65–74.
47. M. E. Rincón, M. E. Trujillo, J. Ávalos and N. Casillas, *J. Solid State Electrochem.*, 2007, **11**, 1287–1294.
48. G. Yu, Z. Chen, Z. Zhang, P. Zhang and Z. Jiang, *Catal. Today*, 2004, **90**, 305–312.
49. S. Zhu, T. Xu, H. Fu, J. Zhao and Y. Zhu, *Environ. Sci. Technol.*, 2007, **41**, 6234–6239.
50. T. Hasobe, S. Hattori, P. V. Kamat and S. Fukuzumi, *Tetrahedron*, 2006, **62**, 1937–1946.
51. V. Apostolopoulou, J. Vakros, C. Kordulis and A. Lycourghiotis, *Colloids Surf., A*, 2009, **349**, 189–194.
52. K.-i. Katsumata, N. Matsushita and K. Okada, *Int. J. Photoenergy*, 2012, **2012**, 256096.
53. M. Grandcolas, J. Ye and K. Miyazawa, *Ceram. Int.*, 2014, **40**, 1297–1302.
54. M.-Q. Yang, N. Zhang and Y.-J. Xu, *ACS Appl. Mater. Interfaces*, 2013, **5**, 1156–1164.

55. Y. Long, Y. Lu, Y. Huang, Y. Peng, Y. Lu, S.-Z. Kang and J. Mu, *J. Phys. Chem. C*, 2009, **113**, 13899–13905.
56. Z.-D. Meng, F.-J. Zhang, L. Zhu, C.-Y. Park, T. Ghosh, J.-G. Choi and W.-C. Oh, *Mater. Sci. Eng., C*, 2012, **32**, 2175–2182.
57. J. Lin, R. Zong, M. Zhou and Y. Zhu, *Appl. Catal., B*, 2009, **89**, 425–431.
58. V. Krishna, N. Noguchi, B. Koopman and B. Moudgil, *J. Colloid Interface Sci.*, 2006, **304**, 166–171.
59. W.-C. Oh, A.-R. Jung and W.-B. Ko, *Mater. Sci. Eng., C*, 2009, **29**, 1338–1347.
60. Z.-D. Meng, L. Zhu, J.-G. Choi, M.-L. Chen and W.-C. Oh, *J. Mater. Chem.*, 2011, **21**, 7596–7603.
61. N. Zhang, Y. Zhang, M.-Q. Yang, Z.-R. Tang and Y.-J. Xu, *J. Catal.*, 2013, **299**, 210–221.
62. J. Lee, J. D. Fortner, J. B. Hughes and J.-H. Kim, *Environ. Sci. Technol.*, 2007, **41**, 2529–2535.
63. C. Chen, W. Zhao, P. Lei, J. Zhao and N. Serpone, *Chem. – Eur. J.*, 2004, **10**, 1956–1965.
64. L. n. Brunet, D. Y. Lyon, E. M. Hotze, P. J. J. Alvarez and M. R. Wiesner, *Environ. Sci. Technol.*, 2009, **43**, 4355–4360.
65. W. Kratschmer, L. D. Lamb, K. Fostiropoulos and D. R. Huffman, *Nature*, 1990, **347**, 354–358.
66. Y. Kajii, T. Nakagawa, S. Suzuki, Y. Achiba, K. Obi and K. Shibuya, *Chem. Phys. Lett.*, 1991, **181**, 100–104.
67. Q. Xie, E. Perez-Cordero and L. Echegoyen, *J. Am. Chem. Soc.*, 1992, **114**, 3978–3980.
68. J. W. Arbogast, C. S. Foote and M. Kao, *J. Am. Chem. Soc.*, 1992, **114**, 2277–2279.
69. P. V. Kamat, M. Haria and S. Hotchandani, *J. Phys. Chem. B*, 2004, **108**, 5166–5170.
70. P. V. Kamat, M. Gevaert and K. Vinodgopal, *J. Phys. Chem. B*, 1997, **101**, 4422–4427.
71. G. Accorsi and N. Armaroli, *J. Phys. Chem. C*, 2010, **114**, 1385–1403.
72. S. Mu, Y. Long, S.-Z. Kang and J. Mu, *Catal. Commun.*, 2010, **11**, 741–744.
73. W.-C. Oh, A.-R. Jung and W.-B. Ko, *J. Ind. Eng. Chem.*, 2007, **13**, 1208–1214.
74. W.-C. Oh and W.-B. Ko, *J. Ind. Eng. Chem.*, 2009, **15**, 791–797.
75. W.-C. Oh, F.-J. Zhang and M.-L. Chen, *J. Ind. Eng. Chem.*, 2010, **16**, 299–304.
76. H. Fu, T. Xu, S. Zhu and Y. Zhu, *Environ. Sci. Technol.*, 2008, **42**, 8064–8069.
77. K. Woan, G. Pyrgiotakis and W. Sigmund, *Adv. Mater.*, 2009, **21**, 2233–2239.
78. S. Z. Kang, Z. Xu, Y. Song and J. Mu, *J. Dispersion Sci. Technol.*, 2006, **27**, 857–859.
79. Y. Yu, J. C. Yu, C.-Y. Chan, Y.-K. Che, J.-C. Zhao, L. Ding, W.-K. Ge and P.-K. Wong, *Appl. Catal., B*, 2005, **61**, 1–11.
80. J. Yu, T. Ma and S. Liu, *Phys. Chem. Chem. Phys.*, 2011, **13**, 3491–3501.
81. W. Wang, P. Serp, P. Kalck and J. L. Faria, *Appl. Catal., B*, 2005, **56**, 305–312.

82. W. Wang, P. Serp, P. Kalck and J. L. Faria, *J. Mol. Catal. A: Chem.*, 2005, **235**, 194–199.
83. W. Wang, P. Serp, P. Kalck, C. G. Silva and J. L. Faria, *Mater. Res. Bull.*, 2008, **43**, 958–967.
84. X.-H. Xia, Z.-J. Jia, Y. Yu, Y. Liang, Z. Wang and L.-L. Ma, *Carbon*, 2007, **45**, 717–721.
85. Y. Yu, J. C. Yu, J.-G. Yu, Y.-C. Kwok, Y.-K. Che, J.-C. Zhao, L. Ding, W.-K. Ge and P.-K. Wong, *Appl. Catal., A*, 2005, **289**, 186–196.
86. Y. Chuan-Yu, L. Yu-Feng, H. Chih-Hung, T. Yao-Hsuan, M. M. Chen-Chi, C. Min-Chao and S. Hsin, *Nanotechnology*, 2008, **19**, 045604.
87. K. Chien-Sheng, T. Yao-Hsuan, L. Hong-Ying, H. Chia-Hung, S. Chih-Yen, L. Yuan-Yao, S. I. Shah and H. Chin-Pao, *Nanotechnology*, 2007, **18**, 465607.
88. B. Ahmmad, Y. Kusumoto, S. Somekawa and M. Ikeda, *Catal. Commun.*, 2008, **9**, 1410–1413.
89. Y. Chen, J. C. Crittenden, S. Hackney, L. Sutter and D. W. Hand, *Environ. Sci. Technol.*, 2005, **39**, 1201–1208.
90. G. Pyrgiotakis, S.H. Lee and W.M. Sigmund, MRS spring meeting, San Francisco, USA, 2005.
91. Y. Luo, Y. Heng, X. Dai, W. Chen and J. Li, *J. Solid State Chem.*, 2009, **182**, 2521–2525.
92. Z. Zhu, Y. Zhou, H. Yu, T. Nomura and B. Fugetsu, *Chem. Lett.*, 2006, **35**, 890–891.
93. Y. Yao, G. Li, S. Ciston, R. M. Lueptow and K. A. Gray, *Environ. Sci. Technol.*, 2008, **42**, 4952–4957.
94. G. An, W. Ma, Z. Sun, Z. Liu, B. Han, S. Miao, Z. Miao and K. Ding, *Carbon*, 2007, **45**, 1795–1801.
95. S. Kim and S. K. Lim, *Appl. Catal., B*, 2008, **84**, 16–20.
96. C. H. Kim, B.-H. Kim and K. S. Yang, *Synth. Met.*, 2011, **161**, 1068–1072.
97. J. Mu, C. Shao, Z. Guo, Z. Zhang, M. Zhang, P. Zhang, B. Chen and Y. Liu, *ACS Appl. Mater. Interfaces*, 2011, **3**, 590–596.
98. B. Pant, H. R. Pant, N. A. M. Barakat, M. Park, K. Jeon, Y. Choi and H.-Y. Kim, *Ceram. Int.*, 2013, **39**, 7029–7035.
99. L. Yang, Y. Zhao, S. Chen, Q. Wu, X. Wang and Z. Hu, *Chin. J. Catal.*, 2013, **34**, 1986–1991.
100. A. Peigney, C. Laurent, E. Flahaut, R. R. Bacsa and A. Rousset, *Carbon*, 2001, **39**, 507–514.
101. K. S. Novoselov, A. K. Geim, S. V. Morozov, D. Jiang, Y. Zhang, S. V. Dubonos, I. V. Grigorieva and A. A. Firsov, *Science*, 2004, **306**, 666–669.
102. K. S. Novoselov, A. K. Geim, S. V. Morozov, D. Jiang, M. I. Katsnelson, I. V. Grigorieva, S. V. Dubonos and A. A. Firsov, *Nature*, 2005, **438**, 197–200.
103. Y. Zhang, Y.-W. Tan, H. L. Stormer and P. Kim, *Nature*, 2005, **438**, 201–204.
104. R. R. Nair, P. Blake, A. N. Grigorenko, K. S. Novoselov, T. J. Booth, T. Stauber, N. M. R. Peres and A. K. Geim, *Science*, 2008, **320**, 1308.
105. G. Williams, B. Seger and P. V. Kamat, *ACS Nano*, 2008, **2**, 1487–1491.
106. E. Bekyarova, M. E. Itkis, P. Ramesh, C. Berger, M. Sprinkle, W. A. de Heer and R. C. Haddon, *J. Am. Chem. Soc.*, 2009, **131**, 1336–1337.

107. J. Du, X. Lai, N. Yang, J. Zhai, D. Kisailus, F. Su, D. Wang and L. Jiang, *ACS Nano*, 2010, **5**, 590–596.
108. H. Zhang, X. Lv, Y. Li, Y. Wang and J. Li, *ACS Nano*, 2010, **4**, 380–386.
109. J. Liu, H. Bai, Y. Wang, Z. Liu, X. Zhang and D. D. Sun, *Adv. Funct. Mater.*, 2010, **20**, 4175–4181.
110. Y. Zhang, Z.-R. Tang, X. Fu and Y.-J. Xu, *ACS Nano*, 2010, **4**, 7303–7314.
111. G. Jiang, Z. Lin, C. Chen, L. Zhu, Q. Chang, N. Wang, W. Wei and H. Tang, *Carbon*, 2011, **49**, 2693–2701.
112. K. Zhou, Y. Zhu, X. Yang, X. Jiang and C. Li, *New J. Chem.*, 2011, **35**, 353–359.
113. T. Xu, L. Zhang, H. Cheng and Y. Zhu, *Appl. Catal., B*, 2011, **101**, 382–387.
114. D.-H. Yoo, T. V. Cuong, V. H. Pham, J. S. Chung, N. T. Khoa, E. J. Kim and S. H. Hahn, *Current Appl. Phys.*, 2011, **11**, 805–808.
115. X.-Y. Zhang, H.-P. Li, X.-L. Cui and Y. Lin, *J. Mater. Chem.*, 2010, **20**, 2801–2806.
116. T.-F. Yeh, J.-M. Syu, C. Cheng, T.-H. Chang and H. Teng, *Adv. Funct. Mater.*, 2010, **20**, 2255–2262.
117. W. Wang, J. Yu, Q. Xiang and B. Cheng, *Appl. Catal., B*, 2012, **119–120**, 109–116.
118. D.-H. Yoo, T. V. Cuong, V. H. Luan, N. T. Khoa, E. J. Kim, S. H. Hur and S. H. Hahn, *J. Phys. Chem. C*, 2012, **116**, 7180–7184.
119. Z. Ren, E. Kim, S. W. Pattinson, K. S. Subrahmanyam, C. N. R. Rao, A. K. Cheetham and D. Eder, *Chem. Sci.*, 2012, **3**, 209–216.
120. I. V. Lightcap, T. H. Kosel and P. V. Kamat, *Nano Lett.*, 2010, **10**, 577–583.
121. J. Yang, X. Zhang, B. Li, H. Liu, P. Sun, C. Wang, L. Wang and Y. Liu, *J. Alloys Compd.*, 2014, **584**, 180–184.
122. S. D. Perera, R. G. Mariano, K. Vu, N. Nour, O. Seitz, Y. Chabal and K. J. Balkus, *ACS Catal.*, 2012, **2**, 949–956.
123. B. Li and H. Cao, *J. Mater. Chem.*, 2011, **21**, 3346–3349.
124. T. Lv, L. Pan, X. Liu, T. Lu, G. Zhu and Z. Sun, *J. Alloys Compd.*, 2011, **509**, 10086–10091.
125. J. Zhang, Z. Xiong and X. S. Zhao, *J. Mater. Chem.*, 2011, **21**, 3634–3640.
126. Y. Yang and T. Liu, *Appl. Surf. Sci.*, 2011, **257**, 8950–8954.

CHAPTER 9

Nanostructured Carbon Materials for Energy Conversion and Storage

9.1 Introduction

Systems for electrochemical energy storage and conversion include batteries, fuel cells, and electrochemical capacitors. Although the energy storage and conversion mechanisms are different, there are a few similarities between these three systems. Common features include the energy-transfer processes taking place at the phase boundary of the electrode/electrolyte interface and electron and ion transport being separated. Batteries, fuel cells, and super-capacitors all consist of two electrodes in contact with an electrolyte solution. In batteries and fuel cells, electrical energy is generated by conversion of chemical energy *via* redox reactions at the anode and cathode. As reactions at the anode usually take place at lower electrode potentials than at the cathode, the terms negative and positive electrode (indicated as minus and plus poles) are used. The more negative electrode is designated the anode, whereas the cathode is the more positive one. The difference between batteries and fuel cells is related to the locations of energy storage and conversion. Batteries are closed systems, with the anode and cathode being the charge-transfer medium and taking an active role in the redox reaction as "active masses". In other words, energy storage and conversion occur in the same compartment. Fuel cells are open systems where the anode and cathode are just charge-transfer media and the active masses undergoing the redox reaction are delivered from outside the cell, either from the environment (*e.g.* oxygen from air), or from a tank (*e.g.* hydrogen and hydrocarbons).

RSC Catalysis Series No. 23
Nanostructured Carbon Materials for Catalysis
By Philippe Serp and Bruno Machado

Published by the Royal Society of Chemistry, www.rsc.org

Energy storage (in the tank) and energy conversion (in the fuel cell) are thus locally separated.

Over the next sections we will address the application of nanostructured carbon materials in energy conversion (fuel and solar cells) and storage (capacitors and batteries).

9.2 Carbon (Nano)materials for Fuel Cells

A fuel cell is a device that converts the chemical energy from a fuel into electricity through a chemical reaction with oxygen or another oxidizing agent. There are many types of fuel cells, but they all consist of an anode, a cathode and an electrolyte. Two chemical reactions occur at the interfaces of the three different segments: at the anode a catalyst oxidizes the fuel (usually hydrogen, methanol, ethanol, *etc.*), turning the fuel into a positively charged ion and an electron; the electrolyte is a substance specifically designed so ions can pass through it, but the electrons cannot. The freed electrons travel through a wire creating the electric current. The ions travel through the electrolyte to the cathode. When they reach the cathode, the ions are reunited with the electrons and the two react with a third chemical, usually oxygen, to create water or carbon dioxide. The net result of the two reactions is that fuel is consumed, water or carbon dioxide is created, and an electric current is generated, which can be used to power electrical devices.

As the main difference among fuel cell types is the nature of the electrolyte, fuel cells are classified by the type of electrolyte they use. Apart from certain specialty types, the five major types of fuel cells are: alkaline fuel cell (AFC), proton exchange membrane fuel cell (PEMFC), phosphoric acid fuel cell (PAFC), molten carbonate fuel cell (MCFC), and solid oxide fuel cell (SOFC) (Table 9.1).

Table 9.1 Different types of fuel cells and their characteristics.

Fuel cell type	Electrolyte	Charge carrier	Operating temperature/K	Fuel	Electric efficiency
AFC	KOH	OH^-	333–393	Pure H_2	35–55%
PEMFC	Solid polymer (Nafion)	H^+	323–373	Pure H_2 (tolerates CO_2)	35–45%
PAFC	Phosphoric acid	H^+	~493	Pure H_2 (tolerates CO_2, 1% CO)	40%
MCFC	Li and K carbonate	CO_3^{2-}	~923	H_2, CO, CH_4, other hydrocarbons (tolerates CO_2)	>50%
SOFC	Solid oxide electrolyte (Yttria Stablized Zirconia, YSZ)	O^{2-}	~1273	H_2, CO, CH_4, other hydrocarbons (tolerates CO_2)	>50%

PEMFCs are one of the most promising energy conversion technologies available today due to their low operation temperature, common fuel sources including hydrogen, alcohols and natural gas, small dimensions and almost no environmental pollution during operation. This makes them as ideal power supplies for portable electronic devices, automobiles, and distributed stationary power sources.[1] The most important part of the design of these fuel cell systems is the preparation of the electrodes, which consist of a cathode for oxygen reduction and an anode for, most commonly, hydrogen or methanol oxidation. The electrochemical reactions are accelerated by the presence of electrocatalysts.[2] Despite intensive research on various non-precious metal catalysts so far, platinum is still considered as the most efficient catalyst in both cathode and anode reactions.[3–5] In order to decrease the amount of metal used, Pt nanoparticles are usually deposited on porous catalyst supports. The ideal catalyst support for PEMFCs should possess all or most of the following properties: (i) high surface area and surface activity for effective dispersion of metal nanoparticles, since high metal loadings are generally necessary, particularly at the cathode; (ii) high conductivity to provide electrical pathways; (iii) high porosity and accessibility for facile mass transportation, and (iv) electrochemical stability for long-term operation.

Porous carbon materials have been considered as promising catalyst supports due to their low cost, designable carbon framework and surface, high chemical and mechanical stability, high conductivity, along with their high surface area and abundant porosity.[6] As seen in the previous chapters, carbon materials have a strong influence on the properties of supported noble metal catalysts, such as metal particle size, morphology, size distribution, degree of alloying, stability and dispersion. On the other hand, these carbon materials can also affect the performance of supported catalysts in fuel cells, such as mass transport and catalyst layer electronic conductivity, electrochemical active area, and metal nanoparticle stability during operation. Commercial catalysts used for fuel cells are prepared by loading Pt onto AC and CB, which can largely decrease the amount of Pt compared to the direct use of common Pt black catalyst.[7] However, these Pt/C catalysts suffer from a lowering of performance during practical operations. Corrosion of the carbon supports has been recognized as the major reason of the catalyst failure. For the cathode catalyst, oxidation of the carbon support can occur in the presence of oxygen, resulting in the detachment of Pt particles and thus, degraded fuel cell performance. For the anode catalyst, the carbon support can also be oxidized in the situation of fuel (*e.g.* hydrogen) starvation.[8] Hence, the optimization of carbon supports is very important in fuel cell electrode development.

9.2.1 Preparation of Carbon-Supported Electrocatalysts

Many methods have been applied to the preparation of carbon-supported fuel cell catalysts. These methods can be divided into two main groups: those based on deposition from the gas-phase (or vacuum), and those based on metal deposition from the liquid-phase. Metal deposition from the gas-phase is well adapted to flat surfaces but less useful for porous substrates such as CBs, because metal precursors often cannot penetrate the pores of carbon particles. For this reason, methods based on metal deposition from the liquid-phase prevail. Amongst the different methods commonly used to prepare electrocatalysts are: impregnation (both wet and incipient wetness), colloidal synthesis, electrodeposition, and vapor-phase deposition (see Chapter 4, Section 4.3 for a detailed description of each method).

Given the normally high loadings necessary to obtain significant activities (typically 20–40 wt%, but can easily increase up to 60–80 wt%) some of these methods are not adequate. As an example, in order to prepare a catalyst using IWI, consecutive depositions might need to be performed in order to achieve the intended metal loading (sometimes there is a solubility issue in dissolving the necessary high amount of metal precursor in the usually small volume corresponding to the pore structure of the carbon). Furthermore, there would be no guarantee that the newly deposited particles would not aggregate to form bigger metal clusters over the old ones.

One methodology that often yields well-dispersed nanoparticles in high loading conditions, involves the use of surfactant species. In this case, the surfactant adsorbs at the metal–solution interface, limiting the growth of metal nanoparticles. The adsorption step is usually followed by the thermal and/or oxidative treatment required for the surfactant degradation.

9.2.2 Characterization of Carbon-Supported Electrocatalysts

Various approaches have been described in the literature for the characterization of carbon-supported metal electrocatalysts. The catalysts are usually analyzed before and after their electrochemical operation with conventional *ex situ* techniques such as X-ray diffraction, transmission electron microscopy, scanning electron microscopy, energy-dispersive X-ray analysis, X-ray photoelectron spectroscopy, and X-ray absorption spectroscopy (Table 9.2).

Table 9.2 Characterization techniques applicable to fuel cell electrocatalysts.

Technique	Information
Gas-phase chemisorption (*e.g.*, CO, H_2)	Metal surface area and dispersion
Temperature programmed reduction	Temperature of surface and bulk reduction
Temperature programmed oxidation	Temperature of surface and bulk oxidation
X-ray diffraction	Metal crystallite size Metal lattice parameter (degree of alloying)
X-ray photoelectron spectroscopy	Atomic composition Oxidation state of components
Scanning electron microscopy	Aggregate morphology
Transmission electron microscopy	Particle size distribution Aggregate morphology

Although *ex situ* analysis provides an important starting point in catalyst characterization, one must keep in mind that significant morphological changes may occur under the operational conditions. Hence, *in situ* characterization techniques capable of discerning information about the state of the catalyst under potential control are of great interest.

The standard tools for the electrochemical characterization of fuel cell catalysts include techniques such as Cyclic Voltammetry (CV), with measurements using a Rotating Disk Electrode (RDE) or Rotating Ring Disk Electrode (RRDE) and Electrochemical Impedance Spectroscopy (EIS). CV records the current drawn from the electrode as it is cycled between chosen high and low potentials. Unlike the majority of physical characterization techniques, voltammetry is surface specific, giving information about the nature of the catalyst surface as a function of potential. From the resulting voltammogram the electrochemical surface area and mass- and area-specific activities of oxygen reduction catalysts can be obtained.[9] In the case of Pt-only catalysts, the hydride adsorption charge is often used as a measure of active metal surface area in the electrochemical environment and compares well with values determined from gas-phase H_2 or CO chemisorption measurements. However, this method becomes less reliable when applied to bi- or tri-metallic systems.

The RDE technique has been used extensively to study electrode kinetics for a variety of oxidation and reduction reactions. Analysis of data from the RDE is commonly done by applying the Koutecky–Levich equation,[10] allowing for the calculation of the number of electrons transferred:

$$1/j = 1/j_k + 1/B\omega^{1/2} \tag{9.1}$$

where j_k is the kinetic current; ω is the electrode rotating rate (the constant 0.2 is adopted when the rotation speed is expressed in rpm); and B is Levich slope which is given by:

$$B = 0.2nF(D_{O_2})^{2/3}\, v^{-1/6} C_{O_2} \tag{9.2}$$

where n represents the number of electrons transferred in the reduction of one oxygen molecule; F is the Faraday constant ($F = 96\,485$ C mol^{-1}); D_{O_2} is the diffusion coefficient of O_2 in 0.1M KOH (1.9×10^{-5} cm^2 s^{-1}); v is the kinetic viscosity of the solution (0.01 cm^2 s^{-1}); C_{O_2} is the bulk concentration of O_2 (1.2×10^{-6} mol cm^{-3}).

A variation of the RDE is a rotating ring disk electrode (RRDE) in which a central ring disk electrode is surrounded by a concentric ring, separated by a coaxial insulating ring. The advantage that the RRDE offers over the RDE, for the ORR, is the ability to detect H_2O_2 produced in the reaction using the outer ring.

Increasingly, *in situ* spectroscopies are also being applied to fuel cell catalysts as electrodes under potential control. One prominent use has been the application of X-ray absorption spectroscopy techniques (XAS) such as extended X-ray absorption fine structure (EXAFS) and X-ray absorption near edge structure (XANES) to fuel cell catalysts. In particular, the use of *in situ* EXAFS is interesting in understanding how the nature of supported catalysts changes as a function of potential. *In situ* XAS has also been widely used to probe the structure of bimetallic catalysts, particularly PtRu catalysts. As well as indicating whether Pt and Ru are mixed at the atomic level, information on the influence of Ru and on the electronic properties of Pt has been gained.

9.2.3 Carbon-Supported Electrocatalysts for PEMFCs

PEMFCs have attracted considerable interest recently because of various advantages, including high power density, zero or low exhaust, ease of recharging, simple structure, and quick start-up at low temperature.[11] In light of these advantages, these fuel cells are being developed as electrical power sources for vehicles and for stationary and portable applications as an alternative to conventional internal combustion engines, secondary batteries, and other conventional power sources. However, PEMFCs are still far from market launch, which is hindered by two main issues: the excessive production cost and poor durability and reliability.[12]

9.2.3.1 Working Principle of PEMFCs

The conversion of chemical energy to electrical energy in a PEMFC occurs through a direct electrochemical reaction. A PEMFC consists of two electrodes in contact with an electrolyte membrane (Figure 9.1). The membrane is designed as an electronic insulator that separates the reactants and allows only the transport of protons/hydroxyl ions and water between the electrodes, serving also as an effective barrier to reactant crossover. The membranes are

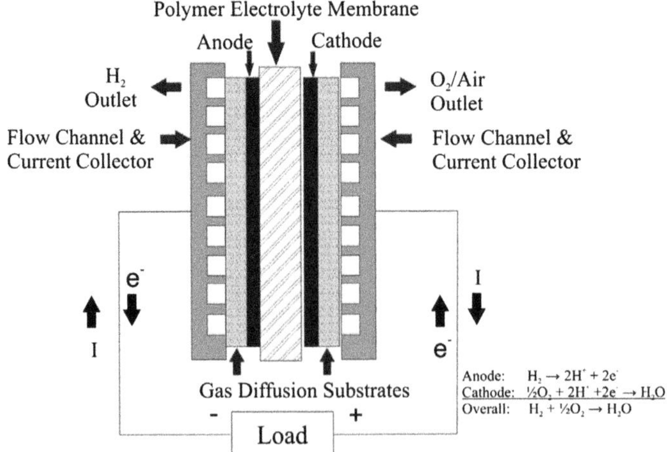

Figure 9.1 Schematic representation of a PEMFC. Reprinted with permission from ref. 13. Copyright 2014 Elsevier Science Ltd.

usually polymers such as Nafion®, modified to include ions, such as sulfonic groups. These hydrophilic ionic moieties are the key for allowing proton transport across the membrane. The electrodes are constituted by a porous gas diffusion layer (GDL) and a catalyst (usually platinum supported on high surface area carbon) containing the active layer. The assembly between membrane and electrodes is known as the membrane electrode assembly (MEA) and it is the heart of a fuel cell (Figure 9.2a). The MEA is composed of catalytic layers (CLs), where electrochemical reactions occur; gas diffusion layers providing access of the fuel and the oxidant to the CLs (carbon paper or cloth of 0.2–0.5 mm thickness that also confers rigidity to the MEA); and a proton exchange membrane such as Nafion. The structure and composition of the MEAs has undergone decisive changes during the past decades.

The triple phase boundary (TPB) of a fuel cell is the area of contact between the three phases necessary for electrochemical reactions at the electrode: ion conducting phase, electron conducting phase, and gas-phase. A good fuel cell maximizes the TPB area, allowing the reaction to occur in more sites, thus maximizing current flow (Figure 9.2b).

The most common method to prepare the MEA of a PEM fuel cell consists on using appropriate techniques to add the carbon-supported catalyst to a porous and conductive material, such as carbon cloth or carbon paper (GDL). Normally, polytetrafluoroethylene (PTFE) and Nafion solution are added (PTFE, which is hydrophobic, can drain the product water to the surface where it can evaporate, and Nafion can help to attach the catalyst layer to the membrane, in addition to increasing the ionic conductivity of the catalyst layer). Then the catalyst-surfaced GDL is hot-pressed together with the pre-treated membrane, which has catalyst on each side.

The CLs of the state-of-the-art fuel cells are multicomponent media that include: (i) a catalyst, accelerating the rates of electrochemical reactions;

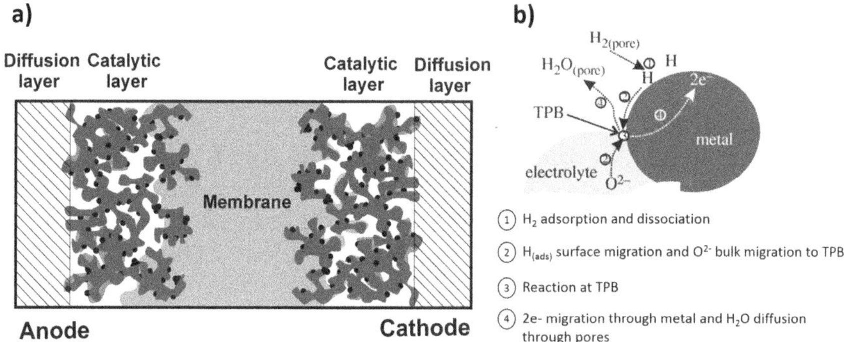

Figure 9.2 (a) Schematic representation of a MEA. Reprinted with permission from ref. 14. Copyright 2009 John Wiley & Sons, Inc. (b) Close-up of anode cermet TPB.

(ii) an ionomer, providing a flow of protons through the MEA, and (iii) gas- and liquid-filled pores, providing access of the reagents to, and products from, the catalyst surface. The MEA is, in turn, sandwiched between two electrically conducting bipolar plates. These plates are a multi-functional components and their primary function is to supply reactant gases to the GDL *via* gas distribution flow channels.

A good electrode has to effectively facilitate the trade-off between enabling high catalytic activity, retaining enough water to guarantee good proton conductivity in the ionomer phase, and having an optimal pore size distribution to facilitate rapid gas transport. All the components of the MEA need to be stable (under both chemical and mechanical stresses) for several thousands of hours in the fuel-cell under the prevailing operating and transient conditions.

When using hydrogen as fuel, at the anode it is decomposed into positively charged protons and negatively charged electrons. Positively charged protons pass through the polymer electrolyte membrane to the cathode, whereas the negatively charged electrons travel along an external circuit to the cathode, creating an electrical current. At the cathode, the electrons recombine with the protons, and together with the oxygen molecules, form pure water as the only reaction byproduct, which flows out of the cell. The splitting of the hydrogen molecule is relatively easy using a platinum catalyst. However, splitting of the oxygen molecule is more problematic, which causes significant activation loss. So far platinum is still the best option for the oxygen reduction reaction (ORR). Another significant source of performance loss is the resistance of the membrane to proton flow, which is minimized by making it as thin as possible (around 50 μm). The thermodynamic voltage for a hydrogen–oxygen fuel cell operating at standard temperature and pressure is 1.229 V. The maximum theoretical energy efficiency of a hydrogen fuel cell is 83% (operating at low power density and using pure hydrogen and oxygen as reactants). This value compares with a maximum theoretical efficiency of 58% for internal combustion engines.

Other fuels (*e.g.*, methanol, ethanol, or other organic molecules), which can be oxidized under moderate conditions on PEMFCs (≤373 K at ambient pressure), are also being used to replace hydrogen gas. Direct-methanol fuel cells (DMFCs) are a subcategory of proton-exchange fuel cells in which methanol is used as the fuel. In this system, chemical energy is converted into electrical energy by oxidizing methanol to CO_2 and H_2O. At the anode, a methanol molecule reacts with a H_2O molecule and releases CO_2, six protons (that are free to migrate through the electrolyte towards the cathode) and six electrons (that can travel through the external load). The CO_2 produced in the reaction is rejected by the acid electrolyte solution. The generally accepted mechanism of methanol oxidation on Pt catalysts proceeds with the electrosorption of methanol, followed by proton and electron stripping (on low index planes of Pt, these initial steps are considered to be rate determining); removal of protons gives rise to Pt-bound CO. Water is co-adsorbed at sites adjacent to the bound CO, and oxygen transfer occurs to give carbon dioxide, which desorbs from the catalyst surface. At potentials below *ca.* 450 mV, the surface of Pt becomes poisoned with near-monolayer coverage of CO, and further adsorption of water or methanol cannot occur. Hence, the methanol oxidation rate drops to an insignificant level. The maximum theoretical voltage attainable from the overall reaction in the methanol–air fuel cell is *ca.* 1.21 V with a theoretical efficiency of 96.5%, but in practice, this voltage is not obtained due to poor electrode kinetics and ohmic losses in the electrolyte.

So, the rate determining reaction is the oxygen reduction reaction. The mechanism of the electrochemical ORR is very complicated and involves many intermediates. In addition, the ORR is highly dependent on the nature of the electrode material, catalyst, and electrolyte solution. Currently, different metals are regarded as active and efficient for ORR, with Pt being by far the most studied. However, many issues including high cost, CO poisoning and sintering tend to severely hinder its commercialization. Hence, much effort has been devoted to developing cheaper and more efficient ORR electrocatalysts, mainly with respect to three aspects: (i) increasing Pt efficiency;[15–17] (ii) developing non-precious-metal electrocatalysts;[18–21] and (iii) developing metal-free ORR electrocatalysts.[22–25]

Oxygen gas molecules can be reduced either in aqueous or in non-aqueous aprotic electrolytes with distinct reaction pathways and products. In aqueous electrolytes, the ORR occurs mainly by two pathways: the 4-electron reduction of O_2 to H_2O in acidic electrolytes (eqn (9.3)) or to OH^- in basic electrolytes (eqn (9.5)), and the 2-electron reduction of O_2 to H_2O_2 in acidic electrolytes (eqn (9.4)) or to HO_2^- in basic electrolytes (eqn (9.6)). In non-aqueous aprotic electrolytes, aside from the 4-electron and 2-electron pathways, O_2 can also be reduced to superoxide (O_2^-) involving one electron transfer. Note that the potentials should depend on the actual electrolyte pH. The practical reaction sequences associated with the ORR are rather complicated, involving many intermediates and elementary steps (electron transfer or chemical reaction) depending on the natures of the catalysts and electrolytes.

$4e^-$ pathway in aqueous acid electrolytes (eqn (9.3))

$$O_2 + 4H^+ + 4e^- \rightarrow 2H_2O \tag{9.3}$$

2-step $2e^-$ pathway in aqueous acid electrolytes (eqn (9.4))

$$O_2 + 2H^+ + 2e^- \rightarrow H_2O_2 \tag{9.4}$$

$$H_2O_2 + 2H^+ + 2e^- \rightarrow 2H_2O$$

$4e^-$ pathway in aqueous alkaline electrolytes (eqn (9.5))

$$O_2 + 2H_2O + 4e^- \rightarrow 4OH^- \tag{9.5}$$

2-step 2e$^-$ pathway in aqueous alkaline electrolytes (eqn (9.6))

$$O_2 + 2H_2O + 2e^- \rightarrow HO_2^- + OH^- \quad (9.6)$$

$$HO_2^- + H_2O + 2e^- \rightarrow 3OH^-$$

To achieve maximum energy capacity, reduction of O_2 must occur *via* the 4e$^-$ pathway. Hence, the four-electron reduction pathway is especially important in fuel cell systems, whereas the two-electron reduction pathway is used in H_2O_2 production.

9.2.3.2 *Carbon as Anode Catalyst for PEMFCs*

The basic function of the electrocatalyst support is to: (i) provide good electrical conductivity with high surface area, (ii) bring the catalyst particles close to the reactants *via* the pore structure and (iii) provide corrosion stability under oxidizing conditions.[26] Hence, the choice of support material to disperse the active phase is vital in determining the performance and durability of the catalyst. CB is the most commonly used support for Pt and Pt-alloy catalysts. Due to its susceptibility to oxidation, alternate materials such as CNTs, CNFs, graphene, and mesoporous carbons have been investigated to replace CB as support, in order to provide both higher corrosion resistance and surface area.[27]

Carbon Blacks. Carbon blacks are commonly used as supports for PEMFC anode catalysts. There are many types of CBs (see Chapter 1, Section 1.2.4), such as Acetylene Black, Vulcan XC-72, Ketjen Black, *etc.*, and these are usually manufactured by pyrolyzing hydrocarbons such as natural gas or oil fractions taken from petroleum processing. Acetylene Black possesses a low surface area, hindering metal dispersion for supported catalysts. High surface area carbon blacks such as Ketjen Black can support highly dispersed catalyst nanoparticles, but this support shows high Ohmic resistance and mass transport limitation during fuel cell operation.[28] Vulcan XC-72 with a surface area of *ca.* 250 m^2 g^{-1} has been widely used as a catalyst support for Pt and Pt-alloy catalysts in fuel cell anode catalyst preparation. However, CB is susceptible to oxidation at low (potential greater than 1.2 V *vs.* SHE) and high temperature (potential lower than 0.9 V *vs.* SHE) fuel cells resulting in active surface area loss,[7] and alteration of pore surface characteristics (generation of surface oxygenated groups).[29]

The contact between the metal nanoparticles and Nafion micelles in the catalyst layer of MEA is also affected by carbon support pore size and distribution. Since the Nafion ionomer forms rather large (>40 nm) micelles, metal nanoparticles residing in carbon pores below 40 nm in diameter have no access to the Nafion ionomer and do not contribute to the electrochemical activity.[30] This is due to the fact that the metal catalyst utilization is determined by an electrochemical accessible active area, rather than the carbon specific surface area.

Mesoporous Carbons. Mesoporous carbons are characterized by high surface area and porosity leading to higher activity and stability than CB.[6,31–35] The performance of the mesoporous carbon supported electrocatalysts is influenced by both the pore structures and sizes. Generally, a high performance PEMFC anode requires an efficient TPB zone at nanoscale, in which the electrochemical reactions occur on the surface of the metal nanoparticle involving electron and proton transport. In addition, it also requires the provision of an efficient transport passage for liquid-phase reactants (*e.g.* CH_3OH, H_2O) and the gas-phase product (CO_2). Too many small micropores (<2 nm) in carbon supports (*e.g.*, Vulcan XC 72) decreases catalyst utilization because the mass transport of reactants and product is poor in these micropores. When the macroporous size is larger than 50 nm, the surface area will become small and the electrical resistance will increase.[36] Mesoporous carbons with tunable pore sizes in the range 2–50 nm are, thus, attractive for use as catalyst supports and have the potential to enhance both the dispersion and utilization of metal catalysts. This represents an equilibrium between high surface area, which allows a higher degree of catalyst dispersion, pore size, which allows for efficient transport of reactants and products, and electric conductivity.[6]

The ordered mesoporous carbon is usually synthesized by a template method starting with either highly ordered mesoporous silica or nanosized silica spheres as templates. Uniform mesoporous carbon can be formed after removing the silica template by HF etching.

Carbon Nanotubes and Nanofibers. The field to which the specific features of nanostructured carbon materials could bring the most significant advancements is perhaps fuel cell electrocatalysis.[2,37] The family of CNTs is probably the most well-known within the nanostructured carbons, showing very promising results in catalyst support for fuel cell applications due to their unique electrical and structural properties.

Pristine CNTs are chemically inert and metal nanoparticles cannot be attached very efficiently. Hence, most research is focused on functionalized CNTs in order to incorporate oxygen groups on their surface that will increase their hydrophilicity and improve the catalyst support interaction. Several studies have shown that CNTs are superior to CB as catalyst supports in PEMFCs.[2,13,38] Both SWCNTs and MWCNTs have been extensively studied, with MWCNTs having been found to be more conductive while SWCNTs provide larger surface areas.[39–41]

CNTs have been used to support a wide variety of Pt-based mono-, bi- (*e.g.* Pt–Ru, Pt–Co, Pt–Fe, Pt–Sn) and even trimetallic (*e.g.* Pt–Ru–Pd, Pt–Ru–Ni, Pt–Ru–Os) catalyst systems.[42–47] Surprisingly, only a few studies have been conducted on the use of cheaper metals such as Pd,[48–54] Ag,[55,56] Au,[57–60] and Co.[61,62]

Among the observed advantages are: (i) CNTs are electrochemically durable in fuel cell conditions; (ii) CNTs provide a high electric conductivity and a specific interaction between Pt and CNT support (delocalized π-electrons of

CNTs and Pt d-electrons), resulting in a high catalytic activity; (iii) CNTs have few impurities, while CB (*e.g.*, Vulcan XC-72) contains a certain impurities, such as sulfur, which are known to poison Pt; (iv) CNTs are free from the deep-crack structure, which is the so-called "dead zone" of electrocatalysts because Pt nanoparticles have no catalytic activity when deposited there due to the absence of electrochemical TPB.

Another nanostructured carbon material that has found application as catalyst support in PEMFCs is CNFs.[63–68] Unlike CNTs, CNFs do not have (or have a very thin) hollow cavities and often present larger diameters than CNTs. However, the main difference between CNTs and CNFs lies on the exposure of active edge planes. A predominant basal plane is exposed in CNTs while only the edge planes with anchoring sites for the electrocatalyst are exposed in CNFs.[69] Bessel *et al.* demonstrated that CNF-supported catalysts showed improved methanol oxidation activity as compared to CB, using 5 wt% Pt supported on "platelet" and "ribbon" type CNFs.[63] The authors found that these catalysts exhibited activities comparable to that displayed by about 25 wt% Pt on Vulcan carbon (Figure 9.3). Furthermore, they observed that the CNF-supported metal particles were significantly less susceptible to CO poisoning than the traditional catalyst systems. This improvement in performance was believed to depend on the fact that the metal particles adopted specific crystallographic orientations when dispersed on the highly tailored CNF structure.

Although it is not possible to compare all these studies mainly because of the different origins of CNT and CNF samples (pointing to the crucial importance of CNT and CNF standardization), the general tendency observed is that the catalysts prepared on CNTs or CNFs are more active and in some cases present a better resistance to poisoning than those prepared

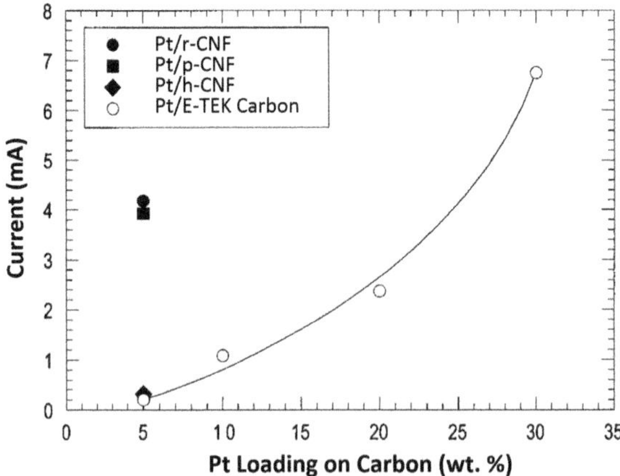

Figure 9.3 Comparison of methanol oxidation currents for platinum particles supported on various types of carbon nanofiber supports (0.5 M methanol in 0.5 M H_2SO_4, 313 K). Reprinted with permission from ref. 63. Copyright 2009 Elsevier Science Ltd.

on conventional carbon supports, such as Vulcan XC-72.[70] In some cases it is also possible to obtain similar or better performances with a significant reduction of Pt loadings.[63,71,72] When compared to the commonly used CB support, the increase in the power density of a single stack is between 20 and 40%, even if 70,[39] 100%,[73] or even higher values have been reported.[74,75] Only a few studies concern systematic comparison of the different types of CNTs and CNFs. If we consider the electronic conductivity and specific surface area of the supports, we would expect the following order: SWCNTs > DWCNTs > MWCNTs > CNFs. In general, and this result could be rationalized on the basis of the electronic conductivity of the support, the activity of metal supported MWCNT systems is superior to CNF-based catalysts. For DWCNTs, which present a higher specific surface area than MWCNTs,[39] better performances have been reported. For SWCNTs contrasting results have been reported, and further work is needed to shed light on the potential of these materials.

The advances made by the use of CNTs and CNFs as supports for fuel cell applications are generally attributed to: (i) the possibility to reach high metal dispersions and high electroactive surface area; for Vulcan XC-72R the catalyst particles can sink into the microporosity, thus reducing the number of three-phase boundary active sites; (ii) the peculiar 3D mesoporous network formed by these materials, which gives an improved mass transport; and (iii) their excellent conducting properties that improve electron transfer.

In addition to CNTs and CNFs, other nanostructured carbons such as carbon nanocoils,[76–81] nanocages,[82] and fullerenes,[83–88] were also explored as catalyst supports for PEMFCs.

Nanodiamonds and Doped Diamonds. An interesting alternative for electrocatalyst support involves diamond powders, which are characterized by a high chemical and mechanical stability, as well as a rich surface chemistry. Depending on the preparation method, the surface of diamond powders can be H-terminated and/or feature a range of functional groups;[89–93] Moore *et al.* demonstrated that Pd nanostructures generated by impregnation on commercially available high-pressure, high-temperature diamond particles could be used as high surface area electrocatalysts for formic acid oxidation.[94] This approach was inspired by seminal studies on boron-doped diamond (BDD) films, demonstrating their suitability as dimensionally stable support for electrodeposited metal nanostructures.[95–97] Undoped diamond is an electrical insulator with a band gap greater than 5 eV. Common dopant species include nitrogen, phosphorous and boron, with the latter being the most popular due to its low charge activation energy (0.37 eV). Although doping improves the conductivity of the diamonds, it can inversely affect its stability. Conductive doped diamonds are intrinsically attractive for application as a durable catalyst support because of its specific properties, such as an extremely wide potential window, a very low background current, and in particular a high chemical and dimensional stability.[96,98–105] However, there are still some problems with doped diamonds as electrocatalyst supports:

the low conductivity, the low surface area, and the poor dispersion of the catalytic metals. In addition, it is still difficult to realize a homogeneous and controllable boron doping level in diamond powders.[106,107]

Various different routes such as co-deposition of Pt during CVD growth,[108] implantation,[109,110] thermal decomposition,[98] electroless deposition,[111] and electrochemical deposition[112–117] have been explored to optimize the Pt deposition process on BDD. Cabrera *et al.* investigated the performance of detonated diamond nanoparticles and boron-doped diamond powders, obtained by ion implantation, as supports for Pt and Pt–Ru catalysts in direct methanol fuel cells.[99,118,119]

Lu *et al.* studied the performance of Pt–Ru alloy nanoparticles electrodeposited on BDD.[117] A potentiostatic method was used to simultaneously deposit Pt and Ru at various working potentials, and the deposits were compared with those from sequential deposition. Pt–Ru electrodeposits from simultaneous deposition showed more stable CVs in sulfuric acid. The same method also resulted in higher Ru content, and could also be controlled by a choice of the electrodeposition potentials (the more negative the deposition potential, the lower the Ru content). These electrodes also exhibited higher activity and CO tolerance for the methanol oxidation reaction (MOR). The authors also observed that while methanol dehydrogenation dominated at lower overpotentials, CO_{ads} oxidation dominated at higher overpotentials.

Graphene-Based Materials. Since the existence of 2D graphene was reported, many researchers have started to shift their research target towards this material due to its outstanding properties (high conductivity and one of the fastest available electron transfer capabilities). As such, it has been widely studied as a fuel cell catalyst support, evidencing both enhanced activity and durability.[120–126] The improvements are mainly related to the fact that graphene exhibits an extraordinary modification of the properties of supported Pt cluster electrocatalysts. In addition, graphene also improves the stability of Pt, as a lower aggregation of Pt nanoparticles on functionalized graphene and higher retained activity of Pt/graphene (mostly in comparison with Pt/CB) has been observed. Due to the high available surface area, smaller Pt nanoclusters supported on graphene nanosheets can be obtained, which have a very positive effect on MOR performance. As an example, Honma *et al.* observed that the current density of a 20 wt% Pt/FLG electrocatalyst was 4 times higher (0.12 mA cm^{-2}) than that of commercial Pt/C (0.03 mA cm^{-2}).[120] In addition, the CO adsorption rate by the Pt/FLG was much smaller (*ca.* 40 times less) compared to that of Pt/CB, revealing an excellent CO poisoning tolerance of Pt/FLG.

The high surface area of graphene also allows for very high Pt loaded catalysts to be prepared. Catalysts with Pt loadings as high as 80 wt% were prepared by Kim *et al.* without the assistance of any surfactants, yielding Pt particles smaller than 3 nm.[127] The Pt/FLG catalysts with 40, 60 and 80 wt% Pt loadings showed electrochemical active surface areas (ECSA) of 53, 51 and 36 $m^2 g^{-1}$, respectively, while a conventional 40 wt% Pt/C catalyst had an ECSA of only 26 $m^2 g^{-1}$, with all Pt/FLG catalysts exhibiting much higher current densities for MOR than the commercial Pt/CB (up to three times higher).

Graphene's oxidized counterpart, *i.e.* GO, has also drawn a lot of interest and attention. Although GO has lower conductivity (a difference of two to three orders of magnitude compared to graphene), it offers a different set of properties (hydrophilicity, high mechanical strength, chemical "tunability") compared to graphene, which makes it suitable for a wide range of applications. Moreover, variable oxygen content enables tunable electronic conductivity for various applications.[128–130] The use of GO as a catalyst support material in PEMFC is one of the latest applications of GO, which has shown very promising results.[131–135] Oxygen groups introduced into the graphene structure during the preparation of GO create defect sites on surface as well as edge planes. These defect sites act as nucleation centers and anchoring sites for the growth of metal nanoparticles. Sharma *et al.* recently reported the synthesis of Pt nanoparticles on GO using the microwave-assisted polyol method.[131] The process allowed simultaneous partial reduction of GO and growth of Pt nanoparticles on the rGO support, producing Pt nanoparticles with good control over particle size and distribution. Apart from high catalytic mass activity and higher electrochemically active surface area, MOR studies also revealed I_F/I_R ratios (forward anodic peak current density to the reverse anodic peak current density) as high as 2.7 compared to 1.3 for Pt/CB, indicating a tolerance against CO poisoning. The latter was attributed to the presence of covalently bonded residual oxygen functional groups on the rGO support.

Bi- (Pt–Ru, Pt–Pd, Pt–Au)[126,136–140] and even trimetallic (Pt–Pd–Au)[141] nanoparticles supported on graphene substrates have attracted considerable interest due to their enhanced catalytic properties relative to individual nanoparticles.

Dong *et al.* demonstrated that Pt–Ru nanoparticles synthesized onto FLG exhibit high electrocatalytic activity towards both methanol and ethanol oxidation.[126] In addition, Pt/FLG revealed a much higher tolerance against CO poisoning as shown by an I_F/I_R ratio of 6.5, which was much higher than that of Pt/CB (1.4) and Pt/graphite (1.0). Furthermore, the methanol oxidation potential of Ru–Pt/graphene was shifted to 0.50 V compared with 0.65 V for Pt/FLG, suggesting that the addition of Ru can improve methanol oxidation compared to Pt alone.

Ternary alloy systems have also been studied due to their superior catalytic activity and stability to binary alloys. Wang *et al.* prepared a Pt–Pd–Au/rGO catalyst *via* simple ethylene glycol-water reduction from GO and Pt, Pd and Au precursor salts.[141] Compared with Pt–Pd/rGO and Pt–Au/rGO, the Pt–Pd–Au/rGO electrode showed: (i) the lowest onset potential and (ii) enormous current density of electro-oxidation (I_F on the forward scan was 1.5 and 2.3 times higher than those of Pt–Pd/rGO and Pt–Au/rGO, respectively) (Figure 9.4). The high catalytic activity of the ternary system

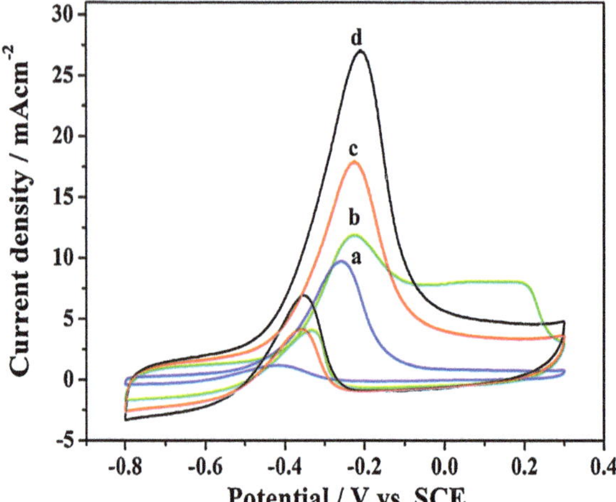

Figure 9.4 Cyclic voltammograms of (a) Pt/rGO, (b) Pt–Au/rGO, (c) Pt–Pd/rGO and (d) Pt–Pd–Au/rGO catalysts (1.0 M KOH containing 1.0 M CH_3OH solutions with a scan rate of 50 mV s^{-1}). Reprinted with permission from ref. 141. Copyright 2001 American Chemical Society.

for MOR in alkaline media was attributed to the combination of the third metal. As the scan rate and methanol concentration increased, the high MOR activity of Pt–Pd–Au/rGO catalysts was proven to be a diffusion controlled process.

Finally, the use of graphene and its derivatives in fuel cells is not limited to that of catalyst support, as we have discussed in this section, but it has also been explored as material for conducting membranes (as a composite with polymers) and also as a bipolar plate material.[142]

9.2.3.3 *Carbon as Cathode Catalyst for PEMFCs*

The preparation of cathode Pt-based electrocatalysts for ORR is practically the same as for the anode electrocatalysts in PEMFCs. Currently, expensive noble metals such as Pt, Au, Ru, and their alloys are among the best performing cathode materials.[123,143–145] Given the limitations of these Pt-based electrodes, the development of active, selective and less expensive non-precious metal and/or metal-free ORR electrocatalysts has thus triggered a great deal of interest.[23]

There are a few options that can be investigated to reduce the costs and to improve the electrocatalytic activity of Pt, especially in the presence of methanol crossover. One is to increase Pt utilization; this can be achieved either by increasing its dispersion on carbon and the interfacial region with the electrolyte.[146–148]

Hayashi *et al.* built an ideal TPB inside the mesopores of carbon support in order to examine the electrochemical reactions occurring at the nanoscale.[146] Depending on the solvent used to dilute Nafion, the reactivity toward oxygen reduction was different. Due to the hydrophobic character of the pores, Nafion diluted by a more hydrophobic solvent (2-propanol) was able to penetrate more easily. As a result, more Pt inside the mesopores was efficiently used and higher ORR reactivity was observed. Liu *et al.* reported a procedure to disperse Pt nanoparticles in the pore walls of an ordered mesoporous carbon (denoted as Pt/OMC) by a one pot organic–organic self-assembly approach.[148] These Pt/OMC electrocatalysts were found to possess not only highly stable and well-dispersed Pt nanoparticles, but also superior catalytic activities and durabilities during ORR, allowing potential applications as cathodic electrocatalysts for DMFCs and PEMFCs.

Nanostructured carbon materials such as CNF have also been studied on ORR activity using Pt as active phase.[149–151] Sebastián *et al.* studied the influence of the CNF support on the ORR for DMFCs.[149] In their study, different temperatures were used for the synthesis of herringbone CNFs with varying textural and crystalline properties. A microemulsion method was used to deposit Pt nanoparticles (2–3 nm) on the CNF support and performance was compared with that of commercial Pt/CB (Vulcan XC-72R) with Pt nanoparticle size between 2 and 3.5 nm. Highly graphitic CNFs displayed better catalytic activity, despite their lower surface area and pore volume compared to CB.

GO has also drawn a significant amount of interest as a support for ORR applications. Liu *et al.* prepared Pt nanoparticles with an average diameter of 2 nm uniformly dispersed on functionalized FLG obtained through the thermal expansion of GO.[123] The 20 wt% Pt/FLG electrocatalyst showed high initial current densities and good retention on both the ECSA and ORR activity compared with a commercial E-TEK Pt/CB (20 wt% Pt loading on Vulcan XC-72 carbon). After a durability test of 5000 cycles, Pt/FLG retained 50% of its initial ORR value, whereas Pt/CB maintained only *ca.* 34%. The improved performance was attributed to the higher stability of Pt nanoparticles. The average particle size of Pt/FLG increased from 2 to 5.5 nm (more than 75% of Pt particles retaining their size below 7 nm), whereas that of Pt/CB increased from 2.8 nm to 5.5 nm (more than 45% of the particles having a size over 7 nm). In another study, Ruoff *et al.* used a modified polyol strategy to prepare high loading (70 wt%) Pt catalysts with small nanoparticle size and uniform particle dispersion (*ca.* 2.9 nm) supported on reduced graphene oxide platelets.[152] The Pt/rGO composite catalyst not only outperformed the catalytic activity for ORR, but also has enhanced fuel cell polarization performance compared with a commercially available Johnson Matthey Pt/C catalyst (75 wt% Pt). The maximum power density of the Pt/rGO composite was about 128 mW cm^{-2}, 11% greater than the Pt/CB commercial catalyst.

Pt–Transition Metal Alloys and Non-Pt Metals. Another successful approach to enhance the electrocatalytic O_2 reduction is by alloying Pt with transition metals.[153] This enhancement in electrocatalytic activity has been interpreted in different ways, and several studies have been conducted to make an in depth analysis of the surface properties of the proposed alloy combinations.[154–156]

Graphene-supported Pt, Pt_3Co and Pt_3Cr alloy nanoparticles were prepared by ethylene glycol reduction followed by pyrolysis at 573 K.[157] The ECSAs were found to be 65, 57 and 55 $m^2 g^{-1}$ for graphene-supported Pt, Pt_3Co and Pt_3Cr catalysts, respectively. The amount of oxide formation on these catalysts follows the order Pt > Pt_3Co > Pt_3Cr, because of the increased donor ability of Co and Cr. The ORR activity of the graphene supported Pt_3Co and Pt_3Cr catalysts was increased by *ca.* 3–4 times compared with graphene supported Pt. The higher activity of Pt alloys was attributed to the suppression of hydroxyl formation on Pt surface.[157] Moreover, the overpotential for ORR of the Pt alloys is 45–70 mV less than that of Pt alone.

Heteroatom-Doped Carbon Materials. Finally, one last approach to improve ORR activity is to develop heteroatom-doped carbon-based electrocatalysts.[158] This concept holds great promise and has drawn much attention due to excellent electrocatalytic activities as well as the low cost, good durability, and environmental friendliness of the catalyst. These materials can be used either as electrocatalysts by themselves (metal-free catalysis) or as a support for different metals to further improve their performance. The use of doped carbons as metal-free catalysts has already been discussed in detail in Chapter 6 (Section 6.4.1), whereas metal-containing doped carbon materials are discussed over the next few paragraphs.

Heteroatom-doped carbon nanomaterials present better conductivity compared with undoped ones and have been used as catalyst supports in fuel cells. The presence of *e.g.* nitrogen (one of the most common dopant species) results in good coverage and dispersion of the *in situ* synthesized Pt nanoparticles compared with the undoped carbonaceous surface (see Section 6.5). This advantage, along with the N-doping induced increase of the electronic conductivity, has led to much enhanced electrocatalytic activities for carbon nanomaterials.[16,159]

Chen *et al.* synthesized CNTs and N-doped CNTs with a floating chemical vapor deposition method under similar conditions for a comparative test.[159] Pt/N-CNTs showed a more uniform dispersion and a smaller particle size than regular Pt/CNTs and exhibited a higher catalytic activity for ORR, as indicated by a larger kinetic current, higher half-wave potential, and higher four-electron transfer efficiency to H_2O. Pt/N-CNTs also showed higher fuel cell performance as a cathode catalyst in a single cell test. Hu *et al.* developed a facile strategy for the preparation of Pt–Co/N-CNT electrocatalysts.[16] The authors observed that Pt-based alloyed nanoparticles with a size of *ca.* 3 nm were highly and homogeneously dispersed over the N-CNTs. Compared with commercial Pt/C and monometallic Pt/N-CNT catalysts, binary Pt–Co/N-CNT catalysts showed much higher electrocatalytic activities and similar stabilities for oxygen reduction in an acidic electrolyte, but with far lower consumption of expensive Pt. It was proposed that the good performance of the catalysts mainly arose from: (i) high dispersion of Pt-based species, (ii) the alloying effect of Pt/Co, and (iii) the intrinsic catalytic capacity of the highly conductive N-CNTs.

The fast electron transport mechanism offered by N-doped graphene can particularly enhance the ORR activity in fuel cells. Jafri *et al.* used N-doped graphene nanoplatelets as a Pt nanoparticle support for electrocatalytic studies.[160] Graphene nanoplatelets were synthesized by thermal exfoliation of GO and further treated in nitrogen plasma to produced N-doped (3 at%) graphene nanoplatelets. MEAs fabricated using Pt/N-G and Pt/G as the ORR catalyst showed a maximum power density of 440 mW cm^{-2} and 390 mW cm^{-2}, respectively. The improved performance of Pt/N-G was attributed to the formation of pyrrolic nitrogen defects that increased the anchoring sites for the deposition of Pt on the surface leading to increased electrical conductivity and carbon–catalyst interaction, leading to an increase in the conductivity of neighboring C atoms.

Non-precious metal catalysts supported on N-graphene have also been studied for the ORR. Wu *et al.* successfully fabricated 3D N-doped graphene aerogel-supported Fe_3O_4 nanoparticles *via* a combined hydrothermal self-assembly, freeze-drying, and thermal treatment process.[161] Due to the 3D macroporous composite structure and high surface area, the resulting Fe_3O_4/N-GA showed excellent electrocatalytic activity for the ORR in alkaline electrolytes, exhibiting higher current density, lower ring current, lower H_2O_2 yield, higher electron transfer number (*ca.* 4), and better durability, potentially making Fe_3O_4/N-GA a non-precious metal cathode catalyst candidate for fuel cells. Liang *et al.* grew Co_3O_4 nanocrystals on reduced mildly oxidized GO (rmGO) both with and without NH_4OH, obtaining Co_3O_4/rmGO and Co_3O_4/N-rmGO catalysts.[162] The authors showed that Co_3O_4 (a material with little ORR activity by itself), when grown on reduced mildly oxidized graphene oxide (rmGO), exhibited surprisingly high performance in both the ORR and oxygen evolution reaction (OER) in alkaline solutions. Moreover, the hybrid exhibited comparable ORR catalytic activity to a commercial carbon-supported Pt catalyst (20 wt% Pt on Vulcan XC-72) and superior stability, thus leading to a new bi-functional catalyst for the ORR and OER.

Research is also focused on non-precious metal catalysts involving conjugated heterocyclic conducting polymers such as polypyrrole (Ppy), polyaniline (Pani), poly-(3-methylthiophene) (P3MT) and poly-(ethyleneimine) with nitrogen-doped carbon hybrids.[163–168]

The ORR activity and stability of carbon-supported materials with or without cobalt based on different heterocyclic polymers was investigated by Smit *et al.*[165] The addition of cobalt resulted in improved electrocatalytic activity of C–P3MT and C-Pani catalysts even though their thermal stability was slightly reduced at lower temperatures due to the degradation of cobalt oxides. The potential at which the ORR occurred was found to be the highest for Ppy-C-Co (−0.2 to 1 V *vs.* SHE), followed by Ppy-C, and P3MT-C-Co. Based on these results, the Ppy-C-Co catalyst is a suitable material for use in PEMFCs, even though the potential range in which the ORR occurs needs to be further improved. Popov *et al.* synthesized N-doped ordered porous carbons *via* a nanocasting process using polyacrylonitrile as the carbon and nitrogen precursor and mesoporous silica as hard template.[169] RDE measurements demonstrated a *ca.* 0.88 V (*vs.* NHE) onset potential for ORR in 0.5 M H_2SO_4, while a current density of 0.6 A cm^{-2} at 0.5 V was obtained in a H_2/O_2 PEMFC using 2 mg cm^{-2} catalyst loading.

9.2.4 Carbon-Supported Electrocatalysts for AFCs

Alkaline fuel cells have one of the longest histories of all fuel cell types, as it was first developed as a working system by fuel cell pioneer F. T. Bacon in the 1930s. This technology was further developed for the Apollo space program that took man to the moon. AFCs allow the use of non-precious metal catalysts as cathode electrodes due to the faster ORR kinetics and better stability of these materials under basic conditions.[170–172] This unique characteristic elevates AFCs to a promising candidate for portable power sources and electric vehicles applications.

9.2.4.1 *Working Principle of AFCs*

In an alkaline fuel cell (Figure 9.5), the overall reaction is the same as that for PEMFC, but the reactions at each electrode are different. Hence, hydroxyl ions are produced at the cathode and migrate to the anode side where they react with hydrogen; some of the water formed at the anode diffuses to the cathode and reacts with oxygen to form hydroxyl ions in a continuous process. The overall reaction produces water and heat as by-products and generates four electrons per mole of oxygen, which travel *via* an external circuit producing the electrical current.

For these reactions to proceed continuously, the HO^- ions must be able to pass through the electrolyte, and there must be an electrical circuit for the electrons to go from the anode to the cathode. Also, comparing

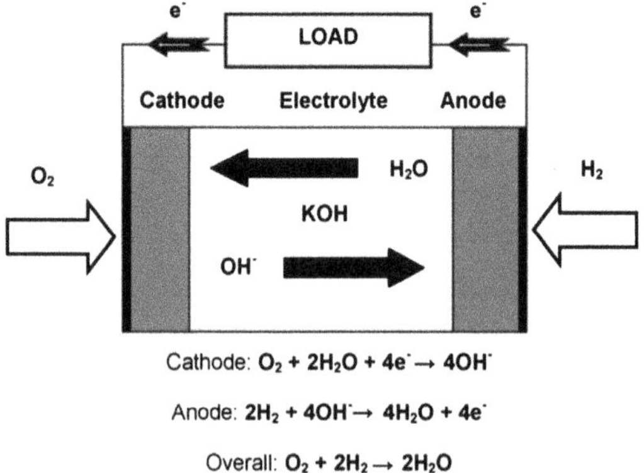

Figure 9.5 Diagram showing the fundamentals of an alkaline fuel cell. Reprinted with permission from ref. 173. Copyright 2011 Elsevier Science Ltd.

the reactions on both electrodes, we see that we need twice the amount of hydrogen compared to oxygen, as with the acid electrolyte. Note that although water is consumed at the cathode, it is created twice as fast at the anode.

The electrolyte obviously needs to be an alkaline solution. Sodium hydroxide and potassium hydroxide solutions being of lowest cost, highly soluble, and not excessively corrosive, are the main candidates. However, potassium hydroxide solution is almost exclusively used as the electrolyte because it has a higher ionic conductivity than the sodium hydroxide one, and potassium carbonate has a higher solubility product that sodium hydroxide, which renders the former less likely to precipitate.

Despite their decline in relative importance, and the lack of high-profile research interest, the outlook for the AFC is not so bleak. The reason is that the AFCs do have a number of clear and fundamental advantages over other types of fuel cells. The first important advantage is that the activation over-voltage at the cathode is generally lower than that with an acid electrolyte. This is the most important voltage loss in low-temperature fuel cells. It is not clearly understood why the reduction of oxygen proceeds more rapidly in an alkaline system, but this fact has long been observed. This allows AFCs to have operating voltages as high as 0.875 V per cell, considerably higher than that with, for example, PEMFCs. Another advantage of great importance is related to the system cost. The electrolyte cost of the AFC is far less expensive than any other type. Also, the electrodes (particularly the cathode) can be made from non-precious metals, and no particularly exotic materials are needed. The electrodes are thus considerably cheaper than other types of fuel cells. The use of a non-noble metal catalyst is possible because the oxygen reduction reaction in alkaline media is more facile than in acidic media. As a consequence, higher voltages at a given current density can be obtained, leading to a higher electrical efficiency. The final advantage is related to the absence of bipolar plates on AFCs, as these play a very significant contribution to the cost of the PEMFC, for example. This then has the benefit of reducing the cost, but it is also an important cause of the lower power density of AFCs, in comparison with the PEMFC.

The main disadvantage of AFCs is that carbon dioxide can react with the electrolyte to form carbonate (eqn (9.7)), decreasing the electrolyte conductivity, oxygen solubility and electrode activity. The formation of precipitated carbonate (eqn (9.8)) can also lead to the blockage of the electrolyte pathways and electrode pores.

$$CO_2 + 2OH^- \rightarrow CO_3^{2-} + H_2O \tag{9.7}$$

$$CO_3^{2-} + 2K^+ \rightarrow K_2CO_3 \tag{9.8}$$

In general, AFC electrodes consist of several PTFE-bonded CB layers, which fulfil different functions. AFC electrodes can be made of different materials with different structures, but modern electrodes tend to use high surface area carbon supported catalysts and PTFE to obtain the necessary three phase boundary. Electrode performance in AFCs depends on catalyst surface area rather than catalyst weight. As with all other fuel cells, the catalyst loading is a critical parameter in determining performance. The nature of the catalyst support is also of prime importance to achieve high catalytic activity.

9.2.4.2 *Carbon as Anode Catalyst for AFCs*

Anodes for AFC applications consist of a composite containing a metal catalyst and PTFE supported by a metal web on the backside of the electrode. The metal web electrically conducts the electrode and guarantees mechanical stability. The PTFE is used as an organic binder to hold the catalyst particles together. Due to its hydrophobic character the PTFE is important for gas transport in the electrode, and yields a hydrophobic pore system that is not flooded by the alkaline electrolyte so that the electrode can be supplied with the fuel gas from the backside.

The anode electrode for AFCs has been less studied than the cathode, where catalysts containing platinum-group metals such as Pt/Pd have shown good performance and stability.[174,175] Nickel has also been studied as a potential low cost catalyst for AFCs anodes.[176,177] Raney® nickel, which offers the advantage of high surface area, has been shown to be one of the most active catalysts for the hydrogen oxidation reaction in alkaline media, but has been reported to suffer from deactivation,[178–180] which could be improved by the addition of copper[181] or treatment with H_2O_2.[182]

The power output and lifetime of alkaline fuel cells are directly linked to the behavior of the cathode, where most of the polarization losses occur. As a consequence, cathode development requires special attention to find the best catalyst and electrode structure to combine performance and stability.

9.2.4.3 *Carbon as Cathode Catalyst for AFCs*

Platinum is the most commonly used and active catalyst for the ORR and all of the platinum-group metals reduce oxygen in alkaline media according to the direct 4-electron process.[183–189] At a very low Pt/C ratio, the overall number of electrons exchanged is approximately two due to the carbon contribution, but increases as the Pt/C ratio increases, reaching four electrons at 60 wt% Pt.[190] Pt-based alloys have been studied and generally exhibit higher activity and stability than Pt-alone.[191,192] The enhanced electrocatalytic activity of Pt-alloy systems has been explained by a number of phenomena, including: (i) reduction in Pt–Pt bond distance, favoring the adsorption of oxygen; (ii) the electron density in the Pt 5d orbital; and (iii) the presence of surface oxide layers.[190,193–196]

Ag and Ag-alloys are commonly used as catalysts in AFCs due to their low cost and similar ORR activities to Pt and Pt-alloys.[197,198] However, the performance of Ag based catalysts is affected by sintering of the Ag nanoparticles and the electrochemical corrosion of the catalyst supports.[56] The incorporation of Ag catalysts on CNTs is a very promising method to increase the stability and ORR activity of these catalysts in alkaline environments.[198] It was reported that an increase of Ag content (from 5 to 9 wt%) and CNT surface area results in a higher amount of active sites per geometric surface area of the electrode and thus faster ORR kinetics. Recently, Ag nanoparticles were deposited directly on MWCNTs by sputter deposition[199] and *in situ* growth.[56] Ag/MWCNTs catalyst exhibited high ORR activity in alkaline media and the reduction process proceeded *via* a four-electron pathway. The high ORR activity of Ag/MWCNTs was attributed to the ability of dispersed Ag nanoparticles on the surface of MWCNTs to further reduce HO_2^- formed at the surface of CNTs. Even though the ORR activity of Ag/MWCNTs is still lower than pristine Pt/CB (20 wt%),[56] Ag/MWCNTs are promising ORR catalysts for AFCs.

Other metals such as Pd and Pd-alloys are also used as non-Pt catalysts since Pd has similar properties to Pt.[200,201] Pd has slightly lower ORR activity than Pt, and by addition of a suitable metal, such as Co or Fe, the ORR activity of Pd can surpass that of Pt.[202] The incorporation of transition metals (Co, Fe, Cr, and Ni) to Pd enhances the ORR activity due to modification of electron configuration and alteration of surface species and composition.[201] Pd has fully occupied d-orbitals while the alloyed transition metal has low occupancy of d orbitals;[203] the d-orbital coupling effect between those two metals decreases the required Gibbs free energy of the electron transfer steps in the ORR, leading to an increase in ORR kinetics. Among the non-Pt alloys tested, Pd–Co alloy exhibited high ORR activity in alkaline media along with Au–Co and Ag–Co alloys.[204] Cobalt is stable in alkaline media even though it dissolves from the metal alloy as cobalt oxide in acidic media.[205]

One class of materials that has attracted some attention towards ORR in alkaline medium is iron phthalocyanines (FePc), cobalt phthalocyanines (CoPc) and other similar M-N4-macrocycles supported on different carbon nanomaterials, due to their ability to promote the $4e^-$ oxygen reduction to water without the formation of peroxide intermediates.[206–208]

Manganese oxide-based catalysts have also captivated many researchers in recent years for the ORR in alkaline media due to the low cost of the material.[209,210] The electrocatalytic properties of MnO_x have been demonstrated to be highly dependent on its chemical composition, texture, morphology, oxidation state, and crystalline structure. As an example, the catalytic activity of MnO_2 depends on the crystallographic structures, namely α-, β- and γ-MnO_2.[211] ORR catalytic activity order in alkaline media was found to be α- > β- > γ-MnO_2 and a 4-electron ORR pathway was achieved when α-MnO_2 nanostructures were used.[209,212] The change was attributed to a combined effect of their intrinsic tunnel (interspace in the stack of $[MnO_6]$ octahedron) size and electrical conductivity. However MnO_2 has low electrical conductivity and it is always incorporated into a conducting carbon material such as CNTs to overcome this issue.[212–219] Previous studies have shown that nanostructured $Ni^{(II)}$-doped MnO_x nanoparticles supported on high area carbon exhibit remarkable catalytic activity for ORR.[216,217] The four-electron

ORR pathway is favored on such materials, probably because the doping transition metal can stabilize an intermediate $Mn^{(III)}/Mn^{(IV)}$ phase, which enhances the oxygen bond splitting.[218] In addition, MnO_x-based catalysts exhibit high tolerance to fuel crossover, including methanol, ethanol, and borohydrides.[219] Despite these promising results, the main obstacle that needs to be addressed when using these materials is the deactivation of MnO_x/C catalysts in concentrated LiOH or KOH electrolytes (*e.g.*, concentration >5 M) because of an insufficient activity of water, thereby limiting the necessary proton insertion into the MnO_x lattice, a prerequisite for ORR activity.[217] In addition, when LiOH electrolyte is used, Li^+ ions may be inserted into the MnO_x lattice and stabilize $Mn^{(III)}$ and the oxygen groups at the carbon surface, which prevents their role as redox-mediating species and further blocks the catalytic process, eventually yielding an increased ORR overpotential.[220]

Nitrogen doped carbon materials were among the first to be studied as electrocatalytic cathode materials for AFCs.[221–225] Nitrogen doping of carbon leads to similar ORR kinetics with commercial Pt/C electrocatalyst, and higher kinetics than commercial Ag/C catalysts.[225,226] Apart from the contribution of nitrogen functionalities (pyridinic, quaternary and pyrrolic nitrogen and pyridine-nitrogen-oxide),[158,224,227,228] there are a few reports in the literature attributing the ORR activity in alkaline media to the number of defects and edge sites of carbon upon nitrogen doping,[229–231] which are also considered as active sites for ORR. Hence, despite the fact that the origin of the ORR activity of N-doped nanocarbons is still unclear, nitrogen-doped carbons are a promising candidate as cathode catalysts in AFCs, both as a metal-free electrocatalyst (see Chapter 6, Section 6.4.1 for more details) or as a support for different ORR-electroactive metals. Tan *et al.* synthesized a novel MnO-containing mesoporous nitrogen-doped carbon (m-NC) nanocomposite (MnO/m-NC) *via* an *in situ* MnOx-template method.[232] Throughout this synthesis process and electrocatalysis, manganese oxides played very important roles, including serving as a template for carbon coating and creating synergetic active sites for the ORR. Due to the synergetic effects between MnO and m-NC, the resulting MnO/m-NC composite catalyst exhibited high ORR activity in alkaline solution. The high surface area of the MnO/m-NC nanocomposite, resulting from its mesoporous structure, also made a remarkable contribution to the high ORR activity observed. In addition to excellent electrocatalytic activity, the as prepared MnO/m-NC composite catalyst exhibited superior stability and methanol tolerance compared to the commercial Pt/C catalyst for ORR, indicating a promising cathode catalyst candidate for alkaline methanol fuel cell applications.

In summary, AFCs offer facile ORR kinetics in alkaline media allowing a wide range of low cost, non-platinum electrocatalysts to be used in the cathode, namely, nitrogen doped carbonaceous materials, Ag and Pd nanoparticles, transition metal oxides and M-N4-macrocycles supported on carbon. Among these materials, nitrogen doped CNTs and nitrogen doped FLG/rGO exhibit the highest ORR activity in alkaline media, even though the activity is still lower when compared to pristine Pt/C.

9.2.5 Corrosion and Stability Issues

One of the more significant problems hindering the large-scale implementation of PEMFCs technology is the loss of performance during extended operation. These fuel cells are operated under extremely harsh conditions.[233–236] The anode catalysts are exposed to a strong reducing H_2 atmosphere; at the cathode, the catalysts are under strong oxidizing conditions: high O_2 concentration, high potential (>0.6 V *vs.* SHE), and sometimes even higher potentials (*e.g.*, >1.2 V) for short periods of time.[237] In addition, both the anode and cathode of PEMFCs are operated under the condition of low pH (<1), high temperature (353 K or higher), and with significant levels of water in both vapor- and liquid-phase.[238]

Various failure modes have been identified in a fuel cell, and include catalyst particle ripening (particle coalescence), preferential alloy dissolution in the catalyst layer, carbon support oxidation (corrosion), catalyst poisoning, membrane thinning and pin-hole formation, loss of sulfonic acid groups in the ionomer phase of the catalyst layer or in the membrane, bipolar plate surface film growth, hydrophilicity changes in the CL and/or GDL, and PTFE decomposition in the CL and/or GDL. Component decay or failure is affected by many internal and external factors, including material properties, fuel cell operating conditions (such as humidity, temperature, cell voltage, *etc.*), impurities or contaminants in the feeds, environmental conditions (*e.g.*, subfreezing or cold start) and operation modes (such as start-up, shutdown, *etc.*).[239–241] In addition, the degradation processes of different components are often interconnected in a fuel cell system. Despite the importance of all of these issues, we will be only addressing those directly related to carbon corrosion in this section.

Carbon is an excellent material for supporting electrocatalysts, allowing facile mass transport of reactants and fuel cell reaction products, and providing good electrical conductivity and stability.[242] However, under prolonged conditions of high temperature, high water content, low pH, high oxygen concentration, and/or high potential,[26] oxidation of carbon (carbon corrosion) is prone to acceleration. This corrosion weakens the attachment of Pt particles to the carbon surface, and eventually leads to structural collapse and the detachment of Pt particles from the carbon support, which results in severe Pt agglomeration and performance degradation during long-term operation. The loss of carbon due to corrosion also has a detrimental effect over the electronic behavior of the catalyst layer, as isolated Pt particles are not able to participate in the electrochemical reactions. In addition, when carbon oxidizes to form surface groups, it changes the hydrophilicity of the electrode, causing the development of wettability and probably slackening mass transport during long-term operation of the PEMFC.[29,243] The properties of carbon can also influence its degradation rate, as carbon supports with a higher degree of graphitization and/or a lower specific surface area exhibit lower degradation.[237]

9.2.5.1 Carbon Corrosion Mechanism

Carbon corrosion may occur *via* a chemical or an electrochemical route.[242,244] More specifically, carbon oxidation takes place along two pathways that are believed to proceed by electron transfer, followed by hydrolysis and CO_2 production: (i) incomplete oxidation leads to the formation of surface groups (eqn (9.9) and (9.10)); (ii) complete oxidation leads to gaseous carbon dioxide (eqn (9.11)).[245]

$$C \rightarrow C^+ + e^- \tag{9.9}$$

$$C^+ + H_2O \rightarrow CO + 2H^+ + e^- \tag{9.10}$$

$$2CO + H_2O \rightarrow CO + CO_2(g) + 2H^+ + 2e^- \tag{9.11}$$

$$\textbf{TOTAL}: C + 2H_2O \rightarrow CO_2 + 4H^+ + 4e^- (E^\circ = 0.207\text{V } \textit{vs.}\ \text{SHE}) \tag{9.12}$$

In fact, carbon corrosion (eqn (9.12)) will not be too severe under normal operating conditions (*i.e.*, potential between 0.4 and 0.7 V). Therefore, within this potential range the impact of carbon instability is limited for PEMFC electrocatalyst durability. During the operation of PEMFC, carbon can react with some transient oxygen radicals, such as $HO^\bullet$ and $HOO^\bullet$,[246] generated by the catalyst and/or water to form oxygen functionalities (*e.g.*, lactones, carbonyls, phenols, carboxylic acids, *etc.*),[29] which then proceed to form gaseous products (CO and CO_2). Furthermore, the presence of Pt is also known to accelerate the oxidation rate, and the corrosion may be more severe at places where Pt particles reside.[238,247]

The corrosion reaction is also markedly accelerated under extreme conditions, especially at abnormal potentials. For example, when voltage reversal occurs during cell operation, carbon support oxidation becomes so serious that it can cause irreversible damage to fuel cells.[248–251] This situation can be observed when the fuel cell experiences fuel starvation resulting from bad flow distribution, gas blockages, or sudden current changes to heavy loads in transient conditions, as well start-up and shut-down processes. According to the literature, during unprotected and frequent start-up as well as prolonged shut-downs of a fuel cell, the local cathode potential can reach 1.5 V due to partial hydrogen coverage in the anodes, which significantly accelerates carbon corrosion.[252,253]

9.2.5.2 Fuel Starvation

The carbon corrosion reaction takes place at the electrodes as a result of gross fuel starvation. When the fuel is insufficient to provide the expected current for the PEMFC, the potential value of the anode continues to increase. With fuel starvation, the cell potential will decrease to a value substantially below normal and drive the cell into reverse operation, with the anode potential higher than the cathode potential. This state involves carbon oxidation (eqn (9.13)) and water electrolysis (eqn (9.14)) at the fuel cell anode in order to provide the required protons and electrons for the ORR happening at the cathode.[250]

$$C + 2H_2O = CO_2 + 4H^+ + 4e^- \quad (E_{298K} = +0.207V\,\textit{vs.}\ SHE) \tag{9.13}$$

$$H_2O = 1/2O_2 + 2H^+ + 2e^- \quad (E_{298K} = +0.207V\,\textit{vs.}\ SHE) \tag{9.14}$$

These reactions can be observed when the fuel cell experiences bad flow distribution, gas blockages, or sudden current changes due to heavy load under transient conditions.

9.2.5.3 Start-Up/Shut-Down Cycling

Fuel cell start-up and shut-down are operational modes that can have a profound influence on fuel cell durability. Another mode of carbon corrosion arises from non-uniform distribution of fuel to the anode and the crossover of reactant gas through the membrane. A fuel cell used in an automotive system is expected to experience 30 000 start-up/shut-down cycles during its life.[248] Both start-up and prolonged shut-down can result in this kind of carbon corrosion.[254] Under conditions of prolonged shut-down, all the hydrogen will eventually cross over from the anode to the cathode, resulting in the anode flow channels being filled with air. In this case, fuel cell start-up will create a transient condition in which fuel exists at the inlet but the outlet is still fuel-starved at the anode side. This localized fuel starvation can induce the local potential at the cathode to be higher than 1.8 V, causing serious deterioration in fuel cell performance and durability.[251]

Figure 9.6 shows what could happen in the case of an unprotected fuel cell shut-down. After the shut-down, there will be unreacted air and hydrogen at the cathode and anode, respectively, resulting in an open-circuit voltage (OCV) of *ca.* 1.0 V. If the anode exhaust port is not closed after shut-down, air will gradually diffuse into the anode side, creating an air/hydrogen boundary, represented by the dotted line, and the boundary moves to the other end of the flow channel. The boundary line creates four distinct regions marked A–D. Both regions B and C are connected with region A through the membrane, and thus they will bear a potential of OCV. At the same time, since the entire cathode is about 1.0 V higher than the entire anode, the potential of region D will be close to the sum of the potentials at regions B and C, which is about two times that of OCV, resulting in a quick corrosion of carbon in that region.

All the potentials mentioned refer to the potential difference between the electrode and the membrane electrolyte; and the potential of the membrane

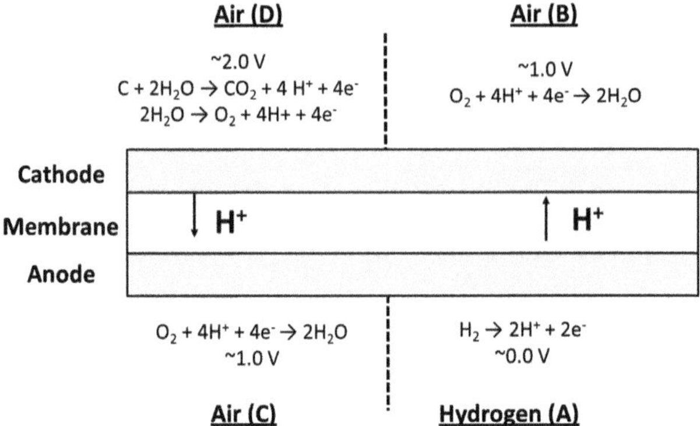

Figure 9.6 Schematic representation of reactions in four distinct regions when an air/fuel boundary is formed at the anode. Adapted from ref. 249.

electrolyte section shared by regions A and B could be largely different from that shared by regions C and D. At such a high potential, both water electrolysis and carbon oxidation could happen in region D at significant rates. Resulting from the reactions occurring in regions A–D, there will be electron flows between regions A and C, and between regions B and D, as well as proton flows between regions A and B, across the dotted line, and between regions C and D, as shown in Figure 9.6 The boundary will disappear after all the hydrogen is consumed or displaced, and the anode will be filled with air eventually, leading to the disappearance of the high potential in region D. However, when hydrogen is fed to the anode during the restart of the fuel cell, a new air/fuel boundary forms, causing the cathode in the region that faces the anode air section to corrode.

Carbon corrosion through start-up/shut-down cycling can also explain why the cathode layer, after start-up and shut-down aging, is normally thinner than the anode layer.[249,255] The sharp voltage change might also cause some damage to Nafion, and the combination of these effects could entirely explain MEA degradation.[249,251]

9.2.5.4 *Contamination of the Electrode/Electrocatalyst*

Contamination can also have adverse effects on the performance and life of PEMFCs. Contamination is the process through which impurities pollute the cell components and/or initiate chemical attack and slow down the actual reactions taking place in the electrode. The contamination products can originate from components inside the cell or be transported into the cell by the reactants. As a result, metal, alkaline metal and ammonium ions, silicon and catalyst particles as well as carbon monoxide, nitrogen oxides (NO_x) or sulfur dioxide (SO_2) can appear in the cell. Even trace amounts of impurities result in considerable degradation of performances.[256,257]

Small amounts of carbon monoxide are sufficient to poison the anode reaction resulting in a lower cell potential output and lower energy conversion efficiency. The basic theory behind CO-poisoning is that CO-molecules are adsorbed on the platinum catalyst sites and block the hydrogen from reaching the platinum particles. Although CO-poisoning is a slow process, it can lead to a significant performance loss, which means voltage loss over time. Hence, the most successful approach to solving electrode poisoning problem is the development of CO tolerant electrocatalysts. In order to avoid, or delay this poisoning effect, Pt is often supported on nanostructured carbons (often carbon nanotubes and more recently graphene), resulting in an increased tolerance towards surface deactivation. Furthermore, the preparation of bimetallic catalysts where Pt can undergo alloy formation with other metals, such as Ru, is also known to inhibit CO chemisorption.

9.2.5.5 GDL Degradation

As mentioned earlier, the GDL is a carbon-based porous substrate situated between the CL and the flow field that enables gas phase transport, water transport, electronic and thermal conduction, in addition to mechanical support. The GDL consists of a macro-porous layer made of carbon fiber paper or carbon cloth that is covered with a micro-porous layer (MPL) made of CB powder and a hydrophobic agent (PTFE). To date, the GDL is the least studied MEA component in terms of durability and degradation. Several GDL degradation mechanisms have been proposed: carbon oxidation,[247,258] PTFE decomposition,[259] and mechanical degradation as a result of compression.[260] The first two mechanisms cause hydrophobicity loss and changes in the GDL pore structure, resulting in an increase in the water content of the GDL and MPL and thus hindering gas-phase mass transport.[261–263] It should be noted that the carbon fibers of the GDL and the CB particles of the MPL are more stable than the CB in the CL due to the absence of Pt that can catalyze the electrochemical oxidation of carbon. However, chemical surface oxidation of carbon by water cannot be excluded.[247,258]

9.3 Carbon (Nano)materials for Solar Cells

Solar cells utilize the energy from the sun by converting solar radiation directly into electricity. There are three basic types of solar cell: (i) crystalline cells; (ii) multi-junction cells and (iii) amorphous or thin-film solar cells. To date this field has been dominated by solid-state junction devices, usually made of silicon, and profiting from the experience and material availability resulting from the semiconductor industry. However, thin film technologies have several advantages based on the fact that they use materials that absorb much more strongly than silicon. This quality allows for the creation of thinner devices, which use less material, and which can be created *via* high throughput deposition schemes such as solution processing, chemical vapor deposition, and sputtering. Nevertheless, these solar cells have low efficiencies, or use rare (*e.g.*, indium, tellurium) or toxic (*e.g.*, cadmium) elements.

Among thin-film solar cell are dye-sensitized (DSSCs), quantum dot (QDSCs) and organic/polymer solar cells. These devices are based on the concept of charge separation at an interface of two materials of different conduction mechanisms. Conventional organic photovoltaic devices use polymers or small molecules to absorb light and separate charge. Rather than an electron/hole pair, an exciton is created, which has a diffusion length on the order of 10 nm. By stacking multiple cells in a tandem structure, each with a different absorption range, devices exhibiting efficiencies over 11% have been fabricated. On the other hand, in DSSC technology the charge generation is done at the semiconductor–dye interface and the charge transport is done by the semiconductor and the electrolyte. Thus, optimization of the spectral properties can be done by modifying the dye alone, while the carrier transport properties can be improved by optimizing the semiconductor and the electrolyte composition. By 1997, efficiencies over 10% were achieved; however, little progress has been made since then, as research has focused on ways to reduce the material and processing costs. QDSCs are based on the DSSC architecture (also named Grätzel cell), but employ low band gap semiconductor nanoparticles, fabricated with crystallite sizes small enough to form quantum dots (such as CdS, CdSe, Sb_2S_3, PbS, *etc.*), instead of organic or organometallic dyes as light absorbers. The QD's size quantization allows for the band gap to be tuned by simply changing the particle size.

Owing to its high efficiency, low cost and ease of fabrication, DSSCs represent an attractive alternative to solid state photovoltaics.[264] The working principle of this device is shown in Figure 9.7. The first step is the absorption of a photon by the sensitizer *S*, leading to the excited sensitizer S^*, which injects an electron into the conduction band of the semiconductor, leaving the sensitizer in the oxidized state S^+. The injected electron flows through the semiconductor network to arrive at the back contact and then through the external load to the counter electrode to reduce the redox mediator, which in turn regenerates the sensitizer to complete the circuit.

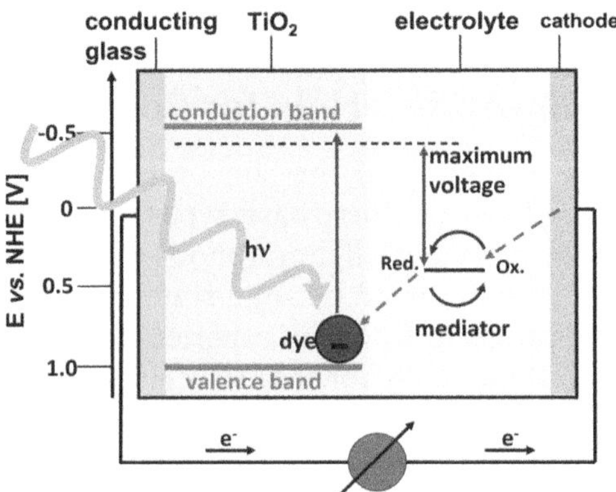

Figure 9.7 Operating principles and energy level diagram of dye-sensitized solar cell. Reprinted with permission from ref. 265. Copyright 2011 Elsevier Science Ltd.

The DSSC key components are the dye-sensitized photoanode, the electrolyte solution with a redox mediator, and the cathode material. The latter is typically an optically transparent film of Pt nanoparticles on F-doped SnO_2 (Pt-FTO) and the former is the I^{3-}/I^- redox couple in an aprotic electrolyte medium. A critical aspect of this type of optoelectronic device is the transparent conductive electrode through which light couples into the device. These typically use transparent indium tin oxide (ITO) or fluorine-doped tin oxide (FTO) as such electrodes. However, the scarcity of indium reserves, intensive processing requirements, and highly brittle nature of metal oxides imposes serious limitations on the use of these materials for applications where cost, physical conformation, and mechanical flexibility are important. There were several attempts to replace Pt-FTO by other materials such as carbons,[266–268] conducting polymers,[269,270] and others,[271,272] but none of the alternative materials was superior to Pt-FTO in terms of optical transparency and electrochemical activity for the I^{3-}/I^- redox couple. However, the redox potential of I^{3-}/I^- (*ca.* 0.35 V *vs.* SHE) is too low to achieve the optimum voltage of the DSSC system and, consequently, the best power conversion efficiency.[273]

Porous carbon-based composite materials are attractive when incorporated as transparent electrodes, counter electrodes, electrolytes, and other parts present in the solar cell. Using various porous carbon materials such as CNTs, graphene, fullerenes, mesoporous carbon, and other fabricated hollow nanoparticles has been found to be beneficial for solar cell performance.[274,275] Various carbon-based composites were synthesized and studied, such as the mixture of graphene, CNTs, and ionic liquid as electrolytes,[276] Pt-decorated CNTs as the catalytic layer on counter electrodes,[277] and TiO_2-modified/CNTs hybridized material as working electrodes,[278,279] as well as polymer/CNTs,[280,281] reduced graphene-CNTs,[282] P25-graphene,[283] CNF–CdSe quantum-dots,[284] and TiN-CNTs composites[285] as counter electrodes. The efficiency of the DSSCs using most composites as counter electrodes is comparable to that with a Pt film electrode. Instead of spinning or depositing CNTs on substrates, ordered CNT arrays grown on FTO glass and graphene paper can also be synthesized and used as counter electrodes,[286,287] as aligned structures would provide faster electron transfer and, thus, more stable performance.

Many researchers focus on the design of CNTs–silicon heterojunction solar cells for achieving high efficiency by enhancing photon absorption, inhibiting charge recombination, and reducing internal resistance.[288,289] Hybrid solar cells composed of a heterojunction cell and a photoelectrochemical cell have also been reported.[290] In this case, a thin DWCNTs film formed a heterojunction with the silicon nanowire array (SiNW) and also functioned as the transparent counter electrode of the photoelectrochemical cell (Figure 9.8). The low reflective SiNW arrays generated most of the charge carriers, and the presence of the redox electrolyte (containing 40% hydrobromic acid and 3% bromine) in the gap between SiNWs facilitated the collection of carrier from the surface of SiNWs. The conversion efficiency obtained with this hybrid cell was higher than those of previously reported SiNW array-based photoelectrochemical solar cells, in which thin platinum

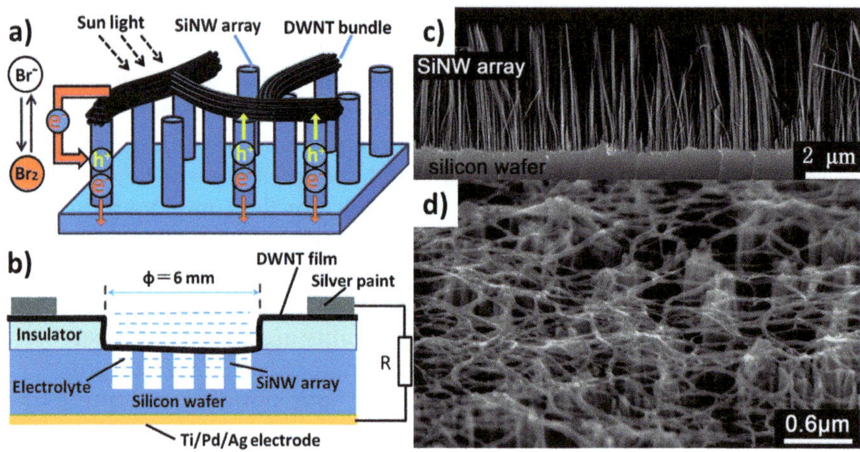

Figure 9.8 (a) Schematic illustration of a hybrid heterojunction/PEC solar cell made with CNTs and SiNW arrays and (b) side view of the structure. (c) Cross section of the obtained SiNW array on the silicon wafer and (d) top view showing CNTs on top of the SiNW array. Reprinted with permission from ref. 290. Copyright 2009 American Chemical Society.

layers were used as counter electrodes, which did not contact directly with the SiNW arrays. Two reasons were accounted for the enhanced overall performance: (i) the strong adhesion between the DWCNT film and the SiNW array which greatly reduced the series resistance; (ii) the highly transparent DWCNT film, which enabled good light transmission onto the SiNW arrays.

With its unique properties, *i.e.*, highly optical transparence, highly electrical conduction, and mechanical flexibility, graphene and its derivatives have been investigated extensively in the field of solar cells. Hence, it is no surprise that graphene materials have quickly made their entry into DSSC applications, having been incorporated into each aspect of a DSSC (Figure 9.9). They were first used in 2008 as a transparent electrode to replace fluorine doped tin oxide (FTO) at the photoanode[291] and have since been used, for example, with the purpose of harvesting light,[292] improving transport through both the titania layer[293–295] and the electrolyte,[276,296] and superseding platinum at the cathode.[268,297]

9.4 Carbon (Nano)materials for Super-Capacitors

Current electrochemical energy storage devices are becoming less appropriate for the ever increasing range of high demand applications utilizing them today. As technology becomes increasingly more advanced and powerful, the requirements for new energy storage systems increase. Thus, in order to produce energy storage devices that can sufficiently meet the mounting demands of consumers, investigation and application of new electrode materials must occur. Electrochemical capacitors are sometimes called super-capacitors, ultra-capacitors, or hybrid capacitors. While the energy

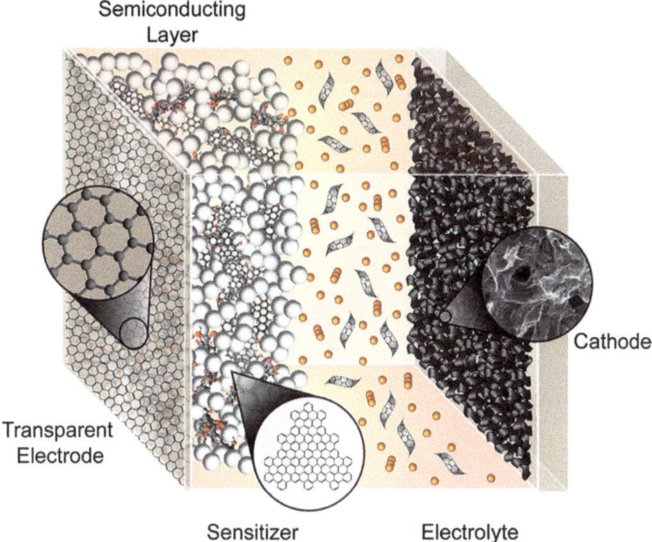

Figure 9.9 Schematic of a DSSC incorporating graphene materials in each part of the device. Adapted with permission from ref. 298. Copyright 2014 American Chemical Society.

density of super-capacitors is much higher than conventional dielectric capacitors, it is still lower than batteries and fuel cells. Most of the commercially available super-capacitor products have a specific energy density less than 10 W h kg^{-1}, which is 3 to 15 times lower than batteries (150 W h kg^{-1} is possible for lithium ion batteries). The term ultra-capacitor or super-capacitor is usually used to describe an energy storage device. The way that super-capacitors store energy is based, in principle, on two types of capacitive behavior: the electrical double layer (EDL) capacitance from the pure electrostatic charge accumulation at the electrode interface, and the pseudo-capacitance developed from fast and reversible surface redox processes at characteristic potentials.

The concept of the EDL was first described and modeled by von Helmholtz in the 19th century.[299] The Helmholtz double layer model states that two layers of opposite charge form at the electrode/electrolyte interface and are separated by an atomic distance. This model was subsequently modified by Gouy and Chapman in order to incorporate a continuous distribution of electrolyte ions (both cations and anions) in the electrolyte solution, driven by thermal motion, which is referred to as the diffuse layer.[300,301] As a result of the combination of these two previous models, Stern recognized the existence of two regions of ion distribution—the inner region called the compact layer or Stern layer and the diffuse layer.[302] In the compact layer, ions (very often hydrated) are strongly adsorbed by the electrode. The diffuse layer region is as what the Gouy and Chapman model defines. Figure 9.10 schematically illustrates an EDL capacitor made of a porous carbon electrode.

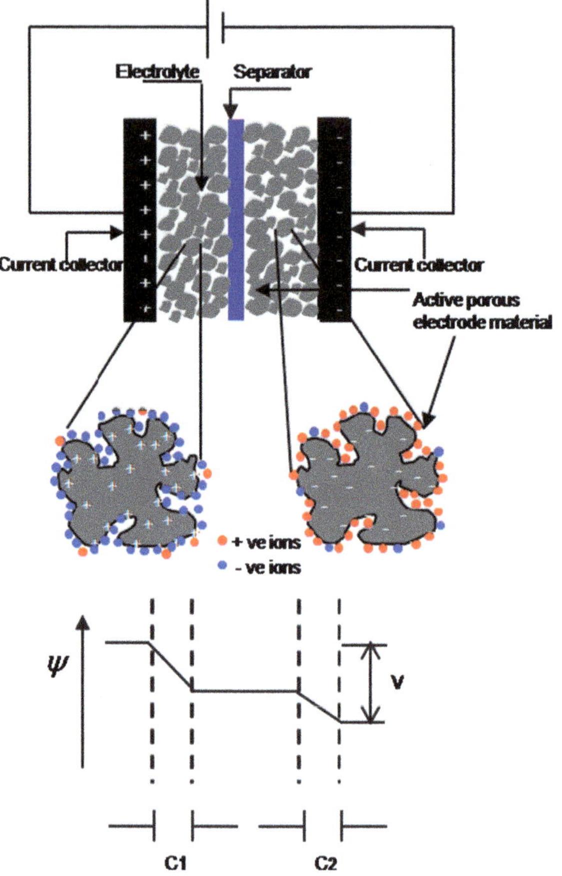

Figure 9.10 Schematic representation of an EDL capacitor based on porous electrode materials. Reprinted with permission from ref. 303.

In contrast to EDL capacitance, pseudo-capacitance arises for thermodynamic reasons and is due to charge acceptance and a change in potential.[304] The main difference between pseudo-capacitance and EDL capacitance lies in the fact that pseudo-capacitance is faradic in origin, involving fast and reversible redox reactions between the electrolyte and electro-active species on the electrode surface. While pseudo-capacitance can be higher than EDL capacitance (10–100 times), it suffers from the drawbacks of a low power density (due to poor electrical conductivity), and lack of stability during cycling.

Nowadays, the use of porous carbon-based materials has seen a huge demand for super-capacitors, in which charge is stored within the double layers at the interfaces between the carbon electrodes and the electrolyte.[305] The energy stored in carbon-based super-capacitors involves charge separation at the carbon/electrolyte interface, which leads to an electric double layer capacitance, and the presence of functional groups or heteroatoms at the surface of carbons leads to a pseudo-capacitance. The common methods for increasing capacitance of the carbon electrode materials focus on the preparation of high surface area carbons with appropriate pore size and surface modification of carbons. The double layer capacitance is directly correlated with the surface area of the carbon, while the pseudo-capacitance arises from sites containing redox active surface functional groups, such as quinone and hydroquinone, which are rapidly oxidized and then reduced, as the potential is changed.[306] For pseudo-capacitors, metal oxides and conjugated conducting polymers, such as polyaniline, polypyrrole and polythiophene, have been investigated for use as electrode materials.[307] The most beneficial metal oxide known to give very high capacitance is RuO_2. Some of the oxides, which have been studied as super-capacitor electrode materials are NiO, $Ni(OH)_2$, MnO_2, Co_2O_3, FeO, TiO_2, V_2O_5 and MoO_x. None of these oxides are used in commercial production of EDLs and they are all still only used in laboratorial-scale research. Manganese oxides are seen to be potentially useful materials for pseudo-capacitors not only due to their low cost but also to their environmental friendliness.[308]

Due to the existence of several different oxidation-state structures of these redox-active species over a potential range, high Faradaic pseudo-capacitance can be obtained. An important parameter in characterizing potential capacitor materials is the specific capacitance (F m^{-2}), which depends on the surface density of redox active functional groups and is independent of surface area. Gravimetric capacitance (F g^{-1}) takes into account both double layer capacitance (surface area) and pseudo-capacitance (functional groups).[309]

EDLs are being studied extensively due to the increasing demand for new energy-storage media with high specific power and improved durability.[310] As compared to conventional capacitors, EDLs have high power densities and relatively high energy densities, and also the high surface area material helps for charge accumulation and creates an interconnected porous network with sufficient pore windows for electrolyte wetting and rapid ionic transport. Hence, mesoporous carbon electrodes, with their more accessible porous infrastructure, are promising materials for EDL applications.

CB and ACs are the most widely used electrode materials because of their large surface area, low cost and easy processability.[305,311,312] However, the limited energy storage capacity (typically below 200 F g^{-1}), low cell voltage ($<$1 V for aqueous and *ca.* 3 V for organic electrolytes), sensitivity to impurities and rate capability have prevented their widespread application in EDL capacitors. Although microtextured carbon materials represent a very attractive material for energy storage, the performance of the capacitor goes beyond the texture of carbon electrodes. It is obvious that electrodes for super-capacitors are multifunctional materials with a strong component of interfacial molecular transport combined with requirements for electrical conductivity and excellent chemical stability in large potential gradients and at high absolute current flows. It is unlikely that any single property, such as microporosity, will dominate the performance profile of such a device. Progress in this area will require some fundamental understanding of the underlying chemical processes under the operating conditions of electrodes.

Table 9.3 Properties and characteristics of various carbon and carbon-based materials as super-capacitors electrode materials. Reprinted with permission from ref. 303. Copyright 2009 Royal Society of Chemistry.

	Specific surface area	Density	Aqueous electrolyte		Organic electrolyte	
Carbon material	/m^2 g^{-1}	/g cm^{-3}	/F g^{-1}	/F cm^{-3}	/F g^{-1}	/F cm^{-3}
ACs	1000–3500	0.4–0.7	<200	<80	<100	<50
Templated porous carbon	500–3000	0.5–1	120–350	<200	60–140	<100
Functionalized porous carbon	300–2200	0.5–0.9	150--300	<180	100–150	<90
Activated carbon fibers	1000–3000	0.3–0.8	120–370	<150	80–200	<120
Carbon cloth	2500	0.4	100–200	40–80	60–100	24–40
Carbon aerogels	400–1000	0.5–0.7	100–125	<80	<80	<40
C_{60}	1100–1400	1.72	—	—	—	—
CNTs	120–500	0.6	50–100	<60	<60	<30
Graphite	10	2.26	—	—	—	—
Graphene	2630	>1	100–205	>100–205	80–110	>80–110

Other carbonaceous materials such as carbon aerogels,[313] glassy carbons,[314,315] carbon nanotubes (both SWCNT[316,317] and MWCNT[318,319]) nanofibers,[320,321] nanodiamonds,[322] nano-onions[323] and graphene[121,122] have also been successfully studied in super-capacitor electrode preparation, because of their exceptional electrical and mechanical properties and unique structures. Table 9.3 summarizes some properties and characteristics of different carbon electrodes materials based on the literature.

As one of the most promising materials in super-capacitors, graphene has already shown immense theoretical and practical advantages, such as a high surface area, and excellent conductivity and capacitance. The preparation method significantly influences the capacitance of the final graphene material. In fact, the current tendency indicates that graphene with lower amounts of oxygen functionalities provide a higher capacitance.[324–326] With regard to the restacking issue of graphene layers, in order to maximize the surface area, many attempts have been made to space out the graphene sheets with nanoparticles (*e.g.*, Au)[327] or CNTs.[328–330] Whereas in some cases, a dramatic increase of capacitance was reported for CNT/graphene hybrid materials;[328–330] in other instances, however, such CNT/graphene hybrid materials exhibited weight-specific capacitance corresponding only to the average capacitances of the individual carbon components.[331]

Further efforts to increase the capacitances of graphene materials include doping graphene with various heteroatoms, such as nitrogen, sulfur, or boron. Despite a large number of research articles published on the increased capacitance of N-doped (n-dopant; the N atom is an electron donor) graphene materials, little is known about the mechanism.[332–334] The effect was suggested to arise from a quantum capacitance effect, as this is closely related to the modification of the electronic structure of graphene.[335] Interestingly, boron (p-dopant) doping on graphene, which leads to the introduction of electron deficient atoms to the graphene structure, also provides an increased capacitance for graphene.[336] As for sulfur-doped graphene, a proposed mechanism suggested that the increased capacitance was a result of sulfur species located in small micropores. These sulfur species affected the charge on the graphene surface and decreased the surface affinity to adsorb water. This resulted in a specific electrosorption of electrolyte ions and increased the capacitance of the sulfur-doped graphene.[337]

In summary, there are several approaches to improve charge storage in carbon super-capacitors. A higher capacitance can be achieved by careful thermal, chemical, or electrochemical treatment to increase the accessible surface area and surface functional groups, or by extending the operating voltage range beyond the limit of an aqueous electrolyte solution. Several critical parameters can contribute to a high capacitance: (i) increasing the surface area is quite important, but significant effort has already been made to maximize the surface area of carbon (the correlation between the surface area and the specific capacitance cannot be detected for many high surface area materials); (ii) inducing pseudo-capacitance can also increase the capacitance. The latter involves voltage dependent Faradaic reactions between the electrode and the electrolyte, either in the form of surface adsorption/desorption of ions, redox reactions with the electrolyte, or doping/undoping of the electrode materials. Surface functionalization proves to be very effective in increasing the pseudo-capacitance arising from oxidation/reduction of surface quinone functional groups generated during the sample treatment.

9.5 Carbon (Nano)materials for Lithium Batteries

The great interest in lithium insertion into different kinds of carbons is connected with the rapid development of lithium-ion batteries in the world market, especially in high-tech devices like laptops and cell phones. Intensive research is still focused on the optimization of anodic carbon materials for getting better performance of this system.

9.5.1 Lithium-Ion

A variety of advanced devices require rechargeable lithium-ion batteries with high energy and high power density. Generally, carbon materials (especially graphite) have been investigated for lithium-ion battery anode materials for quite some time, due to their ability to intercalate lithium ions. Prior to the discovery of graphite anode materials, Li metal had been the main candidate for Li-ion batteries, but it had a serious problem of dendrite formation, which caused a short circuit and presented a safety issue. Commercial lithium ion batteries often employ graphitic carbon anodes, which provide

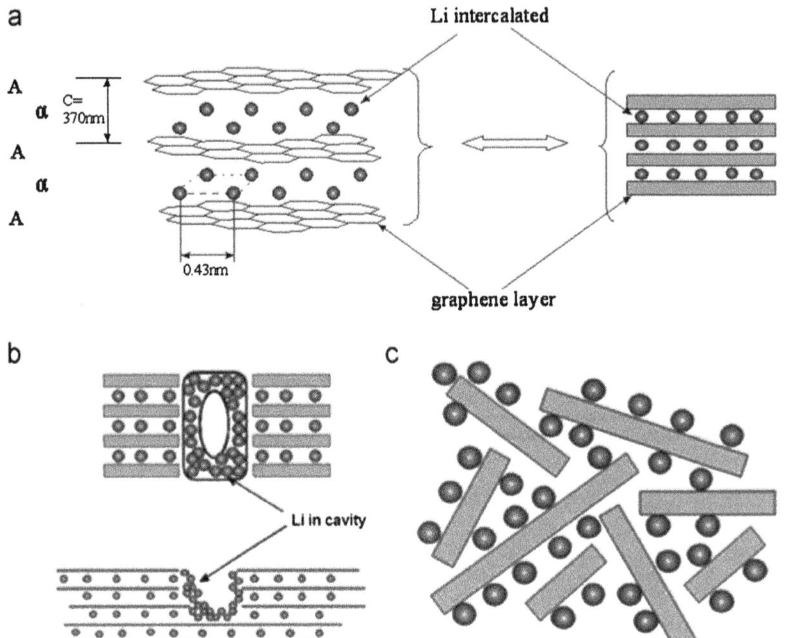

Figure 9.11 Li-ion storage mechanisms in graphite: (a) Li-ion intercalation in graphite; (b) Li-ion storage in micro- or nanocavities; and (c) Li-ion storage on surfaces or at the interfaces of graphite micro- or nanocrystals (house-of-cards model). Reprinted with permission from ref. 340. Copyright 2009 Wiley-VCH.

a theoretical capacity of 372 mA h g^{-1}.[338,339] Graphite is attractive because of its high in-plane electron conductivity due to the π-bond and weak interaction with Li ions, giving rise to high Li ion storage capacity and fast Li ion diffusion (Figure 9.11a). Great efforts have been made to increase the intercalation capacity of Li ions in graphite by altering the carbon structures, increasing lattice disordering, creating pore spaces, and increasing surfaces areas. Important mechanisms for Li storage also include microcavities or nanopores (Figure 9.11b) or the surfaces and interfaces of microcrystalline or nanocrystalline graphite or stacked graphene sheets (the house-of-cards model) (Figure 9.11c). However, new compact and modern portable electronic devices as well as hybrid electric devices require higher energy density power sources with improved cyclability and rate capability.

The electrodes of lithium-ion batteries are based on intercalation materials between which lithium ions are transferred through the electrolytic medium during charge and discharge.[341] The cathodic (positive electrode) materials are lamellar oxides such as $LiCoO_2$, $LiNiO_2$ or $LiMnO_2$, represented by the general formula Li_yMO_2 ($y \approx 1$), while carbon (generally graphite) is used for the negative electrode (Figure 9.12). The storage capacity of a battery is given by the amount of Li that can be stored reversibly in the two electrodes.

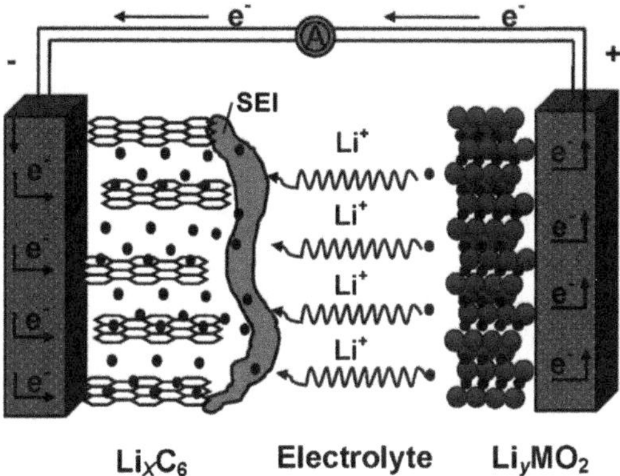

Figure 9.12 Schematic illustration of an intercalation-type Li-ion battery in a charging process with graphite as the anode and Li_yMO_2 as the cathode (M = metal). SEI is the abbreviation for Solid-Electrolyte-Interphase. Adopted from ref. 343. Copyright 2008, Elsevier Science Ltd.

During the charging process, Li ions are extracted from the Li_yMO_2 electrode and simultaneously inserted into the graphitic carbon electrode, coupling with negatively charged electrons to keep overall charge neutrality. During the discharging process, Li ions are reversibly extracted from the negative electrode and simultaneously inserted into the positive electrode. The Li ions have to shift back and forth easily between the storage hosts of the cathode and anode (so called "rocking chair" or "shuttle rock").

Using a graphite anode, such a battery operates at almost constant voltage of about 3.5 V during discharge,[342] that makes this system very attractive for its high energy density. In the case of graphite, lithium penetrates between the graphene layers through an intercalation process with charge transfer to carbon. It is remarkable that for graphite the insertion–extraction proceeds very close to the potential of metallic lithium, allowing the lithium-ion battery to discharge at a high and almost constant voltage.

In the ideal case, the structural features of the host materials should remain unchanged during and after the insertion–extraction of the guests resulting in a long cycling life. In practice, initially a large number of additional irreversible reactions occur. They are associated with structural phase transformations of the electrodes and chemical processes related to the Li ion transport. The number of such additional reactions and therefore the capacity fading caused by such reactions decreases with each cycle, resulting in an important irreversible capacity and a varying voltage during lithium extraction.[341]

An optimal carbon material for lithium-ion batteries should have a higher reversible capacity than graphite, while maintaining a small irreversible capacity and its main discharge (oxidation) below 0.5 V *versus* Li. During the last years, many efforts have been devoted to develop electrodes based on the use of harder carbons, because of the high reversible capacity which can be reached with some of these materials, without the inconvenience of exfoliation during insertion–extraction processes. In general, the reversible Li storage capacity of nanocarbons is higher than that of graphite, at the expense of higher capacity losses (the very high surface area and, consequently, the large solid–electrolyte interface enhance the irreversible capacity). For nanostructured carbons with very high specific surface areas, it is suggested that the storage capacity can be increased through the formation of Li_2 molecules between layers,[344] by the presence of charged Li^+ clusters in the cavities,[345] in the micropores, or in the inner space of CNTs, or through the adsorption of Li ions on the surface and edges of graphite grains. For single-layer graphene, Li may adsorb on both sides of the sheet.

The electrochemical intercalation of lithium in both MWCNTs and SWCNTs has been studied in recent years. MWCNTs exhibit reversible capacities of 80–640 mA h g^{-1} ($Li_{0.2}C_6$–$Li_{1.7}C_6$), while the capacity of the SWCNTs is 450–600 mA h g^{-1} ($Li_{1.2}C_6$–$Li_{1.6}C_6$), which could increase to 790 and 1000 mA h g^{-1} ($Li_{2.1}C_6$ and $Li_{2.7}C_6$) by either mechanical ball-milling or chemical etching.[346,347] However, the very nature of these carbon nanomaterials restricts their capacity. Hence, in order to improve the capacity of the CNT anodes one method is to fabricate composite electrodes of CNTs with other materials. In such hybrid systems, the CNTs function as an effective confining buffer of mechanical stress induced by volume changes in charging and discharging reactions, while the other nanomaterials provide a high capacity. Amongst the different systems already studied, various metals (Sn, Sb, Bi, *etc.*)[348,349] and metal oxides (Sn-, Mn-, Fe-based, *etc.*)[350–352] have showed high charge capacities and good durability.

As the thinnest carbon materials, graphene and graphene-based materials have also shown promising results in Li-ion batteries. The first reversible specific capacity of the prepared graphene sheets (specific surface area of 492.5 m^2 g^{-1}) was as high as 1264 mA h g^{-1} at a current density of 100 mA g^{-1}.[353] After 40 cycles, the reversible capacity was still 848 mA h g^{-1} at a current density of 100 mA g^{-1}, higher than general values of CNT electrodes. The surface oxidation of carbon materials can also improve their electrochemical properties,[354] and Bhardwaj *et al.* confirmed that oxidized graphene nanoribbons (ox-GNRs) outperformed MWCNTs and GNRs, presenting a first charge capacity of *ca.* 1400 mA h g^{-1} and a reversible capacity of *ca.* 800 mA h g^{-1}.[355]

As in the case of CNTs, the preparation of graphene-based composites with different metal oxides (TiO_2, Fe_2O_3, Fe_3O_4, CeO_2, CuO, *etc.*) can also improve the Li-ion insertion properties.[356–360] Three-dimensional graphene-based hybrid structures can also improve the storage capacity of Li-ion batteries by increased specific surface area and more suitable layer spacing between graphene sheets. For example, Yin *et al.* created honeycomb-like electrode materials with hierarchical graphene nanoarchitectures modified by the organic agent dimethyldioctadecylammonium (DODA) with electrostatic interactions. This novel structure simultaneously optimized ion transport and capacity, leading to a high performance of reversible capacity (up to 1600 mA h g^{-1}), and 1150 mA h g^{-1} after 50 cycles.[361]

Hence, the use of nanocarbons as electrodes can provide: (i) a high electrode–electrolyte contact area, (ii) new storage mechanisms that are not possible in bulk materials, and (iii) a shorter pathway both for electron and Li ion transport (this enables a high charge–discharge rate and, thus, higher power).

Although carbon is largely associated with the battery anode for storing Li metal species, it must be understood that the Li-ion battery performance is not limited by this part of the system. While developing anodes with larger capacities and slightly higher potentials than carbon is very demanding, developing cathode materials able to store Li and to exhibit the same electrical transport property as the anode is a much greater challenge.

9.5.2 Lithium–Sulfur

In order to overcome the charge-storage limitations of insertion-compound electrodes, materials that undergo conversion reactions while accommodating more ions and electrons are becoming a promising option. With this in mind, Li–S batteries with high energy are nowadays being intensively pursued.[362,363] Sulfur (one of the most abundant elements on earth) offers a high theoretical capacity of 1672 mA h g^{-1}, which is an order of magnitude higher than those of the transition-metal oxide cathodes, and an energy density of 2600 W h kg^{-1}, which is 3–5 fold higher than those of state-of-the-art Li-ion batteries. The high capacity is based on the conversion reaction of sulfur to form lithium sulfide (Li_2S) by reversibly incorporating two electrons per sulfur atom compared to one or less than one electron per transition metal ion in the insertion-oxide cathodes.

A basic Li–S cell consists of a lithium anode, a carbon–sulfur cathode, and an organic liquid electrolyte that enables Li ions to pass. The overall cell reaction during discharge converts lithium metal in the anode into Li_2S at the surface of the cathode. The flow of two lithium ions from the anode to the cathode is then balanced by the flow of two electrons between the battery contacts, delivering double the current of a Li-ion battery at a voltage between *ca.* 1.7 and 2.5 V, depending on the state of charge of the cell (Figure 9.13). Lithium

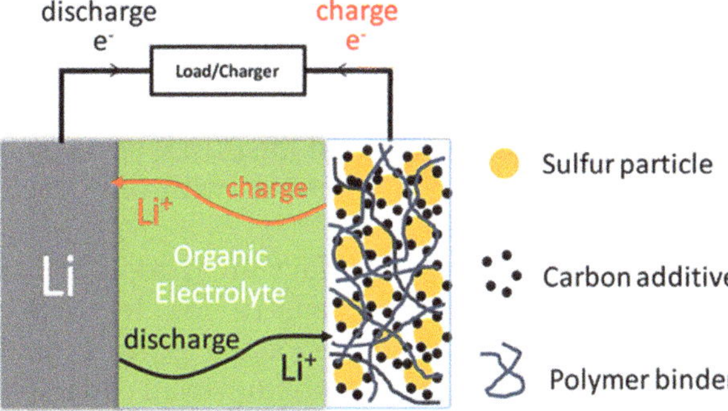

Figure 9.13 Schematic diagram of a Li–S cell with its charge/discharge operation. Reprinted with permission from ref. 363. Copyright 2014 American Chemical Society.

polysulfides (Li_2S_8, Li_2S_6, *etc.*) are formed at intermediate charge levels, which affect the cell voltage. Despite sulfur itself and Li_2S being essentially insoluble in the typical electrolyte used in Li-S cells, these intermediate polysulfides are often soluble, which causes an ongoing and severe loss of sulfur at the cathode. In addition, when the sulfur in the cathode absorbs lithium ions from the electrolyte, the Li_2S has nearly double the volume of the original sulfur, inducing a very large source of mechanical stress on the cathode, which causes mechanical deterioration, reduces the electrical contact between the carbon and the sulfur (the path whereby electrons flow to allow the reaction to occur), and prevents the flow of lithium ions to the sulfur surface.

Carbonaceous materials are extensively used as electronic conductors in the battery industry, and play a particularly crucial role in the sulfur electrode. An ideal carbon matrix for sulfur–carbon composites needs to have (i) high electrical conductivity, (ii) electrochemical affinity for sulfur, (iii) small pores without large outlets to accommodate polysulfides, (iv) accessibility of the liquid electrolyte to the active material, and (v) a stable framework to sustain the strain generated by the volume changes of the active material during cycling.

The frequently used carbon materials include microporous carbon,[364,365] mesoporous carbon,[366,367] hierarchical porous carbon,[368–371] carbon black,[372,373] hollow carbon spheres,[374,375] CNTs,[376–380] CNFs,[381–383] and graphene.[384–387]

Recent progress on the sulfur–carbon composites benefits from their hierarchical design. High carbon content improves conductivity but at the expense of reduced energy density. Carbon spheres with micropores can encapsulate sulfur, but the total pore volume needs to be increased to enhance the sulfur loading.[364] Spherical ordered mesoporous carbon with a large number of inner pores is a superior option to serve as the matrix to accommodate more sulfur without compromising battery performance.[370] Porous hollow carbon–sulfur composites exhibit impressive cyclability over 100 cycles.[374] Graphene oxide has also been used, but the graphene–sulfur composite may need further treatment for immobilizing the intermediate polysulfides.[386,387] Porous CNFs and hollow CNFs or CNTs are also good substrates to be impregnated with sulfur.[378,381,382] Although these carbon-based sulfur composites have shown outstanding progress in Li–S battery performance, high active material loading (>70 wt %), high specific capacity (>1200 mA h g^{-1}), and excellent cycle stability and life (<10% capacity loss over 100 cycles) could not be obtained simultaneously with any single composite material.

Mesoporous carbons with a large surface area, good conductivity, and tunable pores are the ideal conductive matrix for electrical energy storage applications. Furthermore, confining active phases inside the pores leads to composite battery electrodes with improved capacity, cycling stability, and rate capability. Hence, it is only logical that these materials have also found applications in Li–S batteries, yielding the best electrochemical properties that have been reported for carbon–sulfur systems.[368]

This material consists of ordered interwoven carbon–sulfur composites that comprise high pore-volume carbons with 3D-accessible channel nanostructures. The conductive carbon framework constrains the sulfur within its channels and generates the essential electrical contact. Kinetic inhibition to diffusion within the framework, and the sorption properties of the carbon both aid in trapping the intermediate polysulfides. Thus immobilized, their full reduction to Li_2S_2/Li_2S (or oxidation to S_8 on charge) is achieved within the carbon framework. Capacity fading is reduced owing to greatly reduced polysulfide concentration in the electrolyte, and the materials sustained reversible capacities of 1100 mA h g^{-1} over 20 cycles.

9.5.3 Lithium–Air/Oxygen

Similar to the AFCs configuration, metal–air batteries also contain electrodes that need ORR catalysts to reduce oxygen molecules coming from the ambient air for use as the cathode reactant.[388] Among many kinds of metal–air batteries with different metals as anodes, Li–air batteries and Zn–air batteries are considered to be the most promising candidates for application in future electric vehicles over long-distances.[389] Because the overpotentials of the oxidation of metals in metal–air batteries are relatively low, the performance of metal–air batteries, to a great extent, depends on the efficiency of catalysts at oxygen cathodes.

Li–air batteries, which can theoretically deliver super high energy densities, hold greater promise for the transportation applications than lithium-ion batteries. Li–air batteries include four configurations: aprotic, aqueous, solid, and mixed aqueous/aprotic.[390] While all four different architectures assume the lithium metal as the anode material, they include different electrolytes and cathode components. Apart from these differences, all four configurations need a high-efficiency air-breathing cathode with ORR catalysts to suppress the oxygen polarization and improve the round-trip efficiency.[391–393] Although Li–O_2 and Li–S share the same anode, and have active cathode components (O_2 and S) that are nearest neighbors in group 16 of the periodic table, there are important differences related to the different chemistry of O and S and the different states of matter of their cathodes. A notable characteristic of metal–air batteries is their open cell structure, since these batteries use oxygen gas accessed from the air as their cathode material, whilst the cell configuration for conventional rechargeable batteries is a closed system.

The leap forward in theoretical specific energy on migrating from Li-ion to Li–S and then Li–O_2 is clear. It arises because Li_2S, Li_2O_2 and LiOH in the cathode store more Li, and hence charge, than $LiCoO_2$ per unit mass, and Li metal stores more charge per unit mass than a graphite (C_6Li) anode. The theoretical energy density is also greater for Li–O_2 and Li–S than Li-ion but the gain is not as great as for specific energy. A comparison of practical specific energies (W h kg^{-1}) for several rechargeable batteries is presented in Figure 9.14. The values for established technologies are well attested but those obtained for Li–O_2 are at best rough estimates, because so far there are only a few realistic prototypes on which to base such projections. However, the values quoted are in line with those reported by different researchers.[362,390,394]

Several factors lower the energy storage of practical Li–S and Li–O_2 batteries. The cathode in each case consists of a porous conducting matrix (usually carbon), in which the discharge products form, thus adding mass and volume to the cell. More Li metal than is required for the stoichiometric reaction has to be included, to compensate for the inefficiency of Li-metal cycling. Furthermore, packaging, current collectors and gas diffusion channels (for Li–O_2 cells), will also reduce the practical energy storage.

Li–O_2 cells can be based on both aqueous and non-aqueous electrolytes. Regardless of the type of electrolyte, on discharge, the Li-metal anode is oxidized, releasing Li^+ into the electrolyte, and the process is reversed on charge. At the positive electrode, O_2 from the atmosphere enters the porous cathode, dissolves in the electrolyte and is reduced at the electrode surface on discharge. When a suitable non-aqueous electrolyte is employed, O_2^{2-} is formed, which, along with Li^+ from the electrolyte, forms Li_2O_2 as the final discharge product. The peroxide is then decomposed on charging: $Li_2O_2 \leftrightarrow 2Li^+ + O_2 + 2e^-$. Aqueous electrolytes involve the formation of OH^- and then LiOH at the cathode on discharge, according to the equation: $2Li^+ + ½O_2 + H_2O + 2e^- \leftrightarrow 2LiOH$, with LiOH being oxidized on charge. Because H_2O and O_2 are involved, this is sometimes referred to as a Li–water battery.

Because oxygen has to diffuse within the carbon pores during discharge, the design of appropriate porous structures for the carbon electrode becomes

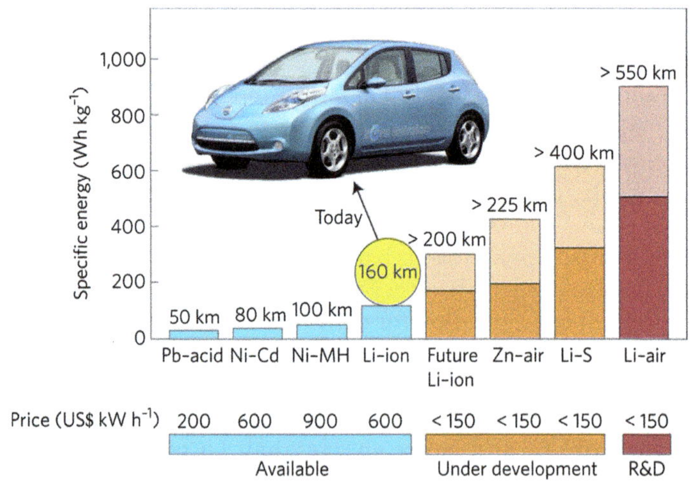

Figure 9.14 Practical specific energies for some rechargeable batteries, along with estimated driving distances and pack prices. Reprinted with permission from ref. 394. Copyright 2012 Macmillan Publishers Limited.

important in this system. The porous carbon electrode not only supplies electrons continuously for the electrochemical process but also provides numerous solid–liquid–gas tri-phase regions for the oxygen reduction. The pores and the surface of the carbon electrode are also the host of the Li_2O_2 discharge product;[395] thus, playing a key role in determining the overall electrochemical performances of the Li–O_2 batteries.

The surface area of porous carbon is directly related to the performance of Li–O_2 batteries since a larger surface area means more active sites for electrochemical reactions and also that more catalysts can be loaded onto the porous carbon. However, detailed studies on the relationship between surface area and the capacity delivered from Li–O_2 batteries reveal that higher surface area does not always lead to a higher capacity. Instead, the discharge capacity is largely determined by the pore volume of the carbon, and especially the amount of mesopores.[396–398] The importance of the mesopore volume of carbon in Li–air batteries can be explained in terms of kinetic process, including oxygen diffusion and reduction. During discharge, oxygen first transports into the whole carbon electrode, either in the gaseous phase or through the electrolyte in the form of dissolved oxygen.[397] However, no matter what the oxygen penetration route, the successful formation of Li_2O_2 depends on the triple junctions where electrolyte, carbon, and oxygen coexist. In other words, the more the tri-phase regions, the more Li_2O_2 will be produced and therefore the higher the capacity obtained from the cell. If the pore size is too small and within the micro-size range, the entrance of the micropores will be quickly blocked by either the electrolyte or Li_2O_2, preventing further access to the interior carbon surface. On the other hand, carbons with a large amount of macropores are not suitable for a Li–O_2 battery because large pores are easily flooded by the electrolyte, reducing the generation of triple junctions. Furthermore, the volumetric energy density of the air electrode will also decrease when using macroporous carbons.[399]

Graphene and its composites have been intensively investigated as ORR catalysts for both fuel cells and metal–air batteries. Zhou *et al.* employed metal-free FLG as an air electrode in a Li-air battery with a hybrid electrolyte (Figure 9.15).[400] The FLG demonstrated very high efficiency, close to that of a commercial Pt/C catalyst. At 0.5 mA cm^{-2}, the Li-air battery with FLG cathode delivered a constant discharge voltage of 3.00 V *vs.* Li/Li^+, only 50 mV lower than that with state-of-the-art commercial Pt/CB catalyst and much higher than that with an acetylene carbon black cathode.

Furthermore, FePc combined with different carbon materials, including FLG, CNTs, and CB, already used for ORR catalysts (see Section 9.2.4.3) were also evaluated as hybrid materials in Li–air batteries.[401] Li–air battery tests demonstrated that the FePc supported on CNTs outperformed that on FLG and CB by showing a higher discharge potential and much more stable cycling performance, indicating a stronger synergistic effect between FePc and CNTs.

In summary, a good carbon candidate for Li–air or Li–O_2 batteries should provide the possibility to accurately control the pore size/volume, which can be implemented during synthesis. The surface area of the ideal carbon

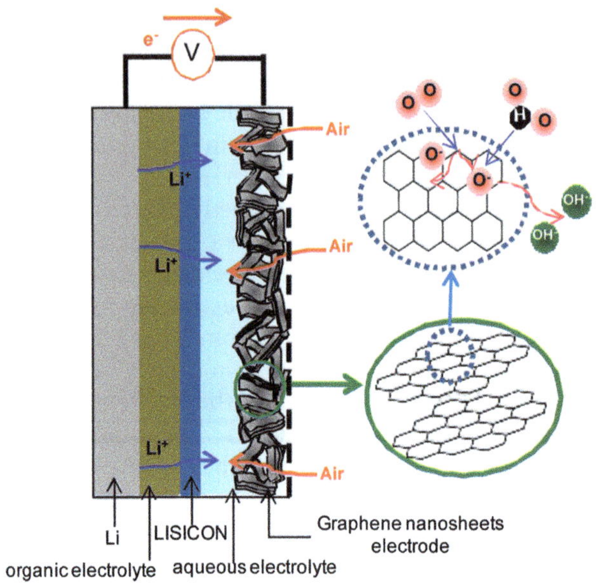

Figure 9.15 Structure of the rechargeable Li-air battery based on FLG as an air electrode. Reproduced with permission from ref. 400. Copyright 2011, American Chemical Society.

should also be tailored to reduce the side reactions and reach an optimized performance for the battery. On the other hand, the porous carbon can also be "produced" from non-porous two-dimensional carbons such as graphene nanosheets. The functional groups and/or defects on different porous carbons (not only in graphene) that can be modified to satisfy different energy requirements and their interactions with Li_2O_2 are an interesting direction that is worth further investigation.

9.6 Concluding Remarks and Future Perspectives

Carbon is nowadays a vital component in both fuel cells and energy storage devices. The unique combination of outstanding chemical stability with electrical conductivity and an endless variability of texture in multiple scales render carbon an element of eminent relevance in both energy conversion and storage technology. Impressive advances have been achieved in the research and development of catalytic layers for fuel cells, with the decisive breakthrough being the change from noble metal blacks to carbon-supported noble metal nanoparticles. This resulted in higher metal dispersions and better particle size distributions of noble metals, meaning higher utilization factors (and increasing PEMFC performances), and lower metal loadings per unit MEA surface area. However, carbon is oxidized at high potentials leading to fuel-cell performance losses. Hence, new carbon support materials such as carbon nanotubes, nanofibers, doped nanodiamonds, as well as mesoporous carbon, few-layer graphene (undoped, doped, and with metal nanoparticles) and reduced graphene oxide (also undoped, doped, and with metal oxides) are being actively investigated. They offer high crystallinity and surface area, leading to increased fuel cell performance and stability.

Platinum is currently used as the electrocatalyst in fuel cells (both anode and cathode), but its high cost and limited resources have led to an extensive research for alternate, non-noble metal electrocatalysts. Among them, nitrogen doping of nanostructured carbon (with or without other metals) demonstrate enhanced catalytic activity and durability. Nitrogen-doped non-precious metal electrocatalysts with conducting polymers (polypyrrole, polyaniline) also exhibit interesting ORR activity and fuel cell performances. Moreover, the use of these doped carbon materials as electrocatalysts is not limited to the oxygen reduction reaction and their application may be extended to other relevant electrochemical reactions. Furthermore, doped carbon materials can also be used as anode materials in Li-ion batteries, as ultra-capacitors, as supports for metal catalysts that are used in a broad set of reactions, and as active catalysts themselves. Interestingly, after several years of intensive research in the domains of energy storage, the most used materials in current industry are still graphite and activated carbons for anodes of lithium-ion batteries and electrodes of super-capacitors, respectively, with cost playing a major role.

In light of all this, more systematic studies are required to unveil the complex influence of the structure and morphology of carbon materials on the performance of fuel cells, but also on storage devices, with the aim of developing a new generation of structurally ordered tailored materials with enhanced catalytic properties, low noble metal contents, and high durability.

References

1. R. F. Service, *Science*, 2002, **296**, 1222–1224.
2. H. Liu, C. Song4, L. Zhang, J. Zhang, H. Wang and D. P. Wilkinson, *J. Power Sources*, 2006, **155**, 95–110.
3. J. Chen, B. Lim, E. P. Lee and Y. Xia, *Nano Today*, 2009, **4**, 81–95.
4. S. Guo and E. Wang, *Nano Today*, 2011, **6**, 240–264.
5. C. Wang, M. Waje, X. Wang, J. M. Tang, R. C. Haddon and Yan, *Nano Lett.*, 2003, **4**, 345–348.
6. H. Chang, S. H. Joo and C. Pak, *J. Mater. Chem.*, 2007, **17**, 3078–3088.
7. E. Antolini, *J. Mater. Sci.*, 2003, **38**, 2995–3005.
8. D. S. Cameron, *Platinum Met. Rev.*, 2009, **53**, 147–154.
9. Y. Garsany, O. A. Baturina, K. E. Swider-Lyons and S. S. Kocha, *Anal. Chem.*, 2010, **82**, 6321–6328.
10. S. Treimer, A. Tang and D. C. Johnson, *Electroanalysis*, 2002, **14**, 165–171.
11. J. N. Tiwari, R. N. Tiwari, G. Singh and K. S. Kim, *Nano Energy*, 2013, **2**, 553–578.

12. C. J. Zhong, J. Luo, B. Fang, B. N. Wanjala, P. N. Njoki, R. Loukrakpam and J. Yin, *Nanotechnology*, 2010, **21**, 062001.
13. P. Trogadas, T. F. Fuller and P. Strasser, *Carbon*, 2014, **75**, 5–42.
14. F. Maillard, P. A. Simonov and E. R. Savinova, *Carbon Materials for Catalysis*, John Wiley & Sons, Inc., 2008, pp. 429–480.
15. V. R. Stamenkovic, B. S. Mun, M. Arenz, K. J. J. Mayrhofer, C. A. Lucas, G. Wang, P. N. Ross and N. M. Markovic, *Nat. Mater.*, 2007, **6**, 241–247.
16. S. Jiang, Y. Ma, G. Jian, H. Tao, X. Wang, Y. Fan, Y. Lu, Z. Hu and Y. Chen, *Adv. Mater.*, 2009, **21**, 4953–4956.
17. D. Wang, H. L. Xin, R. Hovden, H. Wang, Y. Yu, D. A. Muller, F. J. DiSalvo and H. D. Abruña, *Nat. Mater.*, 2013, **12**, 81–87.
18. G. Wu, K. L. More, C. M. Johnston and P. Zelenay, *Science*, 2011, **332**, 443–447.
19. R. B. Levy and M. Boudart, *Science*, 1973, **181**, 547–549.
20. F. Jaouen, E. Proietti, M. Lefevre, R. Chenitz, J.-P. Dodelet, G. Wu, H. T. Chung, C. M. Johnston and P. Zelenay, *Energy Environ. Sci.*, 2011, **4**, 114–130.
21. Z. Chen, D. Higgins, A. Yu, L. Zhang and J. Zhang, *Energy Environ. Sci.*, 2011, **4**, 3167–3192.
22. B. Winther-Jensen, O. Winther-Jensen, M. Forsyth and D. R. MacFarlane, *Science*, 2008, **321**, 671–674.
23. K. Gong, F. Du, Z. Xia, M. Durstock and L. Dai, *Science*, 2009, **323**, 760–764.
24. Y. Zheng, Y. Jiao, M. Jaroniec, Y. Jin and S. Z. Qiao, *Small*, 2012, **8**, 3550–3566.
25. D. Yu, E. Nagelli, F. Du and L. Dai, *J. Phys. Chem. Lett.*, 2010, **1**, 2165–2173.
26. A. L. Dicks, *J. Power Sources*, 2006, **156**, 128–141.
27. H. Huang and X. Wang, *J. Mater. Chem. A*, 2014, **2**, 6266–6291.
28. A. S. Aricò, S. Srinivasan and V. Antonucci, *Fuel Cells*, 2001, **1**, 133–161.
29. K. H. Kangasniemi, D. A. Condit and T. D. Jarvi, *J. Electrochem. Soc.*, 2004, **151**, E125–E132.
30. M. Uchida, Y. Fukuoka, Y. Sugawara, H. Ohara and A. Ohta, *J. Electrochem. Soc.*, 1998, **145**, 3708–3713.
31. S. H. Joo, S. J. Choi, I. Oh, J. Kwak, Z. Liu, O. Terasaki and R. Ryoo, *Nature*, 2001, **412**, 169–172.
32. W. C. Choi, S. I. Woo, M. K. Jeon, J. M. Sohn, M. R. Kim and H. J. Jeon, *Adv. Mater.*, 2005, **17**, 446–451.
33. J. B. Xu and T. S. Zhao, *RSC Adv.*, 2013, **3**, 16–24.
34. F. Su, J. Zeng, X. Bao, Y. Yu, J. Y. Lee and X. S. Zhao, *Chem. Mater.*, 2005, **17**, 3960–3967.
35. G. S. Chai, S. B. Yoon, J.-S. Yu, J.-H. Choi and Y.-E. Sung, *J. Phys. Chem. B*, 2004, **108**, 7074–7079.
36. K.-Y. Chan, J. Ding, J. Ren, S. Cheng and K. Y. Tsang, *J. Mater. Chem.*, 2004, **14**, 505–516.
37. K. Lee, J. Zhang, H. Wang and D. P. Wilkinson, *J. Appl. Electrochem.*, 2006, **36**, 507–522.

38. F. Hasche, M. Oezaslan and P. Strasser, *Phys. Chem. Chem. Phys.*, 2010, **12**, 15251–15258.
39. W. Li, X. Wang, Z. Chen, M. Waje and Y. Yan, *J. Phys. Chem. B*, 2006, **110**, 15353–15358.
40. G. Che, B. B. Lakshmi, E. R. Fisher and C. R. Martin, *Nature*, 1998, **393**, 346–349.
41. P. V. Dudin, P. R. Unwin and J. V. Macpherson, *J. Phys. Chem. C*, 2010, **114**, 13241–13248.
42. S. Liao, K.-A. Holmes, H. Tsaprailis and V. I. Birss, *J. Am. Chem. Soc.*, 2006, **128**, 3504–3505.
43. B. Rajesh, K. Ravindranathan Thampi, J. M. Bonard, N. Xanthopoulos, H. J. Mathieu and B. Viswanathan, *J. Phys. Chem. B*, 2003, **107**, 2701–2708.
44. Z. Liu, X. Lin, J. Y. Lee, W. Zhang, M. Han and L. M. Gan, *Langmuir*, 2002, **18**, 4054–4060.
45. C. H. Wang, H. Y. Du, Y. T. Tsai, C. P. Chen, C. J. Huang, L. C. Chen, K. H. Chen and H. C. Shih, *J. Power Sources*, 2007, **171**, 55–62.
46. L. Gan, R. Lv, H. Du, B. Li and F. Kang, *Electrochem. Commun.*, 2009, **11**, 355–358.
47. E. Antolini, *Appl. Catal., B*, 2007, **74**, 337–350.
48. Z. X. Cai, C. C. Liu, G. H. Wu, X. M. Chen and X. Chen, *Electrochim. Acta*, 2013, **112**, 756–762.
49. M. Noroozifar, M. Khorasani-Motlagh, M. S. Ekrami-Kakhki and R. Khaleghian-Moghadam, *J. Appl. Electrochem.*, 2014, **44**, 233–243.
50. D.-J. Guo and H.-L. Li, *J. Colloid Interface Sci.*, 2005, **286**, 274–279.
51. M. Krishna Kumar and S. Ramaprabhu, *Int. J. Hydrogen Energy*, 2007, **32**, 2518–2526.
52. S. Yang, X. Zhang, H. Mi and X. Ye, *J. Power Sources*, 2008, **175**, 26–32.
53. J. Zhang, H. Y. Chen, H. B. Li, J. T. Di, M. H. Chen, F. X. Geng, Z. G. Zhao and Q. W. Li, *Chem. Mater.*, 2014, **26**, 2789–2794.
54. F. C. Zhu, G. S. Ma, Z. C. Bai, R. Q. Hang, B. Tang, Z. H. Zhang and X. G. Wang, *J. Power Sources*, 2013, **242**, 610–620.
55. D. J. Guo and H. L. Li, *Carbon*, 2005, **43**, 1259–1264.
56. Y. Cheng, W. Li, X. Fan, J. Liu, W. Xu and C. Yan, *Electrochim. Acta*, 2013, **111**, 635–641.
57. R. Singh, T. Premkumar, J.-Y. Shin and K. E. Geckeler, *Chem. – Eur. J.*, 2010, **16**, 1728–1743.
58. P. Santhosh, A. Gopalan and K.-P. Lee, *J. Catal.*, 2006, **238**, 177–185.
59. J. Naruse, H. Le Quynh, Y. Sugano, T. Ikeuchi, H. Yoshikawa, M. Saito and E. Tamiya, *Biosens. Bioelectron.*, 2011, **30**, 204–210.
60. S.-H. Lee, M. F. Philips, S. Komathi, A. Gopalan, K.-P. Lee and Y. Zhu, *J. Nanoelectron. Optoelectron.*, 2011, **6**, 353–356.
61. A. L. Mohana Reddy, N. Rajalakshmi and S. Ramaprabhu, *Carbon*, 2008, **46**, 2–11.
62. D. G. Larrude, P. Ayala, M. E. H. Maia da Costa and F. L. Freire, *J. Nanomater.*, 2012, **2012**, 695453.

63. C. A. Bessel, K. Laubernds, N. M. Rodriguez and R. T. K. Baker, *J. Phys. Chem. B*, 2001, **105**, 1115–1118.
64. E. S. Steigerwalt, G. A. Deluga, D. E. Cliffel and C. M. Lukehart, *J. Phys. Chem. B*, 2001, **105**, 8097–8101.
65. E. S. Steigerwalt, G. A. Deluga and C. M. Lukehart, *J. Phys. Chem. B*, 2002, **106**, 760–766.
66. M. Tsuji, M. Kubokawa, R. Yano, N. Miyamae, T. Tsuji, M.-S. Jun, S. Hong, S. Lim, S.-H. Yoon and I. Mochida, *Langmuir*, 2006, **23**, 387–390.
67. Q. Guo, D. Liu, J. Huang, H. Hou and T. You, *Microchim. Acta*, 2014, **181**, 797–803.
68. G.-H. An and H.-J. Ahn, *ECS Solid State Lett.*, 2014, **3**, M29–M32.
69. S. Sharma and B. G. Pollet, *J. Power Sources*, 2012, **208**, 96–119.
70. L. Li, G. Wu and B.-Q. Xu, *Carbon*, 2006, **44**, 2973–2983.
71. J.-H. Wee, K.-Y. Lee and S. H. Kim, *J. Power Sources*, 2007, **165**, 667–677.
72. J. M. Tang, K. Jensen, M. Waje, W. Li, P. Larsen, K. Pauley, Z. Chen, P. Ramesh, M. E. Itkis, Y. Yan and R. C. Haddon, *J. Phys. Chem. C*, 2007, **111**, 17901–17904.
73. Y. Xing, *J. Phys. Chem. B*, 2004, **108**, 19255–19259.
74. J. M. Liu, H. Meng, J. l. Li, S. j . Liao and J. H. Bu, *Fuel Cells*, 2007, **7**, 402–407.
75. Y. L. Hsin, K. C. Hwang and C.-T. Yeh, *J. Am. Chem. Soc.*, 2007, **129**, 9999–10010.
76. T. Hyeon, S. Han, Y.-E. Sung, K.-W. Park and Y.-W. Kim, *Angew. Chem., Int. Ed.*, 2003, **42**, 4352–4356.
77. K.-W. Park, Y.-E. Sung, S. Han, Y. Yun and T. Hyeon, *J. Phys. Chem. B*, 2003, **108**, 939–944.
78. A. Leela Mohana Reddy, R. I. Jafri, N. Jha, S. Ramaprabhu and P. M. Ajayan, *J. Mater. Chem.*, 2011, **21**, 16103–16107.
79. M. J. Lázaro, V. Celorrio, L. Calvillo, E. Pastor and R. Moliner, *J. Power Sources*, 2011, **196**, 4236–4241.
80. V. Celorrio, L. Calvillo, R. Moliner, E. Pastor and M. J. Lázaro, *J. Power Sources*, 2013, **239**, 72–80.
81. V. Celorrio, J. Flórez-Montaño, R. Moliner, E. Pastor and M. J. Lázaro, *Int. J. Hydrogen Energy*, 2014, **39**, 5371–5377.
82. X. X. Wang, Z. H. Tan, M. Zeng and J. N. Wang, *Sci. Rep.*, 2014, **4**, 4437.
83. N. Koprinarov, M. Konstantinova, G. Pchelarov and M. Marinov, in *Fuel Cell Technologies: State and Perspectives*, ed. N. Sammes, A. Smirnova and O. Vasylyev, Springer, Netherlands, 2005, vol. 202, pp. 81–95.
84. K. Vinodgopal, M. Haria, D. Meisel and P. Kamat, *Nano Lett.*, 2004, **4**, 415–418.
85. B. Xu, X. Yang, X. Wang, J. Guo and X. Liu, *J. Power Sources*, 2006, **162**, 160–164.
86. J. Guo, X. Yang, Y. Yao, X. Wang, X. Liu and B. Xu, *Rare Metals*, 2006, **25**, 305–308.
87. K. Tasaki, R. DeSousa, H. Wang, J. Gasa, A. Venkatesan, P. Pugazhendhi and R. O. Loutfy, *J. Membr. Sci.*, 2006, **281**, 570–580.

88. J.-H. Jung, S. Vadahanambi and I.-K. Oh, *Compos. Sci. Technol.*, 2010, **70**, 584–592.
89. A. Krueger, *Adv. Mater.*, 2008, **20**, 2445–2449.
90. H. Huang, L. Dai, D. H. Wang, L.-S. Tan and E. Osawa, *J. Mater. Chem.*, 2008, **18**, 1347–1352.
91. V. Chakrapani, J. C. Angus, A. B. Anderson, S. D. Wolter, B. R. Stoner and G. U. Sumanasekera, *Science*, 2007, **318**, 1424–1430.
92. Y. Liu, Z. Gu, J. L. Margrave and V. N. Khabashesku, *Chem. Mater.*, 2004, **16**, 3924–3930.
93. A. Krueger and D. Lang, *Adv. Funct. Mater.*, 2012, **22**, 890–906.
94. A. Moore, V. Celorrio, M. M. de Oca, D. Plana, W. Hongthani, M. J. Lazaro and D. J. Fermin, *Chem. Commun.*, 2011, **47**, 7656–7658.
95. J. Wang and G. M. Swain, *J. Electrochem. Soc.*, 2003, **150**, E24–E32.
96. N. Spătaru, X. Zhang, T. Spătaru, D. A. Tryk and A. Fujishima, *J. Electrochem. Soc.*, 2008, **155**, B264–B269.
97. T. Spătaru, M. Anastasescu, N. Spătaru and A. Fujishima, *Electrochem. Commun.*, 2013, **29**, 1–3.
98. F. Montilla, E. Morallón, I. Duo, C. Comninellis and J. L. Vázquez, *Electrochim. Acta.*, 2003, **48**, 3891–3897.
99. L. La-Torre-Riveros, R. Guzman-Blas, A. E. Méndez-Torres, M. Prelas, D. A. Tryk and C. R. Cabrera, *ACS Appl. Mater. Interfaces*, 2012, **4**, 1134–1147.
100. J. Wang, G. M. Swain, T. Tachibana and K. Kobashi, *Electrochem. Solid-State Lett.*, 2000, **3**, 286–289.
101. M. Panizza and G. Cerisola, *Electrochim. Acta*, 2005, **51**, 191–199.
102. V. Celorrio, D. Plana, J. Flórez-Montaño, M. G. Montes de Oca, A. Moore, M. J. Lázaro, E. Pastor and D. J. Fermín, *J. Phys. Chem. C*, 2013, **117**, 21735–21742.
103. G. R. Salazar-Banda, K. I. B. Eguiluz and L. A. Avaca, *Electrochem. Commun.*, 2007, **9**, 59–64.
104. G. Siné, G. Fóti and C. Comninellis, *J. Electroanal. Chem.*, 2006, **595**, 115–124.
105. H. B. Suffredini, V. Tricoli, N. Vatistas and L. A. Avaca, *J. Power Sources*, 2006, **158**, 124–128.
106. A. E. Fischer and G. M. Swain, *J. Electrochem. Soc.*, 2005, **152**, B369–B375.
107. J. B. Zang, Y. H. Wang, H. Huang and W. Tang, *Electrochim. Acta*, 2007, **52**, 4398–4402.
108. J. Wang, G. M. Swain, T. Tachibana and K. Kobashi, *J. New Mater. Electrochem. Syst.*, 2000, **3**, 75–82.
109. T. A. Ivandini, R. Sato, Y. Makide, A. Fujishima and Y. Einaga, *Diamond Relat. Mater.*, 2005, **14**, 2133–2138.
110. T. A. Ivandini, R. Sato, Y. Makide, A. Fujishima and Y. Einaga, *Chem. Lett.*, 2004, **33**, 1330–1331.
111. X. Lyu, J. Hu, J. S. Foord and Q. Wang, *J. Power Sources*, 2013, **242**, 631–637.
112. I. Shpilevaya, W. Smirnov, S. Hirsz, N. Yang, C. E. Nebel and J. S. Foord, *RSC Adv.*, 2014, **4**, 531–537.
113. F. Gao, N. Yang and C. E. Nebel, *Electrochim. Acta*, 2013, **112**, 493–499.

114. J. A. Bennett, Y. Show, S. Wang and G. M. Swain, *J. Electrochem. Soc.*, 2005, **152**, E184–E192.
115. J. Hu, X. Lu, J. S. Foord and Q. Wang, *Phys. Status Solidi A*, 2009, **206**, 2057–2062.
116. F. Gao, N. Yang, W. Smirnov, H. Obloh and C. E. Nebel, *Electrochim. Acta*, 2013, **90**, 445–451.
117. X. Lu, J. Hu, J. S. Foord and Q. Wang, *J. Electroanal. Chem.*, 2011, **654**, 38–43.
118. L. La-Torre-Riveros, E. Abel-Tatis, A. Méndez-Torres, D. Tryk, M. Prelas and C. Cabrera, *J. Nanopart. Res.*, 2011, **13**, 2997–3009.
119. I. González-González, D. A. Tryk and C. R. Cabrera, *Diamond Relat. Mater.*, 2006, **15**, 275–278.
120. E. Yoo, T. Okata, T. Akita, M. Kohyama, J. Nakamura and I. Honma, *Nano Lett.*, 2009, **9**, 2255–2259.
121. H.-J. Choi, S.-M. Jung, J.-M. Seo, D. W. Chang, L. Dai and J.-B. Baek, *Nano Energy*, 2012, **1**, 534–551.
122. D. A. C. Brownson, D. K. Kampouris and C. E. Banks, *J. Power Sources*, 2011, **196**, 4873–4885.
123. R. Kou, Y. Shao, D. Wang, M. H. Engelhard, J. H. Kwak, J. Wang, V. V. Viswanathan, C. Wang, Y. Lin, Y. Wang, I. A. Aksay and J. Liu, *Electrochem. Commun.*, 2009, **11**, 954–957.
124. R. Kou, Y. Shao, D. Mei, Z. Nie, D. Wang, C. Wang, V. V. Viswanathan, S. Park, I. A. Aksay, Y. Lin, Y. Wang and J. Liu, *J. Am. Chem. Soc.*, 2011, **133**, 2541–2547.
125. N. Soin, S. S. Roy, T. H. Lim and J. A. D. McLaughlin, *Mater. Chem. Phys.*, 2011, **129**, 1051–1057.
126. L. Dong, R. R. S. Gari, Z. Li, M. M. Craig and S. Hou, *Carbon*, 2010, **48**, 781–787.
127. S. M. Choi, M. H. Seo, H. J. Kim and W. B. Kim, *Carbon*, 2011, **49**, 904–909.
128. D. A. Dikin, S. Stankovich, E. J. Zimney, R. D. Piner, G. H. B. Dommett, G. Evmenenko, S. T. Nguyen and R. S. Ruoff, *Nature*, 2007, **448**, 457–460.
129. G. Eda, G. Fanchini and M. Chhowalla, *Nat. Nanotechnol.*, 2008, **3**, 270–274.
130. C. Gómez-Navarro, R. T. Weitz, A. M. Bittner, M. Scolari, A. Mews, M. Burghard and K. Kern, *Nano Lett.*, 2007, 7, 3499–3503.
131. S. Sharma, A. Ganguly, P. Papakonstantinou, X. Miao, M. Li, J. L. Hutchison, M. Delichatsios and S. Ukleja, *J. Phys. Chem.*, 2010, **114**, 19459–19466.
132. Y. Li, W. Gao, L. Ci, C. Wang and P. M. Ajayan, *Carbon*, 2010, **48**, 1124–1130.
133. S. Wang, S. P. Jiang and X. Wang, *Electrochim. Acta*, 2011, **56**, 3338–3344.
134. J.-D. Qiu, G.-C. Wang, R.-P. Liang, X.-H. Xia and H.-W. Yu, *J. Phys. Chem., C*, 2011, **115**, 15639–15645.
135. Y. Li, L. Tang and J. Li, *Electrochem. Commun.*, 2009, **11**, 846–849.
136. S. Guo, S. Dong and E. Wang, *ACS Nano*, 2009, **4**, 547–555.

137. S. Zhang, Y. Shao, H.-g. Liao, J. Liu, I. A. Aksay, G. Yin and Y. Lin, *Chem. Mater.*, 2011, **23**, 1079–1081.
138. S. Wang, X. Wang and S. P. Jiang, *Phys. Chem. Chem. Phys.*, 2011, **13**, 6883–6891.
139. N. Kristian, Y. Yan and X. Wang, *Chem. Commun.*, 2008, 353–355.
140. S. Bong, Y.-R. Kim, I. Kim, S. Woo, S. Uhm, J. Lee and H. Kim, *Electrochem. Commun.*, 2010, **12**, 129–131.
141. Y. Zhang, Y.-e. Gu, S. Lin, J. Wei, Z. Wang, C. Wang, Y. Du and W. Ye, *Electrochim. Acta*, 2011, **56**, 8746–8751.
142. Y.-C. Cao, C. Xu, X. Wu, X. Wang, L. Xing and K. Scott, *J. Power Sources*, 2011, **196**, 8377–8382.
143. J. Hernández, J. Solla-Gullón and E. Herrero, *J. Electroanal. Chem.*, 2004, **574**, 185–196.
144. J. Luo, M. M. Maye, V. Petkov, N. N. Kariuki, L. Wang, P. Njoki, D. Mott, Y. Lin and C.-J. Zhong, *Chem. Mater.*, 2005, **17**, 3086–3091.
145. Z.-S. Wu, D.-W. Wang, W. Ren, J. Zhao, G. Zhou, F. Li and H.-M. Cheng, *Adv. Funct. Mater.*, 2010, **20**, 3595–3602.
146. A. Hayashi, H. Notsu, K. i. Kimijima, J. Miyamoto and I. Yagi, *Electrochim. Acta*, 2008, **53**, 6117–6125.
147. S. H. Joo, H. I. Lee, D. J. You, K. Kwon, J. H. Kim, Y. S. Choi, M. Kang, J. M. Kim, C. Pak, H. Chang and D. Seung, *Carbon*, 2008, **46**, 2034–2045.
148. S.-H. Liu, C.-C. Chiang, M.-T. Wu and S.-B. Liu, *Int. J. Hydrogen Energy*, 2010, **35**, 8149–8154.
149. D. Sebastián, J. C. Calderón, J. A. González-Expósito, E. Pastor, M. V. Martínez-Huerta, I. Suelves, R. Moliner and M. J. Lázaro, *Int. J. Hydrogen Energy*, 2010, **35**, 9934–9942.
150. J.-S. Zheng, M.-X. Wang, X.-S. Zhang, Y.-X. Wu, P. Li, X.-G. Zhou and W.-K. Yuan, *J. Power Sources*, 2008, **175**, 211–216.
151. D. Sebastian, M. J. Lazaro, R. Moliner, I. Suelves, A. S. Arico and V. Baglio, *Int. J. Hydrogen Energy*, 2014, **39**, 5414–5423.
152. H.-W. Ha, I. Y. Kim, S.-J. Hwang and R. S. Ruoff, *Electrochem. Solid-State Lett.*, 2011, **14**, B70–B73.
153. J. Wu and H. Yang, *Acc. Chem. Res.*, 2013, **46**, 1848–1857.
154. T. Toda, H. Igarashi and M. Watanabe, *J. Electrochem. Soc.*, 1998, **145**, 4185–4188.
155. M. Watanabe, D. A. Tryk, M. Wakisaka, H. Yano and H. Uchida, *Electrochim. Acta*, 2012, **84**, 187–201.
156. X. Li, L. An, L. Zhang, F. Li, X. Wang and D. Xia, *Prog. Chem.*, 2011, **23**, 501–508.
157. C. V. Rao, A. L. M. Reddy, Y. Ishikawa and P. M. Ajayan, *Carbon*, 2011, **49**, 931–936.
158. Z. Yang, H. Nie, X. a. Chen, X. Chen and S. Huang, *J. Power Sources*, 2013, **236**, 238–249.
159. Y. Chen, J. Wang, H. Liu, M. N. Banis, R. Li, X. Sun, T.-K. Sham, S. Ye and S. Knights, *J. Phys. Chem., C*, 2011, **115**, 3769–3776.

160. R. Imran Jafri, N. Rajalakshmi and S. Ramaprabhu, *J. Mater. Chem.*, 2010, **20**, 7114–7117.
161. Z.-S. Wu, S. Yang, Y. Sun, K. Parvez, X. Feng and K. Müllen, *J. Am. Chem. Soc.*, 2012, **134**, 9082–9085.
162. Y. Liang, Y. Li, H. Wang, J. Zhou, J. Wang, T. Regier and H. Dai, *Nat. Mater.*, 2011, **10**, 780–786.
163. R. Bashyam and P. Zelenay, *Nature*, 2006, **443**, 63–66.
164. R. Othman, A. L. Dicks and Z. Zhu, *Int. J. Hydrogen Energy*, 2012, **37**, 357–372.
165. W. Martínez Millán, T. Toledano Thompson, L. G. Arriaga and M. A. Smit, *Int. J. Hydrogen Energy*, 2009, **34**, 694–702.
166. K. Lee, L. Zhang, H. Lui, R. Hui, Z. Shi and J. Zhang, *Electrochim. Acta*, 2009, **54**, 4704–4711.
167. Z. Mo, H. Peng, H. Liang and S. Liao, *Electrochim. Acta*, 2013, **99**, 30–37.
168. A. Serov, M. H. Robson, K. Artyushkova and P. Atanassov, *Appl. Catal., B*, 2012, **127**, 300–306.
169. G. Liu, X. Li, P. Ganesan and B. N. Popov, *Appl. Catal., B*, 2009, **93**, 156–165.
170. G. Jo and S. Shanmugam, *Electrochem. Commun.*, 2012, **25**, 101–104.
171. G. Wu and P. Zelenay, *Acc. Chem. Res.*, 2013, **46**, 1878–1889.
172. J. Sanetuntikul and S. Shanmugam, *Electrochim. Acta*, 2014, **119**, 92–98.
173. F. Bidault, D. J. L. Brett, P. H. Middleton and N. P. Brandon, *J. Power Sources*, 2009, **187**, 39–48.
174. Y. Kiros, C. Myrén, S. Schwartz, A. Sampathrajan and M. Ramanathan, *Int. J. Hydrogen Energy*, 1999, **24**, 549–564.
175. Y. Kiros and S. Schwartz, *J. Power Sources*, 2000, **87**, 101–105.
176. M. A. Al-Saleh, S. Gültekin, A. S. Al-Zakri and H. Celiker, *J. Appl. Electrochem.*, 1994, **24**, 575–580.
177. M. Schulze, E. Gülzow and G. Steinhilber, *Appl. Surf. Sci.*, 2001, **179**, 251–256.
178. S. Gultekin, M. A. Al-Saleh, A. S. Al-Zakri and K. A. A. Abbas, *Int. J. Hydrogen Energy*, 1996, **21**, 485–489.
179. T. Kenjo, *Bull. Chem. Soc. Jpn.*, 1981, **54**, 2553–2556.
180. K. Schultze and H. Bartelt, *Int. J. Hydrogen Energy*, 1992, **17**, 711–718.
181. M. A. Al-Saleh, S. Gultekin, A. S. Al-Zakri and A. A. A. Khan, *Int. J. Hydrogen Energy*, 1996, **21**, 657–661.
182. M. A. Al-Saleh, R. Sleem Ur, S. M. M. J. Kareemuddin and A. S. Al-Zakri, *J. Power Sources*, 1998, **72**, 159–164.
183. W. M. Vogel, *Electrochim. Acta*, 1968, **13**, 1821–1826.
184. D. B. Sepa, M. V. Vojnovic, M. Stojanovic and A. Damjanovic, *J. Electroanal. Chem. Interfacial Electrochem.*, 1987, **218**, 265–272.
185. R. Pattabiraman, *Appl. Catal., A*, 1997, **153**, 9–20.
186. J. M. Martinovic, D. B. Sepa, M. V. Vojnovic and A. Damjanovic, *Electrochim. Acta*, 1988, **33**, 1267–1272.
187. D. B. Sepa, M. V. Vojnovic, L. M. Vracar and A. Damjanovic, *Electrochim. Acta*, 1987, **32**, 129–134.

188. Y.-F. Yang, Y.-H. Zhou and C.-S. Cha, *Electrochim. Acta*, 1995, **40**, 2579–2586.
189. N. A. Anastasijević, Z. M. Dimitrijević and R. R. Adžić, *J. Electroanal. Chem. Interfacial Electrochem.*, 1986, **199**, 351–364.
190. F. H. B. Lima and E. A. Ticianelli, *Electrochim. Acta*, 2004, **49**, 4091–4099.
191. K. V. Ramesh and A. K. Shukla, *J. Power Sources*, 1987, **19**, 279–285.
192. M. Chatenet, M. Aurousseau, R. Durand and F. Andolfatto, *J. Electrochem. Soc.*, 2003, **150**, D47–D55.
193. G. Couturier, D. W. Kirk, P. J. Hyde and S. Srinivasan, *Electrochim. Acta*, 1987, **32**, 995–1005.
194. M.-k. Min, J. Cho, K. Cho and H. Kim, *Electrochim. Acta*, 2000, **45**, 4211–4217.
195. T. Toda, H. Igarashi and M. Watanabe, *J. Electroanal. Chem.*, 1999, **460**, 258–262.
196. M. B. Vukmirovic, J. Zhang, K. Sasaki, A. U. Nilekar, F. Uribe, M. Mavrikakis and R. R. Adzic, *Electrochim. Acta*, 2007, **52**, 2257–2263.
197. V. Hacker, E. Wallnöfer, W. Baumgartner, T. Schaffer, J. O. Besenhard, H. Schröttner and M. Schmied, *Electrochem. Commun.*, 2005, **7**, 377–382.
198. M. A. Kostowskyj, R. J. Gilliam, D. W. Kirk and S. J. Thorpe, *Int. J. Hydrogen Energy*, 2008, **33**, 5773–5778.
199. L. Tammeveski, H. Erikson, A. Sarapuu, J. Kozlova, P. Ritslaid, V. Sammelselg and K. Tammeveski, *Electrochem. Commun.*, 2012, **20**, 15–18.
200. E. Antolini, *Energy Environ. Sci.*, 2009, **2**, 915–931.
201. M. Shao, P. Liu, J. Zhang and R. Adzic, *J. Phys. Chem., B*, 2007, **111**, 6772–6775.
202. S. Maheswari, S. Karthikeyan, P. Murugan, P. Sridhar and S. Pitchumani, *Phys. Chem. Chem. Phys.*, 2012, **14**, 9683–9695.
203. Y. Wang and P. B. Balbuena, *J. Phys. Chem., B*, 2005, **109**, 18902–18906.
204. J. L. Fernández, D. A. Walsh and A. J. Bard, *J. Am. Chem. Soc.*, 2004, **127**, 357–365.
205. H. T. Duong, M. A. Rigsby, W.-P. Zhou and A. Wieckowski, *J. Phys. Chem., C*, 2007, **111**, 13460–13465.
206. C. Song, L. Zhang, J. Zhang, D. P. Wilkinson and R. Baker, *Fuel Cells*, 2007, **7**, 9–15.
207. A. Morozan, S. Campidelli, A. Filoramo, B. Jousselme and S. Palacin, *Carbon*, 2011, **49**, 4839–4847.
208. R. Chen, H. Li, D. Chu and G. Wang, *J. Phys. Chem. C*, 2009, **113**, 20689–20697.
209. Y. L. Cao, H. X. Yang, X. P. Ai and L. F. Xiao, *J. Electroanal. Chem.*, 2003, **557**, 127–134.
210. L. Mao, D. Zhang, T. Sotomura, K. Nakatsu, N. Koshiba and T. Ohsaka, *Electrochim. Acta*, 2003, **48**, 1015–1021.
211. F. Cheng, Y. Su, J. Liang, Z. Tao and J. Chen, *Chem. Mater.*, 2009, **22**, 898–905.
212. F. H. B. Lima, M. L. Calegaro and E. A. Ticianelli, *Electrochim. Acta*, 2007, **52**, 3732–3738.

213. F. H. B. Lima, M. L. Calegaro and E. A. Ticianelli, *J. Electroanal. Chem.*, 2006, **590**, 152–160.
214. I. Roche, E. Chaînet, M. Chatenet and J. Vondrák, *J. Appl. Electrochem.*, 2008, **38**, 1195–1201.
215. I. Roche and K. Scott, *J. Appl. Electrochem.*, 2009, **39**, 197–204.
216. A. C. Garcia, A. D. Herrera, E. A. Ticianelli, M. Chatenet and C. Poinsignon, *J. Electrochem. Soc.*, 2011, **158**, B290–B296.
217. I. Roche, E. Chaînet, M. Chatenet and J. Vondrák, *J. Phys. Chem. C*, 2006, **111**, 1434–1443.
218. J. Vondrák, B. Klápšte, J. Velická, M. Sedlaríková, J. Reiter, I. Roche, E. Chainet, J. F. Fauvarque and M. Chatenet, *J. New Mater. Electrochem. Syst.*, 2005, **8**, 209–212.
219. A. C. Garcia, F. H. B. Lima, E. A. Ticianelli and M. Chatenet, *J. Power Sources*, 2013, **222**, 305–312.
220. F. Moureaux, P. Stevens and M. Chatenet, *Electrocatalysis*, 2013, **4**, 123–133.
221. L. Qu, Y. Liu, J.-B. Baek and L. Dai, *ACS Nano*, 2010, **4**, 1321–1326.
222. W. Y. Wong, W. R. W. Daud, A. B. Mohamad, A. A. H. Kadhum, K. S. Loh and E. H. Majlan, *Int. J. Hydrogen Energy*, 2013, **38**, 9370–9386.
223. A. Zhao, J. Masa, W. Schuhmann and W. Xia, *J. Phys. Chem. C*, 2013, **117**, 24283–24291.
224. M. Vikkisk, I. Kruusenberg, U. Joost, E. Shulga, I. Kink and K. Tammeveski, *Appl. Catal., B*, 2014, **147**, 369–376.
225. H. Li, H. Liu, Z. Jong, W. Qu, D. Geng, X. Sun and H. Wang, *Int. J. Hydrogen Energy*, 2011, **36**, 2258–2265.
226. S. Maldonado and K. J. Stevenson, *J. Phys. Chem., B*, 2005, **109**, 4707–4716.
227. H. Wang, T. Maiyalagan and X. Wang, *ACS Catal.*, 2012, **2**, 781–794.
228. S. Wang, L. Zhang, Z. Xia, A. Roy, D. W. Chang, J.-B. Baek and L. Dai, *Angew. Chem., Int. Ed.*, 2012, **51**, 4209–4212.
229. P. H. Matter, L. Zhang and U. S. Ozkan, *J. Catal.*, 2006, **239**, 83–96.
230. P. H. Matter, E. Wang, M. Arias, E. J. Biddinger and U. S. Ozkan, *J. Phys. Chem., B*, 2006, **110**, 18374–18384.
231. E. J. Biddinger and U. S. Ozkan, *J. Phys. Chem., C*, 2010, **114**, 15306–15314.
232. Y. Tan, C. Xu, G. Chen, X. Fang, N. Zheng and Q. Xie, *Adv. Funct. Mater.*, 2012, **22**, 4584–4591.
233. J. Xie, D. L. Wood, K. L. More, P. Atanassov and R. L. Borup, *J. Electrochem. Soc.*, 2005, **152**, A1011–A1020.
234. Y. Shao, G. Yin, Y. Gao and P. Shi, *J. Electrochem. Soc.*, 2006, **153**, A1093–A1097.
235. Y. Shao, G. Yin, J. Zhang and Y. Gao, *Electrochim. Acta*, 2006, **51**, 5853–5857.
236. Y. Shao, G. Yin, Z. Wang and Y. Gao, *J. Power Sources*, 2007, **167**, 235–242.
237. D. A. Stevens, M. T. Hicks, G. M. Haugen and J. R. Dahn, *J. Electrochem. Soc.*, 2005, **152**, A2309–A2315.
238. L. M. Roen, C. H. Paik and T. D. Jarvi, *Electrochem. Solid-State Lett.*, 2004, **7**, A19–A22.
239. X.-Z. Yuan, H. Li, S. Zhang, J. Martin and H. Wang, *J. Power Sources*, 2011, **196**, 9107–9116.

240. S. Zhang, X.-Z. Yuan, J. N. C. Hin, H. Wang, K. A. Friedrich and M. Schulze, *J. Power Sources*, 2009, **194**, 588–600.
241. W. Schmittinger and A. Vahidi, *J. Power Sources*, 2008, **180**, 1–14.
242. B. Merzougui and S. Swathirajan, *J. Electrochem. Soc.*, 2006, **153**, A2220–A2226.
243. H. Chizawa, Y. Ogami, H. Naka, A. Matsunaga, N. Aoki and T. Aoki, *ECS Trans.*, 2006, **3**, 645–655.
244. E. Guilminot, A. Corcella, F. Charlot, F. Maillard and M. Chatenet, *J. Electrochem. Soc.*, 2007, **154**, B96–B105.
245. S. C. Ball, S. L. Hudson, D. Thompsett and B. Theobald, *J. Power Sources*, 2007, **171**, 18–25.
246. A. Panchenko, H. Dilger, J. Kerres, M. Hein, A. Ullrich, T. Kaz and E. Roduner, *Phys. Chem. Chem. Phys.*, 2004, **6**, 2891–2894.
247. D. A. Stevens and J. R. Dahn, *Carbon*, 2005, **43**, 179–188.
248. T. Fuller and G. Gray, *ECS Trans.*, 2006, **1**, 345–353.
249. H. Tang, Z. Qi, M. Ramani and J. F. Elter, *J. Power Sources*, 2006, **158**, 1306–1312.
250. P. T. Yu, W. Gu, R. Makharia, F. T. Wagner and H. A. Gasteiger, *ECS Trans.*, 2006, **3**, 797–809.
251. J. P. Meyers and R. M. Darling, *J. Electrochem. Soc.*, 2006, **153**, A1432–A1442.
252. X. Wang, W. Li, Z. Chen, M. Waje and Y. Yan, *J. Power Sources*, 2006, **158**, 154–159.
253. X. Yu and S. Ye, *J. Power Sources*, 2007, **172**, 145–154.
254. Y. Takagi and Y. Takakuwa, *ECS Trans.*, 2006, **3**, 855–860.
255. C. A. Reiser, L. Bregoli, T. W. Patterson, J. S. Yi, J. D. Yang, M. L. Perry and T. D. Jarvi, *Electrochem. Solid-State Lett.*, 2005, **8**, A273–A276.
256. B. Kienitz, H. Baskaran, T. Zawodzinski and B. Pivovar, *ECS Trans.*, 2007, **11**, 777–788.
257. X. Cheng, Z. Shi, N. Glass, L. Zhang, J. Zhang, D. Song, Z.-S. Liu, H. Wang and J. Shen, *J. Power Sources*, 2007, **165**, 739–756.
258. M. Cai, M. S. Ruthkosky, B. Merzougui, S. Swathirajan, M. P. Balogh and S. H. Oh, *J. Power Sources*, 2006, **160**, 977–986.
259. M. Schulze, N. Wagner, T. Kaz and K. A. Friedrich, *Electrochim. Acta*, 2007, **52**, 2328–2336.
260. C. Lee and W. Mérida, *J. Power Sources*, 2007, **164**, 141–153.
261. L. R. Jordan, A. K. Shukla, T. Behrsing, N. R. Avery, B. C. Muddle and M. Forsyth, *J. Power Sources*, 2000, **86**, 250–254.
262. U. Pasaogullari and C.-Y. Wang, *Electrochim. Acta*, 2004, **49**, 4359–4369.
263. A. Z. Weber and J. Newman, *J. Electrochem. Soc.*, 2005, **152**, A677–A688.
264. M. Gratzel, *Nature*, 2001, **414**, 338–344.
265. M. K. Nazeeruddin, E. Baranoff and M. Grätzel, *Sol. Energy*, 2011, **85**, 1172–1178.
266. J. E. Trancik, S. C. Barton and J. Hone, *Nano Lett.*, 2008, **8**, 982–987.
267. P. Joshi, L. Zhang, Q. Chen, D. Galipeau, H. Fong and Q. Qiao, *ACS Appl. Mater. Interfaces*, 2010, **2**, 3572–3577.

268. J. D. Roy-Mayhew, D. J. Bozym, C. Punckt and I. A. Aksay, *ACS Nano*, 2010, **4**, 6203–6211.
269. S. Ahmad, J.-H. Yum, Z. Xianxi, M. Gratzel, H.-J. Butt and M. K. Nazeeruddin, *J. Mater. Chem.*, 2010, **20**, 1654–1658.
270. S. Ahmad, J.-H. Yum, H.-J. Butt, M. K. Nazeeruddin and M. Grätzel, *ChemPhysChem*, 2010, **11**, 2814–2819.
271. T. N. Murakami and M. Grätzel, *Inorg. Chim. Acta*, 2008, **361**, 572–580.
272. M. Wu, X. Lin, A. Hagfeldt and T. Ma, *Angew. Chem., Int. Ed.*, 2011, **50**, 3520–3524.
273. T. W. Hamann and J. W. Ondersma, *Energy Environ. Sci.*, 2011, **4**, 370–381.
274. L. J. Brennan, M. T. Byrne, M. Bari and Y. K. Gun'ko, *Adv. Energy Mater.*, 2011, **1**, 472–485.
275. T. Chen, S. Wang, Z. Yang, Q. Feng, X. Sun, L. Li, Z.-S. Wang and H. Peng, *Angew. Chem., Int. Ed.*, 2011, **50**, 1815–1819.
276. I. Ahmad, U. Khan and Y. K. Gun'ko, *J. Mater. Chem.*, 2011, **21**, 16990–16996.
277. A. Mathew, G. M. Rao and N. Munichandraiah, *Mater. Res. Bull.*, 2011, **46**, 2045–2049.
278. A. K. K. Kyaw, H. Tantang, T. Wu, L. Ke, C. Peh, Z. H. Huang, X. T. Zeng, H. V. Demir, Q. Zhang and X. W. Sun, *Appl. Phys. Lett.*, 2011, **99**, 021107.
279. T. Charinpanitkul, P. Lorturn, W. Ratismith, N. Viriya-empikul, G. Tumcharern and J. Wilcox, *Mater. Res. Bull.*, 2011, **46**, 1604–1609.
280. J. Zhang, X. Li, W. Guo, T. Hreid, J. Hou, H. Su and Z. Yuan, *Electrochim. Acta*, 2011, **56**, 3147–3152.
281. H. Chang, Y. Liu, H. Zhang and J. Li, *J. Electroanal. Chem.*, 2011, **656**, 269–273.
282. G. Zhu, L. Pan, T. Lu, T. Xu and Z. Sun, *J. Mater. Chem.*, 2011, **21**, 14869–14875.
283. H. Zhang, X. Lv, Y. Li, Y. Wang and J. Li, *ACS Nano*, 2009, **4**, 380–386.
284. B. Fang, M. Kim, S.-Q. Fan, J. H. Kim, D. P. Wilkinson, J. Ko and J.-S. Yu, *J. Mater. Chem.*, 2011, **21**, 8742–8748.
285. G.-r. Li, F. Wang, Q.-w. Jiang, X.-p. Gao and P.-w. Shen, *Angew. Chem., Int. Ed.*, 2010, **49**, 3653–3656.
286. P. Dong, C. L. Pint, M. Hainey, F. Mirri, Y. Zhan, J. Zhang, M. Pasquali, R. H. Hauge, R. Verduzco, M. Jiang, H. Lin and J. Lou, *ACS Appl. Mater. Interfaces*, 2011, **3**, 3157–3161.
287. S. Li, Y. Luo, W. Lv, W. Yu, S. Wu, P. Hou, Q. Yang, Q. Meng, C. Liu and H.-M. Cheng, *Adv. Energy Mater.*, 2011, **1**, 486–490.
288. O. Pang-Leen, B. E. William and A. L. Igor, *Nanotechnology*, 2010, **21**, 105203.
289. K. Golap, A. Sudip, A. Hare Ram, A. Rakesh, S. Tetsuo, S. Maheshwar, K. Wakita and U. Masayoshi, *J. Phys. D: Appl. Phys.*, 2009, **42**, 115104.
290. Q. Shu, J. Wei, K. Wang, H. Zhu, Z. Li, Y. Jia, X. Gui, N. Guo, X. Li, C. Ma and D. Wu, *Nano Lett.*, 2009, **9**, 4338–4342.
291. X. Wang, L. Zhi and K. Müllen, *Nano Lett.*, 2007, **8**, 323–327.
292. X. Yan, X. Cui, B. Li and L.-s. Li, *Nano Lett.*, 2010, **10**, 1869–1873.

293. T. Chen, W. Hu, J. Song, G. H. Guai and C. M. Li, *Adv. Funct. Mater.*, 2012, **22**, 5245–5250.
294. Z. Peining, A. S. Nair, P. Shengjie, Y. Shengyuan and S. Ramakrishna, *ACS Appl. Mater. Interfaces*, 2012, **4**, 581–585.
295. H. Wang, S. L. Leonard and Y. H. Hu, *Ind. Eng. Chem. Res.*, 2012, **51**, 10613–10620.
296. M.-H. Jung, M. G. Kang and M.-J. Chu, *J. Mater. Chem.*, 2012, **22**, 16477–16483.
297. D. W. Zhang, X. D. Li, H. B. Li, S. Chen, Z. Sun, X. J. Yin and S. M. Huang, *Carbon*, 2011, **49**, 5382–5388.
298. J. D. Roy-Mayhew and I. A. Aksay, *Chem. Rev.*, 2014, **114**, 6323–6348.
299. H. V. Helmholtz, *Ann. Phys. (Leipzig)*, 1853, **89**, 21.
300. G. Gouy, *J. Phys.*, 1910, **4**, 457.
301. D. L. Chapman, *Philos. Mag.*, 1913, **25**, 475–481.
302. O. Stern, *Z. Elektrochem.*, 1924, **30**, 508.
303. L. L. Zhang and X. S. Zhao, *Chem. Soc. Rev.*, 2009, **38**, 2520–2531.
304. B. E. Conway, *Electrochemical Supercapacitors - Scientific Fundamentals and Technological Applications*, Springer US, New York, 1999.
305. A. G. Pandolfo and A. F. Hollenkamp, *J. Power Sources*, 2006, **157**, 11–27.
306. E. Frackowiak and F. Béguin, *Carbon*, 2001, **39**, 937–950.
307. J.-i. Hong, I.-H. Yeo and W.-k. Paik, *J. Electrochem. Soc.*, 2001, **148**, A156–A163.
308. C.-C. Hu and T.-W. Tsou, *J. Power Sources*, 2003, **115**, 179–186.
309. M. Winter and R. J. Brodd, *Chem. Rev.*, 2004, **104**, 4245–4270.
310. H. Li, H. a. Xi, S. Zhu, Z. Wen and R. Wang, *Microporous Mesoporous Mater.*, 2006, **96**, 357–362.
311. E. Frackowiak, *Phys. Chem. Chem. Phys.*, 2007, **9**, 1774–1785.
312. O. Barbieri, M. Hahn, A. Herzog and R. Kötz, *Carbon*, 2005, **43**, 1303–1310.
313. W. Li, G. Reichenauer and J. Fricke, *Carbon*, 2002, **40**, 2955–2959.
314. A. Braun, M. Bärtsch, B. Schnyder, R. Kötz, O. Haas, H. G. Haubold and G. Goerigk, *J. Non-Cryst. Solids*, 1999, **260**, 1–14.
315. A. Braun, M. Bärtsch, O. Merlo, B. Schnyder, B. Schaffner, R. Kötz, O. Haas and A. Wokaun, *Carbon*, 2003, **41**, 759–765.
316. K. H. An, W. S. Kim, Y. S. Park, J. M. Moon, D. J. Bae, S. C. Lim, Y. S. Lee and Y. H. Lee, *Adv. Funct. Mater.*, 2001, **11**, 387–392.
317. K. H. An, K. K. Jeon, J. K. Heo, S. C. Lim, D. J. Bae and Y. H. Lee, *J. Electrochem. Soc*, 2002, **149**, A1058–A1062.
318. E. Frackowiak, S. Delpeux, K. Jurewicz, K. Szostak, D. Cazorla-Amoros and F. Béguin, *Chem. Phys. Lett.*, 2002, **361**, 35–41.
319. C. Niu, E. K. Sichel, R. Hoch, D. Moy and H. Tennent, *Appl. Phys. Lett.*, 1997, **70**, 1480–1482.
320. Y.-H. Hsu, C.-C. Lai, C.-L. Ho and C.-T. Lo, *Electrochim. Acta*, 2014, **127**, 369–376.
321. C. Ma, Y. Li, J. Shi, Y. Song and L. Liu, *Chem. Eng. J.*, 2014, **249**, 216–225.
322. I. Kovalenko, D. G. Bucknall and G. Yushin, *Adv. Funct. Mater.*, 2010, **20**, 3979–3986.

323. M. E. Plonska-Brzezinska and L. Echegoyen, *J. Mater. Chem. A*, 2013, **1**, 13703–13714.
324. L. Buglione, E. L. K. Chng, A. Ambrosi, Z. Sofer and M. Pumera, *Electrochem. Commun.*, 2012, **14**, 5–8.
325. A. Bonanni and M. Pumera, *Electrochem. Commun.*, 2013, **26**, 52–54.
326. W. Song, X. Ji, W. Deng, Q. Chen, C. Shen and C. E. Banks, *Phys. Chem. Chem. Phys.*, 2013, **15**, 4799–4803.
327. L. Buglione, A. Bonanni, A. Ambrosi and M. Pumera, *ChemPlusChem*, 2012, **77**, 71–73.
328. Q. Cheng, J. Tang, J. Ma, H. Zhang, N. Shinya and L.-C. Qin, *Phys. Chem. Chem. Phys.*, 2011, **13**, 17615–17624.
329. S.-Y. Yang, K.-H. Chang, H.-W. Tien, Y.-F. Lee, S.-M. Li, Y.-S. Wang, J.-Y. Wang, C.-C. M. Ma and C.-C. Hu, *J. Mater. Chem.*, 2011, **21**, 2374–2380.
330. K.-S. Kim and S.-J. Park, *Electrochim. Acta*, 2011, **56**, 1629–1635.
331. L. Buglione and M. Pumera, *Electrochem. Commun.*, 2012, **17**, 45–47.
332. W. Kim, J. B. Joo, N. Kim, S. Oh, P. Kim and J. Yi, *Carbon*, 2009, **47**, 1407–1411.
333. M. Seredych, D. Hulicova-Jurcakova, G. Q. Lu and T. J. Bandosz, *Carbon*, 2008, **46**, 1475–1488.
334. L. Zhao, L.-Z. Fan, M.-Q. Zhou, H. Guan, S. Qiao, M. Antonietti and M.-M. Titirici, *Adv. Mater.*, 2010, **22**, 5202–5206.
335. L. L. Zhang, X. Zhao, H. Ji, M. D. Stoller, L. Lai, S. Murali, S. McDonnell, B. Cleveger, R. M. Wallace and R. S. Ruoff, *Energy Environ. Sci.*, 2012, **5**, 9618–9625.
336. J. Han, L. L. Zhang, S. Lee, J. Oh, K.-S. Lee, J. R. Potts, J. Ji, X. Zhao, R. S. Ruoff and S. Park, *ACS Nano*, 2012, **7**, 19–26.
337. M. Seredych and T. J. Bandosz, *J. Mater. Chem. A*, 2013, **1**, 11717–11727.
338. Y. Idota, T. Kubota, A. Matsufuji, Y. Maekawa and T. Miyasaka, *Science*, 1997, **276**, 1395–1397.
339. M. Noh, Y. Kwon, H. Lee, J. Cho, Y. Kim and M. G. Kim, *Chem. Mater.*, 2005, **17**, 1926–1929.
340. N. A. Kaskhedikar and J. Maier, *Adv. Mat.*, 2009, **21**, 2664–2680.
341. *Lithium Ion Batteries - Fundamentals and Performance,* ed. M. Wakihara and O. Yamamoto, Wiley-VCH Verlag GmbH, Tokyo, Weinheim, 1998.
342. K. Sawai, Y. Iwakoshi and T. Ohzuku, *Solid State Ionics*, 1994, **69**, 273–283.
343. F. Béguin and E. Frackowiakz, in *Adsorption by Carbons*, ed. E. J. Bottani and J. M. D. Tascón, Elsevier, Amsterdam, 2008, pp. 593–629.
344. K. Sato, M. Noguchi, A. Demachi, N. Oki and M. Endo, *Science*, 1994, **264**, 556–558.
345. M. Winter and J. O. Besenhard, in *Lithium Ion Batteries - Fundamentals and Performance*, ed. M. Wakihara and O. Yamamoto, Wiley-VCH Verlag GmbH, 2007, pp. 127–155.
346. B. Gao, C. Bower, J. D. Lorentzen, L. Fleming, A. Kleinhammes, X. P. Tang, L. E. McNeil, Y. Wu and O. Zhou, *Chem. Phys. Lett.*, 2000, **327**, 69–75.

347. S. Y. Chew, S. H. Ng, J. Wang, P. Novák, F. Krumeich, S. L. Chou, J. Chen and H. K. Liu, *Carbon*, 2009, **47**, 2976–2983.
348. W. X. Chen, J. Y. Lee and Z. Liu, *Carbon*, 2003, **41**, 959–966.
349. Y. NuLi, J. Yang and M. Jiang, *Mater. Lett.*, 2008, **62**, 2092–2095.
350. Y. Li, X. Lv, J. Lu and J. Li, *J. Phys. Chem. C*, 2010, **114**, 21770–21774.
351. H. Xia, M. Lai and L. Lu, *J. Mater. Chem.*, 2010, **20**, 6896–6902.
352. Y. Zhao, J. Li, Y. Ding and L. Guan, *Chem. Commun.*, 2011, **47**, 7416–7418.
353. P. Lian, X. Zhu, S. Liang, Z. Li, W. Yang and H. Wang, *Electrochim. Acta*, 2010, **55**, 3909–3914.
354. L. J. Fu, H. Liu, C. Li, Y. P. Wu, E. Rahm, R. Holze and H. Q. Wu, *Solid State Sci.*, 2006, **8**, 113–128.
355. T. Bhardwaj, A. Antic, B. Pavan, V. Barone and B. D. Fahlman, *J. Am. Chem. Soc.*, 2010, **132**, 12556–12558.
356. D. Wang, D. Choi, J. Li, Z. Yang, Z. Nie, R. Kou, D. Hu, C. Wang, L. V. Saraf, J. Zhang, I. A. Aksay and J. Liu, *ACS Nano*, 2009, **3**, 907–914.
357. X. Zhu, Y. Zhu, S. Murali, M. D. Stoller and R. S. Ruoff, *ACS Nano*, 2011, **5**, 3333–3338.
358. G. Zhou, D.-W. Wang, F. Li, L. Zhang, N. Li, Z.-S. Wu, L. Wen, G. Q. Lu and H.-M. Cheng, *Chem. Mater.*, 2010, **22**, 5306–5313.
359. G. Wang, J. Bai, Y. Wang, Z. Ren and J. Bai, *Scr. Mater.*, 2011, **65**, 339–342.
360. Y. J. Mai, X. L. Wang, J. Y. Xiang, Y. Q. Qiao, D. Zhang, C. D. Gu and J. P. Tu, *Electrochim. Acta*, 2011, **56**, 2306–2311.
361. S. Yin, Y. Zhang, J. Kong, C. Zou, C. M. Li, X. Lu, J. Ma, F. Y. C. Boey and X. Chen, *ACS Nano*, 2011, **5**, 3831–3838.
362. X. Ji and L. F. Nazar, *J. Mater. Chem.*, 2010, **20**, 9821–9826.
363. A. Manthiram, Y. Fu, S.-H. Chung, C. Zu and Y.-S. Su, *Chem. Rev.*, 2014, **114**, 11751–11787.
364. B. Zhang, X. Qin, G. R. Li and X. P. Gao, *Energy Environ. Sci.*, 2010, **3**, 1531–1537.
365. H. Ye, Y.-X. Yin, S. Xin and Y.-G. Guo, *J. Mater. Chem. A*, 2013, **1**, 6602–6608.
366. X. Li, Y. Cao, W. Qi, L. V. Saraf, J. Xiao, Z. Nie, J. Mietek, J.-G. Zhang, B. Schwenzer and J. Liu, *J. Mater. Chem.*, 2011, **21**, 16603–16610.
367. X. Tao, X. Chen, Y. Xia, H. Huang, Y. Gan, R. Wu, F. Chen and W. Zhang, *J. Mater. Chem., A*, 2013, **1**, 3295–3301.
368. X. Ji, K. T. Lee and L. F. Nazar, *Nat. Mater.*, 2009, **8**, 500–506.
369. C. Liang, N. J. Dudney and J. Y. Howe, *Chem. Mater.*, 2009, **21**, 4724–4730.
370. J. Schuster, G. He, B. Mandlmeier, T. Yim, K. T. Lee, T. Bein and L. F. Nazar, *Angew. Chem., Int. Ed.*, 2012, **51**, 3591–3595.
371. S.-R. Chen, Y.-P. Zhai, G.-L. Xu, Y.-X. Jiang, D.-Y. Zhao, J.-T. Li, L. Huang and S.-G. Sun, *Electrochim. Acta*, 2011, **56**, 9549–9555.
372. Y.-S. Su and A. Manthiram, *Electrochim. Acta*, 2012, **77**, 272–278.
373. B. Zhang, C. Lai, Z. Zhou and X. P. Gao, *Electrochim. Acta*, 2009, **54**, 3708–3713.
374. N. Jayaprakash, J. Shen, S. S. Moganty, A. Corona and L. A. Archer, *Angew. Chem., Int. Ed.*, 2011, **50**, 5904–5908.

375. C. Zhang, H. B. Wu, C. Yuan, Z. Guo and X. W. Lou, *Angew. Chem., Int. Ed.*, 2012, **51**, 9592–9595.
376. W. Wei, J. Wang, L. Zhou, J. Yang, B. Schumann and Y. NuLi, *Electrochem. Commun.*, 2011, **13**, 399–402.
377. L. Yuan, H. Yuan, X. Qiu, L. Chen and W. Zhu, *J. Power Sources*, 2009, **189**, 1141–1146.
378. J. Guo, Y. Xu and C. Wang, *Nano Lett.*, 2011, **11**, 4288–4294.
379. C. Jia-jia, J. Xin, S. Qiu-jie, W. Chong, Z. Qian, Z. Ming-sen and D. Quan-feng, *Electrochim. Acta*, 2010, **55**, 8062–8066.
380. S.-C. Han, M.-S. Song, H. Lee, H.-S. Kim, H.-J. Ahn and J.-Y. Lee, *J. Electrochem. Soc.*, 2003, **150**, A889–A893.
381. G. Zheng, Y. Yang, J. J. Cha, S. S. Hong and Y. Cui, *Nano Lett.*, 2011, **11**, 4462–4467.
382. L. Ji, M. Rao, S. Aloni, L. Wang, E. J. Cairns and Y. Zhang, *Energy Environ. Sci.*, 2011, **4**, 5053–5059.
383. M. Rao, X. Song and E. J. Cairns, *J. Power Sources*, 2012, **205**, 474–478.
384. C. Zu and A. Manthiram, *Adv. Energy Mater.*, 2013, **3**, 1008–1012.
385. S. Evers and L. F. Nazar, *Chem. Commun.*, 2012, **48**, 1233–1235.
386. N. Li, M. Zheng, H. Lu, Z. Hu, C. Shen, X. Chang, G. Ji, J. Cao and Y. Shi, *Chem. Commun.*, 2012, **48**, 4106–4108.
387. L. Ji, M. Rao, H. Zheng, L. Zhang, Y. Li, W. Duan, J. Guo, E. J. Cairns and Y. Zhang, *J. Am. Chem. Soc.*, 2011, **133**, 18522–18525.
388. R. Cao, J.-S. Lee, M. Liu and J. Cho, *Adv. Energy Mater.*, 2012, **2**, 816–829.
389. J.-S. Lee, S. Tai Kim, R. Cao, N.-S. Choi, M. Liu, K. T. Lee and J. Cho, *Adv. Energy Mater.*, 2011, **1**, 34–50.
390. G. Girishkumar, B. McCloskey, A. C. Luntz, S. Swanson and W. Wilcke, *J. Phys. Chem. Lett.*, 2010, **1**, 2193–2203.
391. Y. Shao, F. Ding, J. Xiao, J. Zhang, W. Xu, S. Park, J.-G. Zhang, Y. Wang and J. Liu, *Adv. Funct. Mater.*, 2013, **23**, 987–1004.
392. Y. Shao, S. Park, J. Xiao, J.-G. Zhang, Y. Wang and J. Liu, *ACS Catal.*, 2012, **2**, 844–857.
393. J. Wang, Y. Li and X. Sun, *Nano Energy*, 2013, **2**, 443–467.
394. P. G. Bruce, S. A. Freunberger, L. J. Hardwick and J.-M. Tarascon, *Nat. Mater.*, 2012, **11**, 19–29.
395. S. S. Zhang, D. Foster and J. Read, *J. Power Sources*, 2010, **195**, 1235–1240.
396. J. Xiao, D. Wang, W. Xu, D. Wang, R. E. Williford, J. Liu and J.-G. Zhang, *J. Electrochem. Soc.*, 2010, **157**, A487–A492.
397. X.-h. Yang, P. He and Y.-y. Xia, *Electrochem. Commun.*, 2009, **11**, 1127–1130.
398. T. Kuboki, T. Okuyama, T. Ohsaki and N. Takami, *J. Power Sources*, 2005, **146**, 766–769.
399. G. Wang, G. Sun, Q. Wang, S. Wang, J. Guo, Y. Gao and Q. Xin, *J. Power Sources*, 2008, **180**, 176–180.
400. E. Yoo and H. Zhou, *ACS Nano*, 2011, **5**, 3020–3026.
401. E. Yoo and H. Zhou, *J. Power Sources*, 2013, **244**, 429–434.

CHAPTER 10

Engineering and Safety Issues

10.1 Introduction

The possibility of using nanostructured carbon materials in catalysis at an industrial scale is of course correlated to: (i) their commercial availability (Figure 10.1), (ii) the environmental exposure assessment, and (iii) the possibility of creating catalyst bodies that can be used in industrial reactors. Thanks to the nature of the raw material (hydrocarbons, biomass, *etc.*) necessary to produce them, as well as the simple processes (pyrolysis, CVD, *etc.*) used for their production, most of the nanostructured carbon materials are potentially cheap if produced on a large scale. Among them, carbon black (containing more than 97% elemental carbon) is in the top 50 industrial chemicals manufactured worldwide, based on annual tonnage. The production capacity of CNTs has increased significantly in the last five years, and nowadays hundreds of tons are produced to meet the market demand. The CNT market is on the upswing with lowering costs and improving performance, availability and end user adaptability.

The use of manufactured carbon nanomaterials in a number of commercial applications, including catalysis, raises questions regarding potential unintended risks to humans and the environment. Indeed, even though there is much knowledge available regarding possible health and environmental effects of traditional chemicals, it is not clear if this knowledge could be transferred directly to nanomaterials. There is a need to collect more information about exposure to carbon nanomaterials in both manufacturing and user scenarios. As the market grows, and as manufacturers switch from the micro- to the nanoscale, the potential for exposure will increase. More research is required to quantify any risks to workers and consumers. Given our current knowledge in this field, it is important to take precautions to minimize exposures and protect safety and health.

RSC Catalysis Series No. 23
Nanostructured Carbon Materials for Catalysis
By Philippe Serp and Bruno Machado

Published by the Royal Society of Chemistry, www.rsc.org

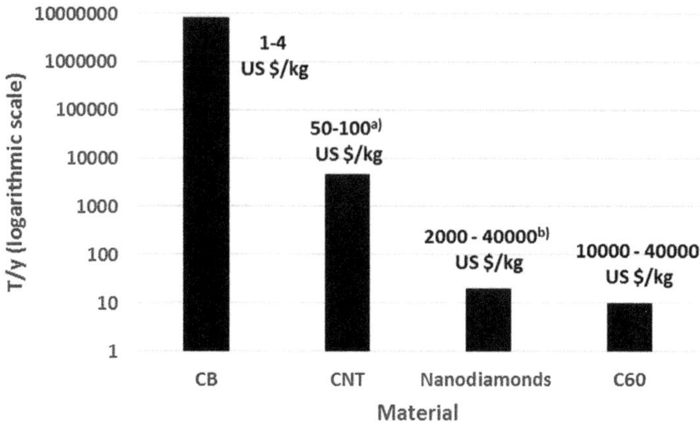

Figure 10.1 Annual worldwide production of nanostructured carbon materials, and their prices. Note: a) For MWCNTs; b) for detonation ND.

Additionally, we have seen in the previous chapters that nanostructured carbon materials have been used at the laboratory scale either as support or directly as catalyst for thermally-, photo- or electro-assisted reactions. Although significant efforts have been dedicated to the understanding of the surface chemistry/functionalization of these materials, there is a very limited amount of work dealing with the formulation and shaping of nanocarbon catalyst bodies for use in a technical reactor. Indeed, conventionally regarded as a secondary unit operation in catalysis science, shaping is often disregarded. To fill the gap between catalysis in academia and industry, shaping is actually a key step in catalyst preparation.

10.2 Macroscopic Shaping of Nanostructured Carbon Materials for Catalytic Applications

Pure CNTs, CNFs or FLG can be obtained after post-synthesis purification steps, such as removal of the catalyst and of undesired carbon forms. At this stage, one can speak of a "fluffy powder" with low bulk density and low mechanical strength, and so far CNTs, CNFs or FLG have found applications as catalyst supports in the form of powders. SEM micrographs of purified MWCNT or FLG powders are shown in Figure 10.2. As far as CB, ND and fullerene powders are concerned they are constituted of aggregates (>100 nm) of primary particles.

Before using these materials in catalysis, the question of their processability should be addressed. This will depend first of the type of catalysis.

For electrocatalysis, where the electrocatalysts are prepared from carbon paste, ink or gel, all the nanostructured carbon materials can be used without restriction.

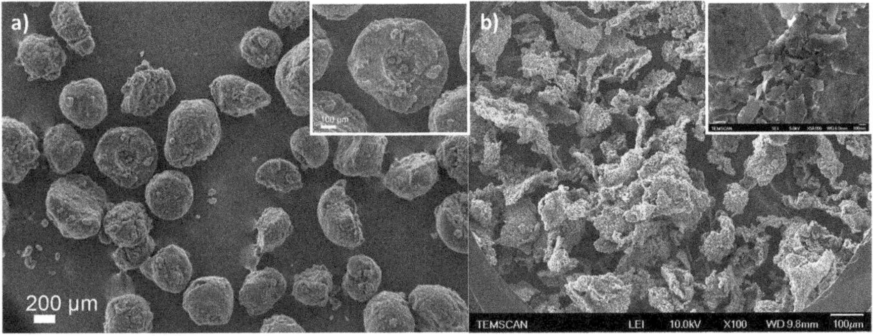

Figure 10.2 SEM micrographs of MWCNT and FLG powders.

For photocatalysis, the small size creates several practical problems such as the separation of nanometric catalyst particles from the suspension once the reaction is complete and possible aggregation of suspended particles. Furthermore, particulate suspensions are not easily applicable in continuous flow systems. Several approaches were taken in order to overcome these problems. One of these approaches was to immobilize the powder on polymeric films either by TiO_2 solution casting and sintering or by mounting them on a polymer nanofiber formed by electrospinning. Thus, composite nanofibers containing nanometric TiO_2 particles and MWCNTs dispersed in poly(acrylonitrile) were prepared by an electrospinning technique.[1]

For thermal catalysis, the type of reactor should be carefully taken into consideration. It is known that whereas fluidized-bed and slurry-type reactors can afford the use of such powdered catalysts, fixed-bed reactors (the most common industrial reaction units) must be provided with formed catalyst bodies to operate. Spheres, pellets, rings, extrudates, and honeycombs are common catalyst shapes for fixed-bed reactors.

Conventional technologies for heterogeneous catalytic reactions involving both liquid- and gas-phases comprise slurry reactors or trickle-bed reactors. In the case where nanostructured carbon materials are applied in the form of a slurry, an efficient downstream process, such as membrane filtration, is needed to retain and recycle the catalyst. Immobilization of the catalyst on reactor surfaces or membrane filters eliminates the need for separation. However, the effectiveness is then limited by the available surface area in the reactor. This is particularly important for water treatment by photocatalysis, where the issue of the access to the light source should be also considered. One obvious research need is better technologies to retain nanostructured carbon materials. Effective and reliable methods are needed to anchor nanostructured carbon materials to reactor surfaces, or to separate and retain suspended nanoparticles in order to reduce costs associated with premature material loss and to prevent human health and environmental impacts. This includes developing better surface coating techniques perhaps through nanocarbon surface functionalization, minimizing membrane fouling by the nanocarbons suspension and impregnating nanocarbons into filter packing materials, such as granular AC.

Reactions in trickle-bed reactors easily end up in diffusion limitations because of the relatively larger catalyst particle sizes used (1–10 mm), implying relatively long diffusion distances inside the catalyst particles. Particle size can be reduced only to a limited extent because this results in increased pressure drop through the reactor. Catalyst particles for slurry reactors are much smaller (30 μm, typically) and thus more suitable for fast reactions; however, the additional cost for catalyst separation is an issue. Nevertheless, even slurry catalysts suffer from mass transfer limitations in the case of very active catalysts.

Considerable research work has been done to apply CNF or CNT aggregates as catalyst supports in gas phase and in a variety of liquid phase reactions (see Chapter 7). Thin layers of CNFs/CNTs on structured materials is an interesting proposition as it would combine the advantages of slurry phase operation (short diffusion length), fixed-bed operation (no catalyst separation is required, no catalyst attrition, and no catalyst agglomeration), and application of CNF aggregates (high porosity and low tortuosity). In addition, it allows for taking advantage of the use of structured reactor packing, optimizing hydrodynamics and especially gas-liquid mass transfer.

It has been shown that CNTs and CNFs can form layers on structured materials. Structured catalysts are promising, as far as the elimination of several drawbacks is concerned: misdistributions of various kinds (including a non-uniform access of reactants to the catalytic surface), high pressure drop in the bed and broad residence time distributions. Two basic kind of structured catalysts can be distinguished:

- Structural packings covered with catalytically active material, similar in design to those used in distillation and absorption columns and static mixers.
- Monolithic catalysts as continuous unitary structures, which contain many channels. A ceramic or metallic support is coated with a layer of material in which active species are dispersed. The catalytically active material is present on, or inside the walls of these channels.

CNTs and CNFs layers were formed on metallic (mostly Ni),[2–11] carbon,[12,13] or SiC[14–16] foams, reticulated vitreous carbon foam (hairy foam),[17,18] fibrous/sintered metal filters,[19–24] carbon,[25] oxide (mostly cordierite)[26–38] or metallic[39] monoliths, or carbon/graphite felts[40–52] (Figure 10.3 and Table 10.1). This helps to keep diffusion distances short.[53] The structured materials of choice obviously will also determine the hydrodynamic behaviour of the reactor. The synthesis of porous CNTs foam composites with a high accessible surface area and tunable porosity has also been reported.[54]

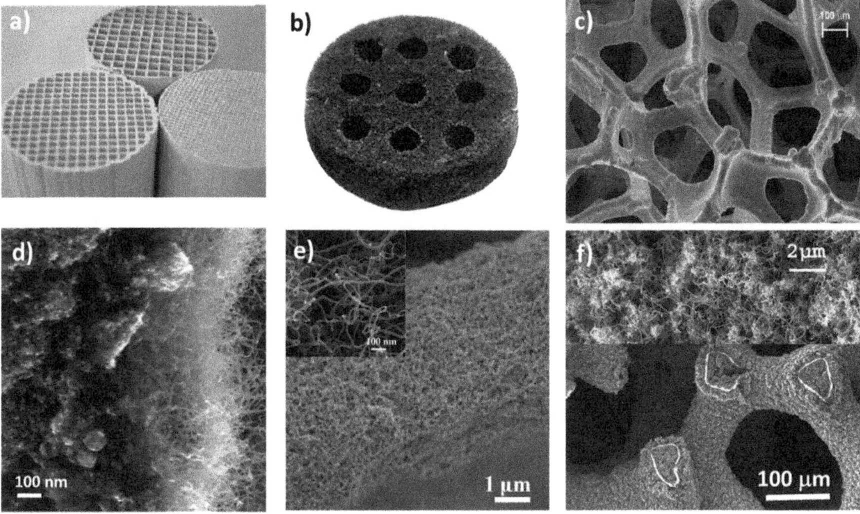

Figure 10.3 Typical structured materials: (a) ceramic monolith, (b) carbon felt, and (c) metal (Ni foam); and SEM micrographs of (d) thin layer of CNFs on the surface of Al_2O_3 wash-coated monolith,[55] (e) CNF composite supported on macroscopic graphite felt, and (f) CNFs-Ni foam.

10.2.1 Carbon Nanostructured Materials on Monoliths

Structured catalyst supports or monoliths have clear advantages over traditional powder catalysts when used in large-scale reactors, including lower pressure drop and better control of heat and gas flow, giving uniform flow distributions and residence times.[56,57] Monoliths are typically made of ceramics, metals or carbon and contain small diameter (0.5–4 mm), uniform channels (Figure 10.3a). Ceramic monoliths like cordierite are commonly used in catalytic converters as the support material. As well as catalytic converters, monolithic reactors have been used in a range of reactions, including NO_x conversion to ammonia for large-scale applications, the Fischer–Tropsch process, synthesis of hydrogen from ammonia, hydrogenation of unsaturated bonds in a number of species, steam methane reforming and biodiesel synthesis.[56] A major limitation of monoliths as catalyst supports is their very low surface area, with a typical bare cordierite monolith having a surface area of about 0.7 $m^2\ g^{-1}$. In order to overcome this problem, monoliths are coated with a thin layer of a high surface area material to allow higher loadings of catalyst onto the support. Commonly, an alumina or silica slurry coating is applied onto the monolith, with the increase in surface area dependent on the thickness of the applied wash-coat, resulting in values of up to 15–30 $m^2\ g^{-1}$ for 0.1–50 μm thick coatings. However, these coatings increase the pressure drop across the monolith, partially reducing their advantage over packed-bed configurations. In addition, achieving a uniform wash-coat on a large scale can be problematic. To further increase the surface area of wash-coated monoliths, growth of CNTs or CNFs on the surface of the wash-coat layer has

Table 10.1 Selected examples of macroscopic shaping of CNFs for catalytic applications.

Type of macro-support	Type of pretreatment	Conditions to grow CNFs	Surface area (SA)/m^2 g^{-1}		
			Composite before CNFs growth	Composite after CNFs growth	CNFs
Ni on γ-Al_2O_3 coated monolith[55]	Reduction in 20% H_2/N_2 at 973 K for 2 h	$T = 843$ K C-source = $CH_4/H_2/N_2$	45	63	190
Ni on SiO_2 coated monoliths[33]	Reduction in H_2 at 773 K for 1 h	C-source = CH_4 in N_2	—	26	—
Ni foam[2]	Oxidation in stagnant air at 973 K for 1 h, reduction in H_2/N_2 at 973 K for 2 h	$T = 723$ K C-source = C_2H_4/N_2	<1	30	90
Ni on SiC foam[14]	Calcination in air at 623 K for 2 h, reduction in H_2 at 673 K for 2 h	$T = 953$ K C-source = C_2H_6/H_2	20	47–100	—
Ni on solid carbon foam[18]	Reduction in H_2 at 773 K for 2 h	$T = 773$ K C-source = $C_2H_6/H_2/N_2$	0.12	146	—
Sintered metal fibers (Inconel), 60.5% Ni	Oxidation in air at 923 K for 3 h, reduction in H_2 at 873 K for 2 h	$T = 928$ K C-source = $C_2H_6/H_2/Ar$	—	22	475
Ni on silica glass fibers	Reduction in H_2/N_2 at 823 K for 3 h	I = 823 K C-source = $CH_4/N;$	<1	19	
Ni on graphite felt	Reduction in H_2 at 673 K for 1 h	$T = 953$ K C-source = $H_2/C_2H;$	<1	90	

been proposed (Table 10.1). Some of these carbon layers have been shown to double the surface area of the wash-coated monoliths. Coating a monolith with CNTs offers a safer and more practical solution to their use in industrial reactors, providing a CNT support with a lower pressure drop directly upon synthesis without the necessity of subsequent treatment, whilst the strong adhesion between the monolith surface and the CNTs makes it unlikely that the CNTs will become airborne.

A uniform layer of CNFs was grown on a cordierite monolith by first coating the monolith with a thin layer of γ-alumina.[28] The nanofibers form a thick, uniform layer on the monolith walls, leading to the formation of a mesoporous and mechanically robust composite. The absence of microporosity in the composite and the ability to tune the thickness of the nanofiber layer suggests that these nanofibers/monolith composites may be useful as catalyst supports for liquid-phase catalytic reactions. CNFs were grown on different macro-structured supports, such as cordierite monoliths, carbon felts and sintered metal fibres.[58] The resulting composites exhibited excellent resistance to attrition/corrosion and their porosity is mainly due to mesoporous structures. The CNF/structured materials were tested in the ozonation of oxalic acid in a conventional semi-batch reactor after being crushed to powder form, and in a newly designed reactor that may operate in semi-batch or continuous operation. The CNFs supported on the different structured materials exhibited high catalytic activity in the mineralization of oxalic acid. CNFs grown on a honeycomb cordierite monolith were also tested in continuous experiments for catalytic ozonation of the herbicide metolachlor.[59] The application of the CNF coated monolith to the continuous ozonation process was shown to have potential, as it improved the TOC removal from 5% to 35%.

CNF/TiO_2/cordierite composite monoliths were prepared and used as a catalyst support in selective hydrogenation of cinnamaldehyde to hydrocinnamaldehyde.[60] The composite was synthesized through TiO_2 coating on the surface of the monolith and the subsequent CNF growth on it. The total BET surface area of the composite was 31 $m^2\ g^{-1}$, and the macro- and mesopore structure dominated the pore space of the material (about 93%). Meanwhile, 94% of the carbon deposit on the surface of the composite was CNFs. The as-prepared CNF/TiO_2/monolith was subsequently employed to prepare its supported palladium catalysts, Pd/CNF/TiO_2/monolith. Attachment strength and acid-resistant properties of the composite have been studied to evaluate its structural stability under severe conditions. TiO_2 and CNF coating was deemed to increase its textural and acid-resistant properties. Although there is a relatively slow reaction rate over as-prepared Pd/CNF/TiO_2/monolith due to its small BET surface area and low Pd dispersion (about 15%), the selectivity to hydro-cinnamaldehyde remained high (about 90%) at 95% cinnamaldehyde conversion.

N-doped CNFs have been grown on cordierite monoliths.[61] Ru has been impregnated on the N-doped CNFs, on undoped CNFs, H_2O_2-treated CNFs, and alumina coated monoliths, and they have been tested in ammonia decomposition. The desired objective for a practical implementation is to uniformly disperse a Ru catalyst on the entire N–CNF/monolith. To this end, incipient wetness impregnation is not a feasible method for a monolith. A suitable catalyst impregnation method is equilibrium adsorption or electrostatic adsorption. It has been found that nitrogen doping contributes to stabilize Ru to a small particle size and in a reduced state. The strong interaction of the precursor with the nitrogen groups enables the preparation of small Ru nanoparticles, uniformly distributed throughout all the CNFs coating the monolith. The testing in NH_3 decomposition of similar Ru loadings supported on CNF monoliths with different functionalization showed an outstanding activity for the catalyst supported on N-doped CNFs with the higher percentage of N doping.

CNT-coated monoliths were prepared by catalytic CVD over deposited cobalt on cordierite.[62] Compared with bare cordierite, the BET surface area and pore volume of CNTs-cordierite increased significantly. CNTs penetrated into the cordierite substrate and led to a remarkable mechanical stability of the CNTs-cordierite monoliths following ultrasonic treatment. The CNT content, BET surface area, pore volume and thermal properties of CNT-cordierite monoliths all could be changed by the variation of the synthesis conditions. Barium promoted ruthenium catalysts supported on the as-synthesized materials showed much higher activity for ammonia synthesis than their counterparts deposited on bare cordierite monoliths. Furthermore, the catalytic activity linearly increased with the BET surface area of CNT-cordierite monoliths. Nitric acid treatment of CNT-cordierite monolith changes the amount of Mg, Si, Al and oxygen-containing functional groups, thereby influencing the surface area and pore size distribution of composite materials.[63] Appropriate treatment of CNT-cordierite with nitric acid increases the surface area and the amount of micropores slightly, but improves the activities for ammonia synthesis noticeably, which might be a consequent of the variation of the amount of Mg, Si, Al and oxygen-containing functional groups.

CNTs and CNFs were also synthesized *via* catalytic CVD of ethane over the surface of stainless steel mesh monoliths.[39] The preparation of such a catalytic coating with good properties, such as good adhesion to substrate, open porosity (mesoporosity) to enhance the diffusion of reactants, uniform thickness of the carbon layer, good mechanical strength and control over the microstructure of CNTs, ensures good activity and durability of the CNT-based structured catalytic reactor.

MWCNT monoliths without any binders were obtained by spark plasma sintering treatment at 2273 K under 800 bar sintering pressure.[64] SEM observation confirmed that these materials maintained the nanosized tube microstructure of raw CNT powder after the plasma treatment.

10.2.2 Carbon Nanostructured Materials on Solid Foams

For heterogeneously catalyzed liquid-phase reactions, liquid–solid mass transfer and/or intra-particle diffusion can be rate limiting. This limits the overall conversion and selectivity. The effects of mass transfer and diffusion can be reduced by using a packing as catalyst support. The packing increases the liquid–solid interfacial area and increases local turbulence in the liquid-phase, thus enhancing mass transfer rates. Structured packings can increase conversion and selectivity when compared to the more conventional packed-bed (*e.g.* porous spherical particles). Many structured packings are discussed in literature, *i.e.* monoliths, Sulzer Katapak elements, cloths, and fibers. The use of these structured packings also overcomes the cumbersome step of catalyst separation often encountered in slurry reactors. Solid foams, a highly porous open celled material, have received ever increasing scientific and industrial interest as catalyst support over the last decade owing to their exceptional properties, namely highly accessible porosity and effective surface area, low pressure drop and good mixing behavior. The structure of the solid foam resembles the inverse of a packed-bed of spherical particles. Solid foams are available in a wide variety of materials (*e.g.* ceramics, metals, plastics, and carbon). The use of the foam structured catalysts instead of slurry catalysts allows one to avoid the costly and time consumption filtration process to separate the catalyst from the final product. Structured catalysts also prevent the problem of fine formation as encountered with slurry reactors due to the attrition of small particles under vigorous stirring, which leads to the plugging of the filtration device. It is expected that the design of a new generation of catalysts containing smaller amounts of the catalytic active phase with higher selectivity, high efficient recovery from reaction medium and high durability will significantly contribute to the development of more environmental friendly catalytic processes with a lower cost-effectiveness. Compared to honeycomb monoliths, foam displays an extremely high degree of radial mixing of the flow passing through its porous matrix, being it gaseous or liquid, which considerably improves the reactant mixing and also the heat transfer for the reaction. In addition, the open structure of the foam also significantly reduces the pressure drop along the catalyst bed. Solid foam employed as a catalytic stirrer has been developed for several gas–liquid phase reactions. The catalytic stirrer developed was based on aluminium metal foam coated with an alumina wash-coat layer. The reactant and product diffusion rates are closely dependent on the thickness of the alumina wash-coat layer. Enhanced liquid–solid mass transfer was reported for CNFs on solid foam as catalyst support.[17] Pd-catalysts on CNFs-covered reticulated vitreous carbon foam (CNF/RVC) were used in the fast oxidation of sodium formate. The reaction was carried out in a rectangular acrylic glass flow reactor (Figure 10.4). The high liquid–solid mass transfer rates of CNFs on solid foam with respect to regular foam is attributed to the hydrodynamic accessibility of the CNFs. The pressure drop of the CNF/RVC foam is significantly lower than for random packings. This shows that higher, or at least equivalent, mass transfer rates can be obtained, but at a considerably lower pressure drop than for random packings.

CNF-foam 'hairy foam' and CNF aggregates-supported Pd catalysts were studied for the reduction of aqueous nitrite solution and compared with

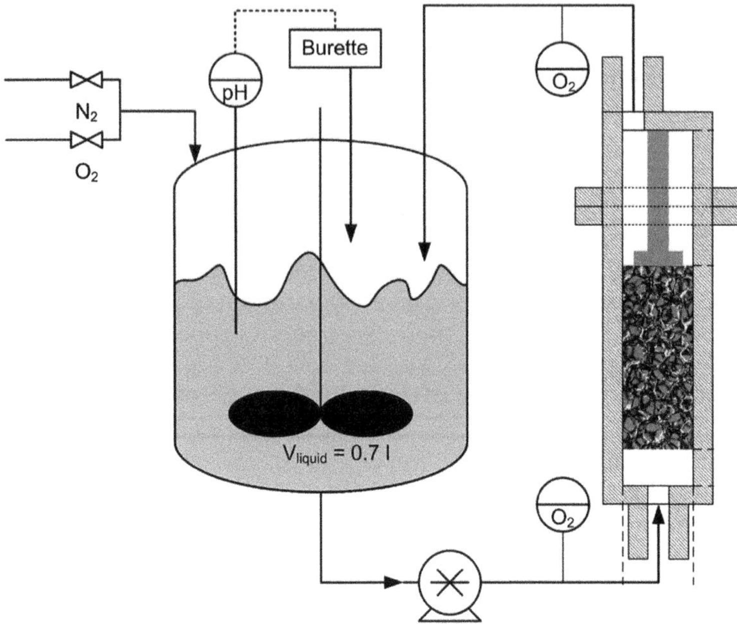

Figure 10.4 Schematic representation of the experimental setup used for the oxidation of sodium formate to carbon dioxide.[17] The liquid in the stirred vessel is saturated with a mixture of nitrogen and oxygen and pumped through the flow reactor with the catalytically active CNF/RVC foam. The oxygen concentration is measured at the inlet and outlet of the flow reactor. The flow exiting the reactor is recycled to the stirred vessel.

conventional catalysts.[65] A series of Pd catalysts has been prepared with limited variation in Pd dispersion, containing relatively large (*ca.* 10 nm) Pd particles, in order to study the effect of the morphology of the support material on the rate of internal mass transfer. It was demonstrated that the limited variation in Pd dispersions in this study resulted in identical TOFs in nitrite hydrogenation for all Ni foam supported catalysts. It was also demonstrated that CNF-based catalysts are highly active for nitrite hydrogenation, resulting in TOFs at least three times larger when compared to conventional catalysts (Figure 10.5). This was attributed to improved mass transfer thanks to the high porosity as well as low tortuosity of CNF aggregates, as well as thin CNF-layers in hairy foam. Surprisingly, the intrinsic activity of graphite and CNF-supported Pd also contributes significantly to the high activity for nitrite reduction. Hairy foam supports provide short diffusion pathways, combined with high porosity and low tortuosity, circumventing the disadvantage of small catalyst particles for liquid-phase catalytic reactions, *i.e.* high pressure drop or the necessity of filtration.

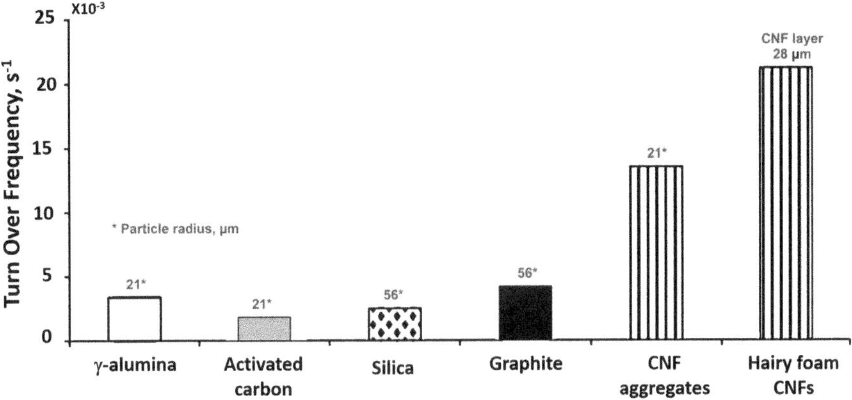

Figure 10.5 Comparison of turn over frequency values for nitrite reduction of conventional catalysts and CNF based catalysts at comparable particle radius and CNF-layer thickness.[65]

10.2.3 Carbon Nanostructured Materials on Fibrous Materials

Fibrous catalysts based on precious metals (Pt, Ru, Ag) have been extensively used in the production of nitric acid, hydrocyanic acid and aldehydes. Since they present the inconvenience of being very expensive, over the last decade cheaper metals (Ni), as well as glass and carbon fibers, were explored as potential material for fibrous catalysts. The small fiber diameter, on the order of 1–80 μm, makes possible the use of support sizes in the same order of magnitude as powders, but without the disadvantages of handling and pressure drop associated with powders. A comparison of some characteristics of fibers with monolith and pellet structures is given in Table 10.2.

The fibrous materials are flexible and can be packed in fixed-beds of different geometries and in different shapes, which would be otherwise impossible to achieve with powdered or granulated materials. Furthermore, fibrous materials possess very high porosities, 70–90%, compared to randomly packed bed porosities of 34–60%. The high porosity leads to low pressure drop, so fibrous materials are seen as attractive alternatives in three-phase systems. However, high porosities are often associated with lower specific surface area and higher risk of channeling. The main advantage of fibrous materials is provided by their immobility and the short diffusion time of reactants through the bed, similar to the one provided by monolithic structures. Other attractive features include the possibility of safer operation and easy scale-up. The most studied and applied classes of materials are: carbon fibers, glass fibers and metallic wires. Activated carbon fibers (ACFs) exhibit an apparent specific surface area (SSA), in the range of 1500–3000 $m^2\ g^{-1}$, but with a complex texture including macro-, meso- and micropores. Microporosity is

Table 10.2 Characteristics comparison for pellet-, monolith- and fibrous-catalysts.

	Pellet	Monolith	Fibrous structure (wire-mesh, cloths)
Particle/channel diameter/mm	3.0–5.0	1.5	0.01–2
Diffusion length/μm	100–2500[a]	25–100[b]	0.15[c]
Specific surface area/m^2 g^{-1}	Up to 1000	2	2000[d]
Geometric surface area/m^2 m^{-3}	1200	1900	Up to 100 000

[a]Depending on catalyst structure.
[b]Washcoat thickness.
[c]Depending on porosity.
[d]Depending on material.

detrimental for the efficiency of the catalysts, because the reactants often have problems diffusing into the micropores where the catalytic active phase can be deposited, giving rise to mass transfer limitations during the catalytic process. Glasses are often used due to their relative chemical inertness, the absence of harmful metal–support interactions and their ability to facilitate the action of promoters. Catalyst based on metallic fibers are mainly used in highly exothermic and fast gas-phase reactions, where their excellent heat transfer properties can help to avoid hot-spot formation and their associated negative effects on the catalyst activity, selectivity and thermal runaway.

Methane, *n*-hexane, benzene, and cyclopentadiene were decomposed at a relatively mild temperature over a Ni catalyst supported on either vapor grown carbon fibers or graphitized carbon fibers.[66] Transmission electron microscopy showed that the morphology of the fibers changed according to the hydrocarbon. Decomposition of methane and *n*-hexane produced fishbone-type fibers. The fibers from *n*-hexane sometimes showed intermittent hollow structures but the diameters of the fibers were widely distributed. Decomposition of benzene and cyclopentadiene mainly produced winding type CNTs of relatively uniform diameters (10–20 nm). Activated carbon fiber fabrics, an excellent adsorbent, were used as catalyst supports to grow CNFs.[67] CNFs with a diameter between 20 and 50 nm were synthesized uniformly and densely on this support, and impregnated by nickel nitrate catalyst precursor, using catalytic chemical vapor deposition. Because of the microporous structure of the activated carbon fibers, the catalysts could be distributed uniformly on the carbon surface. The CNFs grown using a high concentration catalyst solution show a larger diameter and a wider diameter distribution than those using a lower concentration catalyst solution (Figure 10.6). Although the CNFs were not straight but had a crooked morphology, they form a three-dimensional network structure.

CNFs were grown on the surface of micro-scale carbon fibers and glass fibers at low temperature using palladium as a catalyst to create multi-scale fiber structures. MWCNTs were successfully grown on a glass fiber air filter using thermal chemical vapor deposition after the filter was catalytically

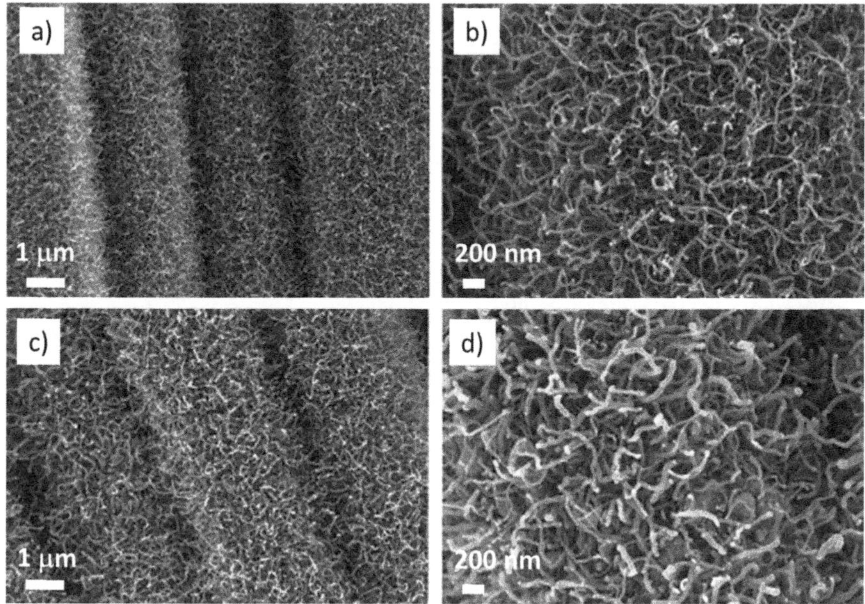

Figure 10.6 SEM images of CNFs grown on activated carbon fiber fabrics using (a) and (b) a low concentration of catalyst impregnation solution; and (c) and (d) a high concentration of catalyst impregnation solution.[67]

activated with a spark discharge.[68] However, up to now nanostructured carbon grown on carbon or glass fibers have not been used for catalytic applications.

Metal supports, such as sintered metal fibers (SMF), present advantages due to high electro- and thermal-conductivity, regular open structures and their filtration properties. These filters consist of a homogenous 3-D structure of metallic micro-filaments. They are characterized by porosity up to 80–90%, high permeability and excellent filtrating properties with a low pressure drop through the catalytic bed during reactor operation. Fibers made of special alloys (stainless steel, Inconel, Fe–Cr alloy) exhibit high mechanical strength, and chemical and thermal stability. The high thermal conductivity of the metal fiber matrix provides a radial heat transfer coefficient two-fold higher compared to randomly packed catalytic beds. This results in nearly isothermal conditions when used as catalytic materials for highly exo/endothermic reactions. This fibrous matrix also acts as a static micro-mixer avoiding channeling. The major drawback associated to metallic fibrous supports is their relatively small specific surface area, which can limit the dispersion of the active catalytic phase. To achieve better performances a non-porous support can be modified by coating with a layer at high surface area.

The use of a CNF/SMF filter as a catalyst support allows the variation of different parameters in order to fit the final material requirement. First of all, its macrostructure can be regulated by the choice of the starting SMF filter. The size of the metallic fibers, the porosity and the thickness of the filter can all be

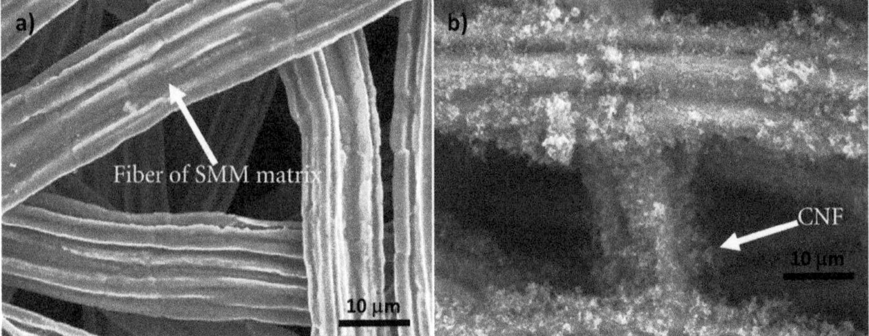

Figure 10.7 3-dimensional deposition of CNFs within 8 μm sintered metal microfibrous (SMM) matrix (Ni): (a) top-view before synthesis, (b) top-view after CNF synthesis.[3]

varied over a broad range. Secondly, the growth of a CNF layer on the metallic fibers can be regulated to obtain the desired carbon layer thickness or the adequate specific surface area. Finally, the deposition of an active component on the CNFs provides the catalytic sites to the composite. This material is designed to control the structural, chemical and physical properties over multiple levels.

Jarrah *et al.* created a novel structured catalyst support modifying the surface of Ni foams by the growth of a layer of CNFs and thus, increasing the SSA.[2] Similarly, Tribolet *et al.* developed a composite material consisting of sintered metal fibers coated with CNFs, directly grown on the metallic surface.[22] The CNFs were firmly anchored to the surface and reached a specific surface area of 310–472 $m^2\ g^{-1}$. Thermal CVD can also be applied for the growth of CNTs on metallic filters.[19] CNFs were also successfully grown at the exterior and the interior of 3-D sintered nickel micro fibrous networks (Figure 10.7).[3]

10.2.4 Carbon Nanostructured Materials on Carbon Felts

Commercial carbon felt is a flexible, cloth-like material available in 3–25 mm thicknesses consisting of an interwoven matrix of extruded amorphous carbon fibers (diameter = 15–20 μm) and is used in diverse applications ranging from high temperature insulation, composite additive, electrode material, adsorbent, and as nanofiber or catalyst supports. A typical example of growing CNFs on a structured support is the immobilization of CNFs on a graphite felt, as proposed by Ledoux *et al.*[40,41,69–71] First, Ni was deposited on the felt and CNFs were synthesized at 953 K using a reaction mixture of ethane and hydrogen. The formed coating was highly porous, consisting of CNFs of about 30 nm in diameter. The CNFs were strongly anchored to the felt as no loss could be detected during ultrasonic treatment in ethanol solution during several minutes. Several authors have followed a similar procedure to produce CNFs or CNTs on graphite felt.[40,42,72] Vieira *et al.* speculated that the anchoring of CNFs was due to penetration of CNFs into the graphite of the graphite felt.[73] Ledoux *et al.* demonstrated the advantage of this material as a catalyst support for Ir in the catalytic decomposition of hydrazine.[70,74] The reaction is extremely fast and exothermic, and is mainly controlled by mass and heat transfer. CNF-based Ir catalysts outperformed the alumina-based catalyst because of the faster mass transport in the former, which allows conversion of all hydrazine immediately leading to higher thrust with a factor of 3. In addition, the high thermal conductivity of the CNF-graphite composite allows faster heating of the catalyst bed, which also contributed to a fast and powerful response to the introduction of hydrazine to the engine. The conductive nature of the CNF composite catalyst allows a rapid and homogeneous dispersion of heat in the catalyst bed, assisting complete decomposition of the reactants. As a result, mass and heat transfer limitations were overcome by using a CNF-based catalyst. Graphite felt supporting 40 nm diameter CNFs was also prepared by the same authors and successfully used as support for a nickel-based catalyst for the low temperature (333 K) selective oxidation of H_2S into elemental sulfur.[75]

Carbon nanofiber/graphite-felt (CNF/GF) composite was also used to support ruthenium (3.0 wt%) for sorbitol hydrogenolysis.[40] The CNF/GF composite was synthesized by *in situ* growing of a CNF layer on graphite felt using Ni as catalyst and ethane as carbon source, as Ru was deposited by incipient wetness impregnation. The monolithic Ru catalyst was then used as stirring blades in an autoclave (Figure 10.8), in which sorbitol hydrogenolysis was carried out. The results showed that, compared with the powder Ru/CNF catalyst, the structured Ru catalyst had a lower activity but a much higher total selectivity, improved from 57.0% to 79.1%, for ethylene glycol, propylene glycol and glycerol.

Hierarchical CNF/carbon felt composites, in which CNFs were directly grown on the surface of microfibers in carbon felt, forming a CNF layer on a micrometre range that completely covers the microfiber surfaces, were tested as a novel support material for cobalt nanoparticles in the highly exothermic Fischer–Tropsch synthesis.[47] A compact, fixed-bed reactor, made of disks of such composite materials, offered the advantages of improved heat and mass transfer, relatively low pressure drop, and safe handling of immobilized CNFs. An efficient 3-D thermal conductive network in the composite provided a relatively uniform temperature profile, whereas the open structure of the CNF layer afforded almost 100% effectiveness for Co nanoparticles in the Fischer–Tropsch synthesis in a fixed-bed.

CNTs were also grown by CVD on different carbon fiber substrates, namely unidirectional carbon fiber tows, bi-directional carbon fiber cloth and three dimensional carbon fiber felt.[76] Nitrogen-doped CNTs have also been grown on graphite felt by a chemical vapor deposition.[77]

However, most of the time, CNFs or CNTs grown directly on commercial carbon felts have poor mechanical stability due to weak binding at the CNF-CNT/carbon felt interface. Even a slight mechanical disturbance leads to extensive CNF mass loss. Knowing that CNFs undergo carbothermal reduction in the presence of silica gel to form silicon carbide, SiC, an opportunity

Figure 10.8 Stirrer with structured Ru catalysts as blades (blade size: 20 mm × 10 mm × 8 mm).[40]

to achieve chemical binding across the CNF/carbon felt interface becomes available. When Fe_2Ni_8 nanocrystals supported on fumed silica particles are dispersed onto commercial carbon felt and then subjected to CNF growth conditions, herringbone CNF/SiO_2/carbon felt composites can be formed, as depicted in Figure 10.9.[45] Due to the physical layering of these phases, this composite also has poor mechanical stability toward loss of CNF material. However, when the as-prepared CNF/SiO_2/carbon felt composite is heated to 1923 K (under Ar atmosphere), carbo-thermal reduction occurs at both carbon/silica interfaces (possibly catalyzed by residual metal growth catalyst) converting the intervening fumed silica phase to SiC. A CNF/SiC/carbon felt hybrid composite is formed having tight binding across both carbon/SiC interfaces. This hybrid composite is now mechanically robust and can survive mechanical rubbing without incurring noticeable CNF mass loss.

10.2.5 Shaping Binder

Activated carbons are normally available in different forms and sizes, ranging from very fine powders to granular and extruded AC. Powdered ACs are used in liquid-phase reactions. Granular AC and extrudates are used in gas-phase reactions. The particle size influences the transport characteristics; the smaller the particles, the better the characteristics, but the finer the particles, the higher the pressure drop or slower the filtration. Granular AC is produced by heating

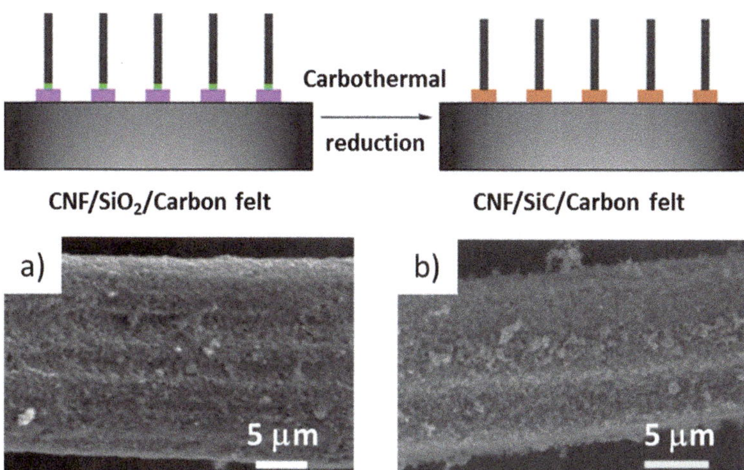

Figure 10.9 Cartoon depictions of CNF/carbon felt nanocomposites. (a) SEM image of the CNF/SiO_2/carbon felt composite; and (b) SEM image of the CNF/SiC/carbon felt composite.

the carbon source (coal, lignite, wood, nutshells or peat) in the absence of air, which produces a high carbon content material. Granular AC has a relatively large particle size compared to powdered activated carbon and, consequently, presents a smaller external surface. Extruded AC (also known as pellet activated carbon) combines powdered AC with a binder, which are fused together and extruded into a cylindrical shaped activated carbon block with diameters from 0.8 to 130 mm. These are mainly used for gas-phase applications because of their low pressure drop, high mechanical strength and low dust content.

Much effort has been devoted to produce CNF or CNT pellets.[78] One approach, conducted by Li *et al.*,[79] was to intentionally synthesize CNF pellets by press molding using polymer binders, resembling the manufacturing process of coal-based activated carbon. The shaping process involves the mixing of CNF powder with a resin mixture as the binder, followed by press molding and thermal treating at high temperature. The CNF composite can be further activated through gas- or liquid-phase oxidation. The main conclusions of this study were the following: (i) a unique fiber morphology remains in the CNF composite and the composite preserves the mesoporous texture of the CNFs; (ii) both the CNFs and the composite have a turbostratic graphite structure and more of the amorphous carbon phase was contained in the composite; and (iii) oxidative treatments led the composites to be more reactive and less massive, but the pellet density and the mechanical strength of the composites were still comparable to those of the activated carbon.

However, using polymer binders usually brings in a large amount of unfavorable micropores, which make the active sites difficult to access because of the pore diffusion resistance. An alternative proposed by Zhou *et al.* consists

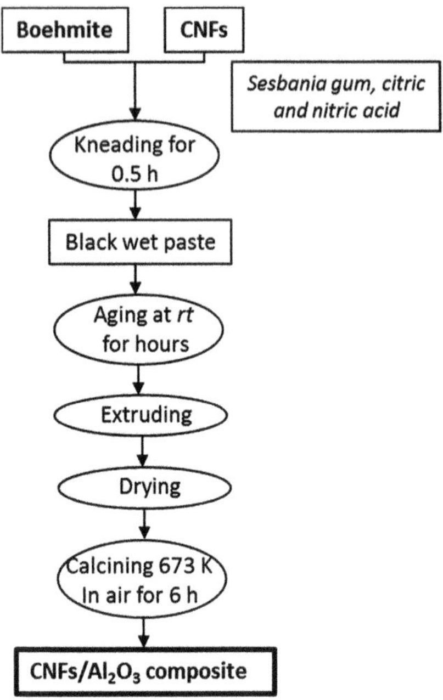

Scheme 10.1 Synthesis scheme of CNF/Al_2O_3 composite by extrusion casting.[80]

of the kneading of CNF powder with boehmite, adding dilute nitric acid as a binder, using citric acid and sesbania gum as additives, and subsequent extrusion and thermal treatment, which is very similar to the conventional alumina shaping by extrusion (Scheme 10.1).[80] A macroscopic CNF/alumina composite was thus synthesized by extrusion casting. It was shown that the CNFs and alumina were evenly and intimately blended in the composite, which displayed a lateral crushing strength greater than 100 N cm^{-1}, thus making it qualified for industrial use. The composite had a mesoporous structure, possessing a surface area greater than 320 m^2 g^{-1} and a narrow pore size distribution.

In 2007, Brying *et al.* were the first to produce aerogels made entirely of CNTs.[81] In their work, powdered CNTs were dissolved in water using the surfactant sodium dodecylbenzenesulfonate, producing a solution with a concentration of CNTs of 5–13 mg mL^{-1}. The solutions were poured into a mold and gelled overnight into an elastic gel. The gels were then solvent exchanged into either water or solutions of 0.25–1 wt% polyvinyl alcohol, then into ethanol, and finally into liquid CO_2 and supercritically dried. The result was CNT aerogels, with less than 5% shrinkage. These aerogels have densities of about 0.01–0.06 g cm^{-3} and can hold 8000 times their weight in applied force. They have remarkably high electrical (and thermal) conductivities for their low densities compared to similar density carbon aerogels; as high as 0.7 Ω^{-1} cm^{-1} for a 0.0075 g cm^{-3} nanotube aerogel, equivalent to a 0.06 g cm^{-3} carbon aerogel. They contain pores ranging from tens of nm to microns, with a unique filamentous structure. Ultralight (4 mg cm^{-3}) MWCNT aerogels were fabricated from a wet gel of well-dispersed pristine MWCNTs.[82] The entangled MWCNTs generate mesoporous structures on the honeycomb walls, creating aerogels with a surface area of 580 m^2 g^{-1}, which is much higher than that of pristine MWCNTs (241 m^2 g^{-1}). CNT–graphene hybrid aerogels have also been fabricated by supercritical CO_2 drying of their hydrogel precursors obtained from heating the aqueous mixtures of graphene oxide and CNTs with Vitamin C.[83] The resulting graphene–CNT hybrid aerogels show light weight, high conductivity (7.5 S m^{-1}), large BET surface area (435 m^2 g^{-1}), and large volume (2.58 cm^3 g^{-1}) with hierarchically porous structure. The high-pressure, high-temperature synthesis of a diamond aerogel from an amorphous carbon aerogel precursor using a laser heated diamond anvil cell has also been reported.[84]

10.2.6 Microreactor

The manufacture of chemicals in catalytic micro-structured reactors has become recently a new branch of chemical reaction engineering focusing on process intensification and safety. The recent developments in micromachining have boosted the use of microreactors for catalytic chemical reactions. Microreactors are structured reactors with some outstanding properties. The main feature of the microreactor is its high surface area to volume ratio, with values ranging from 5000 to 50 000 m^2 m^{-3}, while those of conventional reactors are around 100 m^2 m^{-3}. The dimensions of microreactor channels span from tens to hundreds of microns. The microscopic dimensions of the channels lend microreactors enhanced heat and mass transport properties with respect to conventional reactors. To increase the surface area of microreactor and disperse a catalyst, it is essential to coat the microreactor with a high surface area support material. Thus, the whole preparation process of catalytic microreators needs the involvement of multidisciplinary research teams with backgrounds as different as mechanical engineering, chemistry and catalysis. Concerning the preparation of the catalytic coating, some properties are desired such as good adhesion to substrate, complete coverage and uniform thickness. A frequently used catalyst support is γ-alumina.

After dispersing Ni catalyst on the alumina coating of ultrasonicated microreactors, Cebollada *et al.* have grown CNFs by catalytic CVD of CH_4.[85] The CNFs have diameters smaller than 50 nm and they formed an entangled layer completely covering the channels of microreactor (Figure 10.10). The CNF layer has a thickness between 1.3 and 2.5 μm. The complete alumina coverage after ultrasonic treatment is deemed essential for the growth of a uniform CNF layer from CH_4 decomposition since microreactors without previous alumina coating resulted in negligible CNF yield.

A microstructured catalyst based on aligned MWCNT arrays was synthesized and tested for Fischer–Tropsch synthesis reaction in a microchannel

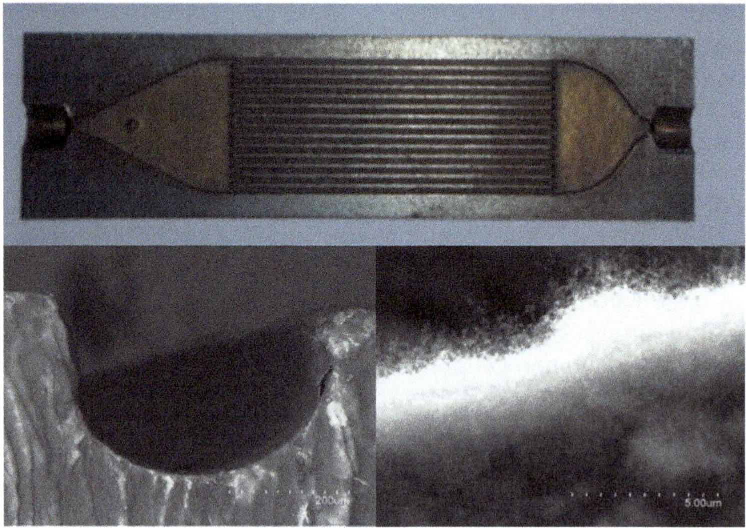

Figure 10.10 Microreactor after CNFs growth.

reactor.[86] Fabrication of such a structured catalyst first involved metal–organic chemical vapour deposition of a dense Al_2O_3 thin film over FeCrAlY foam to enhance the adhesion between the ceramic-based catalyst and metal substrate. Aligned MWCNTs were deposited uniformly over the substrate by controlled catalytic decomposition of ethylene. Another microreactor incorporating vertically-aligned carbon nanotubes supporting Pt nanoparticles was fabricated.[87] It was found that the presence of aligned nanotubes significantly enhances the catalytic octane hydrosylilation reaction and extends the catalyst lifetime as compared with conventional microreactors using a Pt metal film or Pt nanoparticles directly deposited on the channel walls.

A well-adhered layer of CNFs was grown on stainless steel microreactors by decomposition of a hydrocarbon over microreactors previously coated with Ni dispersed on alumina.[88] The mechanical stability of alumina coating was fundamental for the strong attachment of the CNFs grown subsequently. Besides, alumina coating ensured the growth of CNFs with uniform and thin diameter smaller than 50 nm. For the growth with CH_4 or with C_2H_6 at the lowest temperatures, *i.e.* 853 and 873 K, the unique type of carbon present were CNFs smaller than 50 nm, with an average diameter of 15 nm. An *in situ* CVD method was also developed in order to grow CNFs on Ni/alumina and nickel thin film catalyst coated inside a closed channel fused silica microreactor.[89] By directly flowing reactant gases over a catalytic coating inside the microchannels, a mechanically stable and porous CNF–alumina composite was formed with high surface area (160 $m^2\ g^{-1}$).

CNF layers were synthesized from nickel-based thin-films on flat fused silica substrates *via* catalytic chemical vapor deposition of ethylene.[90,91]

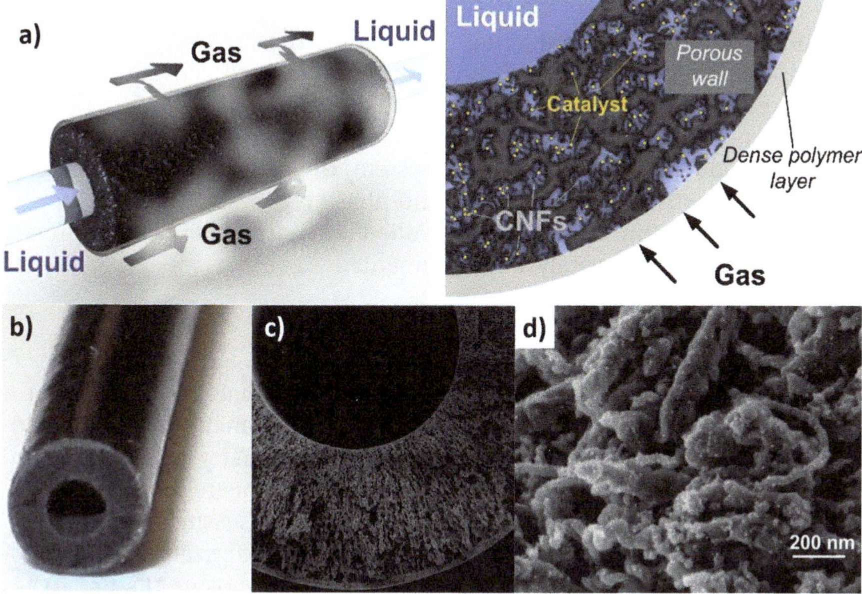

Figure 10.11 (a) Gas–liquid contacting in porous metallic membrane microreactor for multiphase reactions: porous stainless steel membrane decorated with CNFs. The reactor is encapsulated with a selective, gas permeable layer. (b) Porous stainless steel hollow fibers after CNFs growth; (c) SEM of the cross-section of the porous stainless steel hollow fiber; and (d) SEM of Pd particles immobilized on the surface of CNFs.

Different under-layer metal thin films, *viz.* titanium, tantalum and titanium–tungsten, were tested in order to obtain stable and well-attached CNF films on fused silica substrates.[90] The use of tantalum or titanium–tungsten as the adhesion layer resulted in stable and well-adhered CNF films on fused silica substrates, in which the morphology can be controlled by the growth time. Channels containing silicon micropillars covered with these CNFs have a significantly enhanced surface-to-volume ratio compared to bare microreactor channels (3–4 orders of magnitude).[91] Deposition of well-distributed platinum nanoparticles was carried out on these CNF layers, exemplifying their functionality as structured catalyst support to be used in microreactors.

Porous metallic membrane microreactors with CNFs catalyst support have been developed and applied for catalytic nitrite reduction in water.[92] Porous stainless steel hollow fibers with high porosity and mechanical strength were fabricated, and CNFs with high surface area were successfully grown on the porous surface (Figure 10.11). Pd catalyst was then immobilized on the CNFs as a hydrogenation catalyst. All the reactors were encapsulated with a gas permeable polymeric coating. The fabricated microreactors were shown to have a high surface area, mechanical strength and catalytic activity, which was tested for the catalytic reduction of nitrite.

To conclude, it appears that their aforementioned properties and the possibility of macroscopic shaping make nanostructured carbon materials attractive competitive catalyst supports if compared to activated carbons. Indeed, resistance to abrasion, thermal and dimensional stabilities, and specific adsorption properties are important factors controlling the final activity and reproducibility of the catalytic system. In particular, CNTs and CNFs could replace activated carbons in liquid-phase reactions since the properties of the latter cannot be easily controlled, and since their microporosity has often slowed down catalyst development.

10.3 Safety Issues

The amounts of engineered carbon nanostructured materials that are produced are crucial for environmental exposure assessment.[93] The development of risk and safety decision frameworks in industry seems therefore necessary to ensure that the potential risks of engineered nanomaterials are taken into consideration. Fundamentally, risk assessment involves an estimation of the potential for exposure and characterization of hazard. Potential routes of carbon nanostructured material exposure (Figure 10.12) include inhalation, dermal, oral, and in the case of biomedical applications, parenteral (delivered intravenously). Toxicity resulting from nanoparticle exposure could occur at the various portals of entry, such as the lungs and skin, or at distant sites. For prediction of systemic toxicity following non-parenteral exposures, systemic dose is an important parameter to consider. The systemic dose is dependent upon both the barrier function and clearance mechanisms at the portals of entry. Studies addressing the systemic translocation of nanoparticles from sites of deposition are beginning to unravel the dynamics of nanoparticle/organism interaction, and provide the means to relate exposure and hazard data.

Carbon nanomaterials with different physico chemical characteristics exhibit quite different cytotoxicity and bioactivity *in vitro*, although they may not be accurately reflected in the comparative toxicity *in vivo*. The main physico chemical characteristics relevant to safety considerations are: the shape, the typical dimensions, the surface (area, charge, coating), the elasticity/stiffness, the colloidal stability (aqueous dispersibility, aggregation), durability (possible enzymatic degradation), and the presence of impurities such as catalyst residue (Figure 10.13). The cytotoxicity apparently follows a sequence order on a mass basis: SWCNTs > MWCNTs > Graphene > C_{60} ≈ CB ≈ FLG > ND. For example, marked differences in toxicity were observed after short-term inhalation exposure of four carbon-based nanomaterials (CB, MWCNTs, FLG, and graphene).[94] Whereas no relevant toxicity occurred with CB and graphite nanoplatelets, MWCNTs and graphene induced lung toxicity. The toxicity of MWCNT at 2.5 mg m^{-3} was more marked than that of graphene at 10 mg m^{-3}. In the following sections we will briefly discussed some aspects of the toxicity of the main nanostructured carbon materials, and will provide some advice for safely working with these materials.

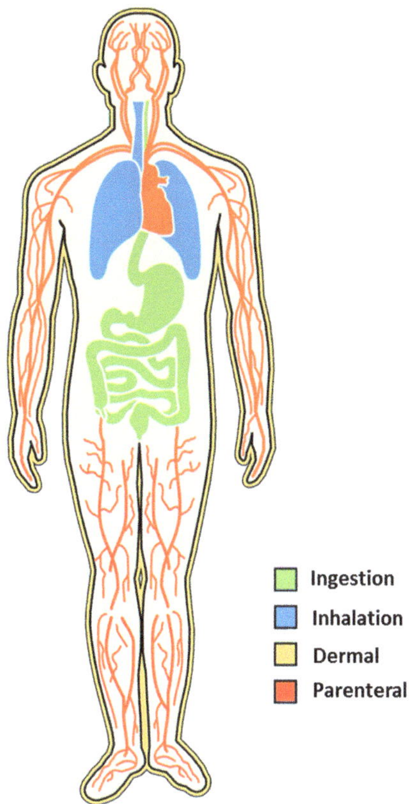

Figure 10.12 Potential routes of nanoparticle exposure.

10.3.1 Carbon Black

CB is a low solubility particle produced industrially from incomplete thermal decomposition of hydrocarbons in which the process is controlled to achieve pre-defined and reproducible particle sizes and properties. The carbon black particles so-formed are complex, with a degenerated graphitic crystallite. The structure of CB is described as nodules, the roughly spherical primary structural elements, aggregates which comprise fused, connected particles, and agglomerates, which are undispersed clusters of aggregates (see Chapter 1). Commercial CB materials generally contain more than 97% elemental carbon with variable amounts of oxygen, hydrogen and sulfur. Less than 1% of CB particles consist of extractable organic materials. Typical classes of organic chemicals adsorbed onto the CB particles surface are polycyclic aromatic hydrocarbons (PAHs), nitro-derivatives of PAHs and sulfur-containing PAHs. Examples of PAHs adsorbed onto carbon black particle surface include benzopyrenes, benzo[*ghi*]perylene, coronene, fluoranthene and pyrene. Several of these compounds are recognized as human carcinogens.

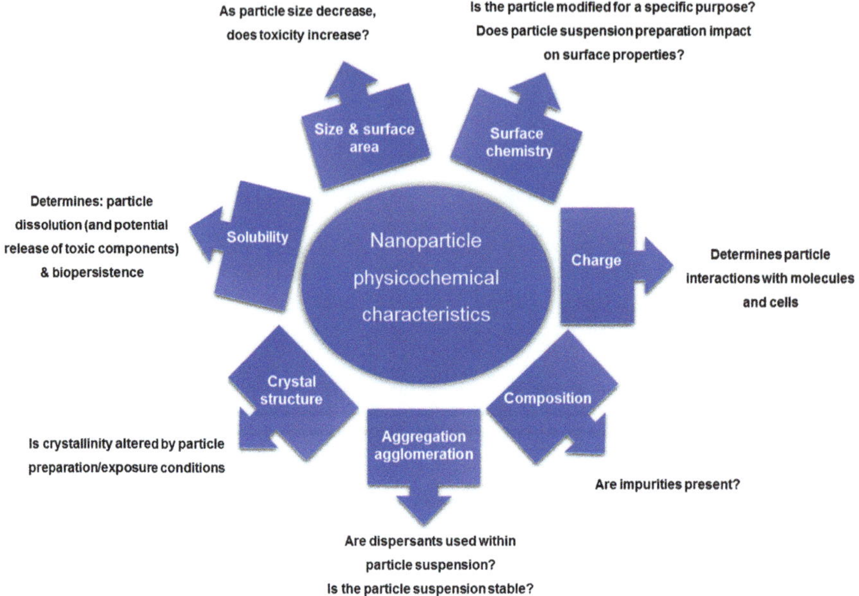

Figure 10.13 Main physico chemical characteristics relevant to safety considerations.

CB has been widely investigated since its use as a benchmark control for *in vivo* toxicological evaluation of diesel exhaust particles and as a model of urban air pollution particulate matter, almost three decades ago.[95–98] Since then, CB has become the focus of numerous toxicity studies as well as an important reference material, especially as an example of a low toxicity, low solubility particle not complicated by harmful levels of toxicologically-relevant organics or metals.[99–101] CB nanoparticles are reactive oxygen species (ROS) generators, as shown in cellular[102,103] and acellular systems.[104] Exposure to CB resulted in the greatest ROS generation followed by SWCNT and C_{60} in both cellular and cell-free particle suspensions. The chemical basis of the ability of CB to cause oxidative stress is unknown, but unlike highly soluble welding fume, ROS production is not related to metal or any other soluble component. Moreover, inhalation or intra-tracheal instillation exposures to CB nanoparticles result in large pulmonary inflammatory responses in rodents,[105–111] which can greatly exacerbate ROS generation *via* activation of polymorphonuclear granulocytes.[111] As such, it is expected that CB nanoparticles can mediate secondary genotoxicity by means of inflammation and oxidative stress. CB nanoparticles are genotoxic *in vitro*, as shown by increases in DNA base oxidation,[112] mutation frequency,[113,114] strand breaks[115,116] and micronucleus frequency in lung epithelial cells[117] as well as increases in strand breaks in fibroblasts.[118] However, not much is known about the genotoxicity of CB nanoparticles *in vivo*. A few studies in rats have demonstrated CB nanoparticle-induced DNA base oxidation[119] and increased mutation frequency.[108] However, rats may not be the most suitable model for exposure to particulates due to their predisposition to particle overload. Studies in mice have demonstrated DNA strand breaks in BAL cells[109,120] and one study has established CB nanoparticle-induced lung DNA strand breaks, but this was found using a high dose immediately post instillation.[117]

In long-term animal studies CB was found to be a carcinogen, although rat lung overload very likely plays a role in this result.[121] Studies of workers exposed to CB revealed respiratory symptoms including impaired lung function arising from deposition of dust in the lungs, abnormalities in chest radiographs and respiratory morbidity.[122–124] Recent studies have shown that nano-sized CB particles are more toxic than large respirable CB particles because of their higher surface area.[125] Agglomeration did not affect the toxicity of nano-sized carbon particles.

Surveys of CB workers have not permitted any firm conclusions on carcinogenicity to be drawn; and up to now industries in which significant airborne CB exposure has occurred were not associated with increased lung cancer risk.

10.3.2 Carbon Nanotubes

There has been a great concern that if CNTs, which are very light, enter the working environment as suspended particulate matter of respirable sizes, they could pose an occupational inhalation exposure hazard.[126] The results of rodent studies collectively showed that regardless of the process by which CNTs were synthesized and the types and amounts of metals they contained, CNTs were capable of producing inflammation, epithelioid granulomas (microscopic nodules), fibrosis, and biochemical/toxicological changes in the lungs. Comparative toxicity studies in which mice were given equal amounts of test materials showed that SWCNTs were more toxic than quartz,[127] which is considered a serious occupational health hazard if it is chronically inhaled; ultrafine CB was shown to produce minimal lung responses. Given that manufactured SWCNTs and MWCNTs were found to elicit pathological changes in the lungs, and SWCNTs (administered to the lungs of mice) were further shown to produce respiratory function impairments, retard bacterial clearance after bacterial inoculation, damage the mitochondrial DNA in aorta, increase the percent of aortic plaque, and induce atherosclerotic lesions in the brachiocephalic artery of the heart, it is speculated that exposure to combustion-generated MWCNTs in fine particulate matter may play a significant role in air pollution-related cardiopulmonary diseases. Therefore, CNTs from manufactured and combustion sources in the environment could have adverse effects on human health. A recent study aimed at investigating the feasibility and challenges associated with conducting a human health risk assessment for CNTs utilizing an approach similar to that of a classical regulatory risk assessment.[128] Results indicate that the main risks for humans arise from chronic occupational inhalation, especially during activities involving high CNT release and uncontrolled exposure. However, it is not yet possible to draw definitive conclusions with regards the potential risk for long, straight MWCNTs to pose a similar risk as asbestos by inducing mesothelioma. The genotoxic potential of CNTs is currently inconclusive and could be either primary or secondary. Possible systemic effects of CNTs would be either dependent on absorption and distribution of CNTs to sensitive organs or could be induced through the release of inflammatory mediators. In conclusion, gaps in the data set in relation to both exposure and hazard do not allow any definite conclusions suitable for regulatory decision-making. In order to enable a full human health risk assessment, future work should focus on the generation of reliable occupational, environmental and consumer exposure data. Data on toxico-kinetics and studies investigating effects of chronic exposure under conditions relevant for human exposure should also be prioritized. The average person consumes an estimated 5000 to 3 000 000 particles per cm^3 daily due to incidental nanoparticles from the ambient environment.[129] The safe systemic dose of CNTs, if it can be made to conform to such numbers, would then make current toxicity reports on biological risk seem overestimated.[130]

10.3.3 Graphene

Even though the research on technical and biomedical applications of graphene and graphene derivative nanomaterials is expanding rapidly, relatively little is known about their interaction with biological systems or intrinsic toxicity.[131,132] However, and because of the potential risk factors associated with the manufacture and use of graphene-related materials, the number of nano-toxicological studies of these compounds has been increasing rapidly in the past decade.[133] These studies have researched the effects of the nanostructural/biological interactions on different organizational levels of the living system, from biomolecules to animals.

The literature published up to now on *in vitro* toxicity of graphene nanomaterials suggest that, analogous to other carbon nanostructured materials, physico-chemical characteristics may play a critical role in the biological activity of this novel class of nanomaterials. Mechanisms that were suggested to underlie the cytotoxic effect include plasma membrane damage, impairment of mitochondrial activity, induction of oxidative stress and DNA damage eventually leading to apoptotic and/or necrotic cell death. The current literature proposes that the generation of reactive oxygen species in target cells is the most important cytotoxicity mechanism of graphene.[133] Further studies are required to better understand the toxicity pathways, in particular those that focus on the investigation of cellular interactions of graphene materials with cell membrane lipids on a molecular level.

Yet, in some cases, results regarding the cytotoxicity of graphene-based nanomaterials obtained by different authors are conflicting (in particular that for GO). These discrepancies may be due to differences in the intrinsic properties of the nanomaterials tested, the availability of the nanomaterial during the assay or the sensitivity of the cell lines used (among other factors).

Furthermore, considering the very high specific surface area of graphene nanomaterials and their chemical nature (conjugated π-electron system, presence of reactive functional surface groups), they can be expected to interfere with most of the commonly used bioassay(s) (*e.g.* physical sorption of assay reagents to the nanomaterial surface, quenching of fluorescent probes, auto-fluorescence of the nanomaterial).

Overall, an important conclusion that can be postulated is that small and hydrophilic graphene nanomaterials (in particular, those capped with biocompatible molecules) tend to form a stable colloid dispersion, avoiding aggregation and, therefore, are more suitable to be internalized and removed/excreted from the application site. Moreover, colloidal dispersions of individualized graphene sheets (or graphene oxide and its derivatives) can be more easily engineered without metallic impurities compared with several types of CNTs, making graphene-based materials promising candidates for biomedical applications. In addition, graphene nanostructures are not fiber-shaped and theoretically offer significant advantages in terms of safety over inhomogeneous dispersions of CNTs. Toxicological studies should consider the purity (quality) of the sample, especially the presence of oxidative debris formed during the early stages of synthesis and/or during the functionalizing process, which largely alters the surface microchemical environment of graphene and GO.

10.3.4 Fullerenes and Nanodiamonds

Pristine fullerenes have shown low toxicity and there is probably no risks expected for humans exposed to fullerenes in the workplace under good hygiene conditions.[134] Indeed, available data clearly shows that pristine C_{60} has no acute or sub-acute toxicity in a large variety of living organisms, from bacterial and fungal to human leukocytes, and also in drosophila, mice, rats and guinea pigs.[135] However, manipulation of surface chemistry and molecular makeup has created a diverse population of fullerenes, which exhibit drastically different behaviors,[136] and some C_{60} derivatives can be highly toxic. Furthermore, under light exposure, C_{60} is an efficient singlet oxygen sensitizer.[137] This shows that environmental conditions, including light exposure and oxygen concentration, have the potential to impact on the generation of toxic ROS by fullerenes. Thus, the studies conducted so far suggest that fullerene toxicity involves an oxidant-driven response, suggesting that toxicity evaluations should assess the potential of fullerenes to cause oxidative stress and related consequences, such as inflammation or genotoxicity.[136] The main concern for consumers is exposure *via* direct dermal application. Available studies do not indicate a short term risk from the tested fullerene types, however no extrapolation to all fullerene types and to chronic exposure can be made. The current dataset on fullerenes in relation to both human exposure and hazard is limited and does not allow reaching any definite conclusions suitable for regulatory decision making. Future work should focus on generating occupational and consumer exposure data, as well as suitable data on toxico-kinetics and potential toxic effects following repeated inhalation and dermal exposure. It seems also relevant to clarify whether certain fullerene types may potentially induce genotoxic and/or carcinogenic effects *via* physiologically relevant routes.

Diamond and glassy carbon are known to be non-toxic, but we cannot assume that carbon nanoparticles are also non-toxic. The cytotoxicity of nanodiamonds ranging in size from 2 to 10 nm has been assessed. Assays of cell viability showed that nanodiamonds were not toxic to a variety of cell types.[138] Furthermore, nanodiamonds did not produce significant reactive oxygen species. Cells can grow on nanodiamond-coated substrates without morphological changes compared to controls. However, it should be noted that, like other nanomaterials such as CNTs and graphene, due to a large surface area and high adsorption capacity, the high adsorption of various components in the environmental media onto NDs may affect their inherent biological effects.[139] Although long-term and *in vivo* toxicity effects remain to be tested, these initial results are encouraging.

10.4 Life Cycle Assessment

The intensity of research interest in nanostructured carbon materials has raised questions regarding the life cycle environmental impact of nanotechnologies, including assessment of: worker and consumer safety, greenhouse gas emissions, toxicological risks associated with production or product emissions and the disposal of nano-products at end of life. However, development of appropriate nanotechnology assessment tools has lagged progress in the nanotechnologies themselves. In particular, current approaches to life cycle assessment (LCA, Figure 10.14) – originally developed for application in mature manufacturing industries such as automobiles and chemicals – suffer from several shortcomings that make applicability to nanotechnologies problematic.[140,141] Among these are uncertainties related to the variability of material properties, toxicity and risk, technology performance in the use phase, nanomaterial degradation and change during the product life cycle and the impact assessment stage of LCA.

The reviewed LCA studies indicate that nanomaterials have higher cradle-to-gate energy demand per functional unit, and thus higher global warming impact than their conventional counterparts. Depending on synthesis method, carbon-based nanoparticles, *i.e.*, CNFs, CNTs and fullerenes require 1900 GJ kg^{-1} of primary energy to produce, compared with *ca.* 200 MJ kg^{-1} for aluminium.[142] This is mainly attributed to the fact that nanomaterials involve an energy intensive synthesis process, or additional mechanical process to reduce particle size.

Early studies suggest that a release of nanostructured carbon materials can occur not only in the production phase, but also in the usage and disposal phases of their applications.[143] The likelihood and form of release is determined by the way nanostructured carbon materials are incorporated into the material. A significant part of all nanostructured carbon materials used may

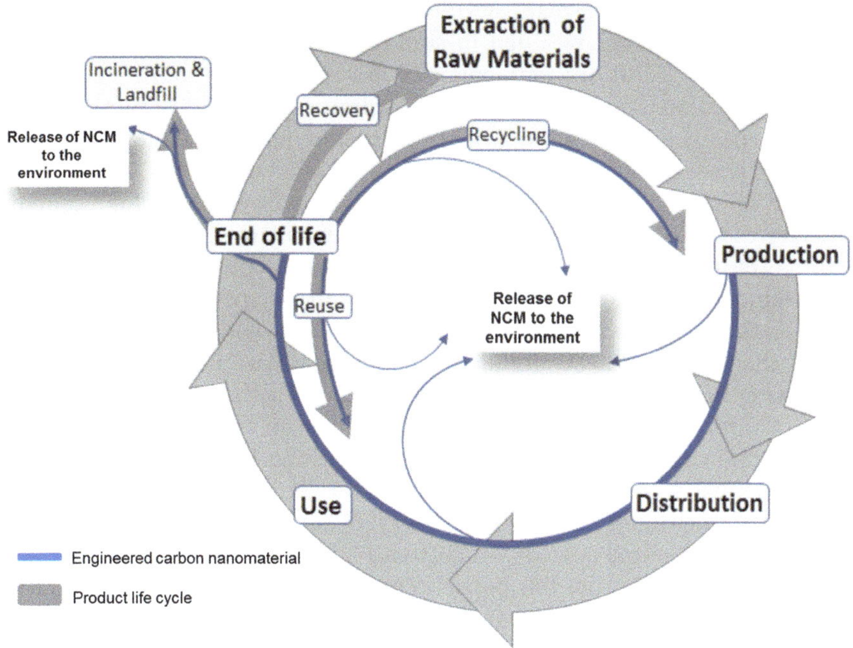

Figure 10.14 Life cycle thinking and carbon nanomaterials.

finally be dispersed somewhere in the technosphere or the environment, *e.g.* by cross-product contamination during recycling.

As long as potential adverse effects of nanostructured carbon materials cannot be ruled out, implementing precautionary measures along the value chain (including the end-of-life treatment) should be recommended in order to reduce the release and possible negative environmental or human health effects of nanostructured carbon materials.

10.5 How to Work Safely with Nanostructured Carbon Materials

The preliminary conclusions to be drawn from the toxicology studies is that some types of carbon nanomaterials can be toxic, if they are not bound up in a substrate and they are available to the body. Researchers must use procedures that prevent inhalation and dermal exposures because at this time toxicological information is limited. Based on particle physics and studies of fine atmospheric pollutants, the nanoparticle size range is the range of minimum settling. This means that once released into air, nanoparticles will remain airborne for considerable periods of time. Nanoparticles can be inhaled and will be collected in all regions of the respiratory tract; about 35% will deposit in the deep region of the lungs. Because they are so small, nanoparticles follow airstreams more easily than larger particles, so they will be easily collected and retained in standard ventilated enclosures such as fume hoods. In addition, nanoparticles are readily collected by High Efficiency Particulate Air (HEPA) filters. Respirators with HEPA filters will be adequate protection for nanoparticles in case of spills of large amounts of material.

Working safely with carbon nanomaterials involves following standard procedures that should be followed for any particulate material with known or uncertain toxicity: preventing inhalation, skin contact, and ingestion. Many nanomaterials are synthesized in enclosed reactors or glove boxes. The enclosures are under vacuum or exhaust ventilation, which prevent exposure during the actual synthesis. Inhalation exposure can occur during additional processing of materials removed from reactors, this processing should be done in fume hoods. In addition, maintenance on reactor parts that may release residual particles in the air should be done in fume hoods. Liquid-phase chemistry involving carbon nanomaterials should be carried out in ventilated fume hoods or glove boxes.

The type of surface coating on carbon nanoparticles often causes them to clump together so that few particles are actually released when particles are removed from reactors. In one of the few workplace industrial hygiene studies of nanoparticles, almost no release of fibers was found when CNTs were removed from a reactor and transferred into a secondary container. The SWCNT clumped together into nanoropes and remained attached to the substrate as it was removed from the reactor.

Since the ability of nanoparticles to penetrate the skin is uncertain at this point, gloves should be worn when handling particulates and solutions containing particles. A glove having good chemical resistance to any solution the particles are suspended in should be used. If working with dry particulate, a sturdy glove with good integrity should be used. Disposable nitrile gloves commonly used in many labs would provide good protection from nanoparticles for most procedures that involve extensive skin contact. Two pairs of gloves can be worn if extensive skin contact is anticipated.

One potential safety concern with nanoparticles is fires and explosions if large quantities of dust are generated during reactions or production. This is expected to become more of a concern when reactions are scaled-up to pilot plant or production levels. Both carbonaceous and metal dusts can burn and explode if an oxidant such as air and an ignition source are present. Nanodusts can be anticipated to have a greater potential for explosivity than larger particles.

There are currently no government occupational exposure standards for nanomaterials. One should also be aware that Material Safety Data Sheets (MSDS) may not have accurate information at this point in time. For example, the MSDSs that are accompanying some commercially available CNTs are referring to the graphite Permissible Exposure Limit (PEL) as a relevant exposure standard. Both graphite and carbon nanotubes are composed of carbon arranged in a honeycomb pattern. However, graphite is composed of coarse particles while CNTs are shaped like fibers and have very different tensile and conductive properties than graphite. CNTs are much more toxic in the short-term animal tests that have been performed to date. Consequently, the graphite PEL and toxicity information is not appropriate for MSDSs of CNTs. CNTs should be treated as potentially toxic fibers, if capable of being released into the air and not bound up in a substrate, and should be handled with appropriate controls. The US Occupational Safety and Health Administration (OSHA) recently published a fact sheet entitled Working Safely with Nanomaterials.[144] OSHA recommends that worker exposure to respirable CNTs and CNFs not exceed 7.0 μg m^{-3} as an 8-hour time-weighted average, based on the National Institute for Occupational Safety and Health (NIOSH) proposed Recommended Exposure Limit (REL). OSHA suggests that, because exposure limits for other nanomaterials do not exist yet, employers should minimize worker exposure by using the hazard control measures and best practices identified in the fact sheet and in the references noted.

References

1. S. Kedem, J. Schmidt, Y. Paz and Y. Cohen, *Langmuir*, 2005, **21**, 5600–5604.
2. N. A. Jarrah, F. Li, J. G. van Ommen and L. Lefferts, *J. Mater. Chem.*, 2005, **15**, 1946–1953.
3. A. N. Karwa, V. A. Davis and B. J. Tatarchuk, *J. Nanotechnol.*, 2012, 14.
4. N. Jeong and J. Lee, *J. Catal.*, 2008, **260**, 217–226.
5. W. Huang, X.-B. Zhang, J. Tu, F. Kong, Y. Ning, J. Xu and G. V. Tendeloo, *Phys. Chem. Chem. Phys.*, 2002, **4**, 5325–5329.
6. A. Sacco Jr, F. W. A. H. Geurts, G. A. Jablonski, S. Lee and R. A. Gately, *J. Catal.*, 1989, **119**, 322–341.
7. W. Wang, S. Guo, M. Ozkan and C. S. Ozkan, *J. Mater. Res.*, 2013, **28**, 912–917.
8. J. K. Chinthaginjala and L. Lefferts, *Carbon*, 2009, **47**, 3175–3183.
9. J. K. Chinthaginjala, D. B. Thakur, K. Seshan and L. Lefferts, *Carbon*, 2008, **46**, 1638–1647.
10. Y. Wang, Y. H. Chin and Y. Gao, *US Pat.*, US6713519 B2, 2006.
11. Y. Wang, Y. H. Chin, Y. Gao, C. L. Aardahl and T. L. Stewart, *US Pat.*, US6824689 B2, 2004.
12. N. Xiao, Y. Zhou, J. Qiu and Z. Wang, *Fuel*, 2010, **89**, 1169–1171.
13. N. Xiao, Y. Zhou, Z. Ling and J. Qiu, *Carbon*, 2013, **59**, 530–536.
14. L. Truong-Phuoc, T. Truong-Huu, L. Nguyen-Dinh, W. Baaziz, T. Romero, D. Edouard, D. Begin, I. Janowska and C. Pham-Huu, *Appl. Catal. A*, 2014, **469**, 81–88.
15. H. Yuan, Z. Sun, H. Liu, B. Zhang, C. Chen, H. Wang, Z. Yang, J. Zhang, F. Wei and D. S. Su, *ChemCatChem*, 2013, **5**, 1713–1717.
16. C. Pham-Huu, C. Pham, P. Nguyen, E. Vanhaecke, S. Ivanova and A. Deneuve, *US Pat*, US20100297428 A1, 2010.

17. P. W. A. M. Wenmakers, J. van der Schaaf, B. F. M. Kuster and J. C. Schouten, *Chem. Eng. Sci.*, 2010, **65**, 247–254.
18. P. W. A. M. Wenmakers, J. van der Schaaf, B. F. M. Kuster and J. C. Schouten, *J. Mater. Chem.*, 2008, **18**, 2426–2436.
19. S. J. Park and D. G. Lee, *Carbon*, 2006, **44**, 1930–1935.
20. A. N. Karwa and B. J. Tatarchuk, *Sep. Purif. Technol.*, 2012, **87**, 84–94.
21. P. Tribolet and L. Kiwi-Minsker, *Catal. Today*, 2005, **102–103**, 15–22.
22. P. Tribolet and L. Kiwi-Minsker, *Catal. Today*, 2005, **105**, 337–343.
23. D. Rosenthal, M. Ruta, R. Schlögl and L. Kiwi-Minsker, *Carbon*, 2010, **48**, 1835–1843.
24. S. J. Park and D. G. Lee, *Curr. Appl Phys.*, 2006, **6**, (Suppl. 1), e182–e186.
25. Y. Gao, G. Hu, J. Zhong, Z. Shi, Y. Zhu, D. S. Su, J. Wang, X. Bao and D. Ma, *Angew. Chem. Int. Ed.*, 2013, **52**, 2109–2113.
26. D. R. Minett, J. P. O'Byrne, M. D. Jones, V. P. Ting, T. J. Mays and D. Mattia, *Carbon*, 2013, **51**, 327–334.
27. M. Pérez-Cadenas, V. Muñoz-Andrés, I. Rodríguez-Ramos, A. Maroto-Valiente and A. Guerrero-Ruíz, *J. Nano Res.*, 2012, **18–19**, 271–279.
28. E. García-Bordejé, I. Kvande, D. Chen and M. Rønning, *Adv. Mater.*, 2006, **18**, 1589–1592.
29. E. García-Bordejé, I. Kvande, D. Chen and M. Rønning, *Carbon*, 2007, **45**, 1828–1838.
30. S. Morales-Torres, A. F. Pérez-Cadenas, F. Kapteijn, F. Carrasco-Marín, F. J. Maldonado-Hódar and J. A. Moulijn, *Appl. Catal. B*, 2009, **89**, 411–419.
31. M. A. Ulla, A. Valera, T. Ubieto, N. Latorre, E. Romeo, V. G. Milt and A. Monzón, *Catal. Today*, 2008, **133–135**, 7–12.
32. S. Armenise, M. Nebra, E. García-Bordejé and A. Monzón, in *Studies in Surface Science and Catalysis*, ed. M. Devillers, S. Hermans, P. A. Jacobs, J. A. Martens, E.M. Gaigneaux and P. Ruiz, Elsevier, 2010, vol. 175, pp. 483–486.
33. K. M. de Lathouder, T. Marques Fló, F. Kapteijn and J. A. Moulijn, *Catal. Today*, 2005, **105**, 443–447.
34. K. M. de Lathouder, D. Lozano-Castelló, A. Linares-Solano, F. Kapteijn and J. A. Moulijn, *Carbon*, 2006, **44**, 3053–3063.
35. C. Moreno-Castilla and A. Pérez-Cadenas, *Materials*, 2010, **3**, 1203–1227.
36. L. Roldan, S. Armenise, Y. Marco and E. Garcia-Bordeje, *Phys. Chem. Chem. Phys.*, 2012, **14**, 3568–3575.
37. K. M. de Lathouder, D. Lozano-Castelló, A. Linares-Solano, S. A. Wallin, F. Kapteijn and J. A. Moulijn, *Microporous Mesoporous Mater.*, 2007, **99**, 216–223.
38. E. Soghrati, M. Kazemeini, A. M. Rashidi and K. J. Jozani, *Procedia Eng.*, 2012, **42**, 1484–1492.
39. V. Martínez-Hansen, N. Latorre, C. Royo, E. Romeo, E. García-Bordejé and A. Monzón, *Catal. Today*, 2009, **147**, (Suppl.), S71–S75.
40. J. H. Zhou, M. G. Zhang, L. Zhao, P. Li, X. G. Zhou and W. K. Yuan, *Catal. Today*, 2009, **147**, Suppl. S225–S229.
41. R. Vieira, C. Pham-Huu, N. Keller and M. J. Ledoux, *Chem. Commun.*, 2002, 954–955.

42. J. J. Delgado, D. S. Su, G. Rebmann, N. Keller, A. Gajovic and R. Schlögl, *J. Catal.*, 2006, **244**, 126–129.
43. M. J. Ledoux, R. Vieira, C. Pham-Huu and N. Keller, *J. Catal.*, 2003, **216**, 333–342.
44. R. Vieira, D. Bastos-Netto, M.-J. Ledoux and C. Pham-Huu, *Appl. Catal. A: Gen.*, 2005, **279**, 35–40.
45. J. Li, S. Sambandam, W. Lu and C. M. Lukehart, *Adv. Mater.*, 2008, **20**, 420–424.
46. N. M. d. A. Coelho, J. L. B. Furtado, C. Pham-Huu and R. Vieira, *Mater. Res.*, 2008, **11**, 353–357.
47. S. Zarubova, S. Rane, J. Yang, Y. Yu, Y. Zhu, D. Chen and A. Holmen, *ChemSusChem*, 2011, **4**, 935–942.
48. T. Zhao, I. Kvande, Y. Yu, M. Ronning, A. Holmen and D. Chen, *J. Phys. Chem. C*, 2010, **115**, 1123–1133.
49. J. Mauricio Rosolen, C. H. Patrick Poá, S. Tronto, M. S. Marchesin and S. R. P. Silva, *Chem. Phys. Lett.*, 2006, **424**, 151–155.
50. J. M. Rosolen, E. Y. Matsubara, M. S. Marchesin, S. M. Lala, L. A. Montoro and S. Tronto, *J. Power Sources*, 2006, **162**, 620–628.
51. L. Jiang, S. R. Vance, L. Yang, L. Weijie, S. F. Timothy and M. L. Charles, *Nanotechnology*, 2007, **18**, 325606.
52. J.-s. Zheng, X.-z. Wang, R. Fu, P. Li, D.-j. Yang, H. Lu and J.-X. Ma, *New Carbon Mater.*, 2011, **26**, 262–270.
53. J. K. Chinthaginjala, K. Seshan and L. Lefferts, *Ind. Eng. Chem. Res.*, 2007, **46**, 3968–3978.
54. Y. Liu, H. Ba, D.-L. Nguyen, O. Ersen, T. Romero, S. Zafeiratos, D. Begin, I. Janowska and C. Pham-Huu, *J. Mater. Chem. A*, 2013, **1**, 9508–9516.
55. N. Jarrah, J. G. van Ommen and L. Lefferts, *Catal. Today*, 2003, **79–80**, 29–33.
56. R. M. Heck, S. Gulati and R. J. Farrauto, *Chem. Eng. J.*, 2001, **82**, 149–156.
57. R. E. Hayes and S. T. Kolaczkowski, *Chem. Eng. Sci.*, 1994, **49**, 3587–3599.
58. J. Restivo, J. J. M. Órfão, M. F. R. Pereira, M. R. E. Vanhaecke, T. Iouranova, L. Kiwi-Minsker, S. Armenise and E. Garcia-Bordejé, *Water Sci. Technol.*, 2012, **65**, 1854–1862.
59. J. Restivo, J. J. M. Órfão, S. Armenise, E. Garcia-Bordejé and M. F. R. Pereira, *J. Hazard. Mater.*, 2012, **239–240**, 249–256.
60. J. Zhu, Y. Jia, M. Li, M. Lu and J. Zhu, *Ind. Eng. Chem. Res.*, 2012, **52**, 1224–1233.
61. S. Armenise, L. Roldán, Y. Marco, A. Monzón and E. García-Bordejé, *J. Phys. Chem. C*, 2012, **116**, 26385–26395.
62. B. Gong, R. Wang, B. Lin, F. Xie, X. Yu and K. Wei, *Catal. Lett.*, 2008, **122**, 287–294.
63. X. Yu, B. Lin, B. Gong, J. Lin, R. Wang and K. Wei, *Catal. Lett.*, 2008, **124**, 168–173.
64. M. Uo, T. Hasegawa, T. Akasaka, I. Tanaka, F. Munekane, M. Omori, H. Kimura, R. Nakatomi, K. Soga, Y. Kogo and F. Watari, *Bio-Med. Mater. Eng.*, 2009, **19**, 11–17.

65. J. K. Chinthaginjala, J. H. Bitter and L. Lefferts, *Appl. Catal. A*, 2010, **383**, 24–32.
66. K. Otsuka, Y. Abe, N. Kanai, Y. Kobayashi, S. Takenaka and E. Tanabe, *Carbon*, 2004, **42**, 727–736.
67. S.-S. Tzeng, K.-H. Hung and T.-H. Ko, *Carbon*, 2006, **44**, 859–865.
68. J. H. Park, K. Y. Yoon, H. Na, Y. S. Kim, J. Hwang, J. Kim and Y. H. Yoon, *Sci. Total Environ.*, 2011, **409**, 4132–4138.
69. B. Louis, R. Vieira, A. Carvalho, J. Amadou, M. J. Ledoux and C. Pham-Huu, *Top. Catal.*, 2007, **45**, 75–80.
70. M.-J. Ledoux and C. Pham-Huu, *Catal. Today*, 2005, **102–103**, 2–14.
71. C. Pham-Huu and M.-J. Ledoux, *Top. Catal.*, 2006, **40**, 49–63.
72. P. Li, T. Li, J.-H. Zhou, Z.-J. Sui, Y.-C. Dai, W.-K. Yuan and D. Chen, *Micropor. Mesopor. Mater.*, 2006, **95**, 1–7.
73. R. Vieira, M.-J. Ledoux and C. Pham-Huu, *Appl. Catal. A*, 2004, **274**, 1–8.
74. V. Ricardo, B. Pierre, L. Marc-Jacques and P.-H. Cuong, *Jpn. J. Appl. Phys.*, 2005, **44**, 4282.
75. J. M. Nhut, R. Vieira, N. Keller, C. Pham-Huu, W. Boll and M. J. Ledoux, in *Studies in Surface Science and Catalysis*, ed. D. E. De Vos, P. Grange, P. A. Jacobs, J. A. Martens, P. Ruiz, E. Gaigneaux and G. Poncelet, Elsevier, 2000, vol. 143, pp. 983–991.
76. R. B. Mathur, S. Chatterjee and B. P. Singh, *Compos. Sci. Technol.*, 2008, **68**, 1608–1615.
77. S. Wang, X. Zhao, T. Cochell and A. Manthiram, *J. Phys. Chem. Lett.*, 2012, **3**, 2164–2167.
78. J. Ma, D. Moy, A. Chishti and J. Yang, *US Pat.*, US20060142149 A1, 2006.
79. P. Li, T.-J. Zhao, J.-H. Zhou, Z.-J. Sui, Y.-C. Dai and W.-K. Yuan, *Carbon*, 2005, **43**, 2701–2710.
80. J.-H. Zhou, C. Chen, R. Guo, X.-C. Fang and X.-G. Zhou, *Carbon*, 2009, **47**, 2077–2084.
81. M. B. Bryning, D. E. Milkie, M. F. Islam, L. A. Hough, J. M. Kikkawa and A. G. Yodh, *Adv. Mater.*, 2007, **19**, 661–664.
82. J. Zou, J. Liu, A. S. Karakoti, A. Kumar, D. Joung, Q. Li, S. I. Khondaker, S. Seal and L. Zhai, *ACS Nano*, 2010, **4**, 7293–7302.
83. Z. Sui, Q. Meng, X. Zhang, R. Ma and B. Cao, *J. Mater. Chem.*, 2012, **22**, 8767–8771.
84. P. J. Pauzauskie, J. C. Crowhurst, M. A. Worsley, T. A. Laurence, A. L. D. Kilcoyne, Y. Wang, T. M. Willey, K. S. Visbeck, S. C. Fakra, W. J. Evans, J. M. Zaug and J. H. Satcher, *Proc. Natl. Acad. Sci. U.S.A.*, 2011.
85. P. A. R. Cebollada and E. Garcia-Bordejé, *Chem. Eng. J.*, 2009, **149**, 447–454.
86. Y.-h. Chin, J. Hu, C. Cao, Y. Gao and Y. Wang, *Catal. Today*, 2005, **110**, 47–52.
87. N. Ishigami, H. Ago, Y. Motoyama, M. Takasaki, M. Shinagawa, K. Takahashi, T. Ikuta and M. Tsuji, *Chem. Commun.*, 2007, 1626–1628.
88. L. Martínez-Latorre, S. Armenise and E. Garcia-Bordejé, *Carbon*, 2010, **48**, 2047–2056.

89. A. Ağıral, L. Lefferts and J. G. E. Gardeniers, *Catal. Today*, 2010, **150**, 128–132.
90. D. B. Thakur, R. M. Tiggelaar, K. Seshan, J. G. E. Gardeniers and L. Lefferts, *Adv. Sci. Technol.*, 2008, **54**, 231–236.
91. D. B. Thakur, R. M. Tiggelaar, J. G. E. Gardeniers, L. Lefferts and K. Seshan, *Chem. Eng. J.*, 2010, **160**, 899–908.
92. H. C. Aran, S. Pacheco Benito, M. W. J. Luiten-Olieman, S. Er, M. Wessling, L. Lefferts, N. E. Benes and R. G. H. Lammertink, *J. Membr. Sci.*, 2011, **381**, 244–250.
93. F. Piccinno, F. Gottschalk, S. Seeger and B. Nowack, *J. Nanopart. Res.*, 2012, **14**, 1–11.
94. V. Strauss, L. Ma-Hock, S. Treumann, K. Küttler, W. Wohlleben, T. Hofmann, S. Gröters, K. Wiench, B. van Ravenzwaay and R. Landsiedel, *Part. Fibre Toxicol.*, 2013, **10**, 1–19.
95. J. Bourdon, A. Saber, N. Jacobsen, K. Jensen, A. Madsen, J. Lamson, H. Wallin, P. Moller, S. Loft, C. Yauk and U. Vogel, *Part. Fibre Toxicol.*, 2012, **9**, 5.
96. A. I. Medalia, D. Rivin and D. R. Sanders, *Sci. Total Environ.*, 1983, **31**, 1–22.
97. Y. Kawabata, K. Iwai, T. Udagawa, K. Tukagoshi and K. Higuchi, *Dev. Toxicol. Environ. Sci.*, 1986, **13**, 213–222.
98. I. A. F. R. O. CANCER, *Printing Processes and Printing inks, Carbon black and Some Nitro Compounds*, 1996.
99. K. J. Nikula, M. B. Snipes, E. B. Barr, W. C. Griffith, R. F. Henderson and J. L. Mauderly, *Fundam. Appl. Toxicol.*, 1995, **25**, 80–94.
100. U. Heinrich, R. Fuhst, S. Rittinghausen, O. Creutzenberg, B. Bellmann, W. Koch and K. Levsen, *Inhalation Toxicol.*, 1995, **7**, 533–556.
101. K. Donaldson, L. Tran, L. A. Jimenez, D. Rodger, D. E. Newby, N. Mills, W. MacNee and V. Stone, *Part. Fibre Toxicol.*, 2005, **2**, 1–14.
102. A. Kroll, C. Dierker, C. Rommel, D. Hahn, W. Wohlleben, C. Schulze-Isfort, C. Gobbert, M. Voetz, F. Hardinghaus and J. Schnekenburger, *Part. Fibre Toxicol.*, 2011, **8**, 9.
103. S. Hussain, S. Boland, A. Baeza-Squiban, R. Hamel, L. C. Thomassen, J. A. Martens, M. A. Billon-Galland, J. Fleury-Feith, F. Moisan, J. C. Pairon and F. Marano, *Toxicology*, 2009, **260**, 142–149.
104. N. R. Jacobsen, G. Pojana, P. White, P. Moller, C. A. Cohn, K. S. Korsholm, U. Vogel, A. Marcomini, S. Loft and H. Wallin, *Environ. Mol. Mutagen.*, 2008, **49**, 476–487.
105. H. Tong, J. K. McGee, R. K. Saxena, U. P. Kodavanti, R. B. Devlin and M. I. Gilmour, *Toxicol. Appl. Pharmacol.*, 2009, **239**, 224–232.
106. D. M. Brown, V. Stone, P. Findlay, W. MacNee and K. Donaldson, *Occup. Environ. Med.*, 2000, **57**, 685–691.
107. X. Y. Li, D. Brown, S. Smith, W. MacNee and K. Donaldson, *Inhalation Toxicol.*, 1999, **11**, 709–731.
108. K. E. Driscoll, J. M. Carter, B. W. Howard, D. G. Hassenbein, W. Pepelko, R. B. Baggs and G. Oberdorster, *Toxicol. Appl. Pharmacol.*, 1996, **136**, 372–380.

109. A. T. Saber, J. Bornholdt, M. Dybdahl, A. K. Sharma, S. Loft, U. Vogel and H. Wallin, *Arch. Toxicol.*, 2005, **79**, 177–182.
110. R. K. Wolff, J. A. Bond, R. F. Henderson, J. R. Harkema and J. L. Mauderly, *Inhalation Toxicol.*, 1990, **2**, 241–254.
111. A. Elder, R. Gelein, J. N. Finkelstein, K. E. Driscoll, J. Harkema and G. Oberdorster, *Toxicol. Sci.*, 2005, **88**, 614–629.
112. A. M. Knaapen, P. J. Borm, C. Albrecht and R. P. Schins, *Int. J. Cancer*, 2004, **109**, 799–809.
113. N. R. Jacobsen, A. T. Saber, P. White, P. Moller, G. Pojana, U. Vogel, S. Loft, J. Gingerich, L. Soper, G. R. Douglas and H. Wallin, *Environ. Mol. Mutagen.*, 2007, **48**, 451–461.
114. N. R. Jacobsen, P. A. White, J. Gingerich, P. Moller, A. T. Saber, G. R. Douglas, U. Vogel and H. Wallin, *Environ. Mol. Mutagen.*, 2011, **52**, 331–337.
115. R. M. Mroz, R. P. Schins, H. Li, E. M. Drost, W. Macnee and K. Donaldson, *J. Physiol. Pharmacol.*, 2007, **58**, Suppl. 5461–470.
116. R. M. Mroz, R. P. Schins, H. Li, L. A. Jimenez, E. M. Drost, A. Holownia, W. MacNee and K. Donaldson, *Eur. Respir. J.*, 2008, **31**, 241–251.
117. Y. Totsuka, T. Higuchi, T. Imai, A. Nishikawa, T. Nohmi, T. Kato, S. Masuda, N. Kinae, K. Hiyoshi, S. Ogo, M. Kawanishi, T. Yagi, T. Ichinose, N. Fukumori, M. Watanabe, T. Sugimura and K. Wakabayashi, *Part. Fibre Toxicol.*, 2009, **6**, 23.
118. H. Yang, C. Liu, D. Yang, H. Zhang and Z. Xi, *J. Appl. Toxicol.*, 2009, **29**, 69–78.
119. J. Gallagher, R. Sams, 2nd, J. Inmon, R. Gelein, A. Elder, G. Oberdorster and A. K. Prahalad, *Toxicol. Appl. Pharmacol.*, 2003, **190**, 224–231.
120. N. Jacobsen, P. Moller, K. Jensen, U. Vogel, O. Ladefoged, S. Loft and H. Wallin, *Part. Fibre Toxicol.*, 2009, **6**, 2.
121. P. A. Valberg and A. Y. Watson, *Regul. Toxicol. Pharmacol.*, 1996, **24**, 155–170.
122. K. Gardiner, M. van Tongeren and M. Harrington, *Occup. Environ. Med.*, 2001, **58**, 496–503.
123. T. Sorahan, L. Hamilton, M. van Tongeren, K. Gardiner and J. M. Harrington, *Am. J. Ind. Med.*, 2001, **39**, 158–170.
124. M. J. van Tongeren, K. Gardiner, C. E. Rossiter, J. Beach, P. Harber and M. J. Harrington, *Eur. Respir. J.*, 2002, **20**, 417–425.
125. C.-H. Lim, M. Kang, J.-H. Han and J.-S. Yang, *Environ. Health Toxicol.*, 2012, **27**, 1–8.
126. C.-w. Lam, J. T. James, R. McCluskey, S. Arepalli and R. L. Hunter, *Crit. Rev. Toxicol.*, 2006, **36**, 189–217.
127. C.-W. Lam, J. T. James, R. McCluskey and R. L. Hunter, *Toxicol. Sci.*, 2004, **77**, 126–134.
128. K. Aschberger, H. J. Johnston, V. Stone, R. J. Aitken, S. M. Hankin, S. A. Peters, C. L. Tran and F. M. Christensen, *Crit. Rev. Toxicol.*, 2010, **40**, 759–790.
129. S. T. Stern and S. E. McNeil, *Toxicol. Sci.*, 2008, **101**, 4–21.
130. C. P. Firme, 3rd and P. R. Bandaru, *Nanomedicine*, 2010, **6**, 245–256.

131. C. Bussy, H. Ali-Boucetta and K. Kostarelos, *Acc. Chem. Res.*, 2012, **46**, 692–701.
132. R. Arvidsson, S. Molander and B. A. Sandén, *Hum. Ecol. Risk Assess.*, 2013, **19**, 873–887.
133. A. B. Seabra, A. J. Paula, R. de Lima, O. L. Alves and N. Durán, *Chem. Res. Toxicol.*, 2014, **27**, 159–168.
134. K. Aschberger, H. J. Johnston, V. Stone, R. J. Aitken, C. L. Tran, S. M. Hankin, S. A. K. Peters and F. M. Christensen, *Regul. Toxicol. Pharmacol.*, 2010, **58**, 455–473.
135. J. Kolosnjaj, H. Szwarc and F. Moussa, *Adv. Exp. Med. Biol.*, 2007, **620**, 168–180.
136. H. J. Johnston, G. R. Hutchison, F. M. Christensen, K. Aschberger and V. Stone, *Toxicol. Sci.*, 2010, **114**, 162–182.
137. L. Kong and R. G. Zepp, *Environ. Toxicol. Chem.*, 2012, **31**, 136–143.
138. A. M. Schrand, H. Huang, C. Carlson, J. J. Schlager, E. Ōsawa, S. M. Hussain and L. Dai, *J. Phys. Chem. B*, 2006, **111**, 2–7.
139. Y. Zhu, J. Li, W. Li, Y. Zhang, X. Yang, N. Chen, Y. Sun, Y. Zhao, C. Fan and Q. Huang, *Theranostics*, 2012, **2**, 302–312.
140. T. P. Seager and I. Linkov, in *Nanomaterials: Risks and Benefits*, ed. I. Linkov and J. Steevens, Springer Netherlands, 2009, pp. 423–436.
141. M. R. Wiesner, G. V. Lowry, P. Alvarez, D. Dionysiou and P. Biswas, *Environ. Sci. Technol.*, 2006, **40**, 4336–4345.
142. H. C. Kim and V. Fthenakis, *J. Ind. Ecol.*, 2013, **17**, 528–541.
143. A. R. Köhler, C. Som, A. Helland and F. Gottschalk, *J. Cleaner Prod.*, 2008, **16**, 927–937.
144. O. S. A. H. Administration, Working Safely with Nanomaterials, https://www.osha.gov/Publications/OSHA_FS-3634.pdf, 2013.

Subject Index